Honda Accord Owners Workshop Manual

Colin Brown

Models covered
Honda Accord 1.6 De Luxe, 1.8 Executive & 1.8 EXR
Saloon & Hatchback; 1598 cc & 1829 cc
Does not cover 1986 model

(1177-9P2)

ABCDE
FGHIJ
KLMNO
PQ/

Haynes Publishing Group
Sparkford Nr Yeovil
Somerset BA22 7JJ England

Haynes Publications, Inc
861 Lawrence Drive
Newbury Park
California 91320 USA

Acknowledgements

Thanks are due to the Champion Sparking Plug Company Limited who supplied the illustrations showing the spark plug conditions. Thanks are also due to Sykes-Pickavant who supplied some of the workshop tools; and all those people at Sparkford who assisted in the production of this manual.

© **Haynes Publishing Group 1986, 1988**

A book in the **Haynes Owners Workshop Manual Series**

Printed by J. H. Haynes & Co. Ltd, Sparkford, Nr Yeovil, Somerset BA22 7JJ, England

All rights reserved. No part of this book may be reproduced or transmitted in any form or by any means, electronic or mechanical, including photocopying, recording or by any information storage or retrieval system, without permission in writing from the copyright holder.

ISBN 1 85010 177 9

British Library Cataloguing in Publication Data
Brown, Colin, 1942–
  Honda Accord ('84 to '85) owner's workshop manual. –
(Owner's Workshop Manuals)
1. Honda Accord automobile
I. Title
II. Series
628.28'722
TL215.H58
ISBN 1-85010-177-9

Whilst every care is taken to ensure that the information in this manual is correct, no liability can be accepted by the authors or publishers for loss, damage or injury caused by any errors in, or omissions from, the information given.

Contents

Introductory pages
    About this manual   5
    Introduction to the Honda Accord   5
    General dimensions, weights and capacities   6
    Jacking, wheel changing and towing   7
    Buying spare parts and vehicle identification numbers   9
    Use of English   11
    General repair procedures   12
    Tools and working facilities   13
    Conversion factors   15
    Safety first!   16
    Routine maintenance   17
    Recommended lubricants and fluids   22
    Fault diagnosis   23

Chapter 1
Engine   27

Chapter 2
Cooling system   63

Chapter 3
Fuel, exhaust and emission control systems   71

Chapter 4
Ignition system   129

Chapter 5
Clutch   142

Chapter 6
Manual transmission   147

Chapter 7
Automatic transmission   175

Chapter 8
Driveshafts   204

Chapter 9
Braking system   208

Chapter 10
Suspension and steering   229

Chapter 11
Bodywork and fittings   254

Chapter 12
Electrical system   296

Index   352

Honda Accord 1.8 Executive (UK model)

Honda Accord 4-door Sedan and LX Hatchback (North American models)

About this manual

Its aim

The aim of this manual is to help you get the best value from your vehicle. It can do so in several ways. It can help you decide what work must be done (even should you choose to get it done by a garage), provide information on routine maintenance and servicing, and give a logical course of action and diagnosis when random faults occur. However, it is hoped that you will use the manual by tackling the work yourself. On simpler jobs it may even be quicker than booking the car into a garage and going there twice, to leave and collect it. Perhaps most important, a lot of money can be saved by avoiding the costs a garage must charge to cover its labour and overheads.

The manual has drawings and descriptions to show the function of the various components so that their layout can be understood. Then the tasks are described and photographed in a step-by-step sequence so that even a novice can do the work.

Its arrangement

The manual is divided into twelve Chapters, each covering a logical sub-division of the vehicle. The Chapters are each divided into Sections, numbered with single figures, eg 5; and the Sections into paragraphs (or sub-sections), with decimal numbers following on from the Section they are in, eg 5.1, 5.2, 5.3 etc.

It is freely illustrated, especially in those parts where there is a detailed sequence of operations to be carried out. There are two forms of illustration: figures and photographs. The figures are numbered in sequence with decimal numbers, according to their position in the Chapter – eg Fig. 6.4 is the fourth drawing/illustration in Chapter 6. Photographs carry the same number (either individually or in related groups) as the Section or sub-section to which they relate.

There is an alphabetical index at the back of the manual as well as a contents list at the front. Each Chapter is also preceded by its own individual contents list.

References to the 'left' or 'right' of the vehicle are in the sense of a person in the driver's seat facing forwards.

Unless otherwise stated, nuts and bolts are removed by turning anti-clockwise, and tightened by turning clockwise.

Vehicle manufacturers continually make changes to specifications and recommendations, and these, when notified, are incorporated into our manuals at the earliest opportunity.

Whilst every care is taken to ensure that the information in this manual is correct, no liability can be accepted by the authors or publishers for loss, damage or injury caused by any errors in, or omissions from, the information given.

Introduction to the Honda Accord

First introduced with the new 1600 or 1800 cc 12 valve engine in 1984 in Hatchback or Saloon form, the Accord range has been further enhanced by the addition of the EXR (UK) and SEi (North America) models in 1985.

Put together, they now form a very impressive range of vehicles which are well engineered, with typical Honda thoroughness and attention to detail, with a host of available extras to satisfy the most discerning motorist.

These vehicles will indeed form formidable opposition to Honda's competitors.

General dimensions, weights and capacities

Dimensions
Overall length:
- Hatchback .... 167.5 in (4255 mm)
- Saloon .... 175.4 in (4455 mm)

Overall width .... 65.6 in (1665 mm)

Overall height:
- UK:
  - Hatchback .... 53.4 in (1355 mm)
  - Saloon .... 54.1 in (1375 mm)
- North America:
  - Hatchback .... 51.1 in (1298 mm)
  - Saloon:
    - Accord .... 51.9 in (1318 mm)
    - Accord LX and SEi .... 51.7 in (1313 mm)

Wheelbase .... 96.53 in (2450 mm)

Ground clearance:
- UK .... 6.5 in (165 mm)
- North America .... 5.7 in (145 mm)

Turning circle .... 34.1 ft (10.4 m)

Kerb weights

	Hatchback	Saloon
UK:		
1600 manual	2117 lb (960 kg)	2183 lb (990 kg)
1600 automatic	2150 lb (975 kg)	2216 lb (1005 kg)
1800 manual	2183 lb (990 kg)	2249 lb (1020 kg)
1800 automatic	2216 lb (1005 kg)	2282 lb (1035 kg)
Models with air conditioning	Add a further 49.6 lb (22.5 kg)	

North America:
- Refer to vehicle FMVSS label

Capacities
Engine oil:
- At assembly .... 7.0 Imp pt (4.2 US qt, 4 litre)
- Oil change:
  - With filter .... 6.2 Imp pt (3.7 US qt, 3.5 litre)
  - Without filter .... 5.2 Imp pt (3.2 US qt, 3.0 litre)

Cooling system (with heater) .... 1.5 Imp gal (1.8 US gal, 6.8 litre)

Fuel tank .... 13.2 Imp gal (15.8 US gal, 60 litre)

Manual transmission:
- At assembly .... 4.4 Imp pt (2.6 US qt, 2.5 litre)
- Oil change .... 4.2 Imp pt (2.5 US qt, 2.4 litre)

Automatic transmission:
- At assembly .... 9.8 Imp pt (5.9 US qt, 5.6 litre)
- Fluid change .... 5.0 Imp pt (3.0 US qt, 2.8 litre)

Power steering:
- Fluid change .... 3.0 Imp pt (1.8 US qt, 1.7 litre)
- Reservoir .... 0.9 Imp pt (0.5 US qt, 0.5 litre)

Jacking, wheel changing and towing

Jacking

The jack supplied with the vehicle should only be used for emergency roadside wheel changing. Before jacking, position the vehicle on firm ground, chock the roadwheels on the opposite side to that being raised, and apply the handbrake. Place the jack under the jacking point nearest to the wheel to be removed; ensuring the jack head engages with the notches in the sill as the jack is raised.

When jacking the vehicle for major repair work, always use a more substantial jack, such as a hydraulic trolley jack, placed under the jacking points. Always supplement the jack with axle stands, again positioned as indicated, especially when working under the vehicle.

To avoid repetition, the jacking procedure is not included before each operation requiring the vehicle to be lifted. It is preferable, and recommended, that the vehicle is placed over an inspection pit, or raised on an hydraulic lift. Where these facilities are not available, use ramps, or jack the vehicle and use axle stands as described above.

Wheel changing

Roadwheels should be removed and refitted using the following procedure:

Prepare the vehicle as described in the previous paragraphs on jacking, but as soon as the jack begins to take the weight of the vehicle, slacken off the wheel nuts a quarter turn. If this is not done before the wheel leaves the ground, difficulty will be experienced in undoing the wheel nuts, as the wheel will turn when force is applied to the nuts. With the wheel nuts slackened, raise the vehicle on the jack until the wheel is clear of the ground. Remove the wheel nuts, wheel trim if fitted, and the wheel.

To refit the wheel, place it back on the wheel studs, fit the wheel trim, and the wheel nuts. Do the wheel nuts up as tight as is possible with the wheel off the ground. Lower the vehicle and, once there is sufficient contact between the wheel and the ground to prevent the wheel turning, fully tighten the wheel nuts. Remove the jack from the jacking position.

Towing

In an emergency, the front and rear towing hooks should be used to tow the vehicle short distances, observing the following precautions:

Turn the ignition key to the I position, and ensure the steering wheel is free to turn.

Place the transmission in neutral.

Do not exceed 35 mph (55 kph) or tow for distances of more than 50 miles.

Where a transmission fault has occurred, a specialist towing agency should be called in, who have the necessary equipment for raising the front wheels off the ground for towing, which is the preferred method of towing the vehicle, especially for long distances.

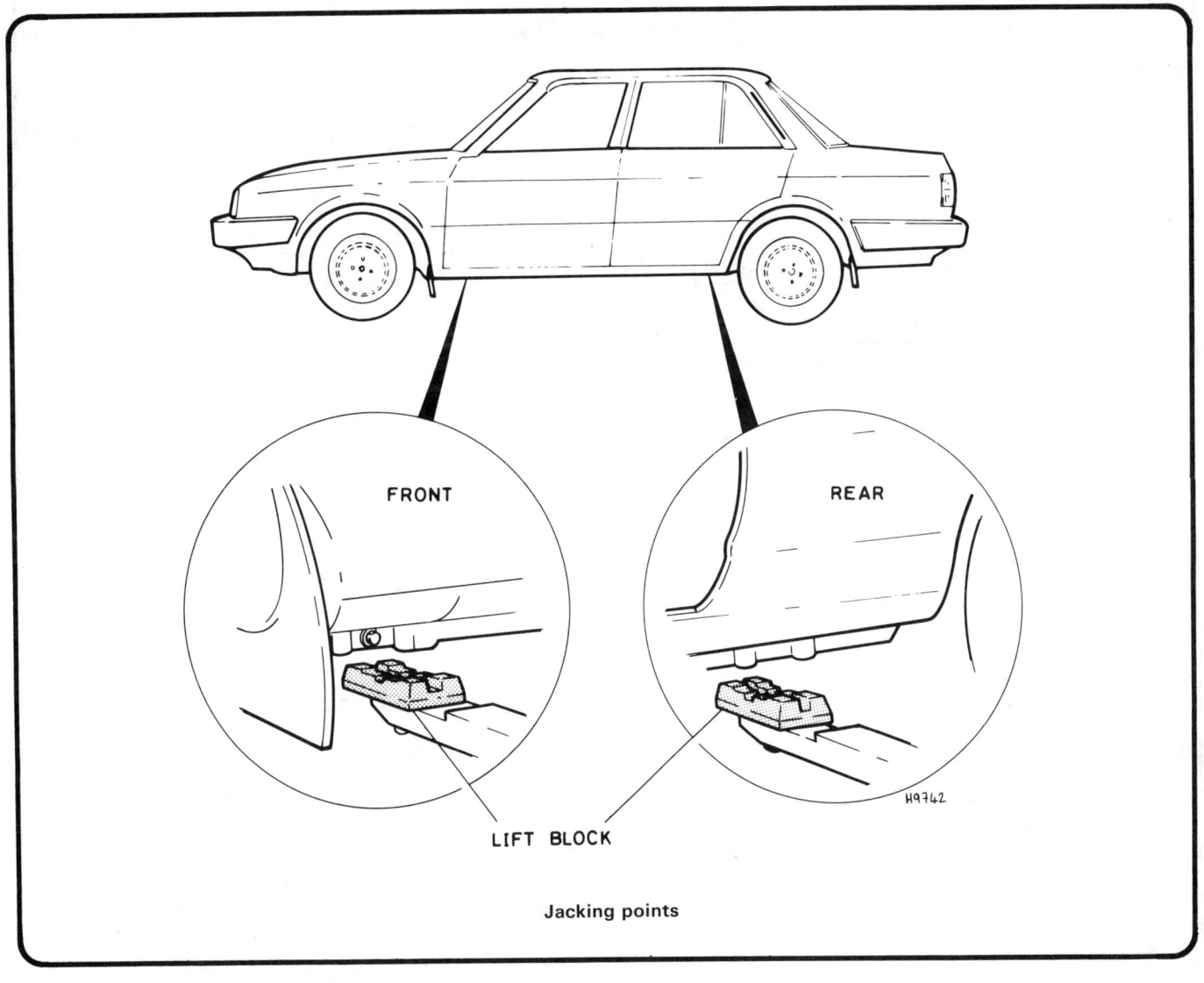

Jacking points

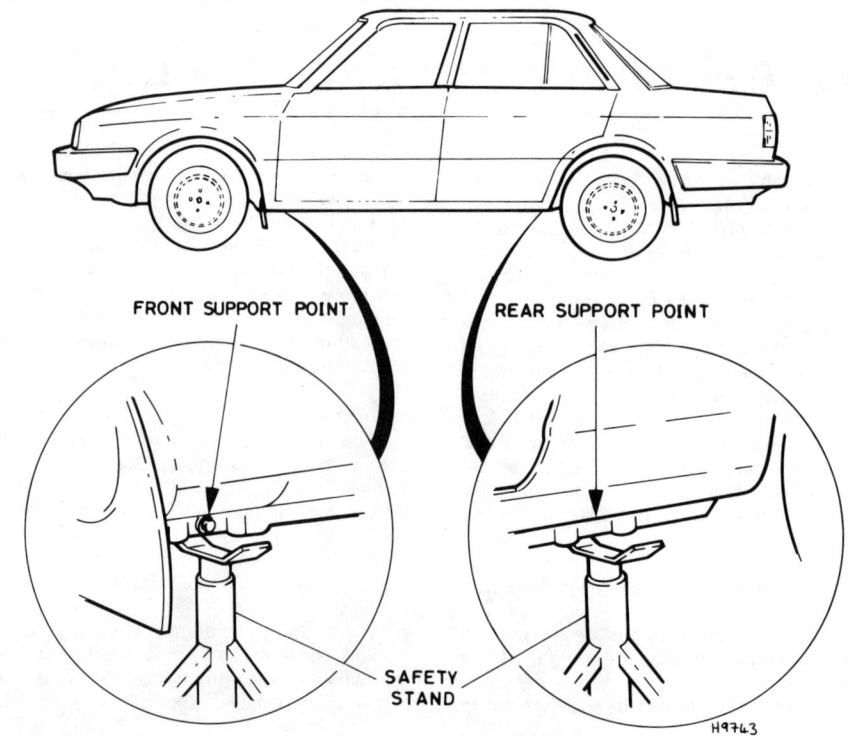

Safety stand points

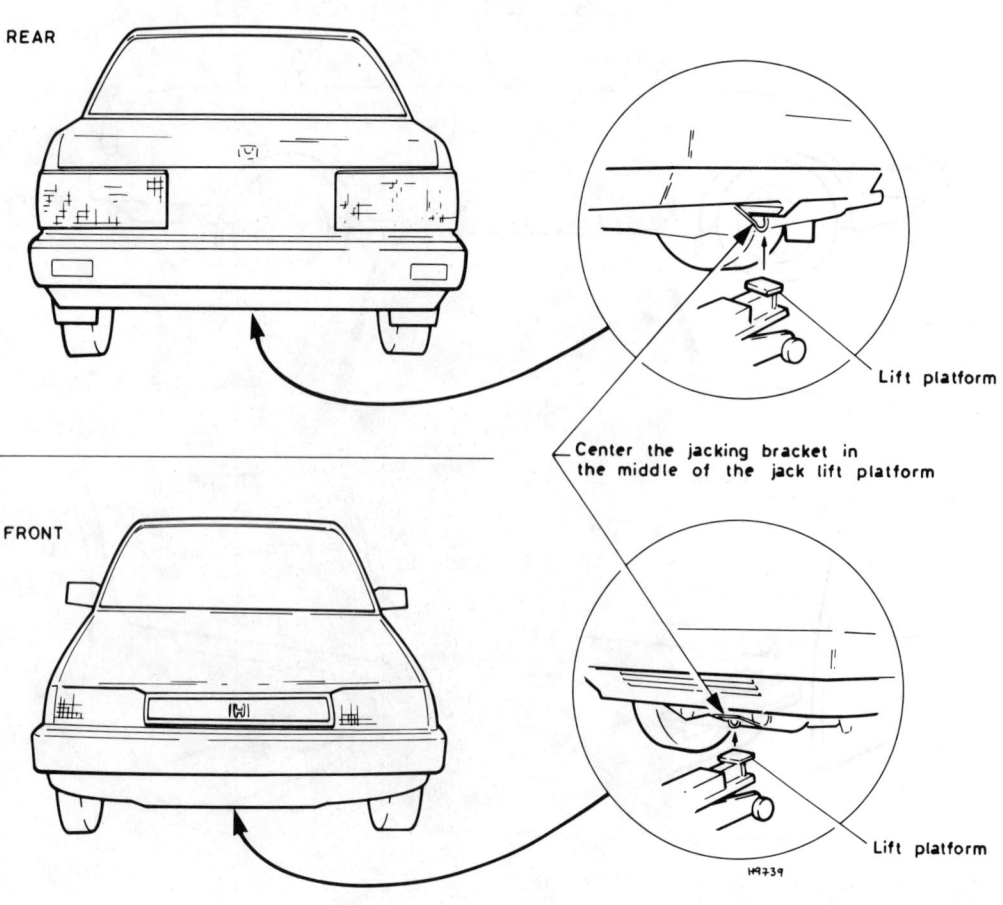

Using a trolley jack

Buying spare parts and vehicle identification numbers

Buying spare parts

Spare parts are available from many sources, for example: Honda dealers, other garages and accessory shops, and motor factors. Our advice concerning spare part sources is as follows:

Official Honda dealers – The best source of parts which are peculiar to your vehicle and which are not generally available (eg, complete cylinder heads, internal gearbox components, badges, interior trim etc). It is also the only place you should buy parts if your vehicle is still under warranty – non-standard components may invalidate the warranty. To ensure the correct parts are obtained, it will always be necessary to give the storeman your vehicle's chassis and engine number. If possible, take the 'old' part with you for further positive identification. Some parts are available on a factory exchange basis, usually the larger or more expensive parts – any parts returned should be clean. It obviously makes good sense to go to the specialists on your type of vehicle, as they are best equipped to provide your needs.

Other garages and accessory shops – These are often good places to buy materials and components needed for the maintenance of your vehicle (eg spark plugs, bulbs, drivebelts, oils and greases, repair materials etc). They also sell general accessories, usually have convenient opening hours, charge lower prices, and generally are not far from home.

Motor factors – Good motor factors will stock all of the more important components which wear out relatively quickly (eg clutch, pistons, valves, exhaust systems, brake cylinders, shoes and pads etc). Motor factors will often provide new or reconditioned components on a part exchange basis – saving a considerable amount of money.

Vehicle identification numbers

Vehicle chassis and engine numbers are located as shown in the illustration. They vary from country to country. There are also various other information labels located around the vehicle.

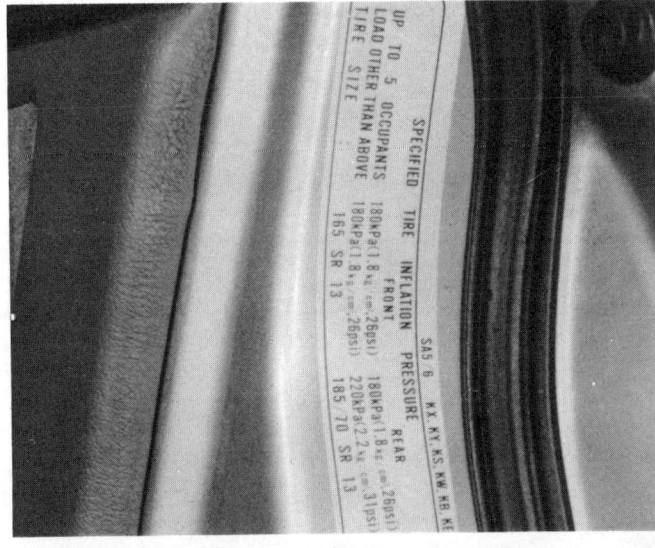

Tyre information label on door pillar

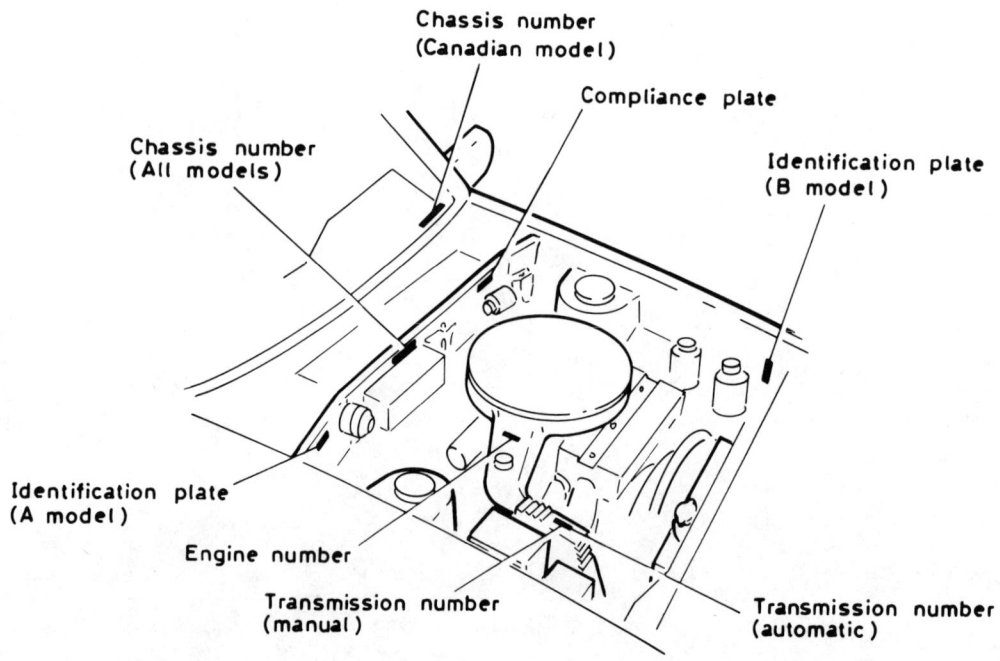

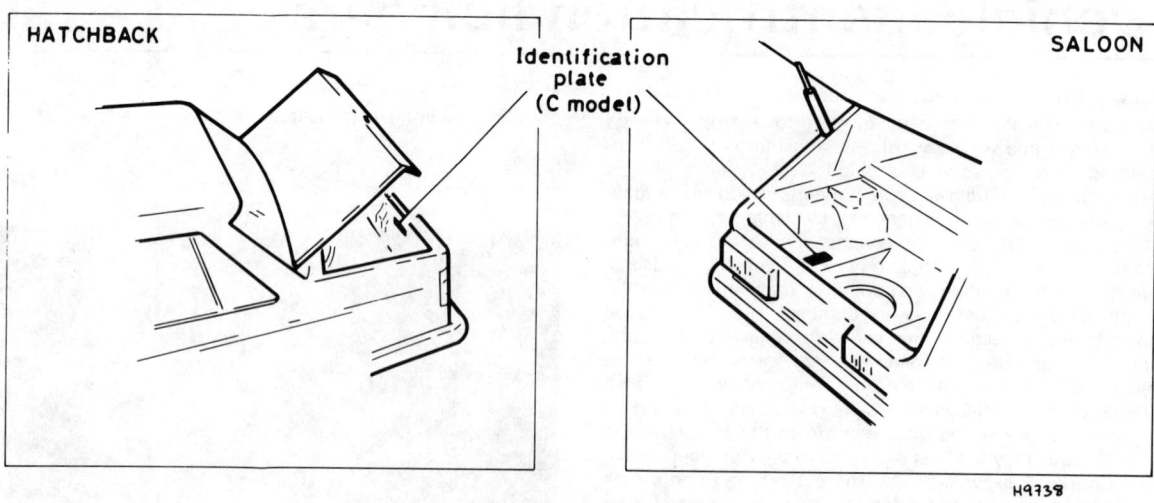

Identification number locations

Use of English

As this book has been written in England, it uses the appropriate English component names, phrases, and spelling. Some of these differ from those used in America. Normally, these cause no difficulty, but to make sure, a glossary is printed below. In ordering spare parts remember the parts list may use some of these words:

English	American	English	American
Accelerator	Gas pedal	Leading shoe (of brake)	Primary shoe
Aerial	Antenna	Locks	Latches
Anti-roll bar	Stabiliser or sway bar	Methylated spirit	Denatured alcohol
Big-end bearing	Rod bearing	Motorway	Freeway, turnpike etc
Bonnet (engine cover)	Hood	Number plate	License plate
Boot (luggage compartment)	Trunk	Paraffin	Kerosene
Bulkhead	Firewall	Petrol	Gasoline (gas)
Bush	Bushing	Petrol tank	Gas tank
Cam follower or tappet	Valve lifter or tappet	'Pinking'	'Pinging'
Carburettor	Carburetor	Prise (force apart)	Pry
Catch	Latch	Propeller shaft	Driveshaft
Choke/venturi	Barrel	Quarterlight	Quarter window
Circlip	Snap-ring	Retread	Recap
Clearance	Lash	Reverse	Back-up
Crownwheel	Ring gear (of differential)	Rocker cover	Valve cover
Damper	Shock absorber, shock	Saloon	Sedan
Disc (brake)	Rotor/disk	Seized	Frozen
Distance piece	Spacer	Sidelight	Parking light
Drop arm	Pitman arm	Silencer	Muffler
Drop head coupe	Convertible	Sill panel (beneath doors)	Rocker panel
Dynamo	Generator (DC)	Small end, little end	Piston pin or wrist pin
Earth (electrical)	Ground	Spanner	Wrench
Engineer's blue	Prussian blue	Split cotter (for valve spring cap)	Lock (for valve spring retainer)
Estate car	Station wagon	Split pin	Cotter pin
Exhaust manifold	Header	Steering arm	Spindle arm
Fault finding/diagnosis	Troubleshooting	Sump	Oil pan
Float chamber	Float bowl	Swarf	Metal chips or debris
Free-play	Lash	Tab washer	Tang or lock
Freewheel	Coast	Tappet	Valve lifter
Gearbox	Transmission	Thrust bearing	Throw-out bearing
Gearchange	Shift	Top gear	High
Grub screw	Setscrew, Allen screw	Trackrod (of steering)	Tie-rod (or connecting rod)
Gudgeon pin	Piston pin or wrist pin	Trailing shoe (of brake)	Secondary shoe
Halfshaft	Axleshaft	Transmission	Whole drive line
Handbrake	Parking brake	Tyre	Tire
Hood	Soft top	Van	Panel wagon/van
Hot spot	Heat riser	Vice	Vise
Indicator	Turn signal	Wheel nut	Lug nut
Interior light	Dome lamp	Windscreen	Windshield
Layshaft (of gearbox)	Countershaft	Wing/mudguard	Fender

General repair procedures

Whenever servicing, repair or overhaul work is carried out on the car or its components, it is necessary to observe the following procedures and instructions. This will assist in carrying out the operation efficiently and to a professional standard of workmanship.

Joint mating faces and gaskets

Where a gasket is used between the mating faces of two components, ensure that it is renewed on reassembly, and fit it dry unless otherwise stated in the repair procedure. Make sure that the mating faces are clean and dry with all traces of old gasket removed. When cleaning a joint face, use a tool which is not likely to score or damage the face, and remove any burrs or nicks with an oilstone or fine file.

Make sure that tapped holes are cleaned with a pipe cleaner, and keep them free of jointing compound if this is being used unless specifically instructed otherwise.

Ensure that all orifices, channels or pipes are clear and blow through them, preferably using compressed air.

Oil seals

Whenever an oil seal is removed from its working location, either individually or as part of an assembly, it should be renewed.

The very fine sealing lip of the seal is easily damaged and will not seal if the surface it contacts is not completely clean and free from scratches, nicks or grooves. If the original sealing surface of the component cannot be restored, the component should be renewed.

Protect the lips of the seal from any surface which may damage them in the course of fitting. Use tape or a conical sleeve where possible. Lubricate the seal lips with oil before fitting and, on dual lipped seals, fill the space between the lips with grease.

Unless otherwise stated, oil seals must be fitted with their sealing lips toward the lubricant to be sealed.

Use a tubular drift or block of wood of the appropriate size to install the seal and, if the seal housing is shouldered, drive the seal down to the shoulder. If the seal housing is unshouldered, the seal should be fitted with its face flush with the housing top face.

Screw threads and fastenings

Always ensure that a blind tapped hole is completely free from oil, grease, water or other fluid before installing the bolt or stud. Failure to do this could cause the housing to crack due to the hydraulic action of the bolt or stud as it is screwed in.

When tightening a castellated nut to accept a split pin, tighten the nut to the specified torque, where applicable, and then tighten further to the next split pin hole. Never slacken the nut to align a split pin hole unless stated in the repair procedure.

When checking or retightening a nut or bolt to a specified torque setting, slacken the nut or bolt by a quarter of a turn, and then retighten to the specified setting.

Locknuts, locktabs and washers

Any fastening which will rotate against a component or housing in the course of tightening should always have a washer between it and the relevant component or housing.

Spring or split washers should always be renewed when they are used to lock a critical component such as a big-end bearing retaining nut or bolt.

Locktabs which are folded over to retain a nut or bolt should always be renewed.

Self-locking nuts can be reused in non-critical areas, providing resistance can be felt when the locking portion passes over the bolt or stud thread.

Split pins must always be replaced with new ones of the correct size for the hole.

Special tools

Some repair procedures in this manual entail the use of special tools such as a press, two or three-legged pullers, spring compressors etc. Wherever possible, suitable readily available alternatives to the manufacturer's special tools are described, and are shown in use. In some instances, where no alternative is possible, it has been necessary to resort to the use of a manufacturer's tool and this has been done for reasons of safety as well as the efficient completion of the repair operation. Unless you are highly skilled and have a thorough understanding of the procedure described, never attempt to bypass the use of any special tool when the procedure described specifies its use. Not only is there a very great risk of personal injury, but expensive damage could be caused to the components involved.

Tools and working facilities

Introduction

A selection of good tools is a fundamental requirement for anyone contemplating the maintenance and repair of a motor vehicle. For the owner who does not possess any, their purchase will prove a considerable expense, offsetting some of the savings made by doing-it-yourself. However, provided that the tools purchased are of good quality, they will last for many years and prove an extremely worthwhile investment.

To help the average owner to decide which tools are needed to carry out the various tasks detailed in this manual, we have compiled three lists of tools under the following headings: *Maintenance and minor repair, Repair and overhaul,* and *Special.* The newcomer to practical mechanics should start off with the *Maintenance and minor repair* tool kit and confine himself to the simpler jobs around the vehicle. Then, as his confidence and experience grow, he can undertake more difficult tasks, buying extra tools as, and when, they are needed. In this way, a *Maintenance and minor repair* tool kit can be built-up into a *Repair and overhaul* tool kit over a considerable period of time without any major cash outlays. The experienced do-it-yourselfer will have a tool kit good enough for most repair and overhaul procedures and will add tools from the *Special* category when he feels the expense is justified by the amount of use to which these tools will be put.

It is obviously not possible to cover the subject of tools fully here. For those who wish to learn more about tools and their use there is a book entitled *How to Choose and Use Car Tools* available from the publishers of this manual.

Maintenance and minor repair tool kit

The tools given in this list should be considered as a minimum requirement if routine maintenance, servicing and minor repair operations are to be undertaken. We recommend the purchase of combination spanners (ring one end, open-ended the other); although more expensive than open-ended ones, they do give the advantages of both types of spanner.

Combination spanners - 10, 11, 12, 13, 14 & 17 mm
Adjustable spanner - 9 inch
Gearbox drain plug key
Spark plug spanner (with rubber insert)
Spark plug gap adjustment tool
Set of feeler gauges
Brake adjuster spanner
Brake bleed nipple spanner
Screwdriver - 4 in long x $1/4$ in dia (flat blade)
Screwdriver - 4 in long x $1/4$ in dia (cross blade)
Combination pliers - 6 inch
Hacksaw (junior)
Tyre pump
Tyre pressure gauge
Grease gun
Oil can
Fine emery cloth (1 sheet)
Wire brush (small)
Funnel (medium size)

Repair and overhaul tool kit

These tools are virtually essential for anyone undertaking any major repairs to a motor vehicle, and are additional to those given in the *Maintenance and minor repair* list. Included in this list is a comprehensive set of sockets. Although these are expensive they will be found invaluable as they are so versatile - particularly if various drives are included in the set. We recommend the ½ in square-drive type, as this can be used with most proprietary torque wrenches. If you cannot afford a socket set, even bought piecemeal, then inexpensive tubular box spanners are a useful alternative.

The tools in this list will occasionally need to be supplemented by tools from the *Special* list.

Sockets (or box spanners) to cover range in previous list
Reversible ratchet drive (for use with sockets)
Extension piece, 10 inch (for use with sockets)
Universal joint (for use with sockets)
Torque wrench (for use with sockets)
'Mole' wrench - 8 inch
Ball pein hammer
Soft-faced hammer, plastic or rubber
Screwdriver - 6 in long x $5/16$ in dia (flat blade)
Screwdriver - 2 in square x $5/16$ in dia (flat blade)
Screwdriver - $1 1/2$ in long x $1/4$ in dia (cross blade)
Screwdriver - 3 in long x $1/8$ in dia (electricians)
Pliers - electricians side cutters
Pliers - needle nosed
Pliers - circlip (internal and external)
Cold chisel - $1/2$ inch
Scriber
Scraper
Centre punch
Pin punch
Hacksaw
Valve grinding tool
Steel rule/straight-edge
Allen keys
Selection of files
Wire brush (large)
Axle-stands
Jack (strong trolley or hydraulic type)

Special tools

The tools in this list are those which are not used regularly, are expensive to buy, or which need to be used in accordance with their manufacturers' instructions. Unless relatively difficult mechanical jobs are undertaken frequently, it will not be economic to buy many of these tools. Where this is the case, you could consider clubbing together with friends (or joining a motorists' club) to make a joint purchase, or borrowing the tools against a deposit from a local garage or tool hire specialist.

The following list contains only those tools and instruments freely available to the public, and not those special tools produced by the vehicle manufacturer specifically for its dealer network. You will find occasional references to these manufacturers' special tools in the text of this manual. Generally, an alternative method of doing the job without the vehicle manufacturers' special tool is given. However, sometimes, there is no alternative to using them. Where this is the case and the relevant tool cannot be bought or borrowed, you will have to entrust the work to a franchised garage.

- Valve spring compressor (where applicable)
- Piston ring compressor
- Balljoint separator
- Universal hub/bearing puller
- Impact screwdriver
- Micrometer and/or vernier gauge
- Dial gauge
- Stroboscopic timing light
- Dwell angle meter/tachometer
- Universal electrical multi-meter
- Cylinder compression gauge
- Lifting tackle
- Trolley jack
- Light with extension lead

Buying tools

For practically all tools, a tool factor is the best source since he will have a very comprehensive range compared with the average garage or accessory shop. Having said that, accessory shops often offer excellent quality tools at discount prices, so it pays to shop around.

Remember, you don't have to buy the most expensive items on the shelf, but it is always advisable to steer clear of the very cheap tools. There are plenty of good tools around at reasonable prices, so ask the proprietor or manager of the shop for advice before making a purchase.

Care and maintenance of tools

Having purchased a reasonable tool kit, it is necessary to keep the tools in a clean serviceable condition. After use, always wipe off any dirt, grease and metal particles using a clean, dry cloth, before putting the tools away. Never leave them lying around after they have been used. A simple tool rack on the garage or workshop wall, for items such as screwdrivers and pliers is a good idea. Store all normal wrenches and sockets in a metal box. Any measuring instruments, gauges, meters, etc, must be carefully stored where they cannot be damaged or become rusty.

Take a little care when tools are used. Hammer heads inevitably become marked and screwdrivers lose the keen edge on their blades from time to time. A little timely attention with emery cloth or a file will soon restore items like this to a good serviceable finish.

Working facilities

Not to be forgotten when discussing tools, is the workshop itself. If anything more than routine maintenance is to be carried out, some form of suitable working area becomes essential.

It is appreciated that many an owner mechanic is forced by circumstances to remove an engine or similar item, without the benefit of a garage or workshop. Having done this, any repairs should always be done under the cover of a roof.

Wherever possible, any dismantling should be done on a clean, flat workbench or table at a suitable working height.

Any workbench needs a vice: one with a jaw opening of 4 in (100 mm) is suitable for most jobs. As mentioned previously, some clean dry storage space is also required for tools, as well as for lubricants, cleaning fluids, touch-up paints and so on, which become necessary.

Another item which may be required, and which has a much more general usage, is an electric drill with a chuck capacity of at least 5/16 in (8 mm). This, together with a good range of twist drills, is virtually essential for fitting accessories such as mirrors and reversing lights.

Last, but not least, always keep a supply of old newspapers and clean, lint-free rags available, and try to keep any working area as clean as possible.

Spanner jaw gap comparison table

Jaw gap (in)	Spanner size
0.250	1/4 in AF
0.276	7 mm
0.313	5/16 in AF
0.315	8 mm
0.344	11/32 in AF; 1/8 in Whitworth
0.354	9 mm
0.375	3/8 in AF
0.394	10 mm
0.433	11 mm
0.438	7/16 in AF
0.445	3/16 in Whitworth; 1/4 in BSF
0.472	12 mm
0.500	1/2 in AF
0.512	13 mm
0.525	1/4 in Whitworth; 5/16 in BSF
0.551	14 mm
0.563	9/16 in AF
0.591	15 mm
0.600	5/16 in Whitworth; 3/8 in BSF
0.625	5/8 in AF
0.630	16 mm
0.669	17 mm
0.686	11/16 in AF
0.709	18 mm
0.710	3/8 in Whitworth; 7/16 in BSF
0.748	19 mm
0.750	3/4 in AF
0.813	13/16 in AF
0.820	7/16 in Whitworth; 1/2 in BSF
0.866	22 mm
0.875	7/8 in AF
0.920	1/2 in Whitworth; 9/16 in BSF
0.938	15/16 in AF
0.945	24 mm
1.000	1 in AF
1.010	9/16 in Whitworth; 5/8 in BSF
1.024	26 mm
1.063	1 1/16 in AF; 27 mm
1.100	5/8 in Whitworth; 11/16 in BSF
1.125	1 1/8 in AF
1.181	30 mm
1.200	11/16 in Whitworth; 3/4 in BSF
1.250	1 1/4 in AF
1.260	32 mm
1.300	3/4 in Whitworth; 7/8 in BSF
1.313	1 5/16 in AF
1.390	13/16 in Whitworth; 15/16 in BSF
1.417	36 mm
1.438	1 7/16 in AF
1.480	7/8 in Whitworth; 1 in BSF
1.500	1 1/2 in AF
1.575	40 mm; 15/16 in Whitworth
1.614	41 mm
1.625	1 5/8 in AF
1.670	1 in Whitworth; 1 1/8 in BSF
1.688	1 11/16 in AF
1.811	46 mm
1.813	1 13/16 in AF
1.860	1 1/8 in Whitworth; 1 1/4 in BSF
1.875	1 7/8 in AF
1.969	50 mm
2.000	2 in AF
2.050	1 1/4 in Whitworth; 1 3/8 in BSF
2.165	55 mm
2.362	60 mm

Conversion factors

Length (distance)
Inches (in)	X	25.4	= Millimetres (mm)	X	0.0394	= Inches (in)
Feet (ft)	X	0.305	= Metres (m)	X	3.281	= Feet (ft)
Miles	X	1.609	= Kilometres (km)	X	0.621	= Miles

Volume (capacity)
Cubic inches (cu in; in^3)	X	16.387	= Cubic centimetres (cc; cm^3)	X	0.061	= Cubic inches (cu in; in^3)
Imperial pints (Imp pt)	X	0.568	= Litres (l)	X	1.76	= Imperial pints (Imp pt)
Imperial quarts (Imp qt)	X	1.137	= Litres (l)	X	0.88	= Imperial quarts (Imp qt)
Imperial quarts (Imp qt)	X	1.201	= US quarts (US qt)	X	0.833	= Imperial quarts (Imp qt)
US quarts (US qt)	X	0.946	= Litres (l)	X	1.057	= US quarts (US qt)
Imperial gallons (Imp gal)	X	4.546	= Litres (l)	X	0.22	= Imperial gallons (Imp gal)
Imperial gallons (Imp gal)	X	1.201	= US gallons (US gal)	X	0.833	= Imperial gallons (Imp gal)
US gallons (US gal)	X	3.785	= Litres (l)	X	0.264	= US gallons (US gal)

Mass (weight)
Ounces (oz)	X	28.35	= Grams (g)	X	0.035	= Ounces (oz)
Pounds (lb)	X	0.454	= Kilograms (kg)	X	2.205	= Pounds (lb)

Force
Ounces-force (ozf; oz)	X	0.278	= Newtons (N)	X	3.6	= Ounces-force (ozf; oz)
Pounds-force (lbf; lb)	X	4.448	= Newtons (N)	X	0.225	= Pounds-force (lbf; lb)
Newtons (N)	X	0.1	= Kilograms-force (kgf; kg)	X	9.81	= Newtons (N)

Pressure
Pounds-force per square inch (psi; lbf/in^2; lb/in^2)	X	0.070	= Kilograms-force per square centimetre (kgf/cm^2; kg/cm^2)	X	14.223	= Pounds-force per square inch (psi; lbf/in^2; lb/in^2)
Pounds-force per square inch (psi; lbf/in^2; lb/in^2)	X	0.068	= Atmospheres (atm)	X	14.696	= Pounds-force per square inch (psi; lbf/in^2; lb/in^2)
Pounds-force per square inch (psi; lbf/in^2; lb/in^2)	X	0.069	= Bars	X	14.5	= Pounds-force per square inch (psi; lbf/in^2; lb/in^2)
Pounds-force per square inch (psi; lbf/in^2; lb/in^2)	X	6.895	= Kilopascals (kPa)	X	0.145	= Pounds-force per square inch (psi; lbf/in^2; lb/in^2)
Kilopascals (kPa)	X	0.01	= Kilograms-force per square centimetre (kgf/cm^2; kg/cm^2)	X	98.1	= Kilopascals (kPa)

Torque (moment of force)
Pounds-force inches (lbf in; lb in)	X	1.152	= Kilograms-force centimetre (kgf cm; kg cm)	X	0.868	= Pounds-force inches (lbf in; lb in)
Pounds-force inches (lbf in; lb in)	X	0.113	= Newton metres (Nm)	X	8.85	= Pounds-force inches (lbf in; lb in)
Pounds-force inches (lbf in; lb in)	X	0.083	= Pounds-force feet (lbf ft; lb ft)	X	12	= Pounds-force inches (lbf in; lb in)
Pounds-force feet (lbf ft; lb ft)	X	0.138	= Kilograms-force metres (kgf m; kg m)	X	7.233	= Pounds-force feet (lbf ft; lb ft)
Pounds-force feet (lbf ft; lb ft)	X	1.356	= Newton metres (Nm)	X	0.738	= Pounds-force feet (lbf ft; lb ft)
Newton metres (Nm)	X	0.102	= Kilograms-force metres (kgf m; kg m)	X	9.804	= Newton metres (Nm)

Power
Horsepower (hp)	X	745.7	= Watts (W)	X	0.0013	= Horsepower (hp)

Velocity (speed)
Miles per hour (miles/hr; mph)	X	1.609	= Kilometres per hour (km/hr; kph)	X	0.621	= Miles per hour (miles/hr; mph)

*Fuel consumption**
Miles per gallon, Imperial (mpg)	X	0.354	= Kilometres per litre (km/l)	X	2.825	= Miles per gallon, Imperial (mpg)
Miles per gallon, US (mpg)	X	0.425	= Kilometres per litre (km/l)	X	2.352	= Miles per gallon, US (mpg)

Temperature
Degrees Fahrenheit = (°C x 1.8) + 32   Degrees Celsius (Degrees Centigrade; °C) = (°F - 32) x 0.56

*It is common practice to convert from miles per gallon (mpg) to litres/100 kilometres (l/100km), where mpg (Imperial) x l/100 km = 282 and mpg (US) x l/100 km = 235

Safety first!

Professional motor mechanics are trained in safe working procedures. However enthusiastic you may be about getting on with the job in hand, do take the time to ensure that your safety is not put at risk. A moment's lack of attention can result in an accident, as can failure to observe certain elementary precautions.

There will always be new ways of having accidents, and the following points do not pretend to be a comprehensive list of all dangers; they are intended rather to make you aware of the risks and to encourage a safety-conscious approach to all work you carry out on your vehicle.

Essential DOs and DON'Ts

DON'T rely on a single jack when working underneath the vehicle. Always use reliable additional means of support, such as axle stands, securely placed under a part of the vehicle that you know will not give way.

DON'T attempt to loosen or tighten high-torque nuts (e.g. wheel hub nuts) while the vehicle is on a jack; it may be pulled off.

DON'T start the engine without first ascertaining that the transmission is in neutral (or 'Park' where applicable) and the parking brake applied.

DON'T suddenly remove the filler cap from a hot cooling system – cover it with a cloth and release the pressure gradually first, or you may get scalded by escaping coolant.

DON'T attempt to drain oil until you are sure it has cooled sufficiently to avoid scalding you.

DON'T grasp any part of the engine, exhaust or catalytic converter without first ascertaining that it is sufficiently cool to avoid burning you.

DON'T allow brake fluid or antifreeze to contact vehicle paintwork.

DON'T syphon toxic liquids such as fuel, brake fluid or antifreeze by mouth, or allow them to remain on your skin.

DON'T inhale dust – it may be injurious to health (see *Asbestos* below).

DON'T allow any spilt oil or grease to remain on the floor – wipe it up straight away, before someone slips on it.

DON'T use ill-fitting spanners or other tools which may slip and cause injury.

DON'T attempt to lift a heavy component which may be beyond your capability – get assistance.

DON'T rush to finish a job, or take unverified short cuts.

DON'T allow children or animals in or around an unattended vehicle.

DO wear eye protection when using power tools such as drill, sander, bench grinder etc, and when working under the vehicle.

DO use a barrier cream on your hands prior to undertaking dirty jobs – it will protect your skin from infection as well as making the dirt easier to remove afterwards; but make sure your hands aren't left slippery.

DO keep loose clothing (cuffs, tie etc) and long hair well out of the way of moving mechanical parts.

DO remove rings, wristwatch etc, before working on the vehicle – especially the electrical system.

DO ensure that any lifting tackle used has a safe working load rating adequate for the job.

DO keep your work area tidy – it is only too easy to fall over articles left lying around.

DO get someone to check periodically that all is well, when working alone on the vehicle.

DO carry out work in a logical sequence and check that everything is correctly assembled and tightened afterwards.

DO remember that your vehicle's safety affects that of yourself and others. If in doubt on any point, get specialist advice.

If, in spite of following these percautions, you are unfortunate enough to injure yourself, seek medical attention as soon as possible.

Asbestos

Certain friction, insulating, sealing, and other products – such as brake linings, brake bands, clutch linings, torque converters, gaskets, etc – contain asbestos. *Extreme care must be taken to avoid inhalation of dust from such products since it is hazardous to health.* If in doubt, assume that they *do* contain asbestos.

Fire

Remember at all times that petrol (gasoline) is highly flammable. Never smoke, or have any kind of naked flame around, when working on the vehicle. But the risk does not end there – a spark caused by an electrical short-circuit, by two metal surfaces contacting each other, by careless use of tools, or even by static electricity built up in your body under certain conditions, can ignite petrol vapour, which in a confined space is highly explosive.

Always disconnect the battery earth (ground) terminal before working on any part of the fuel or electrical system, and never risk spilling fuel on to a hot engine or exhaust.

It is recommended that a fire extinguisher of a type suitable for fuel and electrical fires is kept handy in the garage or workplace at all times. Never try to extinguish a fuel or electrical fire with water.

Fumes

Certain fumes are highly toxic and can quickly cause unconsciousness and even death if inhaled to any extent. Petrol (gasoline) vapour comes into this category, as do the vapours from certain solvents such as trichloroethylene. Any draining or pouring of such volatile fluids should be done in a well ventilated area.

When using cleaning fluids and solvents, read the instructions carefully. Never use materials from unmarked containers – they may give off poisonous vapours.

Never run the engine of a motor vehicle in an enclosed space such as a garage. Exhaust fumes contain carbon monoxide which is extremely poisonous; if you need to run the engine, always do so in the open air or at least have the rear of the vehicle outside the workplace.

If you are fortunate enough to have the use of an inspection pit, never drain or pour petrol, and never run the engine, while the vehicle is standing over it; the fumes, being heavier than air, will concentrate in the pit with possibly lethal results.

The battery

Never cause a spark, or allow a naked light, near the vehicle's battery. It will normally be giving off a certain amount of hydrogen gas, which is highly explosive.

Always disconnect the battery earth (ground) terminal before working on the fuel or electrical systems.

If possible, loosen the filler plugs or cover when charging the battery from an external source. Do not charge at an excessive rate or the battery may burst.

Take care when topping up and when carrying the battery. The acid electrolyte, even when diluted, is very corrosive and should not be allowed to contact the eyes or skin.

If you ever need to prepare electrolyte yourself, always add the acid slowly to the water, and never the other way round. Protect against splashes by wearing rubber gloves and goggles.

When jump starting a car using a booster battery, for negative earth (ground) vehicles, connect the jump leads in the following sequence: First connect one jump lead between the positive (+) terminals of the two batteries. Then connect the other jump lead first to the negative (–) terminal of the booster battery, and then to a good earthing (ground) point on the vehicle to be started, at least 18 in (45 cm) from the battery if possible. Ensure that hands and jump leads are clear of any moving parts, and that the two vehicles do not touch. Disconnect the leads in the reverse order.

Mains electricity

When using an electric power tool, inspection light etc, which works from the mains, always ensure that the appliance is correctly connected to its plug and that, where necessary, it is properly earthed (grounded). Do not use such appliances in damp conditions and, again, beware of creating a spark or applying excessive heat in the vicinity of fuel or fuel vapour.

Ignition HT voltage

A severe electric shock can result from touching certain parts of the ignition system, such as the HT leads, when the engine is running or being cranked, particularly if components are damp or the insulation is defective. Where an electronic ignition system is fitted, the HT voltage is much higher and could prove fatal.

Routine maintenance

The routine maintenance instructions listed are basically those recommended by the vehicle manufacturer. They are sometimes supplemented by additional maintenance tasks proven to be necessary.

The maintenance intervals recommended are those specified by the manufacturer. They are necessarily something of a compromise, since no two vehicles operate under the same conditions. The DIY mechanic, who does not have labour costs to consider, may wish to shorten the service intervals.

Where the vehicle is operated under severe climatic conditions (extremes of heat or cold, dusty or muddy conditions etc), or excessive stop-start driving, then more frequent oil changes and greasing operations are desirable. If in doubt, consult your dealer.

Idle control solenoid valve – see 18.7 p.82
Sight glass A/C fluid

Engine compartment (air cleaner removed)

1. Bonnet latch
2. Windscreen wiper motor
*3. Cruise control solenoid
4. Suspension tower
5. Brake booster vacuum tube
6. Fuel inlet hose
7. Carburettor
8. Air conditioning hose
9. Cruise control unit
10. Cruise control cable
11. Vacuum tank
12. Power steering fluid reservoir
13. Power steering pump
14. Radiator (cooling system) filler cap
15. Distributor
16. Thermostat housing
17. Engine oil filler
18. Cooling system reservoir tank
19. Battery
20. Windscreen wash reservoir
21. Main electrical control box
22. Coil
23. Brake master cylinder reservoir
24. Brake booster unit
25. Engine torque rod
26. Throttle cable
27. Choke cable
28. Cruise control cable
29. Idle boost solenoid (air conditioner models)
30. Chassis identification plate

* Idle control solenoid valve – see 18.7 p.82.

Front underbody view

1. Front chassis crossmember
2. Air conditioning pump adjuster
3. Radius road
4. Power steering and air conditioning drivebelts
5. Engine oil sump drain plug
6. Brake disc caliper
7. Suspension lower arm
8. Tie-rod
9. Exhaust
10. Driveshaft
11. Stabiliser bar
12. Steering rack
13. Floorpan drain plugs
14. Gearshift rod
15. Gearshift torque rod
16. Power steering valve cover
17. Transmission unit
18. Transmission unit drain plug
19. Clutch control lever
20. Engine support beam

Rear underbody view

1. Floorpan drain plugs
2. Exhaust silencer
3. Rear stabiliser bar
4. Suspension lower arm
5. Radius rod
6. Exhaust flange
7. Fuel tank
8. Fuel pump cover
9. Brake system load sensing valve
10. Fuel tank support straps
11. Fuel tank drain plug
12. Handbrake cable

UK MODELS

Weekly or before a long journey

Check engine oil level
Check operation of all lights, direction indicators, horn, wipers and washers
Check coolant system level
Check battery electrolyte level (if applicable)
Check windscreen washer fluid level (photo)
Check tyre pressures (cold) including the spare (photo)
Check brake fluid level
Check transmission oil/fluid level
Check power steering fluid level (if applicable)

Every 6000 miles (10 000 km) or six months, whichever comes first

Change the engine oil and oil filter (Chapter 1)
Inspect the front brake pads for wear (Chapter 9)

Every 12 000 miles (20 000 km) or 12 months, whichever comes first

Check idle speed and idle CO content (Chapter 3)
Check and adjust valve clearances (Chapter 1)
Renew air cleaner filter element (Chapter 3)
Renew sparking plugs (Chapter 1)
Change automatic transmission fluid (Chapter 7)*
Inspect brake hoses and hydraulic lines. (Including ALB components, if fitted) (Chapter 9)
Inspect front brake discs and calipers (Chapter 9)
Inspect handbrake (Chapter 9)
Check clutch release arm travel (Chapter 5)
Check engine exhaust silencer (Chapter 3)
Inspect suspension mounting bolts (Chapter 10)
Check front wheel alignment (Chapter 10)
Check operation of steering, tie-rod ends, steering gearbox and protective rubber boots (Chapter 10)
Check ALB system operation if applicable (Chapter 9)
Check power steering system, if applicable (Chapter 10)
Check load sensing valve (Chapter 10)
Check the constant velocity joints (Chapter 8)

* **Note**: *thereafter change the automatic transmission fluid every 24 000 miles (40 000 km)*

Every 24 000 miles (40 000 km) or 24 months, whichever comes first

In addition to those items listed under the 12 000 mile service, carry out the following:

Inspect alternator drivebelt (Chapter 2)
Renew the manual transmission oil, if applicable (Chapter 6)
Inspect cooling system hoses and connections (Chapter 2)
Renew fuel filter (Chapter 3)
Inspect fuel tank, fuel lines and connections (Chapter 3)
Inspect throttle control system (manual transmission only) (Chapter 3)
Inspect choke mechanism (Chapter 3)
Inspect ignition timing and control system (Chapter 3)
Inspect distributor cap and rotor, and ignition wiring (Chapter 4)
Renew PCV valve (Chapter 3)
Renew blow-by filter (Chapter 3)
Renew brake fluid (including ALB systems) (Chapter 9)
Inspect rear brakes (Chapter 9)
Renew ALB system high pressure hoses (Chapter 9)
Inspect power steering pump belt (Chapter 10)

Every 30 000 miles (48 000 km) or 24 months

Renew the cooling system fluid (Chapter 2)

NORTH AMERICAN MODELS

Weekly, or before a long journey

Check engine oil level
Check operation of lights, direction indicators, horn, wipers and washers
Check coolant system level
Check battery electrolyte level (if applicable)
Check windscreen washer fluid level
Check tyre pressures (cold) including the spare
Check brake fluid level
Check transmission oil/fluid level
Check power steering fluid level (if applicable)

Spare tyre stowage in luggage compartment floor

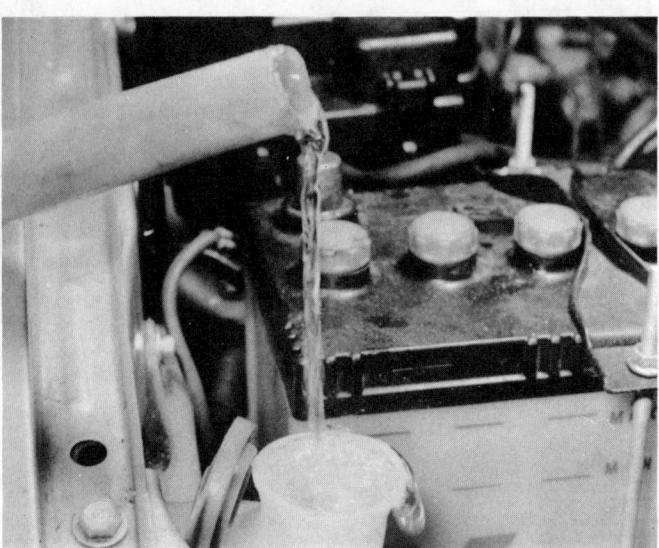

Filling the windscreen washer reservoir

Routine maintenance

Every 7500 miles (12 000 km) or six months, whichever comes first

Change the engine oil and filter (Chapter 1)
Inspect the front brake pads for wear (Chapter 9)

Every 15 000 miles (24 000 km) or 15 months, whichever comes first

Check idle speed and idle CO content (Chapter 3)
Check and adjust valve clearances (Chapter 1)
Renew air filter element (Chapter 3)
Change automatic transmission fluid (Chapter 7)*
Inspect brake hoses and hydraulic lines (including ALB components, if fitted) (Chapter 9)
Inspect front brake discs and calipers (Chapter 9)
Inspect handbrake (Chapter 9)
Check clutch release arm travel (Chapter 5)
Check engine exhaust silencer (Chapter 3)
Inspect suspension mounting bolts (Chapter 10)
Check front wheel alignment (Chapter 10)
Check operation of steering, tie-rod ends, steering gearbox and protective rubber boots (Chapter 10)
Check ALB system operation, if applicable (Chapter 9)
Check the power steering system, if applicable (Chapter 10)
Check load sensing valve (Chapter 10)
Check the constant velocity joints (Chapter 8)

***Note:** *thereafter change the automatic transmission fluid every 30 000 miles (48 000 km)*

Every 30 000 miles (48 000 km) or 30 months, whichever comes first

In addition to those items listed under the 15 000 miles service, carry out the following

Inspect alternator drivebelt (Chapter 2)
Renew manual transmission oil, if applicable (Chapter 6)
Inspect cooling systems hoses and connections. Renew the coolant (Chapter 2)
Renew fuel filter (Chapter 3)
Inspect fuel tank, fuel lines and connections (Chapter 3)
Inspect throttle cable system (manual transmission only) (Chapter 3)
Renew spark plugs (Chapter 4)
Inspect choke mechanism (Chapter 3)
Inspect ignition timing and control system (Chapter 3)
Inspect distributor cap and rotor, and ignition wiring (Chapter 4)
Renew PCV valve (Chapter 3)
Renew brake fluid (including ALB system) (Chapter 9)
Inspect rear brakes (Chapter 9)
Renew ALB system high pressure hoses (Chapter 9)
Inspect power steering pump belt (Chapter 10)

Every 60 000 miles (96 000 km) or every 48 months, whichever comes first

Renew the rear wheel bearing grease (Chapter 10)
Check all fuel line connections, and renew the fuel filter and hoses (and auxiliary filter) on both carburettor and fuel injected models (Chapter 3)
Inspect the carburettor throttle and choke controls, air intake system, and associated controls (Chapter 3)
Inspect the EGR system (Chapter 3)
Inspect the evaporative control system (Chapter 3)
Inspect the catalytic converter heat shield (Chapter 3)

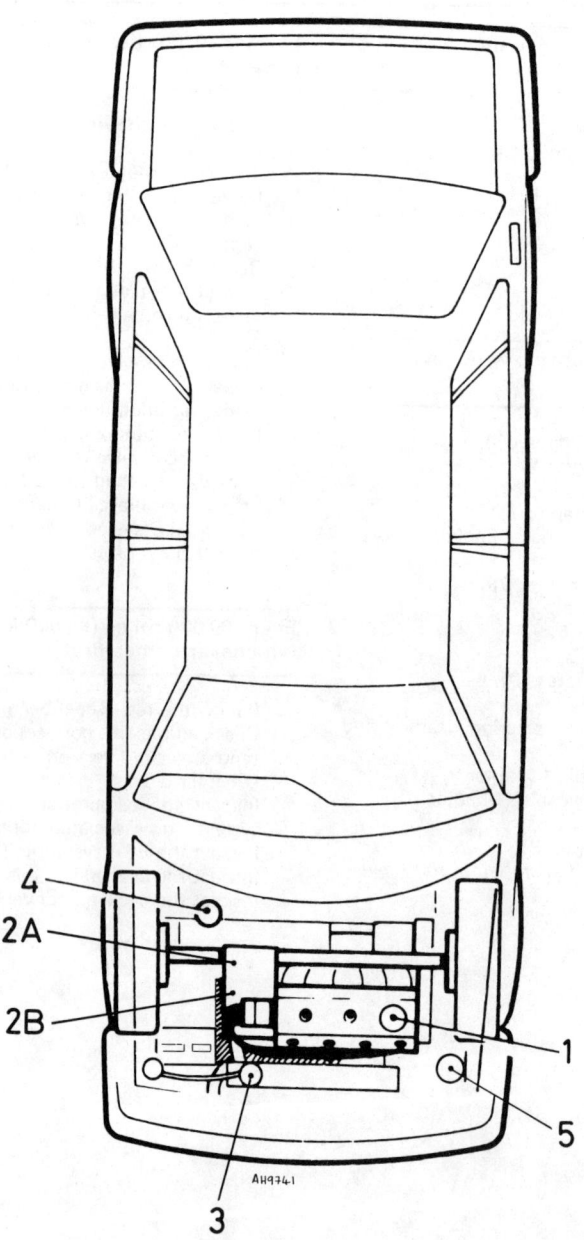

Recommended lubricants and fluids

Component or system	Lubricant type or specification
Engine (1)	Engine oil to API SE or SF – SAE 10W/30, 10W/40 or 10W/50
Manual transmission (2A)	SAE 30, 10W/30, 10W/40 or 20W/40
Automatic transmission (2B)	Dexron R fluid
Cooling system (3)	Ethylene glycol based antifreeze mixture
Brake fluid (4)	DOT 3 or DOT 4
Power-assisted steering (5)	Honda power steering fluid
Wheel bearings	Multi-purpose grease

Note: *The above are general recommendations only. Lubrication requirements vary from territory to territory and depend on vehicle usage. If in doubt, consult the operator's handbook supplied with the vehicle, or your nearest dealer.*

Fault diagnosis

Introduction

The vehicle owner who does his or her own maintenance according to the recommended schedules should not have to use this section of the manual very often. Modern component reliability is such that, provided those items subject to wear or deterioration are inspected or renewed at the specified intervals, sudden failure is comparatively rare. Faults do not usually just happen as a result of sudden failure, but develop over a period of time. Major mechanical failures in particular are usually preceded by characteristic symptoms over hundreds or even thousands of miles. Those components which do occasionally fail without warning are often small and easily carried in the vehicle.

With any fault finding, the first step is to decide where to begin investigations. Sometimes this is obvious, but on other occasions a little detective work will be necessary. The owner who makes half a dozen haphazard adjustments or replacements may be successful in curing a fault (or its symptoms), but he will be none the wiser if the fault recurs and he may well have spent more time and money than was necessary. A calm and logical approach will be found to be more satisfactory in the long run. Always take into account any warning signs or abnormalities that may have been noticed in the period preceding the fault – power loss, high or low gauge readings, unusual noises or smells, etc – and remember that failure of components such as fuses or spark plugs may only be pointers to some underlying fault.

The pages which follow here are intended to help in cases of failure to start or breakdown on the road. There is also a Fault Diagnosis Section at the end of each Chapter which should be consulted if the preliminary checks prove unfruitful. Whatever the fault, certain basic principles apply. These are as follows:

Verify the fault. This is simply a matter of being sure that you know what the symptoms are before starting work. This is particularly important if you are investigating a fault for someone else who may not have described it very accurately.

Don't overlook the obvious. For example, if the vehicle won't start, is there petrol in the tank? (Don't take anyone else's word on this particular point, and don't trust the fuel gauge either!) If an electrical fault is indicated, look for loose or broken wires before digging out the test gear.

Cure the disease, not the symptom. Substituting a flat battery with a fully charged one will get you off the hard shoulder, but if the underlying cause is not attended to, the new battery will go the same way. Similarly, changing oil-fouled spark plugs for a new set will get you moving again, but remember that the reason for the fouling (if it wasn't simply an incorrect grade of plug) will have to be established and corrected.

Don't take anything for granted. Particularly, don't forget that a 'new' component may itself be defective (especially if it's been rattling round in the boot for months), and don't leave components out of a fault diagnosis sequence just because they are new or recently fitted. When you do finally diagnose a difficult fault, you'll probably realise that all the evidence was there from the start.

Electrical faults

Electrical faults can be more puzzling than straightforward mechanical failures, but they are no less susceptible to logical analysis if the basic principles of operation are understood. Vehicle electrical wiring exists in extremely unfavourable conditions – heat, vibration and chemical attack – and the first things to look for are loose or corroded connections and broken or chafed wires, especially where the wires pass through holes in the bodywork or are subject to vibration.

All metal-bodied vehicles in current production have one pole of the battery 'earthed', ie connected to the vehicle bodywork, and in nearly all modern vehicles it is the negative (–) terminal. The various electrical components – motors, bulb holders etc – are also connected to earth, either by means of a lead or directly by their mountings. Electric current flows through the component and then back to the battery via the bodywork. If the component mounting is loose or corroded, or if a good path back to the battery is not available, the circuit will be incomplete and malfunction will result. The engine and/or gearbox are also earthed by means of flexible metal straps to the body or subframe; if these straps are loose or missing, starter motor, generator and ignition trouble may result.

Assuming the earth return to be satisfactory, electrical faults will be due either to component malfunction or to defects in the current supply. Individual components are dealt with in Chapter 12. If supply wires are broken or cracked internally this results in an open-circuit, and the easiest way to check for this is to bypass the suspect wire temporarily with a length of wire having a crocodile clip or suitable connector at each end. Alternatively, a 12V test lamp can be used to verify the presence of supply voltage at various points along the wire and the break can be thus isolated.

If a bare portion of a live wire touches the bodywork or other earthed metal part, the electricity will take the low-resistance path thus formed back to the battery: this is known as a short-circuit. Hopefully a short-circuit will blow a fuse, but otherwise it may cause burning of the insulation (and possibly further short-circuits) or even a fire. This is why it is inadvisable to bypass persistently blowing fuses with silver foil or wire.

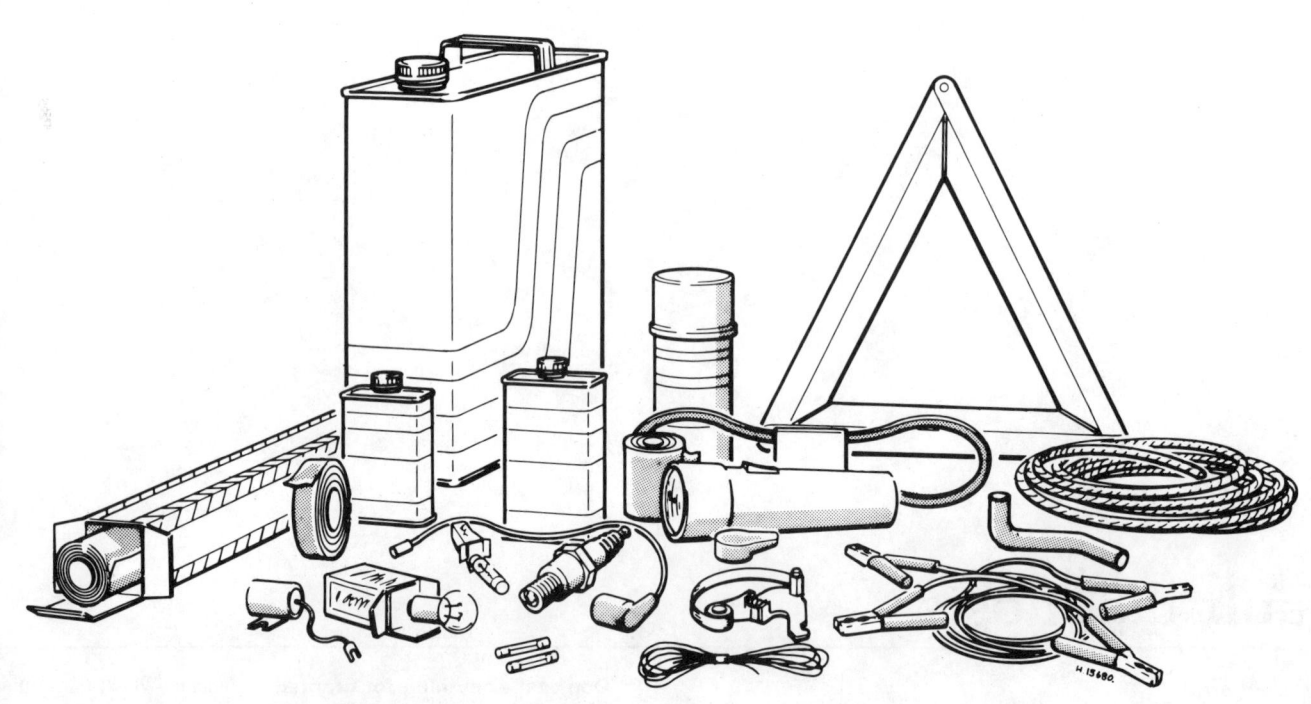

Carrying a few spares may save you a long walk

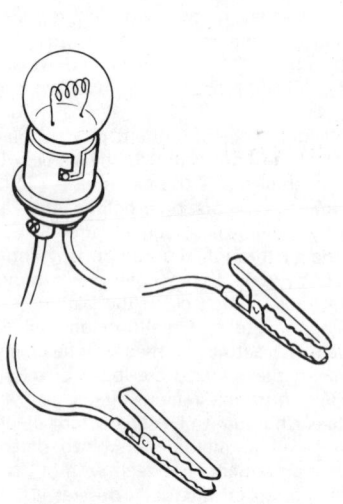

A simple test lamp is useful for tracing electrical faults

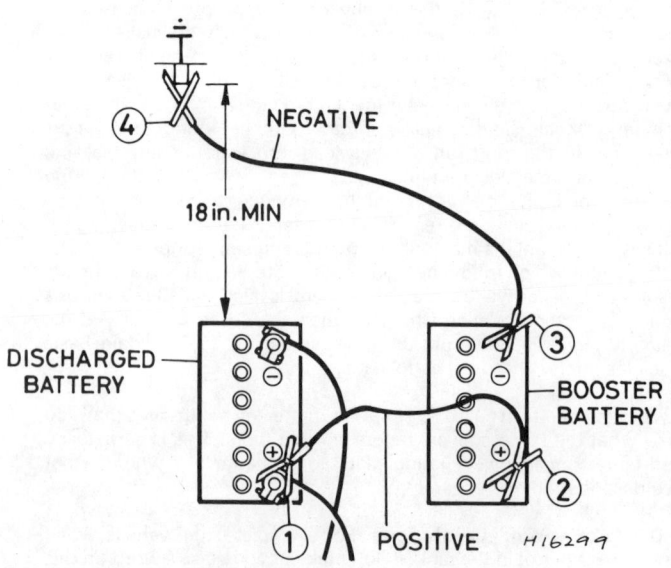

Jump start lead connections for negative earth vehicles – connect leads in order shown

Fault diagnosis

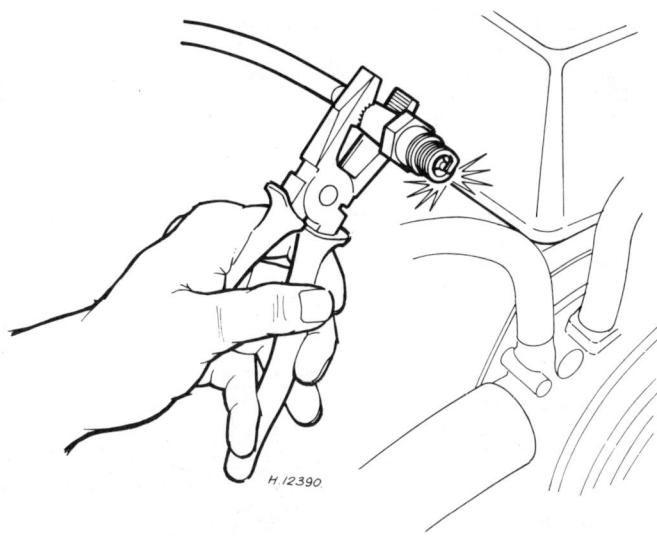

Crank engine and check for spark. Note use of insulated tool to hold plug lead

Spares and tool kit

Most vehicles are supplied only with sufficient tools for wheel changing; the *Maintenance and minor repair* tool kit detailed in *Tools and working facilities*, with the addition of a hammer, is probably sufficient for those repairs that most motorists would consider attempting at the roadside. In addition a few items which can be fitted without too much trouble in the event of a breakdown should be carried. Experience and available space will modify the list below, but the following may save having to call on professional assistance:

Spark plugs, clean and correctly gapped
HT lead and plug cap – long enough to reach the plug furthest from the distributor
Distributor rotor
Drivebelt(s) – emergency type may suffice
Spare fuses
Set of principal light bulbs
Tin of radiator sealer and hose bandage
Exhaust bandage
Roll of insulating tape
Length of soft iron wire
Length of electrical flex
Torch or inspection lamp (can double as test lamp)
Battery jump leads
Tow-rope
Ignition waterproofing aerosol
Litre of engine oil
Sealed can of hydraulic fluid
Emergency windscreen
Worm drive clips
Tube of filler paste

If spare fuel is carried, a can designed for the purpose should be used to minimise risks of leakage and collision damage. A first aid kit and a warning triangle, whilst not at present compulsory in the UK, are obviously sensible items to carry in addition to the above.
When touring abroad it may be advisable to carry additional spares which, even if you cannot fit them yourself, could save having to wait while parts are obtained. The items below may be worth considering:

Clutch and throttle cables
Cylinder head gasket
Alternator brushes
Fuel pump repair kit
Tyre valve core

One of the motoring organisations will be able to advise on availability of fuel etc in foreign countries.

Engine will not start

Engine fails to turn when starter operated
    Flat battery (recharge, use jump leads, or push start)
    Battery terminals loose or corroded
    Battery earth to body defective
    Engine earth strap loose or broken
    Starter motor (or solenoid) wiring loose or broken
    Automatic transmission selector in wrong position, or inhibitor switch faulty
    Ignition/starter switch faulty
    Major mechanical failure (seizure)
    Starter or solenoid internal fault (see Chapter 12)

Starter motor turns engine slowly
    Partially discharged battery (recharge, use jump leads, or push start)
    Battery terminals loose or corroded
    Battery earth to body defective
    Engine earth strap loose
    Starter motor (or solenoid) wiring loose
    Starter motor internal fault (see Chapter 12)

Starter motor spins without turning engine
    Flat battery
    Starter motor pinion sticking on sleeve
    Flywheel gear teeth damaged or worn
    Starter motor mounting bolts loose

Engine turns normally but fails to start
    Damp or dirty HT leads and distributor cap (crank engine and check for spark)
    No fuel in tank (check for delivery at carburettor)
    Excessive choke (hot engine) or insufficient choke (cold engine)
    Fouled or incorrectly gapped spark plugs (remove, clean and regap)
    Other ignition system fault (see Chapter 4)
    Other fuel system fault (see Chapter 3)
    Poor compression (see Chapter 1)
    Major mechanical failure (eg camshaft drive)

Engine fires but will not run
    Insufficient choke (cold engine)
    Air leaks at carburettor or inlet manifold
    Fuel starvation (see Chapter 3)
    Ignition fault (see Chapter 4)

Engine cuts out and will not restart

Engine cuts out suddenly – ignition fault
    Loose or disconnected LT wires
    Wet HT leads or distributor cap (after traversing water splash)
    Coil failure (check for spark)
    Other ignition fault (see Chapter 4)

Engine misfires before cutting out – fuel fault
    Fuel tank empty
    Fuel pump defective or filter blocked (check for delivery)
    Fuel tank filler vent blocked (suction will be evident on releasing cap)
    Carburettor needle valve sticking
    Carburettor jets blocked (fuel contaminated)
    Other fuel system fault (see Chapter 3)

Engine cuts out – other causes
    Serious overheating
    Major mechanical failure (eg camshaft drive)

Engine overheats

Ignition (no-charge) warning light illuminated
    Slack or broken drivebelt – retension or renew (Chapter 2)

Ignition warning light not illuminated
Coolant loss due to internal or external leakage (see Chapter 2)
Thermostat defective
Low oil level
Brakes binding
Radiator clogged externally or internally
Electric cooling fan not operating correctly
Engine waterways clogged
Ignition timing incorrect or automatic advance malfunctioning
Mixture too weak

Note: *Do not add cold water to an overheated engine or damage may result*

Low engine oil pressure

Gauge reads low or warning light illuminated with engine running
Oil level low or incorrect grade
Defective gauge or sender unit
Wire to sender unit earthed
Engine overheating
Oil filter clogged or bypass valve defective
Oil pressure relief valve defective
Oil pick-up strainer clogged
Oil pump worn or mountings loose
Worn main or big-end bearings

Note: *Low oil pressure in a high-mileage engine at tickover is not necessarily a cause for concern. Sudden pressure loss at speed is far more significant. In any event, check the gauge or warning light sender before condemning the engine.*

Engine noises

Pre-ignition (pinking) on acceleration
Incorrect grade of fuel
Ignition timing incorrect
Distributor faulty or worn
Worn or maladjusted carburettor
Excessive carbon build-up in engine

Whistling or wheezing noises
Leaking vacuum hose
Leaking carburettor or manifold gasket
Blowing head gasket

Tapping or rattling
Incorrect valve clearances
Worn valve gear
Worn timing chain or belt
Broken piston ring (ticking noise)

Knocking or thumping
Unintentional mechanical contact (eg fan blades)
Worn drivebelt
Peripheral component fault (generator, water pump etc)
Worn big-end bearings (regular heavy knocking, perhaps less under load)
Worn main bearings (rumbling and knocking, perhaps worsening under load)
Piston slap (most noticeable when cold)

Chapter 1 Engine

Contents

Ancillary components – removal	6
Auxiliary valve – removal, inspection and refitting	12
Bearing selection	21
Camshaft – removal and inspection	9
Camshaft drivebelt – removal, refitting and adjusting	15
Crankshaft – removal and inspection	20
Cylinder block – dismantling	17
Cylinder block – inspection	22
Cylinder block and crankshaft – reassembly	26
Cylinder head – inspection	13
Cylinder head – reassembly and refitting	14
Cylinder head – removal	8
Engine – decarbonising	25
Engine – final assembly and refitting	27
Engine and transmission unit – removal and refitting	4
Engine and transmission unit – separation	5
Engine dismantling – general	7
Engine removal – general	3
Fault diagnosis – engine	28
General description	1
Intake and exhaust valves – removal, inspection and refitting	11
Oil filter and pressure relief valve – removal and refitting	19
Oil pump – removal, inspection and refitting	18
Piston rings – removal, refitting and inspection	24
Pistons and connecting rods – inspection	23
Rocker shaft assembly – dismantling, inspection and reassembly	10
Routine maintenance	2
Valve clearances – adjustment	16

Specifications

General

Type	Transverse, 4-cylinder, in-line, overhead cam, water cooled
Designation:	
ET1	UK and Canada 1800
EZ	UK and Canada 1600
ES2	North America, with carburettor
ES3	North America, with fuel injection
Capacity:	
EZ	1598 cc
ET1, ES2 and ES3	1829 cc
Bore and stroke:	
EZ	80 x 79.5 mm
ET1, ES2 and ES3	80 x 91 mm
Compression ratio:	
ET1 and EZ	9.1 to 1
ES2 and ES3	9.0 to 1
SEi models	8.8 to 1
Firing order	1-3-4-2 (No 1 at timing belt end)

Crankshaft and bearings

Main journal diameter	1.9685 to 1.9694 in (50.000 to 50.024 mm)
Maximum journal taper	0.0002 in (0.005 mm)
Connecting rod journal:	
Diameter:	
1600	1.6526 to 1.6535 in (41.976 to 42.000 mm)
1800	1.7707 to 1.7717 in (44.976 to 45.000 mm)
Taper	0.0002 in (0.005 mm)
Endfloat	0.004 to 0.014 in (0.10 to 0.35 mm)
Run-out	0.0012 in (0.03 mm)
Main bearing-to-journal oil clearance	0.0010 to 0.0022 in (0.026 to 0.055 mm)
Connecting rod bearing-to-journal oil clearance	0.0008 to 0.0015 in (0.020 to 0.038 mm)

Connecting rod
Gudgeon pin interference	0.0005 to 0.0013 in (0.013 to 0.032 mm)
Big-end bore diameter (nominal)	1.89 in (48.0 mm)
Endplay installed on crankshaft	0.006 to 0.012 in (0.15 to 0.30 mm)

Cylinder block
Surface warpage	0.003 in (0.08 mm)
Bore diameter:	
A	3.1500 to 3.1504 in (80.01 to 80.02 mm)
B	3.1496 to 3.1500 in (80.00 to 80.01 mm)
Bore taper	0.0003 to 0.0005 in (0.007 to 0.012 mm)
Maximum rebore limit	0.02 in (0.5 mm)

Pistons
Skirt diameter:	
A	3.1488 to 3.1492 in (79.98 to 79.99 mm)
B	3.1484 to 3.1488 in (79.97 to 79.98 mm)
Piston ring-to-land clearance:	
1600	0.0008 to 0.0022 in (0.020 to 0.055 mm)
1800	0.0008 to 0.0018 in (0.020 to 0.045 mm)
Clearance in cylinder	0.0008 to 0.0016 in (0.020 to 0.040 mm)
Piston ring end gap:	
Top and second ring	0.008 to 0.014 in (0.20 to 0.35 mm)
Oil control ring	0.008 to 0.035 in (0.20 to 0.90 mm)

Cylinder head
Surface warpage (max)	0.002 in (0.05 mm)
Height	3.54 in (90.0 mm)

Camshaft
Endfloat	0.002 to 0.006 in (0.05 to 0.15 mm)
Oil clearance of journals:	
No 1, 3 and 5	0.002 to 0.004 in (0.05 to 0.09 mm)
No 2 and 4	0.005 to 0.007 in (0.13 to 0.17 mm)
Run-out (max)	0.001 in (0.03 mm)
Cam lobe height (UK models):	
1600 manual:	
Inlet	1.4793 in (37.575 mm)
Exhaust	1.4796 in (37.581 mm)
1800 manual:	
Inlet	1.5174 in (38.541 mm)
Exhaust	1.5130 in (38.430 mm)
1600 automatic:	
Inlet	1.4844 in (37.705 mm)
Exhaust	1.4872 in (37.776 mm)
1800 automatic:	
Inlet	1.5174 in (38.541 mm)
Exhaust	1.5200 in (38.607 mm)
Cam lobe height (North American models):	
Manual (carburettor):	
Inlet A	1.492 in (37.899 mm)
Inlet B	1.497 in (38.028 mm)
Exhaust	1.487 in (37.776 mm)
Auxiliary	1.446 in (36.738 mm)
Automatic (carburettor):	
Inlet A	1.484 in (37.705 mm)
Inlet B	1.502 in (38.157 mm)
Exhaust	1.487 in (37.776 mm)
Auxiliary	1.446 in (36.738 mm)
Manual and automatic (fuel injected models):	
Inlet A	1.5297 in (38.855 mm)
Inlet B	1.5200 in (38.608 mm)
Exhaust	1.5274 in (38.796 mm)

Rocker assembly
Rocker arm-to-shaft clearance	0.0003 to 0.0021 in (0.008 to 0.054 mm)

Valves

Valve clearances (cold):
- Inlet ... 0.005 to 0.007 in (0.12 to 0.17 mm)
- Exhaust ... 0.010 to 0.012 in (0.25 to 0.30 mm)
- Auxiliary ... 0.005 to 0.007 in (0.12 to 0.17 mm)

Valve stem, outside diameter:
- Inlet ... 0.2591 to 0.2594 in (6.58 to 6.59 mm)
- Exhaust ... 0.2732 to 0.2736 in (6.94 to 6.95 mm)
- Auxiliary ... 0.2587 to 0.2593 in (6.572 to 6.587 mm)

Stem-to-guide clearance:
- Inlet ... 0.001 to 0.002 in (0.02 to 0.05 mm)
- Exhaust (UK) ... 0.002 to 0.003 in (0.06 to 0.08 mm)
- Exhaust (North America) ... 0.002 to 0.004 in (0.06 to 0.09 mm)
- Auxiliary ... 0.001 to 0.002 in (0.023 to 0.058 mm)

Valve stem installed height:
- Inlet ... 1.913 in (48.59 mm)
- Exhaust ... 1.876 in (47.66 mm)
- Auxiliary ... 1.448 in (36.78 mm)

Valve seat width:
- Inlet and exhaust ... 0.049 to 0.061 in (1.25 to 1.55 mm)
- Auxiliary ... 0.014 to 0.019 in (0.353 to 0.494 mm)

Valve guide, inside diameter:
- Inlet ... 0.260 to 0.261 in (6.61 to 6.63 mm)
- Exhaust ... 0.276 to 0.277 in (7.01 to 7.03 mm)
- Auxiliary ... 0.260 to 0.261 in (6.61 to 6.63 mm)

Lubrication system

- Oil pump displacement (at 5500 rev/min) ... 8.9 Imp gal (10.6 US gal, 40.3 litre) per minute
- Inner-to-outer rotor radial (tip) clearance ... 0.006 in (0.15 mm)
- Pump body-to-rotor radial clearance ... 0.004 to 0.007 in (0.10 to 0.18 mm)
- Pump body-to-rotor axial clearance ... 0.001 to 0.004 in (0.030 to 0.108 mm)
- Relief valve pressure setting (80°C at 3000 rev/min) ... 54 to 65 lbf/in² (3.8 to 4.6 kgf/cm²)

Torque wrench settings

	lbf ft	Nm
Oil sump drain plug	33	45
Engine centre beam mounting bolts	35	47
Engine mountings (tighten in sequence:)		
1	Finger tight initially	Finger tight initially
2	40	54
3	14	19
4	14	19
5	28	38
6	Finger tight initially	Finger tight initially
7	54	73
8	54	73
9	14	19
Valve cover	7	9
Timing belt upper cover	7	9
Rocker shaft pedestal bolts	16	22
Rocker shaft end cap	9	12
Camshaft drive wheel	27	37
Cylinder head bolts:		
Stage 1 (in sequence)	22	30
Stage 2	49	66
Timing belt adjustment bolt	31	42
Crankshaft pulley bolt	83	113
Timing belt lower cover	7	9
Intake and exhaust valve locknuts	14	19
Connecting rod bearing cap bolts	23	31
Main bearing cap bolts	48	65
Oil filter base mounting bolts	9	12
Oil pump mounting bolts	9	12
Oil pick-up assembly bolts	9	12
Oil sump bolts	9	12
Flywheel bolts (manual transmission)	76	103
Converter driveplate bolts (automatic transmission)	54	73

1 General description

The engine is a water-cooled, 4-cylinder, OHC type with a displacement of 1598 cc or 1829 cc.

The camshaft is belt-driven and mounted directly into machined journals in the cylinder head, with white metal bearing liners. The two inlet and one exhaust valve are operated by rocker arms from the camshaft. North American models have an auxiliary valve in each cylinder for increased combustion efficiency.

The crankshaft is a one piece, hardened carbon steel forging, using five main bearings with renewable white metal liners. Endfloat is determined by thrust washers either side of No 3 main bearing.

Lubrication is from a camshaft belt driven oil pump.

The cylinder head is of aluminium alloy and the cylinder block is cast iron.

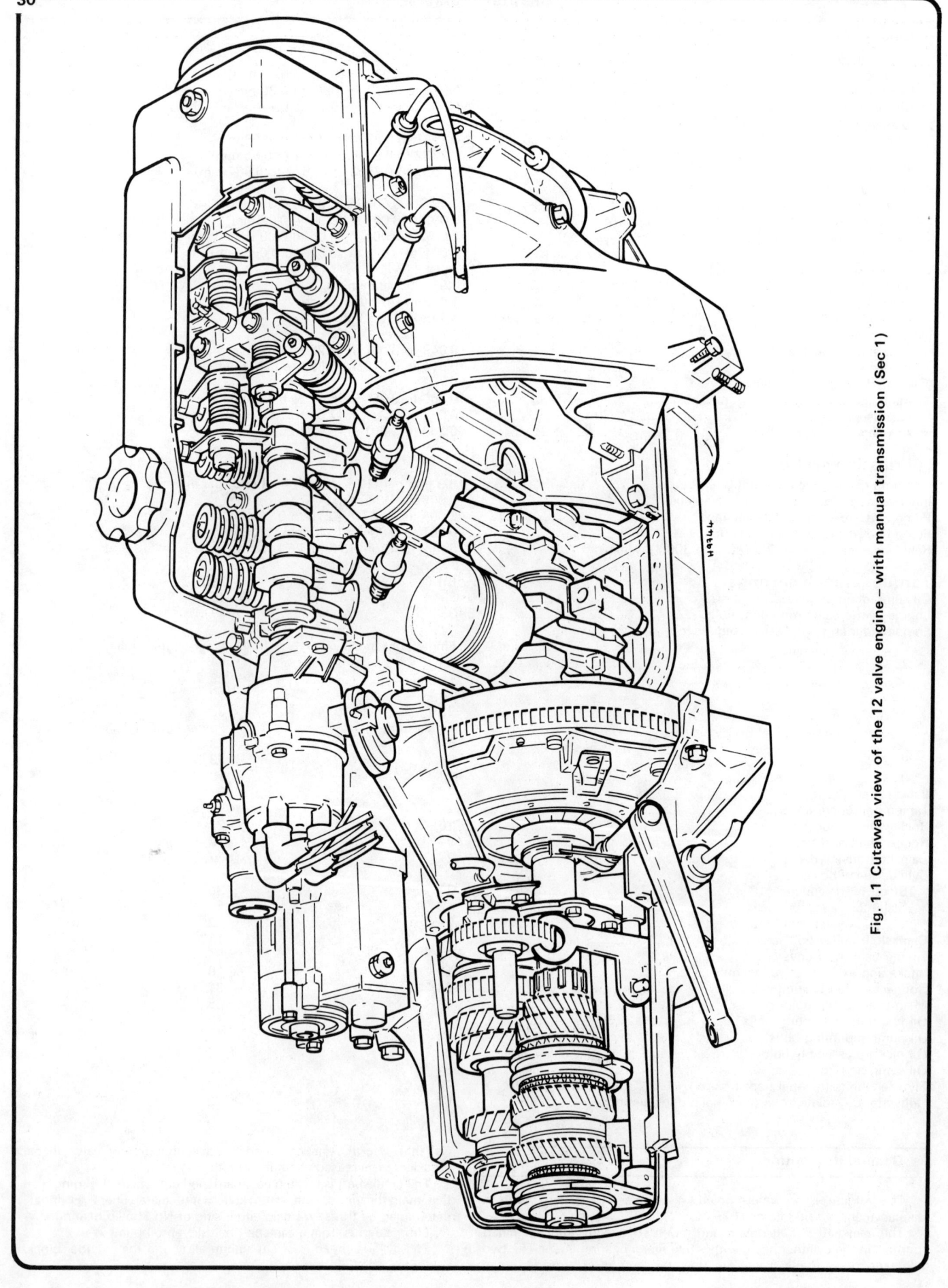

Fig. 1.1 Cutaway view of the 12 valve engine – with manual transmission (Sec 1)

Chapter 1 Engine

2 Routine maintenance

At the intervals given in Routine Maintenance (at the front of the book) undertake the following service tasks.
1 Check the engine oil level. Allow the oil to settle if the engine has been running, and with the vehicle on level ground remove the dipstick from the tube on the side of the engine block (photo).
2 Wipe the dipstick clean and reinsert it in the tube, ensuring it is pushed all the way down, then remove it and check the oil level disclosed on the dipstick. This should be maintained in the cross-hatched area. Do not overfill.
3 Renew the engine oil and oil filter. With the engine hot to ensure complete draining and the vehicle on level ground, remove the engine oil sump drain plug (photo) and drain the oil into a suitable container.
4 Refit the drain plug using a new sealing washer and tightening to the specified torque.
5 Refill the engine with the specified oil through the filler on the valve cover (photo).
6 Oil filter renewal is dealt with in Section 19.
7 Check and adjust the valve clearances.

3 Engine removal – general

1 The engine is normally removed complete with the transmission unit, using an engine hoist and lifting sling.
2 The transmission unit can be removed on its own (see Chapters 6 and 7).
3 The cylinder head may be removed with the engine *in situ*, for work on the camshaft and valves.
4 The sump can be removed for access to the crankshaft, but if major overhaul work is needed, it is better to remove the engine.
5 When stripping internal components, such as the valve train or the bearings etc, keep all parts in order, so that they are not mixed up, and ensure they are returned to their original positions.
6 Oil all parts liberally in clean engine oil on reassembly, and always use new gaskets and O-ring seals.
7 Blank off open pipes/tubes/hoses and cover the driveshaft ends, and power steering pump and air conditioning pump with plastic bags once the engine is removed, to prevent the ingress of dirt and moisture.
8 There are several brackets and cable clips bolted to various parts of the engine and engine mountings, and these should be removed as necessary to do the job in hand; a note being taken of where each one fits. This varies from engine to engine, according to equipment fitted.
9 Reference should be made to the relevant Chapter where instructions to remove a component from another system are given (eg remove the air cleaner – refer to Chapter 3).

4 Engine and transmission unit – removal and refitting

1 The engine is removed complete with the transmission unit, although the transmission unit may be removed by itself (See Chapter 6 or 7).
2 Disconnect the battery negative terminal and then the positive. For absolute safety remove the battery from the vehicle.
3 Remove the bonnet hinge bolts and lift the bonnet away, storing it in a safe place.
4 Remove the splash guard from underneath the engine compartment (two plastic studs at front and six screws/bolts).
5 Remove the sump plug and drain the engine oil.
6 Refer to Chapter 2 and drain the cooling system.
7 Refer to Chapter 6 for manual and Chapter 7 for automatic transmission and drain the oil from the transmission unit.

Carburettor models

8 Refer to Chapter 3 and remove the air cleaner assembly and associated ducting.

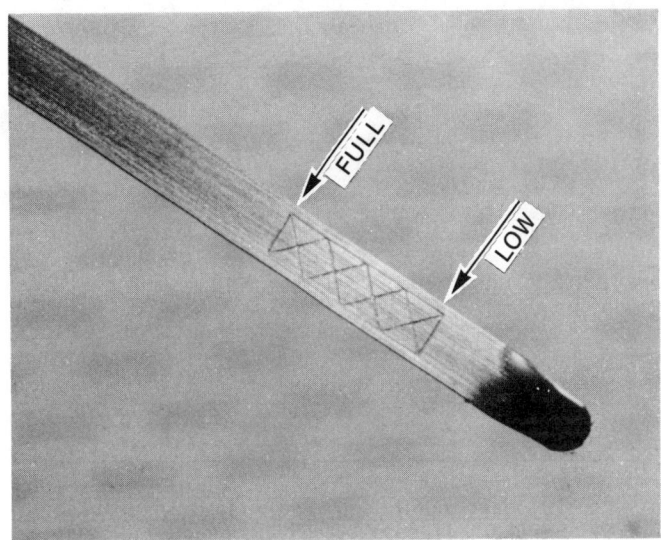

2.1 Engine oil dipstick

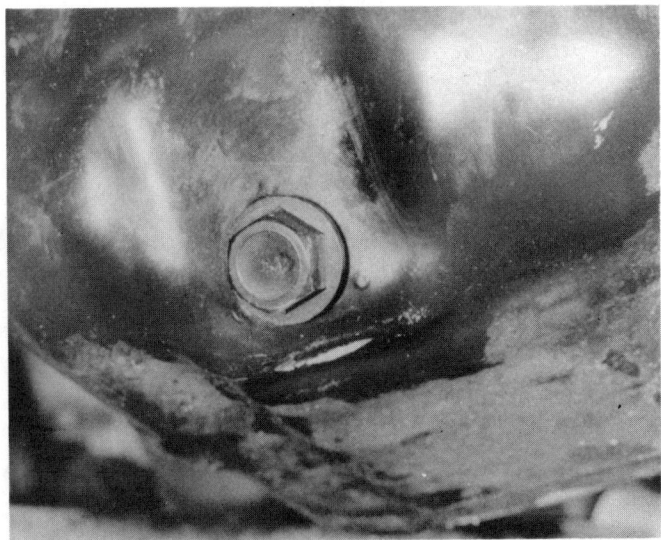

2.3 Engine oil sump drain plug

2.5 Filling the engine with oil

4.9A Disconnecting the engine sub-harness

4.9B Engine secondary earth strap

9 Disconnect the engine compartment sub-harness and the earth strap on the valve cover (photos).
10 Disconnect the LT and HT ignition cables from the coil.
11 Disconnect the vacuum hose from the charcoal canister, and the emission control vacuum hoses from the control box.
Note: When disconnecting vacuum hoses, take note of where they connect so that they are replaced in their correct positions, especially on the more complex systems.
12 Refer to Chapter 3 and disconnect the throttle cable and choke cable on manual choke vehicles.
13 Disconnect the fuel inlet hose at the carburettor.

Fuel injection models

14 Remove the air intake duct.
15 Disconnect the cruise control vacuum tube from the air intake duct.
16 Remove the resonator tube from the air intake duct.
17 Remove the engine earth lead from the valve cover.
18 Disconnect the air duct between the air cleaner and throttle valve.
19 Remove the duct clamp bolt.
20 Disconnect the emission control vacuum hoses.
21 Remove the air cleaner assembly.
22 Remove the throttle cable.
23 Disconnect the earth cable at the fusebox.
24 Disconnect the engine compartment sub harness.
25 Disconnect the HT and LT leads from the coil.
26 Refer to Chapter 3 and relieve the fuel pressure.
27 Disconnect the fuel return hose from the pressure regulator, and the fuel inlet hose.
28 Disconnect the emission vacuum hoses from the multi-connector.
29 Disconnect the electrical cables to No 2 control box and remove the control box mounting bolts. Lift the box from its bracket and allow it to hang down next to the engine.

All models

30 Disconnect the brake booster vacuum hose (photo).
31 On vehicles equipped with the auto-levelling suspension system, remove the hoses and banjo union from the compressor.
32 Disconnect the radiator and heater hoses.
33 On manual transmission vehicles, disconnect the clutch cable.
34 On vehicles equipped with automatic transmission, disconnect the oil cooler hoses and tie the hoses out of the way.
35 Remove the clip and pull the speedometer cable out of the holder. Do not remove the holder or the speedometer gear may fall into the transmission.
36 On vehicles with power steering remove the complete speed sensor, without disconnecting the hoses, and tie it back out of the way.
37 Remove the power steering pump mounting bolt and adjuster bolt, slip off the drivebelt, and tie the pump back out of the way. There should be no need to disconnect the hydraulic hoses (photo).
38 Raise the front end of the vehicle onto stands.
39 Remove the engine centre beam from under the engine bay (photos).
40 Remove the driveshafts, as described in Chapter 8.
41 Remove exhaust pipe from the exhaust manifold and remove the exhaust support bracket (photo).
42 On vehicles equipped with air conditioning, disconnect the compressor clutch electrical lead and loosen the belt adjusting nut.
43 Remove the belt.
44 Remove the compressor mounting bolts and tie the compressor back out of the way. **Note:** there is no need to have the system discharged, but do not disconnect any hoses.
45 Disconnect the cable control solenoid valve vacuum hoses.
46 On vehicles with manual transmission, disconnect the gearchange lever torque rod and the gearchange rod yoke attachment bolt.
47 On automatic transmission vehicles, remove the centre console, select reverse and remove the lockpin from the shift cable.

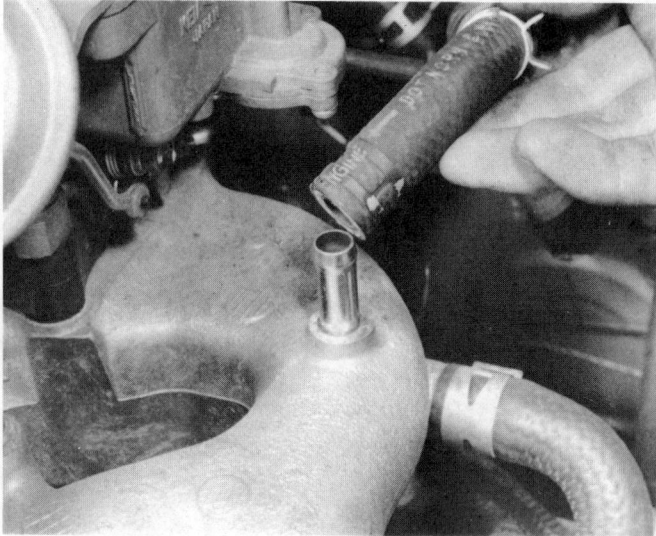

4.30 Disconnecting the brake booster vacuum hose

Chapter 1 Engine

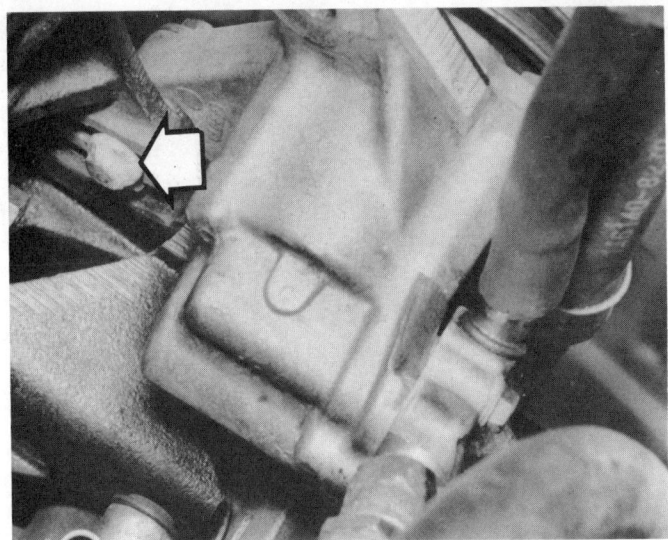

4.37 Power steering pump, showing lower adjustment bolt (arrowed)

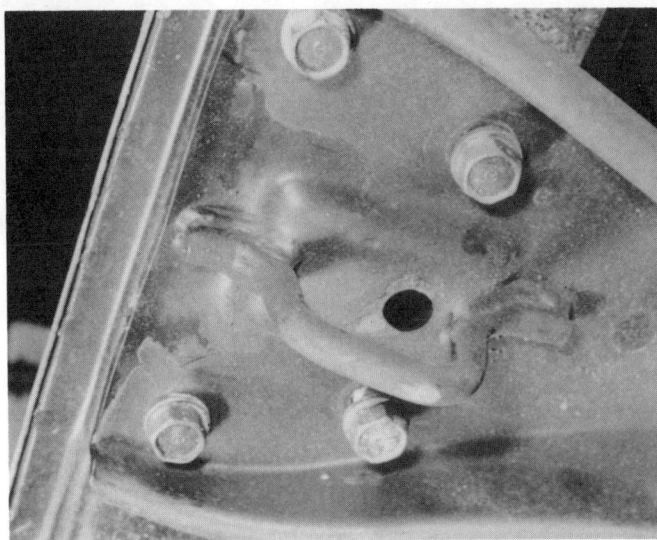

4.39A Engine centre beam front mounting bolts ...

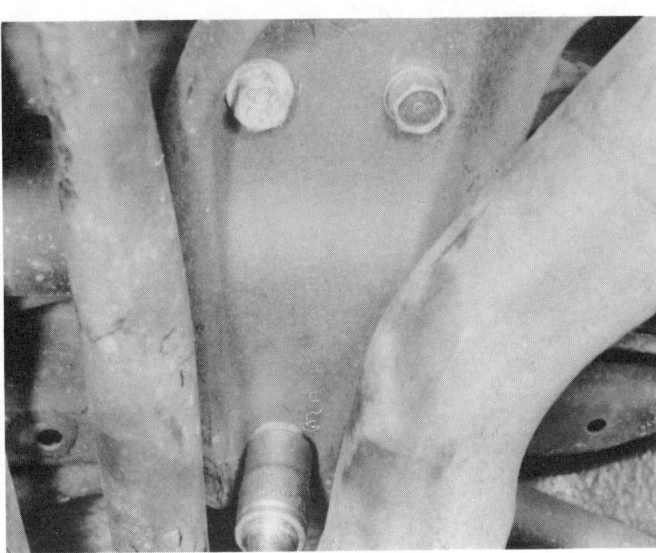

4.39B ... and rear mounting bolts

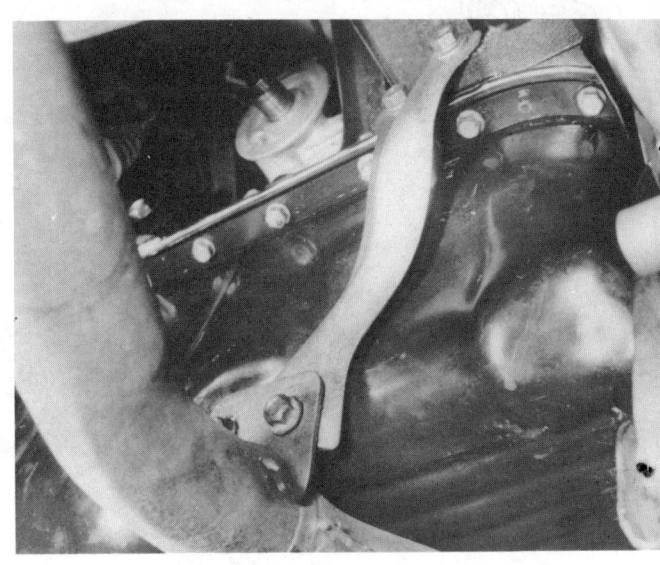

4.41 Exhaust pipe support bracket

48 Remove the cable guide and shift cable holder mounting bolts.
49 Refer to Chapter 3 and remove the throttle control bracket.
50 Disconnect the cruise control vacuum hoses at the intake manifold, and the cruise control cable.

51 Attach a lifting sling to the engine lifting eyes and raise the hoist sufficiently to just take the weight of the engine (photos).
52 Remove the three bolts from the left-hand engine mount and slide the mount back into the housing.

4.51A Engine lifting eye on cylinder head ...

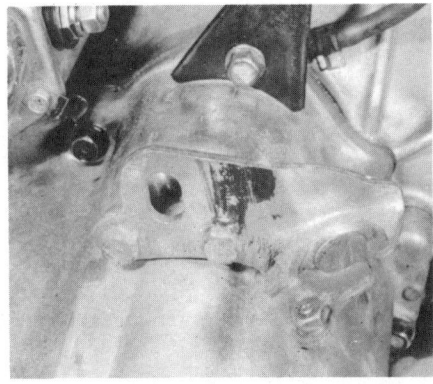

4.51B ... and transmission unit

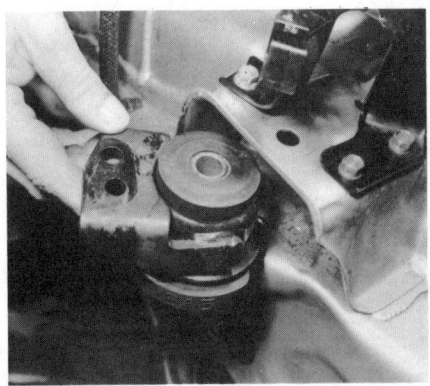

4.52 Left-hand engine mounting (engine removed)

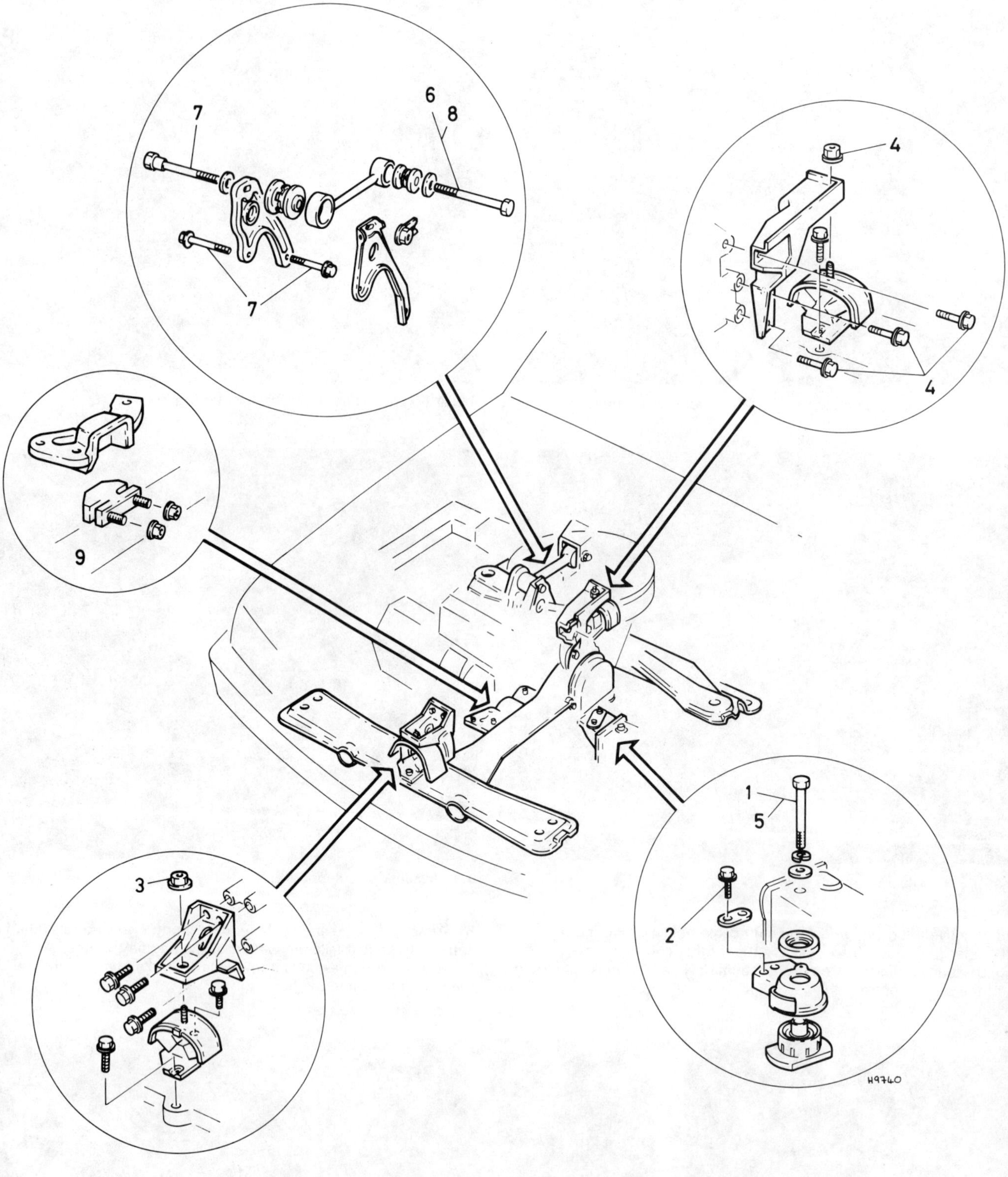

34

Fig. 1.2 Engine mounting tightening sequence (Sec 4)

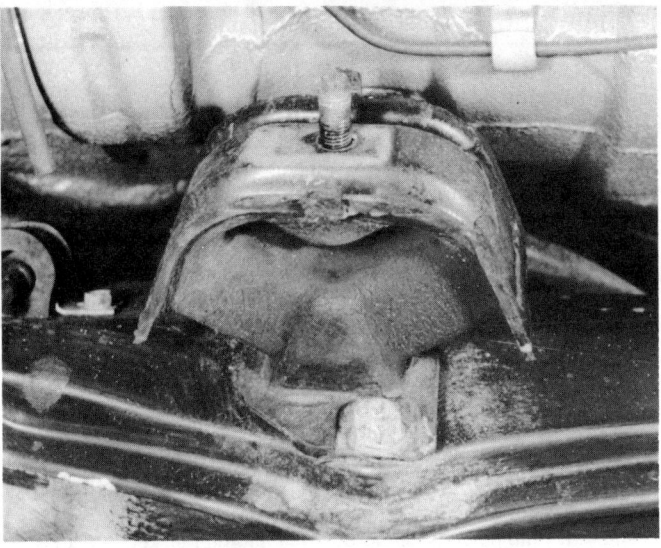
4.53A Rear engine mounting (engine removed)

4.53B Front engine mounting (engine removed)

4.55A Alternator multi-block connector ...

4.55B ... and terminal bolt

4.56A Engine rear torque rod bolt (arrowed). Disconnect from engine ...

4.56B ... and swing it up out of the way

Chapter 1 Engine

4.58 Lifting the engine and transmission unit

53 Remove the front engine mounting nut (photo), and the rear engine mounting nut (photo).
54 Now remove the rear engine mount bracket.
55 Disconnect the alternator electrical leads (photos).
56 Remove the engine rear torque rod from the engine, loosen the frame side bolt, and swing the torque rod up out of the way (photos).
57 Check that all engine systems have been drained, and all disconnections made.
58 Slowly begin to raise the engine, which needs to come out tilted toward the transmission end (photo).
59 Check all round frequently to make sure that no part of the engine catches under brake pipes or the radiator etc.
60 Once clear of the car, swing the engine away and lower it to the ground, supporting it on blocks of wood.
61 Refitting the engine is a reversal of removal.

5 Engine and transmission unit – separation

1 Remove the torque rod brackets from the housing (photo).
2 Remove the damper bracket from the transmission housing (photo).
3 Remove the bolts from the front engine mounting and remove the bracket.

4 Remove the flywheel/driveplate cover.
5 Remove the starter (photo).
6 Remove the remaining transmission housing bolts and then pull the housing clear of the locating dowels, being careful not to damage the input shaft.
7 On automatic transmission models, remove the eight bolts securing the torque converter to the driveplate before removing the transmission housing, and lift the converter from the housing when it is removed.

6 Ancillary components – removal

1 Before work on the engine can begin the ancillary components bolted to the engine must be removed.
2 Work from the top down, in logical order.
3 The carburettor may be removed from the intake manifold, or left in place and removed with the intake manifold.
4 Remove the power steering pump mounting bracket (photos).
5 Remove the exhaust manifold cover and the exhaust manifold (photos).
6 There are several cable clips and other brackets around the engine, remove them and the engine cable harness, taking note of where and how each bracket fits (photos).
7 Remove the spark plugs, distributor, and thermostat housing from the cylinder head.
8 Remove the cylinder block drain plug and drain the remaining coolant from the block (photo).
9 Remove the oil filter, using a strap wrench around the body of the filter.
10 Remove the bolt from the support bracket of the coolant transfer pipe (photo) and remove the pipe.
11 Remove the crankcase breather tank from the side of the crankcase.
12 The remaining components will be dealt with in the following Sections.

7 Engine dismantling – general

1 It is best to mount the engine on a dismantling stand, but if this is not available, stand the engine on a strong bench at a comfortable working height. Failing this, it will have to be stripped down on the floor.
2 During the dismantling process, the greatest care should be taken to keep the exposed parts free from dirt. As an aid to achieving this thoroughly clean down the outside of the engine, first removing all traces of oil and congealed dirt.
3 A good grease solvent will make the job much easier, for, after the solvent has been applied and allowed to stand for a time, a vigorous jet of water will wash off the solvent and grease with it. If the dirt is thick

5.1 The torque rod brackets

5.2 Engine damper bracket (engine *in situ*)

5.5 Remove the starter

6.4A Remove the power steering pump upper bracket ...

6.4B ... and the lower

6.5A Remove the exhaust manifold cover ...

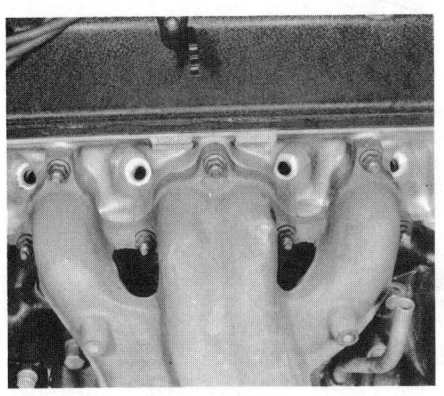

6.5B ... and exhaust manifold

6.6A Engine harness bolted to the bracket near the PCV tank ...

6.6B ... and to the left of the exhaust cover

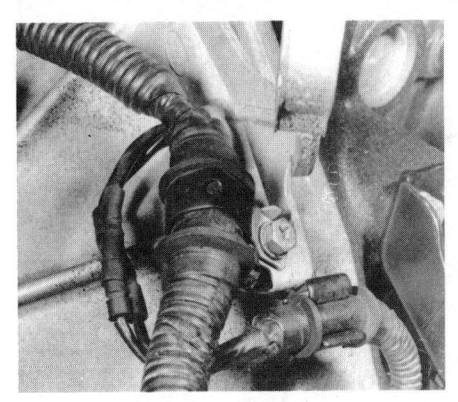

6.6C Cable harness clip on the clutch bellhousing ...

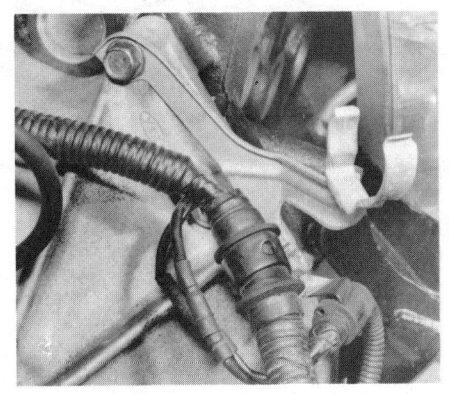

6.6D ... and the support bracket

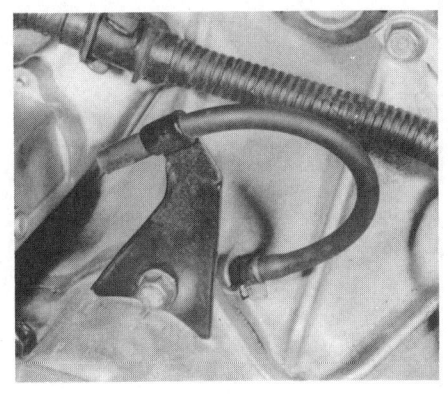

6.6E Transmission housing breather tube

6.8 Cylinder block coolant drain plug

6.10 Coolant transfer pipe support bracket

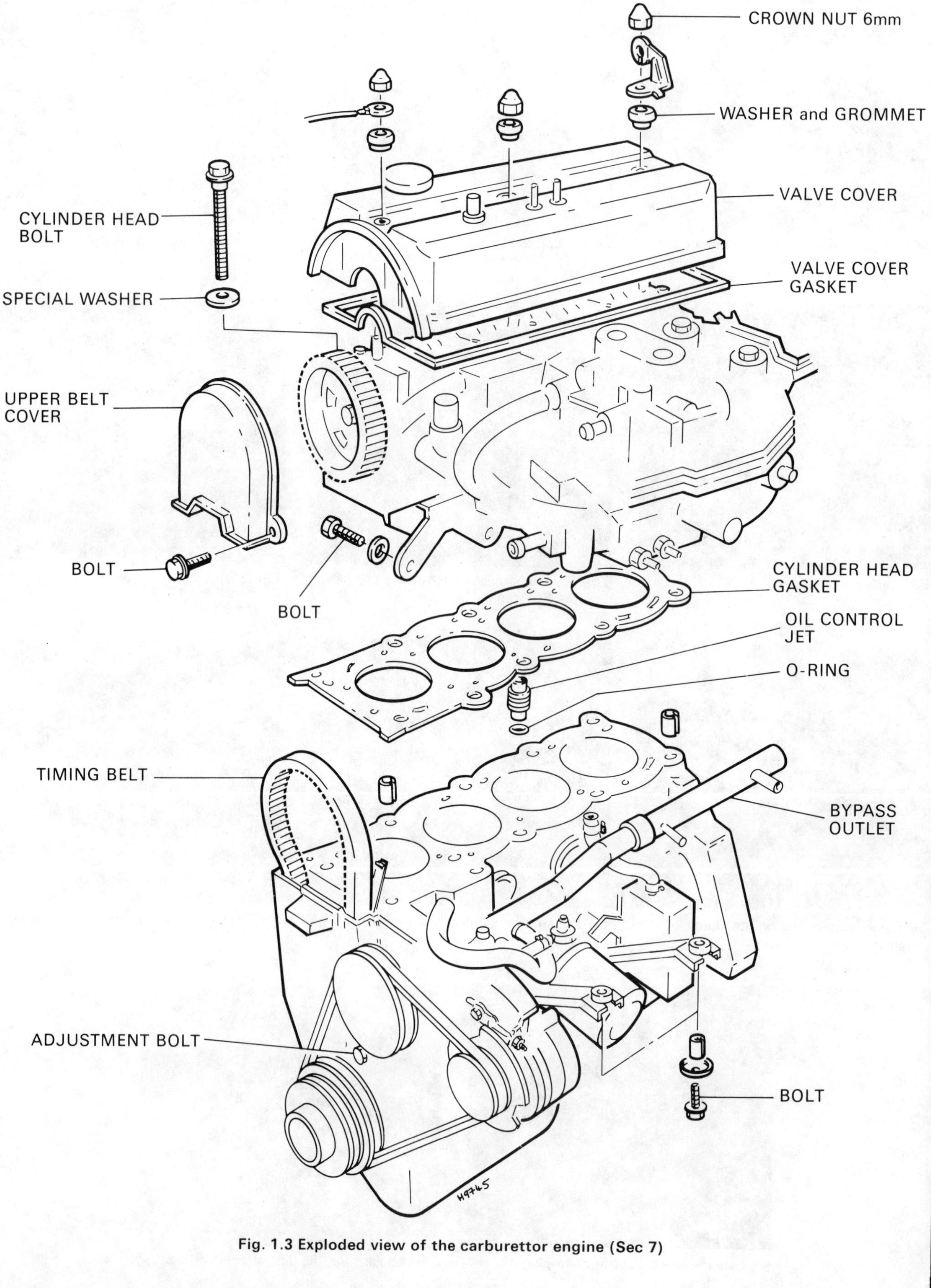

Fig. 1.3 Exploded view of the carburettor engine (Sec 7)

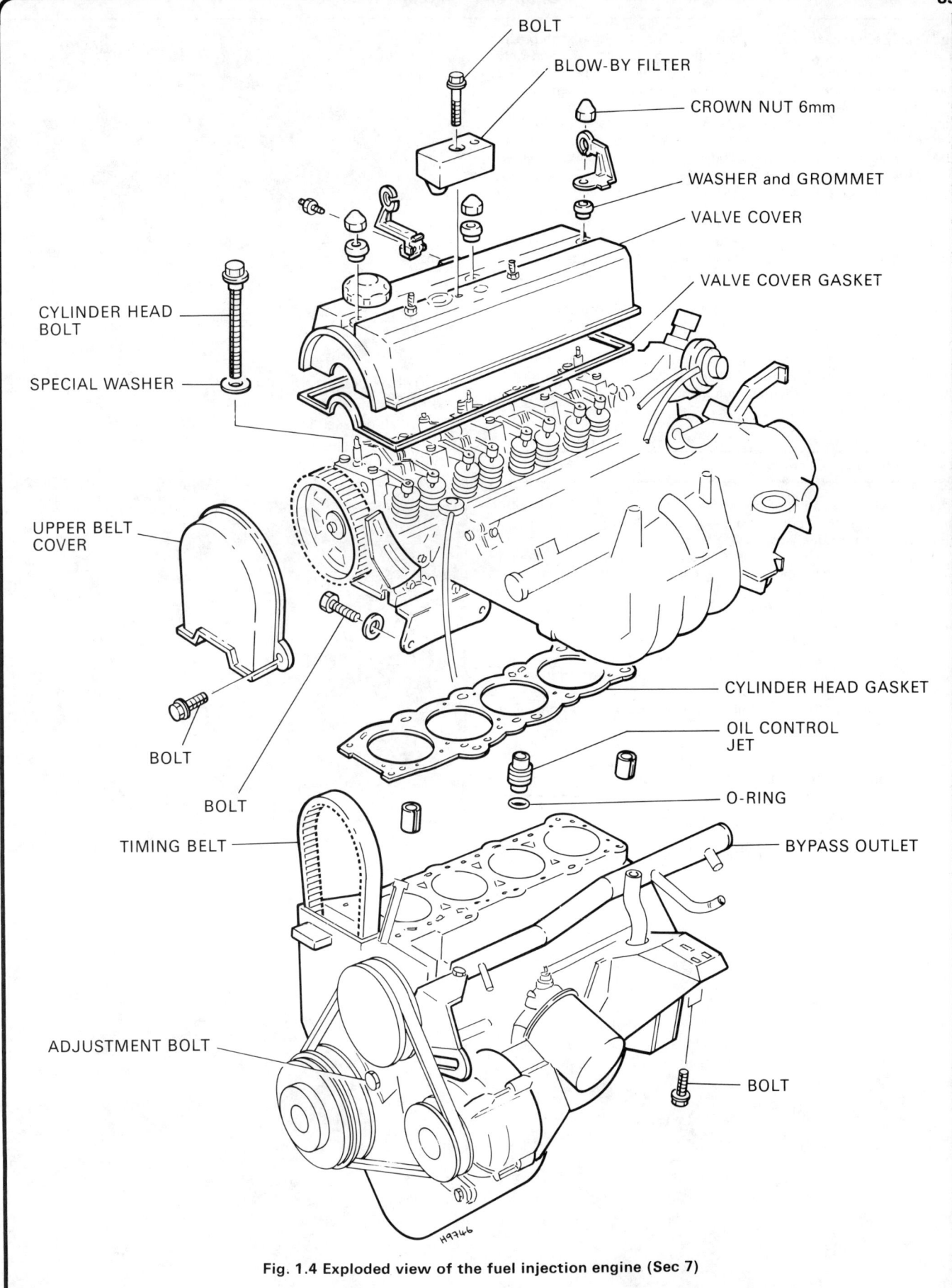

Fig. 1.4 Exploded view of the fuel injection engine (Sec 7)

Chapter 1 Engine

and deeply embedded, work the solvent into it with a strong stiff brush.
4   Finally, wipe down the exterior of the engine with a rag and only then, when it is quite clean, should the dismantling process begin. As the engine is stripped, clean each part with a suitable solvent.
5   Never immerse parts with oilways in paraffin (eg, crankshaft and camshaft). To clean these parts, wipe down carefully with a petrol-dampened rag. Oilways can be cleaned out with wire. If an air line is available, all parts can be blown dry and the oilways blown through as an added precaution.
6   Re-use of old gaskets is false economy. To avoid the possibility of trouble after the engine has been reassembled **always** use new gaskets throughout.
7   Do not throw away the old gaskets, for sometimes it happens that an immediate replacement cannot be found and the old gasket is then very useful as a template. Hang up the gaskets as they are removed.
8   To strip the engine, it is best to work from the top down. When the stage is reached where the crankshaft must be removed, the engine can be turned on its side and all other work carried out with it in this position.
9   Wherever possible, refit nuts, bolts and washers finger tight from wherever they were removed. This helps to avoid loss and muddle. If they cannot be refitted then arrange them in a fashion that it is clear from whence they came.

8.5 Camshaft drivebelt upper cover

8 Cylinder head – removal

1   If the cylinder head is being removed with the engine still *in situ*, then ensure that the coolant temperature is below 38°C (100°F) to prevent warping of the head.
2   Turn the engine so that the No 1 cylinder is on top dead centre on the compression stroke and the timing marks on the flywheel and the crankcase are in alignment.
3   Remove the valve cover (on Fuel injected versions remove also the blow-by filter mounted on the cover).
4   Take note of any brackets fitted under the valve cover nuts.
5   Remove the camshaft drive belt upper cover (photo).
6   Loosen the camshaft drive belt tensioner bolt (photo) and then slip the belt off the camshaft drive wheel.
7   Remove the cylinder head bolts in the reverse order of tightening (see Section 14) and then lift off the cylinder head.

9 Camshaft – removal and inspection

1   Remove the camshaft drive wheel, retaining the Woodruff key (photos).
2   Before removing the rocker arm assembly, check the endfloat of the camshaft using a dial test indicator.

8.6 Camshaft drivebelt tensioner bolt (arrowed)

9.1A Camshaft drive wheel bolt (arrowed) ...

9.1B ... retaining the Woodruff key (arrowed)

Chapter 1 Engine

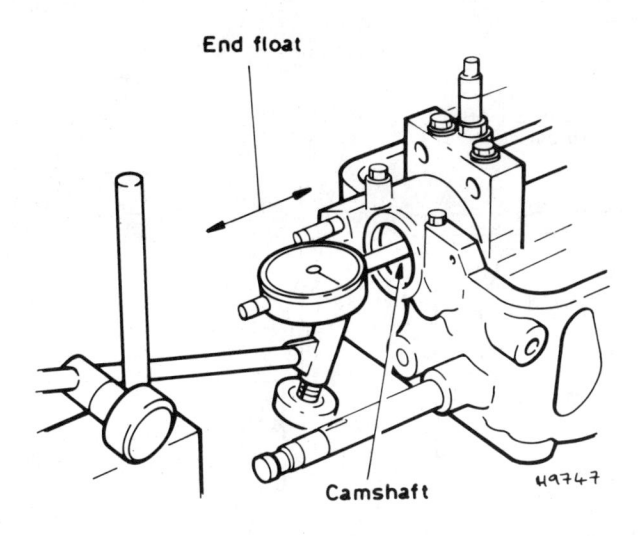

Fig. 1.5 Measuring camshaft endfloat (Sec 9)

9.3 Rocker pedestal bolts (1) and valve cover stud (2)

3 Remove the rocker pedestal bolts by unscrewing them two flats at a time; using a diagonal pattern, to even out valve spring pressure (photo). Remove the rocker arm assemblies.
4 Lift the camshaft from the cylinder head.
5 Clean the camshaft and then inspect the lobes.
6 Renew the camshaft if the lobes are worn, pitted or scored.
7 Clean the camshaft bearing surfaces in the cylinder head, then refit the camshaft.
8 Insert a Plastigage strip across each journal.
9 Refit the rocker arm assembly, and torque load the bolts as described in Section 14.
10 Remove the rocker assembly once more and measure the widest portion of Plastigage on each journal.
11 If the camshaft bearing tolerance (oil clearance) is out of limits (see Specifications) and the camshaft has already been renewed, the cylinder head must be renewed.
12 If the camshaft has not been removed, or this cannot be ascertained, check the total run-out of the camshaft, using V-blocks and a dial test indicator.
13 If the run-out check is within tolerance, then renew the cylinder head.
14 If out of tolerance, renew the camshaft and recheck the bearing clearance as before.
15 If still out of tolerance, renew the cylinder head.
16 Measure the lobes of the cam using a micrometer (photo). If the lobes are worn, renew the camshaft.
17 Refit the camshaft as described in Section 14.

10 Rocker shaft assembly – dismantling, inspection and reassembly

1 Keep all parts in their original positions, ensuring that they are replaced in the same position from which they came.

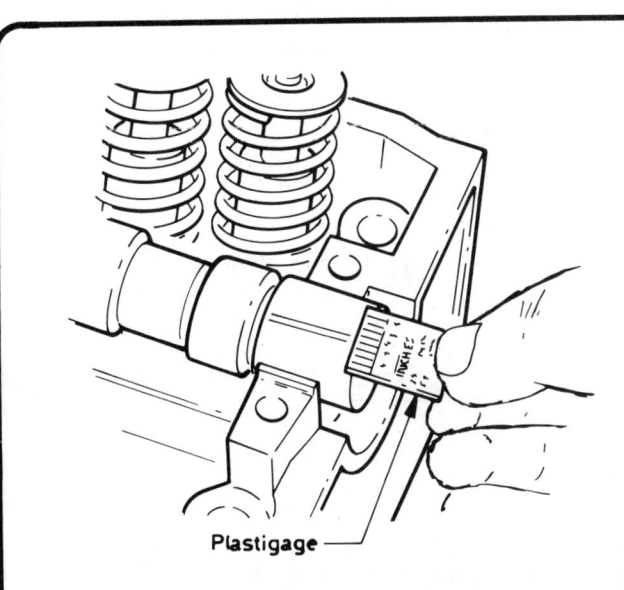

Fig. 1.6 Using a Plastigage strip (Sec 9)

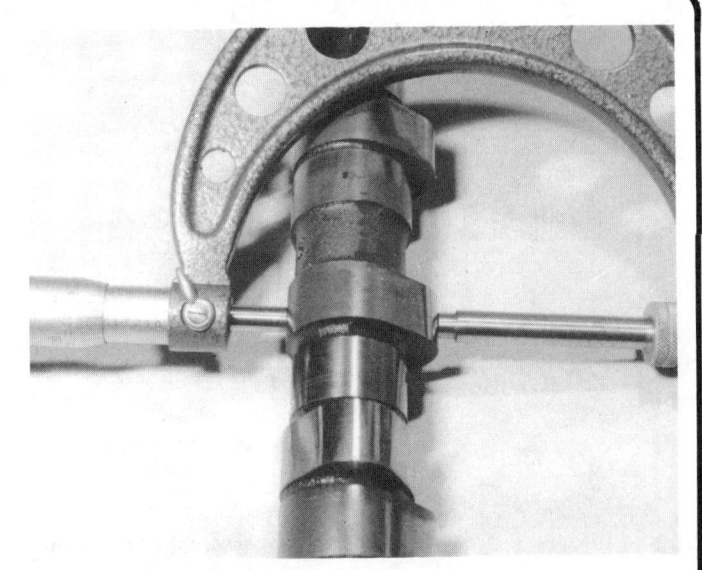

9.16 Measuring camshaft lobes

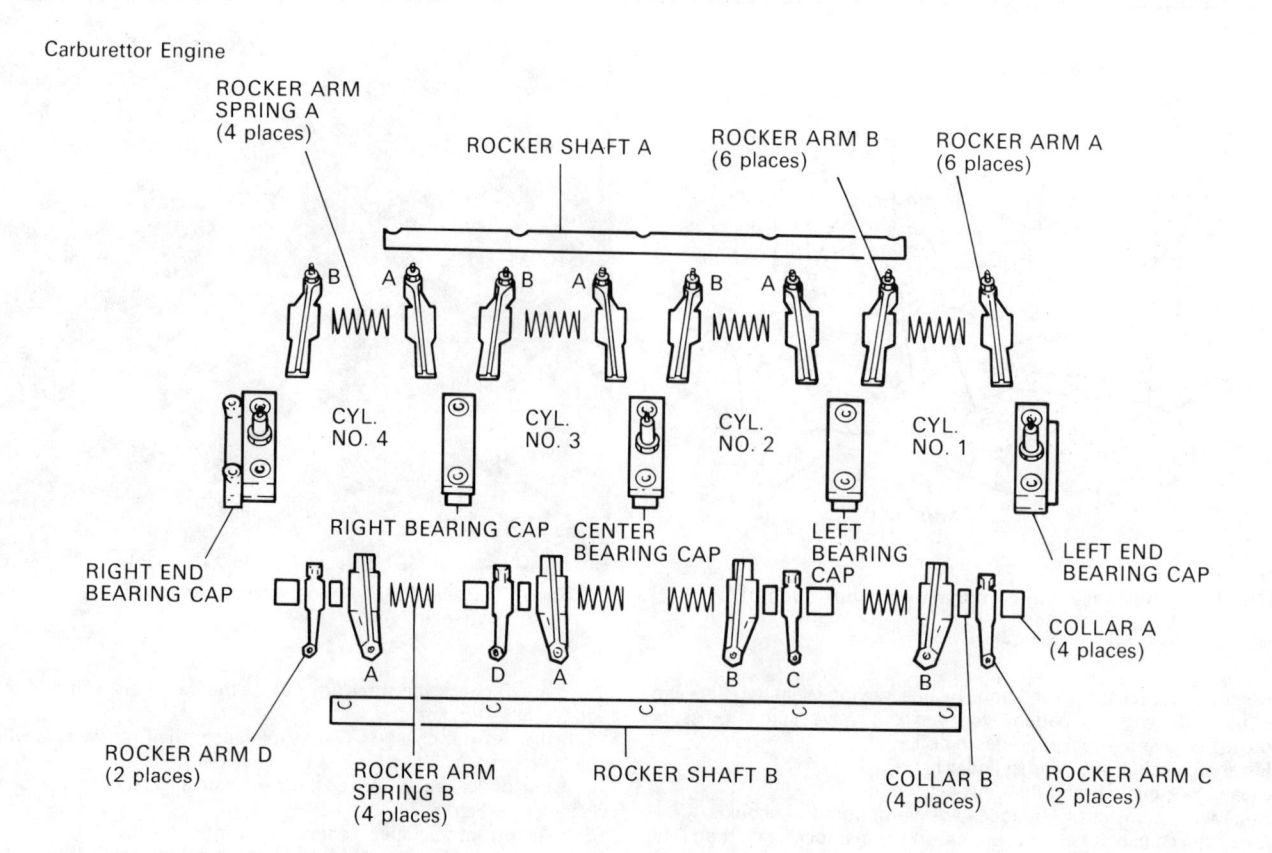

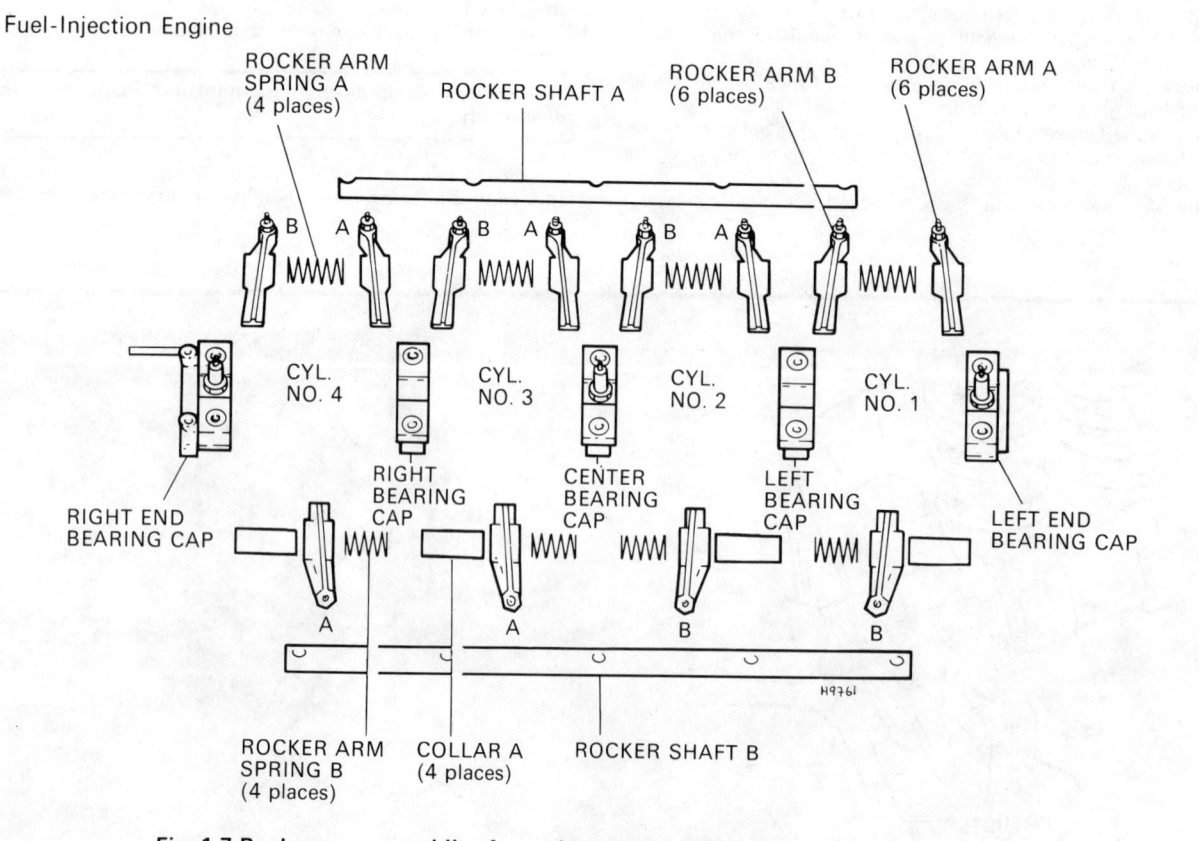

Fig. 1.7 Rocker arm assemblies for carburettor and fuel injection models (Sec 10)

Note: UK carburettor same as fuel injected model

Chapter 1 Engine 43

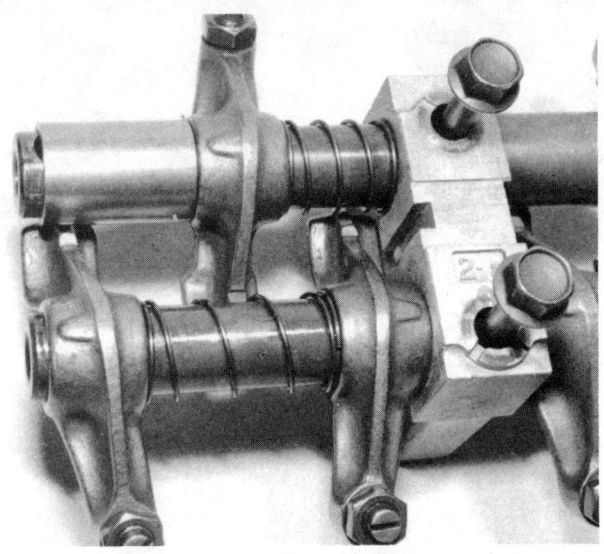

10.4A Rocker assemblies slide off the shaft after removal of the bolts

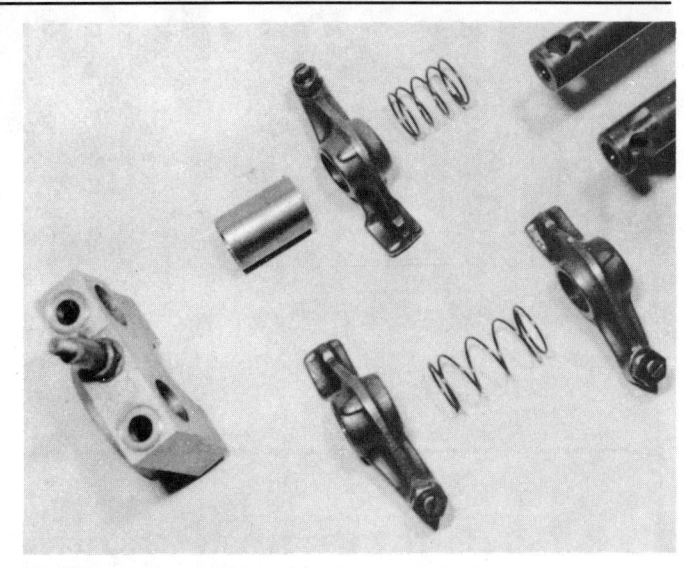

10.4B Keep all components in order

2 Remove the camshaft pedestal bolts using a diagonal pattern and unscrewing the bolts 1/3 of a turn at a time, until all tension from the valve springs has been relieved.
3 The rocker pedestals also serve as the camshaft bearing caps.
4 Lift the rocker shaft assembly from the cylinder head and then remove the bolts, which will allow the rockers and springs to be removed (photos).
5 Both the shafts and all rockers should be examined for wear as follows:
6 Measure the diameter of the shaft at first rocker location, using a micrometer gauge.
7 Use the gauge to zero an internal dial test indicator.
8 Now measure the inside bore of the rocker shaft at the first position.
9 Repeat for all rocker shaft positions and rockers.
10 Renew worn, out of tolerance shafts and rockers.
11 Inspect the face of the rocker where it contacts the valve stem, and renew any rocker which is badly worn or fitted.
12 Coat all parts liberally in clean engine oil before reassembling the rocker shafts, ensuring each component goes back in its original position.
13 Refit the rocker shaft assembly, as described in Section 14.

11 Intake and exhaust valves – removal, inspection and refitting

1 Tap each valve stem with a plastic mallet to loosen the valve keepers.

2 Use a valve spring compressor with a deep reach to compress the valve springs, and remove the valve keepers (photo).
3 Release the tension of the valve spring compressor and remove the compressor.
4 Remove the components of the valve spring assembly (photos).
5 Discard the valve stem seals, always fit new stem seals on reassembly. **Caution:** *Although intake and exhaust components may look the same, there are size differences. Do not mix components.*
6 The valve springs should be tested for length and force, but, as this requires a special gauge, if you suspect a valve spring of being weak, either renew it, or have your dealer test it for you.
7 Remove the valves (photo) and inspect each one as follows:
8 Firstly inspect the valves for obvious wear – burning, pitting, cracking and erosion.
9 The valve stem-to-valve guide clearance should be measured as follows:
10 Using an internal micrometer, measure the internal diameter of the valve guide.
11 Now measure the diameter of the valve stem using a micrometer.
12 Subtract the valve stem measurement from the valve guide measurement.
13 The difference between the largest measurement in the guide and the smallest measurement on the valve stem should not exceed the tolerances given in the Specifications. **Note:** *These measurements should be taken at three places along the valve stem and three places on the inside of the valve guide.*
14 If the specification is exceeded, renew the valve and check again.
15 If the specification is still exceeded, the valve guide should be renewed, as described in Section 13.

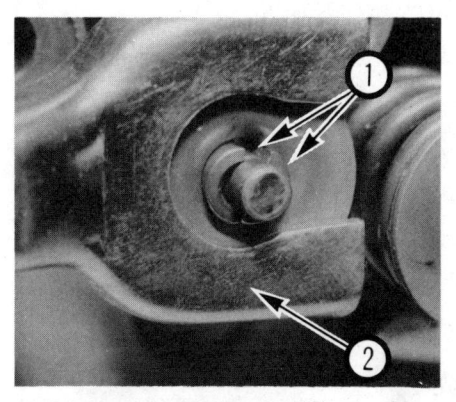

11.2 Valve spring compressor (2) releasing the valve keepers (1)

11.4A Lift off the spring retainer ...

11.4B ... springs ...

11.4C ... seal ...

11.4D ... and valve spring seat

11.7 Removing a valve

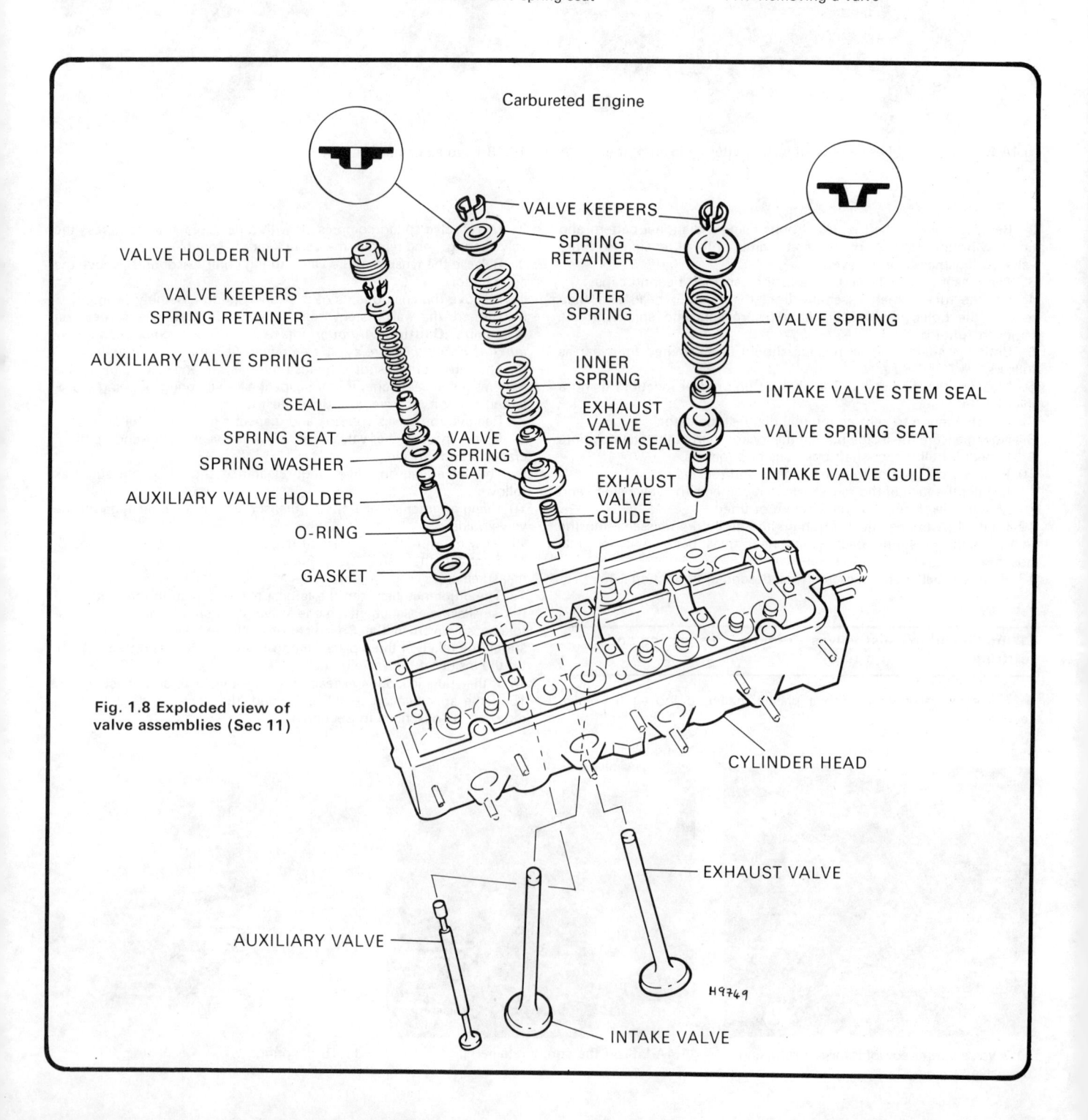

Fig. 1.8 Exploded view of valve assemblies (Sec 11)

Carbureted Engine

- VALVE KEEPERS
- SPRING RETAINER
- VALVE HOLDER NUT
- VALVE KEEPERS
- SPRING RETAINER
- AUXILIARY VALVE SPRING
- SEAL
- SPRING SEAT
- SPRING WASHER
- AUXILIARY VALVE HOLDER
- O-RING
- GASKET
- OUTER SPRING
- INNER SPRING
- VALVE SPRING SEAT
- EXHAUST VALVE STEM SEAL
- EXHAUST VALVE GUIDE
- VALVE SPRING
- INTAKE VALVE STEM SEAL
- VALVE SPRING SEAT
- INTAKE VALVE GUIDE
- CYLINDER HEAD
- EXHAUST VALVE
- AUXILIARY VALVE
- INTAKE VALVE

H9769

Chapter 1 Engine   45

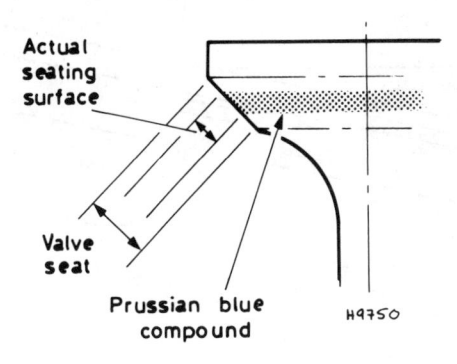

Fig. 1.9 Valve seating area (Sec 11)

16 Inspect the valve seating surface where it contacts the valve seats. The actual seating area can be determined by applying engineers blue to the valve face and inserting it in its valve guide then snapping it closed several times. The valve seat should be centered on the valve, as shown in Fig. 1.9.
17 Slight imperfections to this, and light pitting, can be removed by grinding in the valve seats. Heavier pitting will require the valves to be changed.
18 Valve grinding is carried out as follows. Smear a trace of coarse carborundum paste on the seat face and apply a suction grinding tool to the valve head. With a semi-rotary motion, grind the valve head to its seat, lifting the valve occasionally to redistribute the grinding paste. When a dull matt even surface is produced on both the valve seat and the valve, wipe off the paste and repeat the process with fine carborundum paste, lifting and turning the valve to redistribute the paste as before. A light spring placed under the valve head will greatly ease this operation. When a smooth unbroken ring of light grey matt finish is produced on both valve and valve seat faces, the grinding operation is complete. Carefully clean away every trace of grinding compound, take great care to leave none in the ports or in the valve guides. Clean the valves and valve seats with a solvent-soaked rag, then with a clean rag, and finally, if an air line is available, blow the valves, valve guides and valve ports clean.
19 Valve seats which are worn can be re-cut, but this is best left to your dealer or local engine repair specialist, as a wrong cut will mean renewing the cylinder head.
20 The valve stem guide procedure is dealt with under cylinder head inspection.
21 Once all parts have been inspected, washed free of grinding paste and renewed, as necessary, the valves can be replaced.
22 Oil all components with clean engine oil.
23 Insert the valve into the valve guide.
24 Refit the valve spring assembly with reference to Fig. 1.8. Note the differences between the valve guide seals and the fitting of the springs with the most closely wound coils toward the cylinder head.
25 Compress the springs and insert the valve keepers, then remove the valve compressor.
26 On completion, tap each stem with a plastic mallet to seat the keepers.

12 Auxiliary valve – removal, inspection and refitting

1 The auxiliary valve fitted to some carburettor engines in North America can be removed by using a long box spanner to undo the nut which holds it to the cylinder head.
2 The valve can be dismantled for renewal of the seals by using a valve spring compressor to compress the spring and then removing the valve keepers.
3 If the valve is suspected of being worn, it should be renewed as an assembly.
4 Always renew the O-ring seal and gasket when refitting an auxiliary valve.

5 Note that these valves are a complete unit in themselves, the valve seat being integral in the valve holder, and the assembly being held in the cylinder head by the valve holder nut.

13 Cylinder head – inspection

1 Remove the camshaft and valves, as described in Sections 9, 10 and 11.
2 Clean the cylinder head free from carbon deposits in the valve pockets. Do not use metal scrapers on the soft aluminium of the head.
3 Inspect the valve seats for wear, pitting, burning, cracks and scoring.
4 Refer to Section 11 on valve regrinding and check the valve seat, along with the valve. As explained before, the valve seats can be recut, but this is best done by an engine repair specialist.
5 If the valve guides have been identified as being worn, they can be renewed as follows:
6 Heat the cylinder head to 150°C (300°F).
7 Use heavy gloves when handling the hot cylinder head, and, using a suitable mandrel, drive out the old valve guides from the combustion chamber up.
8 Drive in a new valve guide. Note: this must be done using the special tool provided by Honda, as it sets the valve guide at the correct depth in the cylinder head. A special tool is also available for removing valve guides.
9 After fitting new guides, they must be reamed out again using a Honda special tool, which is of the correct size, in the following manner.
10 Coat the reamer and valve guide with cutting oil.
11 Rotate the reamer, clockwise only, the full length of the valve guide bore.
12 Continue turning the reamer clockwise whilst removing it.
13 Thoroughly wash the valve guide in detergent and water, removing every trace of cutting oil.
14 Check the valve stem-to-guide clearances, as described in Section 11.

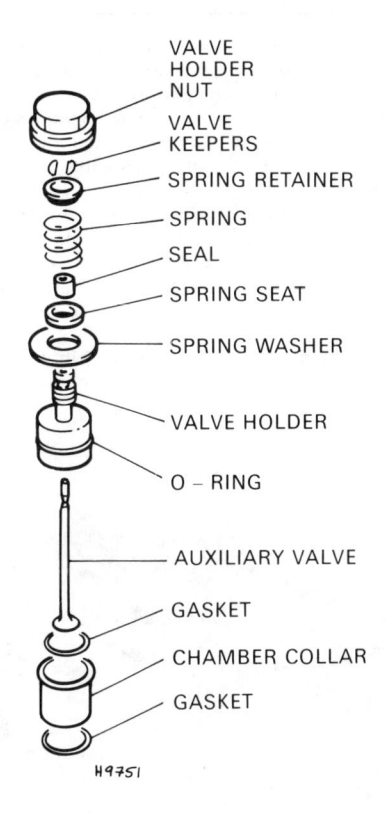

Fig. 1.10 View of the auxiliary valve (Sec 12)

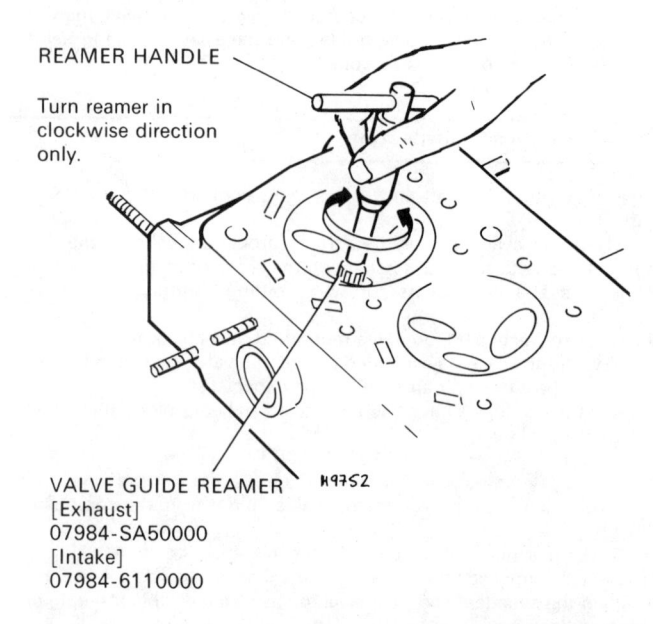

Fig. 1.11 Reaming the valve guides (Sec 13)

15 Refer to Section 9 on camshaft bearing clearances and if these are satisfactory, check the cylinder head for warpage.
16 Take the measurements across all four edges, along the centre line and from corner to corner.
17 Provided the warpage does not exceed the maximum given in the Specifications, the cylinder head may be resurfaced; a job best left to your engine specialist.

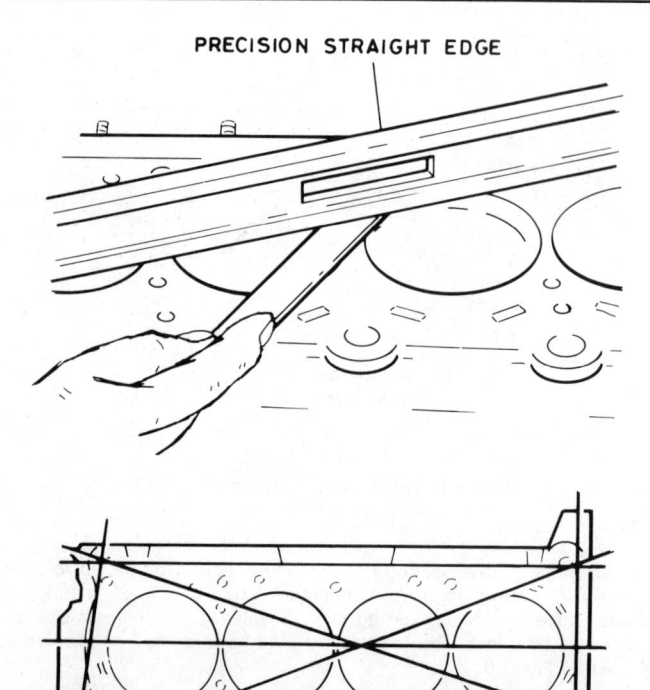

Fig. 1.12 Measuring cylinder head warpage (Sec 13)

14 Cylinder head – reassembly and refitting

1 Refit the valves to the cylinder head, as described in Section 11.
2 Oil the camshaft and camshaft bearing surfaces and set the camshaft in the cylinder head (photo).
3 Turn the camshaft until its keyway is facing upwards.
4 Fit the camshaft oil seal, spring side facing inwards (photo).
5 Refit the rocker arm and shaft assembly and loosely install the retaining bolts (photo).
6 Use a socket of suitable diameter to drive home the camshaft oil seal. **Note:** The seal housing surface should be dry and a light coat of oil applied to the inner lip of the seal and the camshaft.
7 Tighten the rocker shaft bolts to their correct torque figure given in the Specifications, two flats at a time and in the sequence shown in Fig. 1.13.
8 Install the Woodruff key into the shaft (photo).
9 Fit the camshaft drive wheel, checking that the alignment hole is at the top, and the timing marks on each side are parallel to the cylinder head (photos).
10 Tighten the camshaft drive wheel bolt to its specified torque.
11 Fit the oil jet guide to its orifice in the cylinder block (photo).
12 Fit the locating dowels (photo) and place a new cylinder head gasket onto the cylinder block.
13 Back off the valve adjusters on the rocker arms fully, and check that the timing marks are still in alignment.
14 Carefully lift the cylinder head into position on the cylinder block (photo) ensuring the gasket is not disturbed, and that the cylinder head fits over the locating dowels.
15 Fit the cylinder head bolts just more than finger tight (photo) (final tightening is done after installing ancillaries).
16 Fit the inlet and exhaust manifolds, using new gaskets (photo), and tighten the nuts in a diagonal pattern.
17 Adjust the valve clearances (see Section 16).
18 Tighten the cylinder head bolts in two stages as follows:
19 Stage 1 – using the sequence shown in Fig. 1.14 tighten the bolts to the Stage 1 specified torque.

14.2 Fitting the camshaft

14.4 Camshaft oil seal being fitted

14.5 Refitting the rocker assembly

14.8 Fit the Woodruff key and drive wheel

14.9A Drive wheel with the alignment hole uppermost ...

14.9B ... and index marks lined up (arrow)

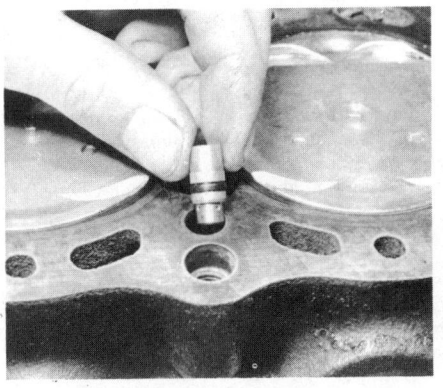

14.11 Fitting the oil control jet

14.12 Locating dowels and cylinder head gasket on the cylinder head

14.14 Fitting the cylinder head

14.15 Just nip the cylinder head bolts

14.16 Use new gaskets under the manifolds

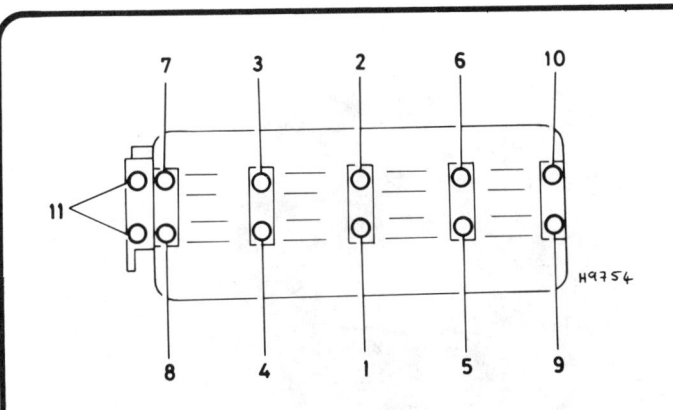

Fig. 1.13 Rocker pedestal bolt tightening sequence (Sec 14)

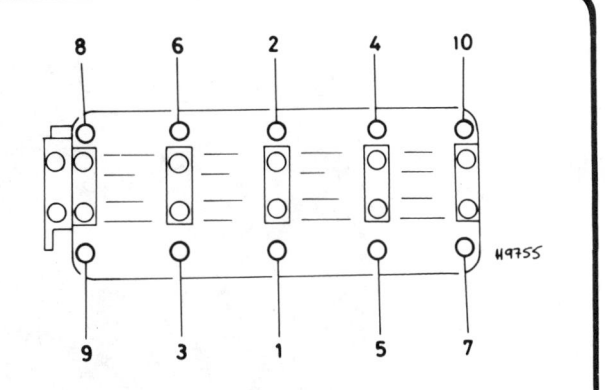

Fig. 1.14 Cylinder head bolt tightening sequence (Sec 14)

48   Chapter 1 Engine

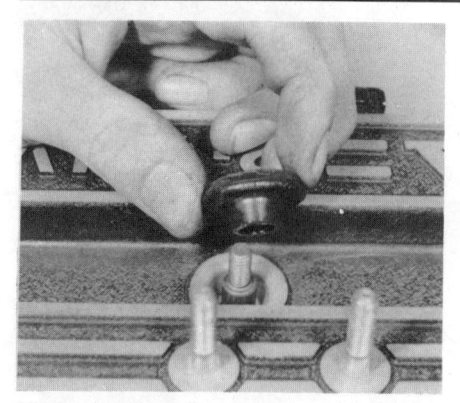

14.22A Fit the sealing ring ...

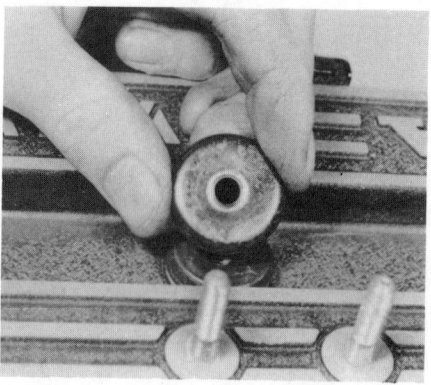

14.22B ... steel washer ...

14.22C ... and tighten the nut on the valve cover

20 Stage 2 – using the same sequence, tighten the bolts to their final torque figure.
21 Refit the timing belt and covers (Section 15).
22 Refit the valve cover using a new gasket and renew the oil seals if they show signs of perishing, not forgetting the brackets which fit under the nuts (photos).
23 The engine is now ready for installation as described in Section 27, or if the head was removed *in situ*, for the final connection.

15 Camshaft drivebelt – removal, refitting and adjusting

Note: *The procedure given here assumes that the engine is in the process of being stripped for overhaul. The timing belt could be changed in situ, after removal of the alternator and air conditioning/power steering sumps and drivebelts, and the left-hand engine mount, with the engine supported on a jack from below.*

1 Remove the crankshaft pulley bolt and remove the pulley, which is keyed to the crankshaft (photos).
2 Remove the water pump pulley wheel (photo).
3 Remove the camshaft belt lower cover.
4 Slacken off the camshaft belt tensioner bolt and slip the belt off the pulleys, marking its direction of rotation (photo).
5 Remove the crankshaft sprocket (photos).
6 Remove the camshaft belt adjuster bolt and remove the adjuster and spring (photo).
7 If the adjuster bearing appears worn, renew as a complete unit (photo).
8 Similarly, if the tensioner spring has stretched, renew it.

15.1A Undoing the crankshaft pulley bolt ...

15.1B ... removing it ...

15.1C ... to reveal the key

15.2 Water pump pulley retaining bolts (arrowed)

15.4 Camshaft drivebelt rotation

Chapter 1 Engine

15.5A Crankshaft sprocket outer plate ...

15.5B ... sprocket ...

15.5C ... and inner plate

15.6 Remove the bolt from the adjuster

15.7 The adjuster and spring unit

9 Inspect the timing belt for wear, especially on the thrust side of the teeth, and cracks, splits, fraying and oil contamination. Renew any belt showing signs of deterioration.
10 Refit the timing belt in the reverse order of removal, ensuring it goes back in the same direction of rotation as it was removed.
11 Before fitting the belt, ensure the UP hole on the camshaft drive wheel is at the top and its two index marks are in line with the cylinder head.
12 Turn the crankshaft pulley so that the timing marks on the flywheel and pointer on the crankcase are in alignment.
13 Fit the timing belt, and temporarily fit the crankshaft drive pulley.
14 Ensure the timing belt adjuster is slack.
15 Turn the engine anti-clockwise a distance of three teeth on the camshaft pulley, to create tension on the timing belt.
16 Tighten the adjuster bolt.
17 Remove the flywheel pulley.
18 Refit the covers, as described in final assembly.
19 If the timing belt has been renewed *in situ*, refit the left-hand engine mount, the crankshaft pulley, power steering, air conditioning and alternator and water pump pulleys and drivebelts, and the valve cover and copper timing belt cover.

16 Valve clearances – adjustment

1 The valves should be adjusted cold, with the cylinder head temperature less than 38°C (100°F).
2 Adjustment for intake and exhaust valves is the same.
3 Remove the valve cover. Set No 1 piston at TDC, the UP hole on the camshaft drive wheel should be at the top, and the index marks aligned with the cylinder head. The distributor rotor should be pointing to No 1 cylinder HT lead.
4 Adjust the valves for No 1 cylinder as follows:
5 Loosen the adjuster nut and back off the adjuster.
6 Insert a feeler gauge of correct thickness (corresponding to the clearance gap) between the valve stem and the adjuster (photo).
7 Tighten the adjuster until it just begins to pinch the feeler blade.
8 Tighten the locknut while holding the adjuster with a screwdriver. Recheck the clearance on completion.
9 Rotate the crankshaft through 180° (the camshaft wheel will turn 90°), and repeat for No 3 cylinder, then turn a further 180° for No 4 and again for No 2.
10 On completion, refit the valve cover.
11 **Note:** Auxiliary valves are adjusted in the same way as the main valves.

16.6 Adjusting the valve clearances

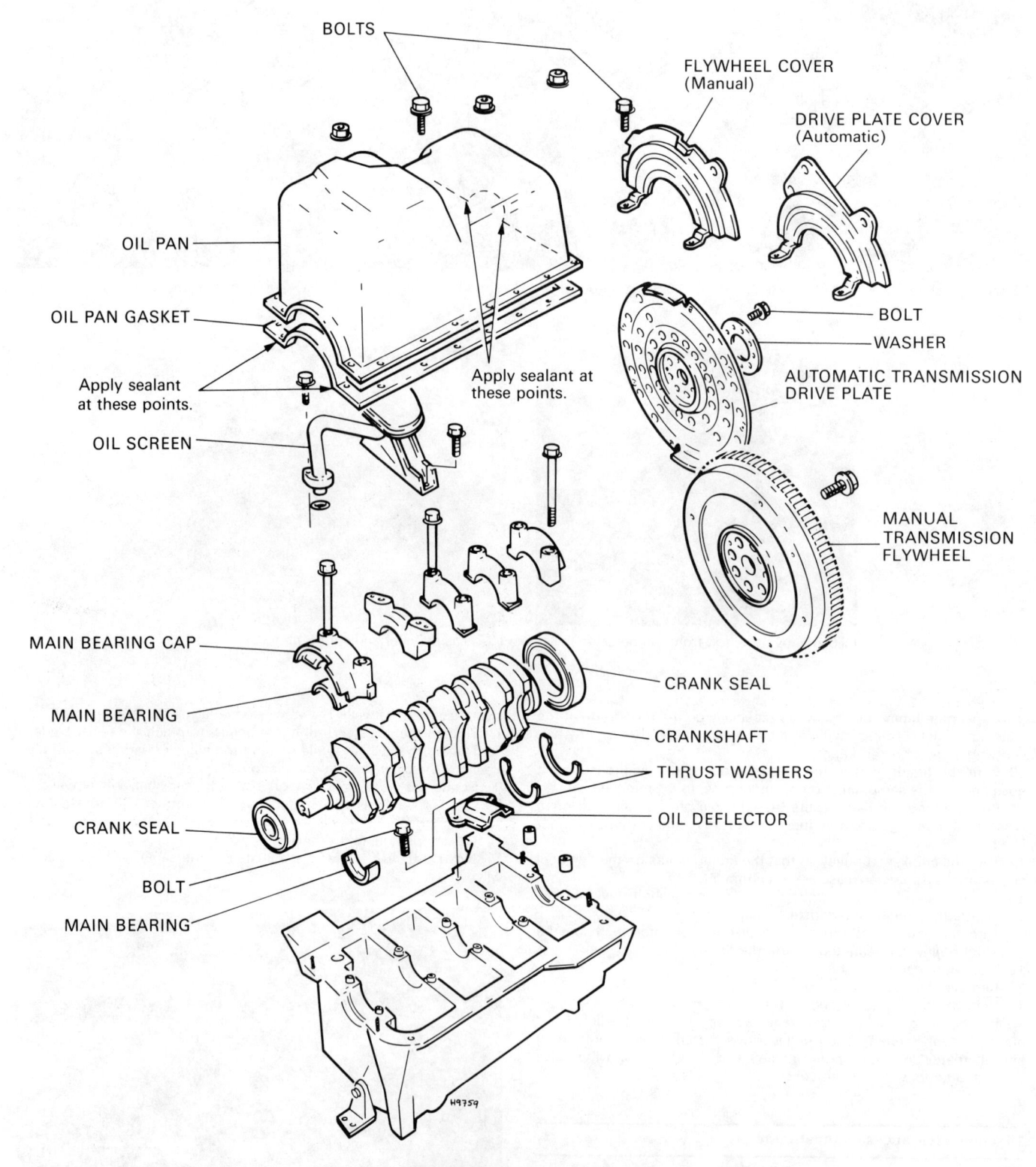

Fig. 1.15 Exploded view of the cylinder block (Sec 17)

17 Cylinder block – dismantling

1 It is assumed that the cylinder head and associated ancillary components have been removed.
2 Place the engine block upside down on the bench, supported on blocks of wood.
3 Remove the oil filter and dipstick.
4 Remove the oil pump.
5 Remove the water pump.
6 Remove the clutch from the flywheel (Chapter 5).
7 Remove the flywheel or driveplate covers.
8 Unbolt and remove the flywheel or driveplate from the crankshaft.
9 Remove the bolts from the oil sump and remove the sump.

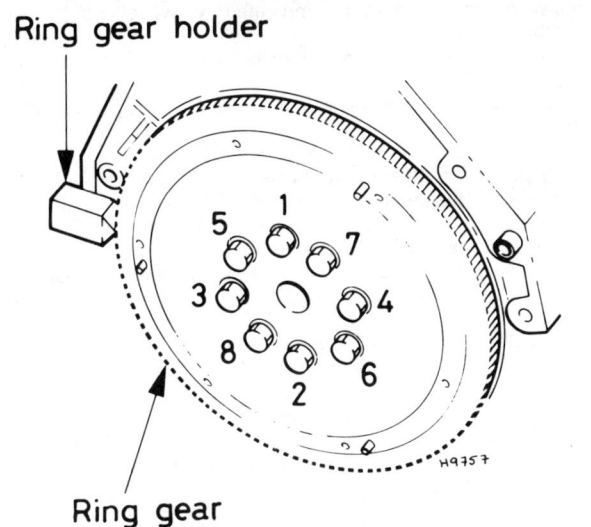

Chapter 1 Engine 51

Fig. 1.16 Flywheel bolt tightening sequence (Sec 17)

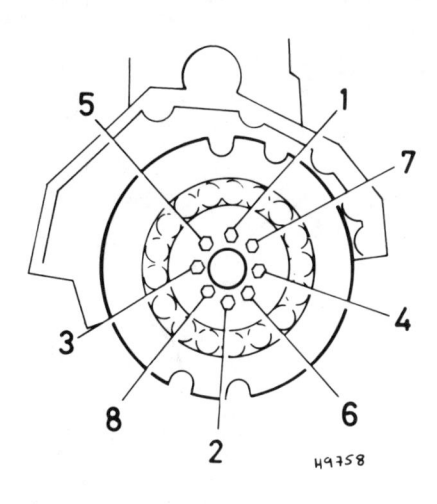

Fig. 1.17 Torque converter bolt tightening sequence (Sec 17)

10 Remove the three bolts securing the oil screen in position, and remove the oil suction pipe (photo).
11 Before proceeding any further, measure the crankshaft endfloat to establish any wear in the crankshaft or thrust washer.
12 Also measure the endplay of the connecting rods.
13 Procedures for both these operations are described in the relevant Sections.
14 The crankshaft, con rods and pistons may now be removed for inspection, as described in the following Sections.

18 Oil pump – removal, inspection and refitting

1 The oil pump is driven by the camshaft drivebelt and is bolted to the cylinder block.

17.10 Three bolts secure the oil suction pipe (arrows)

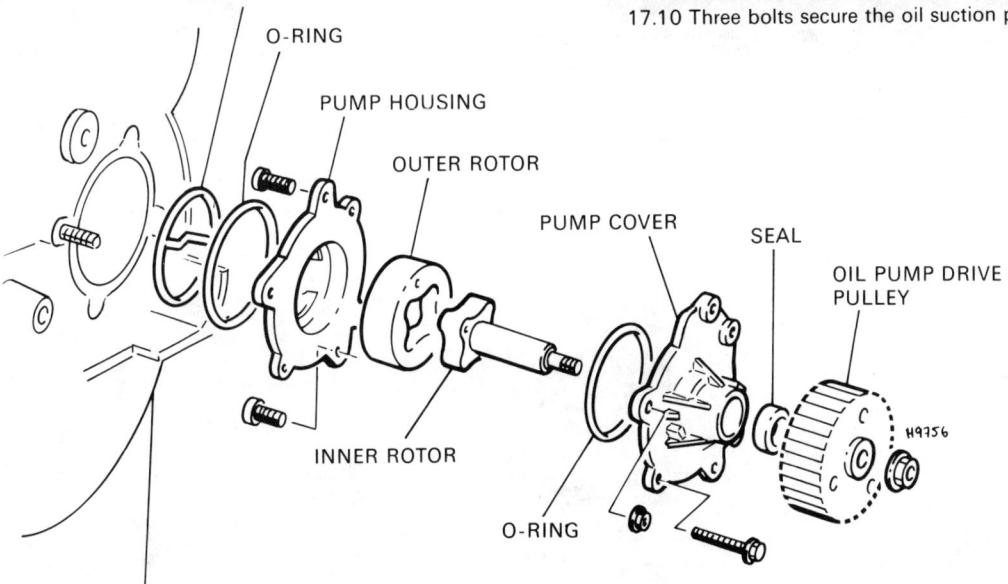

Fig. 1.18 Earlier type oil pump components (Sec 18)

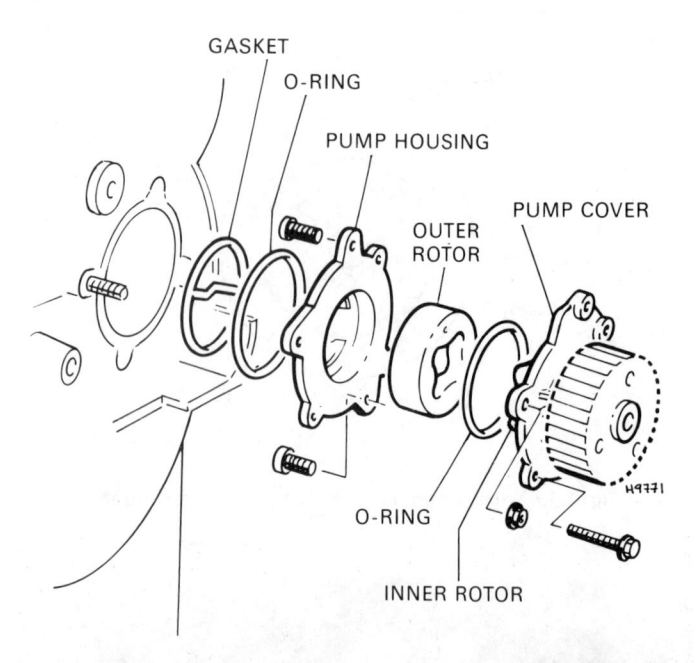

Fig. 1.19 Later type oil pump components (Sec 18)

integral with it, and on earlier pumps the drive wheel is bolted to the inner rotor shaft.
5 In either case, if any parts are worn, replace the pump as a complete unit.
6 Remove the pump housing retaining screws and lift off the housing (photo).
7 Lift out the outer rotor (photo).
8 Wash all parts in paraffin and blow dry.
9 Reassemble the pump and measure the housing-to-rotor radial clearance (photo), rotor tip clearance (photo) and the pump axial clearance between the pump housing and outer rotor.
10 If any clearances are beyond that specified, renew the pump.
11 Apply thread locking compound to the pump housing screws on final assembly.
12 Renew the O-ring seals between the housing and cover (photos), and the gasket between the pump and cylinder block; applying sealant to the gasket.
13 Refit the pump to the cylinder block, which is a reversal of removing. **Caution**: if you remove the drive sprocket from the inner rotor shaft on earlier type pumps, it has a left-hand thread. Later types cannot be dismantled.

19 Oil filter and pressure relief valve – removal and refitting

1 The oil filter will have been removed during the engine dismantling process.
2 Remove the bolts from the filter base/oil pressure relief valve housing and remove the base from the cylinder block (photos).
3 Remove the oil pressure transmitter (photo).
4 Use an Allen key to remove the relief valve, being careful of the spring pressure behind it (photo).
5 Wash all parts in paraffin and blow dry.
6 Renew the pressure transmitter and relief valve assembly if they are suspected of malfunction.

2 Remove the bolts and lift the pump away (photo).
3 There are two types of pump which may be fitted, the difference being in the inner rotor.
4 On later pumps, the inner rotor is keyed to the drive wheel, being

18.2 Removing the oil pump retaining bolts

18.6 Remove the screws and lift off the housing

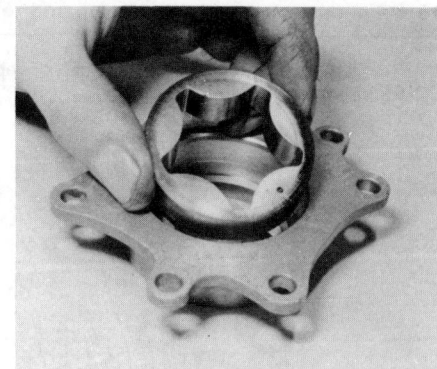

18.7 Lift out the outer rotor

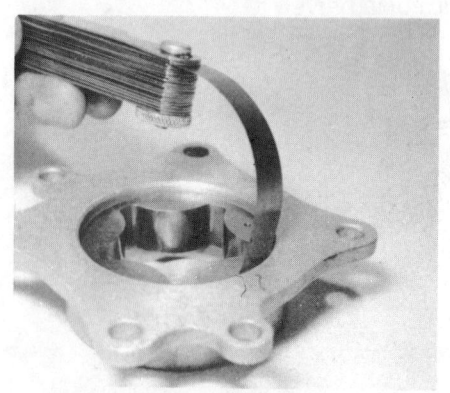

18.9A Housing-to-rotor radial clearance

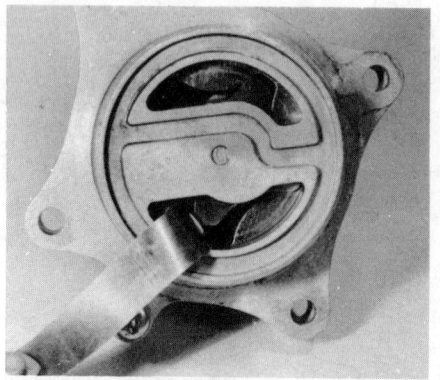

18.9B Rotor tip clearance

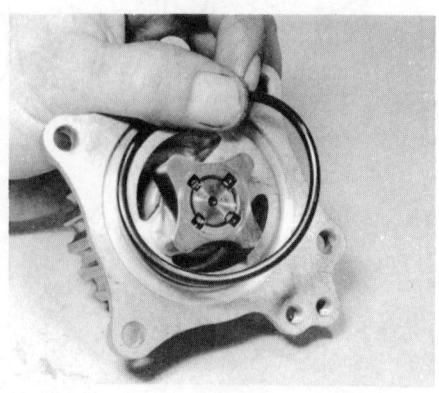

18.12A Fit a new seal ...

Chapter 1 Engine

18.12B ... and gasket

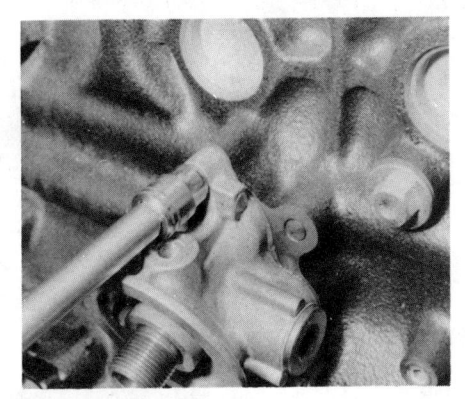

19.2 Removing the bolts from the filter base

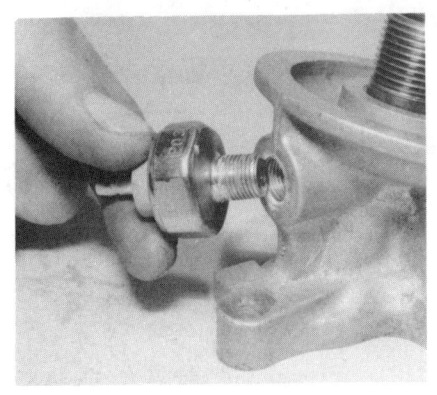

19.3 Removing the oil pressure transmitter

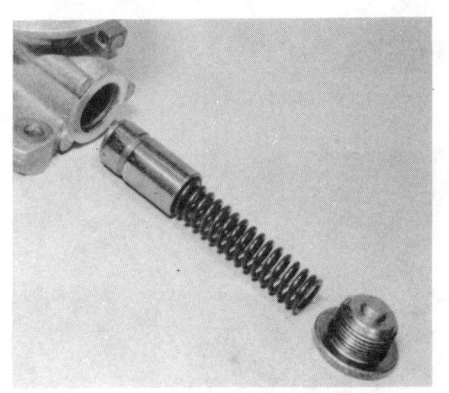

19.4 Components of the relief valve

19.7 Fit a new gasket on reassembly

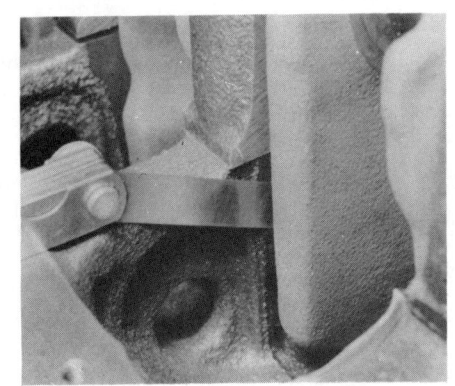

20.2 Measuring crankshaft endfloat with feeler gauges

7 Fit a new gasket to the filter base and refit it to the cylinder block, which is a reversal of removal (photo).
8 When renewing an oil filter in a service, it can be removed with a strap or chain wrench, or by thrusting it through with a screwdriver and using the screwdriver to turn the filter if it is stuck fast. Apply grease to the seal of a new filter and screw it on just more than hand tight. Start the engine and inspect for leaks.

20 Crankshaft – removal and inspection

1 Before removing the crankshaft, measure the endfloat to determine any wear.
2 Use either a dial test indicator erected on the crankshaft pulley end of the crankshaft, or use feeler gauges, inserted between the thrust washers and centre bearing web (photo).
3 Lever the crankshaft to the flywheel end, zero the dial indicator, or take feeler gauge measurement, then lever the crankshaft toward the pulley end and read the indicator.
4 If the tolerance is exceeded, renew the thrust washers.
5 The thrust washers must not be ground down or shimmed. The thickness is fixed, and if the endfloat cannot be brought into tolerance, then the crankshaft must be renewed.
6 Measure the connecting rod endplay with a feeler gauge inserted between the con rod cap and the crankshaft web.
7 If out of tolerance, renew the con rod.
8 If still out of tolerance, renew the crankshaft.
9 Remove the main bearing caps and bearing halves.
10 Each cap is numbered and arrowed, to indicate its position in the block (photo).
11 Clean each main bearing journal and bearing half, then insert a strip of Plastigage across each main journal.
12 Refit the main bearings and caps and torque load the nuts to their specified torque.

13 If this operation is beng done with the engine *in situ*, then support the crankshaft on a jack, or the added weight of the crankshaft on the bearings will give a false reading.
14 Remove the caps and bearings and measure the Plastigage.
15 Compare the reading with the tolerances in the Specifications.
16 Bring the bearing oil clearance into limits by fitting new bearing halves (of the same colour code).
17 If this cannot be achieved with the same colour code, try the next larger or smaller bearings.

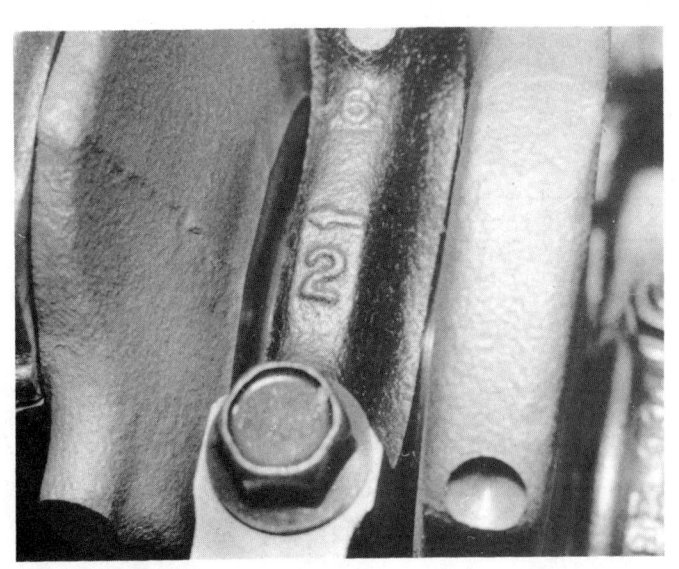

20.10 Bearing cap identification marks

Chapter 1 Engine

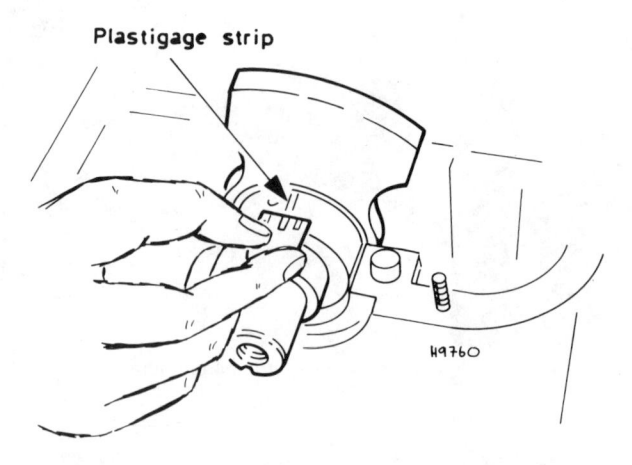

Fig. 1.20 Using Plastigage strip for bearing clearance (Sec 20)

18 If clearances are still out of tolerance, renew the crankshaft.
19 This same routine must be applied to the four connecting rod bearings, and again, if the tolerances cannot be met, the crankshaft must be renewed.
20 Having established wear limits, remove the con rod caps and sealing halves, and the main bearing caps and halves.
21 Remove the crankshaft from the engine. **Note:** To remove the crankshaft the engine must be removed from the vehicle. Before removing the pistons, as described in the next paragraphs, feel around the top of each cylinder for a ridge of hardened carbon. If this ridge is present, it must be removed with a ridge reamer or careful use of fine emery cloth before the pistons are pushed out.
22 Turn the engine on to its side, and using a hammer handle or piece of wood, push each piston up out of the cylinders.
23 Support the crankshaft in a lathe or on V-blocks and, using a dial test indicator, measure the run-out on each main bearing journal.
24 The difference between each measurement must not be more than that specified.
25 Measure the main bearing journals for out-of-round at the middle of each journal in two opposite places.
26 Measure each main bearing journal for taper at the edges of each journal, again the maximum difference must not exceed the service limit.
27 Repeat the main bearing journal measurements on the connecting rod journals.
28 The same tolerances apply equally to the con rod journals as to the main bearing journals.
29 Renew any crankshaft beyond service limits.

21 Bearing selection

1 The bearing code is stamped on the engine web at the flywheel end of the engine (photo).
2 The number nearest the top is for the No 5 journal and the bottom one for No 1 (crank pulley end).
3 These numbers indicate the size of the main journal bores.
4 Stamped on the crankshaft counter weights are another set of letters and numbers (photo).
5 The numbers refer to the crankshaft main bearing size.
6 The letters refer to connecting rod big-end bearing size.
7 The con rod number is stamped across the cap and con rod (photo).
8 Use these letters and the tables in Figs. 1.21 and 22 to select the bearing size.
9 The colour code refers to the colour on the edge of the bearing.

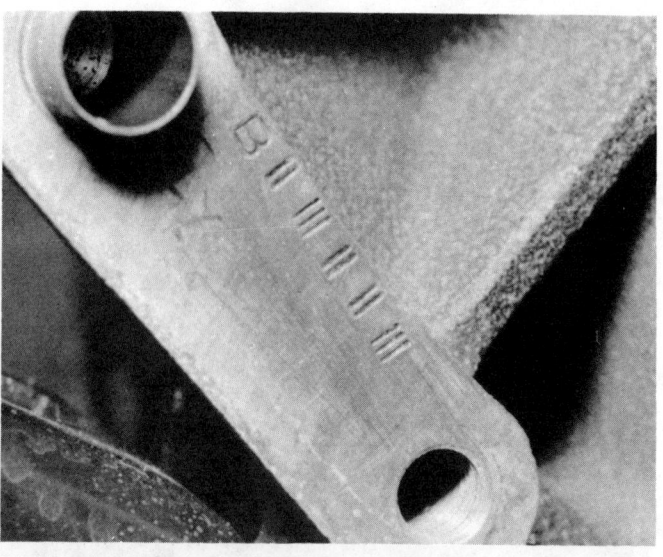

21.1 Bearing identification marks on crankcase web

21.4 Bearing identification marks on the crankshaft

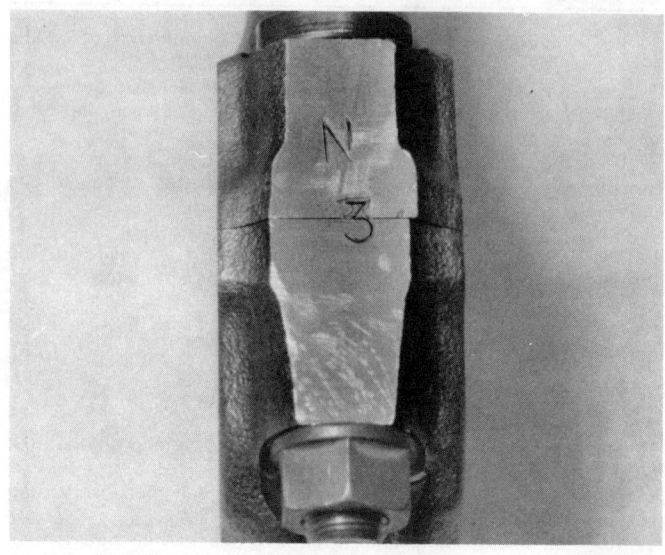

21.7 Identification marks on the connecting rod

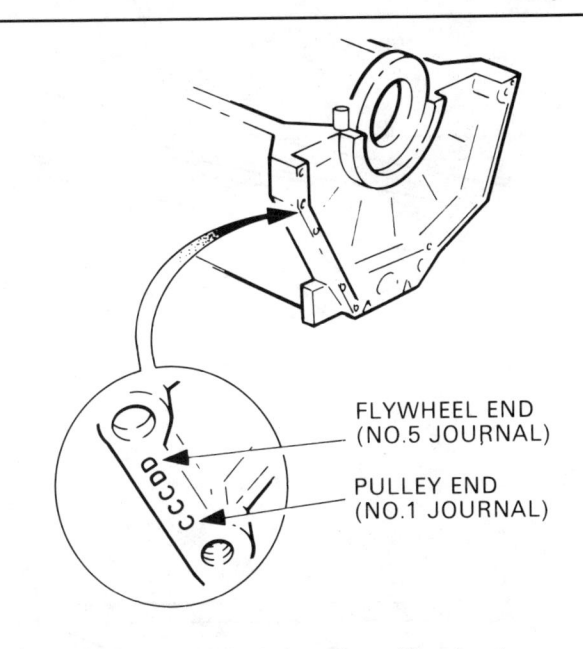

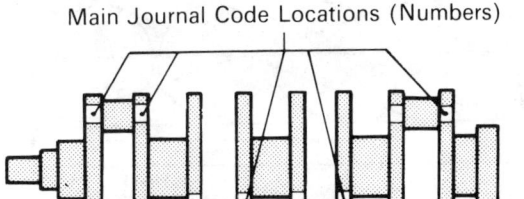

Chapter 1 Engine

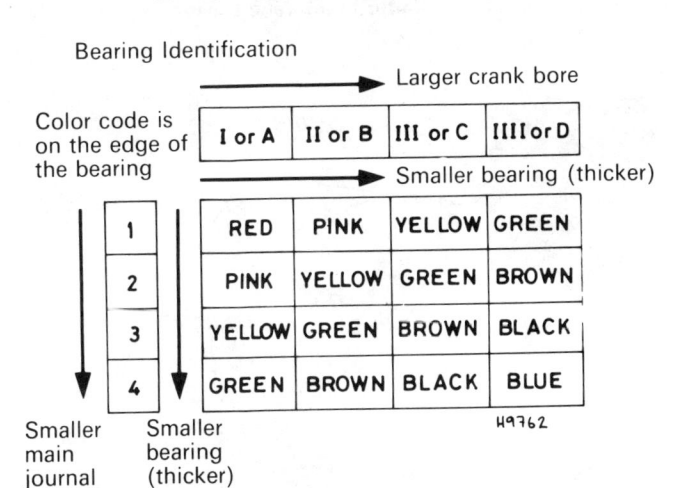

Fig. 1.21 Main bearing identification (Sec 21)

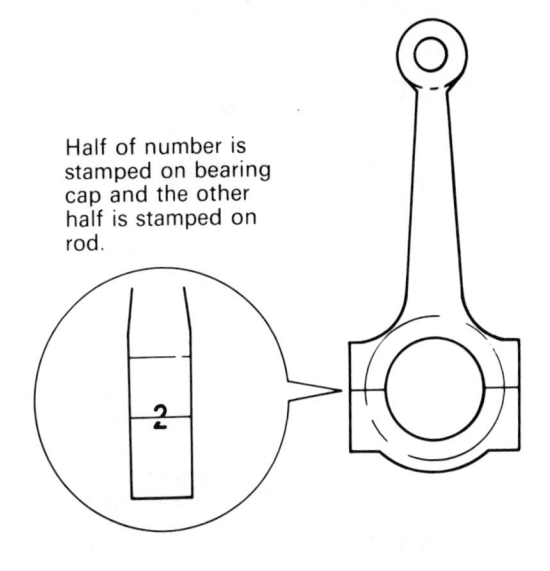

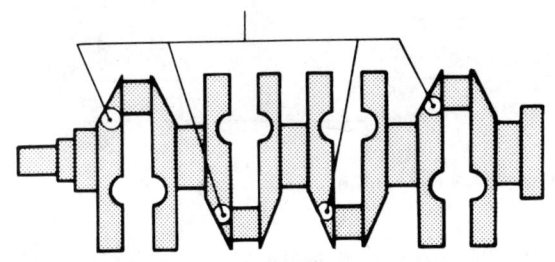

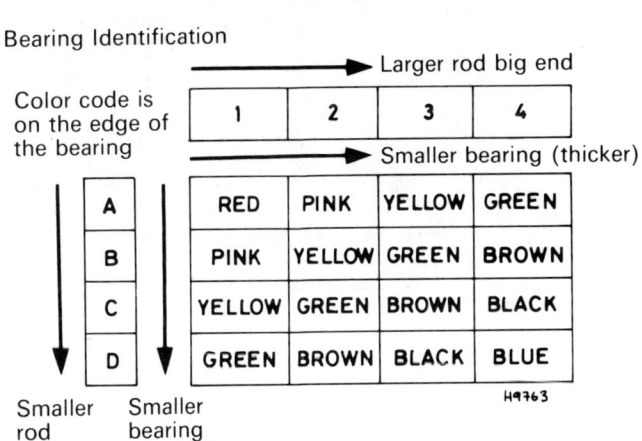

Fig. 1.22 Connecting rod bearing identification (Sec 21)

22 Cylinder block – inspection

1 Measure the bores in the cylinder block to determine any wear.
2 Check for wear and taper in direction X and Y at three levels in each cylinder as shown in Fig. 1.23.
3 Note the cylinder bore size identification letters, which should correspond to the piston size letter stamped on the crown of the piston.
4 If the bores are worn, the cylinders may be rebored, which is a job for a specialist engine repairer.
5 If the difference in taper measurement between the first measurement and the third exceeds the bore taper limit, the block must be renewed.
6 After reboring, refer to the Section 23 on piston clearances.
7 Scored or scratched cylinder bores should be honed as follows. This should also be carried out to ensure fast bedding in of new pistons or rings, as it removes the hard glaze formed on the cylinder walls in service.
8 Hone the cylinder bores using fine grade (400) emery cloth and use a 60° criss-cross pattern. Do not rub straight up and down, or around the bores.
9 Inspect the cylinder block face for warpage using a straight-edge across all four edges and down the centre line.
10 If the service limit in the Specifications is exceeded, the cylinder block should be renewed.
11 If the warpage is below the service limit, it may be possible to have the cylinder block face skimmed; again a job for an engine repair specialist.
12 On completion of all inspections and rectification, wash the cylinder block in solvent, and then wash again with hot soapy water.
13 Clean all oil and waterways and ensure they are free from obstruction.

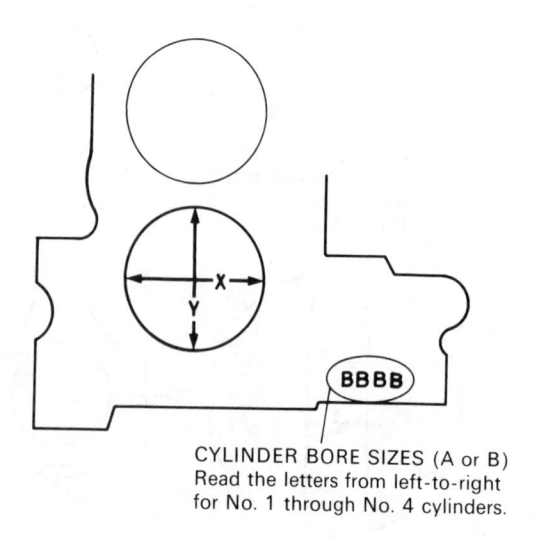

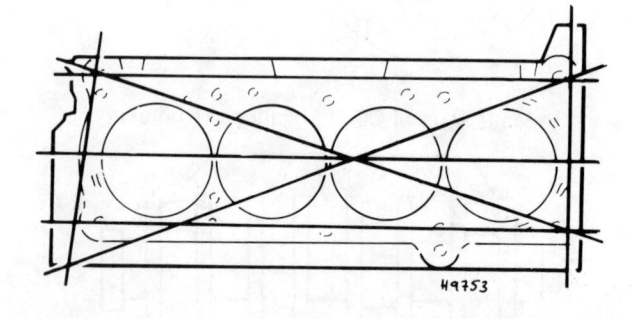

CYLINDER BORE SIZES (A or B)
Read the letters from left-to-right
for No. 1 through No. 4 cylinders.

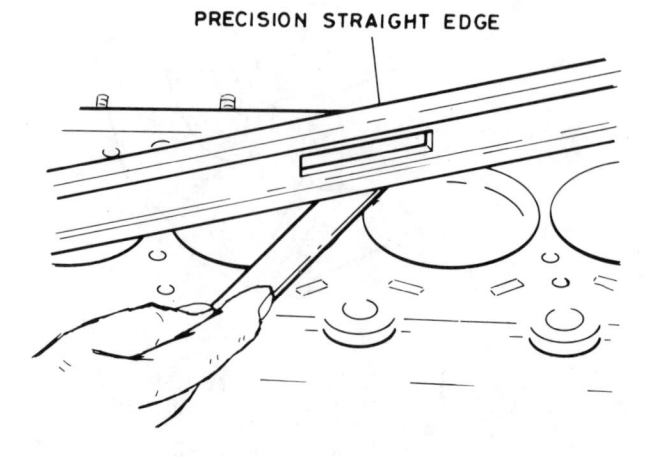

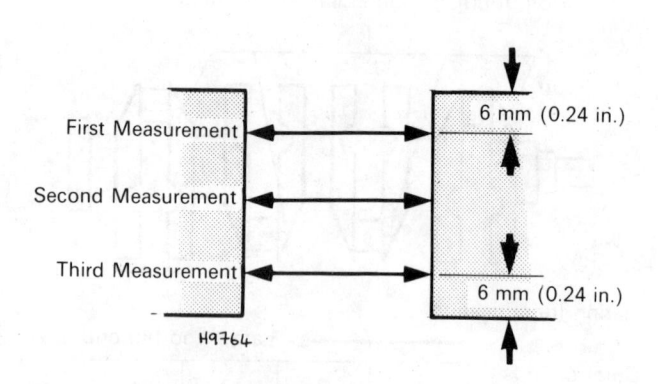

Fig. 1.23 Cylinder bore measurement (Sec 22)

Fig. 1.24 Checking cylinder block face for warpage (Sec 22)

14 Blow out all orifices, passageways and waterways and dry the block thoroughly, before giving it a coat of clean engine oil.

23 Pistons and connecting rods – inspection

1 The pistons and connecting rods will have been removed, as described in Section 20.
2 The numbers stamped over the con rod and bearing caps do not refer to its cylinder number, but to the con rod bore size, so mark the caps and con rods in some way to ensure they are not mixed up.
3 If the piston pin appears worn, or new pistons are to be fitted, this will have to be done by your dealer or engine specialist, as the pin is an interference fit in the piston and a hydraulic press is required to press the pin out and in again.
4 Make sure that the pistons are installed with their index mark facing the con rod oil hole (photo).
5 The big-end inspection and bearing selection are described in Sections 20 and 21.
6 If the cylinders have been rebored, or new pistons have been fitted, the piston-to-cylinder clearance must be established and brought into service limits.
7 Measure the piston skirt at a point 0.83 in (21 mm) from the bottom of the skirt.
8 Piston sizes are given in the Specifications.
9 Measure the cylinder bore, as described in Section 22.
10 Calculate the difference between cylinder bore diameter and piston diameter, and select a piston size that will give the specified piston-to-cylinder clearance.
11 Inspect the connecting rods for signs of burning through overheating, cracks and distortion.

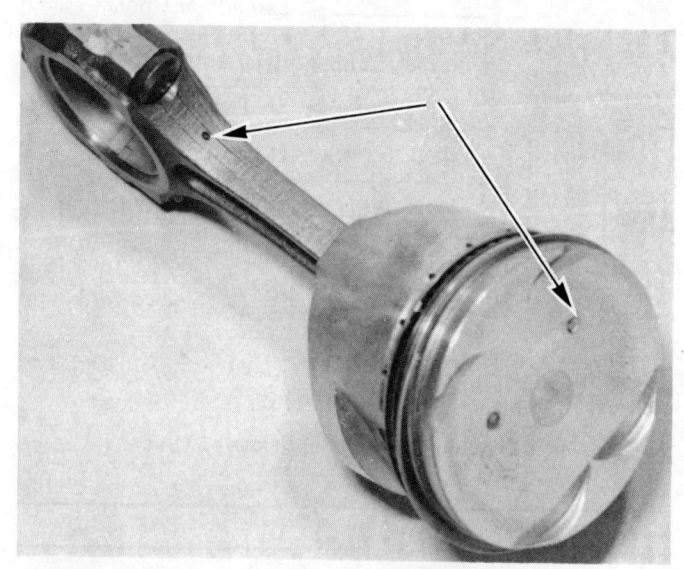

23.4 Piston-to-conrod orientation

24 Piston rings – removal, refitting and inspection

1 The piston rings can be removed by inserting two or three old feeler blades underneath them, spaced around the piston, and sliding the rings off.
2 Old rings should not be re-used.
3 Before fitting the rings to the piston, the ring end gap must be

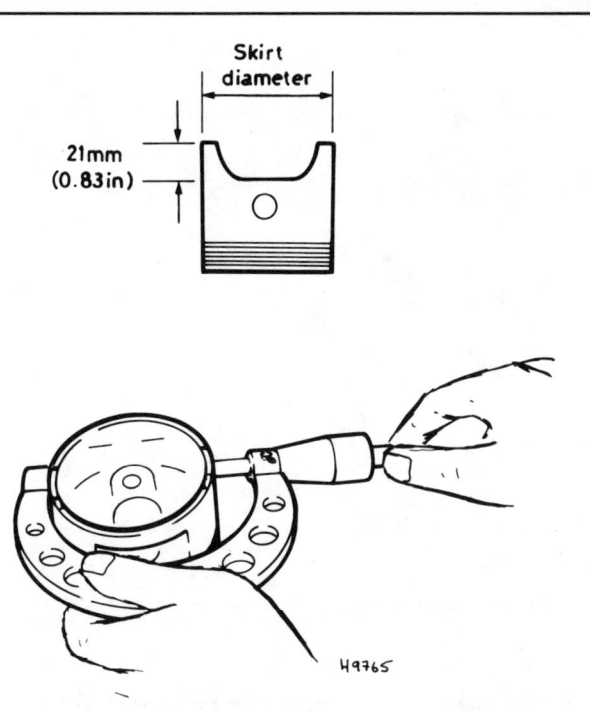

Fig. 1.25 Measuring the piston skirt (Sec 23)

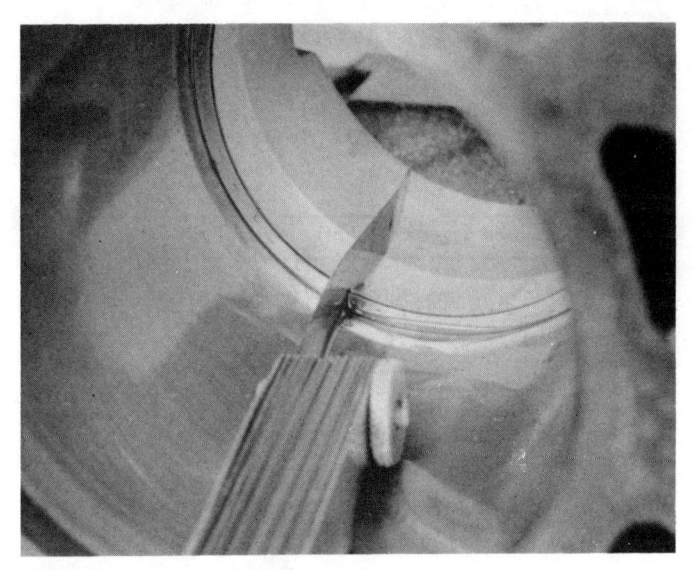

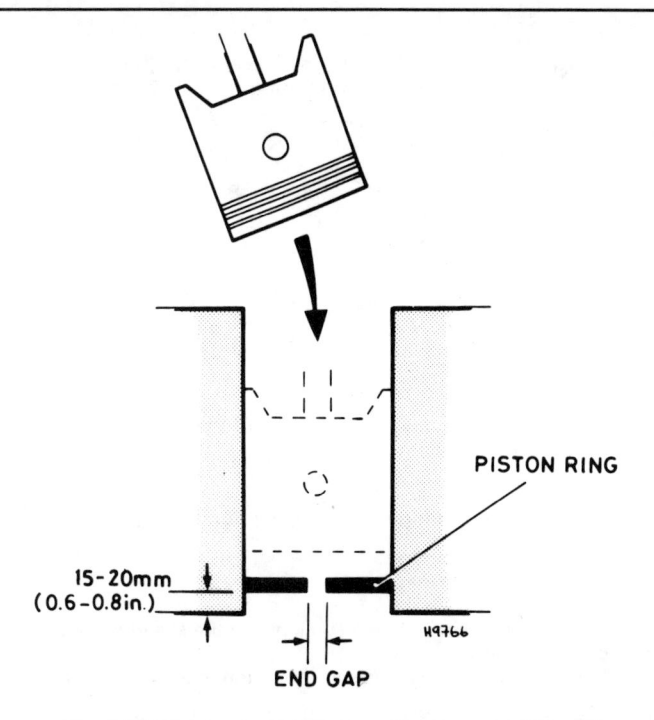

Fig. 1.26 Piston ring end gap measurement (Sec 24)

24.4 Measuring piston ring end gap

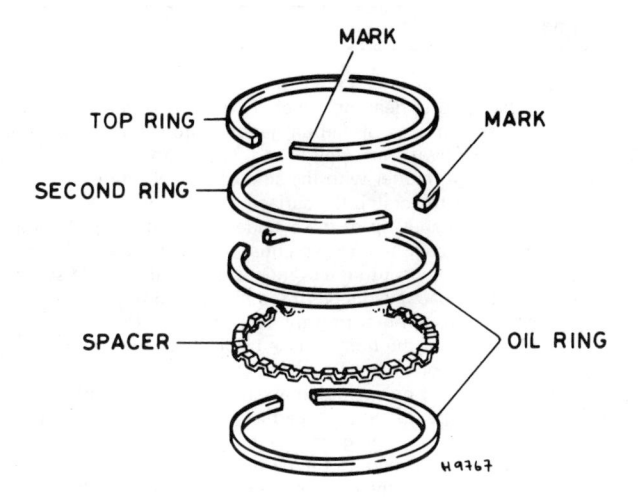

Fig. 1.27 Piston ring fitting order (Sec 24)

established by pushing each ring into its cylinder (use an old piston) until it is positioned as shown in Fig. 1.26.
4 Measure the end gap (photo).
5 If the gap is too small, ensure you have the correct rings for your engine.
6 If the gap is too large, recheck the cylinder bore diameter against the Specifications. If the bore is too large, the cylinder block must be renewed.
7 If new piston rings are to be fitted to old pistons, then the inside top periphery of the new ring should be stepped or chamfered, so that it will clear the wear ridge left in the piston ring groove by the old ring.
8 Clean the piston ring grooves thoroughly.
9 Using a ring expander, or the feeler gauge blade method described earlier, fit the new rings to the pistons.
10 Take note of the piston ring fitting order shown in Fig. 1.27.
11 Identify top and second rings by their different profile, and position the rings as shown in Figs. 1.28 and 1.29.

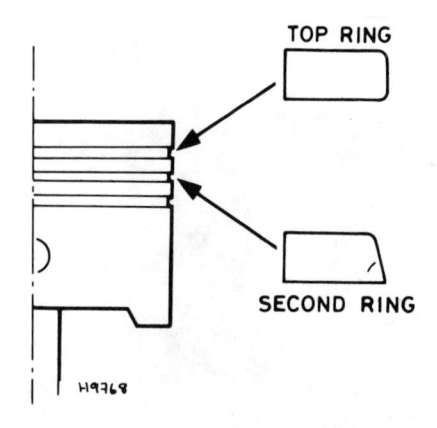

Fig. 1.28 Identifying top and second rings (Sec 24)

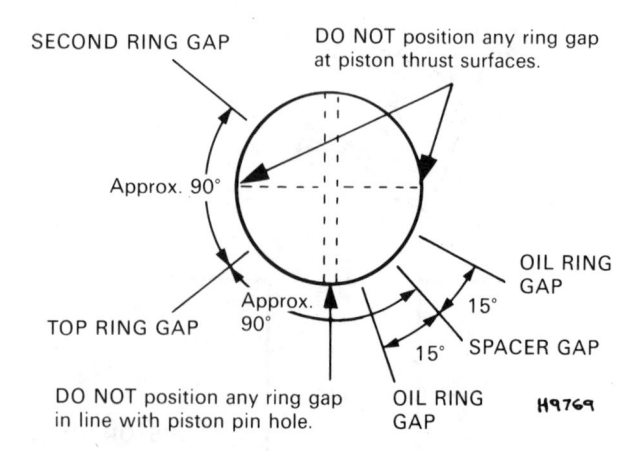

Fig. 1.29 Positioning the ring gaps correctly (Sec 24)

24.13 Measuring piston ring-to-land clearance

12 Do not position any ring gap at the piston thrust surface or above the piston pin holes.
13 After the rings are fitted measure the piston ring-to-land clearance (photo).
14 If out of service limits, renew the piston.

25 Engine – decarbonising

1 This operation can be carried out either with the engine in, or out of the car. With the cylinder head off, carefully remove with a wire brush and blunt scraper all traces of carbon deposits from the combustion spaces and brush down with petrol and scrape the cylinder head surface of any foreign matter with the side of a steel rule or a similar article. Take care not to scratch the surface.
2 Clean the pistons and top of the cylinder bores. If the pistons are still in the cylinder bores then it is essential that great care is taken to ensure that no carbon gets into the cylinder bores as this could scratch the cylinder walls or cause damage to the piston and rings. To ensure that this does not happen first turn the crankshaft so that two of the pistons are at the top of the bores. Place clean non-fluffy rag into the other two bores or seal them off with paper and masking tape. The waterways and pushrod holes should always be covered with a small piece of masking tape to prevent particles of carbon entering the cooling system and damaging the water pump or entering the lubrication system and damaging the oil pump or bearings.
3 Press some grease into the gap between the cylinder walls and the two pistons which are being worked upon. With a blunt scraper carefully scrape away the carbon from the piston crowns taking care not to scratch the aluminium surface. Also scrape the carbon ring from the top of the bores.
4 Remove the rags and masking tape and wipe away the rings of grease which will now be mixed with carbon particles.
5 The crankshaft can now be turned to bring the other two pistons to the top of their strokes and the operations previously described can be repeated.
6 Wipe away every trace of carbon and pour a little thin oil around the pistons to lubricate the rings and to help flush out any remaining carbon particles from the piston grooves.
7 Clean out any holes in the cylinder head and examine for cracks. Any studs which have stripped their threads should be removed and new thread inserts installed.

26 Cylinder block and crankshaft – reassembly

1 Coat all parts liberally with clean engine oil on reassembly.
2 Position the engine block on its side on the bench.
3 Check each piston and con rod assembly is assembled correctly and then use a ring compressor to compress the piston rings on the piston before inserting them and the piston and con rod assemblies into the cylinders (photo). **Note:** It is a good idea to slip short pieces of rubber or plastic tube over the con rod bearing cap studs to prevent them scratching the cylinder bore.
4 Position all pistons at top dead centre.
5 Turn the engine onto the cylinder head face (upside down).
6 Insert the main bearing top halves into the crankshaft block journals (photo).
7 Fit the crankshaft (photo) so that the journals for No 2 and 3

26.3 Inserting a piston into the cylinder

26.6 Fitting the bearing halves

26.7 Fitting the crankshaft

Chapter 1 Engine

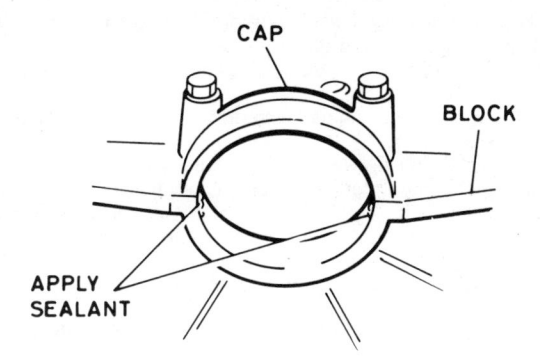

Fig. 1.30 Areas to apply sealant to crankshaft oil seals (Sec 26)

cylinders are facing downward, and seating these journals in their con rods as the crankshaft is lowered.
8 Fit the bearing halves to the con rod caps and fit the caps and retaining nuts finger tight.
9 Rotate the crankshaft clockwise, seat 1 and 4 con rods on their journals and fit bearing halves and caps, again finger tight.
10 Install the thrust washers on each side of No 3 main bearing.
11 Fit the main bearing caps finger tight only.
12 Do not tighten the bearing caps for final assembly until the crankshaft oil seals have been installed.
13 Apply non hardening sealant to the areas shown in Fig. 1.30.
14 The seal surface on the block should be clean and dry, but apply a film of oil to the crankshaft and oil seal lip.
15 Drive both seals into their housing until they bottom against the block, using drivers of suitable diameter. **Note:** Seals are installed with part numbers facing out.

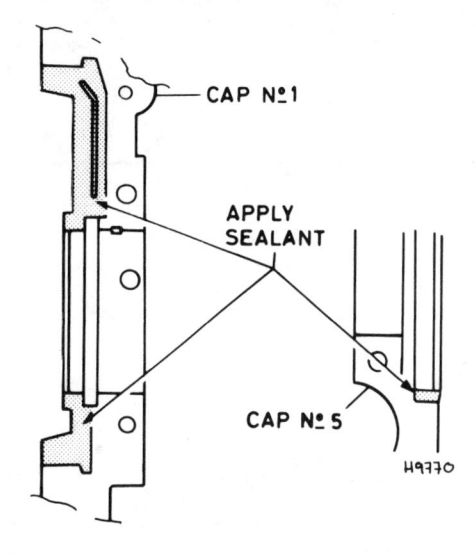

16 Tighten the main and con rod bearing caps to their specified torque.
17 Refit the oil shield over the crankcase breather chamber (photo).
18 Fit the oil pick-up assembly, using a new O-ring on the pick-up pipe end (photo).
19 Fit a new sump gasket (photo), and apply sealant to the four corners of the sump where it fits over the main bearings (photo).
20 Fit and tighten the oil sump, and oil drain plug, bolts to the specified torque.
21 Fit the flywheel or torque converter driveplate and tighten the bolts to their specified torque (photo).

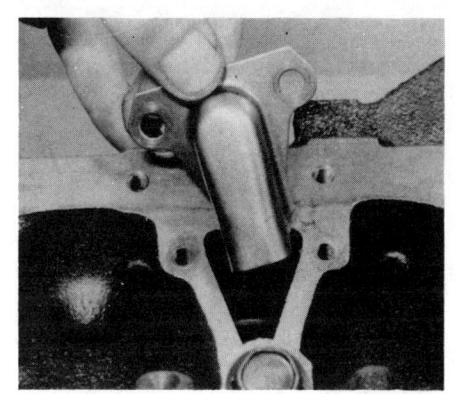

26.17 Install the oil shield

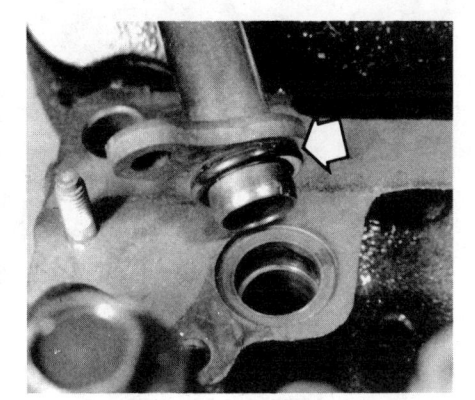

26.18 Use a new O-ring on oil pick-up pipe (arrowed)

26.19A Fitting a new sump gasket

26.19B Fitting the engine oil sump, arrows indicate areas to apply sealant

26.21 Tightening the flywheel bolts

Chapter 1 Engine

22 Fit the flywheel/converter covers. **Note:** on some models there are support straps fitted between the oil sump and transmission housing, do not forget to fit them after installation of the transmission housing and flywheel/torque converter driveplate covers.

27 Engine – final assembly and refitting

The final assembly of the engine components is dealt with under the respective Sections, and refitting of the engine is largely a reversal of removal. The following text is given as a guide only.

1 Turn the engine upright and support it on blocks.
2 Refit the oil system components.
3 Refit the PCV chamber, using a new O-ring (photos).
4 Fit the left-hand engine mount, if it has been removed (photo).
5 Refit the cylinder head.
6 Fit the water pump and alternator mounting bracket (photo).
7 Fit the coolant transfer tube, using a new O-ring (photo) and secure in place with the bracket (photo).
8 Fit and tension the camshaft drivebelt (photo).
9 Fit the camshaft covers, ensuring the gasket is correctly located (photo), and do not forget the seal over the tensioner bolt (photo), or the rubber plug (photo).
10 Fit the water pump drive pulley.
11 Fit the crankshaft pulley.
12 Fit the alternator, then fit and tension the drivebelt.
13 Fit the inlet and exhaust manifolds, and carburettor if it has been removed.
14 Fit the thermostat housing.
15 Fit the distributor.
16 Fit the transmission unit (photo).
17 Work around the engine and refit the various brackets, engine sub harness cable clips, and connect up the coolant hose from the manifold to coolant transfer pipe and the PCV hose.
18 Check the engine block drain plug and oil sump drain plug are fitted.
19 Attach a sling to the engine lifting brackets, and prepare the site for refitting of the engine.
20 Lower the engine slowly into the engine bay, checking frequently that it does not catch on any components or pipework. **Note:** to give sufficient clearance, it may be necessary to fit the rear engine mounting bracket after the engine has been lowered into the engine bay).
21 Work round the engine fitting the mounting brackets, but do not torque load the bolts until the weight of the engine is fully back on the mountings.
22 The mountings should be tightened in the sequence shown in Fig. 1.2.
23 Remove the engine lifting sling.
24 Work around the engine connecting up the various emission control hoses, radiator and heater hoses and brake booster hose.
25 Refit the speedometer cable or speed sensor.
26 Refit the spark plugs and connect the distributor and coil HT and LT cables.
27 Connect the choke and throttle cables and, on automatics, the throttle control cable.
28 Reconnect the gearshift torque rod and shift rod and, on automatics, the shift cable.
29 Reconnect the fuel line to the carburettor.
30 Reconnect the cruise control and suspension levelling system if fitted.
31 Refit the power steering pump and air conditioning pump, and tension the drivebelts (as applicable).
32 Connect up the electrical leads to all pressure and temperature transmitters.
33 Refit the driveshafts and exhaust downpipe.
34 Remove the vehicle from stands.

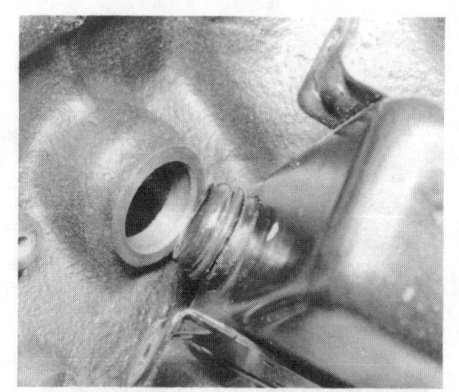

27.3A Fit a new O-ring to the PCV chamber ...

27.3B ... and tighten the bolts

27.4 Left-hand engine mount

27.6 Water pump and alternator bracket

27.7A The coolant transfer tube O-ring (arrowed) ...

27.7B ... and mounting bracket (arrowed)

Chapter 1 Engine 61

27.8 Camshaft drivebelt and components in place

27.9A Ensure the cover gasket is correctly located ...

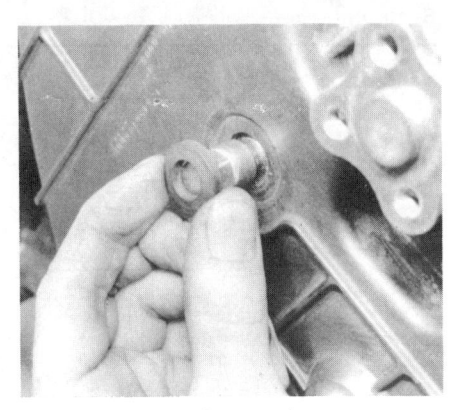

27.9B ... fit the tensioner seal ...

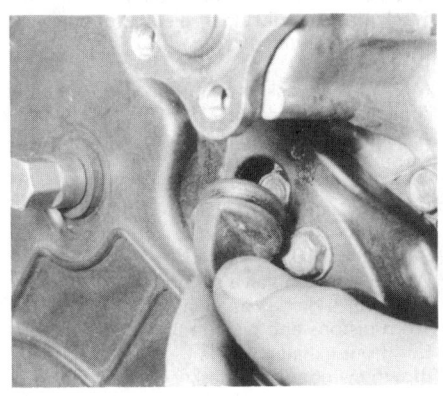

27.9C ... and rubber plug

27.16 Transmission unit ready for refitting

35 Refit the battery and connect the battery and engine sub-harness.
36 Fit the bonnet.
37 Check all round to ensure all connections/components have been fitted.
38 Fill the engine and transmission unit with the recommended lubricant, and the cooling system with coolant.
39 On initial start-up, there will be a smell of burning, accompanied by smoke, especially from the exhaust. Provided this is not excessive, there is no need to be alarmed.
40 Allow the engine to warm up at idle and then let the engine idle for a further 15 minutes. Do not race the engine during this period.
41 Check all round for leaks before fitting the engine splash guard.
42 Switch off the ignition, allow the engine to cool and then recheck the fluid levels.

28 Fault diagnosis – engine

Symptom	Reason(s)
Engine fails to turn when starter operated	Flat or defective battery
	Loose battery leads
	Defective starter solenoid or switch, or broken wiring
	Engine earth strap disconnected
	Defective starter motor
Engine turns on starter but will not start	Ignition damp or wet
	Ignition leads to spark plugs loose
	Shorted or disconnected low tension leads
	Defective ignition switch
	Ignition leads connected wrong way round
	Faulty coil
	No petrol in petrol tank
	Vapour lock in fuel line (in hot conditions or at high altitude)
	Blocked float chamber needle valve
	Fuel pump filter blocked
	Choked or blocked carburettor jets
	Faulty fuel pump

Symptom	Reason(s)
Engine stalls and will not restart	Too much choke allowing too rich a mixture to wet plugs Float damaged or leaking or needle not seating Float level incorrectly adjusted Ignition failure – sudden Ignition failure – misfiring precludes total stoppage Ignition failure – in severe rain or after traversing water splash No petrol in petrol tank Sudden obstruction in carburettor Water in fuel system
Engine misfires or idles unevenly	Ignition leads loose Battery leads loose on terminals Battery earth strap loose on body attachment point Engine earth lead loose Low tension leads to coil terminals loose Low tension lead from coil to distributor loose Dirty, or incorrectly gapped plugs Tracking across inside of distributor cover Ignition too retarded Faulty coil Mixture too weak Air leak in carburettor Air leak at inlet manifold to cylinder head, or inlet manifold to carburettor
Lack of power and poor compression	Incorrect valve clearances Burnt out exhaust valves Sticking or leaking valves Weak or broken valve springs Worn valve guides or stems Worn pistons and piston rings Burnt out exhaust valves Blown cylinder head gasket (accompanied by increase in noise) Worn or scored cylinder bore Ignition timing wrongly set; too advanced or retarded Incorrectly set spark plugs Carburation too rich or too weak Fuel filter blocked causing poor top end performance through fuel starvation Distributor automatic balance weights or vacuum advance and retard mechanisms not functioning correctly Faulty fuel pump giving top end fuel starvation
Excessive oil consumption	Excessively worn valve stems and valve guides Worn piston rings Worn piston and cylinder bores Excessive piston ring gap allowing oil to bypass Piston oil return holes choked Leaking oil filter gasket Leaking rocker cover gasket Leaking sump gasket Loose sump plug
Unusual noises from engine	Worn valve gear Worn big-end bearing (regular heavy knocking) Worn main bearings (rumbling and vibration)

Chapter 2 Cooling system

Contents

Coolant mixture	4	Routine maintenance	2
Cooling system – draining, flushing and refilling	3	Temperature gauge and temperature sender unit – testing, removing and refitting	8
Drivebelts – removing, refitting and adjusting	10	Thermostat – removing, testing and refitting	5
Fault diagnosis – cooling system	11	Water pump – removing and refitting	9
General description	1		
Radiator – removing, repair and refitting	6		
Radiator cooling fan, thermoswitch and thermosensor – removing, testing and refitting	7		

Specifications

General
System type .... Pressurised with belt-driven pump, radiator, thermostat and electric cooling fan
Radiator cap pressure .... 11 to 15 lbf/in² (0.75 to 1.05 kgf/cm²)

Thermostat
Starts to open:
    Primary .... 82 ± 2°C (180 ± 3°F)
    Secondary .... 85 ± 2°C (185 ± 3°F)
Fully open .... 95°C (203°F)
Valve lift at fully open .... 0.31 in (8 mm)

Cooling fan
Thermoswitch closes (fan starts) .... 87 to 93°C (188 to 199°F)
Thermoswitch opens (fan stops) .... 83 to 85°C (181 to 185°F)
System capacity .... 1.5 Imp gal (1.8 US gal, 6.8 litre)

Drivebelt tension
With an applied force of 22 lbs (10 kg):
    Alternator/water pump belt deflection ..... 0.2 to 0.4 in (6 to 9 mm)
    Power steering pump belt deflection .... 0.7 to 0.9 in (18 to 22 mm)
    Air conditioner pump belt deflection .... 0.4 to 0.5 in (10 to 12 mm)

Torque wrench settings

	lbf ft	Nm
Radiator shroud mounting bolts	7	9
Thermoswitch in radiator	17	23
Water pump pulley bolts	9	12
Thermostat housing mounting nuts	16	22
Thermostat housing and cover bolts	9	12
Thermostat housing bleed nipple	7	9
Cylinder block drain plug	23	31
Cylinder block thermoswitch	20	27
Water pump mounting bolts	9	12
Coolant transfer tube mounting bolt	9	12
Alternator mounting bolt	33	45
Belt tensioner adjusting bolt	16	22

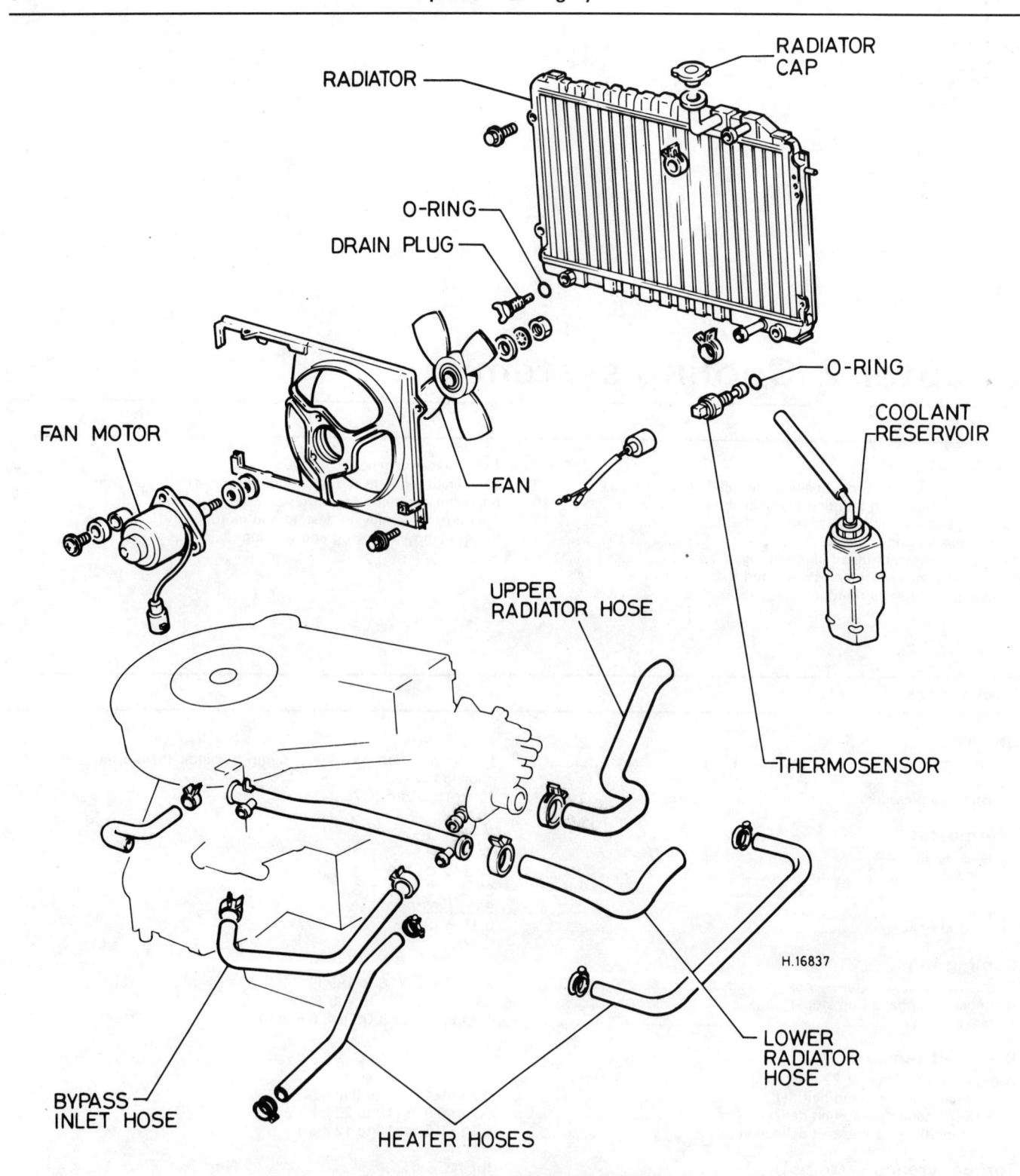

Fig. 2.1 Components of the cooling system (Sec 1)

1 General description

The cooling system is of conventional design, consisting of a front-mounted radiator; a water pump driven by belt from the crankshaft pulley (which also drives the alternator); a thermostat, mounted in a housing bolted to the cylinder head; and an electric cooling fan mounted on the radiator and controlled by a thermoswitch. A reserve tank, mounted in the engine compartment, allows for expansion and contraction of the coolant at different temperatures preventing loss of fluid.

Coolant in the cylinder block is circulated by the pump around the cylinder block, inlet manifold and car interior heater until, at a predetermined temperature, the thermostat opens allowing coolant to pass to the radiator to be cooled.

A thermoswitch, fitted in the bottom right-hand corner of the radiator, will turn the cooling fan on above a given temperature to provide additional cooling, especially when the vehicle is stationary

Chapter 2 Cooling system

with the engine running. A secondary thermal control is also fitted in the form of a thermosensor mounted on the fan motor. This thermosensor will prevent the electric fan motor operating below a predetermined level, and ensures quicker warm-up in cold weather, and greater engine temperature stability.

On vehicles fitted with air conditioning, two cooling fans are mounted behind the radiator.

2 Routine maintenance

At the intervals given in Routine Maintenance (at the front of the book) undertake the following service tasks.
1 Inspect the cooling system, hoses and connections for security, deterioration and leaks. Repair/renew as necessary.
2 Check the drivebelt tension. Retension if necessary.
3 Drain the cooling system. Refill with fresh antifreeze solution.
4 Inspect the radiator for damage and leaks. Brush or blow out any debris from the fins. Repair/renew as necessary.

3 Cooling system – draining, flushing and refilling

1 The cooling system should preferably be drained when the engine is cool, but after the engine has recently been operated at normal temperature. This will ensure that sediment, which has built-up in the system, will be drained off with the coolant.
2 Set the heater temperature control lever to the HOT position.
3 Remove the radiator cap, and then the radiator drain plug (photos), and drain the old coolant.
4 If the coolant is badly contaminated with rust and sediment, flush the system through using a cold water hose inserted in the radiator filler neck.
5 Refit the radiator drain plug.
6 Remove the reserve tank, flush it out and refit it (photo).
7 Refer to Section 4 and mix sufficient coolant to refill the system.
8 Fill the reserve tank, with coolant mixture, to the MAX mark.
9 Loosen the air bleed nipple on the thermostat housing and fill the system to the bottom of the radiator filler neck with coolant (photos).
10 Tighten the bleed nipple as soon as an air free flow of coolant issues from the nipple.
11 With the radiator cap removed, start the engine and run it until it has reached normal operating temperature.
12 If necessary, add more coolant mixture to maintain the level at the bottom of the filler neck.
13 Refit the radiator cap, and check for leaks, with the engine still running.
14 Switch off the engine.

4 Coolant mixture

1 The cooling system should be kept filled with the recommended mixture of antifreeze and water all year round, to protect not only against frost damage, but also against corrosion.
2 The recommended mixture is 50% antifreeze to 50% water. Anything less in antifreeze content may not provide sufficient protection.
3 Concentrations of antifreeze above 60% may impair cooling efficiency and are not recommended.
4 Do not mix different kinds of antifreeze, and do not use additional rust inhibitors, they may not be compatible with the antifreeze.
5 Use only top quality ethylene glycol based antifreeze. Cheaper brands tend to use chemicals which evaporate and soon become ineffective.
6 When topping the system up during periodic servicing, use the same 50/50 mixture.
7 If the vehicle is operated where antifreeze is not necessary, use a corrosion inhibitor.

3.3A The radiator cap ...

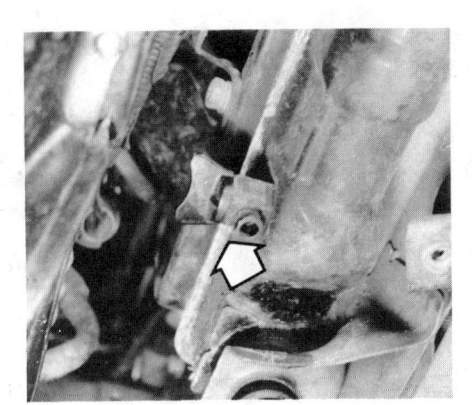

3.3B ... and drain plug (arrowed)

3.6 The coolant reserve tank

3.9A Cooling system air bleed nipple on the thermostat housing (arrowed)

3.9B Filling the radiator

Chapter 2 Cooling system

5.2 Coolant hose retaining spring clip

5.3A Undoing the thermostat housing nuts

5.3B Removing the housing

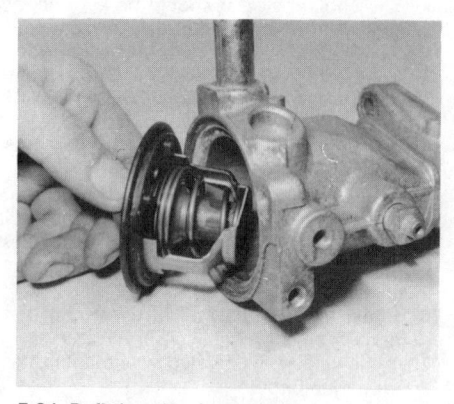

5.8A Refitting the thermostat ...

5.8B ... and end cover

5 Thermostat – removal, testing and refitting

1 Drain off sufficient coolant to bring the coolant level below the thermostat housing.
2 Disconnect the coolant hoses from the thermostat housing (photo).
3 The thermostat end cap may be removed and the thermostat lifted out, or the whole thermostat housing removed by undoing the appropriate nuts or bolts (photos).
4 Discard the gasket.
5 If the thermostat is stuck, do not attempt to lever it out, but cut around its periphery using a sharp knife.
6 To test the thermostat, suspend it in boiling water and check that it opens, then allow it to cool and check that it closes.
7 If the thermostat is seized, fit a replacement.

8 Refit the thermostat into its housing, using a new gasket, and ensure that the air bleed pin is uppermost (photos).
9 Use a new O-ring seal when refitting the housing to the cylinder head, and tighten all nuts and bolts to the specified torque.
10 Reconnect the hoses, fill the system with coolant and run the engine to check for leaks.

6 Radiator – removal, repair and refitting

1 Drain the cooling system, as described in Section 3.
2 Disconnect the radiator top and bottom hoses, and the reserve tank pipe line (photo).
3 Disconnect the electric fan lead at the connector (photo), and the thermoswitch connection.

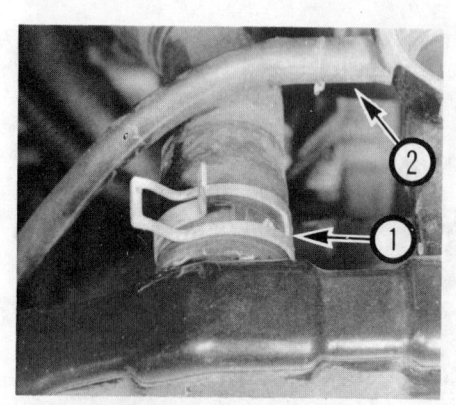

6.2 The radiator top hose clip (1) and reserve tank pipe line (2)

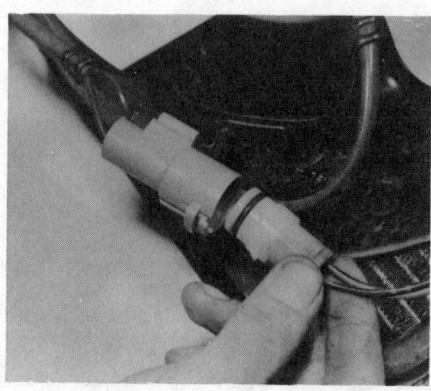

6.3 Disconnecting the radiator fan electrical connector

6.6A Remove the bolts from the mounting blocks ...

Chapter 2 Cooling system

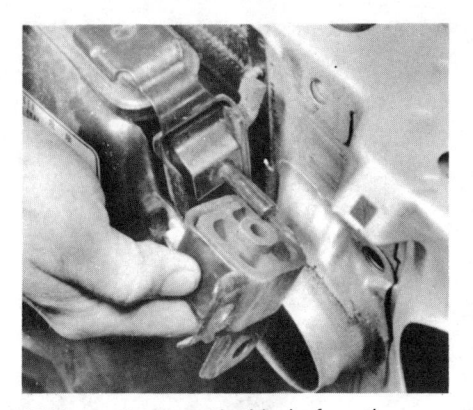

6.6B ... and remove the blocks from the spigot

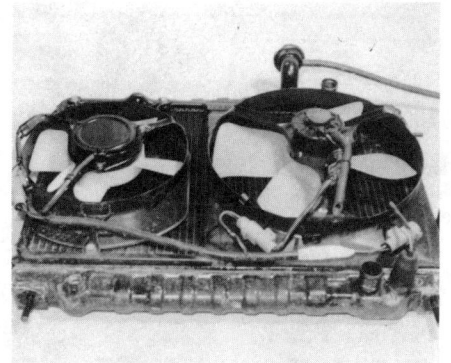

6.7 The radiator and fan assembly

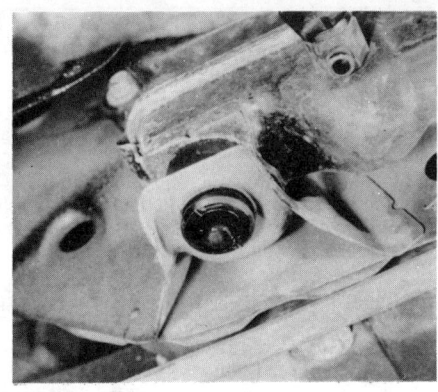

6.10 Radiator lower mounting assembly

4 On models equipped with air conditioning, there are two cooling fans, disconnect both.
5 On models with automatic transmission, disconnect the oil cooler pipe lines and blank off the pipe lines.
6 Remove the bolts from the mounting blocks at each side of the radiator, and remove the blocks from the spigots (photos).
7 Carefully lift out the radiator, complete with cooling fan(s) (photo).
8 The cooling fan(s) may be removed from the radiator by unscrewing the four retaining bolts which hold the support cowling, and lifting the whole assembly away from the radiator.
9 If the radiator was removed because of a leak, it is best to take it to a repair specialist. Pressure testing of the radiator and filler cap can be done by your local dealer, who has the test equipment.
10 Refitting is a reversal of removal, but ensure that the lower mounting rubbers are located correctly in their housings (photo).
11 Refill the radiator, as described in Section 3, on completion.

7 Radiator cooling fan, thermoswitch and thermosensor – removal, testing and refitting

1 Remove the fan assembly from the radiator, as described in Section 6.
2 The fan blades may be removed from the motor by undoing the retaining screw which holds them to the motor driveshaft.
3 Remove the fan from the support cowling by undoing the retaining screws and disconnecting the wiring harness at the connector.
4 Little can be done by way of repair to the fan motor, and if it is defective, it is best to replace it with a new motor.
5 The fan is actuated by a thermoswitch fitted in the radiator bottom right-hand corner which is the primary control.
6 If the fan fails to operate, check the operation at the thermoswitch as follows.
7 Disconnect the black and blue wires to the switch at the connector block.
8 Use a short piece of wire to connect these two wires together, thus shorting them out.
9 Switch on the ignition.
10 The cooling fan should start to run.
11 If it does run, renew the thermoswitch and retest.
12 If the motor fails to start, check the battery is fully charged, and that there is a supply available from the blue wire (positive) to the black wire (negative) at the thermoswitch.
13 If there is no supply, check for a blown fuse, loose terminals or connectors.
14 If supply is available to the thermoswitch, then suspect the motor itself of malfunction.
15 When renewing the thermoswtch use a new sealing washer.
16 Refitting of the fan assembly is a reversal of removing.
17 A thermosensor fitted on the cooling fan motor acts as a secondary control, its purpose being to keep the fan switched off during cold weather to allow for quicker warm-up of the engine.

8 Temperature gauge and temperature sender unit – testing, removal and refitting

1 The temperature gauge sender unit is situated on the inlet manifold, and transmits water temperature to the temperature gauge (photo).
2 If the temperature and fuel gauges read 'H' and 'F' respectively, suspect the gauge unit itself of being faulty and refer to Chapter 12 for details of instrument removal.
3 To test the sender unit, disconnect the yellow/green wire and connect it to a good earth.

8.1 Temperature sender unit in the inlet manifold

4 Switch on the ignition.
5 Temperature gauge should read 'H'.
6 If it does not, check the fuse, wiring and connections, and if these prove to be all right, renew the sender unit.
7 Use a new sealing washer when refitting the sender unit.
8 Refit the pump to the cylinder block and tighten the bolts to the specified torque (photo).
9 Refit the pulley wheel to the pump.
10 Refit and tension the drivebelt, as described in Section 10, and refill the system, as described in Section 3.

9 Water pump – removal and refitting

1 The water pump is mounted on the front face of the cylinder blocks, and is driven by belt from the crankshaft pulley.
2 To remove the pump, drain the cooling system, as described in Section 3.
3 Slacken the alternator mounting bolt and adjustment bolt and push the alternator inward towards the engine to release the tension on the drivebelt, and slip the drivebelt off the pulleys (photos).
4 Remove the bolts securing the water pump drive pulley to the water pump, and remove the pulley (photo).
5 Remove the bolts securing the water pump to the cylinder block, and remove the pump.
6 Little can be done by the way of repair to the water pump, and if it is defective, it should be renewed.
7 Fit a new O-ring seal to the pump housing (photo).
8 Refit the pump to the cylinder block and tighten the bolts to the specified torque (photo).
9 Refit the pulley wheel to the pump.
10 Refit and tension the driveshaft, as described in Section 10, and refill the system, as described in Section 3.

9.3A The alternator mounting bolt (arrowed) ...

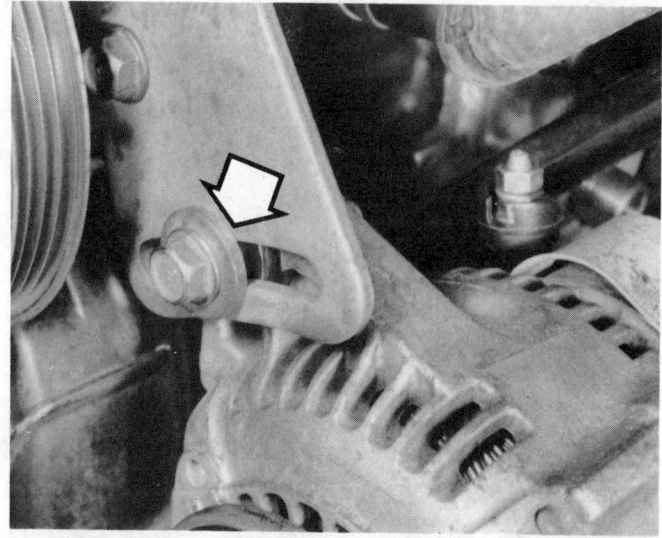

9.3B ... and adjustment bolt (arrowed)

9.4 The water pump with drive pulley removed

9.7 Fit a new O-ring seal to the pump housing

9.8 Refitting the pump to the cylinder block

Chapter 2 Cooling system

pulley wheel results in a deflection of the belt as given in the Specifications.
7 On completion, tighten the alternator mounting bolt and adjusting bolt.

Power steering pump drivebelt
8 Slacken the pump mounting bolt and adjusting nut and push the pump inward towards the engine.
9 The belt may now be slipped off the pulleys.
10 Refit the belt and, using a wooden lever, push the pump away from the engine.
11 Tighten the adjusting bolt.
12 The belt is correctly tensioned when a force applied to the belt midway between the two pulleys results in a deflection as given in the Specification.
13 Tighten the pump mounting bolt and adjusting bolt.

Air conditioner pump drivebelt
14 Loosen the air conditioner pump pivot bolt and then slacken off the adjusting nut sufficiently to give enough play to slip the belt off the pulleys (photo).

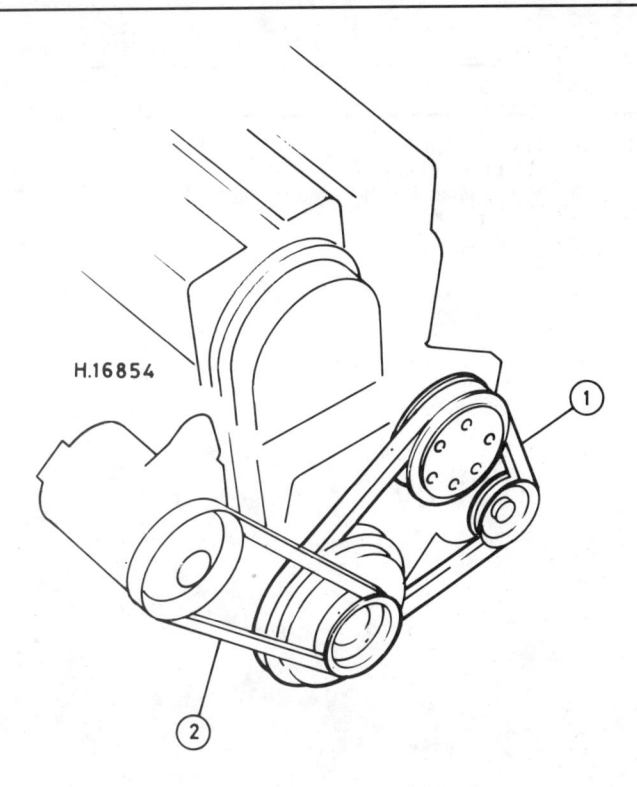

Fig. 2.2 Alternator/water pump drive belt (1) and power steering belt (2) (Sec 10)

10.14 The air conditioner pump adjusting bracket

1 Pivot bolt   2 Adjusting nut

15 On replacement, screw down the adjusting nut to tension the belt so as a force applied midway between the two pulleys results in a belt deflection as given in the Specifications.
16 Tighten the pivot bolt.

10 Drivebelts – removal, refitting and adjusting

1 There will be up to three drivebelts fitted, depending upon model and equipment fitted.
2 All models have an alternator/water pump drivebelt, and additionally there may be a power steering drivebelt and an air conditioner drivebelt.

Alternator/water pump drivebelt
3 The procedure for removing the belt is described in Section 9. If power steering and air conditioning are fitted, these drivebelts will have to be removed first, before the alternator belt can be removed.
4 To adjust the belt, loosen the alternator mounting bolt and the adjusting bolt.
5 Using a wooden lever, push the alternator outwards, applying tension to the belt.
6 The belt is correctly tensioned when a force applied to the belt midway between the alternator pulley wheel and the water pump

Fault diagnosis overleaf

11 Fault diagnosis – cooling system

Symptom	Reason(s)
Overheating	Coolant loss due to leakage Faulty electric cooling fan or switch Alternator/water pump drivebelt slack or broken Faulty thermostat Radiator matrix clogged
Overcooling	Faulty thermostat Faulty electric cooling fan switch
Coolant loss	External leakage (hose, joint etc) Internal leakage (head gasket) Overheating (see above)

Chapter 3
Fuel, exhaust and emission control systems

Contents

Accelerator pump – adjustment	19
Air cleaner assembly – removal and refitting	4
Air cleaner element – removal and refitting	3
Air intake system (FI) – removal, refitting and testing of components	58
Air jet controller	55
Air vent cut-off diaphragm	43
Anti-afterburn valve	51
Automatic choke – general discription	22
Automatic choke (Canadian models) – adjustment	37
Automatic choke (Canadian models) fast idle – adjustment	38
Automatic choke, choke opener – inspection and renewal	25
Automatic choke coil heater – testing and renewal	26
Automatic choke fast idle – adjustment	27
Automatic choke fast idle unloader – inspection	24
Automatic choke linkage – adjustment	23
Automatic transmission throttle control cable – adjustment	36
Automatic transmission throttle control bracket – adjustment	35
Carburettor – general description	15
Carburettor – overhaul	31
Carburettor – removal and refitting	30
Carburettor adjustments – general	16
Catalytic converter	67
Charcoal cannister	44
Choke cable – removal, refitting and adjustment	34
Choke relief valve – adjustment	20
Crankcase controls (FI) – general	63
Cranking opener solenoid valve	48
Cut-off valves – testing and renewing	29
Dashpot system	46
Electronic control unit (ECU) – general description	59
Emission controls – general description	40
Emission controls, (FI) – general description	62
Evaporative controls (carburettor model) – general description	41
Evaporative controls (FI) – general	64
Exhaust gas recirculation system – general description	49
Exhaust gas recirculation system (FI) – general	66
Exhaust system – general	68
Fault diagnosis – fuel, exhaust and emission control systems	70
Feedback control system	53
Float level – inspection and adjustment	32
Fuel cut-off relay	56
Fuel gauge – testing	7
Fuel filter – removing and refitting	5
Fuel injection system – general description	57
Fuel injection system components – removal, refitting and testing	60
Fuel pump – removal and refitting	10
Fuel pump cut-off relay – general	11
Fuel tank – removal and refitting	6
Fuel tank sender unit – removal, testing and refitting	8
General description	1
High altitude reduced emission – adjustment	52
Idle controller (air conditioning equipped models) – adjustment	18
Idle speed and mixture – adjusting	17
Ignition control system	45
Ignition timing controls (FI) – general	65
Intake air control systeem – inspection	14
Intake air temperature sensor – testing	28
Intake and exhaust manifolds – removal and refitting	69
Low fuel level warning light – testing	9
Manual choke fast idle – adjustment	21
PCV valve – checking, removal and refitting	13
Positive crankcase ventilation (PVC) filter – removal and refitting	12
Power valve (North American models) – description	39
Routine maintenance	2
Secondary air supply system	50
Speed sensor	54
Throttle cable – removal, refitting and adjusting	33
Throttle cable (FI) – removal and adjustment	61
Throttle controller	47
Two-way valve	42

Specifications

General
System type:
   UK models .... Electric fuel pump, downdraught carburettor
   North American models .... Electric fuel pump, downdraught carburettor or fuel injection
Fuel tank capacity .... 13.2 Imp gal (15.9 US gal, 60 litre)
Fuel pump pressure (carburettor) .... 2.4 to 3.1 lbf/cm² (0.17 to 0.22 kg/cm²)
Carburettor .... Keihin, twin choke downdraught
Fuel injection .... Honda programmed fuel injection

Carburettor settings – UK models
Venturi diameter:
   Primary .... 1.26 in (32.0 mm)
   Secondary .... 1.34 in (34.0 mm)
Idle speed:
   Manual transmission .... 750 ± 50 rev/min
   Automatic transmission .... 700 ± 50 rev/min
Idle CO content .... Below 2%
Accelerator pump travel .... 0.73 to 0.77 in (18.5 to 19.5 mm)
Manual choke fast idle .... 1500 to 2500 rev/min
Float level (from gasket to bottom face of float) .... 1.39 to 1.47 in (35.4 to 37.4 mm)

Clearance between needle valve and float lever .... 0.0 to 0.004 in (0.0 to 0.1 mm)
Throttle cable deflection .... 0.16 to 0.40 in (4.0 to 10.0 mm)
Choke cable deflection .... 0.20 to 0.24 in (5.0 to 6.0 mm)
Automatic transmission throttle control cable adjustment
(cable end-to-locknut A) .... 3.366 in (85.5 mm)
Throttle cable end in lever free play .... 0.078 to 0.157 in (2.0 to 4.0 mm)

Carburettor settings – North American models (where different to UK models)

Idle speed (high altitude):
    Manual transmission .... 700 ± 50 rev/min
    Automatic transmission .... 650 ± 50 rev/min
Idle CO content 0.1%
Accelerator pump travel (Canadian models are as UK) .... 0.45 to 0.47 in (11.5 to 12.0 mm)
Automatic choke linkage adjustment (not Canadian models):
    1st stage clearance:
        49 ST and HI ALT:
            Manual transmission .... 0.051 ± 0.003 in (1.29 ± 0.07 mm)
            Automatic transmission .... 0.045 ± 0.003 in (1.15 ± 0.07 mm)
        California
            Manual transmission .... 0.048 ± 0.003 (1.22 ± 0.07 mm)
            Automatic transmission .... 0.043 ± 0.003 in (1.08 ± 0.07 mm)
    2nd stage clearance:
        Manual transmission .... 0.067 ± 0.004 in (1.71 ± 0.09 mm)
        Automatic transmission .... 0.074 ± 0.004 in (1.89 ± 0.09 mm)
    3rd Stage clearance:
        Manual transmission .... 0.181 ± 0.008 in (4.60 ± 0.20 mm)
        Automatic transmission .... 0.189 ± 0.008 in (4.80 ± 0.20 mm)
    Fast idle speed .... 2500 ± 500 rev/min
Automatic choke linkage adjustment (Canadian models):
    Fast idle speed .... 1000 to 2000 rev/min
    1st stage clearance .... 0.102 ± 0.004 in (2.6 ± 0.1 mm)
    2nd stage clearance .... 0.146 ± 0.005 in (3.7 ± 0.12 mm)
    3rd stage clearance .... 0.217 ± 0.010 in (5.5 ± 0.26 mm)

Fuel injection
Idle speed .... 750 ± 50 rev/min
Fast idle speed .... 1000 to 1800 rev/min
Idle CO content 0.1%
Fuel pump pressure .... 33 to 39 lbf/in² (2.35 to 2.75 kgf/cm²)
Throttle cable deflection .... 0.15 to 0.40 in (4.0 to 10.0 mm)

Torque wrench settings

	lbf ft	Nm
Fuel tank drain bolt	36	49
Carburettor mounting nuts	14	19
Fuel pump-to-fuel pump mount bolts	7	9
Fuel pump mounting bolts	7	9
Fuel tank retaining strap nuts	16	22
Fuel injection throttle body mounting nuts	16	22
Fast idle valve mounting bolts (FI)	9	12
Fuel filter fuel bleed bolt (FI)	9	12
Banjo union bolt	16	22
Special banjo bolt	18	24
Filter mounting bolts	9	12
Fuel injection manifold nuts	9	12
Earth cable bolts	9	12
Pressure regulator mounting bolts	7	9
Catalytic converter-to-exhaust nuts	25	34
Converter heat shield bolts	16	22
Support strap bolts	16	22
Exhaust heat shield bolts	7	9
Exhaust downpipe-to-manifold nuts	40	54
Exhaust flange nuts/bolts	16	22
Intake manifold nuts	16	22
Intake manifold support brackets	16	22
EGR valve nuts	16	22
Temperature sender unit	6	8
Exhaust manifold nuts	22	30
Exhaust manifold shroud bolts	18	24
Exhaust manifold bracket to engine (FI)	18	24
Exhaust manifold to bracket (FI)	16	22
Exhaust manifold bracket to engine (carburettor)	20	27
Oxygen sensor probe	33	45
EGR tube upper flanged union (FI)	7	9
EGR tube union in exhaust manifold	43	58
EGR tube upper union (carburettor)	36	49
Air suction tube unions (carburettor)	50	68

Chapter 3 Fuel, exhaust and emission control systems

1 General description

The fuel system is either of downdraught carburettor type with manual or automatic choke or fuel injection.

The fuel tank is mounted at the rear of the vehicle, and the fuel is pumped to the carburettor by an electric fuel pump.

All models have exhaust emission control system, from simple crankcase breathing to the very comprehensive systems for use in North America.

It should be noted that the type of carburettor fitted depends upon several factors such as country of use, manual or automatic transmission etc, which also governs the amount of ancillary carburettor controls fitted, to comply with the emission controls of the country in which the vehicle is to be operated.

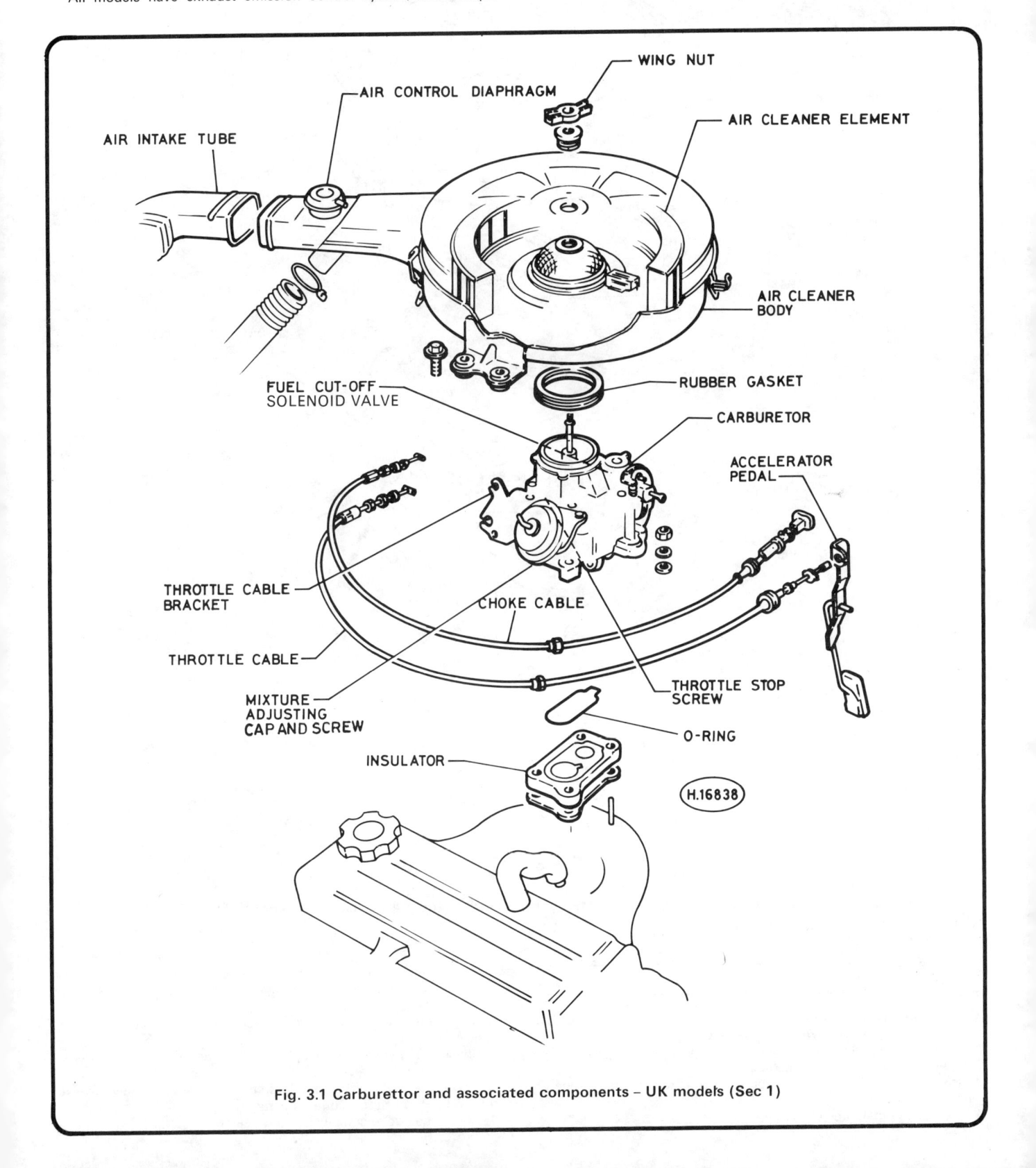

Fig. 3.1 Carburettor and associated components – UK models (Sec 1)

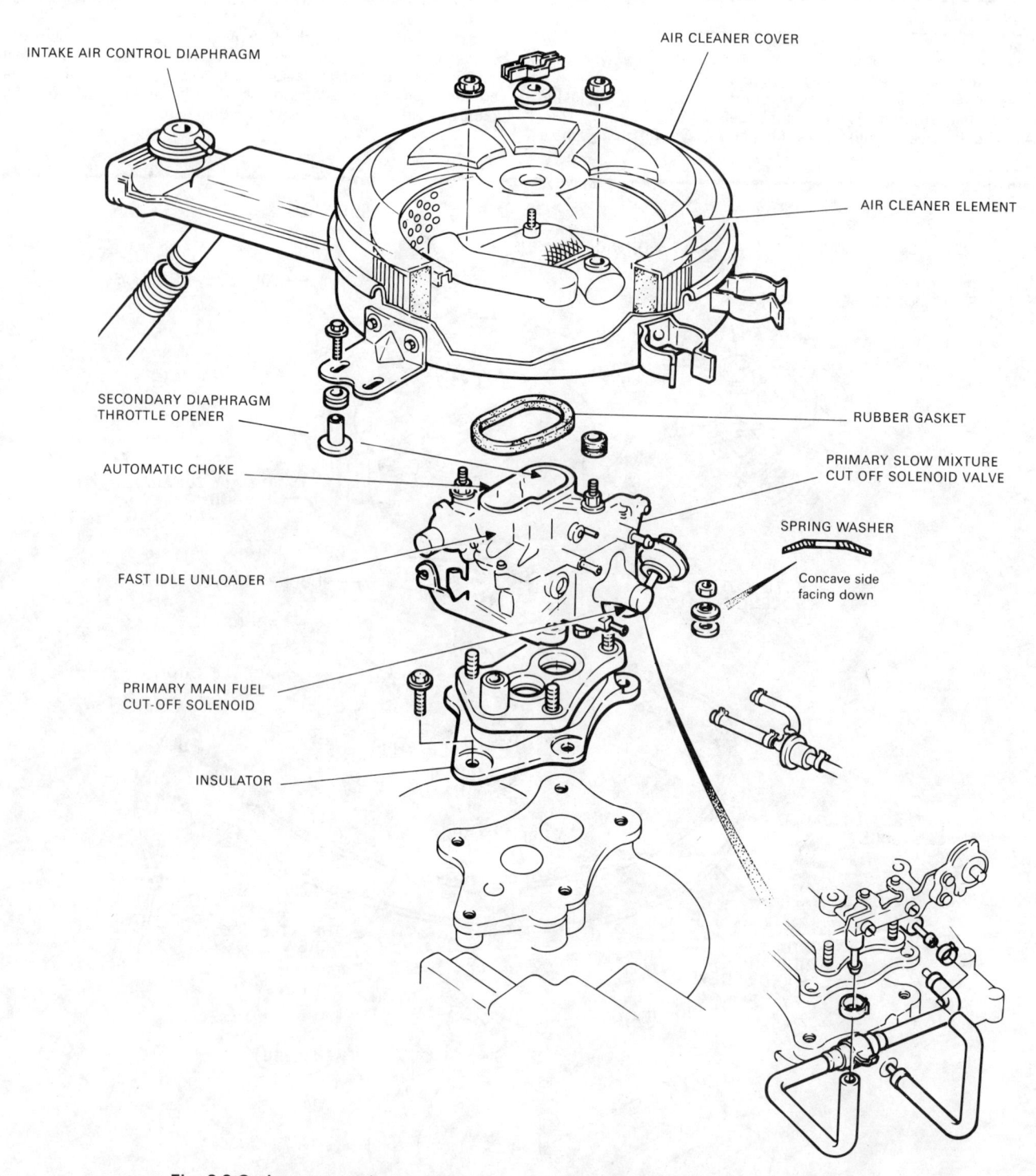

Fig. 3.2 Carburettor and associated components – North American models (Sec 1)

2 Routine maintenance

At the intervals given in Routine Maintenance (at the front of the book) undertake the following service tasks.
1 Check and, if necessary, adjust the idle speed and CO level.
2 Renew the air cleaner element.
3 Inspect the exhaust system for security and condition. Renew/repair as necessary.
4 Renew the fuel filter(s) and hoses.
5 Inspect the fuel tank and fuel lines for leaks, corrosion and security.
6 Inspect the throttle cable system for correct operation (manual transmission).
7 Check the choke opener for correct operation. Clean the mechanism.
8 Check the ignition control system for correct operation (if fitted).
9 Renew the PCV valve and blow-by filter (if fitted).
10 Check the intake air temperature control system, the thermovalve, and the EGR system for correct operation.
11 Check the choke coil tension and heater.

Chapter 3 Fuel, exhaust and emission control systems

12 Check the operation of the throttle control units and the idle control systems.
13 Inspect the catalytic converter heat shield for security.
14 Inspect the charcoal cannisters and two-way valve for condition.

3 Air cleaner element – removal and refitting

1 On carburettor models, remove the wing nut and washer from the centre of the cleaner cover.
2 Undo the three clips around the edge of the air cleaner unit.
3 Lift off the cover and remove the air cleaner element (photo).
4 On fuel injection models, remove the top cover, undo the wing nut and lift out the element.
5 Refitting is a reversal of removal.

4 Air cleaner assembly – removal and refitting

1 Remove the air cleaner element, as described in Section 3.
2 Disconnect the hot air intake and the cold air intake ducting (photo).
3 Disconnect the crankcase breather hose (photo).

4 Remove the bolts/nuts securing the air cleaner to the carburettor and valve cover.
5 On UK models lift the air cleaner, disconnect the vacuum tube to the intake air temperature control and remove the air cleaner.
6 On North American models, remove the air control valve before removing the air cleaner.
7 Refit in the reverse order.

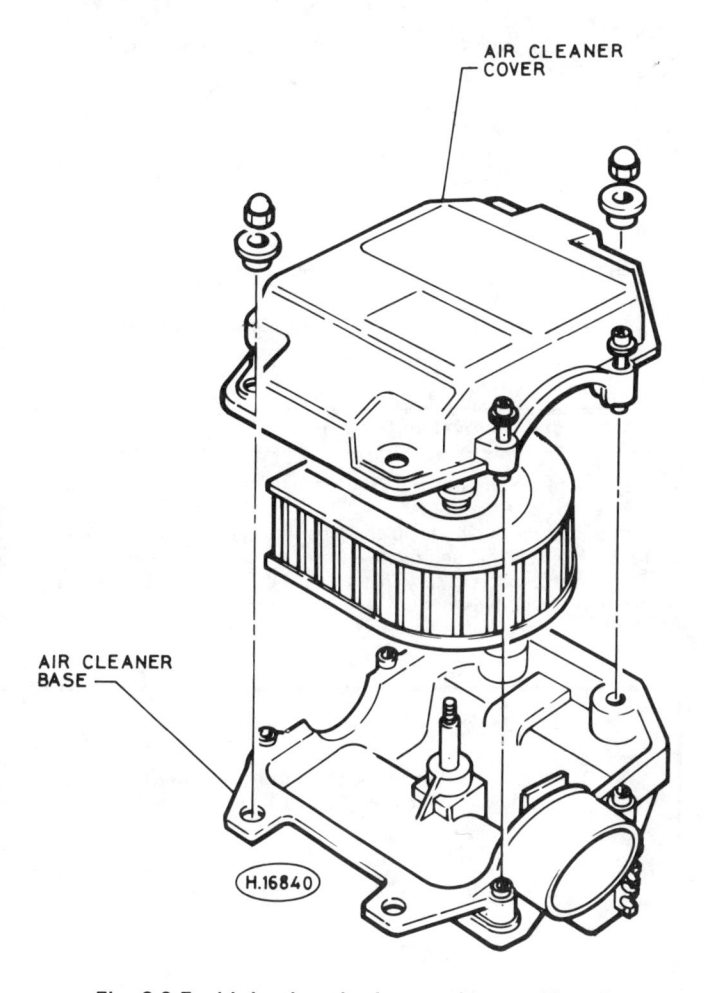

Fig. 3.3 Fuel injection air cleaner element (Sec 3)

3.3 Lift out the air cleaner

4.2 Cold air intake duct (arrowed)

4.3 Crankcase breather hose connection (arrowed)

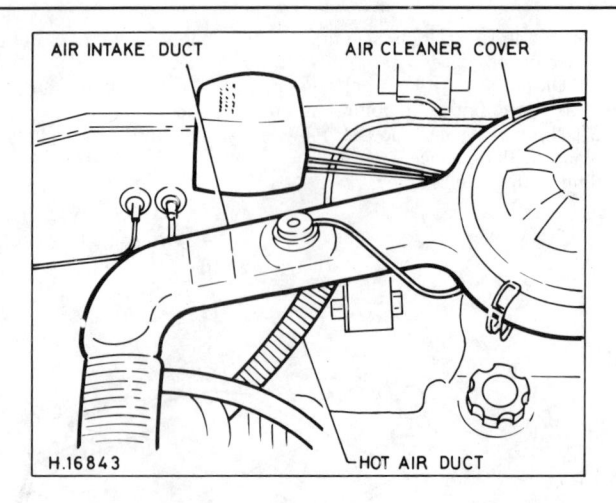

Chapter 3 Fuel, exhaust and emission control systems

Fig. 3.4 Air intake ducts to air cleaner (Sec 4)

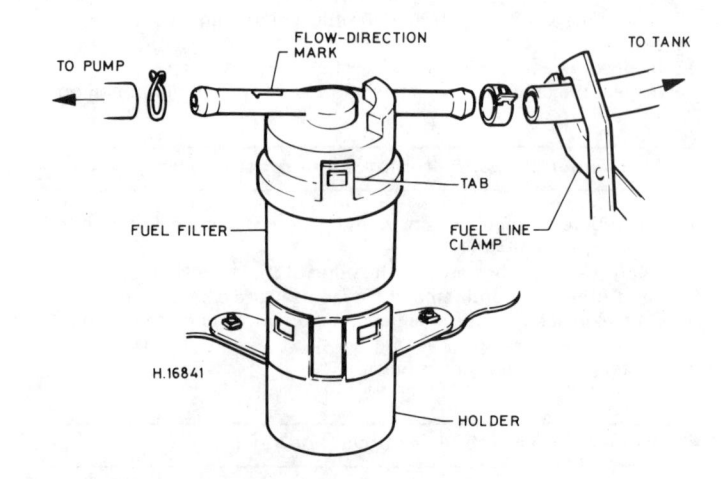

Fig. 3.5 Fuel filter housing assembly (Sec 5)

5 Fuel filter – removal and refitting

1 The fuel filter is mounted on the left-hand side of the fuel tank.
2 It is a clip fit in the holder.
3 To remove the filter, clamp the inlet and outlet hoses to prevent excessive fuel spillage, and remove the hoses from the filter.
4 Push in the locating tab on the fuel filter to release it from the holder, and lift the filter out.
5 Refit in the reverse order.

6 Fuel tank – removal and refitting

Caution: *Fuel tanks are in an enhanced explosive state when empty – it is the vapour which is more dangerous than the fuel itself. Take every fire precaution when working on fuel tanks.*
1 Disconnect the battery.
2 Raise the rear of the car onto stands. Chock the front wheels.
3 Remove the fuel tank drain bolt and drain the fuel into a suitable container.

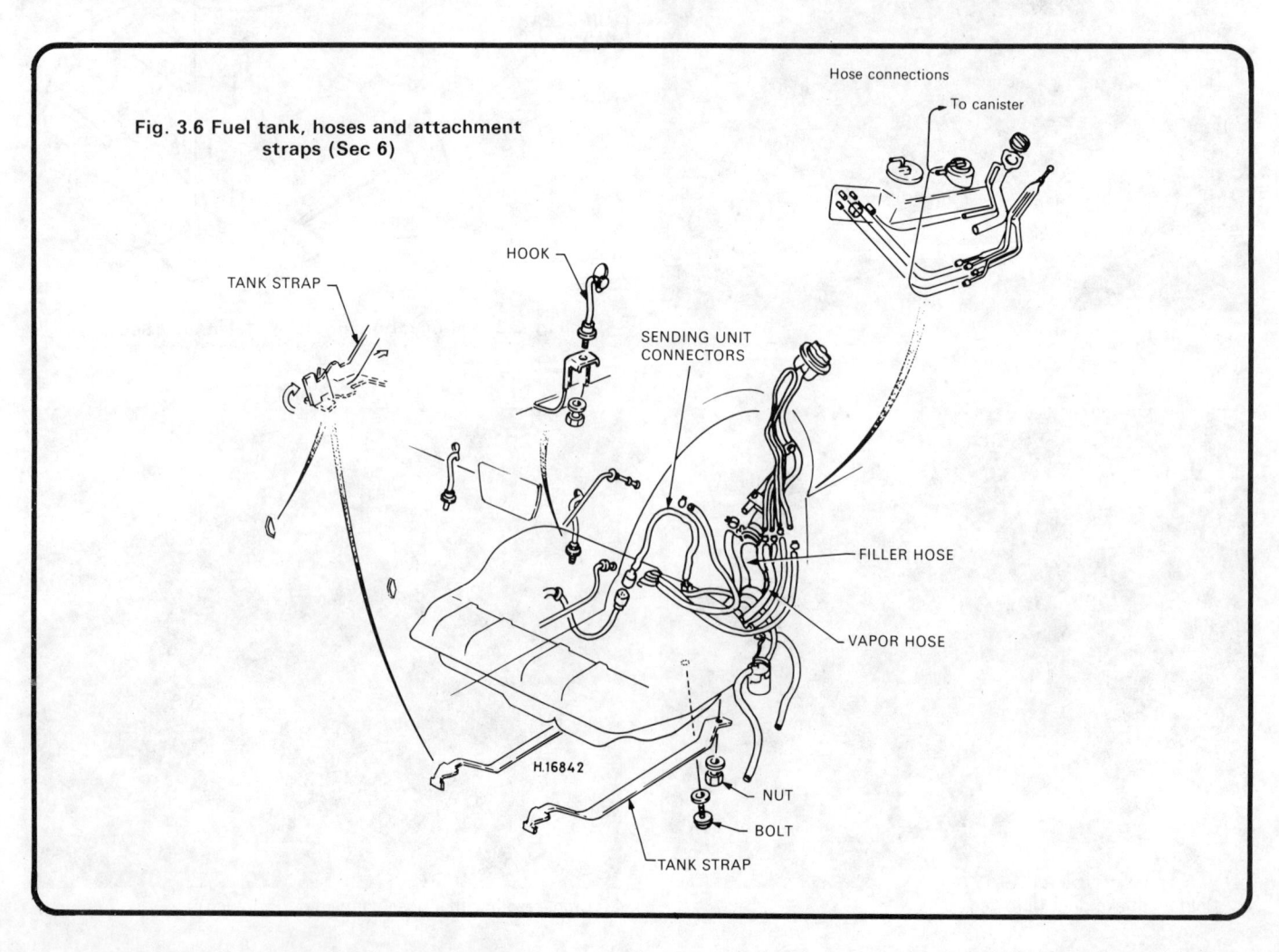

Fig. 3.6 Fuel tank, hoses and attachment straps (Sec 6)

Chapter 3 Fuel, exhaust and emission control systems

4 Disconnect the fuel tank sender unit connectors.
5 Disconnect the fuel hoses to the fuel filter and two-way valve, the vent hoses and fuel filler hose.
6 Position a jack under the tank, with a pad between the jack head and the tank, and just take the weight of the tank.
7 Remove the tank support straps.
8 Slowly lower the jack, steadying the tank as you do so and remove the tank from under the vehicle. **Note:** the tank may have become stuck to the underside under-seal, and may require levering to remove it.
9 Refitting is a reverse of removal.

7 Fuel gauge – testing

1 Jack the vehicle and remove the left rear wheel.
2 Switch the ignition off.
3 Disconnect the fuel gauge sender unit.
4 Connect the yellow/white wire to the black wire with a jumper lead.
5 Switch on the ignition and watch the fuel gauge.
6 Switch the ignition off as soon as the needle stops moving; it should have moved to the full mark.
Caution: *do not leave the ignition switched on for longer than 5 seconds, or damage to the fuel gauge will result.*
7 If the fuel gauge moved to the full mark, it is serviceable, and it can be assumed that if there is a fault, then it is in the sender unit.
8 If the fuel gauge did not move to the full mark, then check the fuse, wiring and connectors.
9 Renewal of the fuel gauge will be found in Chapter 12.
10 Reconnect the electrical connection and refit the roadwheel before removing the vehicle from jacks.

8 Fuel tank sender unit – removal, testing and refitting

1 Drain and remove the fuel tank, as described in Section 6.
2 Remove the fuel tank sender unit from the tank, using the special wrench or a peg spanner.
3 Connect an ohmmeter to the yellow/white wire and the black wire, and measure the resistance at the three float positions indicated in Fig. 3.10.
4 The resistance should be as follows:

Empty	105 to 110 ohm
1/2 full	25.5 to 39.5 ohm
Full	2 to 5 ohm

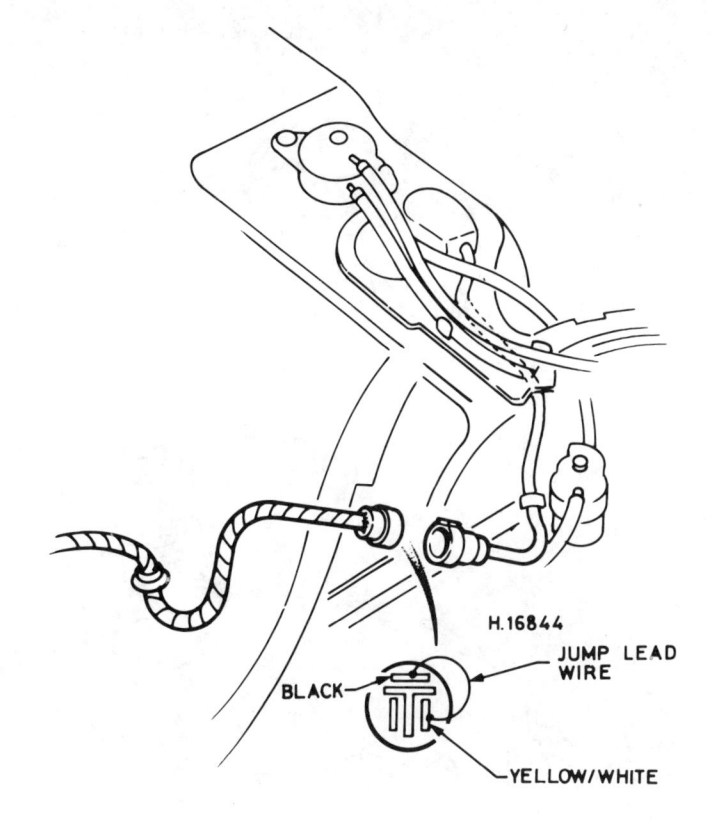

Fig. 3.7 Fuel gauge testing (Sec 7)

5 If the resistance is correct, then suspect the fuel gauge of being faulty.
6 If the resistance is incorrect, renew the sender unit, testing it as described above before installation.
7 Use a new sealing washer under the sender unit when refitting to the tank.
8 Refit the tank as described in Section 6.

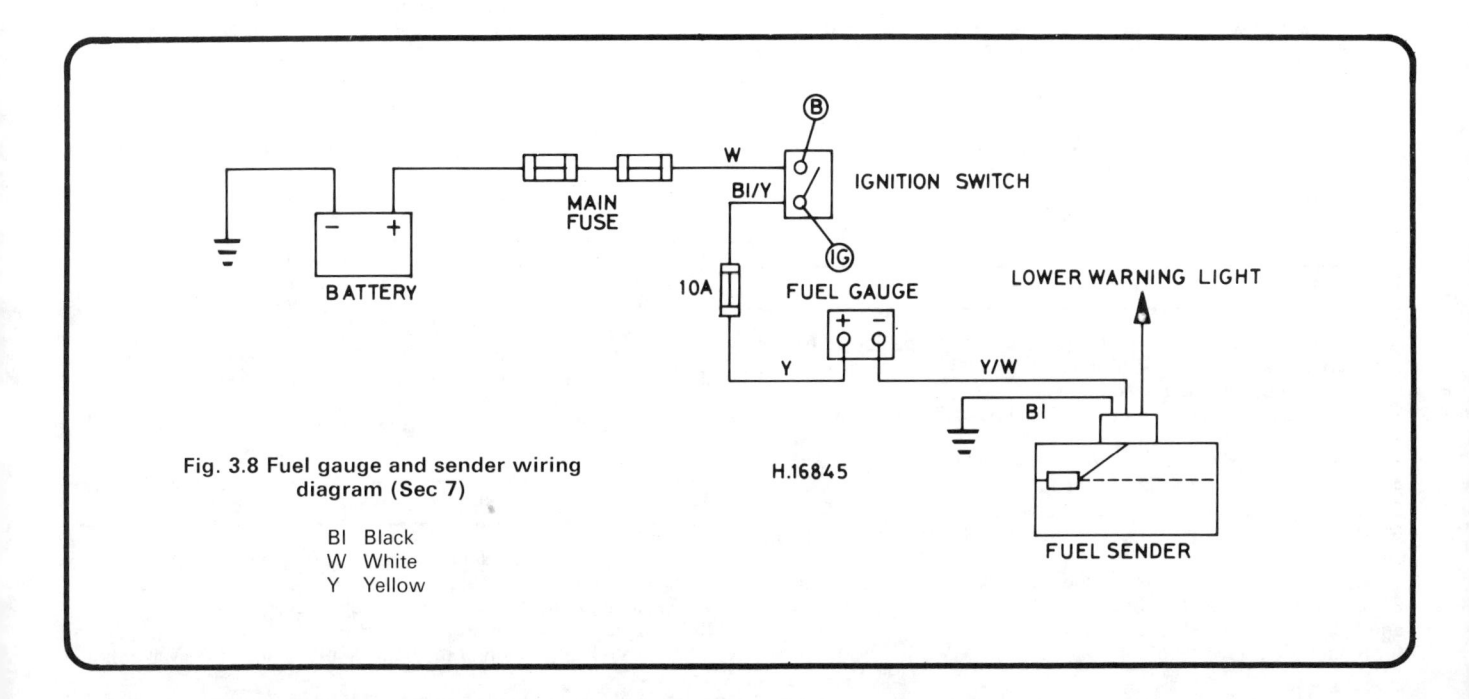

Fig. 3.8 Fuel gauge and sender wiring diagram (Sec 7)

Bl Black
W White
Y Yellow

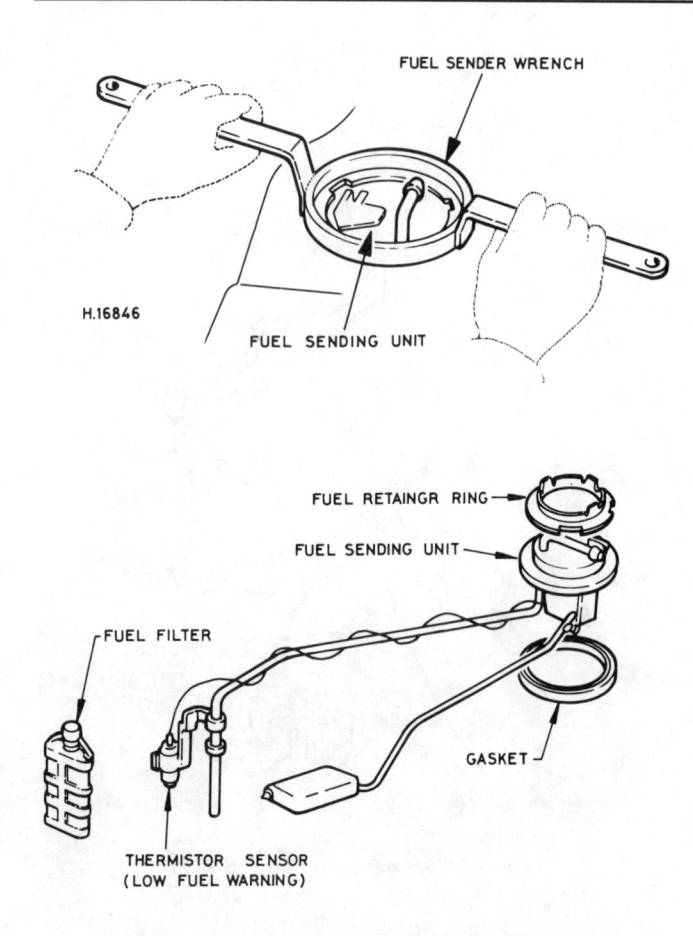

Fig. 3.9 Components of the fuel sender unit (Sec 8)

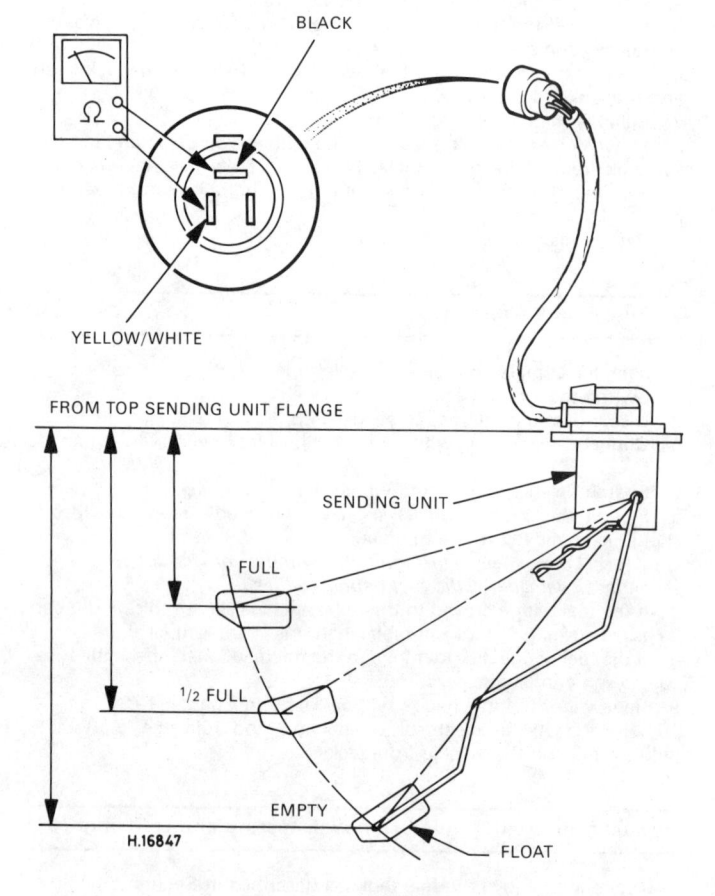

Fig. 3.10 Checking positions for resistance of fuel sender unit (Sec 8)

9 Low fuel level warning light – testing

1 With the fuel tank empty and the vehicle parked on level ground, fill the tank with 2.2 Imp gal (2.6 US gal, 10 litres) of fuel.
2 Switch on the ignition.
3 The low fuel warning light should come on within 3 minutes.
4 If the light comes on, add at least 2.5 Imp gal (3.0 US gal, 11.5 litre) of fuel and check that the light goes off.
5 If the light does not go out, check for a short circuit in the wiring harness or the printed circuit of the instrument board.
6 If no fault can be found here, renew the sender unit.
7 If the light did not come on initially (paragraph 3), check the wiring and the bulb of the warning light.
8 If no fault found renew the sender unit.

10 Fuel pump – removal and refitting

1 The testing of the fuel pump to assess its correct operating pressure and delivery rate is best left to your dealer, as it requires the use of pressure gauges and graduated beakers.
2 If it is proved or suspected that the pump is defective, remove it as follows:
3 Jack the rear of the vehicle and remove the left-hand rear wheel.
4 Remove the protective shield from the fuel pump (photo).
5 Clamp the fuel lines to the pump to prevent excessive fuel leakage.
6 Disconnect the fuel lines and electrical leads from the pump (photo).
7 Remove the fuel pump mounting bolts and remove the pump.
8 The fuel pump and fuel flow meter, if fitted, cannot be overhauled and should be renewed as a unit.
9 Refitting is a reversal of removal.

11 Fuel pump cut-off relay – general

1 The fuel pump cut-off relay is mounted into the back of the fusebox panel.
2 If it is suspected of malfunction, your local dealer can test the cut-off relay circuit, but it is easier to renew the relay with a new item; testing by substitution.
3 This is a wise step to take before renewing the pump.

12 Positive crankcase ventilation (PCV) filter – removal and refitting

1 The PCV filter is fitted to the air cleaner casing.
2 Remove the air cleaner element, as described in Section 3.
3 Disconnect the breather hose from the filter housing.
4 Remove the two bolts securing the filter housing to the air cleaner case (photo).
5 Remove the filter housing (photo), discard the element and wash the filter housing in solvent, drying it off afterwards.
6 Fit a new element to the housing.
7 Refit the filter and housing in the reverse order of removal.

13 PCV valve – checking, removal and refitting

1 With the engine cold, check the PCV breather hose and valve for cracks, disconnection or blockage.
2 Remove the PCV valve from the valve cover.
3 Start the engine and allow it to idle.
4 Repeated application of the finger over the end of the valve should

10.4 Fuel pump shield retaining bolts (arrowed)

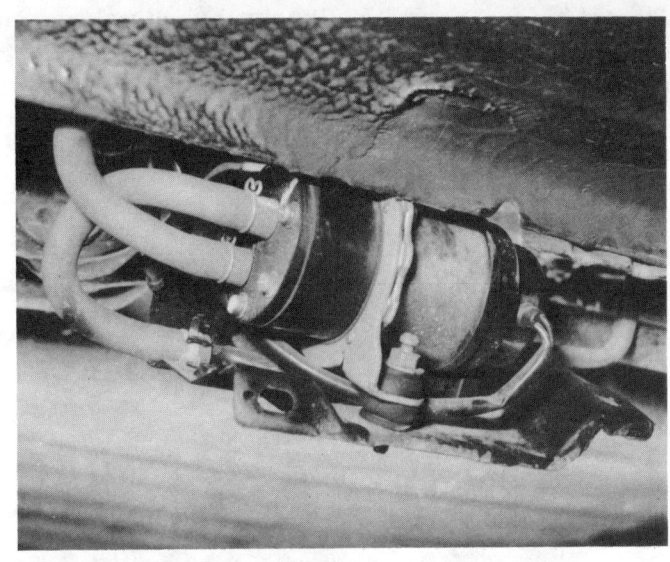

10.6 General view of the fuel pump

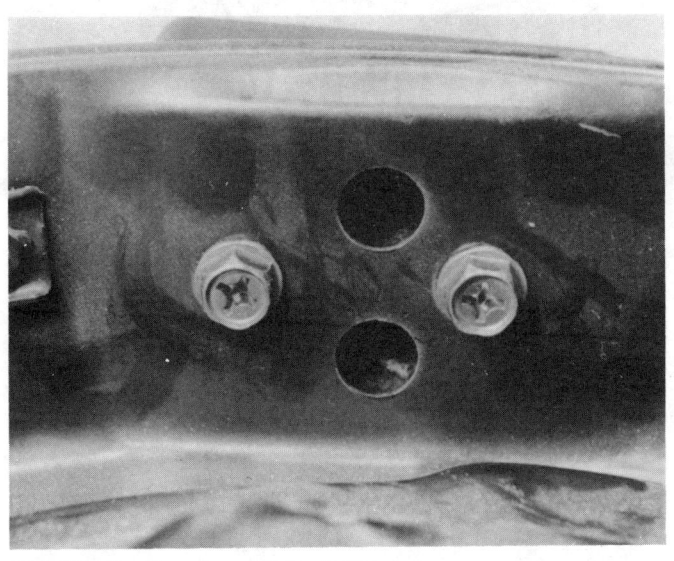

12.4 PCV filter housing retaining bolts

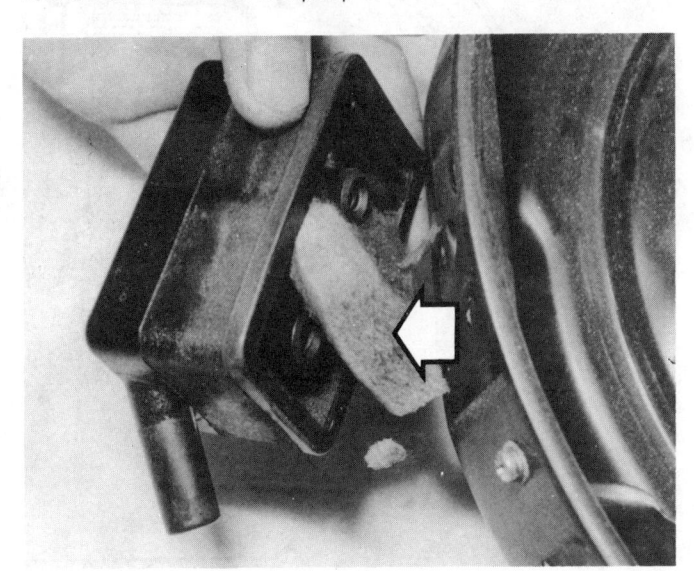

12.5 PCV filter (arrowed) and housing

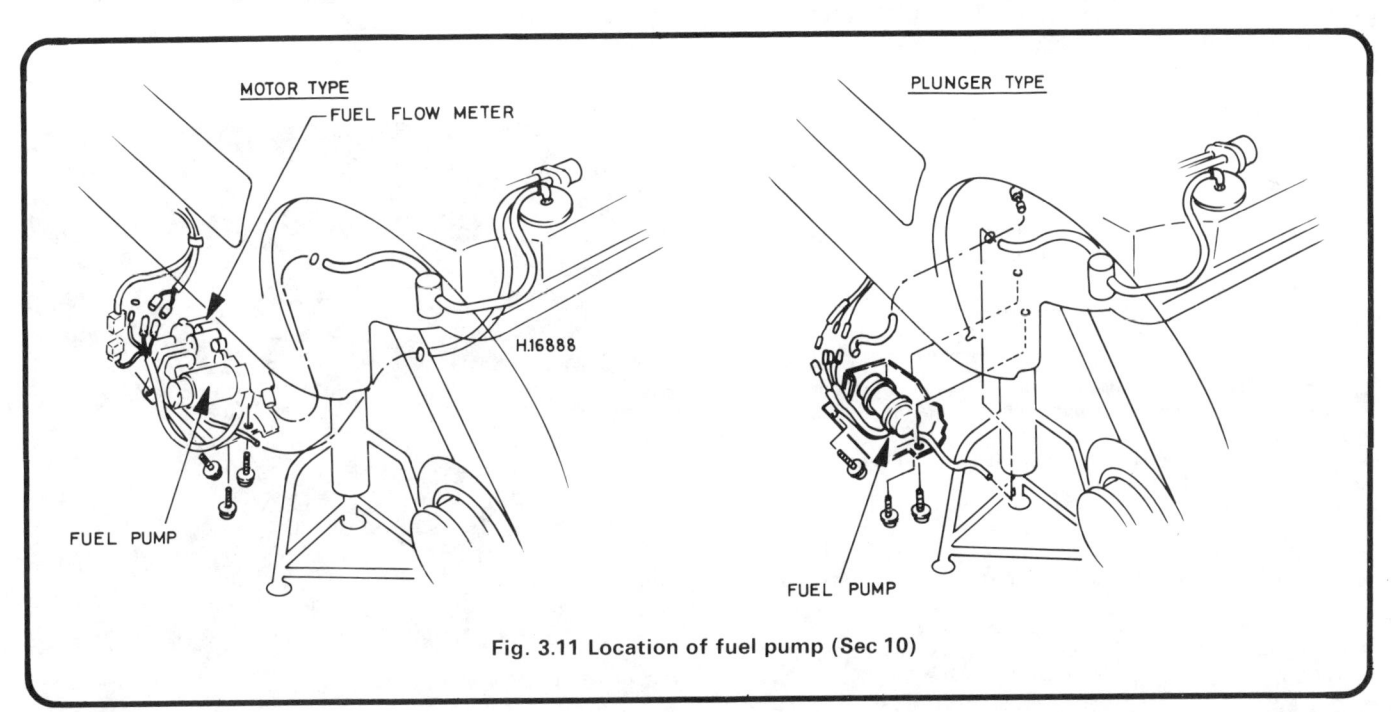

Fig. 3.11 Location of fuel pump (Sec 10)

80   Chapter 3 Fuel, exhaust and emission control systems

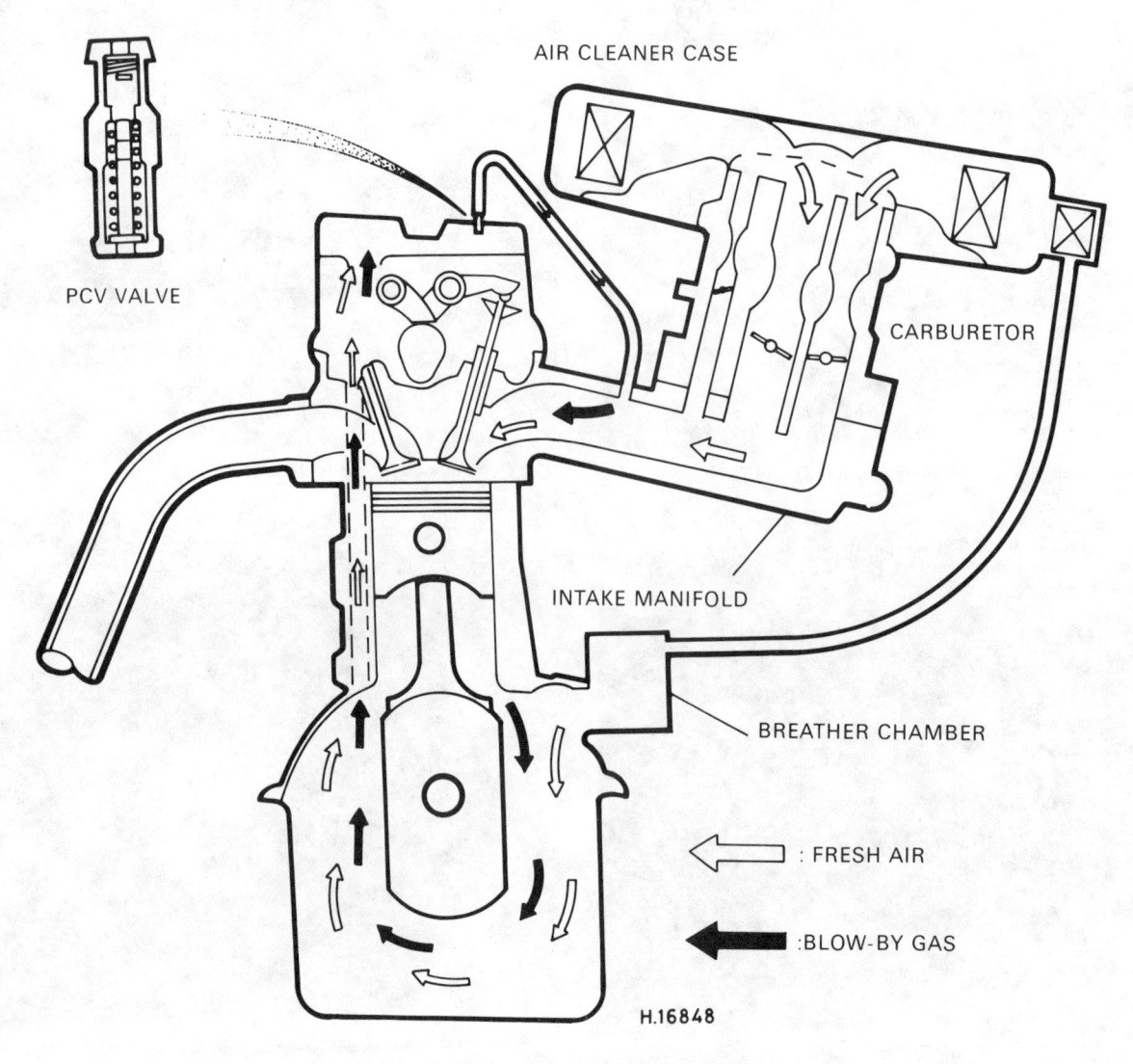

Fig. 3.12 Positive crankcase ventilation (PCV) system (Sec 12)

result in a clicking sound being heard, showing that the valve is working.
5 If not, renew the valve, which is a simple matter of disconnecting it from the hose.
6 Switch off the engine.
7 Reconnect the PCV valve into the valve cover, inspecting the rubber grommet into which it plugs for integrity.

14 Intake air control system – inspection

1 Remove the air cleaner element and disconnect the air inlet duct.
2 With the engine cold, start and run the engine for about 5 seconds and then stop it.
3 The air control door should rise on start-up and remain fully open for at least 3 seconds after the engine is stopped.
4 If these conditions are met, the system is working correctly, if not, proceed as follows:
5 Check that the air door can operate freely and is not binding.
6 Disconnect and plug the hose leading to the air intake temperature sensor.
7 Repeat the test procedure given above.
8 If the door now operates correctly, renew the air temperature sensor (photo), connect the hose to it and retest.

14.8 Intake air temperature sensor

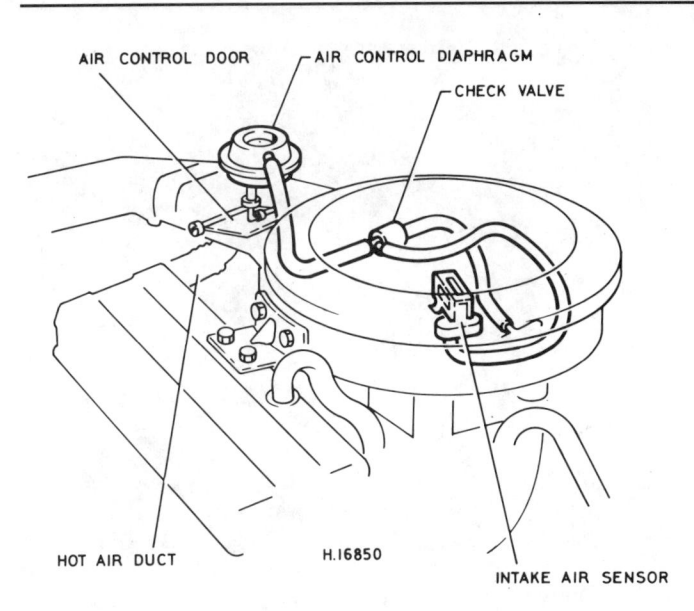

Chapter 3 Fuel, exhaust and emission control systems

Fig. 3.13 Air intake control system (Sec 14)

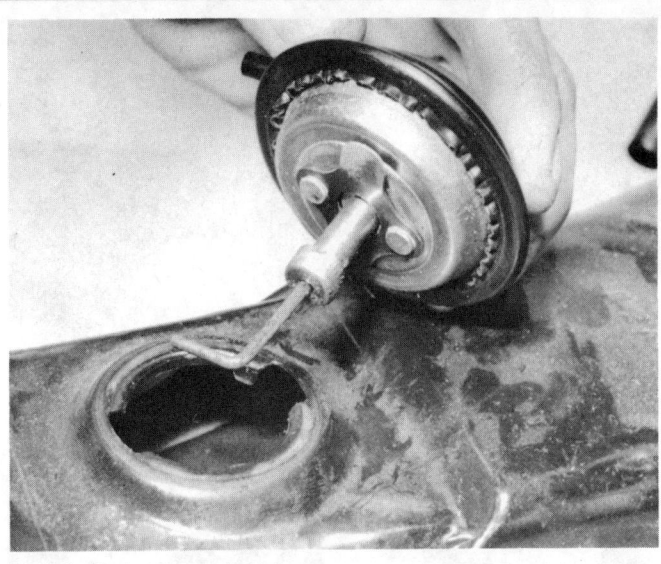

(Labels: AIR CONTROL DOOR, AIR CONTROL DIAPHRAGM, CHECK VALVE, HOT AIR DUCT, H.16850, INTAKE AIR SENSOR)

14.12A Air door control valve

14.12B Vacuum hose connection to air door control valve

9 If the door still malfunctions, disconnect the hose to the air door control valve.
10 Raise the door by hand, then block the inlet pipe to the valve and release the door.
11 If the door stays up, renew the check valve and retest.
12 If the door drops, renew the air door control valve (photos).

15 Carburettor – general description

1 The carburettors fitted to the vehicles covered by this manual are all Keihin downdraught types, but vary according to model and country of use.
2 The degree of additional carburettor control also depends upon the law of the country in which the vehicle is to be operated regarding exhaust emission controls.
3 Either a manual of automatic choke is fitted, again depending on model.

16 Carburettor adjustments – general

1 Before attempting to adjust the carburettor, make sure that the ignition system is in good order, the air cleaner element is clean, and the engine itself is in good mechanical order.
2 Certain adjustment screws or orifices may be protected with a 'tamperproof' plug or seal. This is both to discourage and to detect adjustment by unqualified operators.
3 In some countries, though not yet in Britain, it is an offence to drive a vehicle without these tamperproof devices fitted.
4 Be sure, therefore, that current local legislation permits the removal of tamperproof plugs, and fit new plugs on completion of adjustment.

17 Idle speed and mixture – adjusting

Note: For accurate adjustment of the carburettor, a tachometer, vacuum gauge and exhaust gas analyser will be required.

CO meter method

1 Warm up both the engine and the CO meter, in accordance with the manufacturer's instructions.
2 Insert the gas probe into the exhaust tail pipe by at least 16 in (400 mm).
3 Connect up a tachometer to the ignition coil, in accordance with the maker's instructions.
4 Check the idle speed with cooling fan, air conditioner and headlights off.
5 Adjust by turning the throttle stop screw (photo).
6 If the correct idle speed cannot be obtained, check the throttle cable adjustment.
7 With the idle speed correct, and again with cooling fan, air conditioner and headlights off, check that the exhaust gas CO content is within limits.
8 If adjustment is required, remove the plug or limiter cap from the mixture adjusting screw and adjust the screw to give the desired reading (photo).
9 Turning the screw clockwise decreases CO content, and anti-clockwise increases CO content.
10 On completion, refit the plug or limiter cap, remove the tachometer and CO meter.

Note: On cars fitted with an exhaust catalytic converter, CO meters do not work. The adjustment will have to be entrusted to your dealer, who has the propane enrichment equipment to do the adjusting accurately, or use the idle drop method described here.

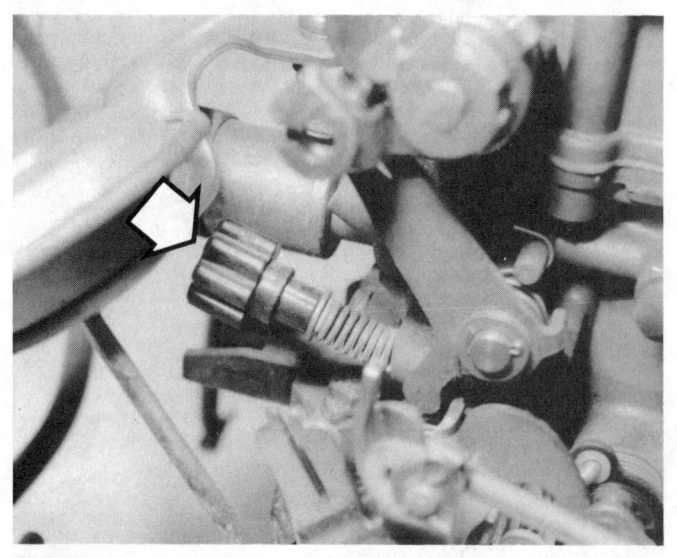

Chapter 3 Fuel, exhaust and emission control systems

17.5 Throttle stop screw (arrowed)

17.8 Idle mixture adjusting screw (arrowed)

Idle drop method
11 Warm up the engine to normal operating temperature.
12 Remove the plug or limiter from the mixture screw.
13 With the cooling fan, air conditioner and headlights switched off, adjust the engine speed and mixture to give the best idle at 800 rev/min (manual transmission) or 780 rev/min (automatic in gear).
14 Once satisfactory idling has been obtained, turn the mixture screw clockwise to decrease the engine speed to that specified.
15 Switch off the engine and refit the tamperproof plug.

18 Idle controller (air conditioning equipped models) – adjustment

1 On vehicles equipped with air conditioning an additional idle control is fitted, designed to prevent the idle speed from dropping when the A/C is switched on (photo).
2 After having set idle and mixture, as previously described in Section 17, switch on the A/C.
3 Idle speed should stay within limits.
4 Adjust by removing the rubber cap on the idle controller adjustment screw, and turning the screw in or out as required.
5 If the idle speed cannot be brought within limits by the adjusting screw, disconnect the vacuum hose from the idle controller diaphragm and check for vacuum.
6 If there is vacuum, renew the diaphragm and retest.
7 If there is no vacuum, check for voltage at the idle control solenoid valve (photo).
8 If there is no voltage, check wiring and fuse.
9 If there is voltage, check the vacuum line to the intake manifold for leaks or blockage, and if found unsatisfactory, renew.
10 If there is still a problem, renew the idle control solenoid valve.

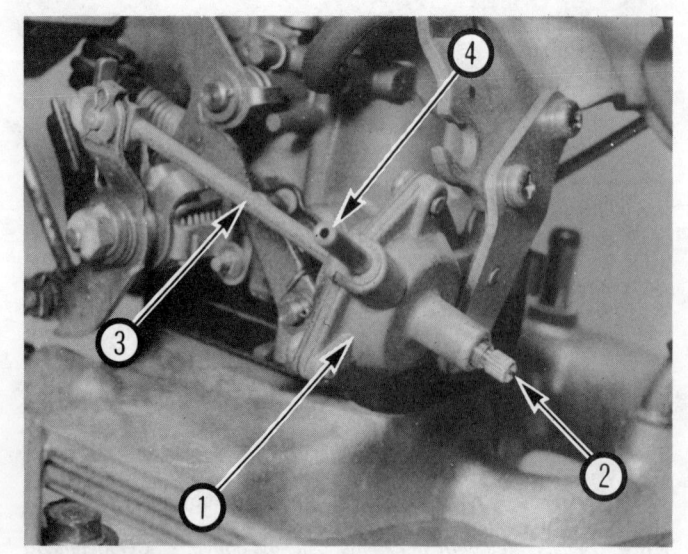

18.1 Idle control diaphragm fitted to models with air conditioning
1 Diaphragm
2 Adjuster screw
3 Control rod
4 Vacuum connection

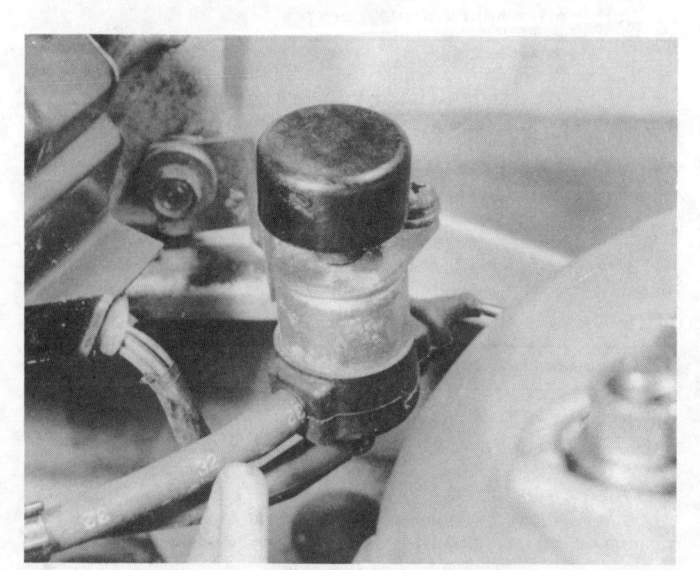

18.7 Idle control solenoid valve fitted to cars with air conditioning

Chapter 3 Fuel, exhaust and emission control systems

83

19.1 Accelerator pump linkage
1 Pump lever
2 Accelerator pump
3 Adjustment point

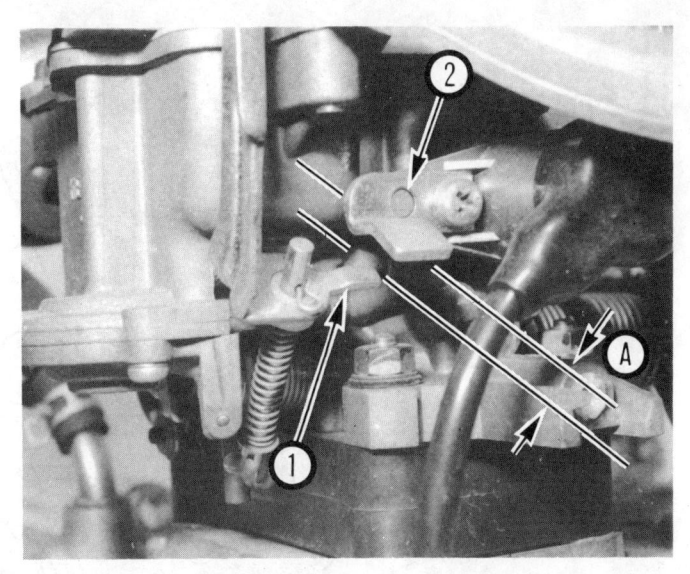

19.3 Accelerator pump clearance
1 Pump lever tang
2 Stop bracket
A = clearance to be measured

19 Accelerator pump-adjustment

1 Ensure the accelerator pump linkage and lever are free to move, and that the lever is just contacting the pump shaft (photo).
2 Measure the distance between the tang on the pump lever and the stop bracket on the carburettor.
3 If out of specification, bend the tang on the pump lever (photo).
4 On carburettors fitted to North American models the measurement is taken between the accelerator pump lever and the throttle control bracket.
5 Bend the tang on the throttle control end of the lever – do not bend the lever where it contacts the accelerator pump.

20 Choke relief valve – adjustment

1 If, during cold weather, the engine becomes hard to start, or it is impossible to drive the car with the choke operating, move the position of the spring to the next notch on the valve plate (Fig. 3.14).

21 Manual choke fast idle – adjustment

1 Connect a tachometer to the coil in accordance with the manufacturer's instructions.
2 Start the engine and allow it to warm up.
3 Position the choke control knob to the first position.
4 Check that the reference arm on the choke flap shaft and the reference boss on the carburettor line up.
5 Check the fast idle speed is as specified.
6 If the fast idle speed is too high, narrow the slot in the fast idle adjusting link, using pliers (photo).
7 If the fast idle speed is too low, widen the slot.
8 Make all adjustments in small increments.
9 On completion, switch off the engine and remove the tachometer.

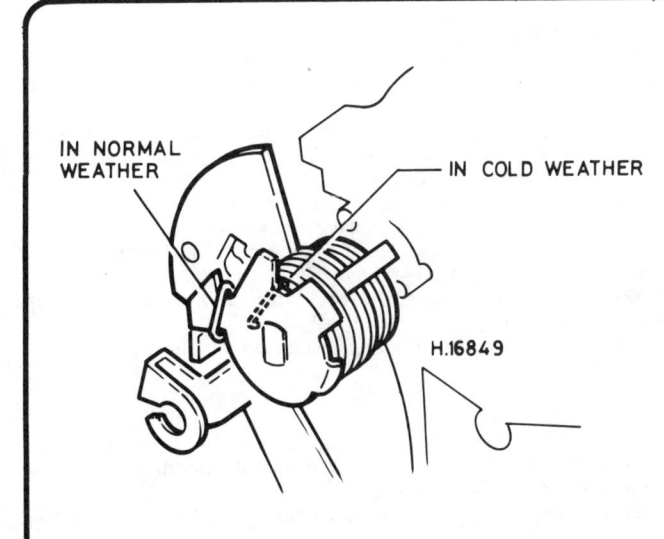

Fig. 3.14 Choke relief valve spring (Sec 20)

21.7 Manual choke fast idle adjustment link (arrowed)

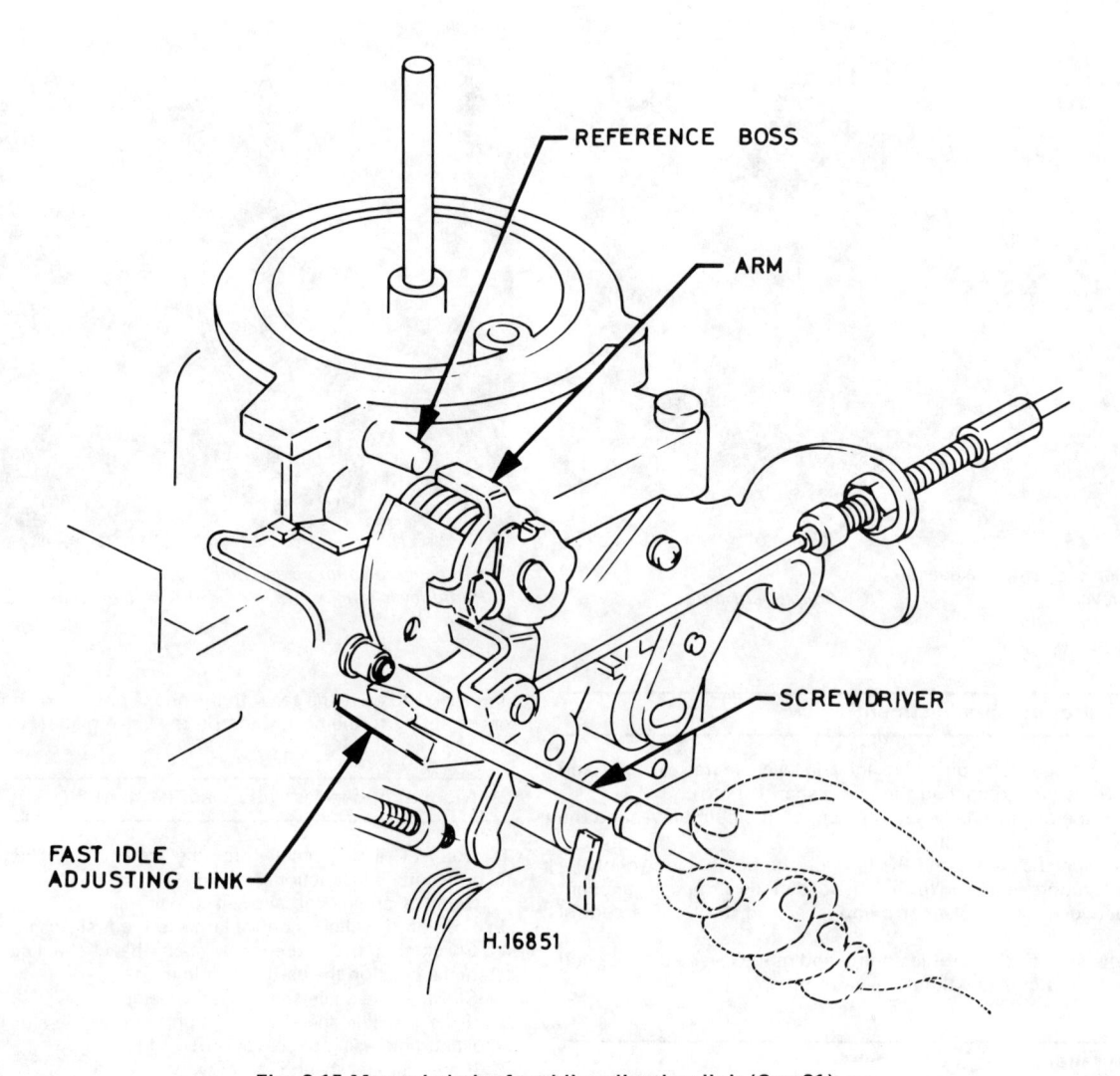

Fig. 3.15 Manual choke fast idle adjusting link (Sec 21)

22 Automatic choke – general description

1 The automatic choke is designed to control both the choke valve setting and fast idle position during warm-up.
2 The mechanism consists of a bi-metal coil and its heater, a thermistor, intake air sensor, external resistor, choke opener, thermovalves A and B, and a fast idle unloader.
3 The coil, heater and thermistor are in the choke cover, the choke opener and fast idle unloader are mounted externally on the curburettor, the intake air sensor is in the air cleaner, the resistor is on the engine bulkhead next to the emission box, and the two thermovalves are screwed into the intake manifold.
Note: the automatic choke fitted to Canadian models is different, and is described in Section 37 and 38.

23 Automatic choke linkage – adjustment

1 Remove the choke cover by drilling out the rivets securing the choke cover retaining ring to the housing.
2 Use a $^5/_{32}$ in (4.1 mm) drill, and ensure that no drillings fall into the carburettor.
3 Do not lose the two gears in the cover and housing.
4 Hold the choke blade closed and open and close the throttle fully to engage the choke and fast idle linkage.
5 Disconnect the vacuum hose from thermovalve A and pressurise the choke opener with air to hold the bleed valve closed: 15 to 85 lbf/in² (1.1 to 6.0 kgf/cm²) is sufficient.
6 Gently push the choke opener lever towards the opener until it stops, then pull the choke drive lever down against the opener lever, taking all free play out of the linkage.
7 Measure the clearance between the choke blade and carburettor intake wall, which should be as specified (1st stage).
8 Adjust by bending tab D in Fig. 3.19.
9 Remove the air pressure and reconnect the vacuum hose.
10 Hold the choke drive and choke opener lever together and push them both towards the diaphragm until they stop. Tab A on the opener lever should be seated against the carburettor (Fig. 3.20).
11 Measure the clearance at the choke blade, which should be as specified (2nd stage).
12 Keep the choke opener lever against its seat and release the choke drive lever.
13 Again measure the choke blade clearance, which should be as specified (3rd stage).
Note: Tab C on the drive lever should stay seated against the spring loop (Fig. 3.21). If it does not, open and close the throttle as described in paragraph 4, and recheck.
14 With all checks and clearances completed and within tolerance, refit the choke cover.
15 Adjust the cover so that the index marks line up.
16 Refit the gears into the cover and housing.
17 Install the retainer ring and secure with rivets.

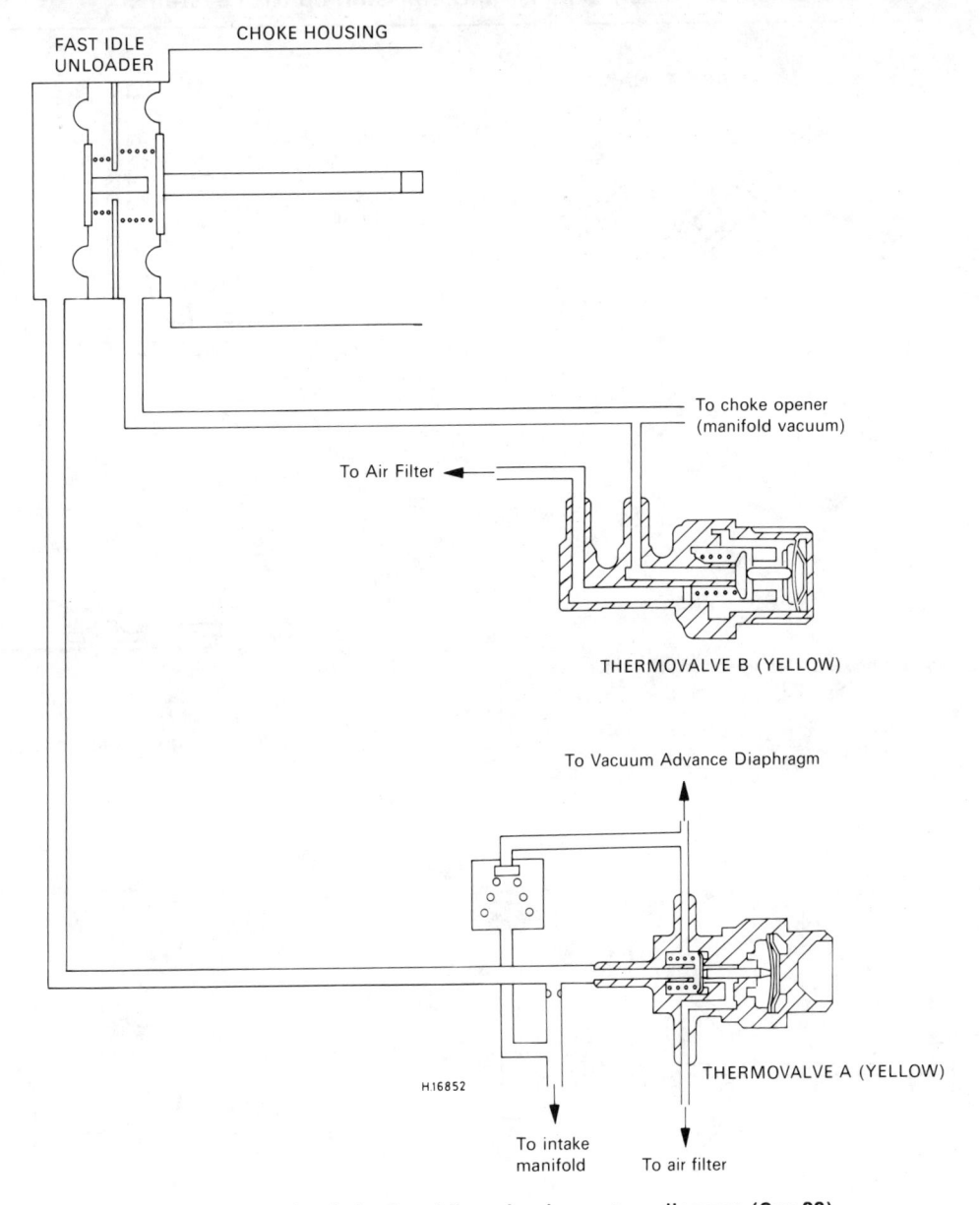

Fig. 3.16 Automatic choke fast idle unloader system diagram (Sec 22)

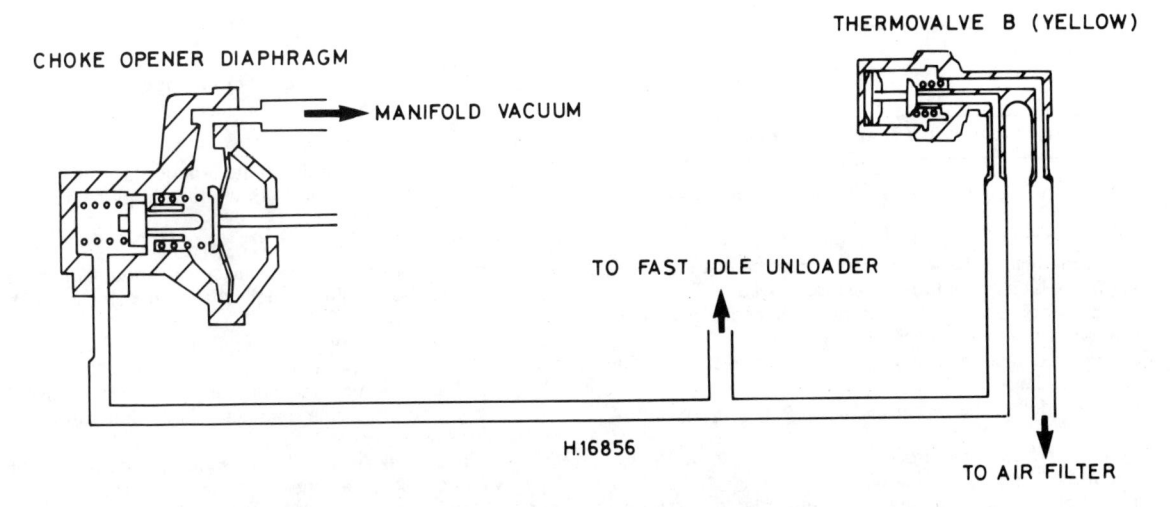

Fig. 3.17 Automatic choke – choke opener system diagram (Sec 22)

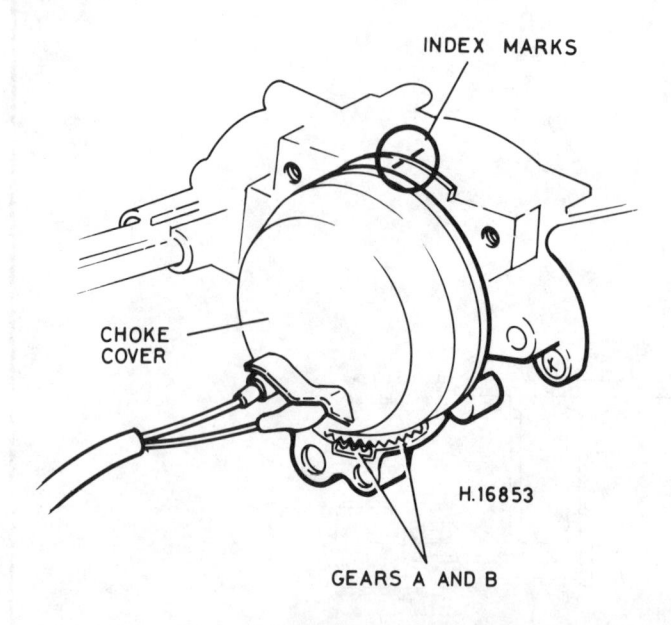

86　　　　　　　　　　　Chapter 3 Fuel, exhaust and emission control systems

Fig. 3.18 Automatic choke cover index marks and gears (Sec 23)

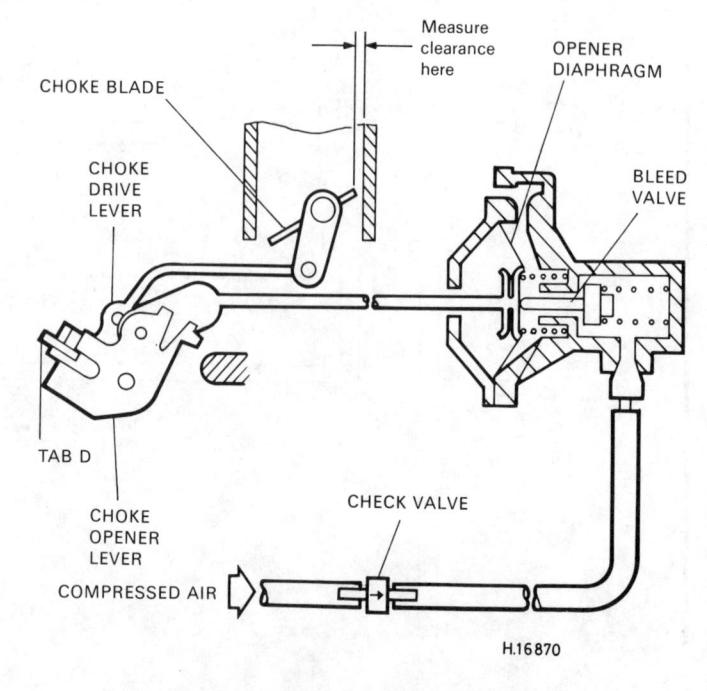

Fig. 3.19 Stage 1 clearance (Sec 23)

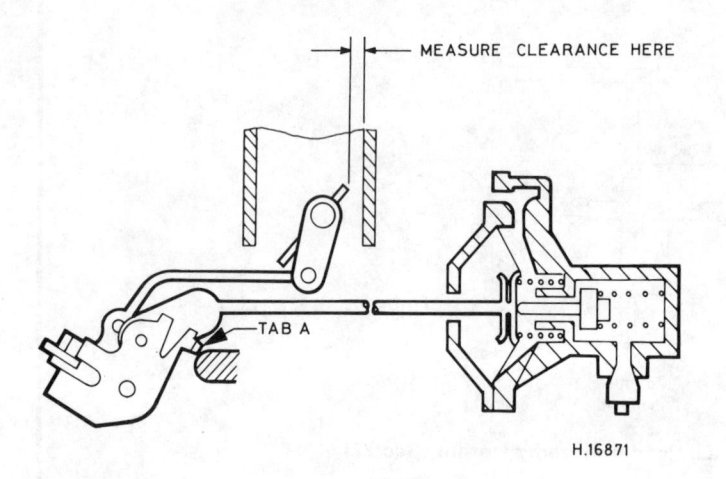

Fig. 3.20 Stage 2 clearance (Sec 23)

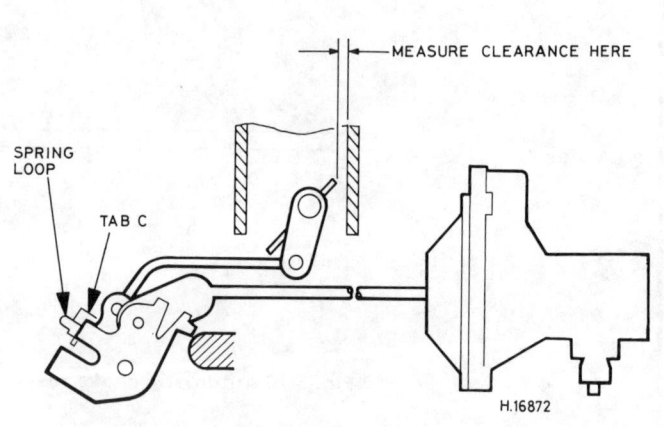

Fig. 3.21 Stage 3 clearance (Sec 23)

24 Automatic choke fast idle inloader – inspection

1 To perform this inspection, the engine coolant temperature must be below 30°C (86°F).
2 Disconnect the two hoses from the fast idle unloader.
3 Open fully and then close the throttle to engage the fast idle cam.
4 Start the engine, which should run at fast idle speed.
5 If the engine does not run at fast idle, remove the choke cover and check the operation of the fast idle cam (see Section 23).
6 Connect a hand held vacuum pump to the inner connection on the fast idle unloader and draw vacuum.
7 With the engine running, the fast idle speed should drop when the vacuum is applied.
8 If the fast idle speed does not drop, check the unloader for leaks, blockage or a split diaphragm.
9 Remove the choke cover and check the unloader rod for free movement.
10 Renew defective parts as necessary.
11 Once satisfactory operation of the choke unloader is achieved with the engine in the cold condition, its function should be checked with the engine hot, as described in the following paragraphs.

Hot engine

12 As the engine warms up, idle speed should drop below 1400 rev/min as the choke unloader pulls the internal linkage off the fast idle cam.
13 If idle speed does not decrease, disconnect the vacuum hoses and check that vacuum is present in each hose.
14 If so, check the unloader for leaks or blockage, and remove the choke cover and check the unloader rod for full and free movement. Renew parts as necessary.
15 If no vacuum exists, check for vacuum at the thermovalves.
16 If vacuum exists at the inlet, but not the outlet, renew the thermovalve. If there is no vacuum at the inlet, check all hoses for kinking, leaks or blockage, and repair/renew as necessary.
17 Carry out the test procedure once more to check satisfactory operation of the unloader.

Chapter 3 Fuel, exhaust and emission control systems

25 Automatic choke, choke opener – inspection and renewal

1 If the operation of choke opener is suspect, check the diaphragm for slits or cracks, the linkage for damage and distortion, and vacuum hoses for blockage or splits.
2 To renew the opener, disconnect the vacuum hose and the choke opener operating rod, and remove the screws securing the opener to the carburettor.
3 Remove the opener.
4 Refit in the reverse order, and carry out the linkage adjustment checks in Section 23.

26 Automatic choke coil heater – testing and renewal

1 If the choke blade does not open when the engine has reached normal operating temperature, first check the fast idle unloader for correct operation.
2 If the unloader is serviceable, then suspect the choke coil heater, which is integral with the choke cover.
3 Renew the choke cover, as described in Section 23.

27 Automatic choke fast idle – adjustment

1 Fit a tachometer to the ignition coil in accordance with the manufacturer's instructions.
2 Disconnect and plug the inner vacuum hose of the fast idle unloader.
3 Disconnect the intake air temperature sensor.
4 Engage the fast idle cam by holding the choke flap closed and operating the throttle.
5 Start the engine.
6 Adjust fast idle speed to that specified by turning the adjuster screw in or out as required (Fig. 3.22).

28 Intake air temperature sensor – testing

1 Disconnect the temperature sender and check the continuity across its terminals.
2 If air temperature is above 23°C (73°F) there should be continuity.
3 If air temperature is below 4.5°C (40°F) there should be no continuity.
4 Renew the temperature sensor if these conditions are not met.

29 Cut-off valves – testing and renewing

1 The primary slow mixture cut-off valve cuts off fuel supply to the primary idle circuit, and the main cut-off valve cuts fuel supply to the main jet when the engine is swiched off.
2 If the engine will not start, or stay running, suspect the primary cut-off valve, and if the engine will not idle or if idle is erratic, suspect the slow mixture cut-off valve.
3 Both valves are removed by disconnecting their electrical leads and removing their retaining screws.
4 Refit in the reverse order.

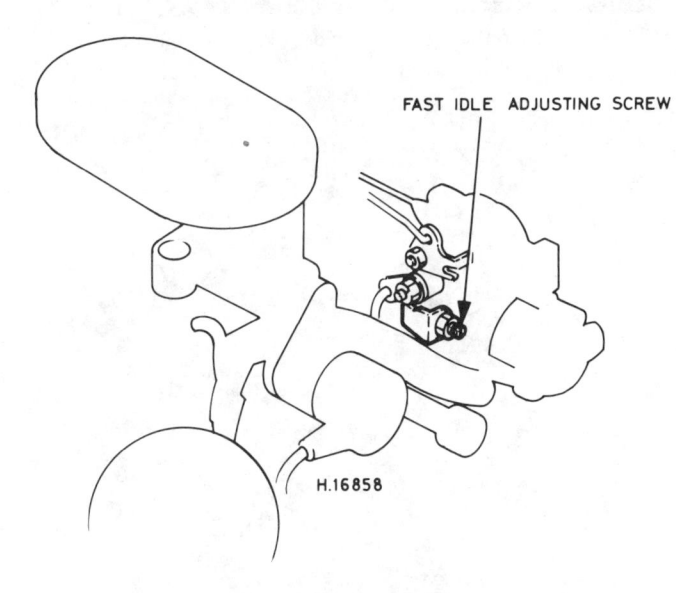

Fig. 3.22 Automatic choke fast idle adjustment screw (Sec 27)

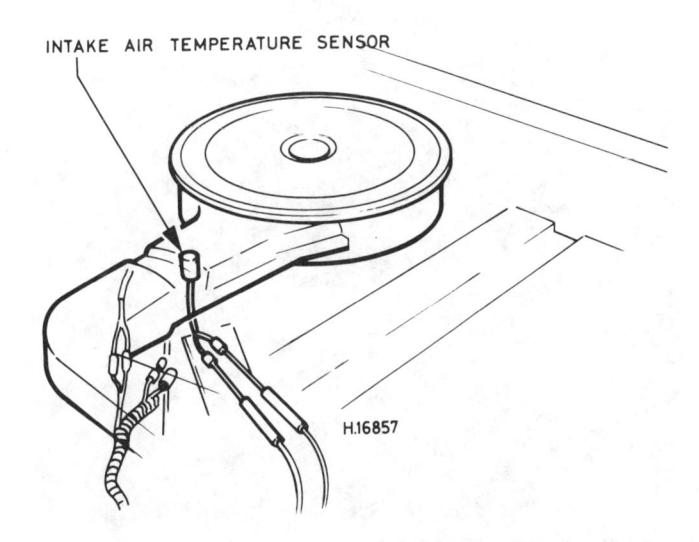

Fig. 3.23 Intake air temperature sensor (Sec 28)

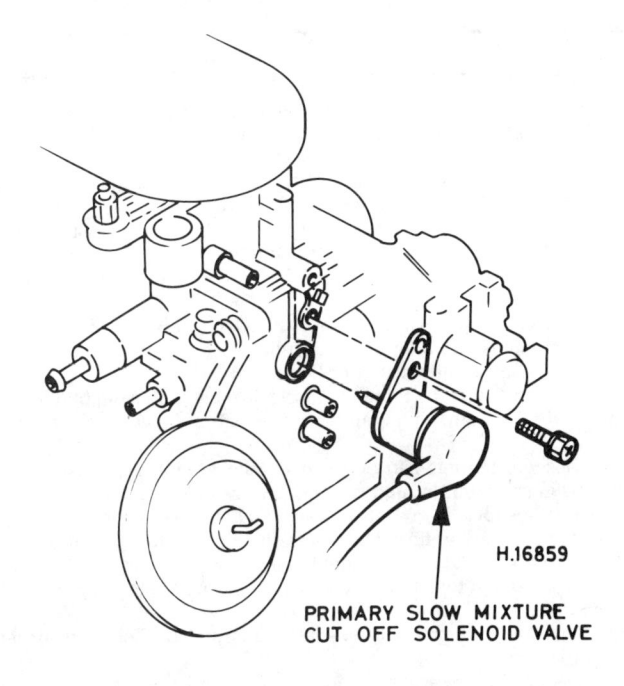

Fig. 3.24 Primary slow mixture cut-off valve (Sec 29)

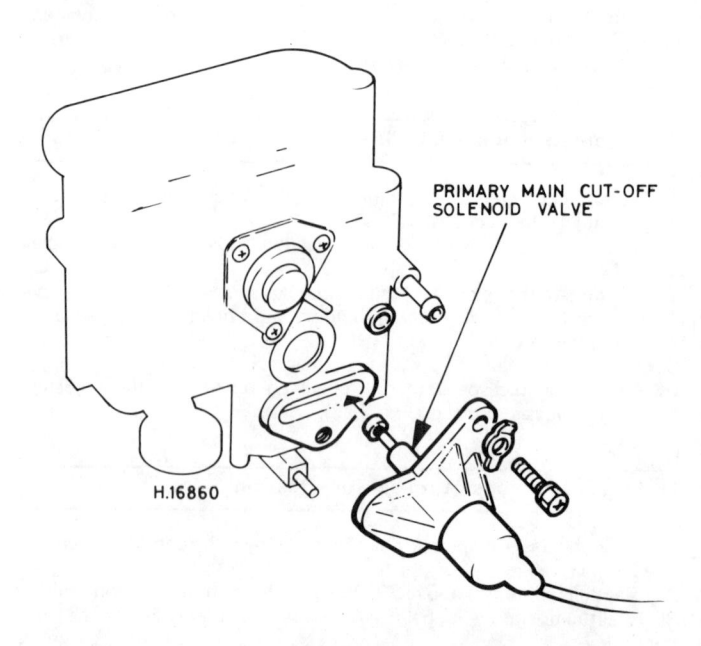

88   Chapter 3 Fuel, exhaust and emission control systems

Fig. 3.25 Primary main cut-off solenoid valve (Sec 29)

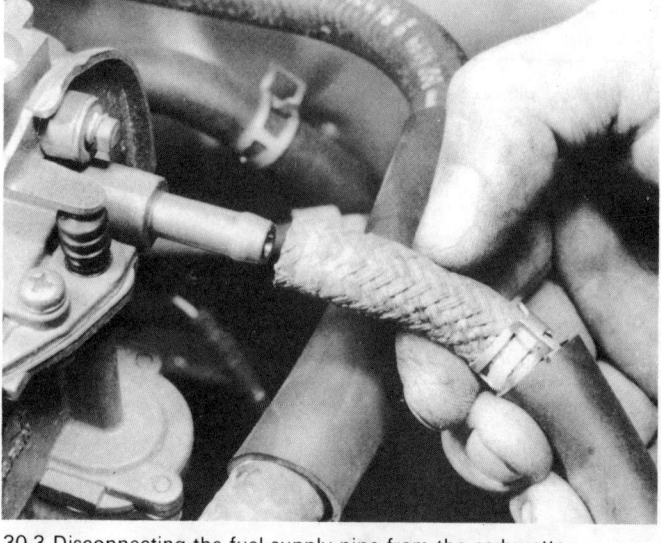

30.3 Disconnecting the fuel supply pipe from the carburettor

30 Carburettor – removal and refitting

1   Remove the air cleaner assembly.
2   Disconnect all electrical leads, vacuum hoses and the throttle control and choke cables.
3   Disconnect the fuel supply pipe to the carburettor (photo).
4   Remove the carburettor-to-manifold mounting nuts (photo).
5   Lift the carburettor from the intake manifold.
6   Refitting is a reversal of removal, but use a new gasket under the insulator block, and new seal between the insulator and the carburettor.
7   Ensure the insulator block is not cracked before refitting.

31 Carburettor – overhaul

1   With the carburettor removed from the engine, clean the external surfaces thoroughly with solvent, and blow dry.
2   The need for complete dismantling of a carburettor seldom occurs; it is usually adequate to clean out the float chamber and check and clean the jets, and to carry out internal adjustment as described in later paragraphs. Where a carburettor has seen long service it is more beneficial to fit a new or reconditioned unit.
3   Disconnect the throttle control and secondary diaphragm operating rod ends.
4   Remove the carburettor top cover securing screws and lift away the two diaphragms and their brackets (photo).
5   Lift off the carburettor top cover.
6   Remove the float by tapping out the float pin with a pin punch – do not tap the float arm – and tap the pin out from the float support long leg side.
7   If desired, the jets can be removed and all orifices, passageways and drillings blown through with compressed air.
8   Before removing any jets, obtain a carburettor overhaul kit from your dealer to ascertain which parts the kit contains, as all O-rings, sealing washers and gaskets should be renewed on assembly.
9   Never attempt to clean out a jet with a piece of wire; they are easily damaged.
10 Clean and lubricate all linkages shafts and bushes, and inspect them for wear.
11 Reassemble the carburettor in the reverse order of dismantling, and refit the carburettor to the intake manifold.
12 Carry out the checks and adjustments described elsewhere in this Chapter.

30.4 Carburettor manifold mounting nuts (arrowed)

31.4 Carburettor top cover screws
1   Support bracket screws   2   Top cover screws

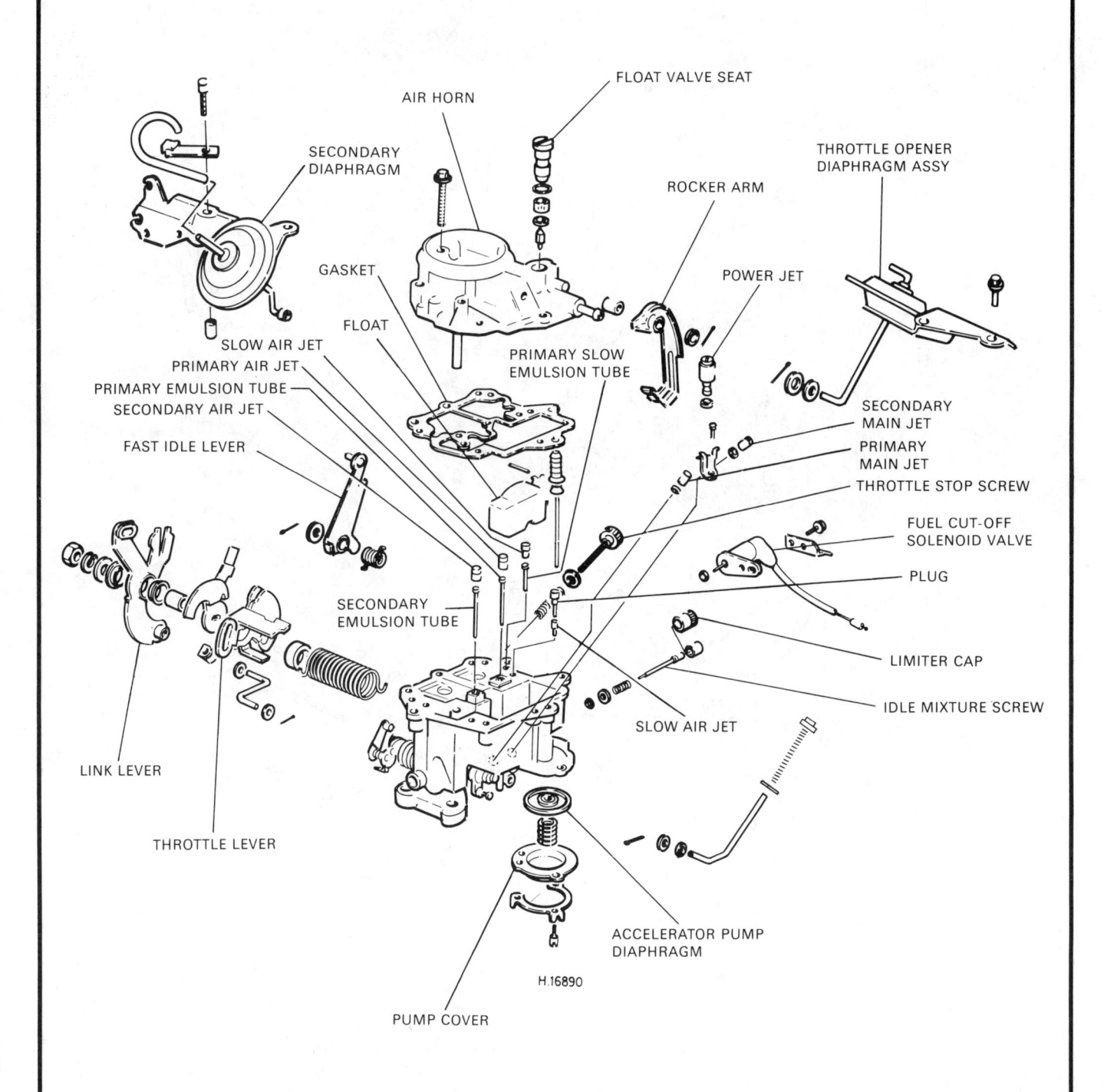

Fig. 3.26 Components of the carburettor – UK models
(Sec 31)

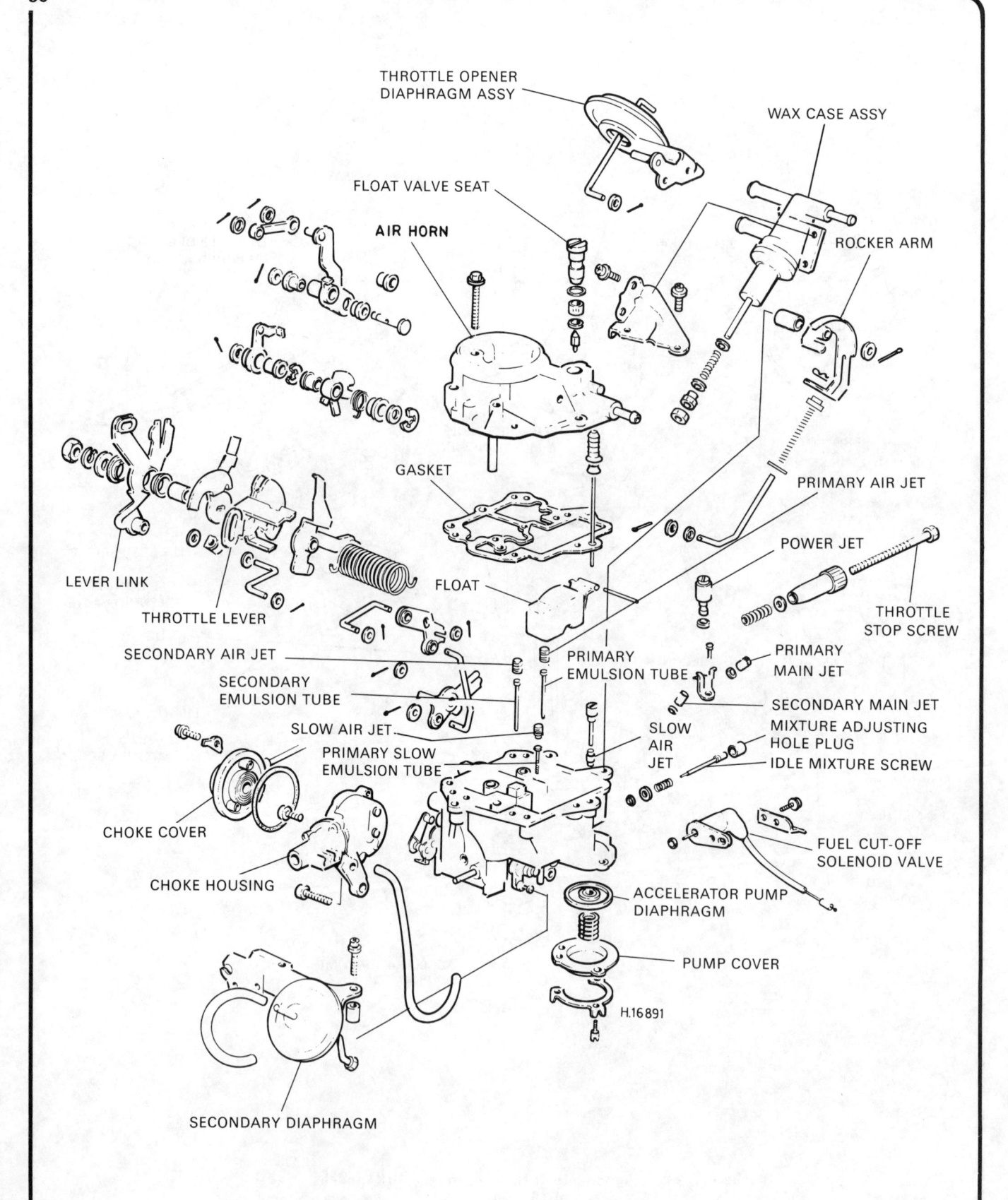

Fig. 3.27 Components of the carburettor – North American models (Sec 31)

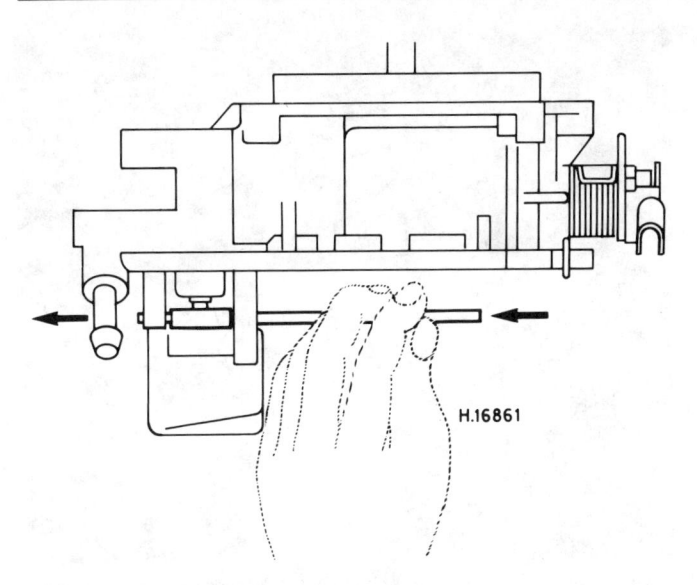

Chapter 3 Fuel, exhaust and emission control systems 91

Fig. 3.28 Tapping out the float pin (Sec 31)

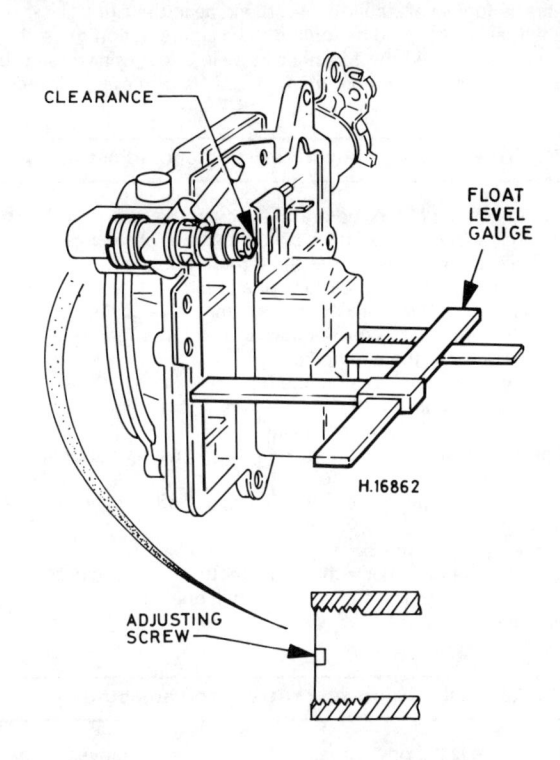

Fig. 3.29 Measuring float level – UK models (Sec 32)

32 Float level-inspection and adjustment

UK models
1 Remove the carburettor top cover as described in Section 31.
2 Tilt the cover so that the float is hanging down.
3 Using a float level gauge, or steel rule, measure the distance from the face of the cover gasket to the face of the float, which should be as specified.
4 Adjust with the float adjusting screw.
5 After setting the float level, back off the float adjusting screw to give a clearance between the float needle and float arm as specified.

North American models (with sight glass)
6 Position the vehicle on level ground.
7 Warm up the engine, blip the throttle several times between idle and 3000 rev/min, then allow it to idle.

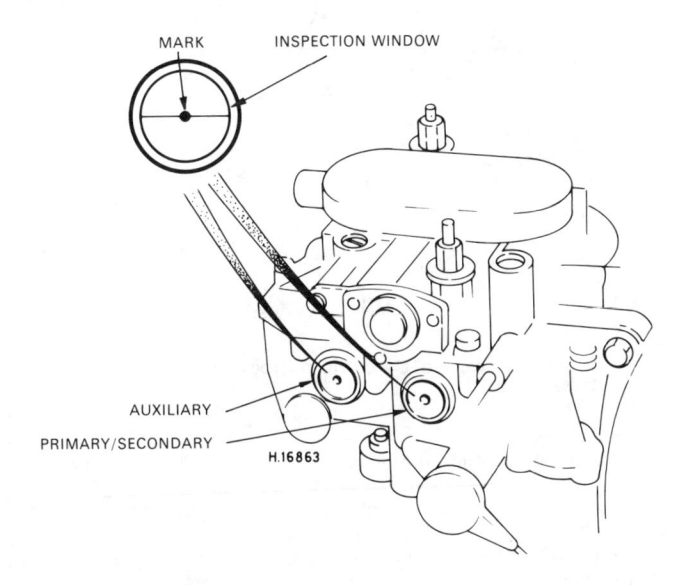

Fig. 3.30 Float level sight glass on North American models (Sec 32)

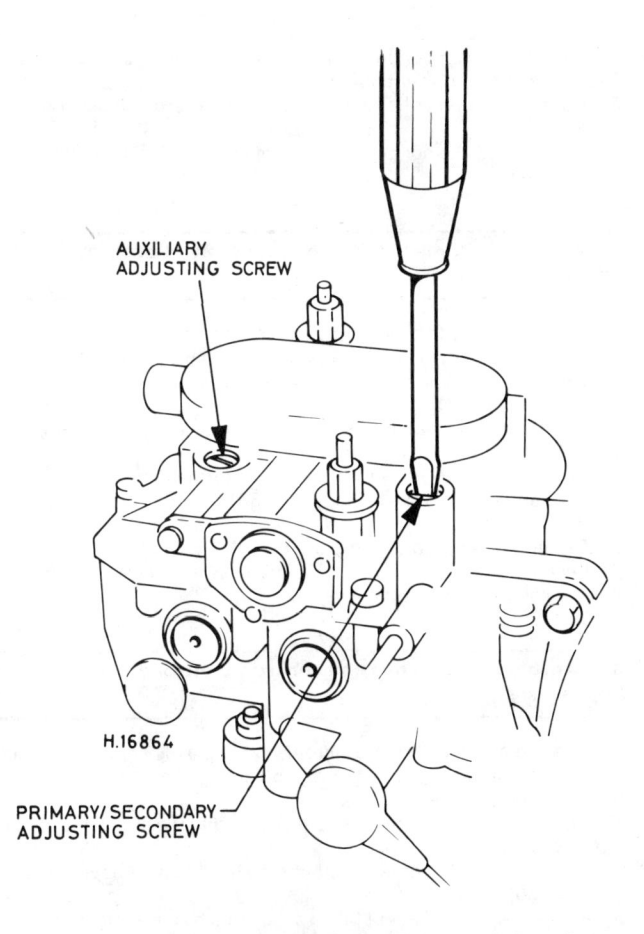

Fig. 3.31 Float level adjusting screws on North American models (Sec 32)

8 Let the fuel level stabilise, and then check the sight glass.
9 The fuel level should be touching the centre dot on the sight glass.
10 Adjust by turning the adjuster screws in or out as necessary, but do not turn the screws by more than an 1/8 of a turn every 15 seconds.

33 Throttle cable – removal, refitting and adjusting

1 Push back the rubber boot and unscrew the locknut (photo).
2 Pull the cable housing back and slide it out of the throttle bracket.
3 Unhook the cable end from the throttle quadrant.
4 Unhook the other end of the cable from the accelerator pedal.
5 Remove the cable from the stay on the valve cover.
6 Turn the grommet in the engine bulkhead through 90° and pull the cable through into the engine bay.
7 Remove the cable from engine bay.
8 Fit a new cable in the reverse order, but apply sealant to the grommet in the bulkhead after installation.
9 Adjust the cable by first tightening the adjusting nut until the cable is taut; without pulling on the throttle lever.
10 Loosen the adjuster until the specified deflection of the cable is obtained.
11 Tighten the locknut.
12 On completion; check that the throttle is fully closed with the accelerator pedal released, and that full opening is achieved when pedal is depressed.

34 Choke cable – removal, refitting and adjustment

1 The choke cable on manual choke versions is removed in the same way as the throttle cable; see Section 33.
2 On completion of refitting the cable, adjust in the same way as the throttle cable, but refer to Specifications for cable deflection.
3 Check that the choke flap is fully closed when the choke is pulled out, and fully open when choke is pushed in.
4 If the flap does not fully open, turn the adjusting nut until the choke flap moves off the positioning stop tab.
5 Continue turning the adjuster until there is no deflection in the cable.
6 Now loosen the adjuster so that the specified cable deflection can be obtained.

35 Automatic transmission throttle control bracket – adjustment

1 Disconnect the throttle control cable from the throttle control lever.
2 Remove the bolts and discard the lockplate from the control bracket.
3 Refit the bracket loosely using a new lockplate.
4 Position the special tool between the throttle control lever and the control bracket.
  Note: a gauge may have to be made up to the dimensions shown if the special tool is not available (Fig. 3.35).
5 Position the bracket with the gauge.
6 Tighten the two bolts and bend over the locking tabs.
7 Recheck the gauge fit, and then remove the gauge and refit the throttle cable to the throttle lever.

36 Automatic transmission throttle control cable – adjustment

1 Check that the carburettor throttle cable play is adjusted correctly.
2 The engine should be at operating temperature, the cooling fan having come on at least twice.
3 Check that the idle speed is correct, and that the automatic choke is working correctly.
4 Check that the throttle control cable bracket is correctly adjusted (Section 35).
5 Attach a weight of about 3lb (1.5 kg) to the accelerator pedal, raise and release the pedal. This will take up the normal free play in the cable.

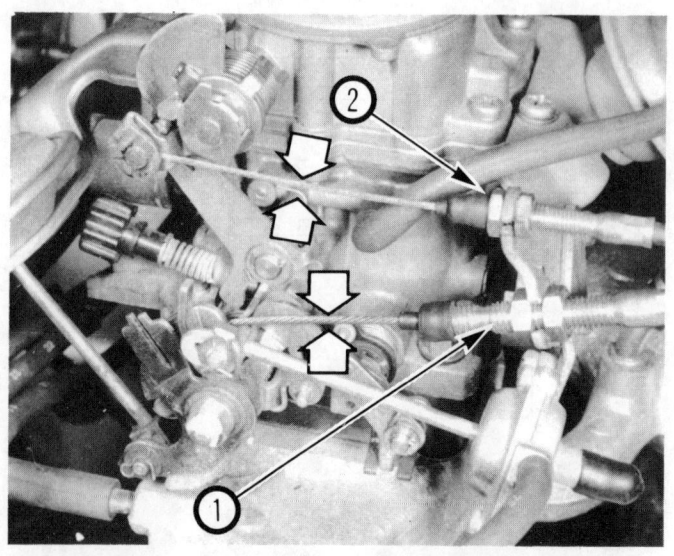

33.1 View of the throttle and choke cable attachment points at the carburettor
1 Throttle cable adjuster
2 Choke cable adjuster
Arrows indicate deflection measurement points

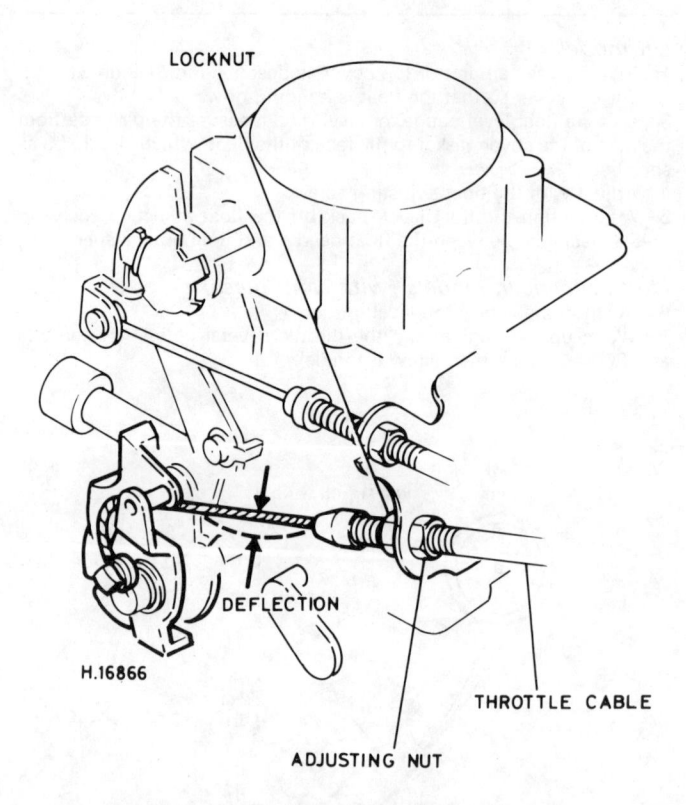

Fig. 3.32 Throttle cable deflection (Sec 33)

6 Secure the throttle control cable with clamps, as shown in Fig. 3.36.
7 Adjust the distance between the throttle control cable end and the locknut A to the specified distance.
8 Insert the end of the throttle control cable in the groove of the throttle control lever.
9 Insert the throttle control cable in the bracket and secure with locknut B. Ensure the cable is not kinked or twisted.
10 Check that the cable moves freely by depressing the accelerator.

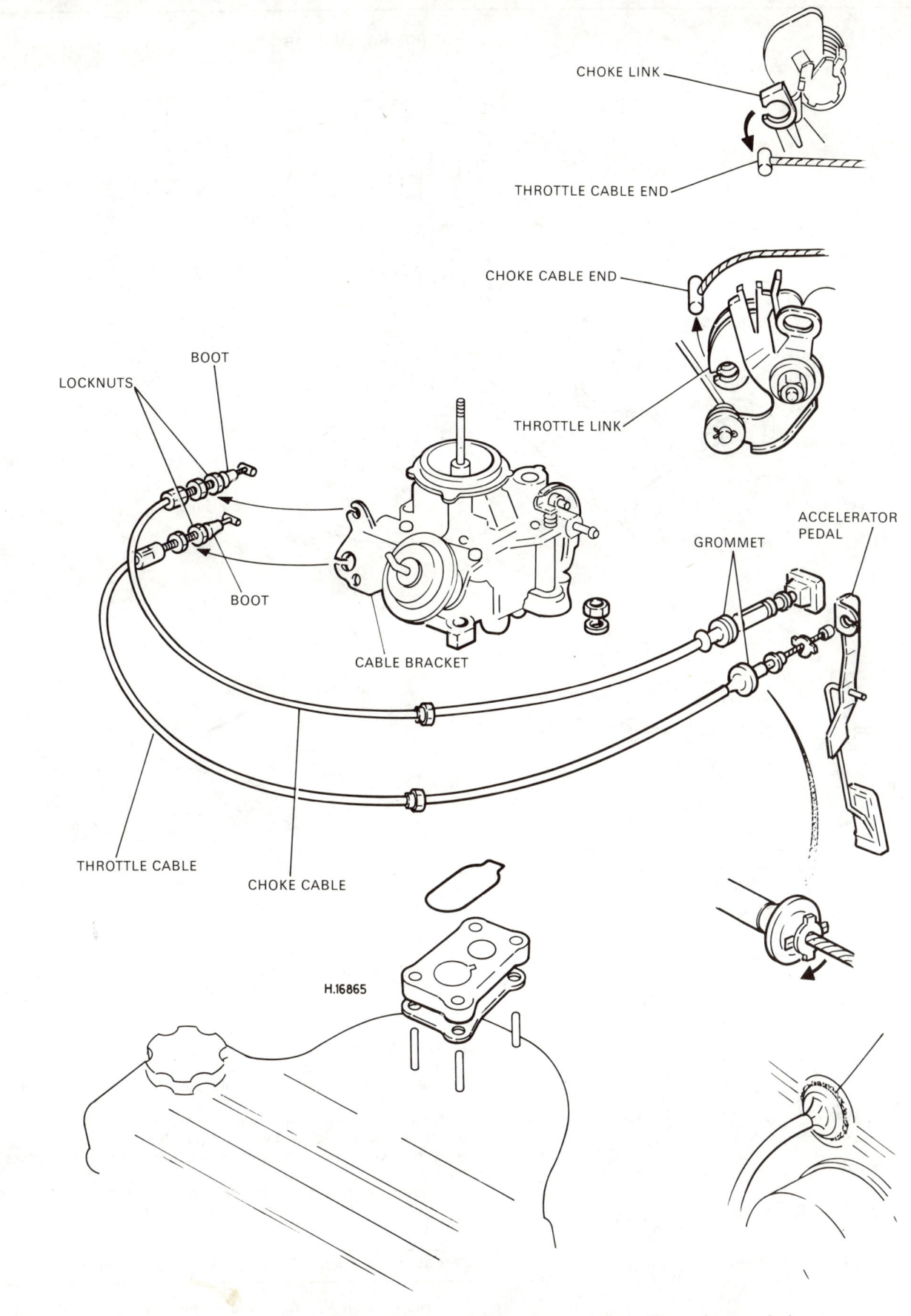

Fig. 3.33 Throttle and choke cable assemblies on manual choke/manual transmission models (Sec 33)

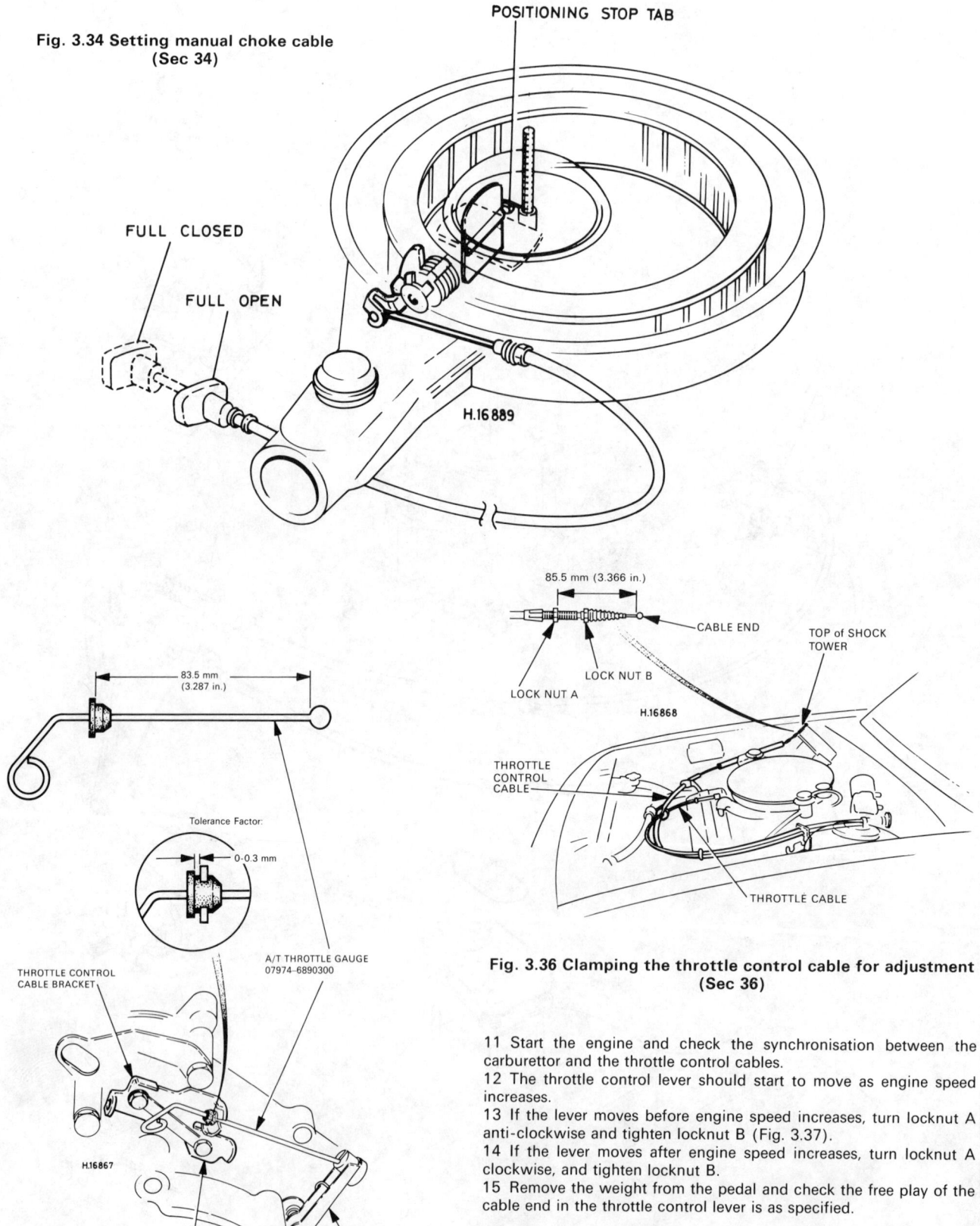

94　　　　　　　　　　Chapter 3 Fuel, exhaust and emission control systems

Fig. 3.34 Setting manual choke cable (Sec 34)

Fig. 3.35 Automatic transmission throttle cable control bracket and setting gauge (Sec 35)

Fig. 3.36 Clamping the throttle control cable for adjustment (Sec 36)

11 Start the engine and check the synchronisation between the carburettor and the throttle control cables.
12 The throttle control lever should start to move as engine speed increases.
13 If the lever moves before engine speed increases, turn locknut A anti-clockwise and tighten locknut B (Fig. 3.37).
14 If the lever moves after engine speed increases, turn locknut A clockwise, and tighten locknut B.
15 Remove the weight from the pedal and check the free play of the cable end in the throttle control lever is as specified.

37 Automatic choke (Canadian models) – adjustment

1 Remove the choke cover and disconnect the heater lead.
2 Disconnect the vacuum hose to the choke opener and apply air pressure to the opener to close the bleed valve.
3 Turn the choke opener lever clockwise until it stops, then turn the

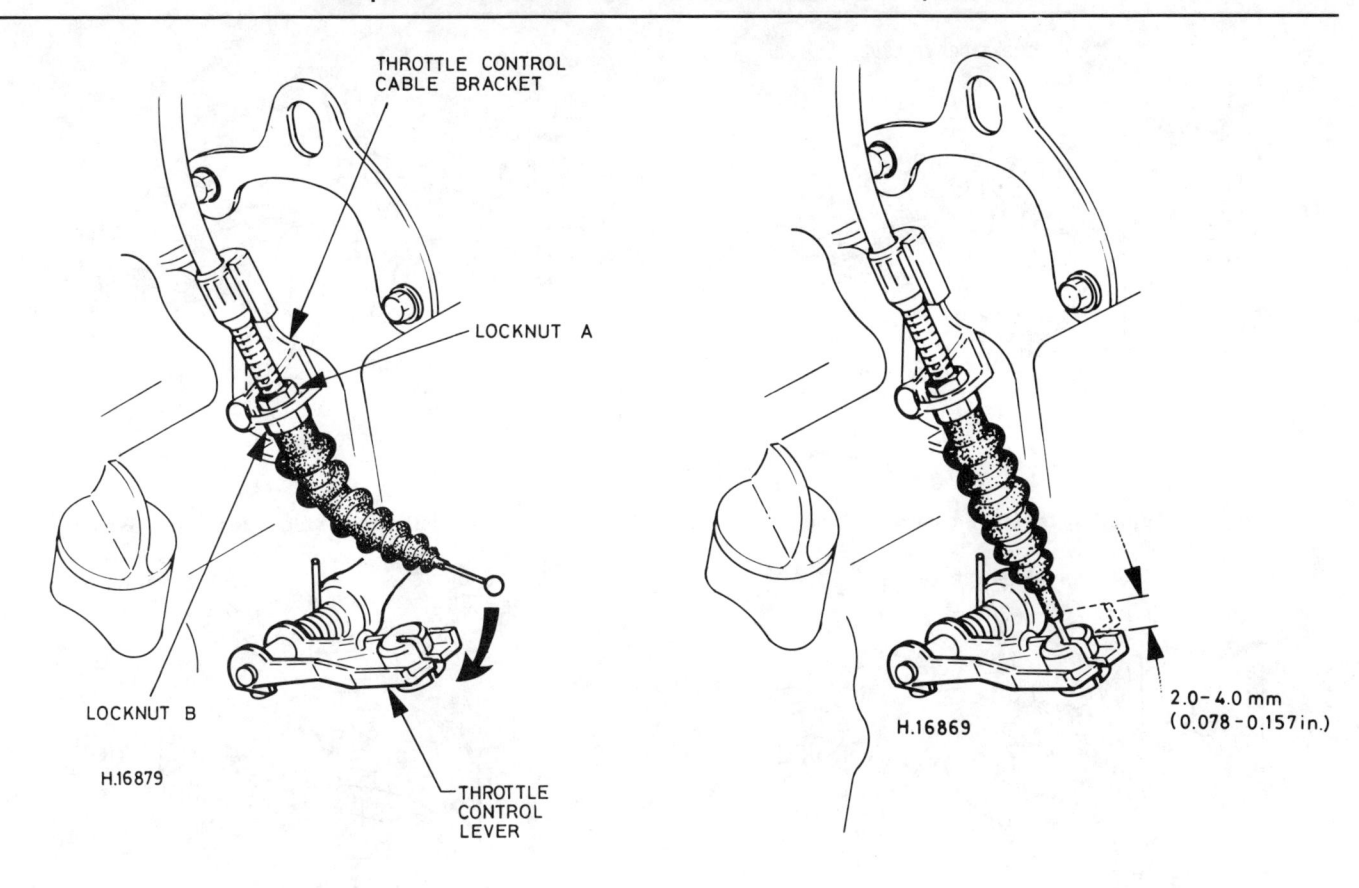

Chapter 3 Fuel, exhaust and emission control systems

Fig. 3.37 Throttle control cable adjustment – automatic transmission (Sec 36)

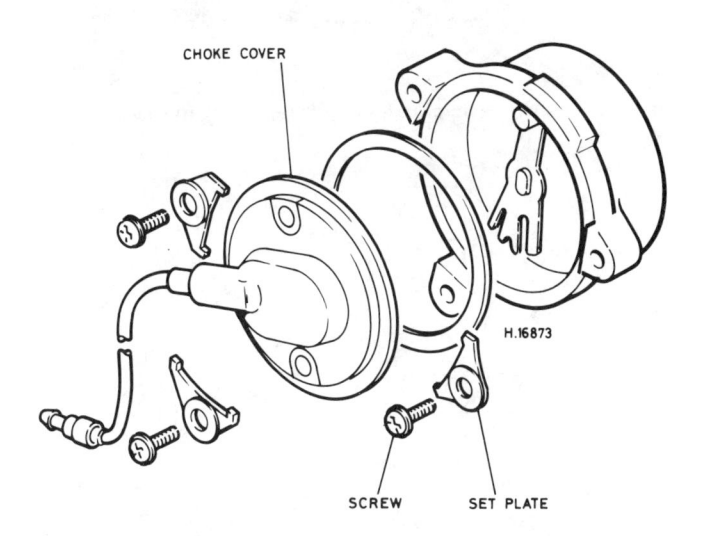

Fig. 3.38 Throttle control cable free play (Sec 36)

Fig. 3.39 Removing the automatic choke cover (Sec 37)

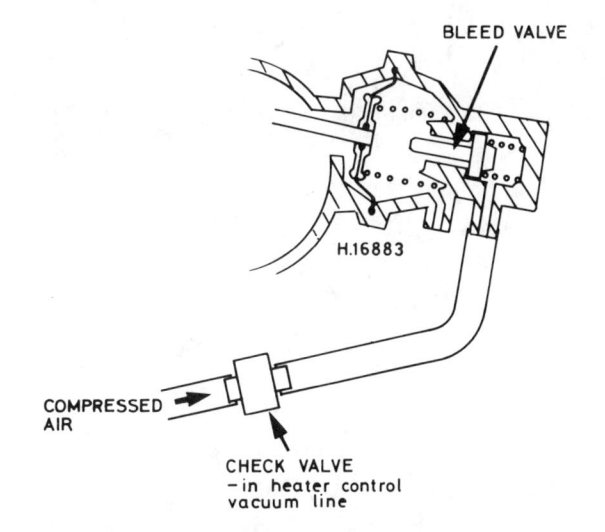

Fig. 3.40 Applying air pressure to the choke opener (Sec 37)

choke drive lever anti-clockwise until it touches the choke opener lever.
4 Measure the clearance between the choke flap and carburettor wall.
5 If the clearance is not as specified, (1st stage clearance) bend tab A (Fig. 3.41).
6 Remove the air pressure and reconnect the vacuum hose.
7 Turn the choke opener lever clockwise until tab B rests against the boss (Fig. 3.42), then turn the choke drive lever anti-clockwise to seat tab A against the choke opener lever.
8 Measure the clearance between the choke flap and carburettor wall, which should be as specified (2nd stage clearance).
9 Adjust by bending tab B.
10 Keep the opener lever against the tab, and turn the choke drive lever until the tab C touches the spring (Fig. 3.43).
11 Again measure the clearance at choke flap which should be as specified (Stage 3 clearance) adjust by bending tab C.
12 On completion refit the choke cover by fitting the choke drive lever boss to the choke bi-metal end hook.
13 Align the index marks by turning the cover clockwise, fit and tighten the retainers and screws.
14 Reconnect the heater cable.

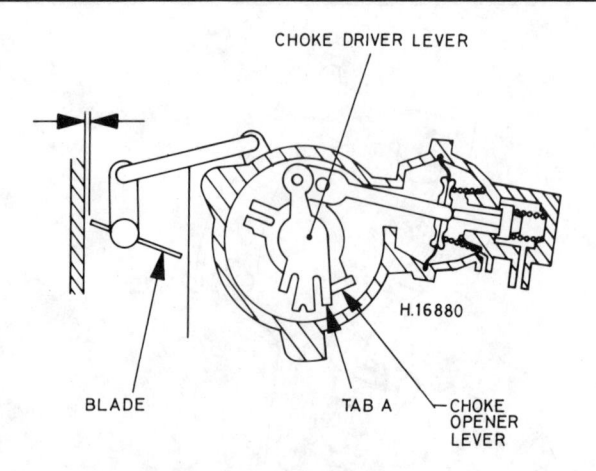

Fig. 3.41 Adjusting 1st stage clearance (Sec 37)

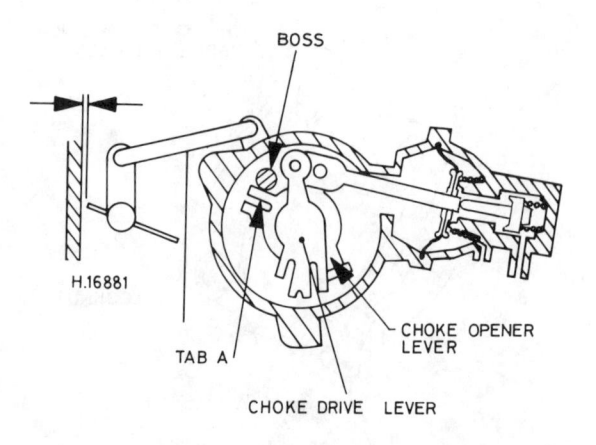

Fig. 3.42 Adjusting 2nd stage clearance (Sec 37)

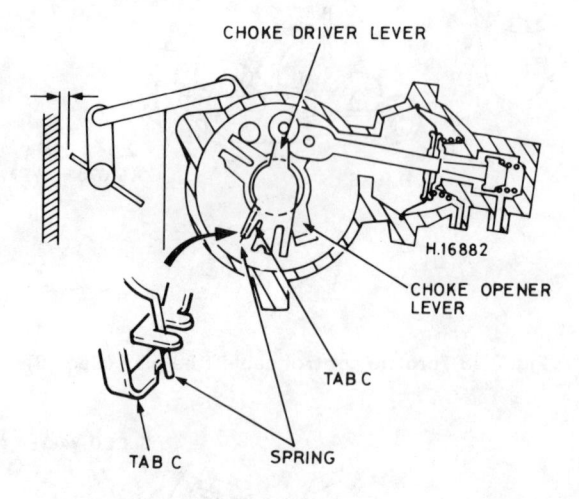

Fig. 3.43 Adjusting 3rd stage clearance (Sec 37)

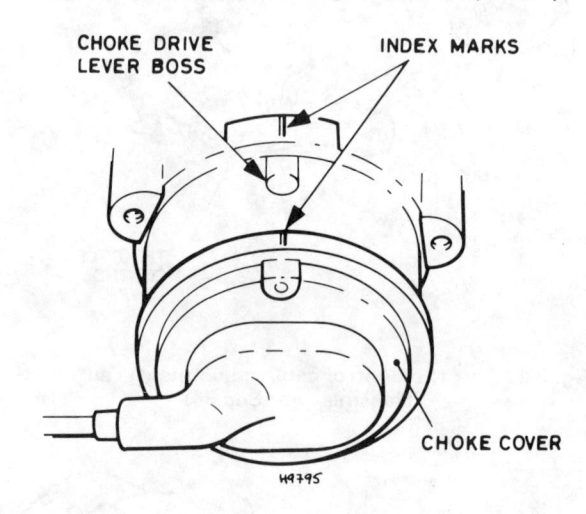

Fig. 3.44 Automatic choke cover alignment marks (Sec 37)

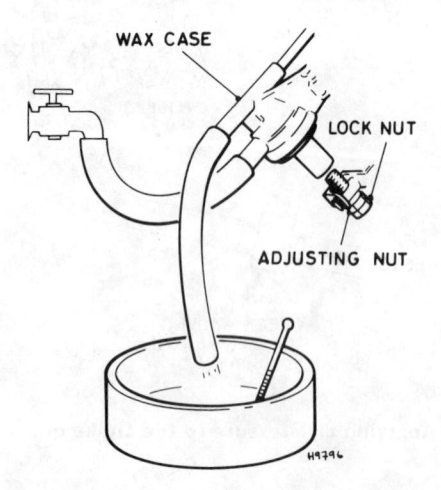

Fig. 3.45 Connecting water hoses to the wax case (Sec 38)

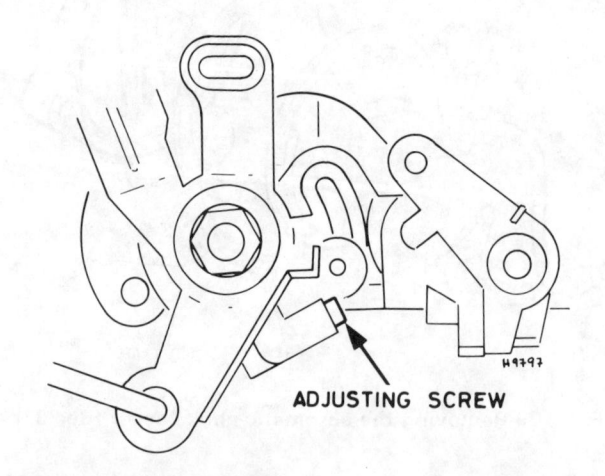

Fig. 3.46 Fast idle adjusting screw (Sec 38)

38 Automatic choke (Canadian models) fast idle – adjustment

1 Disconnect the hoses from the carburettor wax case and connect the bypass hose.
2 Connect up hoses to the wax case so that pure cool water may be fed through the wax case.
3 Start the engine and check the fast idle speed, which should be as specified, with the fast idle lever on the second step of the cam.
4 Adjust on the adjuster screw.
5 To adjust the position of the lever on the cam, measure the temperature of the water and compare it with the graph in Fig. 3.47.

Chapter 3 Fuel, exhaust and emission control systems

6 Adjust the cam position by turning the adjuster nut.
7 If the correct lever-to-cam step position cannot be obtained, renew the wax case by removing the adjuster nut and retaining screws.
8 On refitting, reconnect the coolant hoses.

39 Power valve (North American models) – description

1 The system is provided to supply supplementary fuel to the primary main fuel passage when the vehicle is operated under power.

2 Under normal driving the power valve is held closed by manifold vacuum.
3 On sudden acceleration, the power valve is opened through lack of vacuum in the inlet manifold, allowing additional fuel to the primary fuel passage.
4 Additional control is achieved through the intake air temperature sensor, coupled to the power valve solenoid through the control unit.

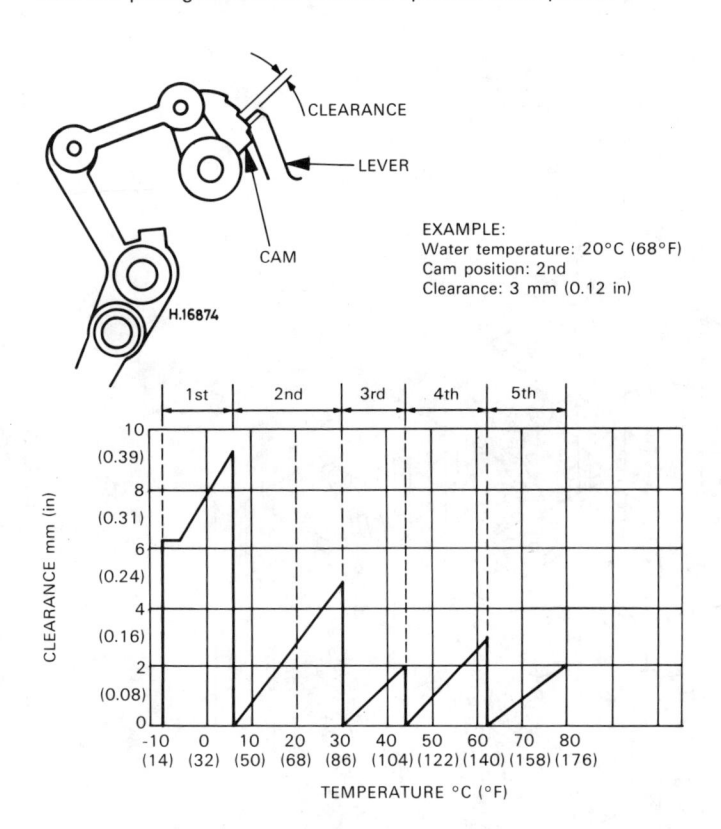

EXAMPLE:
Water temperature: 20°C (68°F)
Cam position: 2nd
Clearance: 3 mm (0.12 in)

Fig. 3.47 Cam-to-lever position and graph (Sec 38)

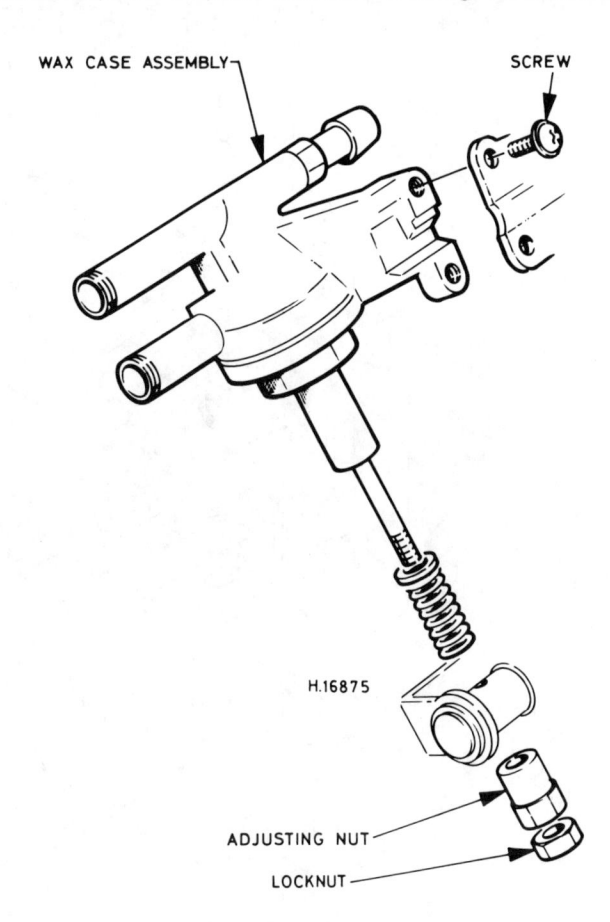

Fig. 3.48 Wax case attachment points (Sec 38)

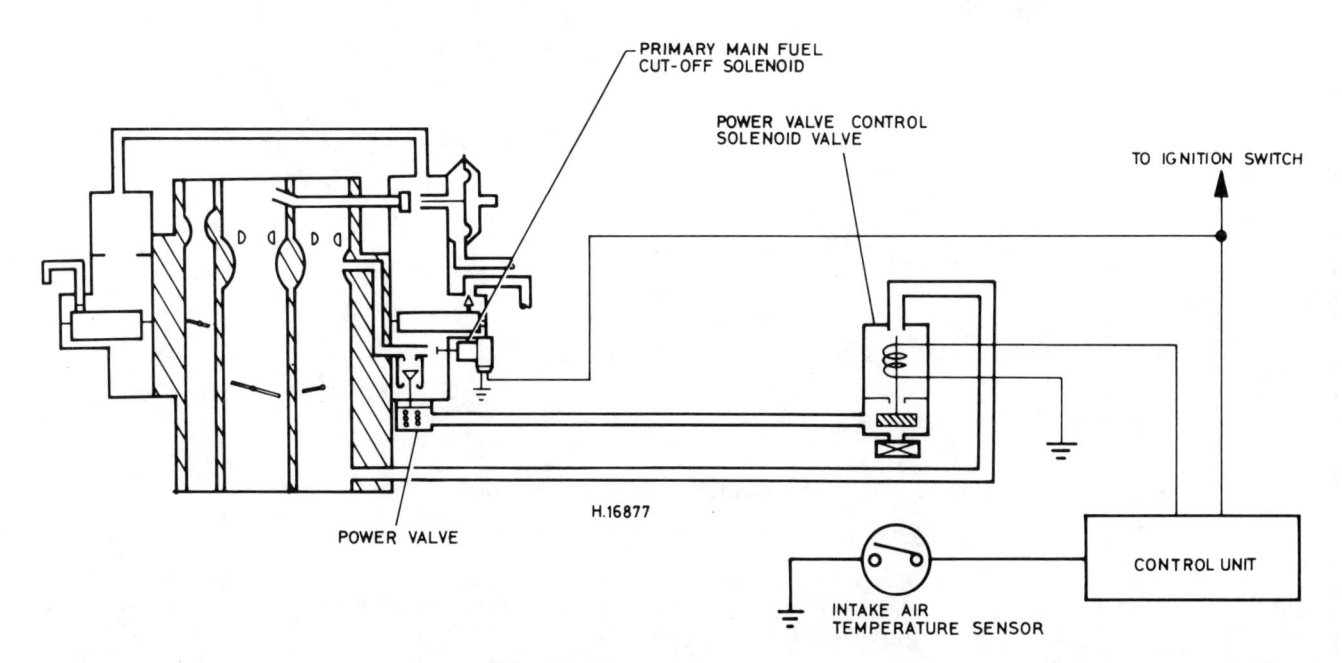

Fig. 3.49 Power valve system diagram (Sec 39)

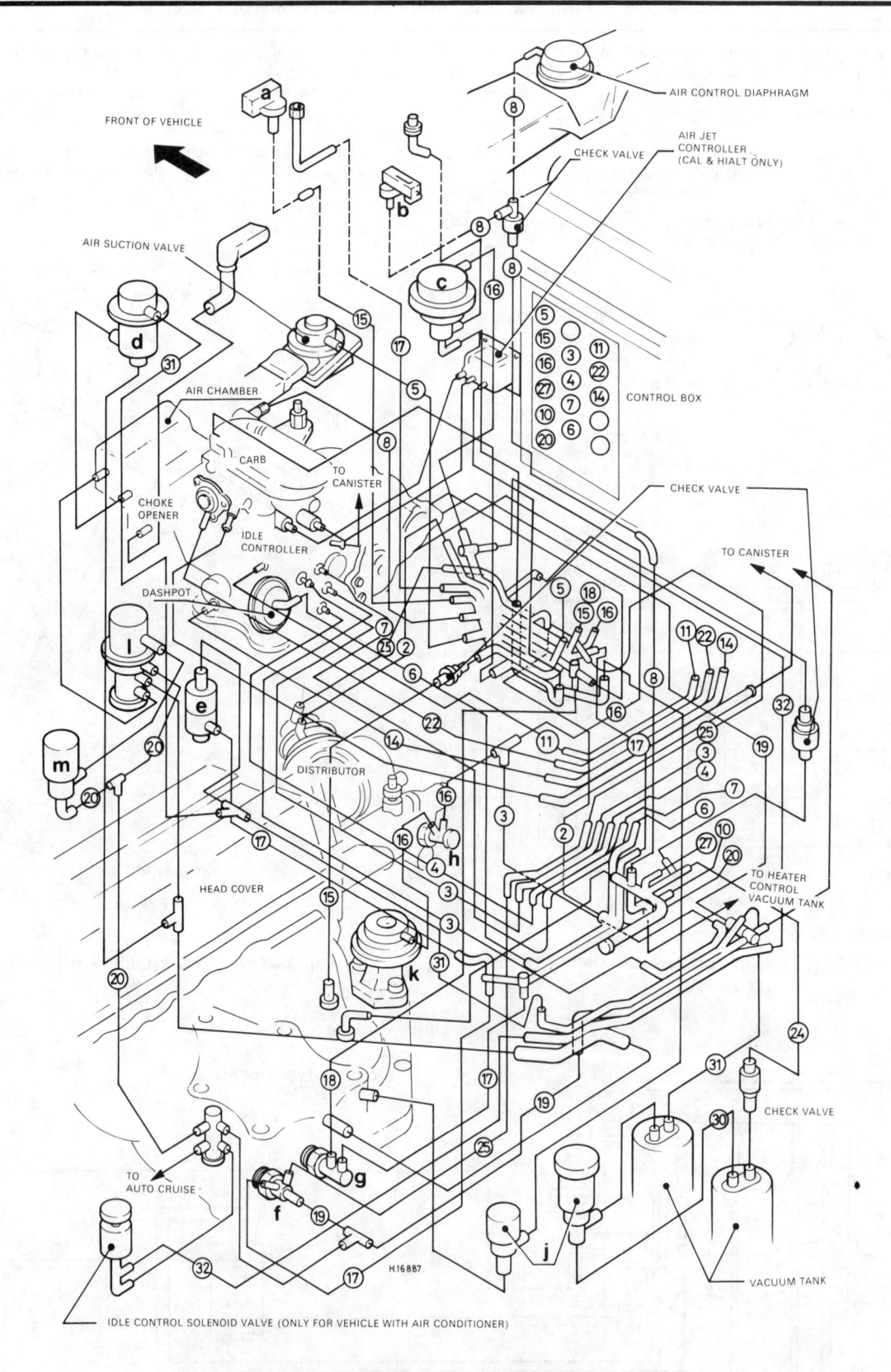

Fig. 3.50 Emission control interconnect diagram (Sec 40)

The numbers are imprinted on the hoses

a	Air bleed valve A	e PX leak solenoid valve	h Thermo valve C	l Shot air valve
b	Air bleed valve B	f Thermo valve A	j Frequency solenoid valve	m Shot air valve control
c	Air control valve A	g Thermo valve B	k EGR valve	solenoid
d	Air control valve B			

Chapter 3 Fuel, exhaust and emission control systems

40 Emission controls – general description

1 The emission control system, designed to prevent as much as is possible of harmful gases caused by vehicle operation entering the atmosphere.
2 The system can be divided into three main groups which are:
3 **Evaporative emission controls** – to dispense with fuel vapour given off from fuel storage and supply areas.
4 **Crankcase vapour emission control** – to dispense with blowby gases created in the crankcase.
5 **Exhaust emission control** – to reduce the gases given off by the exhaust system after combustion has taken place.
6 The control system is mainly vacuum operated from the depression in the manifolds, and by electric solenoid valves.
7 Although the complete system is complicated, it becomes much simpler if each sub-system is broken down to its own particular function; but bearing in mind their relationship with each other, especially when fault finding.
8 Each vacuum hose has a number imprinted on it, which corresponds to the numbers in the interconnect diagram, making it easier to trace hoses between components.
9 It should also be remembered that test procedures can be long and laborious and sometimes require special equipment, and that your dealer is best equipped to keep the system in good order.
10 The simpler tests and component replacement are dealt with in the following Sections.

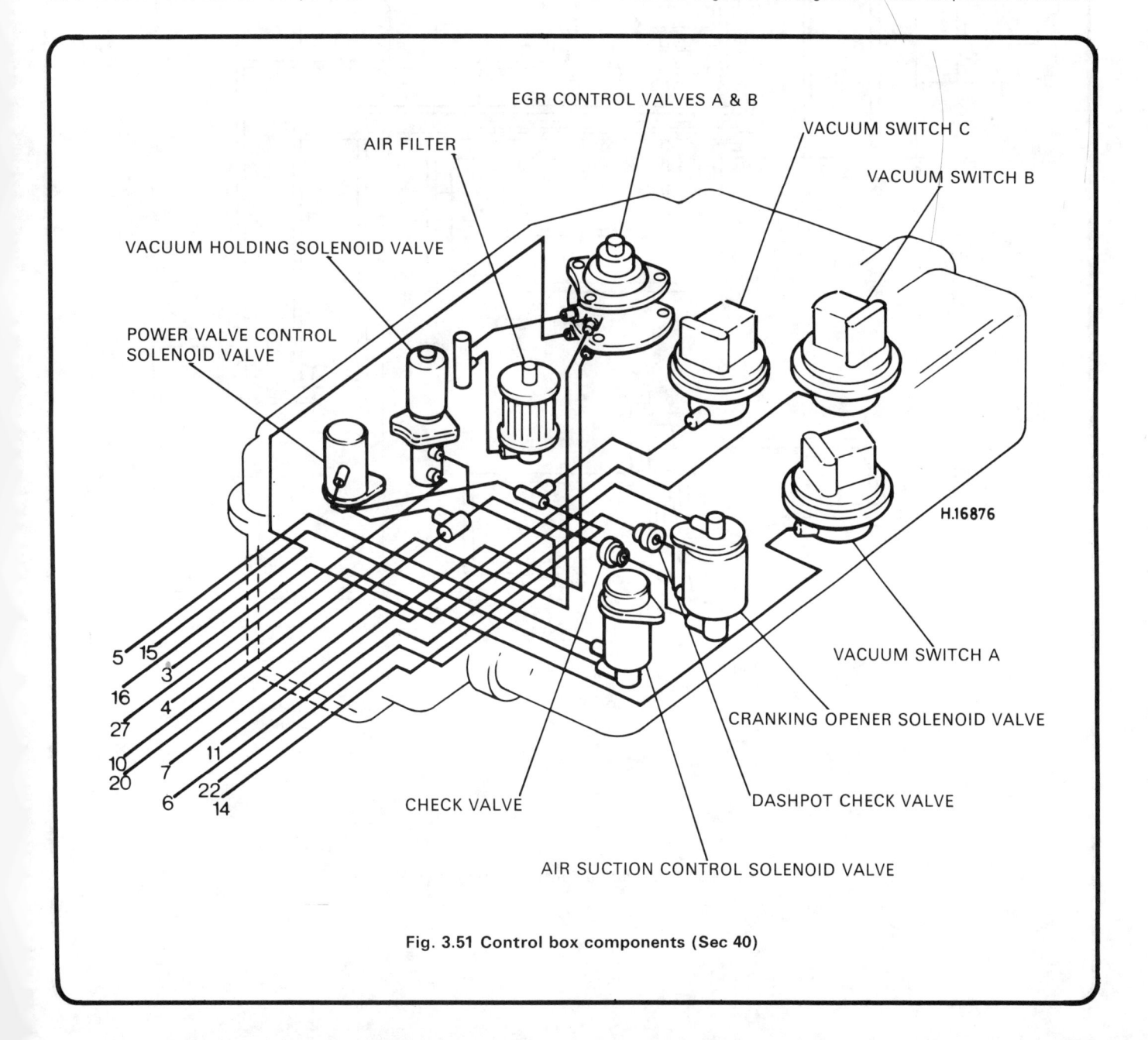

41 Evaporative controls (carburettor models) – general description

1 A two-way valve mounted on the fuel tank regulates the pressure or vacuum caused by changes in temperature and full level inside the tank.
2 A liquid/vapour separator allows liquid fuel to return to the tank via a return line.
3 Vapour is carried out by the vent line to the charcoal cannister, where it is absorbed by the activated charcoal.
4 An air vent cut-off diaphragm is mounted on the carburettor and vents fuel vapour from the float chambers to the charcoal cannister when the engine is not running.
5 When the engine is running, and coolant temperature is above a

Fig. 3.51 Control box components (Sec 40)

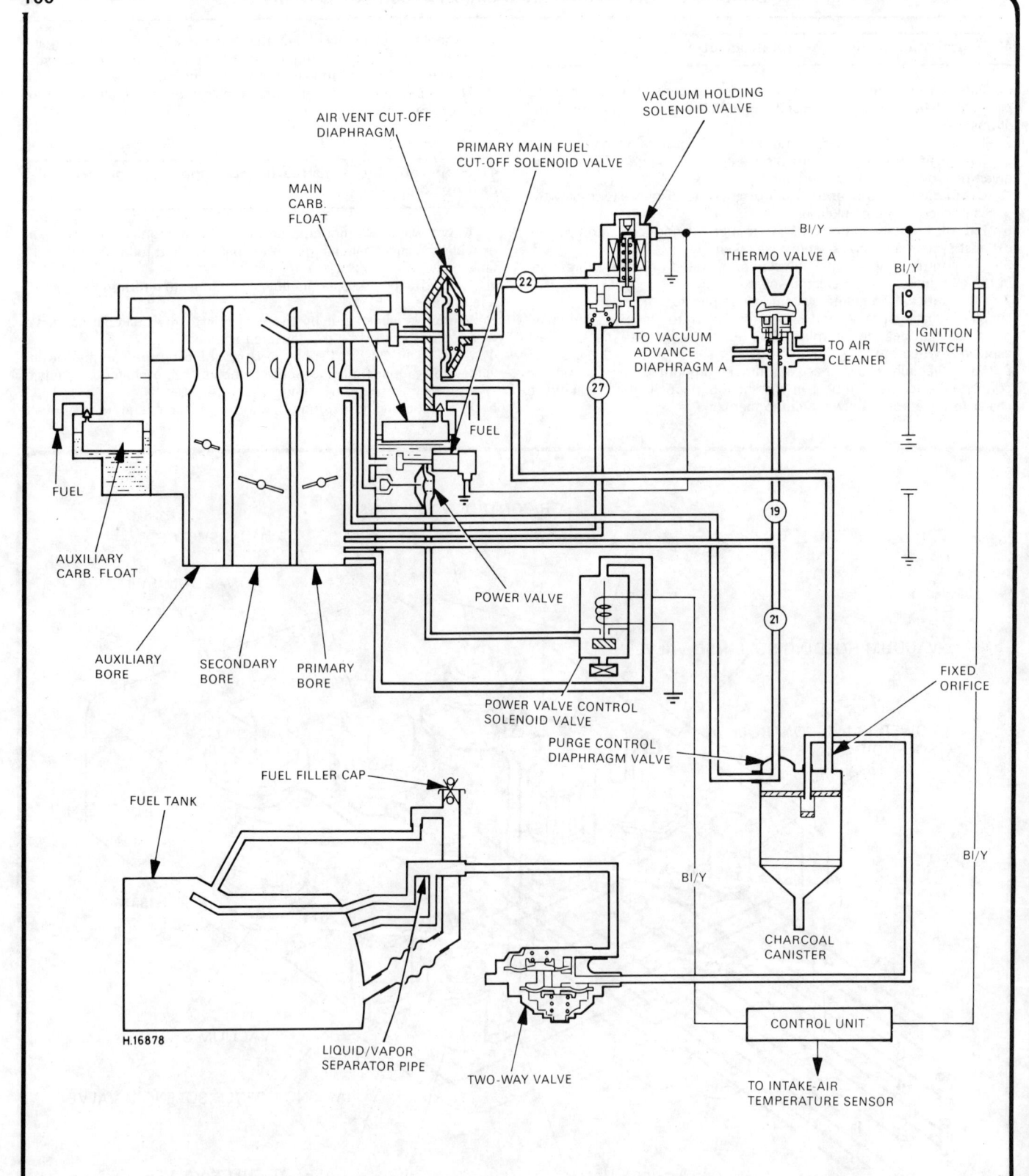

Fig. 3.52 Evaporative control system (Sec 41)

Chapter 3 Fuel, exhaust and emission control systems

preset temperature, a thermovalve in the system opens to allow vapour in the charcoal cannister to be drawn into the intake manifold.
6 The fuel filler cap is fitted with a two-way valve as a safety device should the evaporative system fail.
7 Fuel cut-off solenoid valves are also incorporated in the system (see Section 56).

42 Two-way valve

General description
1 The two-way valve is mounted on the top of the fuel tank.
2 Its function is to relieve pressure of vacuum inside the tank as fuel level and temperature changes occur.

Checking
3 Remove the fuel filler cap.
4 Remove the hose from the liquid/vapour separator pipe and connect a vacuum gauge and pump to the two-way valve.
5 Apply vacuum, which should stabilise at 0.2 to 0.6 in (5 to 15 mm) Hg.
6 If this condition is not met, renew the valve.
7 Remove the vacuum equipment and replace it with a pressure gauge and pump.
8 Pressurise the system, which should stabilise at 1.0 to 2.2 in (25 to 55 mm) Hg.
9 If this condition is not met, renew the valve.

Component replacement
10 Remove the fuel tank, as described in Section 6.
11 Remove the two-way valve from the tank.
12 Refit in the reverse order.

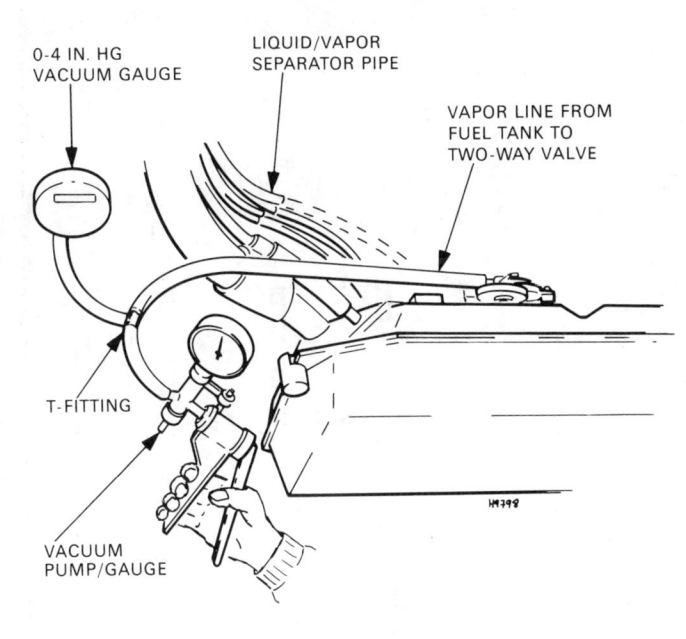

Fig. 3.53 Vacuum pump and gauge connected to the two-way valve (Sec 42)

43 Air vent cut-off diaphragm

General description
1 The air vent diaphragm is mounted on the carburettor.
2 Its function is to vent fuel vapour from the float chambers back to the charcoal cannister when the engine is not running.
3 With the engine running, manifold vacuum holds the diaphragm open, allowing vapour to bleed into the carburettor venturi.
4 The vacuum holding solenoid valve stabilises manifold vacuum at the diaphragm.

Checking
5 Disconnect the hose to the air vent diaphragm and connect up a vacuum pump to the hose.
6 You should not be able to draw vacuum from the hose.
7 Turn on the ignition.
8 Vacuum should now be available. If not, suspect the vacuum holding solenoid.
9 Transfer the vacuum pump to the air vent diaphragm.
10 Draw a vacuum, which should remain steady, renew the diaphragm if the vacuum drops.

Component replacement
11 Removal and refitting of the components is self-evident. The diaphragm being mounted on the carburettor, and the solenoid valve in the control box.

44 Charcoal cannister

General description
1 The charcoal cannister is mounted on the engine bulkhead in the engine compartment.
2 The activated charcoal element inside the cannister absorbs fuel vapour drawn to the cannister from the fuel tank and float chambers.
3 Testing of the cannister is best done by substitution, removal and refitting being self-evident.

45 Ignition control system

General description
1 The additional distributor control by this system optimises ignition timing during and after engine warm-up; controlling emission exhaust levels.
2 There are two inputs to the system, engine intake manifold vacuum and engine coolant temperature.
3 Basically, the distributor has two vacuum diaphragms, (A and B) both operating from intake vacuum, but A has a thermovalve in line.
4 With engine coolant temperature below a preset valve the thermovalve remains open, allowing vacuum to diaphragm A, bringing additional advance to the distributor.

Checking
5 Test the components for vacuum integrity, referring to Chapter 4 for details of the distributor.
6 Check the continuity of the thermovalve above and below its preset value.

Component replacement
7 Refer to Chapter 4 for details of the distributor.
8 Thermovalves are screwed into the manifold. Use a new sealing washer when replacing.

46 Dashpot system

General description
1 The dashpot system is fitted in conjunction with the cranking opener solenoid valve and throttle controller.
2 It improves combustion by holding the throttle slightly open, admitting additional air during periods of gear changing and deceleration.
3 Above idle, vacuum is applied to the throttle controller, through the dashpot check valve. On deceleration, the vacuum is bled off, allowing the throttle controller to slowly close the throttle.
4 Throttle closing speed is determined by the size of the dashpot check valve orifice, tension of the throttle return spring, and the amount of vacuum available.

Checking
5 The dashpot check valve is best tested by substituting a known serviceable valve into the system.

Component replacement
6 The dashpot check valve is in the control box.

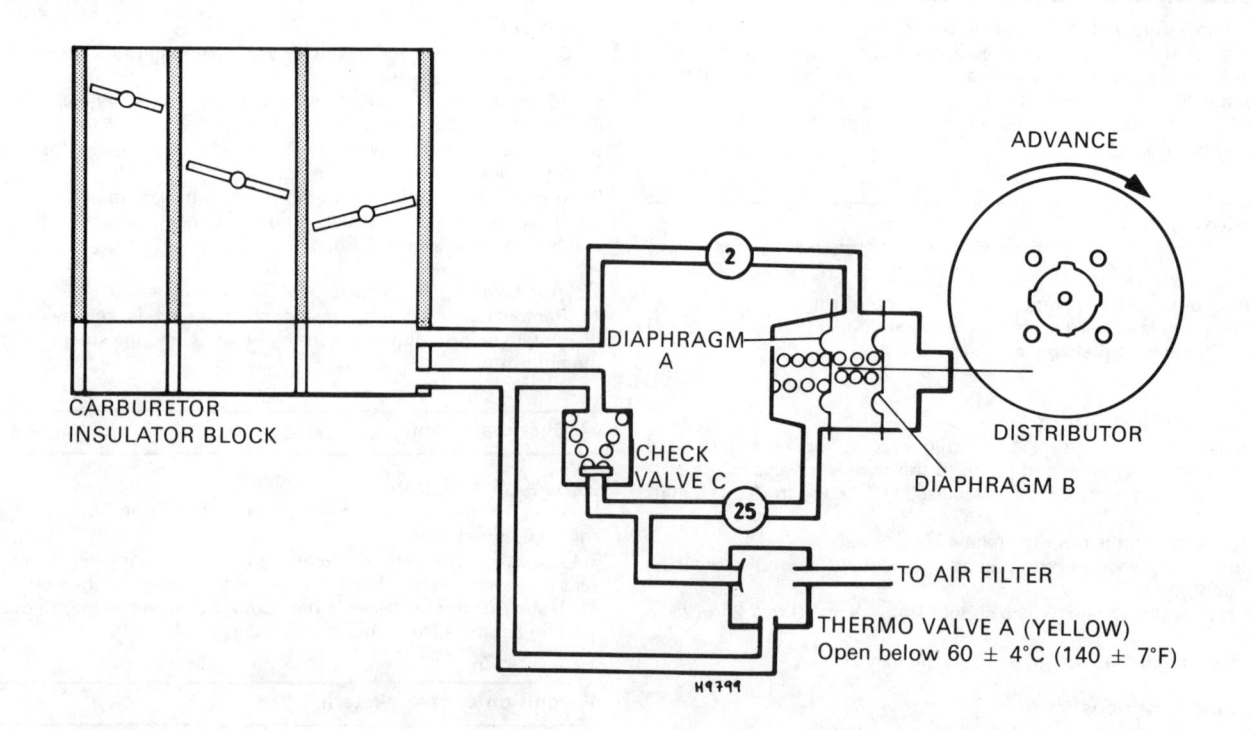

Fig. 3.54 Ignition control system diagram (Sec 45)

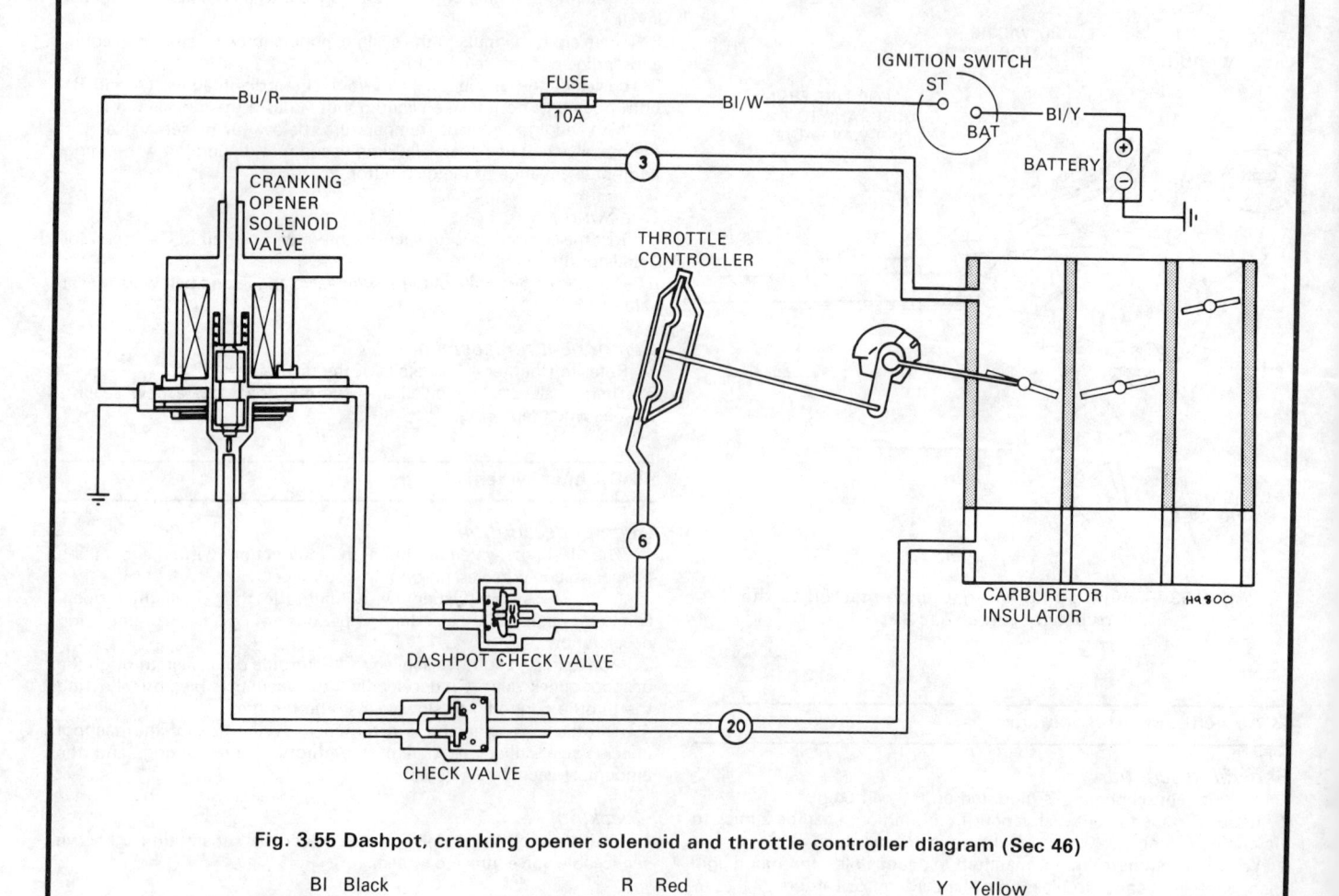

Fig. 3.55 Dashpot, cranking opener solenoid and throttle controller diagram (Sec 46)

| Bl Black | R Red | Y Yellow |
| Bu Blue | W White | |

Chapter 3 Fuel, exhaust and emission control systems

47 Throttle controller

General description
1 The system is fitted to position the throttle valve to its optimum position during the starting cycle.
2 This is achieved by activation of the cranking opener solenoid when the starter is operated, which allows manifold vacuum to the throttle controller via the dashpot check valve.

Checking
3 Start the engine and bring it to normal operating temperature.
4 Introduce a vacuum pump to the hose to the throttle controller.
5 Apply a vacuum of 8 in (200 mm) Hg.
6 Engine speed should rise to 2000 ± 500 rev/min within 1 minute.
7 Adjust by altering the width of the slot in the throttle controller lever (photo).

47.7 Throttle controller adjustment slot (arrowed)

8 Widen the slot to increase and narrow the slot to decrease the rev/min.
9 If the rev/min cannot be adjusted, or the throttle controller will not hold vacuum, renew the throttle controller.
10 Reconnect the vacuum hose to the throttle controller.
11 Increase engine speed to 3500 rev/min for 2 to 3 seconds.
12 Release the throttle and check the time taken for the throttle control arm to fully extend, which should be 1 to 4 seconds.
13 If more than 4 seconds, renew the dashpot check valve.
14 If less than 1 second, disconnect the hose to the throttle controller and fit a vacuum gauge to the hose.
15 With the engine running at 4000 rev/min, vacuum should be 1.2 in (30 mm) Hg.
16 If vacuum is at or above this value, renew the throttle controller, if less, the problem is elsewhere.

Component replacement
17 Remove the retaining screws from the throttle controller, and disconnect the throttle control arm and vacuum hose.
18 Remove the throttle controller.
19 Refit in the reverse order and retest.

48 Cranking opener solenoid valve

General description
1 The cranking opener solenoid valve is activated on starting, allowing vacuum to the throttle controller and positioning the throttle at the optimum angle for starting.

Checking
2 Complete the throttle controller checks (Section 47).
3 Earth the coil secondary wire to prevent the engine starting and turn the ignition key to the start position.
4 The throttle controller arm should retract.
5 If the controller does not work, check for voltage at the solenoid valve.
6 If no voltage, renew the solenoid valve, after checking the wiring and fuse.
7 If voltage is present, check for vacuum at the check valve while cranking the engine.
8 If vacuum is present renew the cranking solenoid valve, if no vacuum, renew the check valve.

Component replacement
9 The check valve is situated in the control box, its renewal being self-evident.
10 Similarly, the cranking solenoid can be removed from the control box by disconnecting the vacuum and electrical leads, and removing its retaining screws.
11 Refit components in the reverse order and retest.

49 Exhaust gas recirculation system (EGR) – general description

1 The EGR system is designed to reduce oxides of nitrogen emission (NOx) by recirculating exhaust gas through the EGR valve and intake manifold into the combustion chambers.
2 The EGR valve, operated by vacuum from a tapping above the throttle idle valve to eliminate gas recirculation at idle, provides gas recirculation proportional to engine load, by the operation of the EGR control valves A and B (Fig. 3.56).
3 Additional control is by thermovalve C, which is open when coolant temperature is below a preset value, bleeding the vacuum off and so keeping the EGR valve closed; keeping exhaust gas out of the manifold in cold weather.
4 EGR valve B is normally closed, and only opens when vacuum taken from the carburettor insulator block reaches a preset level, allowing venturi vacuum to enter control valve A and the EGR valve.

Checking
5 With the engine coolant temperature below the preset value of thermovalve C, so that the thermovalve is open, connect a vacuum gauge to the EGR valve hose.
6 Start the engine and increase speed to 4500 to 5000 rev/min.
7 There should be no vacuum. If there is vacuum, renew the thermovalve C and retest.
8 Allow the engine to warm up, and connect the vacuum gauge to the EGR valve.
9 Apply 6 in (150 mm) Hg vacuum to the EGR valve.
10 If vacuum remains steady and engine dies, EGR valve is working properly, if not renew the EGR valve.
Note: more comprehensive testing of the system should be undertaken by your dealer.

Component replacement
11 Removal and replacement of the EGR valve is self-evident.
12 The EGR control valves A and B are mounted in the control box.
13 The thermovalve C is screwed into the thermostat housing.

50 Secondary air supply system

General description
1 The system is designed to improve emission control performance by introducing fresh air from the air cleaner into the exhaust manifold, through the air suction valve.
2 An air bleed valve is incorporated to prevent overheating of the exhaust system which could be caused by catalyst reaction.

Checking
3 Checking of this system is best left to your dealer.

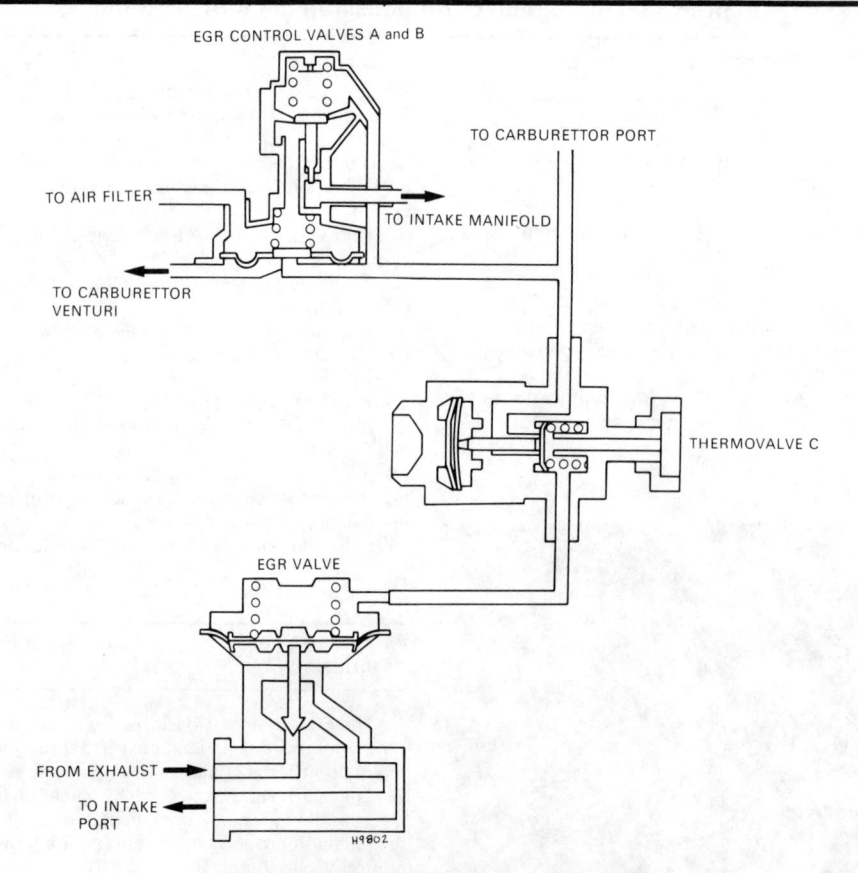

Fig. 3.56 EGR system diagram (Sec 49)

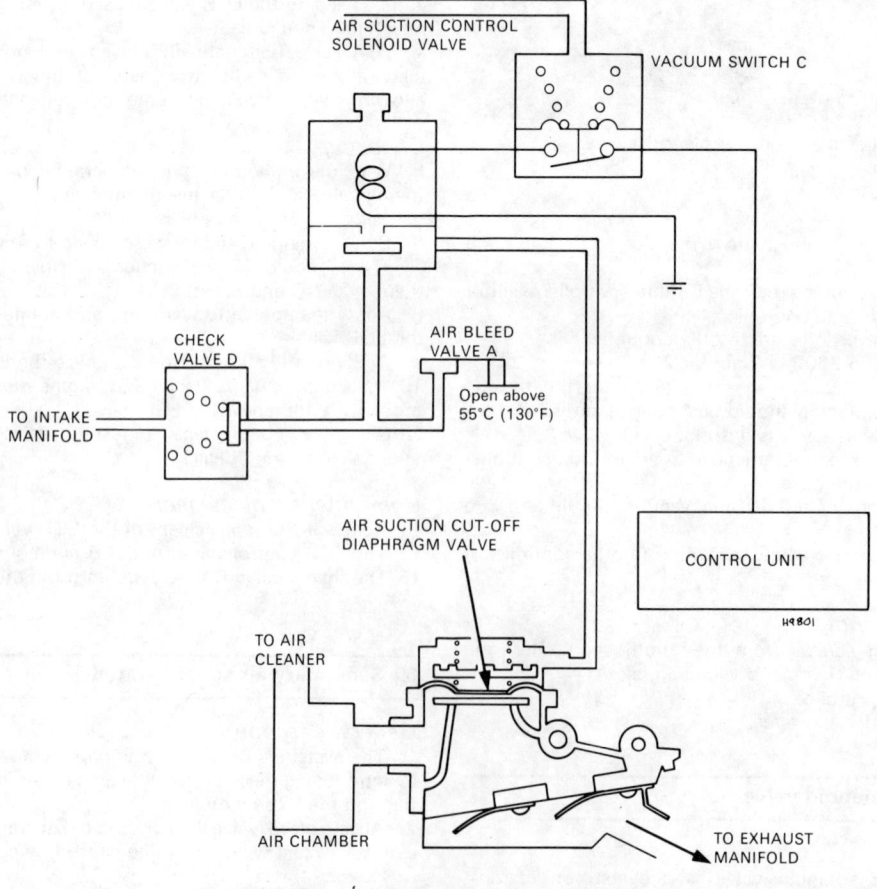

Fig. 3.57 Secondary air supply system (Sec 50)

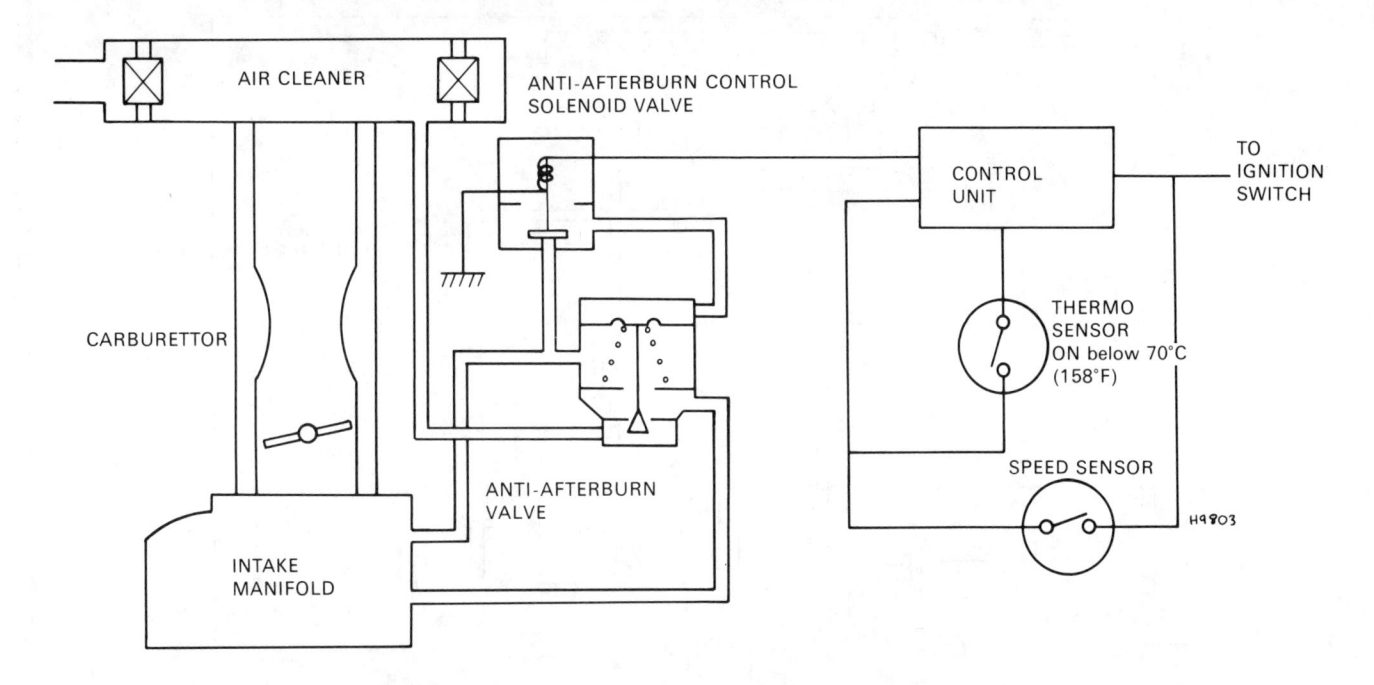

Chapter 3 Fuel, exhaust and emission control systems

Fig. 3.58 Anti-afterburn valve system (Sec 51)

Component replacement
4 The air suction valve is mounted on the air chamber by the air cleaner.
5 Air bleed valve is located in the air cleaner.
6 The remaining components are in the control box.

51 Anti-afterburn valve

General description
1 The system is fitted to supply fresh air to the intake manifold, to prevent an over rich mixture during periods of gear changing and deceleration.
2 Additional control is by means of a thermosensor and speed sensor.
3 Testing and component replacement are best left to your dealer.

52 High altitude reduced emission – adjustment

1 In accordance with EPA regulations the following adjustment may be made to any vehicle normally driven above 4000ft (1220 m).
2 This adjustment is only applicable to vehicles sold at low altitude, outside of California.
3 California cars should not be adjusted, and cars originally sold at high altitude need not be adjusted for low altitude use.
4 Remove the carburettor mounting bolts and tilt the carburettor so the throttle opener bracket screw may be removed.
5 Remove the bracket and mixture screw cap, and turn the mixture screw one half turn clockwise; the end result should be that the screw is no more than 1 full turn from the seated position.
6 Refit the carburettor.
7 Check idle speed, which should be as specified.

53 Feedback control system

General description
1 The feedback control system is designed to provide a stoichiometric air/fuel ratio, making the most of the three-way catalyst's performance, to give a simultaneous reduction of hydrocarbons, carbon monoxide and oxides of nitrogen.

Checking
2 The system is best tested by your local dealer who has the necessary test equipment.

Component replacement
3 Again, best left to your dealer.

54 Speed sensor

General description
1 The speed sensor is essentially a photo-interrupter, and is mounted in the speedometer.

55 Air jet controller

General description
1 The air jet controller is an atmospheric pressure sensing device which controls the amount of air flow into the slow and main air jets of the primary carburettor throat, and the slow air jet of the secondary throat.
2 As atmospheric pressure is reduced by increasing altitude, the bellows expand to open the valve in the air jet controller, increasing airflow to the jets to maintain optimum air/fuel ratio.

Checking
3 The valve is best checked by substitution.

Component replacement
4 The valve is mounted on the engine bulkhead by the control box.

56 Fuel cut-off relay

General description
1 The fuel cut-off relay is an electrically-operated solenoid valve controlled by the ignition switch.
2 As soon as the igniton is switched off the valve closes, shutting off supply to the slow mixture jet.

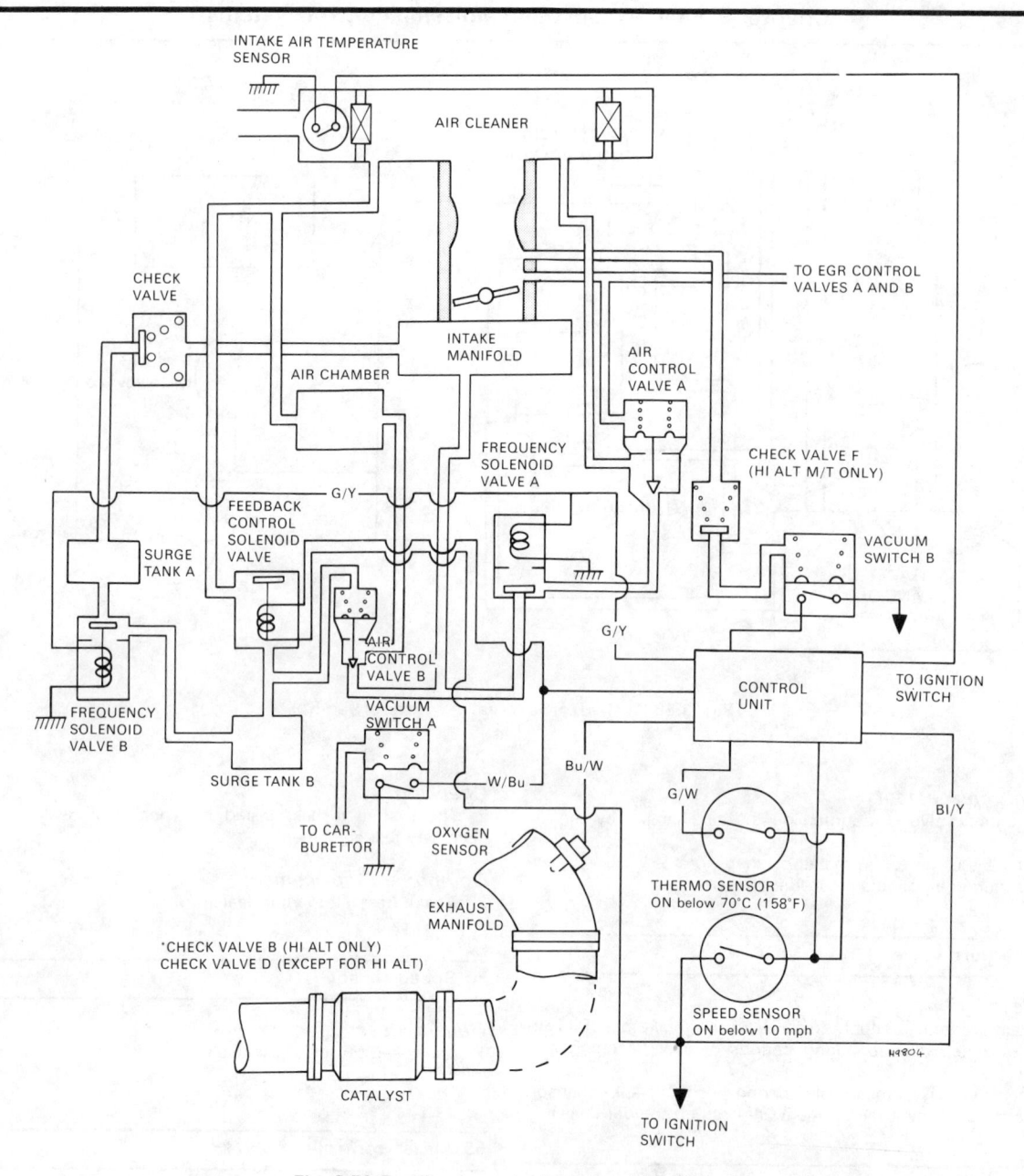

Fig. 3.59 Feedback control system (Sec 53)

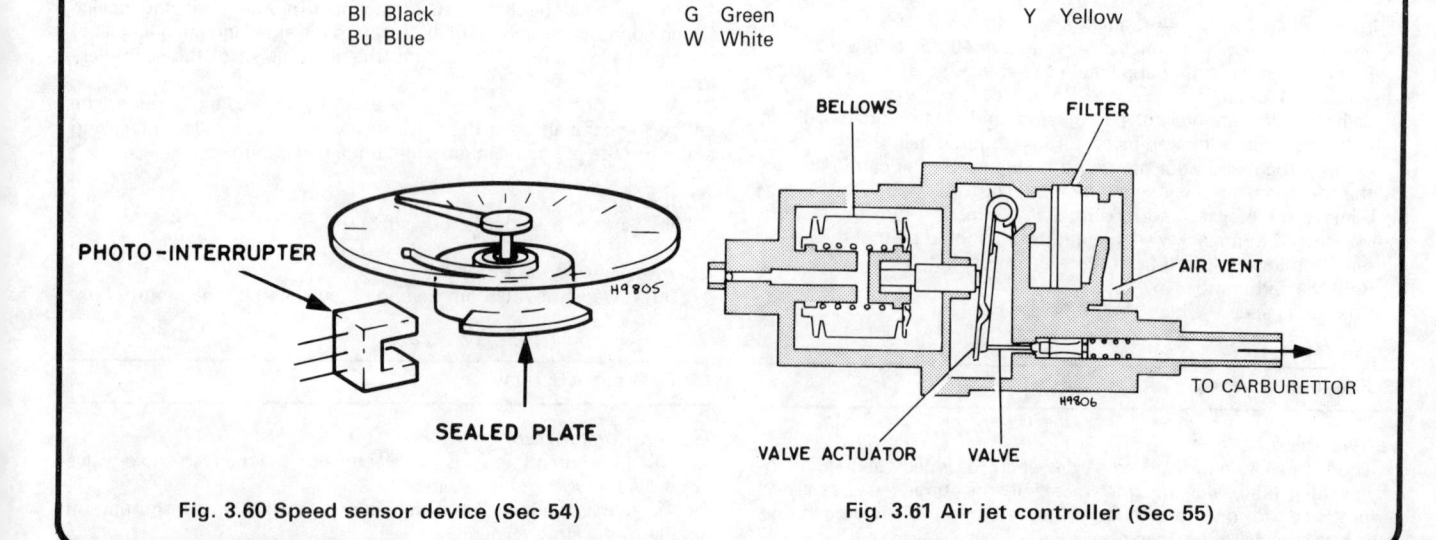

Bl Black
Bu Blue
G Green
W White
Y Yellow

Fig. 3.60 Speed sensor device (Sec 54)

Fig. 3.61 Air jet controller (Sec 55)

Chapter 3 Fuel, exhaust and emission control systems

3 Some carburettors have an additional cut-off valve, which cuts supply to the main jet.

Checking

4 Remove the valve from the carburettor by undoing the retaining screws (photo).
5 Switch the ignition on and off and check that the solenoid operates, and the needle retracts with the ignition on and protrudes when switched off.

Component replacement

6 Remove the valve as described above, and disconnect the electrical lead.

57 Fuel injection system – general description

1 The programmed fuel injection system (PGM-FI), consists of three sub-systems which are: air intake control, electronic control and fuel.
2 To provide the correct fuel/air ratio, engine speed and absolute pressure in the manifold are used to determine the amount of fuel to be injected. This is known as the Speed-Density method.
3 An eight bit microcomputer and various sensors provide extremely accurate control of the fuel/air mixture under all operating conditions.

56.4 Fuel cut-off solenoid valve (arrowed)

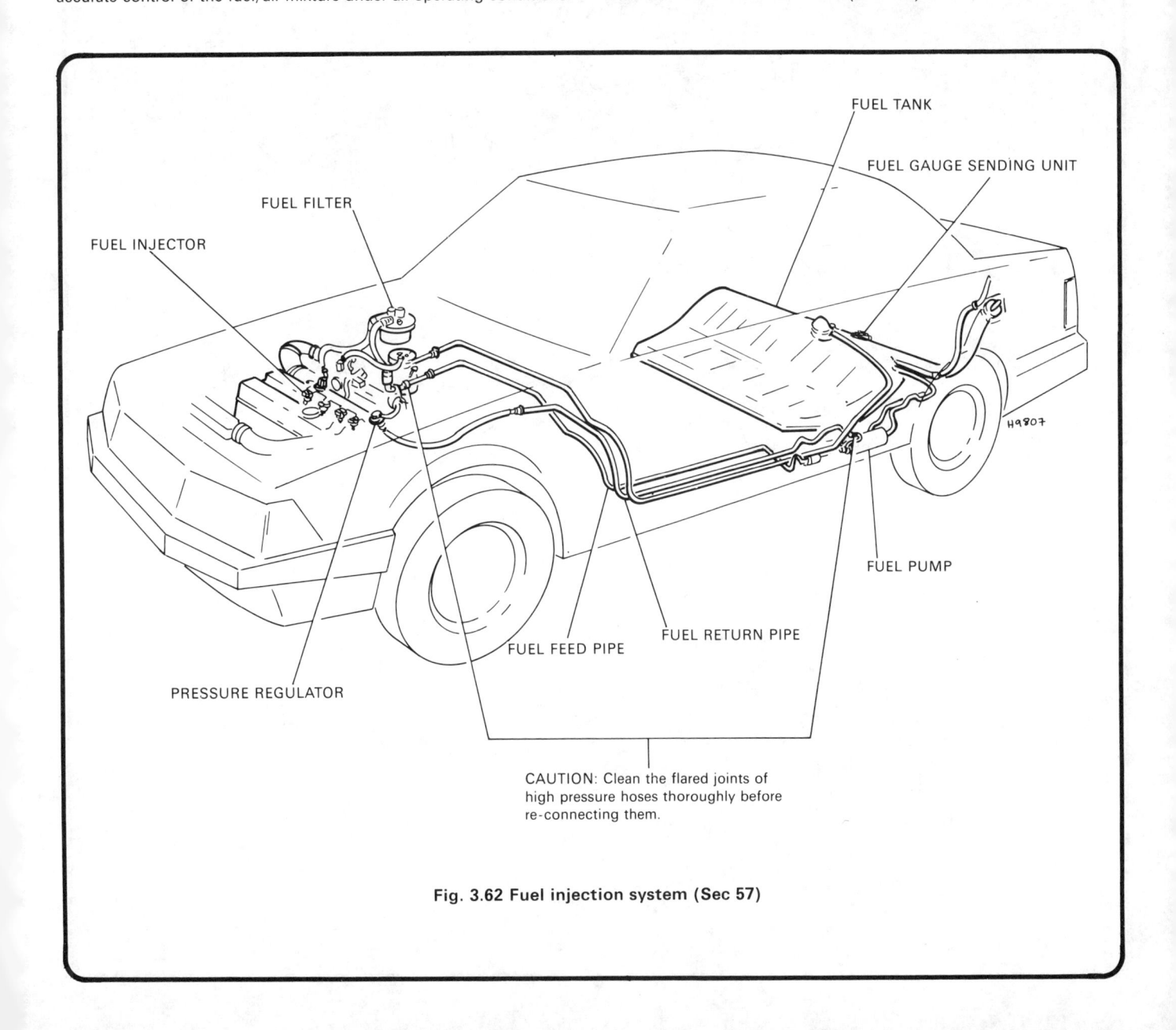

Fig. 3.62 Fuel injection system (Sec 57)

CAUTION: Clean the flared joints of high pressure hoses thoroughly before re-connecting them.

Chapter 3 Fuel, exhaust and emission control systems

4 Fuel is injected into the intake manifold of the cylinder on induction stroke, known as sequential injection.
5 The following sections describe the various components and their function, and the servicing and removal/refitting procedures for the simpler components.
6 Fault diagnosis can be a long and complex affair, and is best left to your dealer, although testing by substitution is an easy way of checking whether or not a component is working correctly.
7 The electronic control system, will be found very reliable in service, although periodic application of water repellant fluid to the electrical components will be found beneficial.
8 Before beginning any fault diagnosis, check all electrical connections for integrity and freedom from corrosion, which may give rise to spurious signals being received by the ECU.

58 Air intake system (FI) – removal, refitting and testing of components

Sensors
1 The fitting and removal of the air control system sensor devices is straightforward, after having located them from Fig. 3.63.

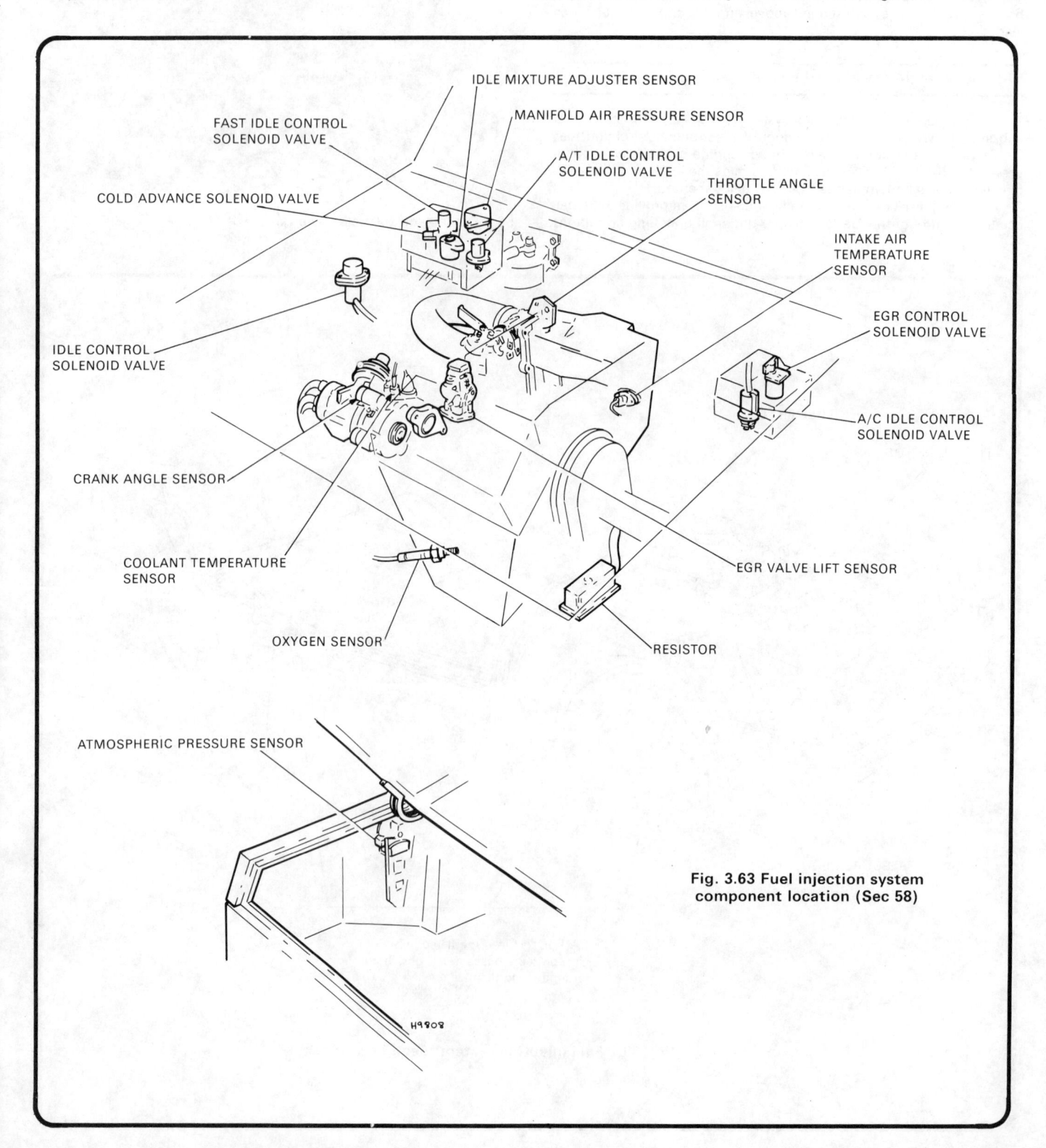

Fig. 3.63 Fuel injection system component location (Sec 58)

Chapter 3 Fuel, exhaust and emission control systems 109

2 Generally, the testing of these sensors should be left to your dealer, or test by substituion.
3 Some items, such as the throttle angle sensor, are fitted using shear screws, which will have to be drilled out to remove the sensor.

Throttle body
4 Disconnect the throttle control cable.
5 Disconnect the electrical leads and vacuum hoses to the throttle body.
6 Disconnect the air intake ducting.
7 Remove the four mounting nuts and remove the throttle body.
8 Refitting is a reversal of removing, using a new gasket.
9 Adjust the throttle cable as described in Section 61, and the throttle, as described in the following paragraphs.

Idle speed/mixture – adjustment
10 Start and warm up the engine to normal operating temperature.
11 Connect up a tachometer to the ignition coil, in accordance with the manufacturer's instructions.
12 Check idle speed, with headlights, heater, heated rear window, cooling fan and air conditioner off.
13 Prevent the idle control system from operating by disconnecting and plugging the vacuum hose from the idle control diaphragm.
14 Adjust the idle speed to its specified valve by turning the idle speed adjusting screw in to decrease idle speed and out to increase (Fig. 3.67).
15 If idle speed cannot be adjusted correctly by the adjusting screw, check the fast idle valve.
16 On vehicles with automatic transmission, after adjusting idle speed, check that idle remains in limits when shifted into gear.
17 Reconnect the idle control diaphragm and check that the idle speed is within limits when air conditioning is selected, adjusting on screw B if necessary (Fig. 3.68).
18 Using a CO meter, check exhaust emission which should be as specified.

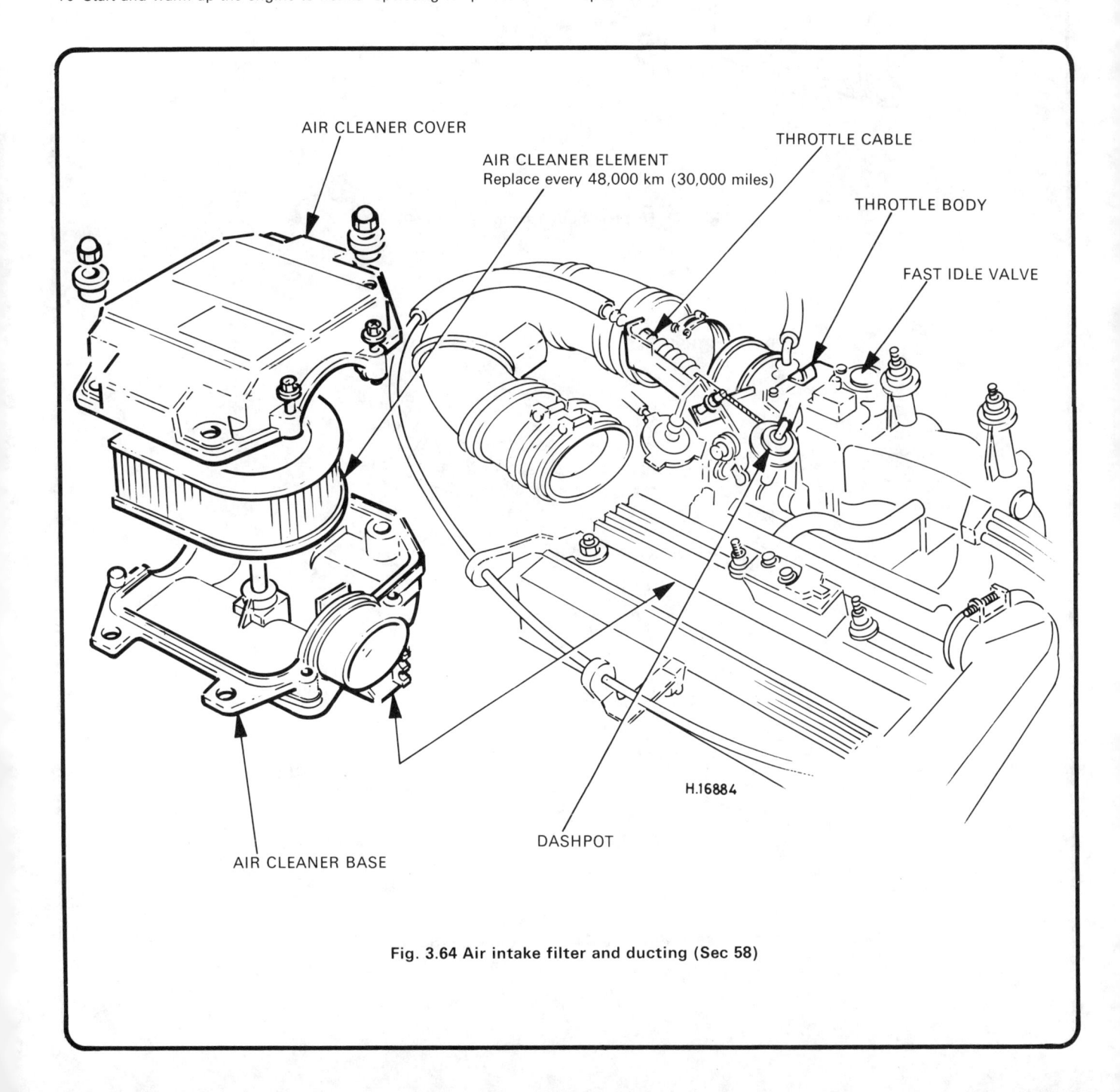

Fig. 3.64 Air intake filter and ducting (Sec 58)

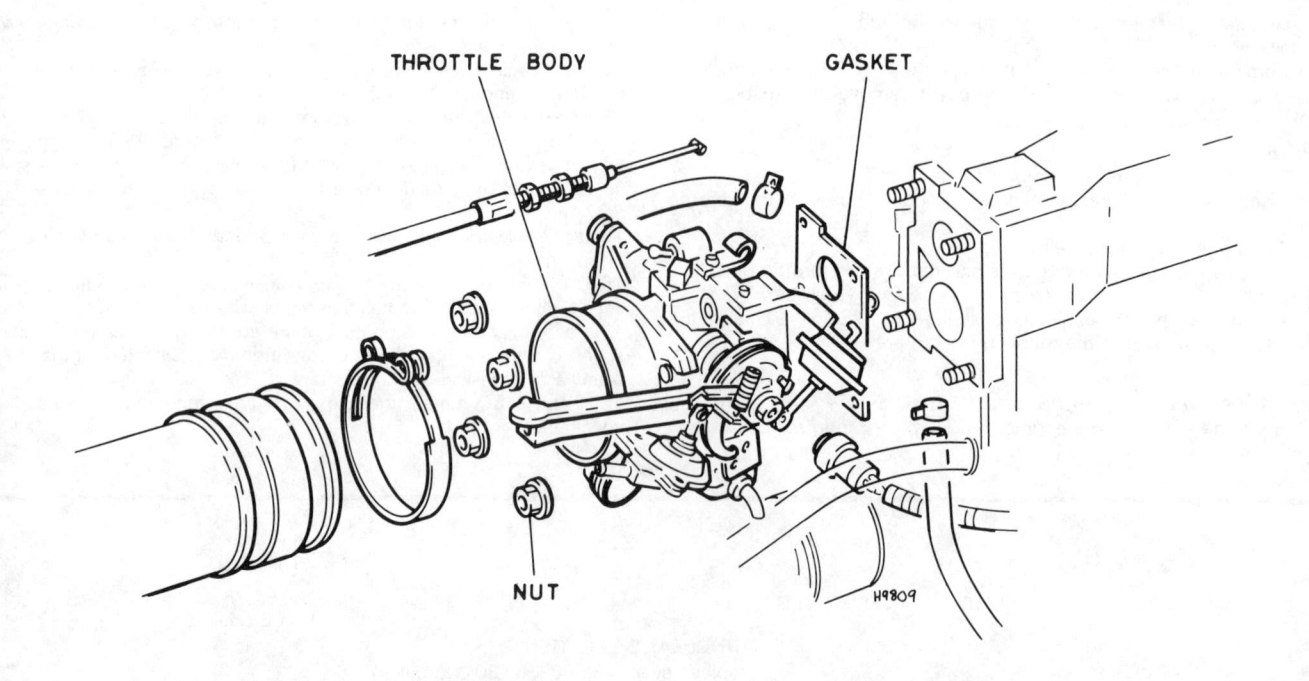

Fig. 3.65 Throttle body assembly (Sec 58)

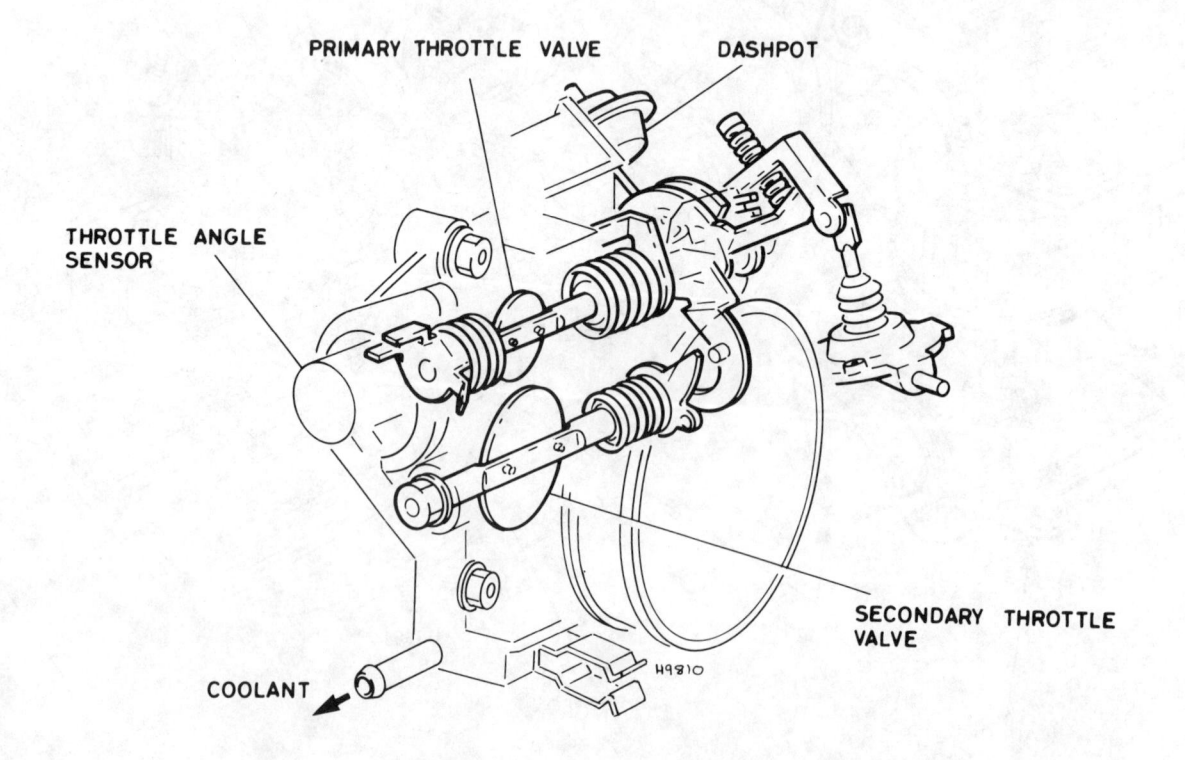

Fig. 3.66 Internal view of throttle body (Sec 58)

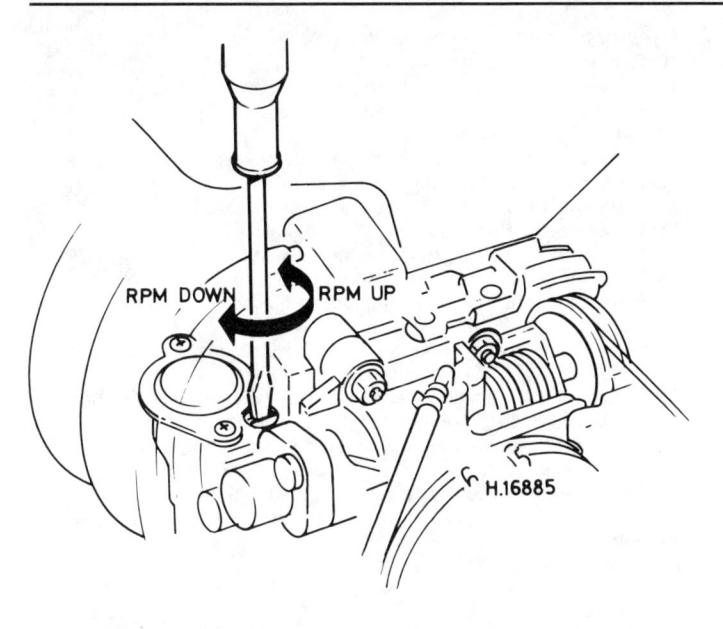

Chapter 3 Fuel, exhaust and emission control systems

Fig. 3.67 Idle speed adjustment screw (Sec 58)

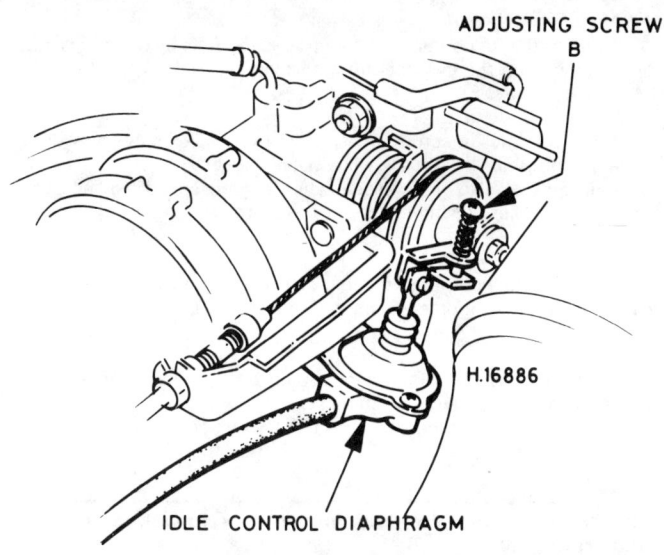

Fig. 3.68 Idle control diaphragm and adjusting screw B (Sec 58)

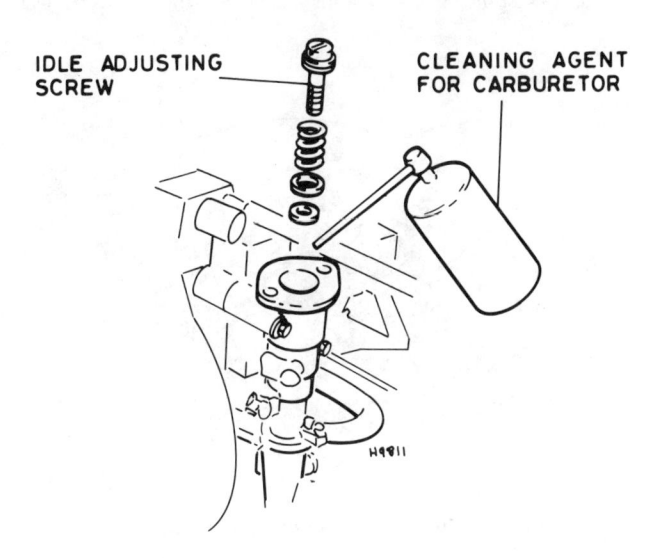

Fig. 3.69 Fast idle valve assembly (Sec 58)

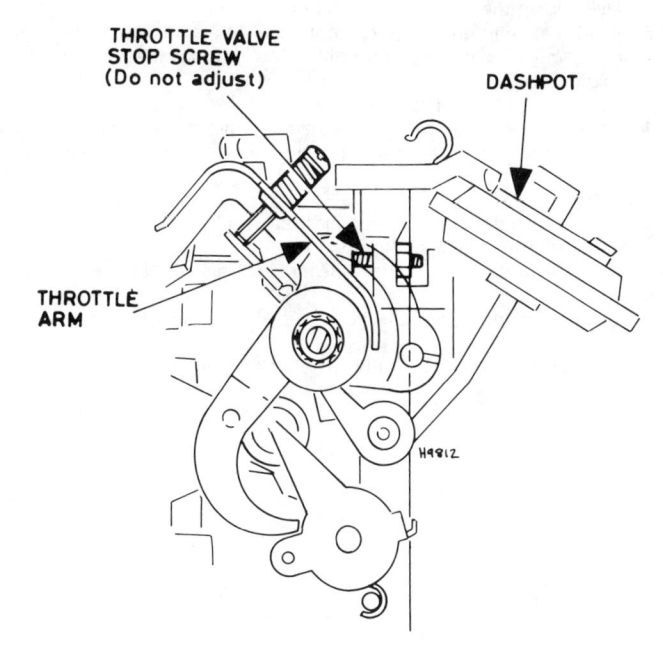

Fig. 3.70 Dashpot assembly (Sec 58)

Fast idle valve
19 The fast idle valve is factory set and should not be tampered with.
20 To test the valve, check that the PCV valve and associated tubing is in serviceable condition, and that the throttle valves are fully closed.
21 Check that the idle control function is normal.
22 Warm up the engine.
23 Remove the cover of the fast idle valve, and check that the fast idle valve is completely closed.
24 If it is not, and air can be felt being sucked past the valve seat, the valve should be renewed.
25 Note that the valve operates from coolant temperature, so if removing the valve, then drain the cooling system at least below the level of the valve before removing it.
26 Use new O-ring seals on reassembly.

Dashpot system
27 Slowly open the throttle arm until the dashpot rod is raised as far as it will go.
28 Release the throttle arm and measure the time taken for the throttle arm to contact the stop screw, which should be less than 2 seconds.
29 If longer, then renew the dashpot.

59 Electronic control unit (ECU) – general description

1 In order to inject fuel into the cylinders at the correct volume and time, the control system must perform various separate functions.
2 The ECU is the 'brain' of the programmed fuel injection system, using an eight bit microcomputer, and consists of a central processing unit (CPU), memories, and input/output ports.
3 Basic data stored in the memories are compensated by the signals sent from the various sensors to provide the correct air/fuel mixture for all engine needs.
4 The ECU is mounted under the passenger seat inside the vehicle, its basic functions are as follows:

Starting control
5 Varies the fuel/air mixture according to information received from the starter switch, engine speed and coolant temperature sensors, providing the extra fuel needed for starting.

Chapter 3 Fuel, exhaust and emission control systems

Fuel pump control
6 Cuts off the fuel pump when engine speed falls below a prescribed limit, preventing the injectors from discharging fuel.

Fuel cut-off control
7 Cuts off the injectors at speeds over 1200 rev/min during deceleration with the throttle valve nearly closed.
8 Fuel cut-off also takes place when engine speed exceeds 6600 rev/min regardless of the position of the throttle valve.

Safety
9 The system incorporates a fail-safe mode ensuring safe driving even when one or more sensors fail, or if the ECU malfunctions.

Checking
10 Fault diagnosis of the system should be restricted to the simpler tasks given in the ensuing sections.
11 Your dealer is best equipped to test the system competently.

60 Fuel injection system components – removal, refitting and testing

Caution: *before working on any part of the fuel system, the pressure in the system must be relieved as follows:*
1 Disconnect the negative terminal from the battery.
2 Hold the banjo union on top of the fuel filter with an open-jawed spanner, and loosen the bleed bolt with a ring spanner, slowly, one complete turn.
3 Be prepared for fuel spillage.
4 Allow full pressure to dissipate, before removing the bolt.
5 Use a new washer on refitting, when work is complete.

Injectors – testing
6 With the engine idling, disconnect each injector electrical lead in turn and note the idle speed drop.
7 If the idle drop is about the same for each injector, the injectors are serviceable.
8 Using a stethoscope, check the clicking sound of each injector while the engine is idling.
9 If the clicking sound is not heard, check all wiring before renewing the injector.

Injectors – removal and refitting
10 Disconnect the battery.
11 Relieve the fuel pressure in the system.
12 Disconnect each injector electrical lead.
13 Remove the air cleaner case.
14 Disconnect the vacuum and fuel return hose from the pressure regulator, being prepared for fuel spillage.
15 Disconnect the two earth leads from the intake manifold.
16 Disconnect the fuel lines.
17 Remove the injectors from the intake manifold.
18 On refitting, first slide the cushion ring onto the injectors.
19 Place an O-ring onto the injectors and insert the injectors into the fuel pipe.
20 Position the sealing ring into each injector position in the intake manifold.
21 Install the injectors and fuel pipe assembly, using new O-rings and seals, lubricated with clean engine oil. Line up the index marks on the injector and intake manifold before tightening the nuts.
22 Tighten the retaining nuts to the specified torque.
23 Reconnect the electrical earth leads to the manifold.
24 Reconnect the vacuum and fuel return hose to the pressure regulator.
25 Connect the electrical leads to the injectors.
26 After fitting new injectors, switch on the ignition but do not start the engine.
27 Allow the fuel pump to run for about 2 seconds to build up pressure.
28 Repeat this procedure two or three times and then check for leaks.

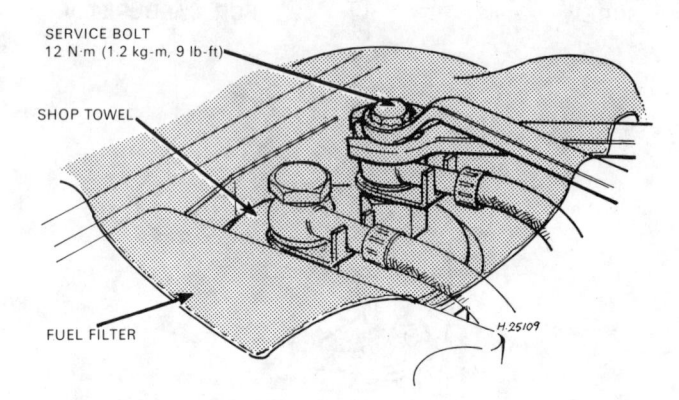

Fig. 3.71 Fuel pressure vent union (Sec 60)

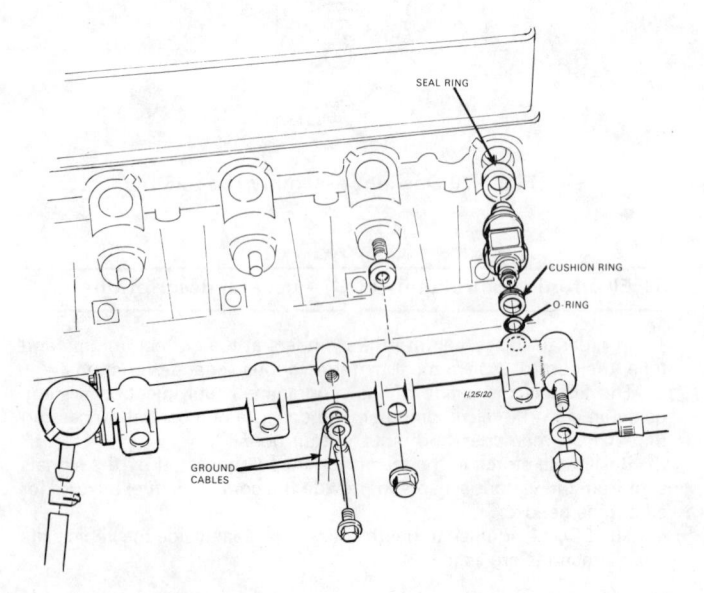

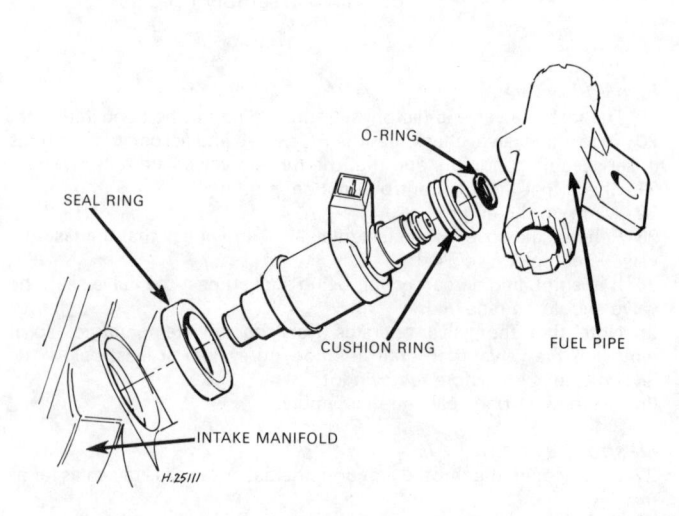

Fig. 3.72 Fuel injector-to-manifold arrangement (Sec 60)

Fig. 3.73 Injector sealing rings (Sec 60)

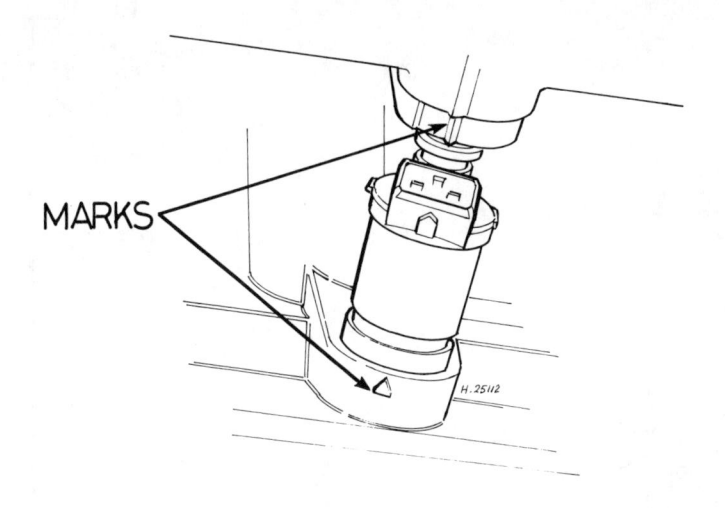

Chapter 3 Fuel, exhaust and emission control systems

Fig. 3.74 Injector alignment index marks (Sec 60)

Resistor
29 The resistance between the resistor terminals and power supply terminals should be 5 to 7 ohm. Renew the resistor if out of specification.

Fuel pressure testing
30 Relieve fuel pressure as described earlier.
31 Attach a suitable pressure gauge to the bleed bolt orifice on the banjo union of the fuel filter.
32 Start the engine and then check fuel pressure with the vacuum hose on the pressure regulator disconnected.
33 If the fuel pressure is not as specified, check the fuel pump, all fuel lines for kinking or leaks or blockages, and the pressure regulator for correct operation.

Pressure regulator
34 Test the regulator by fitting a pressure gauge to the fuel filter, as previously described.
35 Disconnect the vacuum hose to the regulator.
36 Fuel pressure should rise.
37 If it does not, a further check is to pinch the fuel return hose: pressure should rise.
38 Renew the pressure regulator if it fails the above tests.
39 Disconnect the battery negative terminal.
40 Disconnect the vacuum and fuel return hoses to the regulator, being ready for fuel spillage.
41 Remove the regulator retaining bolts and remove the regulator.
42 Refitting is a reversal of removal, but use new O-ring seals, lubricated with clean engine oil, and tighten the bolts to the specified torque.

Fuel filter
43 Renew the fuel filter at the specified intervals in the Routine Maintenance section, or when fuel pressure drops below its specified value (with the vacuum hose disconnected) and having ensured that the fuel pump and pressure regulator are serviceable.
44 Relieve fuel pressure as described previously.
45 Remove the bolts from the two fuel line banjo connections.
46 Remove the filter clamp bolt and filter.
47 Refit the new filter in the reverse order, using new sealing washers on the banjo unions, and tightening all bolts to their specified torque.
Note: new sealing washers should be used whenever the banjo unions are disturbed.

Fuel pump – testing
48 The fuel pump is mounted just in front of the left rear wheel under the car.
49 Test the pump by switching on the ignition and listening at the pump. You should be able to hear it running.
50 If not, switch off the ignition, jack up the rear of the vehicle and remove the left rear wheel.

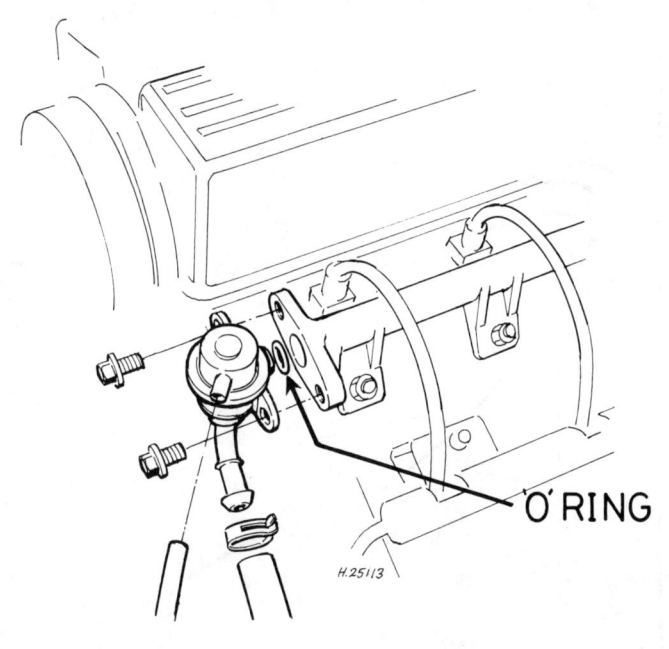

Fig. 3.75 Pressure regulator (Sec 60)

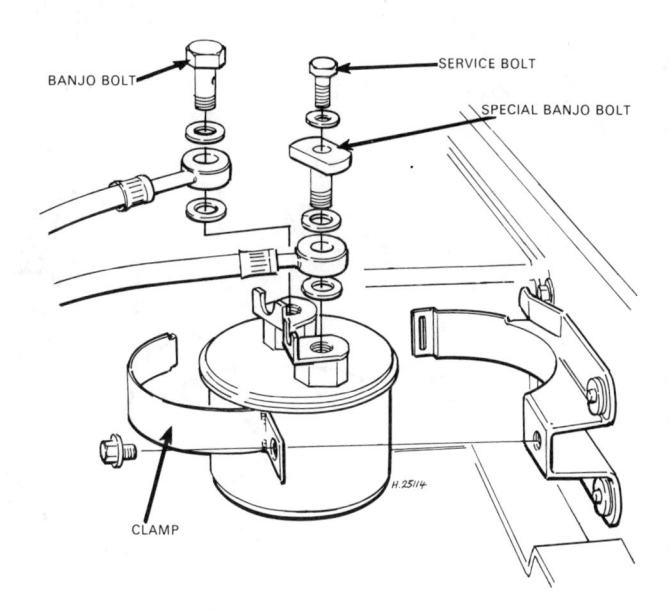

Fig. 3.76 Fuel filter, bracket and pipelines (Sec 60)

51 Remove the fuel pump cover and disconnect the electrical lead.
52 Switch on the ignition and check that battery voltage is available between the black wire and the black/yellow wire.
53 If voltage is available, renew the pump, if no voltage, check the main relay and wire harness.

Fuel pump – removal and refitting
54 Carry out paragraphs 50 and 51.
55 Disconnect the fuel lines at the connectors.
56 Remove the clamp and remove the fuel pump.
57 Remove the fuel line and silencer from the pump.
58 Fit the new pump to the mounting tray.
59 Clean the flare nut sealing surfaces and fit and tighten the flare nut.

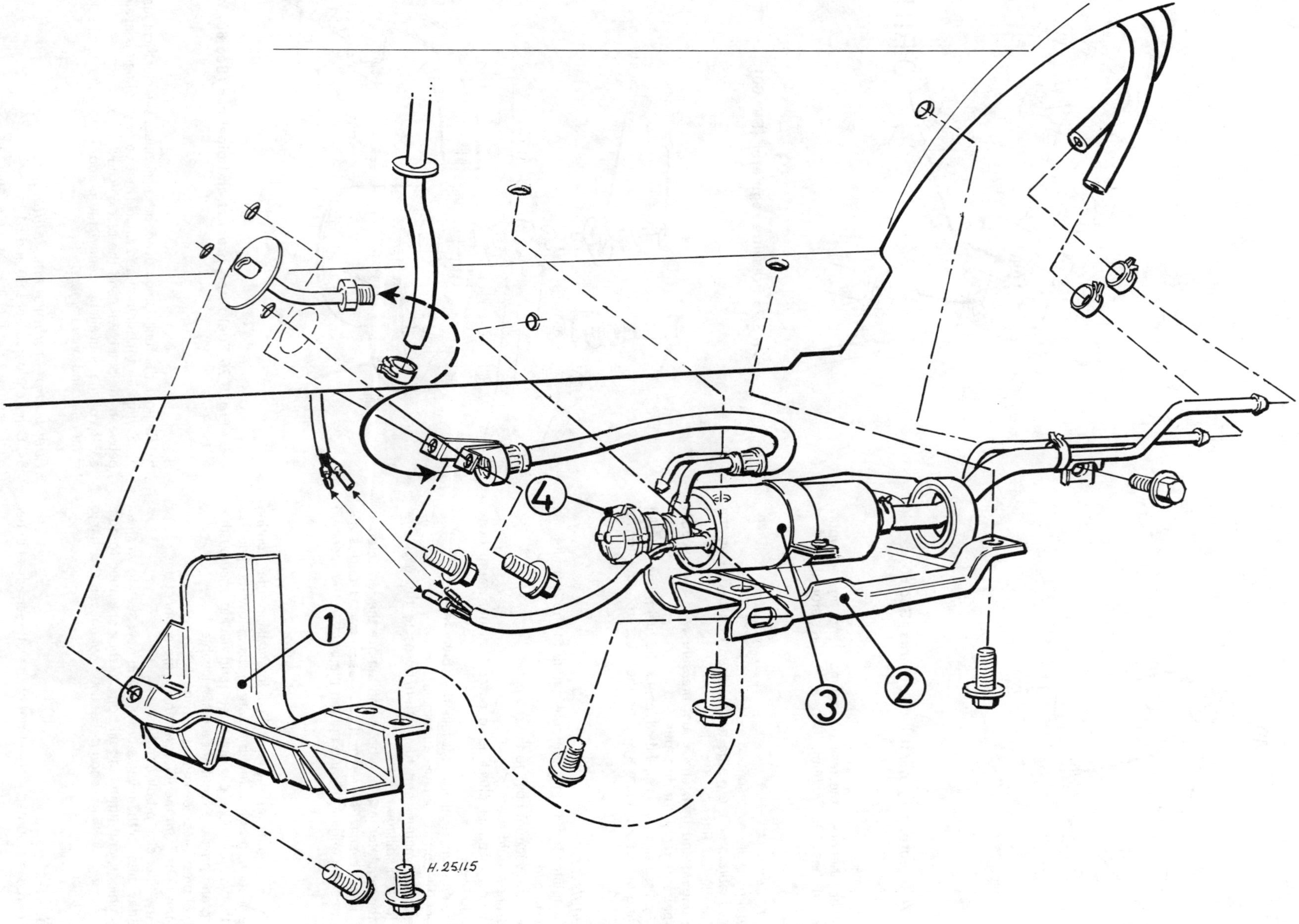

Fig. 3.77 Fuel pump assembly (Sec 60)

1 Cover   2 Mount   3 Pump   4 Silencer

Chapter 3 Fuel, exhaust and emission control systems   115

69 While listening to an injector, raise engine speed to 3000 rev/min, and then release the throttle.
70 The injectors should cease momentarily while the throttle is released.

61 Throttle cable (FI) – removal and adjustment

1 Loosen the locknut and remove the throttle cable from its support bracket.
2 Remove the cable from the throttle linkage.
3 Disconnect the cable from the accelerator pedal.
4 Remove the cable, feeding it through the engine bulkhead.
5 Refit a new cable in the reverse order and carry out the following adjustments.
6 Holding the cable sheath, remove all slack from the cable.
7 Turn the adjuster nut until it is 0.118 in (3 mm) away from the cable bracket.
8 Fit the adjuster into the bracket and tighten the locknut.
9 After adjusting the throttle cable, adjust the automatic transmission control cable and cruise control cable.
10 Adjust the toe board to the dimensions shown in Fig. 3.80.
11 On completion, check the throttle cable deflection which should be as specified.

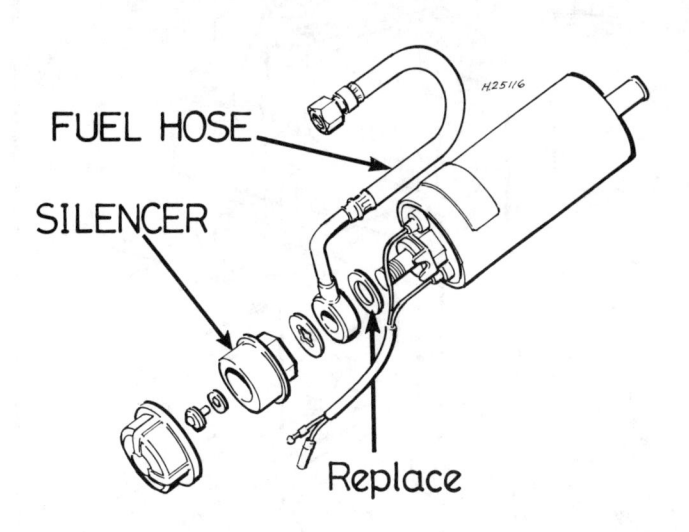

Fig. 3.78 Fuel pump silencer assembly (Sec 60)

60 Fit the fuel hose and silencer to the other end of the pump, using a new sealing washer, and tighten the union.
61 Refit the pump and connect up the fuel lines and electrical leads.
62 Switch on the ignition and check that the pump runs and there are no fuel leaks.
63 Repeat the fuel leak check several times before refitting the cover, rear wheel and removing the vehicle from jacks.
64 The fuel pump cannot be dismantled for repair and must be renewed if found defective.

Fuel pump main relay
65 The fuel pump relay is bolted to a bracket under the fusebox. Testing is best done by substitution.

Fuel cut-off system
66 Start the engine and bring it to normal operating temperature.
67 On manual transmission vehicles, disconnect the vacuum hose from the dashpot on the throttle body.
68 Using a stethoscope, confirm that the injectors are working.

62 Emission control (FI) – general description

1 The emission controls fitted to injected engines basically follow the same pattern as for carburettor versions, in that there are evaporative, crankcase and exhaust controls.
2 A description of each system and its components is given in the following Sections.
3 As the system is controlled by the electronic control unit, testing and subsequent renewal of the various components is best left to your dealer.

63 Crankcase controls (FI) – general

1 The Positive crankcase ventilation (PCV) system is basically the same as is fitted to carburettor models, except that the blow-by filter is mounted on the valve cover. The PCV valve is surrounded by engine coolant to prevent icing in cold weather. Testing of the PCV valve and renewal of the blow-by filter are as for carburettor versions.

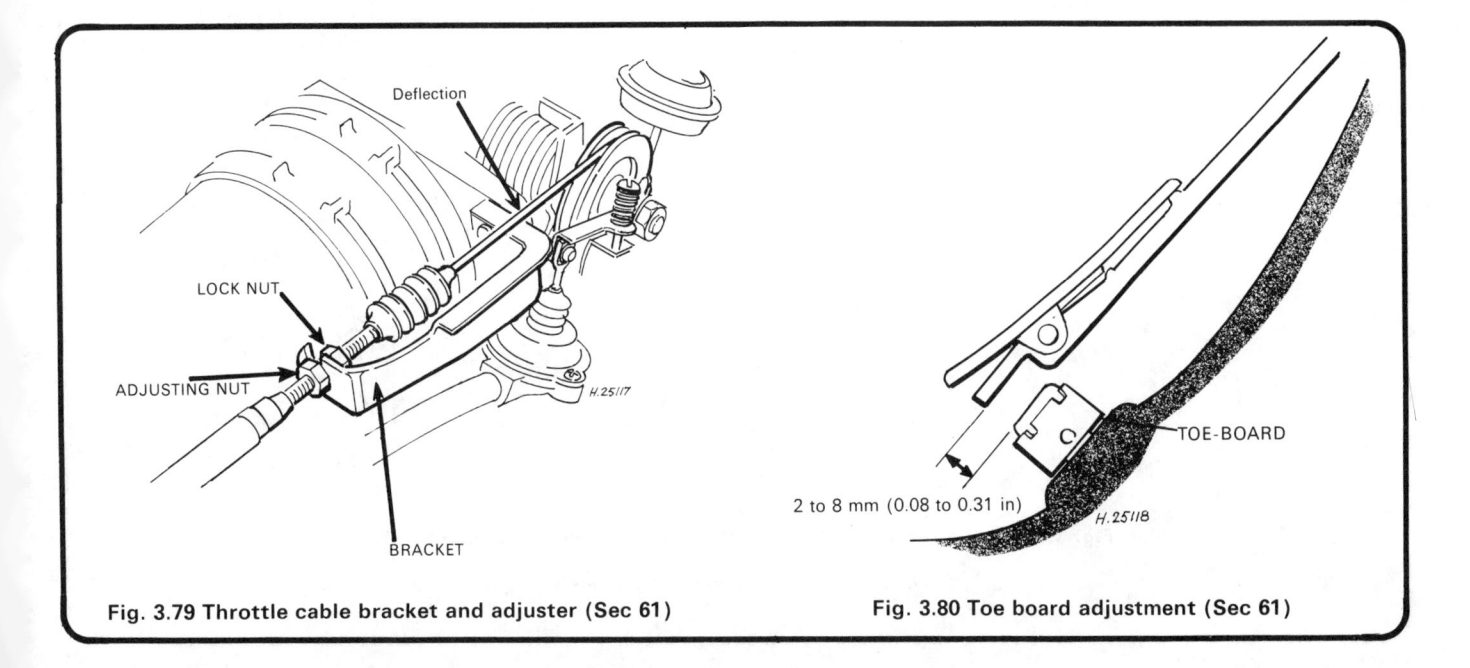

Fig. 3.79 Throttle cable bracket and adjuster (Sec 61)

Fig. 3.80 Toe board adjustment (Sec 61)

116

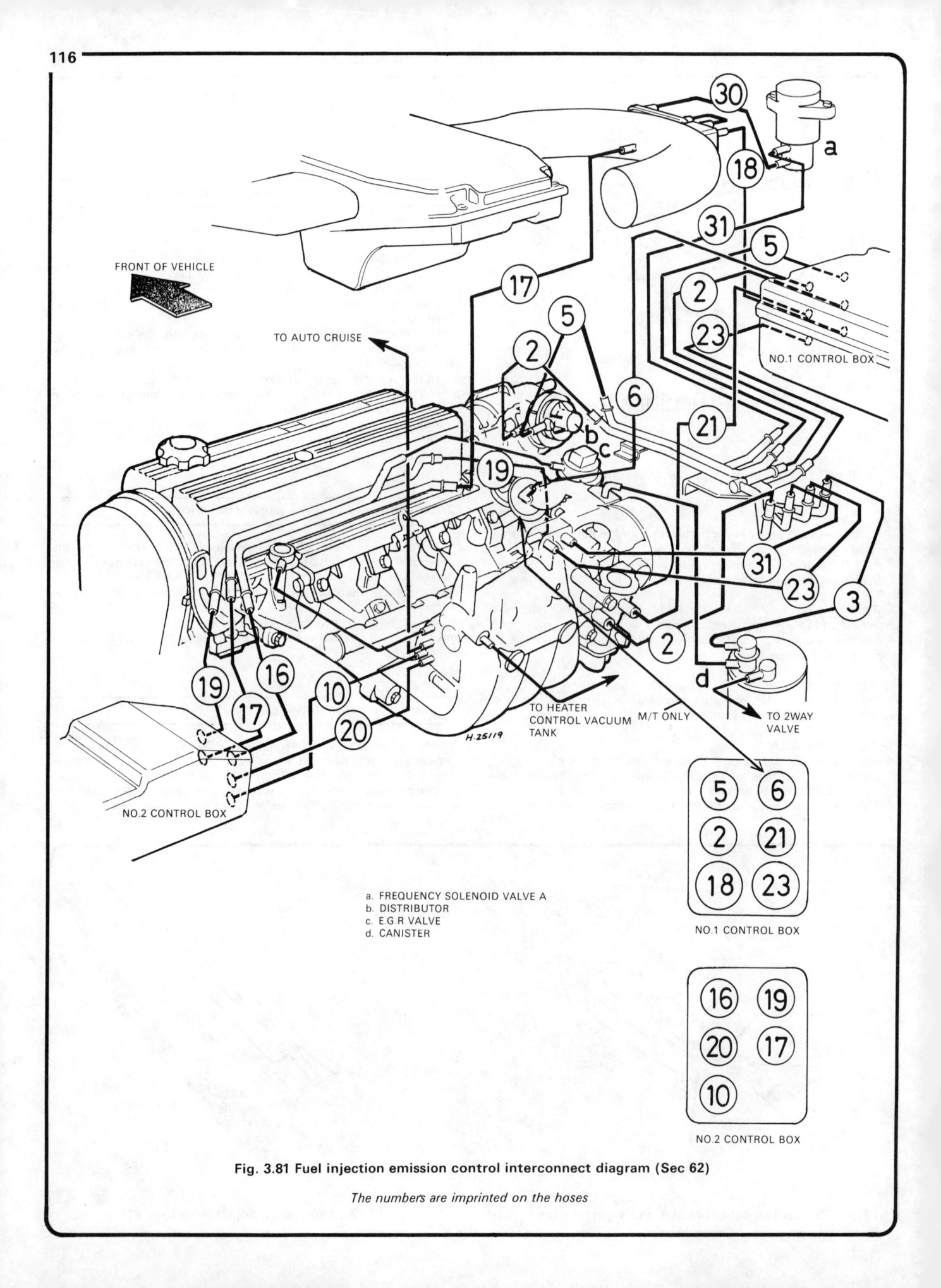

FRONT OF VEHICLE

TO AUTO CRUISE

TO HEATER CONTROL VACUUM TANK

M/T ONLY

TO 2WAY VALVE

NO.2 CONTROL BOX

NO.1 CONTROL BOX

a. FREQUENCY SOLENOID VALVE A
b. DISTRIBUTOR
c. E.G.R VALVE
d. CANISTER

5	6
2	21
18	23

NO.1 CONTROL BOX

16	19
20	17
10	

NO.2 CONTROL BOX

Fig. 3.81 Fuel injection emission control interconnect diagram (Sec 62)

The numbers are imprinted on the hoses

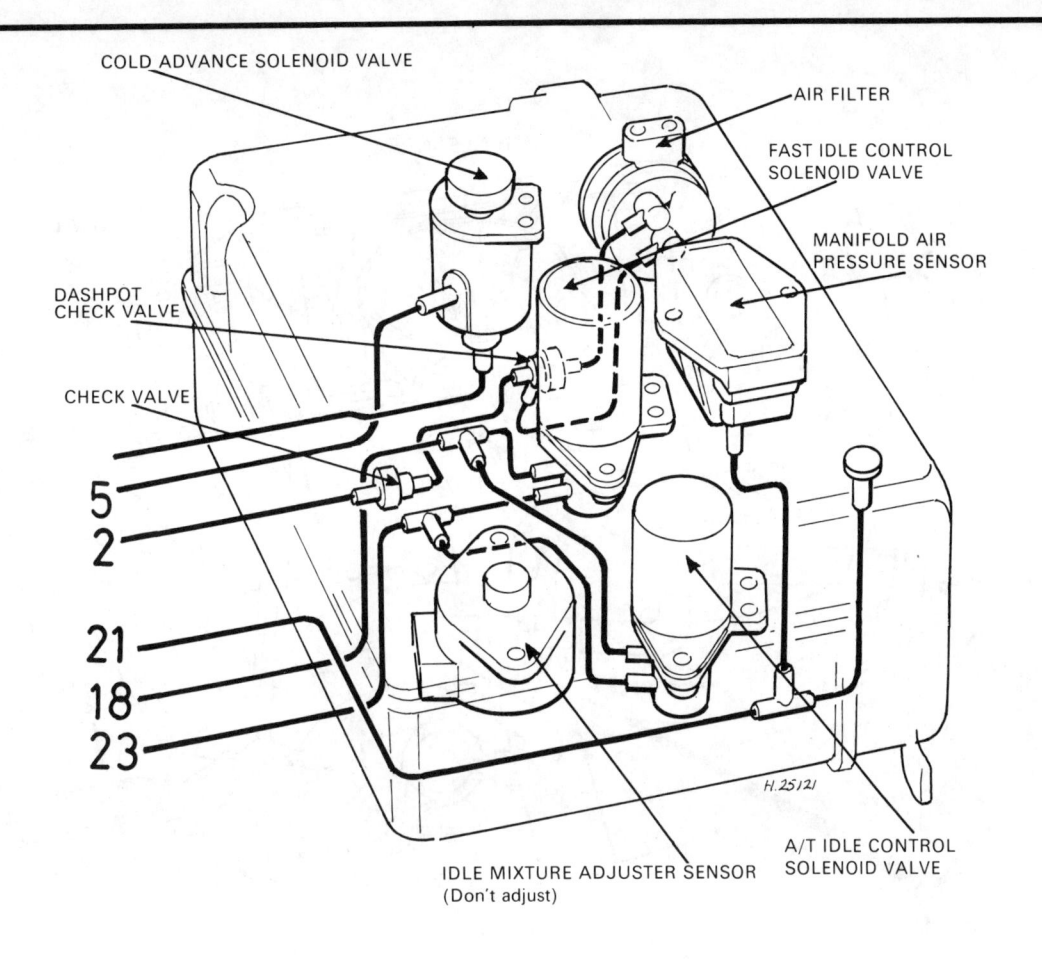

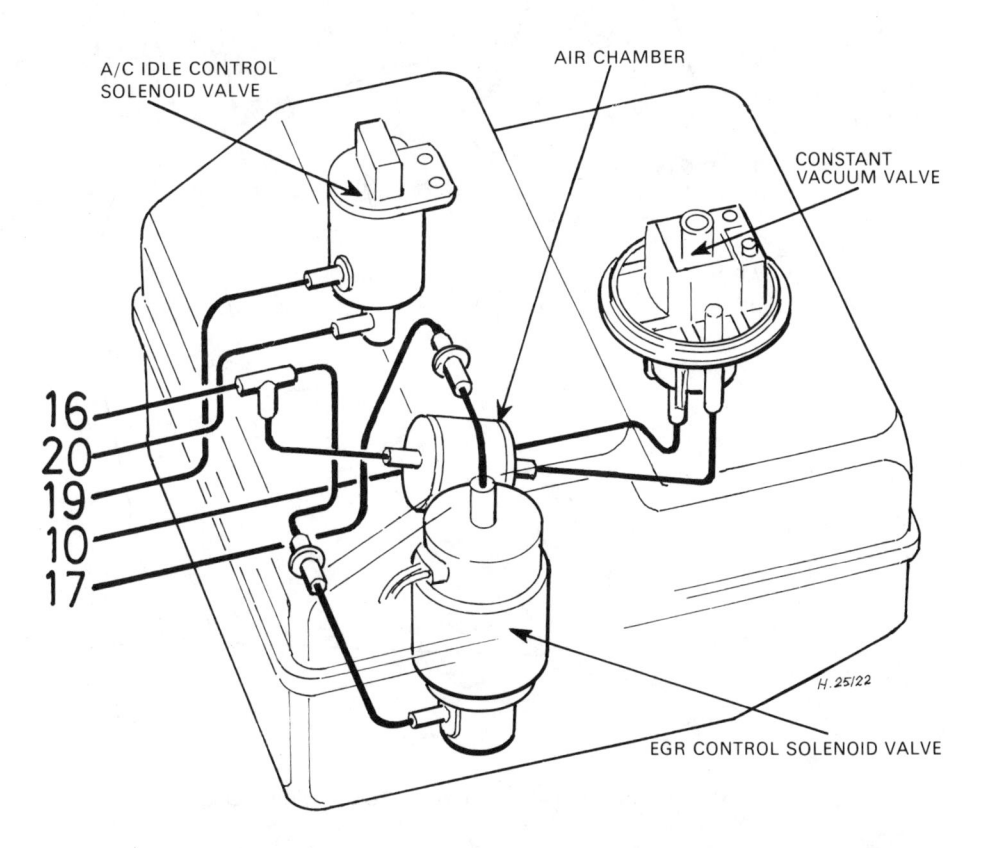

Fig. 3.82 Components of No 1 and No 2 control boxes (Sec 62)

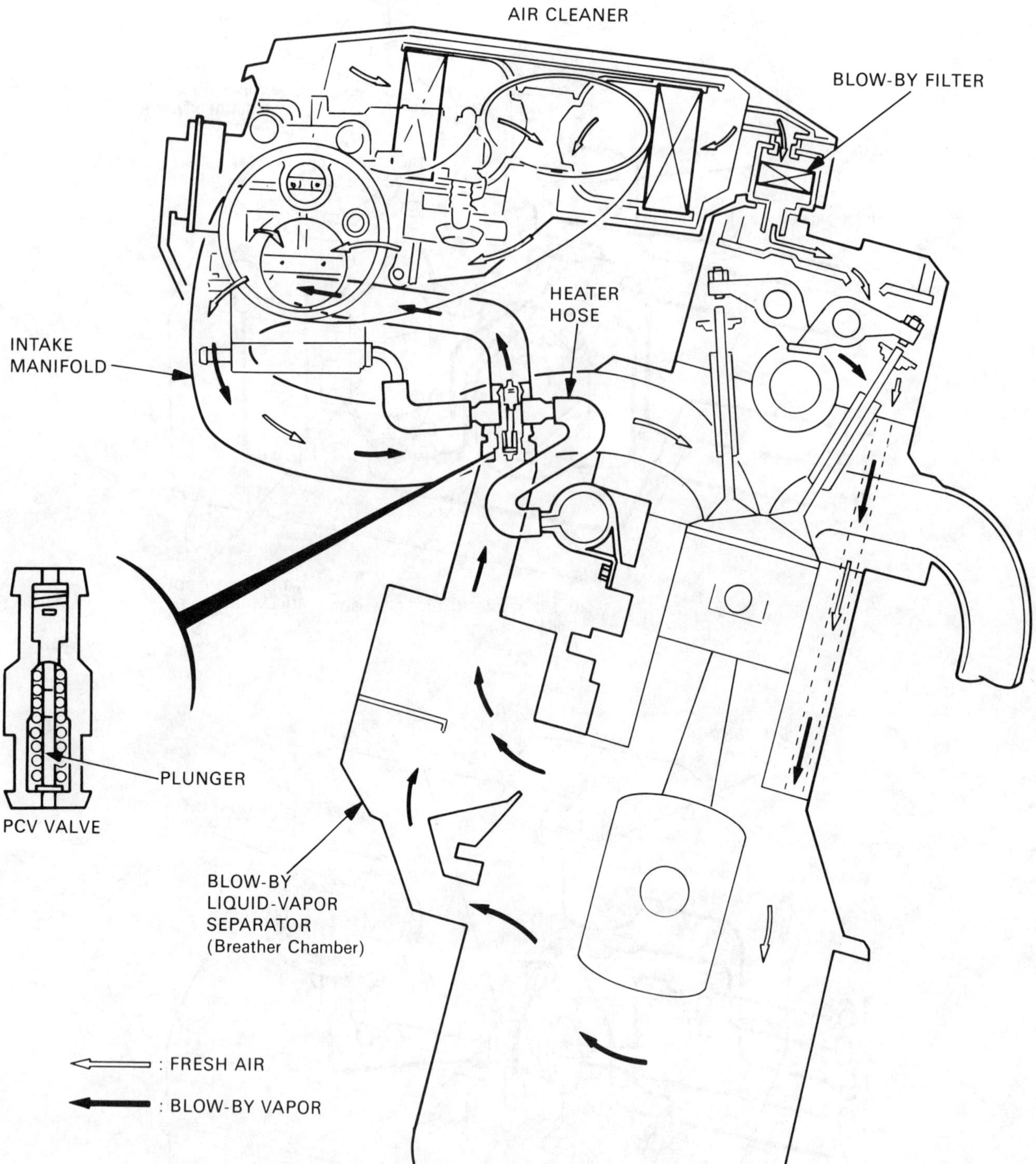

Fig. 3.83 Positive crankcase ventilation (PCV) system (Sec 63)

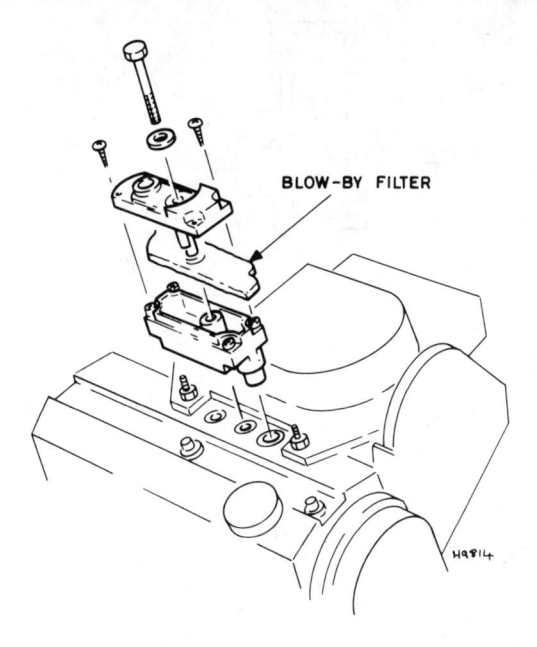

Chapter 3 Fuel, exhaust and emission control systems 119

Fig. 3.84 Blow-by filter assembly (Sec 63)

64 Evaporative controls (FI) – general

1 Again very similar to the carburettor versions.
2 A diagram of the system appears in Fig. 3.85.
3 Testing of the various controls and valves is the same as for carburettor versions.

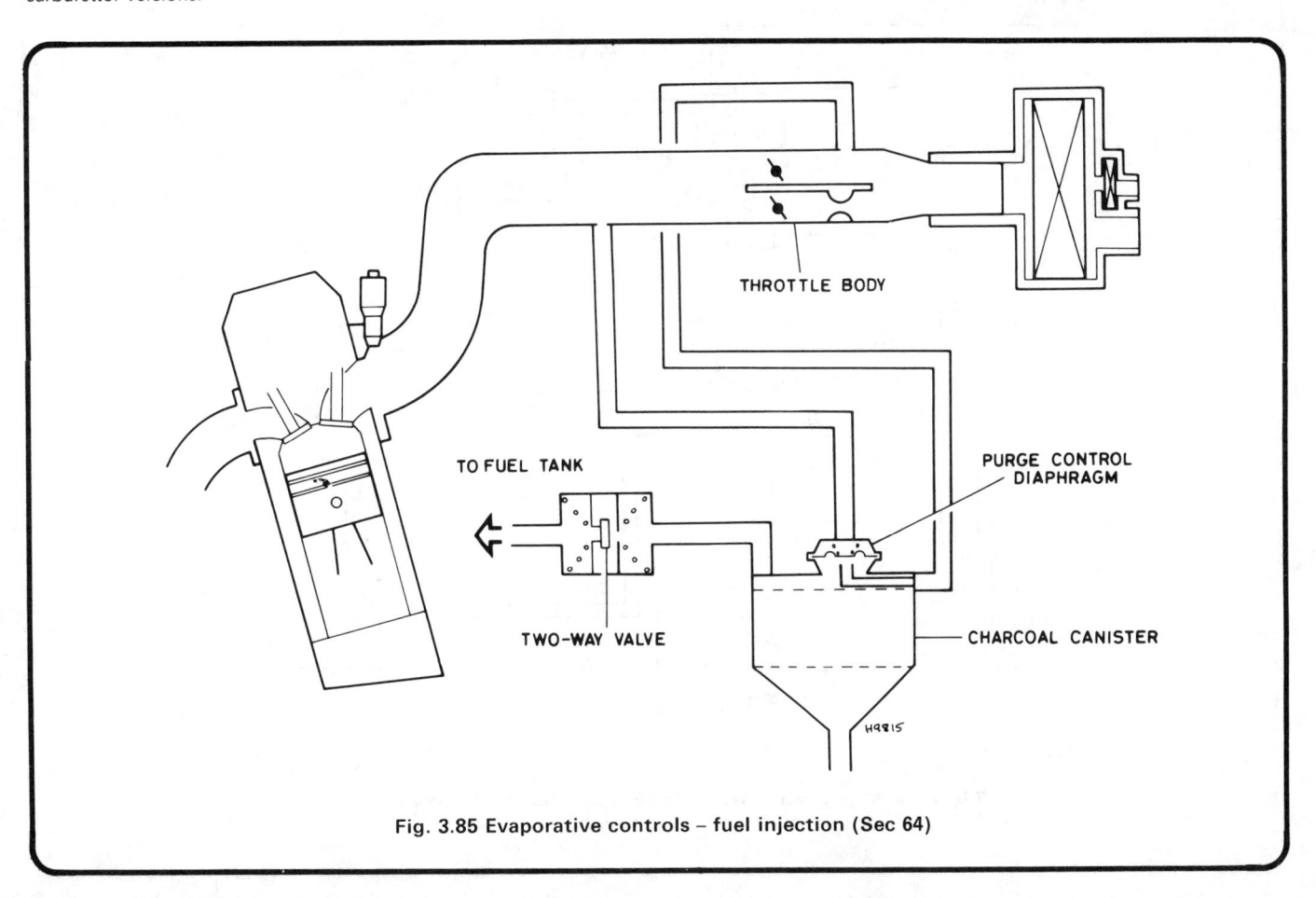

65 Ignition timing controls (FI) – general

1 Again basically the same as carburettor versions, with the cold advance solenoid valve taking the place of the thermovalve.
2 Also fitted to the distributor is the crank angle sensor, which senses the position of number 1 cylinder as a base for sequential fuel injection and detects engine speed to control the basic discharge duration under different operating conditions (see also Chapter 4).

66 Exhaust gas recirculation system (FI) – general

1 The EGR system is shown in Fig. 3.87.
2 Inputs from the various sensors fed to the Electronic Control Unit (ECU) are analysed by the ECU which then controls the EGR valve, keeping it at its optimum control position for all driving conditions.

67 Catalytic converter

General description
1 The catalytic converter converts hydrocarbons (HC), carbon monoxide (CO) and oxides of nitrogen (NOx) in the exhaust gas to carbon dioxide (CO_2), dinitrogen (N_2) and water vapour.
2 These converters can get extremely hot, and care should be exercised when parking the vehicle on grass, dry leaves, etc. for this reason, never operate the vehicle without the heat shields fitted. Allow the exhaust system to cool before beginning any work on it.

Checking
3 Inspect the heat shield for cracks and corrosion.
4 Remove the heat shield to inspect the converter for cracks and damage.
5 If build-up of exhaust pressure in the exhaust system is suspected, remove the converter and inspect it for blockage.

Fig. 3.85 Evaporative controls – fuel injection (Sec 64)

120

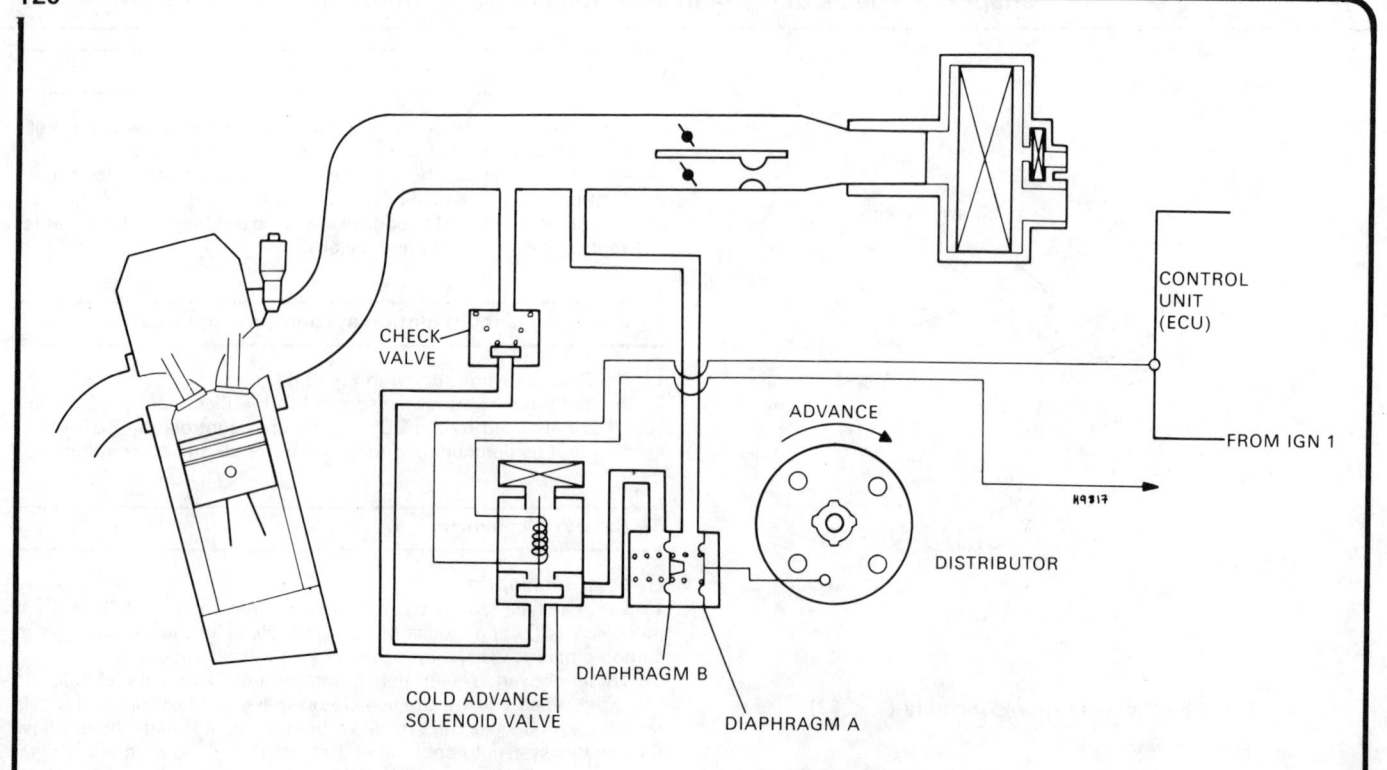

Fig. 3.86 Ignition timing control – fuel injection (Sec 65)

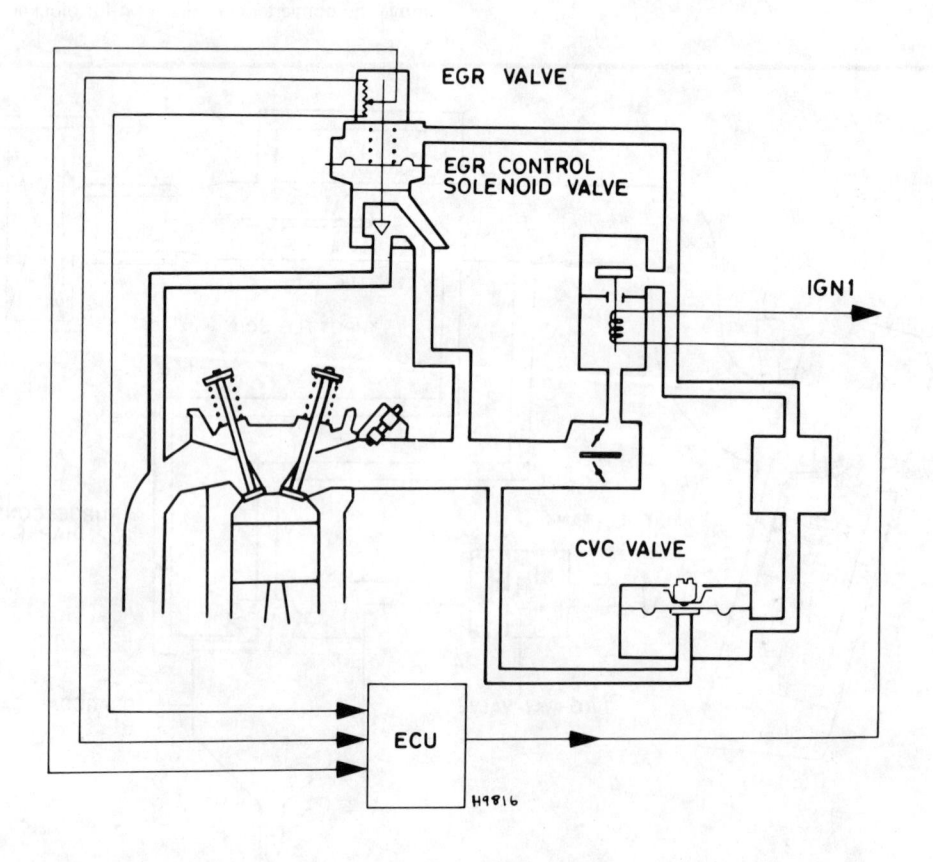

Fig. 3.87 Exhaust gas recirculation (EGR) controls – fuel injection (Sec 66)

121

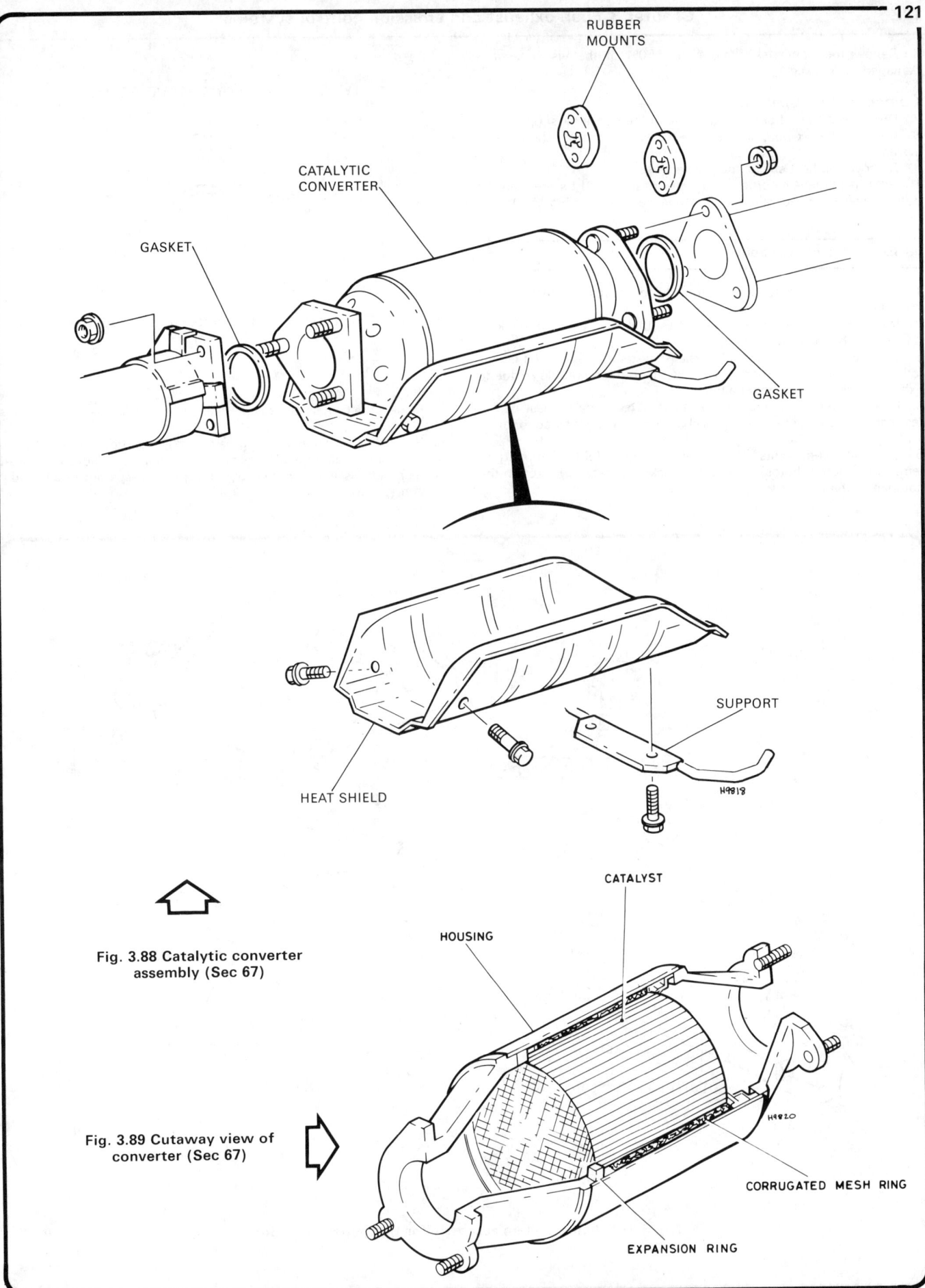

Fig. 3.88 Catalytic converter assembly (Sec 67)

Fig. 3.89 Cutaway view of converter (Sec 67)

6 Renew the converter if more than 50% of the visible area is damaged or blocked.

Component replacement
7 Disconnect the rubber mountings from the converter support.
8 Remove the six nuts and remove the converter from the exhaust system.
9 Remove the heat shield and converter support.
10 Refit in the reverse order, using new gaskets and torque load the nuts to their specified value in the sequence shown in Fig. 3.90.

68 Exhaust system – general

1 Inspect the exhaust system regularly for corrosion cracking and security of mountings.
2 It is generally accepted these days that any exhaust work is undertaken by exhaust specialists.
3 Beware, however, of cheaper exhaust systems, they do not last as long as original equipment and can cause carburation problems due to exhaust back pressure being altered.
4 If you decide to change an exhaust yourself, the hardest part is removing the old exhaust, which will have rusted and welded together with age.
5 Soak all flange bolts with releasing fluid before attempting to remove the nuts (photos). If they are stuck fast, hacksaw through the complete exhaust.

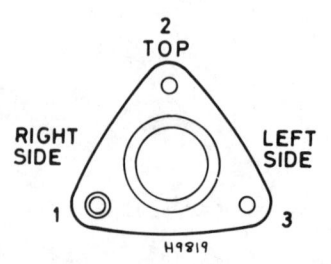

Fig. 3.90 Converter mounting nuts torque loading sequence (Sec 67)

6 Inspect the heat shield for cracks and corrosion while the exhaust is off.
7 Refit the new exhaust, using new gaskets, and do not tighten the securing bolts/nuts until all tension in the different sections is relieved as far as is possible and the rubber support straps are under equal tension (photos).

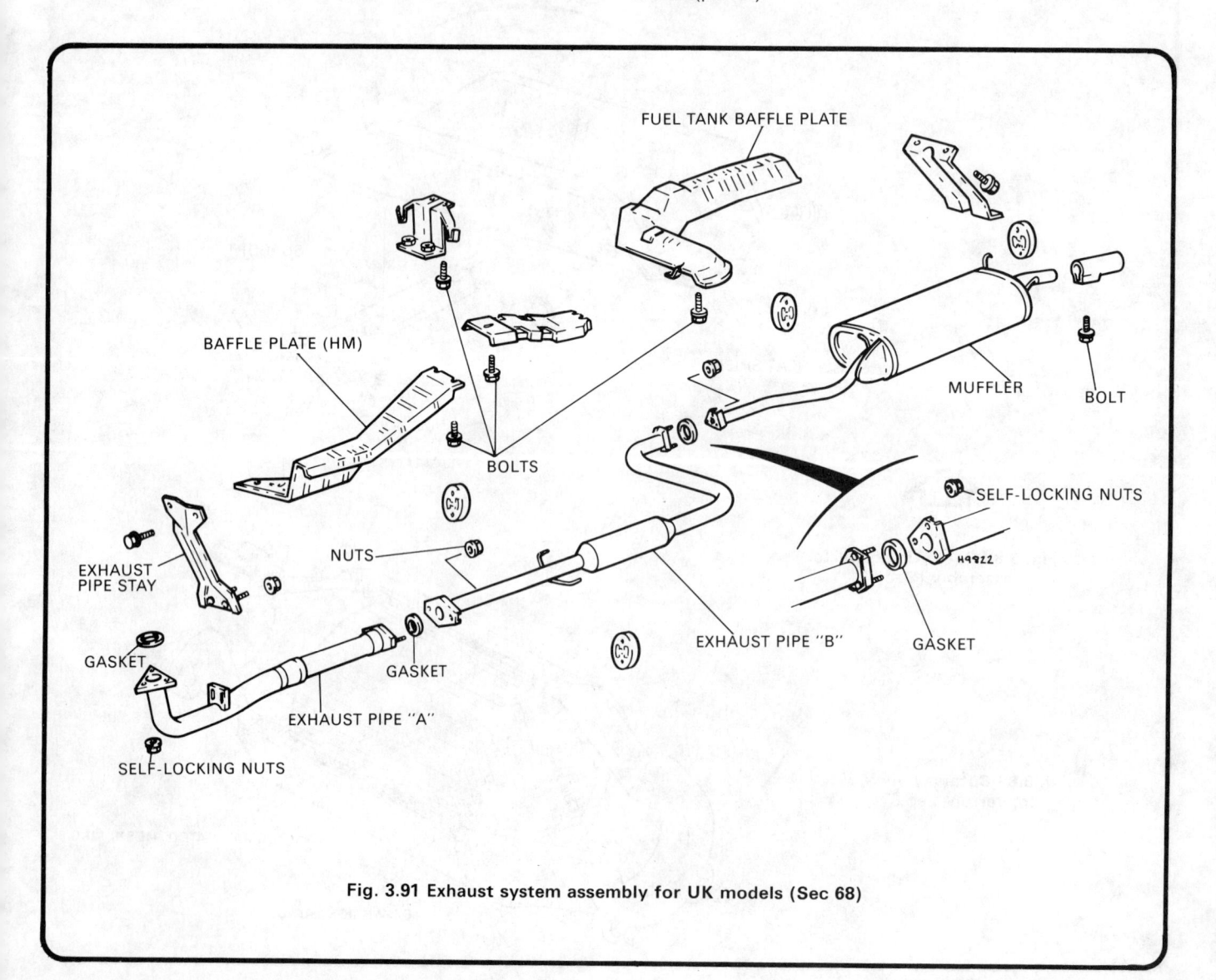

Fig. 3.91 Exhaust system assembly for UK models (Sec 68)

123

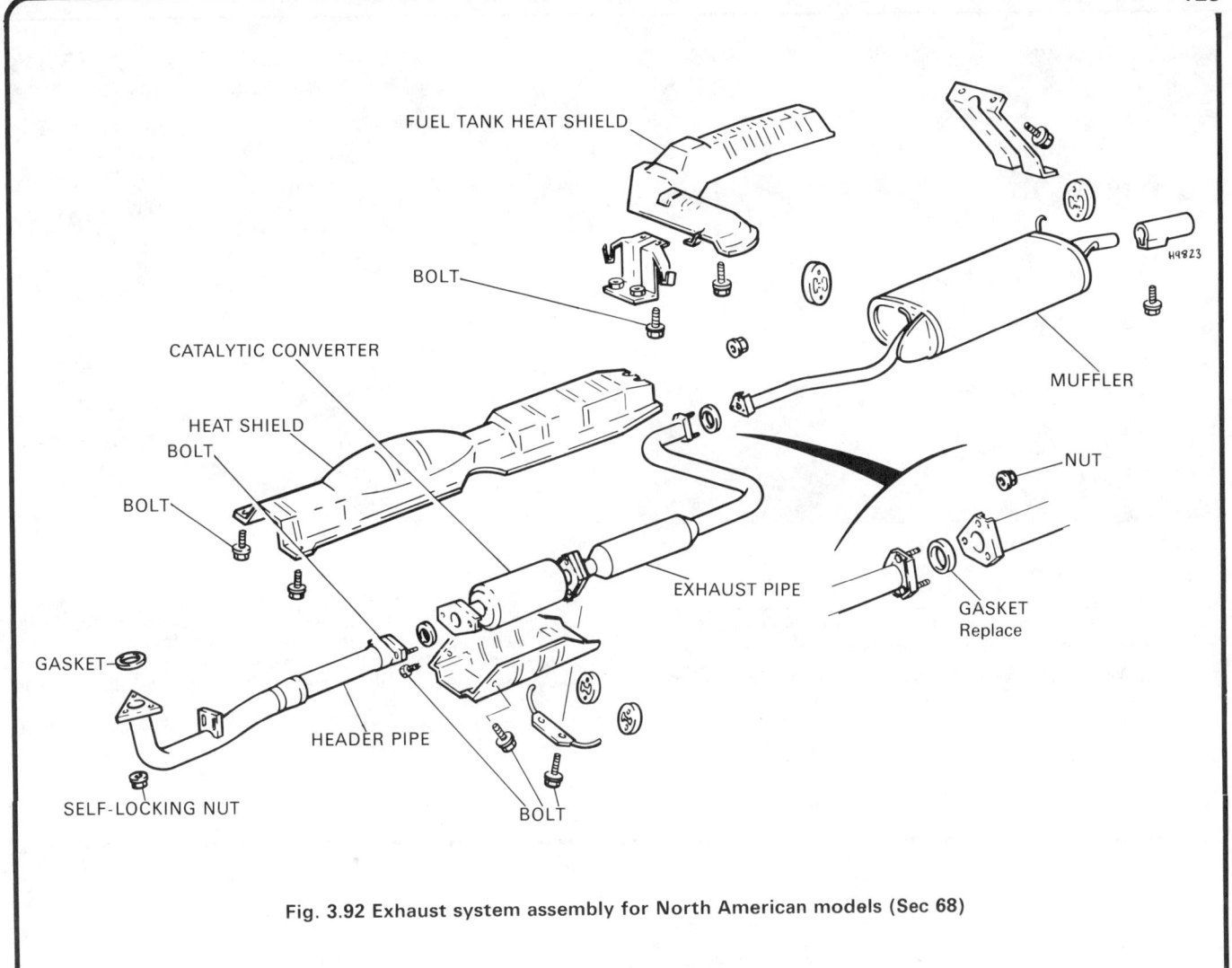

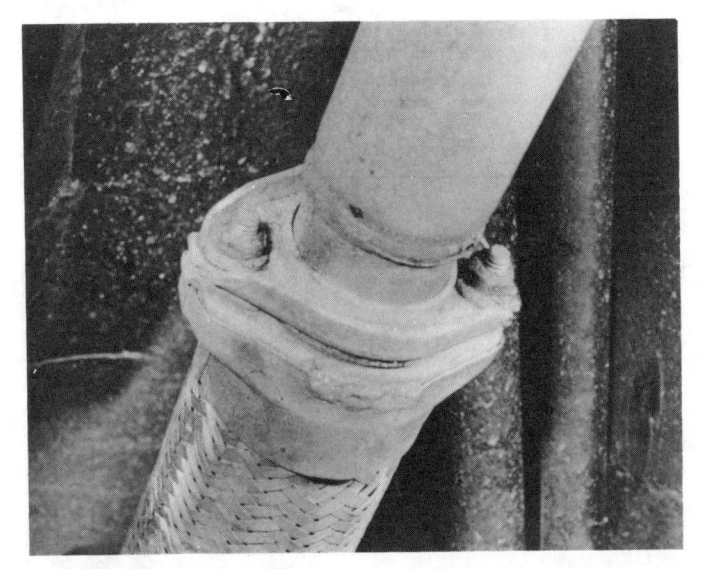

Fig. 3.92 Exhaust system assembly for North American models (Sec 68)

68.5A Exhaust pipe front flange ...

68.5B ... and rear flange

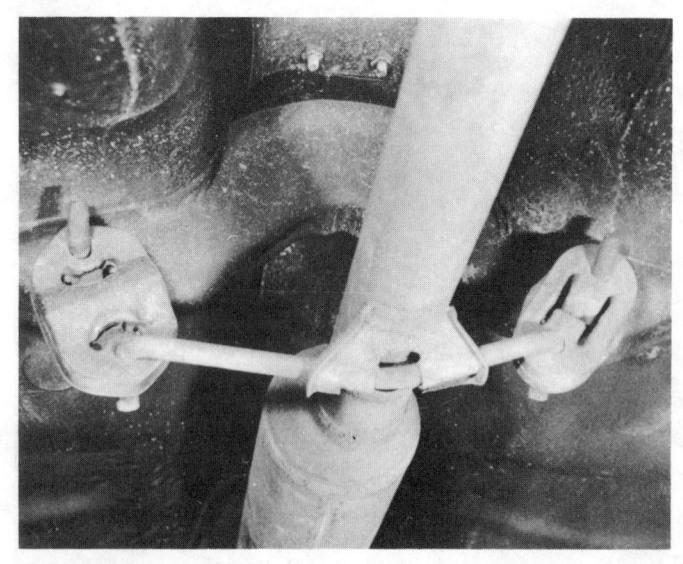

124 Chapter 3 Fuel, exhaust and emission control systems

68.7A Exhaust pipe rubber support strap

68.7B Tail pipe support

69 Intake and exhaust manifolds – removal and refitting

1 The intake and exhaust manifolds can be removed after disconnecting all the ancillary equipment attached to them.
2 Note that when the manifolds are being removed for access, it is not necessary to remove the carburettor first, but disconnect all inputs to the carburettor, and remove the manifold with the carburettor in place.
3 Similarly, where sensor probes are fitted, leave them in the manifolds and disconnect at their electrical connections.
4 Inspect the manifolds and their covers for cracking, and for signs of leakage around their mating surfaces with the engine, which may be an indication of warping.
5 On refitting, always use new gaskets and new self-locking nuts on the exhaust downpipe connection.
6 Tighten all nuts to their specified torque.

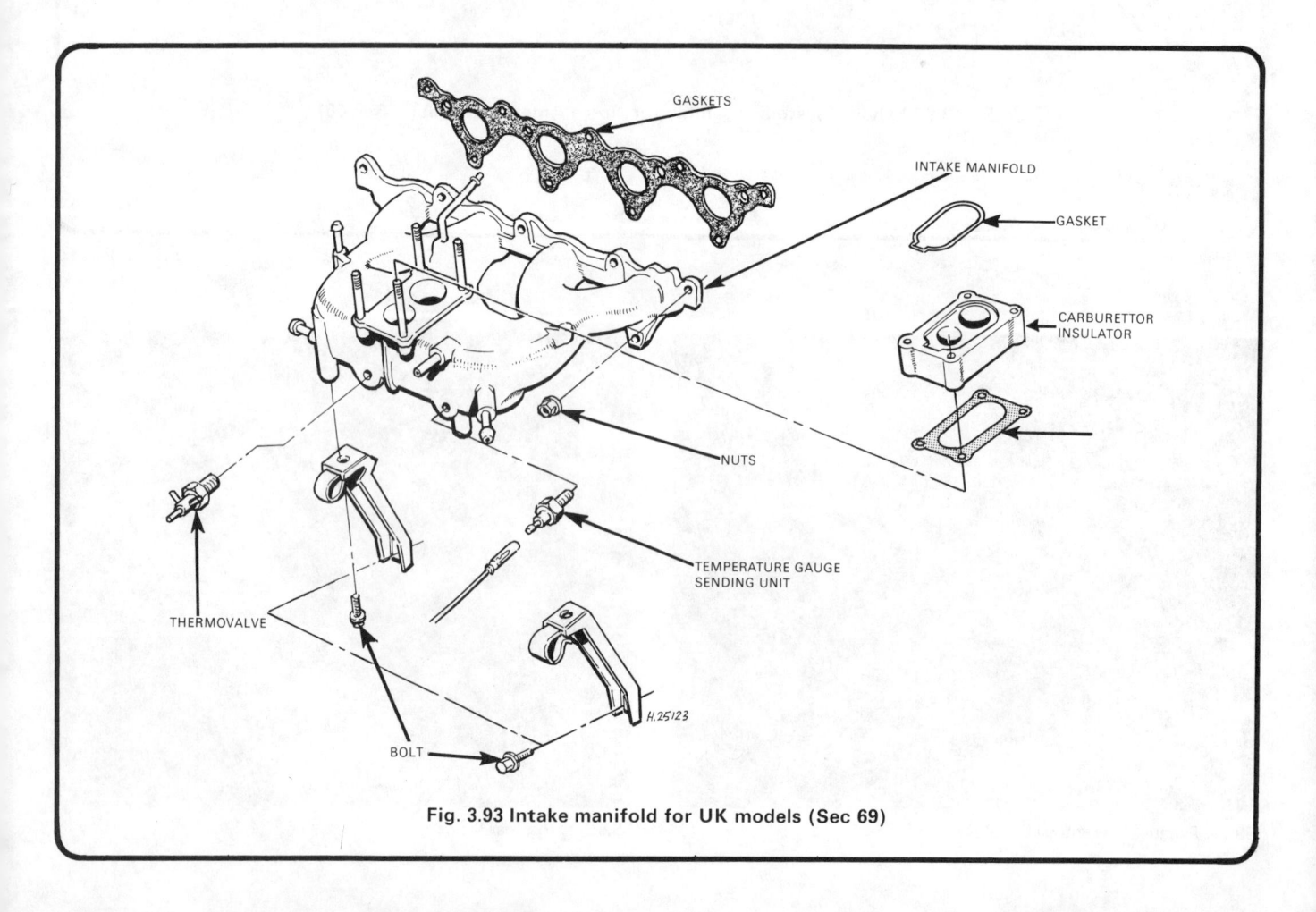

Fig. 3.93 Intake manifold for UK models (Sec 69)

125

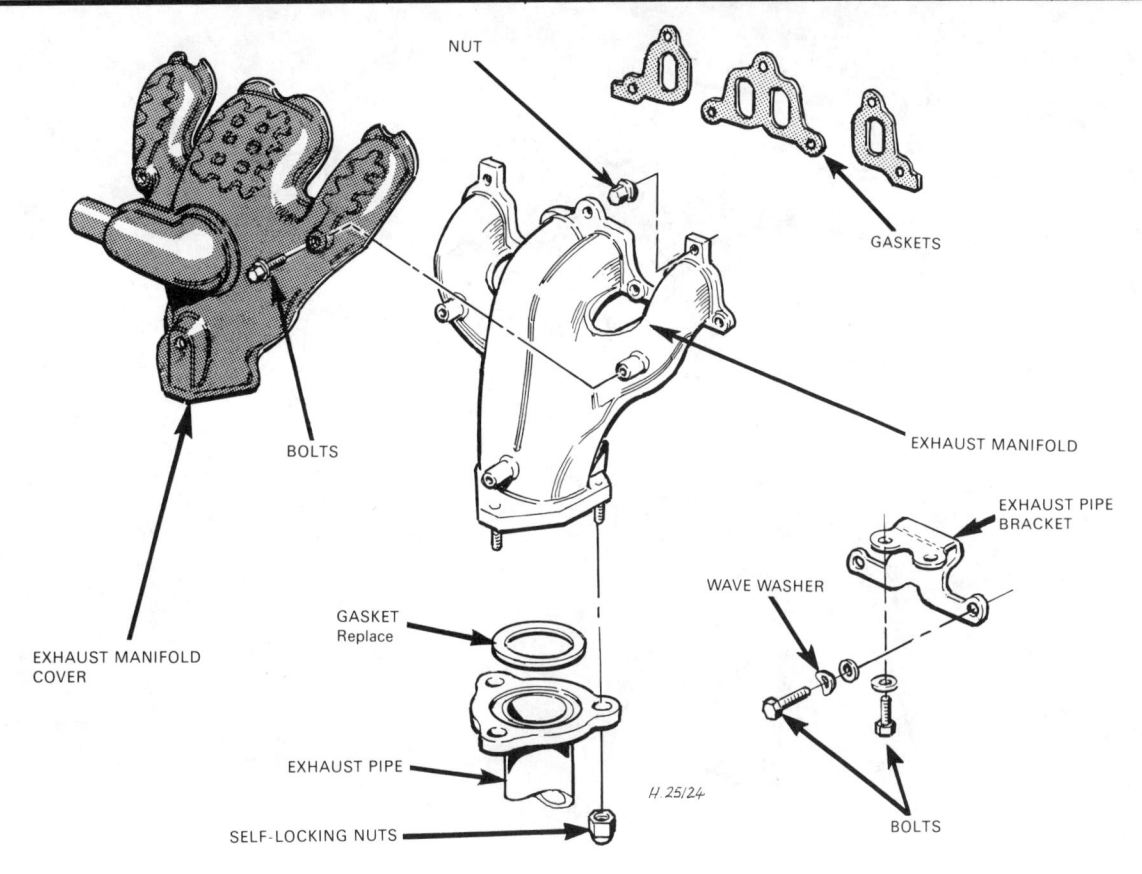

Fig. 3.94 Exhaust manifold for UK models (Sec 69)

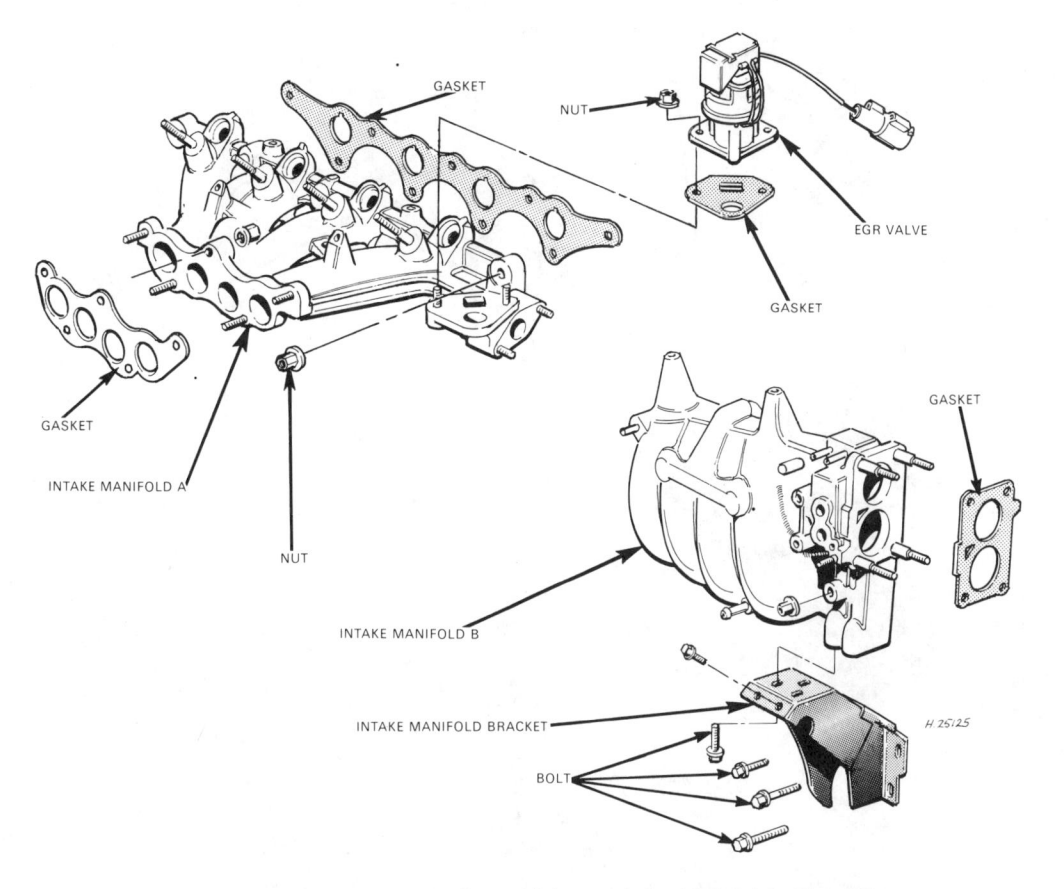

Fig. 3.95 Intake manifold for vehicles with fuel injection (Sec 69)

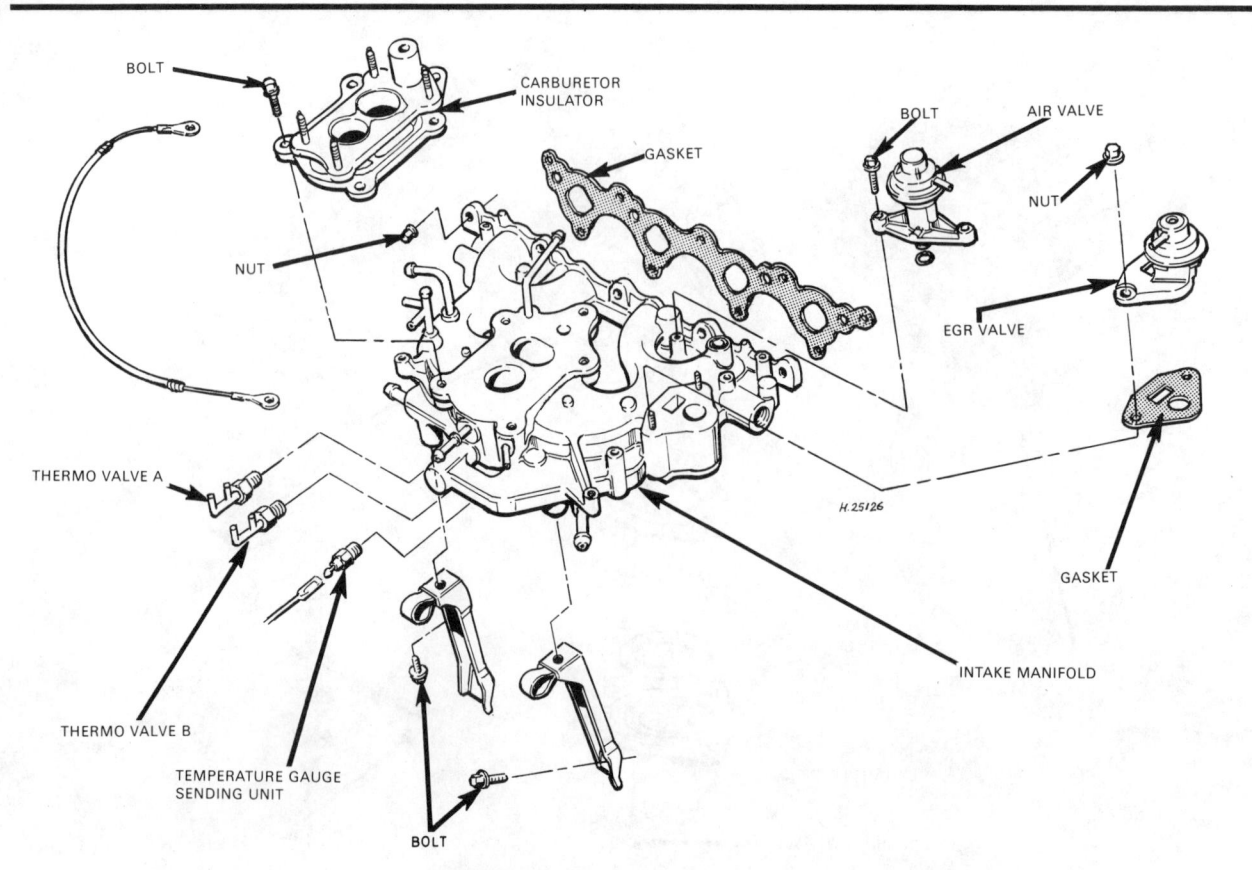

Fig. 3.96 Intake manifold for North American models – carburettor versions (Sec 69)

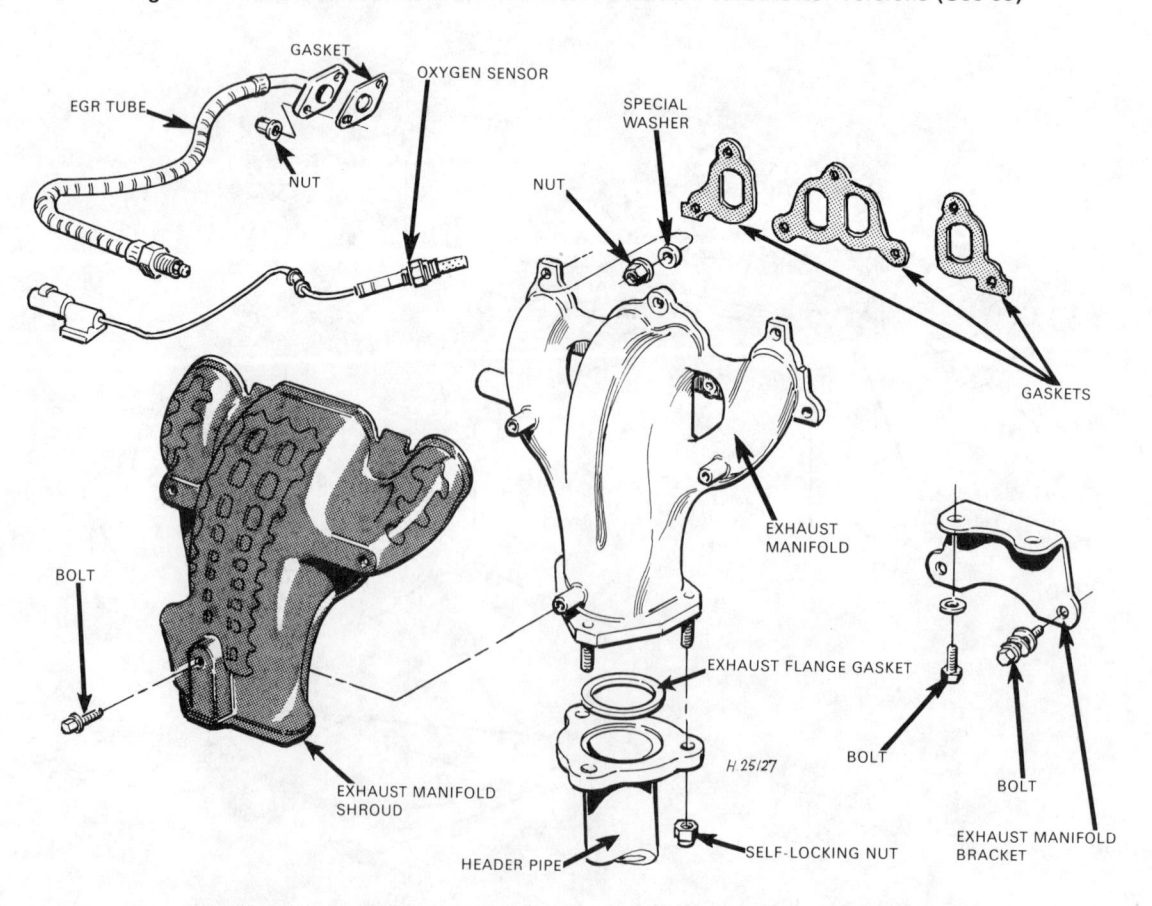

Fig. 3.97 Exhaust manifold for vehicles with fuel injection (Sec 69)

Fig. 3.98 Exhaust manifold for North American models – carburettor versions (Sec 69)

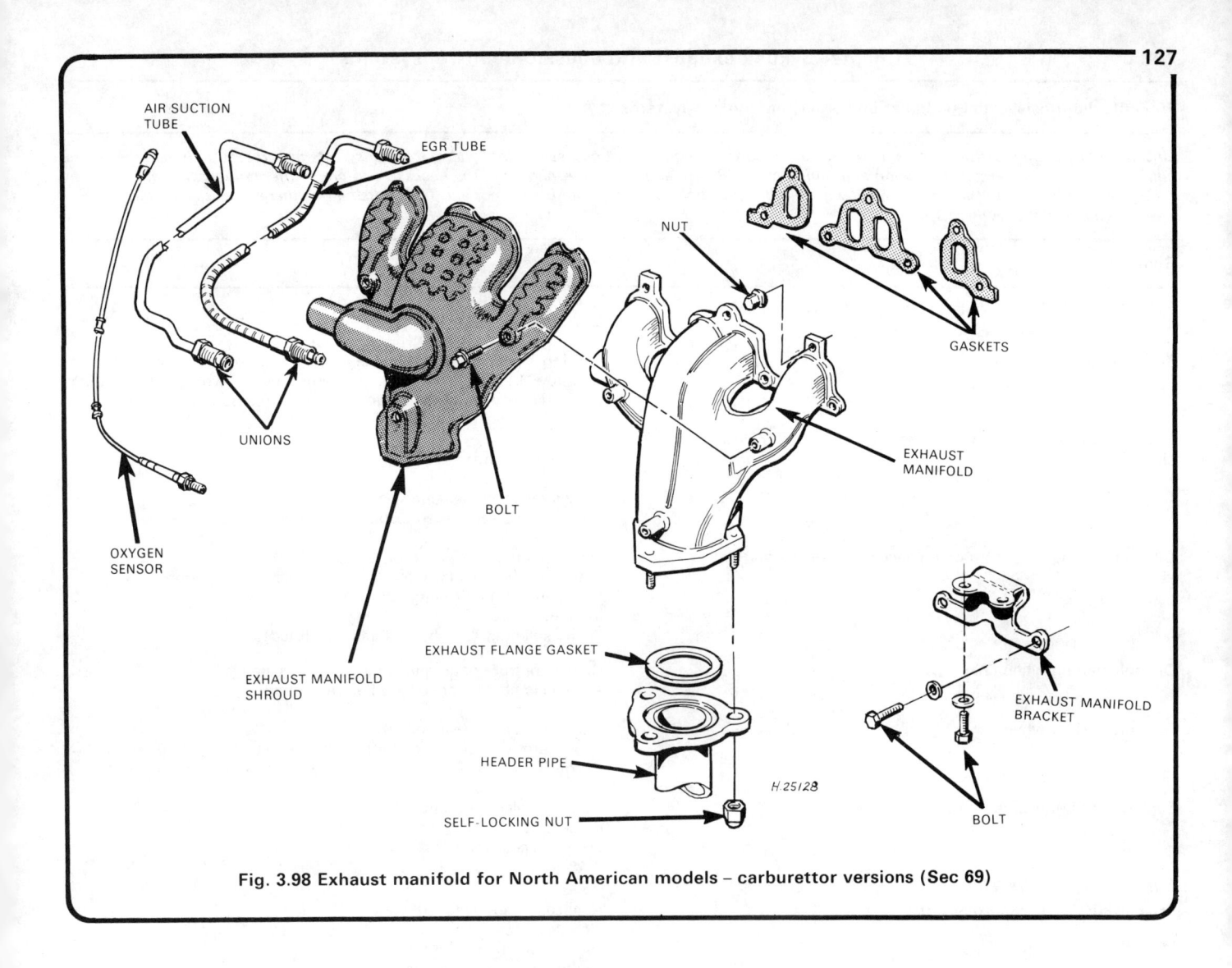

Fault diagnosis overleaf

70 Fault diagnosis – fuel, exhaust and emission control systems

Unsatisfactory engine performance and excessive fuel consumption are not necessarily the fault of the fuel system or carburettor. In fact they more commonly occur as a result of ignition and timing faults. Before acting on the following it is necessary to check the ignition system first. Even though a fault may lie in the fuel system it will be difficult to trace unless the ignition is correct. The faults below, therefore, assume that this has been attended to first (where appropriate).

Symptom	Reason(s)
Smell of fuel when engine is stopped	Leaking fuel lines or unions Leaking fuel tank
Smell of fuel when engine is idling	Leaking fuel line unions between pump and carburettor/injectors Overflow of fuel from float chamber due to wrong level setting, ineffective needle valve or punctured float (carburettor models)
Excessive fuel consumption for reasons not covered by leaks or float chamber faults	Worn jets Over-rich setting Sticking mechanism Dirty air cleaner element Sticking air cleaner thermostatic mechanism
Difficult starting, uneven running, lack of power, cutting out	One or more jets blocked or restricted Float chamber fuel level too low or needle valve sticking Fuel pump not delivering sufficient fuel Induction leak Faulty solenoid fuel shut-off valve (if fitted)
Difficult starting when cold	Choke control or automatic choke maladjusted Automatic choke not cocked before starting
Difficult starting when hot	Automatic choke malfunction Accelerator pedal pumped before starting Vapour lock (especially in hot weather or at high altitude)
Engine does not respond properly to throttle	Faulty accelerator pump Blocked jet(s) Slack in accelerator cable

Emission control system

Symptom	Reason(s)
Excessive HC or CO in exhaust gas	Air cleaner clogged Float level too high Faulty spark control system Faulty throttle opener control system Leaking intake manifold gasket
Excessive HC, CO and NOx in exhaust gas	Worn piston rings Incorrect valve clearances Faulty thermostat Blown cylinder head gasket Clogged PCV valve Incorrect idle mixture Clogged fuel filter Faulty idle compensator Choke not fully off Incorrect ignition settings Malfunction of emission control system component

HC   Hydrocarbons
CO   Carbon monoxide
NOx Oxides of nitrogen

Chapter 4 Ignition system

Contents

Advance and retard mechanism – testing	12
Coil – description and testing	4
Coil – removing and refitting	5
Distributor – removal and refitting	7
Distributor overhaul – carburettor models	8
Distributor overhaul – fuel injection models	9
Fault diagnosis – ignition system	14
General description	1
Ignition (HT) cables – removal and testing	3
Ignition timing – adjusting	13
Igniter unit – testing	10
Reluctor air gap – adjustment	11
Routine maintenance	2
Spark plugs – removal, inspection and refitting	6

Specifications

System type
Breakerless electronic

Distributor
Firing order .... 1-3-4-2 (No 1 at timing belt end)
Rotor rotation .... Anti-clockwise (viewed from cap)

Coil
Primary resistance .... 1.06 to 1.24 ohm at 20°C (70°F)
Secondary resistance .... 7400 to 11 000 ohm at 20°C (70°F)
Condenser capacity .... 0.47 ± 0.09 microfarad
Ignition (HT) cable resistance .... 25 000 ohm (maximum)

Ignition timing (at idle)
UK models:
  Manual .... 18° ± 2° BTDC
  Automatic (in gear) .... 18° ± 2° BTDC
Canadian models:
  Manual .... 2° ± 2° BTDC
  Automatic (in gear) .... 2° ± 2° BTDC
North American models:
  Carburettor:
    Manual .... 18° BTDC
    Automatic .... 18° BTDC
  Fuel injection:
    Manual 49 ST .... 22° BTDC
    Manual CAL and HI ALT .... 18° BTDC
    Automatic .... 18° BTDC

Spark plugs
UK models .... NGK BPR 6EY-11 or BPR 6ES-11
                                                  ND W20 EXR-U11 or W20 EPR-U11
                                                  Champion RN 9 YC
North American models:
  Carburettor .... NGK BUR 5EB-11
                                                  ND W16 EKR-S11
  Fuel injection .... NGK BPR 6EY-11 or BPR 6ES-11
                                                  ND W20 EXR-U11 or W20 EPR-U11
Spark plug gap .... 0.039 to 0.043 in (1.0 to 1.1 mm)

Torque wrench settings
	lbf ft	Nm
Spark plugs	13	18

Chapter 4 Ignition system

1 General description

The ignition system is of the breakerless (no contact points) electronic type. The electronic control unit is located in the distributor, incorporating a magnetic stator, reluctor and igniter unit.

Conventional centrifugal weights and a vacuum diaphragm are used to advance and retard the ignition, with additional external controls more closely associated with the fuel system and electronic control system (see Chapter 3), fitted according to model and country of operation.

In order for the engine to run correctly, the spark which ignites the fuel/air mixture in the combustion chambers must be delivered at precisely the right time, and this is achieved by the ignition system.

Battery voltage is fed to the coil, which converts this Low Tension (LT) voltage, to High Tension (HT) voltage, which is fed to the distributor.

The distributor, which is run directly from the camshaft, distributes the HT voltage to the spark plugs at the right time.

The spark plugs cause the HT voltage to jump the gap in their nose, thus causing the spark which ignites the fuel/air mixture.

Warning: *transistorised electronic ignition systems can generate considerably higher HT voltage than conventional systems which, in certain circumstances, can prove lethal.*

2 Routine maintenance

At the intervals given in Routine Maintenance (at the front of the book) undertake the following tasks.
1 Renew the spark plugs.
2 Inspect the distributor cap and rotor, clean/renew where applicable (Section 8 or 9).
3 Inspect the ignition system wiring and the HT leads (Section 3).
4 Inspect the control system.
5 Check the ignition timing, reset if necessary.
6 On vehicles with fuel injection, inspect the throttle control diaphragm.

3 Ignition (HT) cables – removal and testing

1 Carefully remove each cable from between the distributor and the spark plugs, and the distributor and coil.
2 Do this by pulling on the rubber boot, not on the lead itself, and do not bend the cables too sharply, as conductivity may be impaired.
3 Inspect the terminals for damage and corrosion.
4 Connect an ohmmeter between the two terminals of each cable in turn and check the resistance, which should not exceed that figure given in the Specifications.
5 Renew all cables which are deteriorated.
6 Refit the cables, again being careful not to bend them excessively.

4 Coil – description and testing

1 The coil is located on the right-hand side of the engine bay, next to the suspension strut tower.
2 To test the coil, switch the ignition off, pull back the rubber cap (photo) and connect an ohmmeter between the two LT terminals (positive and negative).
3 The resistance should be as specified.
4 Carry out the same test between one of the LT terminals and the HT terminal (the central terminal which goes to the distributor), again the resistance should be as specified.
5 Note that these tests are carried out with the ignition off.
6 Renew the coil if resistance is outside that specified.
7 Note also that resistance will vary with the temperature of the coil.

5 Coil – removing and refitting

1 Switch off the ignition and disconnect the battery.
2 Remove the rubber cap from the coil and disconnect the primary (LT) wires, and secondary (HT) leads.
3 Retain the noise suppression condenser.
4 Undo the clamp bolt sufficiently to slide the coil out from the clamp.
5 Refit in the reverse order, not forgetting the noise suppression condenser, which is connected between the positive terminal and earth (failure of the noise suppression condenser can cause the engine to stop running).

6 Spark plugs – removal, inspection and refitting

1 Disconnect the HT leads from the spark plugs, as described in Section 3, and brush or blow out any dirt in the spark plug recesses in the cylinder head.

4.2 Pulling back the rubber cap on the coil

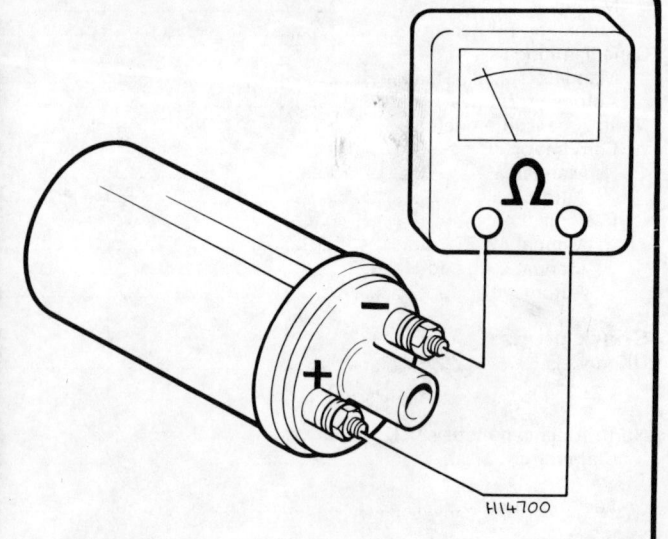

Fig. 4.1 Positive and negative LT terminals on the coil (Sec 4)

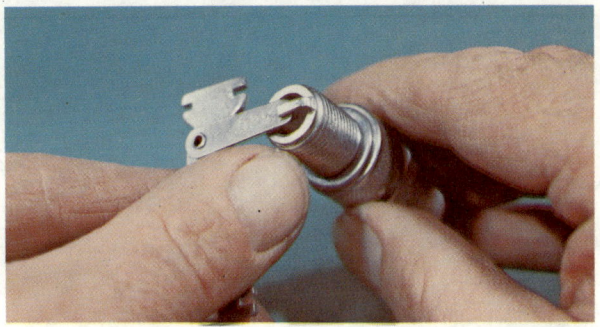

Measuring plug gap. A feeler gauge of the correct size (see ignition system specifications) should have a slight 'drag' when slid between the electrodes. Adjust gap if necessary

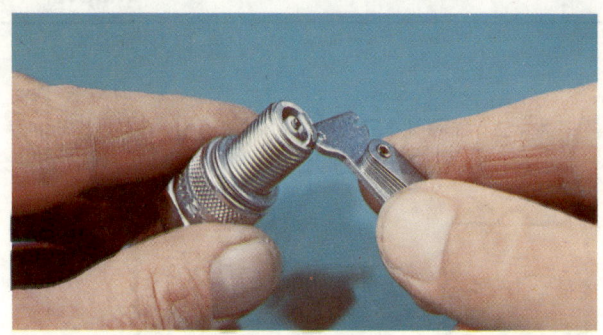

Adjusting plug gap. The plug gap is adjusted by bending the earth electrode inwards, or outwards, as necessary until the correct clearance is obtained. Note the use of the correct tool

Normal. Grey-brown deposits, lightly coated core nose. Gap increasing by around 0.001 in (0.025 mm) per 1000 miles (1600 km). Plugs ideally suited to engine, and engine in good condition

Carbon fouling. Dry, black, sooty deposits. Will cause weak spark and eventually misfire. Fault: over-rich fuel mixture. Check: carburettor mixture settings, float level and jet sizes; choke operation and cleanliness of air filter. Plugs can be re-used after cleaning

Oil fouling. Wet, oily deposits. Will cause weak spark and eventually misfire. Fault: worn bores/piston rings or valve guides; sometimes occurs (temporarily) during running-in period. Plugs can be re-used after thorough cleaning

Overheating. Electrodes have glazed appearance, core nose very white – few deposits. Fault: plug overheating. Check: plug value, ignition timing, fuel octane rating (too low) and fuel mixture (too weak). Discard plugs and cure fault immediately

Electrode damage. Electrodes burned away; core nose has burned, glazed appearance. Fault: pre-ignition. Check: as for 'Overheating' but may be more severe. Discard plugs and remedy fault before piston or valve damage occurs

Split core nose (may appear initially as a crack). Damage is self-evident, but cracks will only show after cleaning. Fault: pre-ignition or wrong gap-setting technique. Check: ignition timing, cooling system, fuel octane rating (too low) and fuel mixture (too weak). Discard plugs, rectify fault immediately

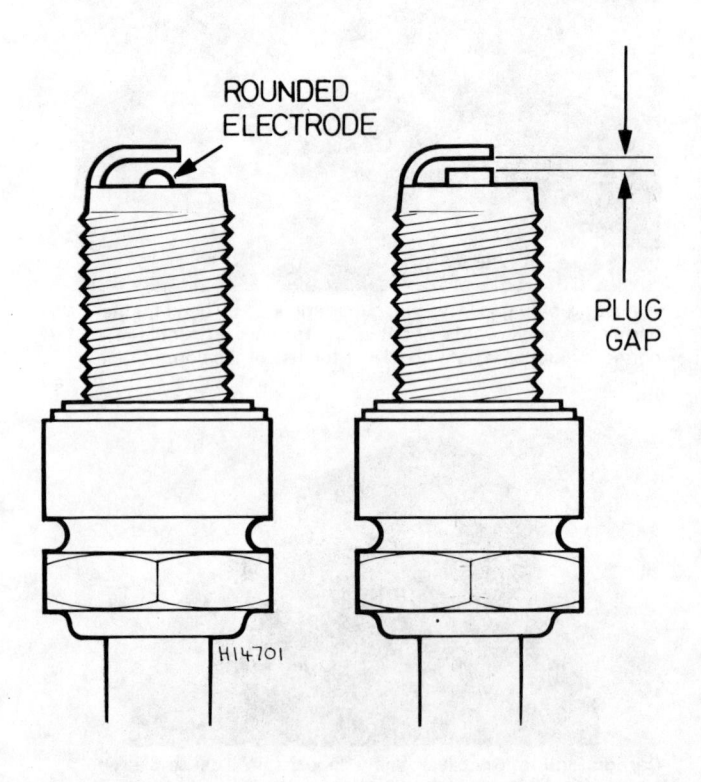

132   Chapter 4 Ignition system

Fig. 4.2 Spark plug gap (Sec 6)

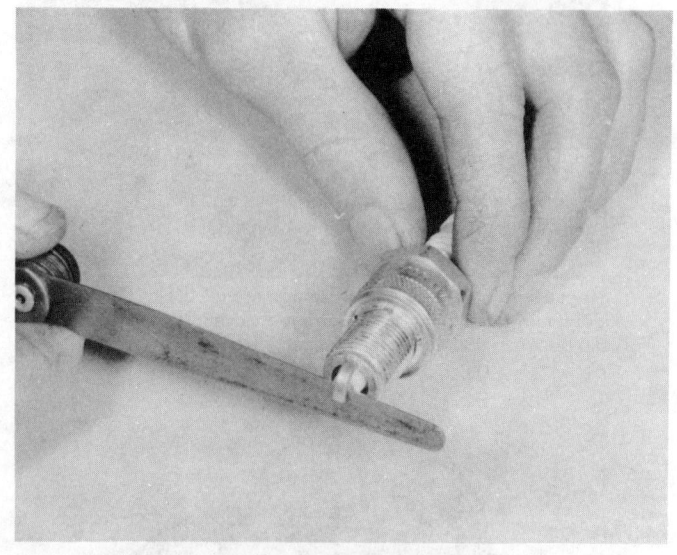

6.7 Measuring a spark plug gap with feeler gauges

2 Remove the plugs, using a double depth socket or box spanner.
3 Inspect each plug for cracked or broken ceramic insulators, worn electrodes, and contamination with oil.
4 The condition of the plugs is a reliable indication of general engine condition.
5 Clean each plug using a wire brush or plug cleaning machine, and inspect the electrodes for wear. If the shoulders of the centre electrode have become worn and rounded off, the plug should be renewed.
6 The spark plugs should also be renewed, regardless of external condition, at the intervals shown in Routine Maintenance.
7 Before refitting new plugs, or plugs which have been removed for cleaning, check and reset the electrode gap (photo). Use a gapping tool to bend the electrode to the desired gap, being careful not to damage the ceramic insulator.

8 Refit the plugs, starting them off in their threads by hand; the soft aluminium of the cylinder head threads are easily stripped.
9 Tighten to the specified torque with a torque spanner.
10 If no torque spanner if available, tighten the plug until the sealing washer contacts the surface of the cylinder head, then tighten by a further half of one turn.

7 Distributor – removal and refitting

1 The distributor is driven directly from the camshaft and is mounted on the right-hand side of the engine.
2 To remove it from the engine, disconnect the battery.
3 Disconnect the spark plug and coil HT leads.
4 Disconnect the hose or hoses from the vacuum advance diaphragm (there may be one or two diaphragms, depending on model).
5 Disconnect the HT and LT leads at the coil.
6 Remove the distributor adjusting bolts, and remove the distributor from the engine (photos).
7 To refit the distributor, fit a new O-ring seal to the housing.
8 The locating lugs on the distributor shaft and camshaft (photo) are offset to prevent the distributor being fitted 180° out.

7.6A Removing the distributor bolts 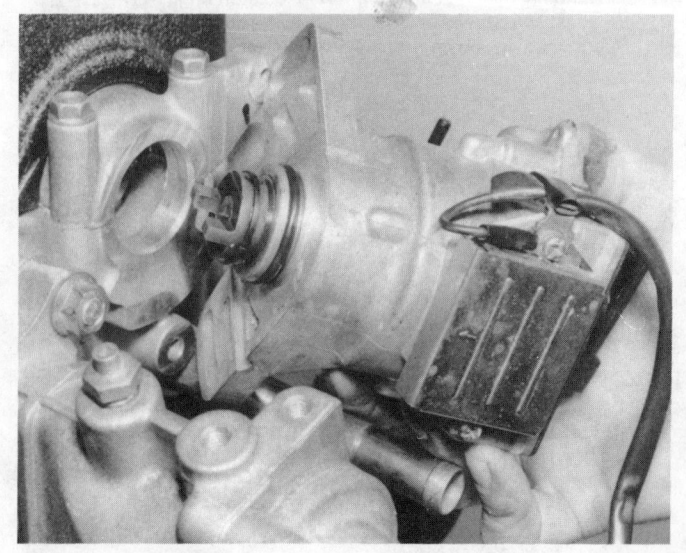 ...   7.6B ... and distributor

Chapter 4 Ignition system

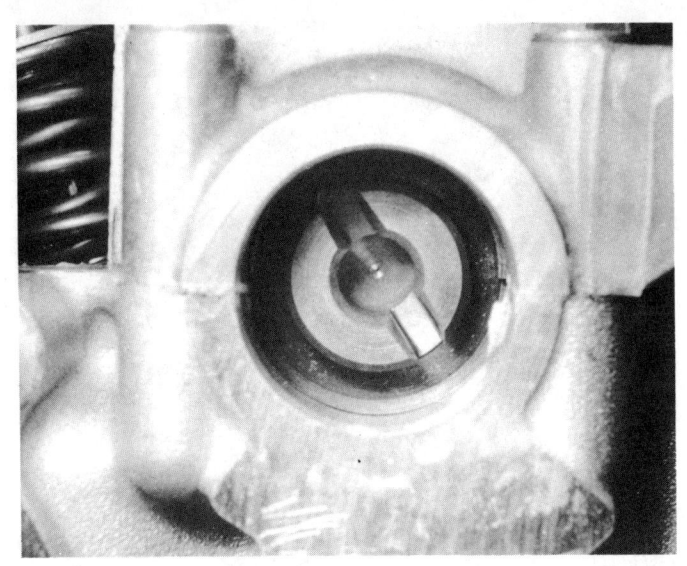

7.8 Offset locating lugs in the camshaft

9 Fit the distributor to the engine housing, turning the distributor to locate the lugs, and then fit the adjusting bolts finger tight.
10 If ignition timing has been lost it can be reset approximately by turning the engine by hand until No 1 piston is on compression stroke and the timing mark on the flywheel and the pointer line up, then turning the distributor body until the reluctor peak is aligned with the stator post. **Note:** *the distributor caps have an index mark on them which indicates Number 1 cylinder. The rotor arm should be pointing to this mark when No 1 cylinder is on compression stroke and the timing marks are in alignment.*
11 Refit the HT and LT leads and the vacuum hose(s).
12 Reconnect the battery.
13 Carry out the strobe timing in accordance with Section 13.

8 Distributor overhaul – (carburettor models)

1 There are two different kinds of distributor fitted according to model, all of them are basically the same and the overhaul procedure

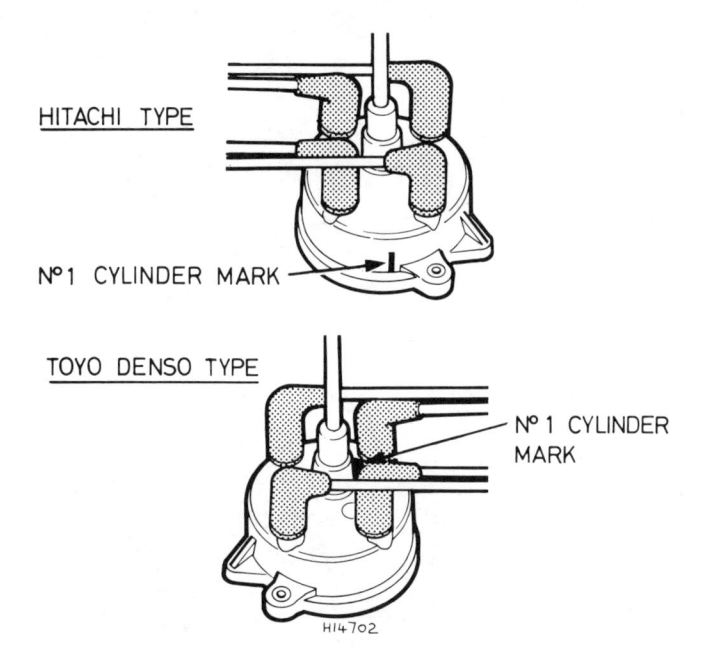

Fig. 4.3 Distributor cap No 1 cylinder index mark (Sec 7)

will suffice for either; any major differences being pointed out in the text.
2 With the distributor removed from the engine, remove the two distributor cap retaining screws and remove the distributor cap.
3 Pull off the rotor (photo).
4 Inspect the inside of the cap for damage and cracks and clean the four contact posts free from deposits (photo).
5 If they are badly worn or pitted, renew the cap.
6 Clean the central carbon brush with 600 grade sand paper, and check it is free to move in and out on its spring.
7 Clean the rotor tip and remove deposits with a fine file; renewing the rotor if it is badly worn.
8 Prise out the reluctor using two screwdrivers as shown in Fig. 4.6.
9 Remove the screw or C-clip holding the vacuum diaphragm shaft

8.3 Pulling off the rotor

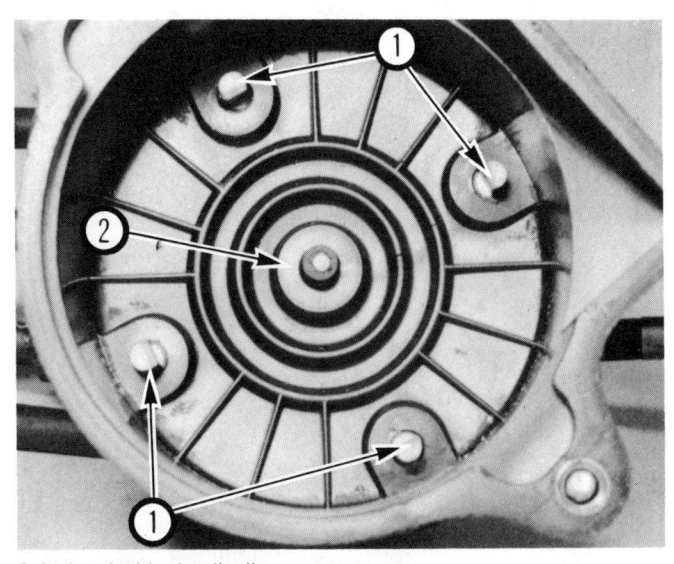

8.4 View inside the distributor cap
1 Contact posts   2 Carbon brush

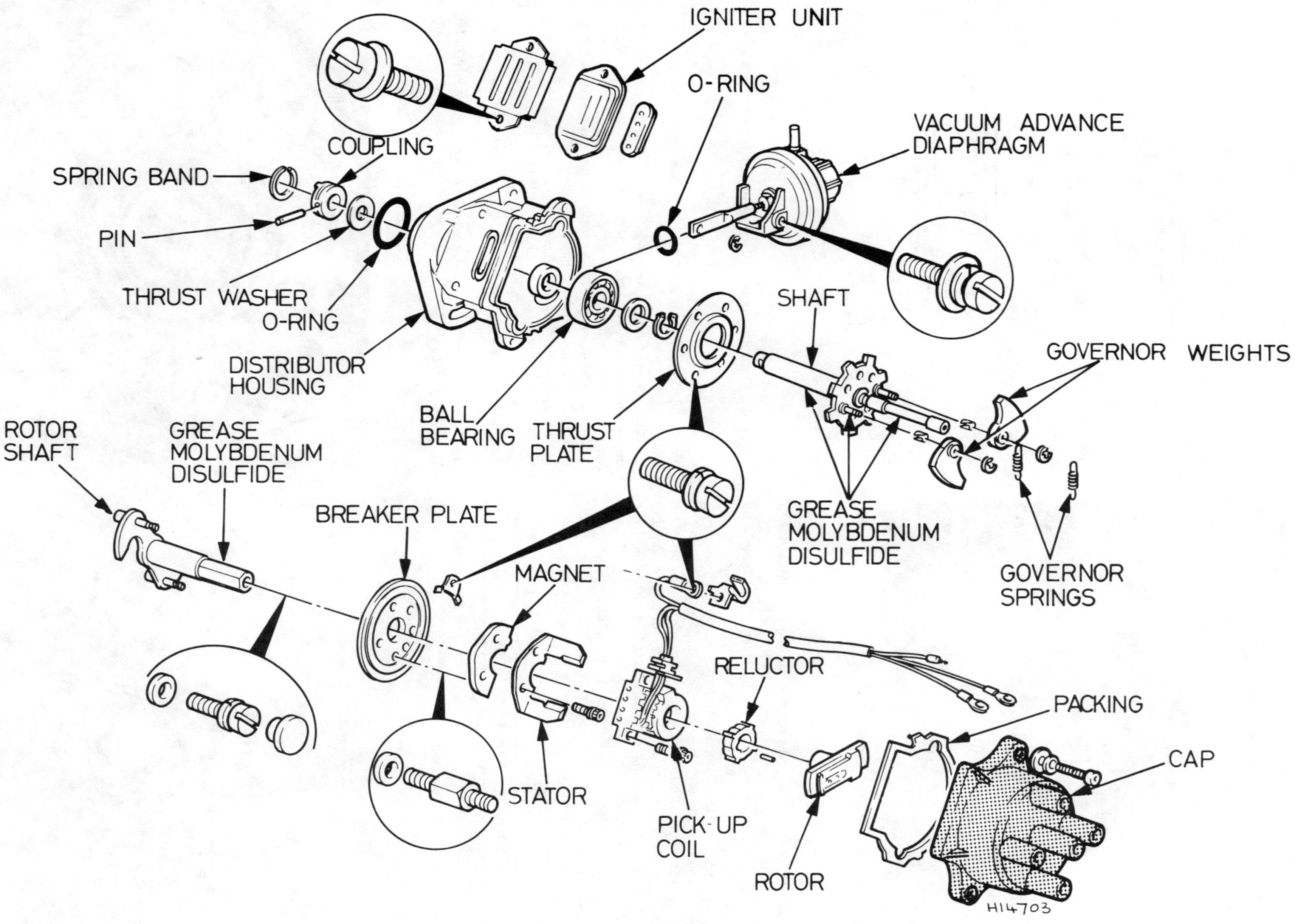

Fig. 4.4 Components of the Toyo Denso distributor (Sec 8)

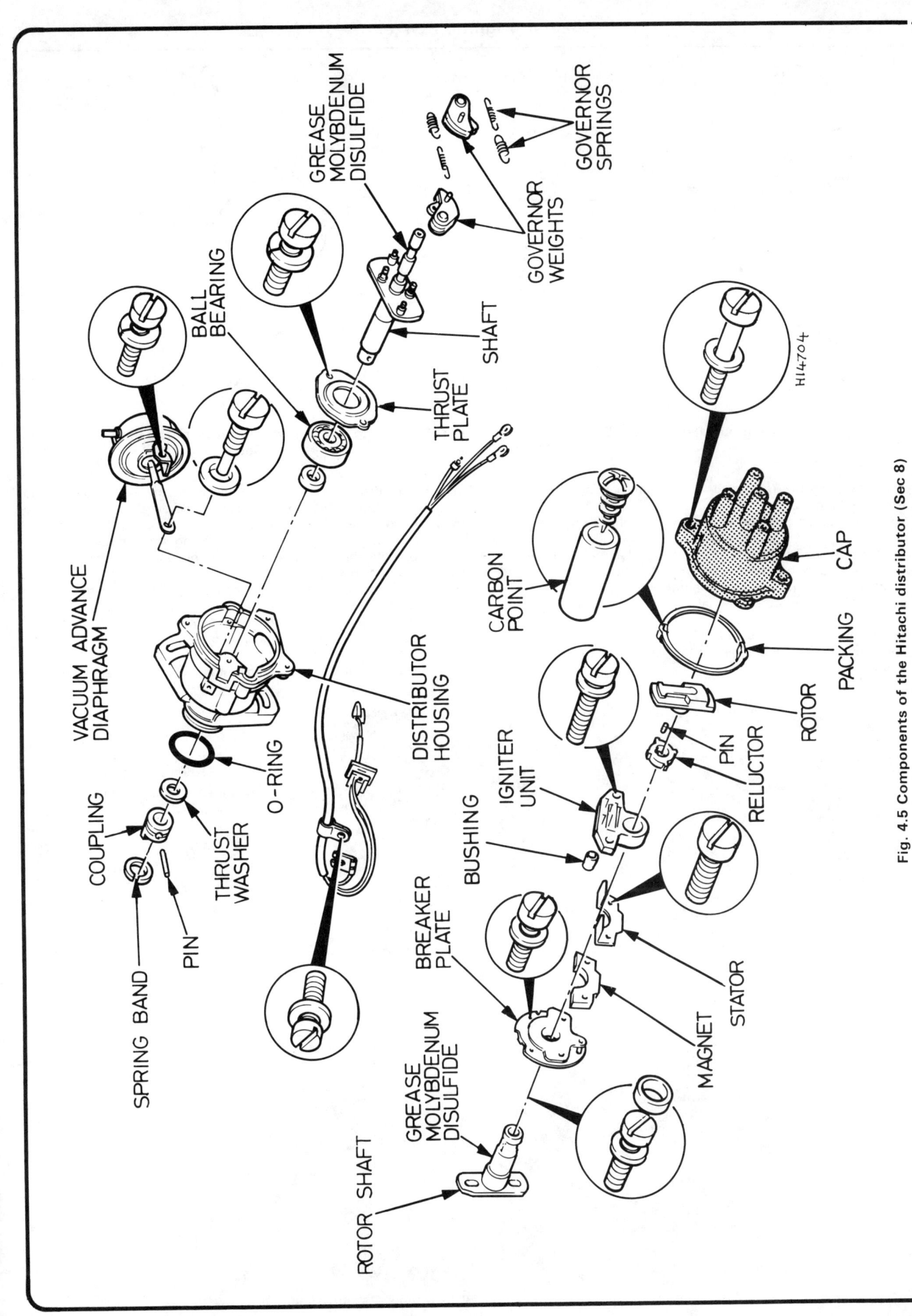

Fig. 4.5 Components of the Hitachi distributor (Sec 8)

Chapter 4 Ignition system

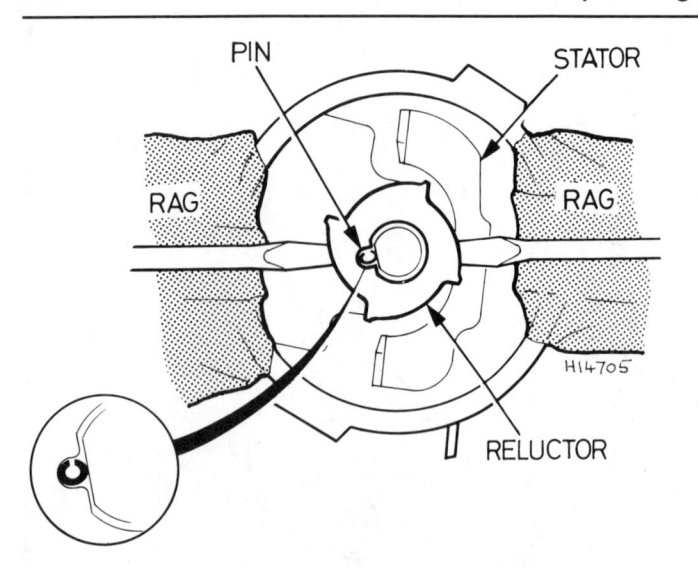

Fig. 4.6 Prising out the reluctor (Sec 8)

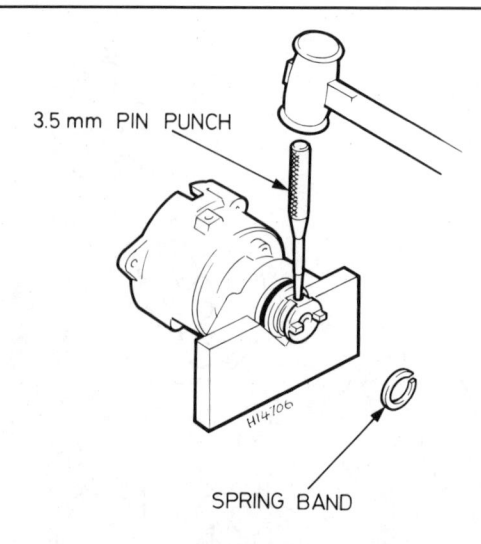

Fig. 4.7 Driving out the spring pin from the distributor shaft (Sec 8)

to the baseplate (photo), remove the diaphragm securing screws and lift out the diaphragm unit.

10 Remove the screws from the igniter cover (photo) lift off the cover (photo) and pull off the igniter (photo).

11 On Hitachi type distributors, the igniter is internal and can be removed by extracting the retaining screws, lifting up the igniter unit to disconnect the leads, and removing the unit.

12 Remove the screws from the stator and lift out the stator and magnet.

13 Similarly, remove the breaker plate.

14 Remove the rubber cap and extract the screw from the rotor shaft.

15 Lift out the rotor shaft.

16 Place the distributor in a padded vice and remove the spring band from the distributor shaft. (This is used as the pin retainer – try not to expand it more than is necessary).

17 Support the shaft and drive out the spring pin using a pin punch.

18 Remove the shaft and then the thrust plate and ball-bearing from the distributor housing.

19 The governor weights may be removed by extracting the C-clips and lifting them off their shafts.

20 Inspect all parts for wear, renewing as necessary.

21 The governor weight springs should be renewed, as they tend to stretch with age.

22 Lubricate all parts with molybdenum grease before reassembly.

8.9 Vacuum unit retaining C-clip (arrowed)

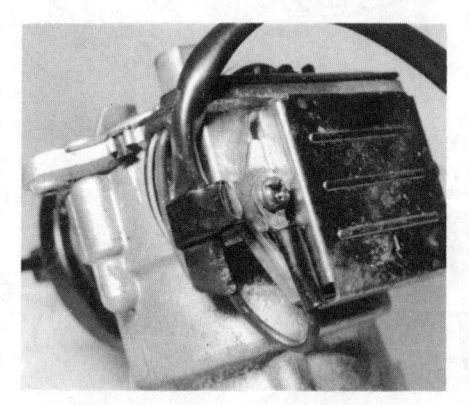

8.10A Remove the screws from the igniter ...

8.10B ... lift off the cover ...

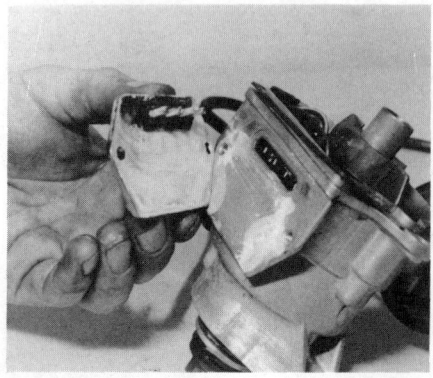

8.10C ... and pull off the igniter

8.25 Distributor shaft O-ring seal (arrowed)

Chapter 4 Ignition system

23 Fit the governor weights, thrust plate and ball-bearing to the distributor shaft.
24 Grease the shaft and fit it to the distributor housing.
25 Install a new O-ring (photo), washer and coupling on the lower end of the shaft line up the holes in the shaft and coupling and drive in a new roll pin.
26 Turn the coupling so that the index marks line up.
27 Install the rotor shaft onto the distributor shaft, as shown in Fig. 4.9.
28 Check that the rotor shaft arms are correctly located over the governor weight pins, then install the lock washer and screw to the top of the shaft.
29 Slide the spring band back on to the coupling.
30 Refit the breaker plate, magnet and stator.
31 Refit the igniter pick up coil (Toyo Denso) or igniter unit (Hitachi).
32 Refit the reluctor, driving in the spring pin with its gap facing away from the rotor shaft.
33 Adjust the gaps between the reluctor and stator, as described in Section 11.
34 Refit the diaphragm unit, using a new O-ring on the Toyo Denso type.
35 Refit the rotor arm and cap, using a new seal under the cap.

9 Distributor overhaul – fuel injection models

1 The distributor fitted to fuel injected models is the same as for carburettor models except that the crank angle sensor is fitted under the distributor (between the distributor and cylinder head).

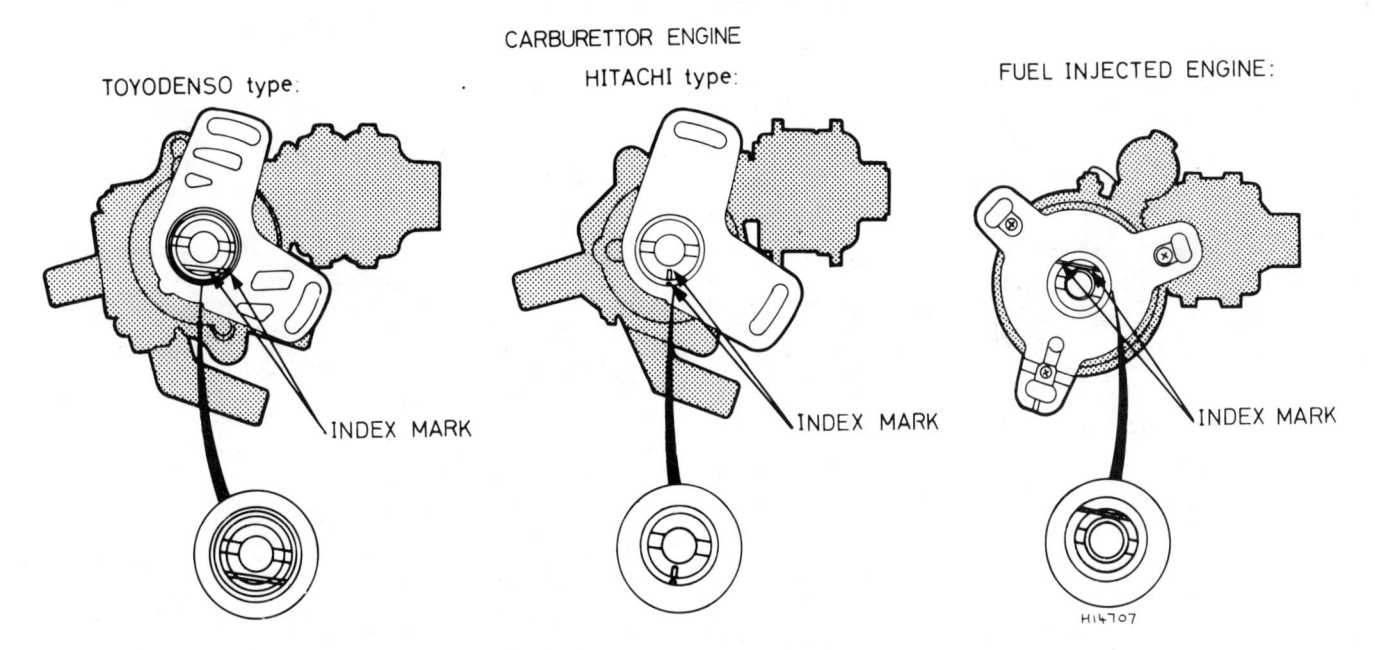

Fig. 4.8 Coupling-to-distributor housing index marks (Sec 8)

Fig. 4.9 Correct orientation of rotor arm (Sec 8)

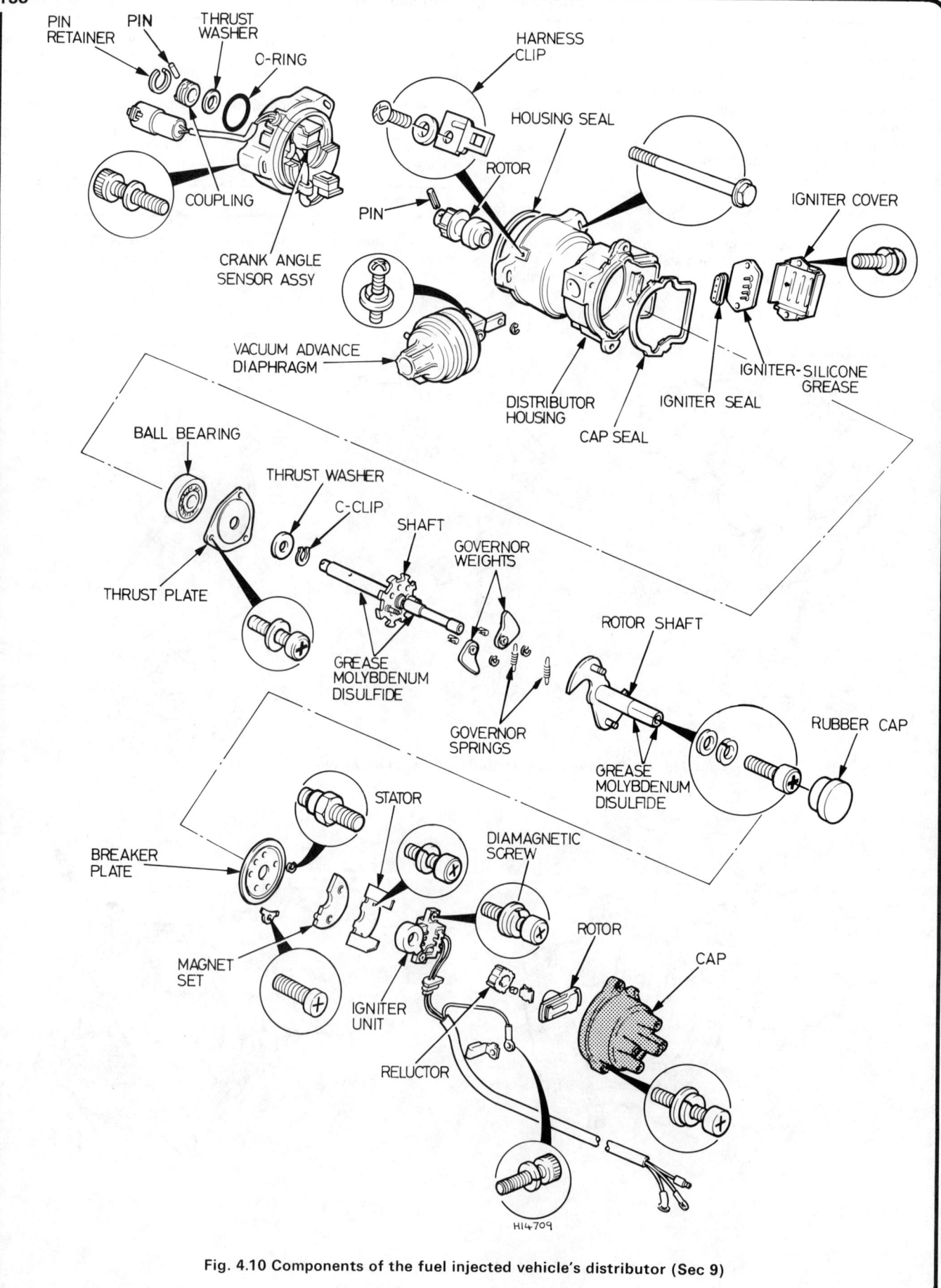

Fig. 4.10 Components of the fuel injected vehicle's distributor (Sec 9)

Chapter 4 Ignition system

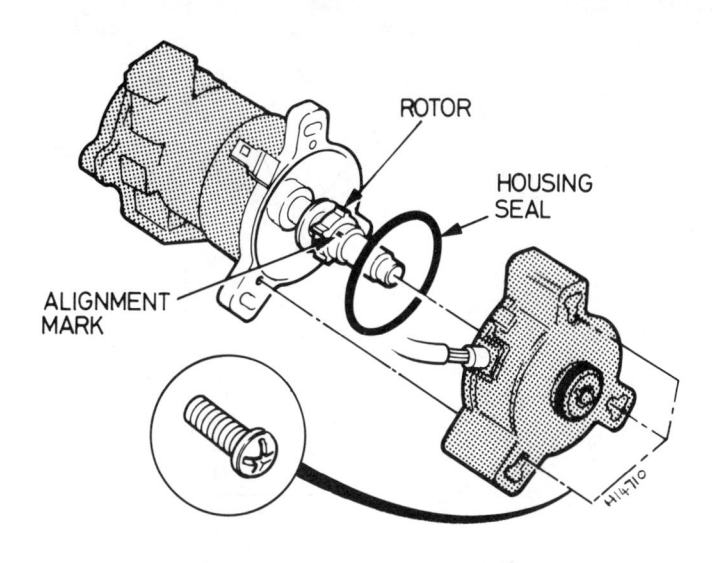

Fig. 4.11 The crank angle sensor assembly (Sec 9)

2 To remove the sensor, first remove the coupling, as described in Section 8, and remove the three retaining screws from the sensor, and lift it off the shaft.
3 The sensor rotor is held to the distributor shaft by another roll pin. Scribe an alignment mark between the rotor and shaft before removing it.
4 Do not attempt to remove the sensors from the sensor housing.
5 Refitting is a reversal of removal, using a new O-ring seal.

10 Igniter unit – testing

Igniter unit test
1 Disconnect the electrical wires from the igniter.
2 Switch on the ignition and check for battery voltage between the blue wire and earth, and the black/yellow wire and earth.
3 With the leads still disconnected, check the continuity of the igniter unit, first with the positive probe of the ohmmeter on the black/yellow and negative probe on the blue terminal, when there should be no continuity. Swap the probes over, when there should be continuity.
Note: fit a jumper wire between the blue and green terminals on the Toyo Denso type during the continuity test.
4 Renew defective igniter units.

Pick-up coil test
5 Check the resistance of the pick-up coil with an ohmmeter between the blue and green terminals on the igniter unit, which should be approximately 750 ohm.
6 When refitting the igniter, pack the terminals with silicone grease.

11 Reluctor air gap – adjustment

1 Remove the distributor cap and rotor arm.
2 Using a feeler gauge, check that the air gaps are all equal.
3 On the Hitachi type, loosen the stator screws and move the stator accordingly. Tighten the screws and recheck.
4 On Toyo Denso types, if the air gaps are not equal, check for damage to the stator or reluctor.

12 Advance and retard mechanism – testing

Vacuum advance
1 Disconnect the vacuum advance hose(s).
2 Connect a vacuum pump to each diaphragm connection in turn and draw a vacuum.

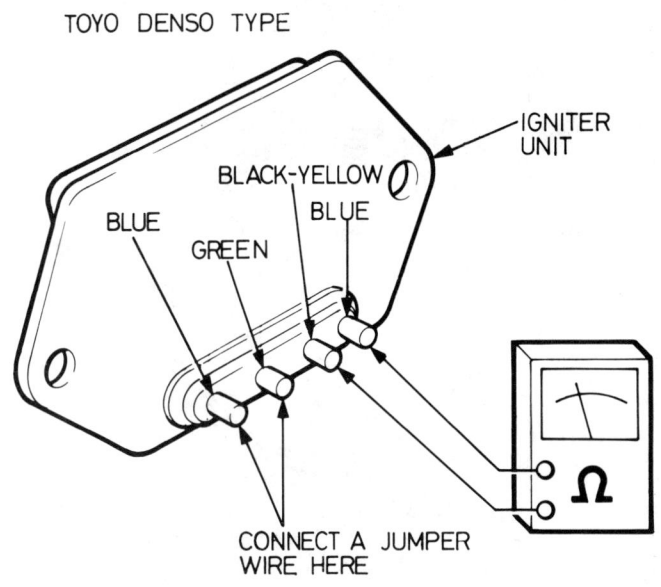

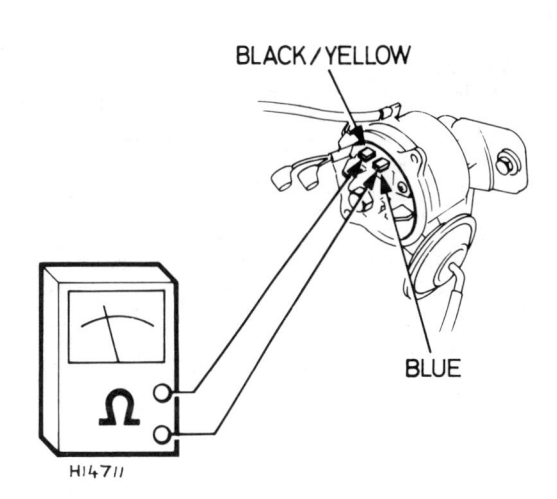

Fig. 4.12 Testing the igniter unit (Sec 10)

3 The breaker plate should move.
4 Hold the vacuum and check that there is no loss of vacuum, which would indicate a defective diaphragm.
5 Renew defective diaphragm units, as described in Section 8.

Mechanical advance
6 Disconnect the vacuum hose(s) from the distributor and clamp their ends.
7 Connect up the strobe light to the coil.
8 Start the engine and increase its speed.
9 Observe the timing mark, which should appear to move as engine speed increases.
10 If not, check the centrifugal weights for correct operation.

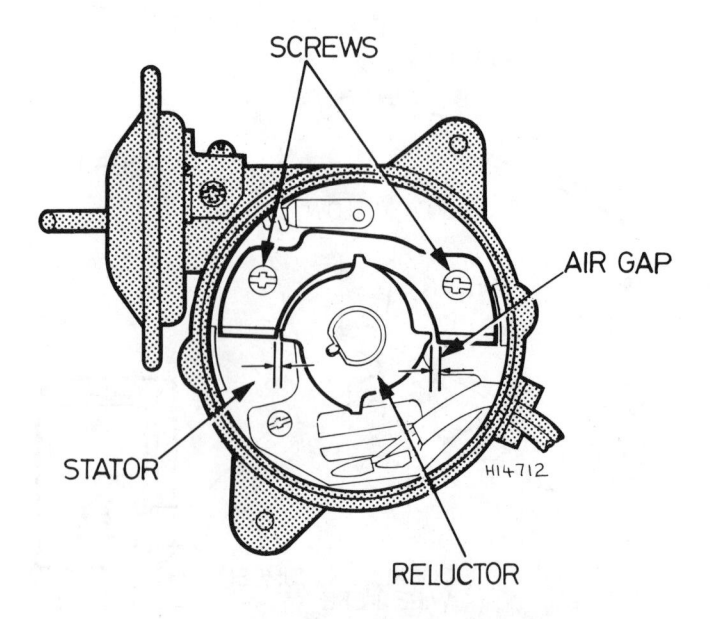

Chapter 4 Ignition system

Fig. 4.13 Reluctor air gap (Hitachi) (Sec 11)

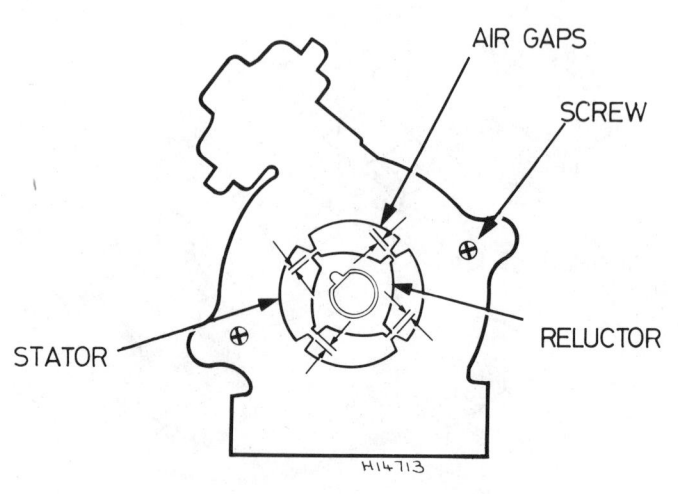

Fig. 4.14 Reluctor air gap (Toyo Denso) (Sec 11)

13 Ignition timing – adjustment

1 Bring the engine to normal operating temperature.
2 Stop the engine and connect up a strobe light, in accordance with the maker's instructions.
3 Remove the rubber cap from the viewing 'window' in the cylinder block.
4 Start the engine, and point the tuning light through the window and onto the flywheel or driveplate (automatics).
5 The relevant timing mark on the flywheel or driveplate should appear directly opposite the pointer on the engine casing if the timing is correct (photos).
6 To adjust the timing, loosen the distributor adjustment bolts, and turn the distributor housing clockwise to retard the ignition and anti-clockwise to advance. Refer to the Specifications for timing values.
7 On completion tighten the adjusting bolts and recheck the timing.
8 Some upper adjusting bolts have a cover fitted to them, refit it after tightening the bolt.
9 Remove the strobe light and refit the rubber cover to the viewing window.

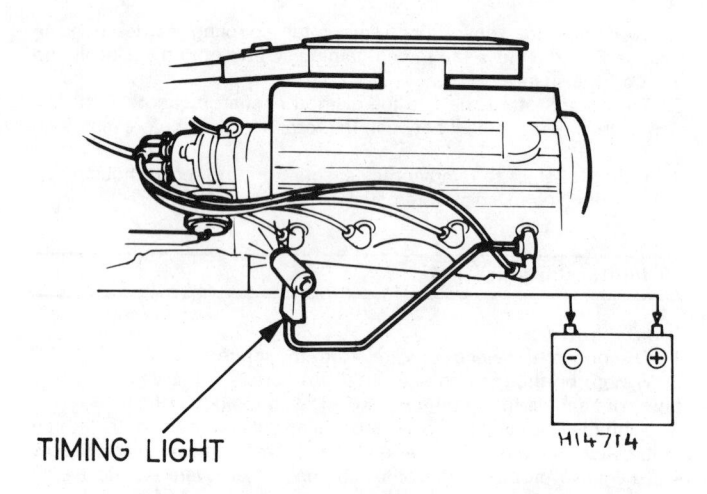

Fig. 4.15 Using a strobe light (Sec 13)

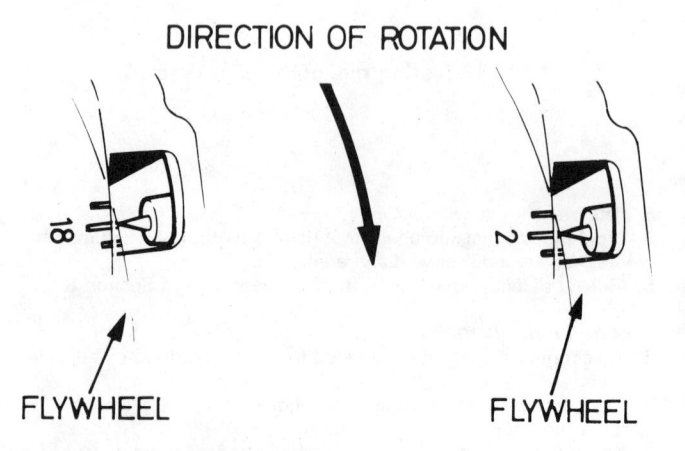

Fig. 4.16 Rotational direction of flywheel (Sec 13)

13.5A Timing marks on the flywheel

Chapter 4 Ignition system

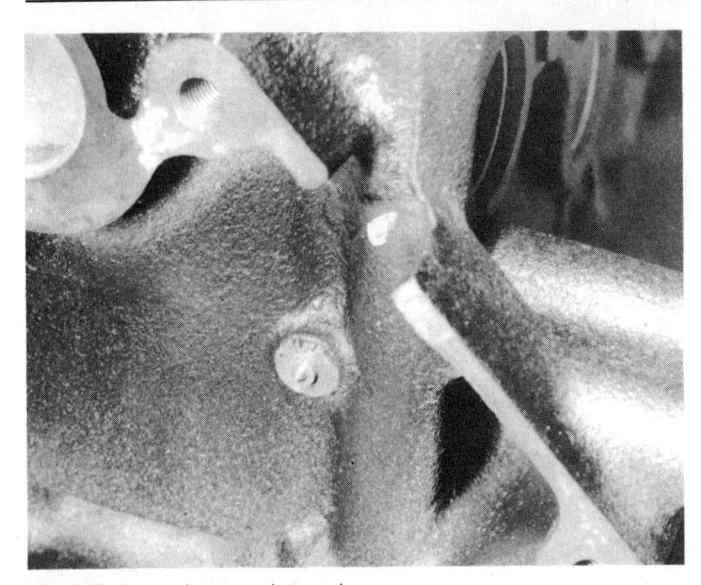

13.5B Timing pointer on the crankcase

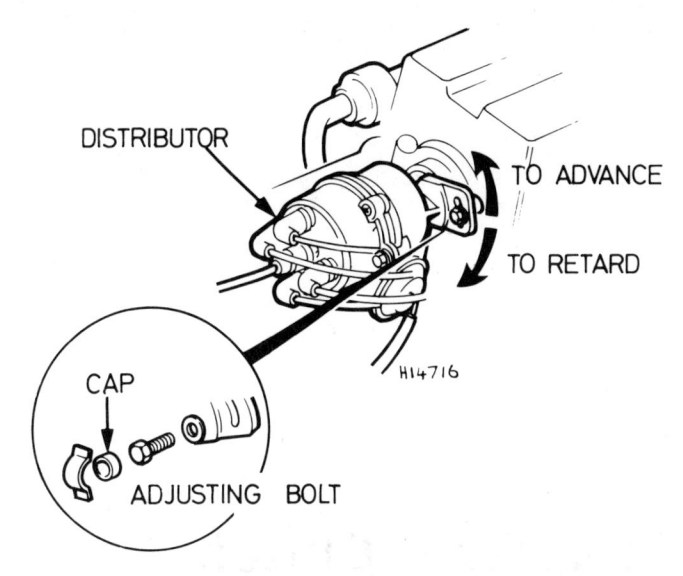

Fig. 4.17 Adjusting the distributor (Sec 13)

14 Fault diagnosis – ignition system

Engine fails to start

1 If the engine fails to start and the car was running normally when it was last used, first check there is fuel in the tank. If the engine turns over normally on the starter motor and the battery is evidently well charged, then the fault may be in either the high or low tension circuits. First check the secondary high tension (HT) circuit. **Note:** If the battery is known to be fully charged; the ignition light comes on, and the starter motor fails to turn the engine check the tightness of the leads on the battery terminals and also the security of the earth lead to its connection to the body. It is quite common for the leads to have worked loose even if they look and feel secure. If one of the battery terminal posts gets very hot while operating the starter motor this is a sure indication of a faulty connection to that terminal.
2 One of the commonest reasons for bad starting is wet or damp spark plug leads and distributor. Remove the distributor cap. If condensation is visible internally, dry the cap with a rag and also wipe over the leads. Refit the cap.
3 If the engine still fails to start, check that current is reaching the plugs, by disconnecting each plug lead in turn at the spark plug end, and hold the end of the cable about 3/16th inch (5.0 mm) away from the cylinder block. Spin the engine on the starter motor.
4 Sparking between the end of the cable and the block should be fairly strong with a regular blue spark. (Hold the lead with rubber to avoid electric shocks). If current is reaching the plugs, then remove them and clean and regap them. The engine should now start.
5 If there is no spark at the plug leads take off the secondary high tension (HT) lead from the centre of the distributor cap and hold it to the block as before. Spin the engine on the starter once more. A rapid succession of blue sparks between the end of the lead and the block indicate that the coil is in order and that the distributor cap is cracked, the rotor arm faulty, or the carbon brush in the top of the distributor cap is not making good contact with the rotor arm.

Engine misfires

6 If the engine misfires regularly, run it at a fast idling speed. Pull off each of the plug caps in turn and listen to the note of the engine. Hold the plug cap in a dry cloth or with a rubber glove as additional protection against a shock from the HT supply.
7 No difference in engine running will be noticed when the lead from the defective circuit is removed. Removing the lead from one of the good cylinders will accentuate the misfire.
8 Remove the plug lead from the end of the defective plug and hold it about 3/16th inch (5.0 mm) away from the block. Restart the engine. If the sparking is fairly strong and regular the fault must lie in the spark plug.
9 The plug may be loose, the insulation may be cracked, or the points may have burnt away giving too wide a gap for the spark to jump. Worse still, one of the points may have broken off. Either renew the plug, or clean it, reset the gap, and then test it.
10 If there is no spark at the end of the plug lead, or if it is weak and intermittent, check the ignition lead from the distributor to the plug. If the insulation is cracked or perished, renew the lead. Check the connections at the distributor cap.
11 If there is still no spark, examine the distributor cap carefully for tracking. This can be recognised by a very thin black line running between two or more electrodes, or between an electrode and some other part of the distributor. These lines are paths which now conduct electricity across the cap thus letting it run to earth. The only answer is a new distributor cap.
12 Apart from the ignition timing being incorrect, other causes of misfiring have already been dealt with under the section dealing with the failure of the engine to start. To recap – these are that:

(a) The coil may be faulty giving an intermittent misfire
(b) There may be a damaged wire or loose connection in the low tension circuit
(c) The condenser may be short circuiting
(d) There may be a mechanical fault in the distributor

13 If the ignition timing is too far retarded, it should be noted that the engine will tend to overheat, and there will be a quite noticeable drop in power. If the engine is overheating and the power is down, and the ignition timing is correct, then the carburettor should be checked, as it is likely that this is where the fault lies.

Chapter 5 Clutch

Contents

Clutch – adjustment	3	Clutch pedal – removal and refitting	5
Clutch – inspection	7	Clutch release bearing – removal, inspection and refitting	8
Clutch – refitting	9	Fault diagnosis – clutch	10
Clutch – removal	6	General description	1
Clutch cable – renewal	4	Routine maintenance	2

Specifications

Type .... Single dry plate, diaphragm spring, pressure plate, cable operated

Driven plate
Rivet head depth (minimum) .... 0.008 in (0.2 mm)
Surface run-out (maximum) .... 0.04 in (1.0 mm)
Overall plate thickness (minimum) .... 0.22 in (5.7 mm)
Radial play in splines (maximum) .... 0.16 in (4.0 mm)

Clutch cover
Diaphragm spring finger maximum height variation .... 0.04 in (1.0 mm)
Clutch release bearing holder:
    Inside diameter (maximum) .... 1.224 in (31.09 mm)
    Holder-to-guide sleeve clearance (maximum) .... 0.0087 in (0.22 mm)

Pedal
Height from floor:
    UK models .... 7.0 in (178 mm)
    North American models .... 7.1 in (180 mm)
Stroke:
    UK models .... 5.2 to 5.7 in (133 to 145 mm)
    North American models .... 5.2 to 5.6 in (133 to 143 mm)
Free play (all models) .... 0.9 to 1.1 in (23 to 28 mm)

Clutch release arm free play
All models .... 0.20 to 0.25 in (5.2 to 6.4 mm)

Torque wrench settings   lbf ft   Nm
Clutch pressure plate bolts   19   26
Release bearing shaft bolt   19   26

1 General description

The clutch is of single dry plate type and is cable operated from a pendant pedal.

When the pedal is depressed, the cable pulls the release lever which forces the ball-bearing type release bearing against the 'fingers' of the clutch cover pressure plate assembly. This releases the pressure plate from the driven plate, and there is no longer any drive between them, thus disengaging the engine from the transmission.

When the pedal is released, the opposite happens and engine torque is again fed to the transmission through the input shaft.

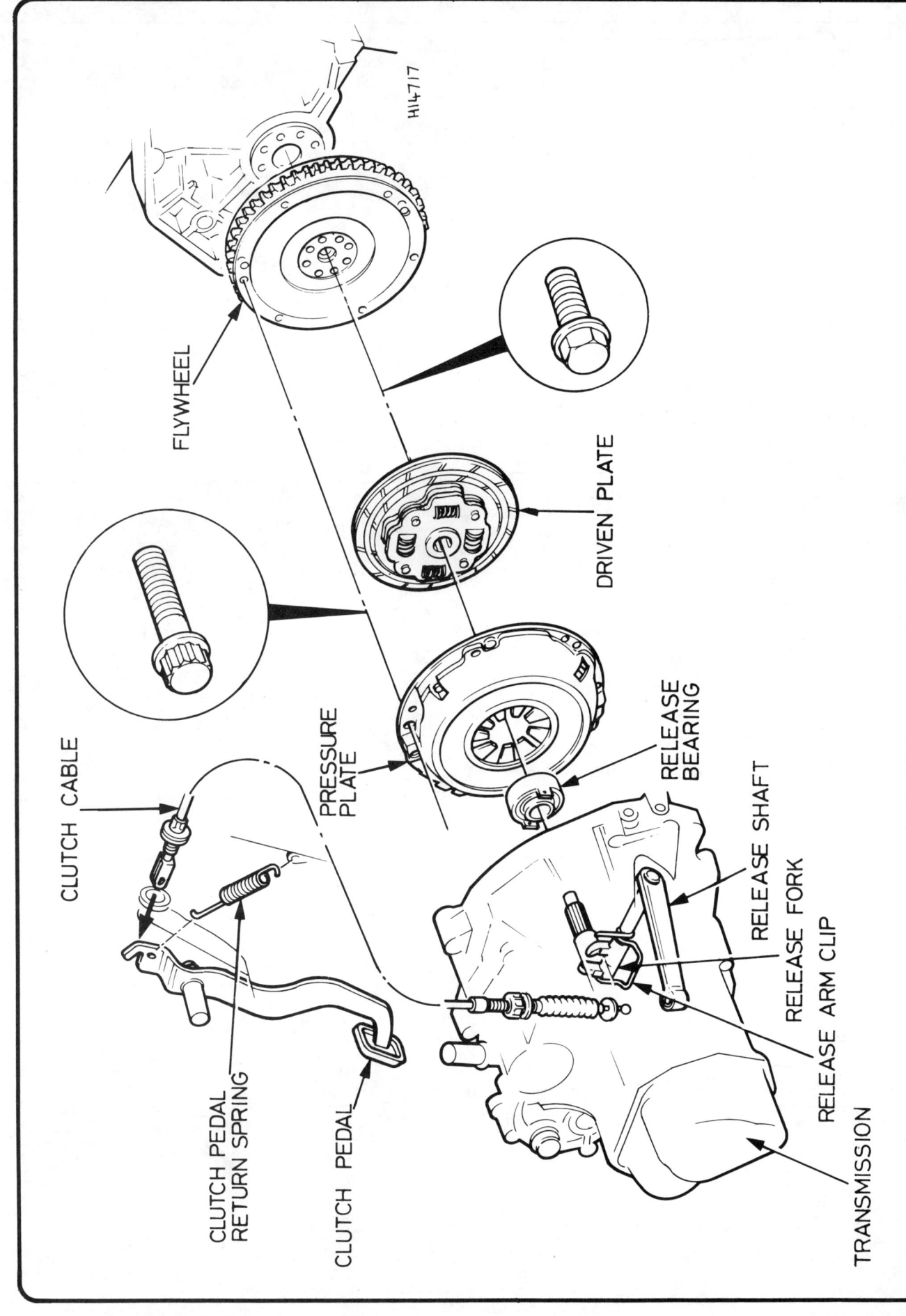

Fig. 5.1 Exploded view of the clutch components (Sec 1)

Chapter 5 Clutch

2 Routine maintenance

At the intervals given in Routine Maintenance (at the front of the book) undertake the following service tasks.
1 Check the clutch release arm travel, and adjust as necessary.
2 Lightly oil the clutch pedal pivot point, and check that the pedal operates smoothly. Check the pedal height and adjust if necessary.

3 Clutch – adjustment

1 Working inside the car, measure the distance from the upper surface of the clutch pedal pad to the floorpan, with the carpet removed.
2 If it is not as specified, adjust on the pedal stop-bolt.
3 From inside the engine compartment, check the clutch release arm free play at the cable end (photo).
4 If it is out of adjustment (see Specifications) adjust on the knurled wheel at the support bracket (photo).
5 Operate the clutch several times and recheck.
6 Finally, check that the clutch pedal free play is as specified.

4 Clutch cable – renewal

1 Unhook the cable from the release arm (photo).
2 Loosen the knurled adjuster and remove the cable from the support bracket.
3 Loosen the locknuts on the cable grommet where it passes through the bulkhead.
4 Unhook the cable from the top end of the clutch pedal.
5 Feed the cable through the bulkhead, and out from the engine compartment.
6 Refitting is a reversal of this procedure. On completion adjust as described in Section 3.

5 Clutch pedal – removing and refitting

1 Loosen the cable at the adjuster sufficiently to allow the cable to be unhooked from the top of the pedal.
2 Remove the E-clip or nut which secures the pedal to the pivot cross-shaft.
3 Unhook the pedal return spring.
4 Slide the pedal off the pivot cross-shaft.
5 Refitting is a reversal of this procedure, but apply some grease to the pivot bushes and cross-shaft, and adjust as described in Section 3.

6 Clutch – removal

1 Remove the transmission, as described in Chapter 6.
2 Remove the pressure plate bolts progressively, a little at a time, to avoid distorting the pressure plate (photo).
3 Remove the pressure plate, catching the driven plate as you do so.

7 Clutch – inspection

1 Inspect the driven plate, for scoring and cracking and signs of burning or oil contamination (photo).
2 Examine the hub for cracks and check the splines are in good condition.
3 Refer to the Specifications and measure the lining thickness. Replace the driven plate if necessary.
4 Inspect the pressure plate for cracks and the diaphragm fingers for wear, especially where the release bearing contacts it.
5 Check the spring fingers for height to ensure none are bent out of tolerance (photo).
6 If there is any sign of oil contamination in either component, suspect either the crankshaft rear oil seal or the transmission input shaft oil seal of being faulty.

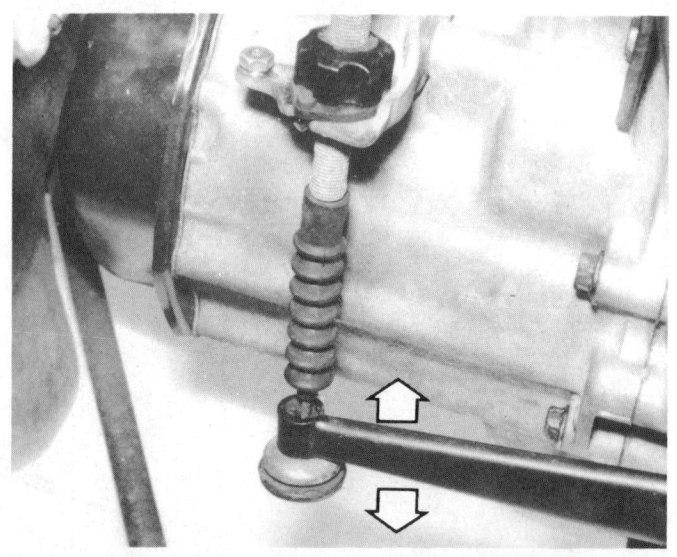

3.3 Move the release arm up and down to measure its free play

3.4 Clutch cable adjuster wheel

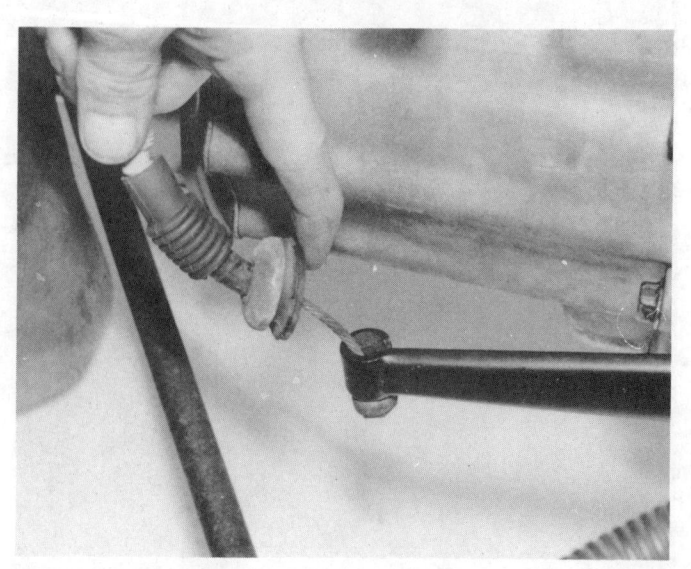

4.1 Clutch cable-to-release arm fitting

Chapter 5 Clutch

6.2 Removing the clutch pressure plate bolts

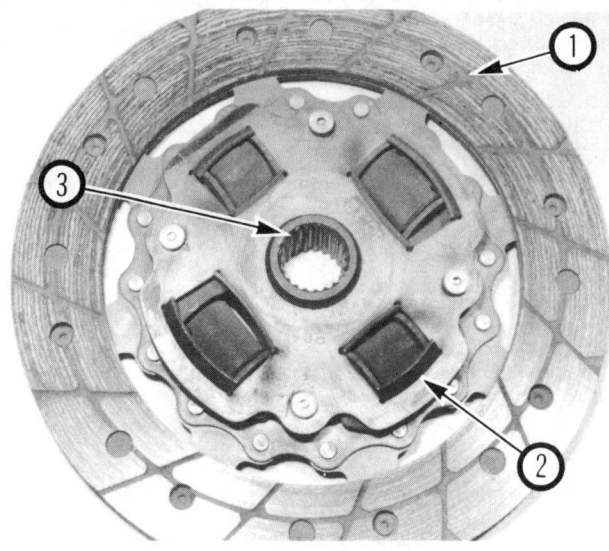

7.1 The driven plate
1 Lining
2 Torsion springs
3 Splined hub

7 Check the flywheel for scoring. If not too badly damaged it can be resurfaced – consult your dealer.

8 Clutch release bearing – removal, inspection and refitting

1 Examine the clutch release bearing in the transmission housing.
2 The bearing is pre-packed so do not wash it in petrol or the grease will be washed out.
3 If the transmission has been dismantled, it is normal practice to renew the release bearing at the same time.
4 To remove the bearing, undo the securing bolt (photo) and slide the release shaft out, releasing the bearing assembly.
5 Slide the bearing assembly off the input shaft.
6 Remove the bearing from the release arm by pulling out the spring from each side (photo).
7 Check the bearing for excessive play or 'rattle' by spinning it, renew if worn.

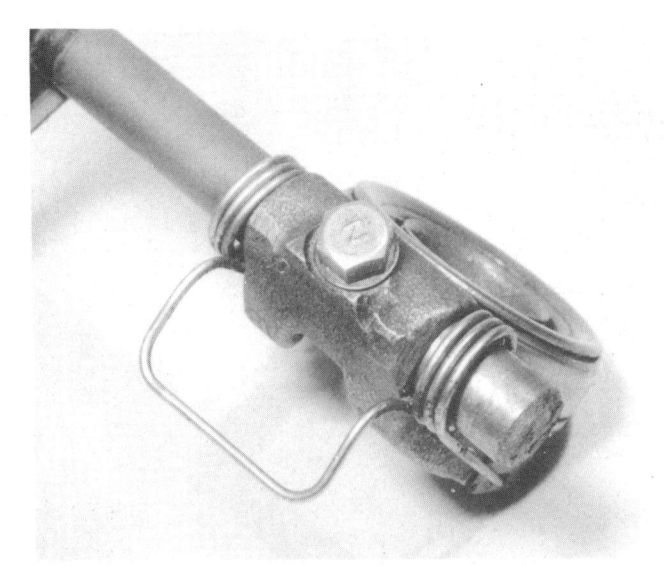

7.5 Clutch pressure plate. The diaphragm spring fingers are arrowed

8.4 The clutch release bearing securing bolt

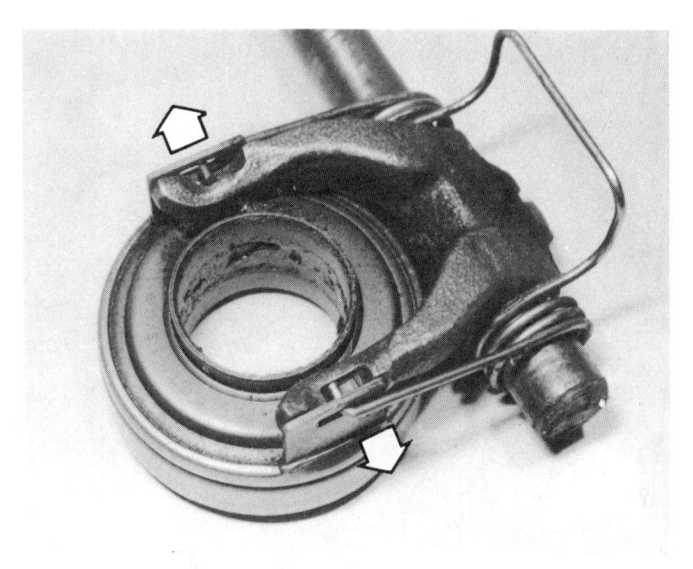

8.6 Pull the spring out in the direction shown to release the bearing

9 Clutch – refitting

1 Clean the flywheel and pressure plate, removing any protective grease, and ensure the driven plate is free from oil or grease.
2 Offer the driven plate up to the flywheel with the torsion springs facing away from the flywheel, then loosely bolt the pressure plate into position (photo).
3 The driven plate must now be centralised in order that the input shaft on the transmission unit will enter it and allow the transmission unit to line up.
4 This can be done by inserting an old shaft through the driven plate hub and into the flywheel or by making up an improvised tool.
5 Once the driven plate is centralised, tighten the pressure plate bolts evenly, in sequence a bit at a time to avoid distorting the plate. Torque figures are in the Specifications.
6 The transmission unit may now be refitted, with reference to Chapter 6.
7 If difficulty is found when attempting to align the input shaft, check the centralisation of the driven plate.

8 Refitting is a reversal of this procedure, applying molybdenum disulphide grease to the release arm shaft and to the inner face of the bearing where the input shaft passes through it. Use a new spring washer under the bolt.

9.2 Positioning the driven plate and pressure plate

10 Fault diagnosis – clutch

Symptom	Reason(s)
Judder when taking up drive	Loose engine/transmission mountings or worn flexible mountings Badly worn friction surfaces or driven plate contaminated with oil deposits Worn splines in the driven plate hub or on the transmission input shaft, or damaged plate Badly worn flywheel spigot bearing Worn or loose input shaft
Clutch drag (or failure to disengage) so that gears cannot be meshed	Clutch clearance too great Clutch driven plate sticking because of rust on splines (usually apparent after standing idle for some length of time) Damaged or misaligned pressure plate assembly
Clutch slip – (Increase in engine speed does not result in increase in car speed – especially on hills)	Clutch clearance too small resulting in partially disengaged clutch at all times Clutch friction surfaces worn out (beyond further adjustment) Clutch surfaces oil soaked
Noise from clutch	Worn release bearing Worn or loose components in driven plate or pressure plate

Chapter 6 Manual transmission

Contents

Countershaft bearing and mainshaft oil seal – removal, inspection and refitting	7
Differential oil seals – removing and refitting	8
Differential unit – removal, inspection and reassembly	9
Fault diagnosis – manual transmission	14
Gearchange lever, shift rod and torque rod – removal and refitting	12
Gearshift selector mechanism – removal, inspection and refitting	10
General description	1
Mainshaft and countershaft – reassembly	6
Mainshaft, countershaft and components – inspection	5
Reversing light switch – testing	13
Routine maintenance	2
Transmission – dismantling into components	4
Transmission – reassembly	11
Transmission – removal and refitting	3

Specifications

Type .... Five forward and one reverse gear. Synchromesh on all forward gears

Ratios
UK models:
  1600 cc:
    1st .... 3.181 : 1
    2nd .... 1.842 : 1
    3rd .... 1.200 : 1
    4th .... 0.870 : 1
    5th .... 0.676 : 1
    Reverse .... 3.000 : 1
  1800 cc:
    1st .... 3.181 : 1
    2nd .... 1.944 : 1
    3rd .... 1.250 : 1
    4th .... 0.933 : 1
    5th .... 0.757 : 1
    Reverse .... 3.000 : 1
North American models:
    1st .... 3.181 : 1
    2nd .... 1.842 : 1
    3rd .... 1.200 : 1
    4th .... 0.870 : 1
    5th .... 0.676 : 1
    Reverse .... 3.000 : 1

Oil capacity .... 4.2 Imp pt (2.5 US qt, 2.4 litre)

Chapter 6 Manual transmission

Clearances

Fifth gear-to-shoulder on spacer collar:
- New ... 0.001 to 0.005 in (0.03 to 0.13 mm)
- Wear limit ... 0.01 in (0.25 mm)

Reverse fork-to-shift shaft:
- New ... 0.002 to 0.014 in (0.05 to 0.35 mm)
- Wear limit ... 0.02 in (0.5 mm)

Reverse idler gear-to-reverse idler gear shift fork:
- New ... 0.008 to 0.04 in (0.2 to 1.0 mm)
- Wear limit ... 0.07 in (1.7 mm)

Reverse shift fork slot (new) ... 0.278 to 0.285 in (7.05 to 7.25 mm)
Reverse shift fork fingers (new) ... 0.46 to 0.48 in (11.8 to 12.1 mm)

Synchroniser ring-to-gear:
- New ... 0.033 in to 0.043 in (0.85 to 1.10 mm)
- Wear limit ... 0.016 in (0.4 mm)

Shift fork-to-synchroniser sleeve:
- New ... 0.014 to 0.026 in (0.35 to 0.65 mm)
- Wear limit ... 0.039 in (1.0 mm)

Mainshaft:
- 3rd gear shoulder-to-2nd gear shoulder:
  - New ... 0.0012 to 0.0071 in (0.03 to 0.18 mm)
  - Wear limit ... 0.012 (0.3 mm)
- 4th gear shoulder-to-spacer collar:
  - New ... 0.0012 to 0.0071 in (0.03 to 0.18 mm)
  - Wear limit ... 0.012 in (0.3 mm)
- 5th gear shoulder-to-spacer collar:
  - New ... 0.0012 to 0.0051 in (0.03 to 0.13 mm)
  - Wear limit ... 0.01 in (0.25 mm)
- Mainshaft ball-bearing-to-spacer washer (new) ... 0.0 to 0.004 in (0.0 to 0.1 mm)
- Replacement washers:
  - A ... 0.074 to 0.075 in (1.88 to 1.92 mm)
  - B ... 0.076 to 0.078 in (1.94 to 1.98 mm)
  - C ... 0.079 to 0.080 in (2.00 to 2.04 mm)
  - D ... 0.081 to 0.082 in (2.06 to 2.10 mm)
  - E ... 0.083 to 0.085 in (2.12 to 2.16 mm)

Countershaft:
- 1st gear-to-thrust washer (new) ... 0.0012 to 0.0031 in (0.03 to 0.08 mm)
- Replacement washers:
  - A ... 0.119 to 0.120 in (3.02 to 3.04 mm)
  - B ... 0.118 to 0.119 in (3.00 to 3.02 mm)
  - C ... 0.117 to 0.118 in (2.98 to 3.00 mm)
  - D ... 0.116 to 0.117 in (2.96 to 2.98 mm)
- 3rd gear shoulder-to-2nd gear:
  - New ... 0.0012 to 0.004 in (0.03 to 0.1 mm)
  - Wear limit ... 0.007 in (0.18 mm)

Transmission housing bearing snap-ring (dimension 'A'):
- Mainshaft ... 0.118 to 0.314 in (3.0 to 8.0 mm)
- Countershaft ... 0.276 to 0.279 in (7.0 to 7.1 mm)

Differential pinion gear backlash (new) ... 0.002 to 0.006 in (0.05 to 0.15 mm)
Differential thrust washer available thickness ... 0.028 in (0.70 mm)
0.030 in (0.75 mm)
0.031 in (0.80 mm)
0.033 in (0.85 mm)
0.035 in (0.90 mm)
0.037 in (0.95 mm)
0.039 in (1.00 mm)

Gearshift selector mechanism:
- Selector arm collar-to-shim ... 0.0004 to 0.0079 in (0.01 to 0.2 mm)
- Available shims:
  - A ... 0.031 in (0.8 mm)
  - B ... 0.039 in (1.0 mm)
  - C ... 0.047 in (1.2 mm)
  - D ... 0.055 in (1.4 mm)
  - E ... 0.063 in (1.6 mm)
- Shift arm-to-shift guide:
  - New ... 0.004 to 0.012 in (0.1 to 0.3 mm)
  - Wear limit ... 0.024 in (0.6 mm)
- Shift guide slot (new) ... 0.311 to 0.315 in (7.9 to 8.0 mm)
- Selector arm-to-interlock:
  - New ... 0.002 to 0.010 (0.05 to 0.25 mm)
  - Wear limit ... 0.03 in (0.7 mm)
- Selector arm finger gap (new) ... 0.396 to 0.400 (10.05 to 10.15 mm)
- Shift arm-to-shift rod guide:
  - New ... 0.002 to 0.010 in (0.05 to 0.25 mm)
  - Wear limit ... 0.03 in (0.8 mm)
- Slot in shift rod guide ... 0.465 to 0.472 in (11.8 to 12.0 mm)

Chapter 6 Manual transmission

Gear selector arm-to-shift rod guide:
  New .... 0.002 to 0.010 in (0.05 to 0.25 mm)
  Wear limit .... 0.002 in (0.05 mm)
Tab on selector arm .... 0.469 to 0.472 in (11.9 to 12.0 mm)

Mainshaft
Outside diameter:
  New:
    A .... 1.0238 to 1.0243 in (26.004 to 26.017 mm)
    B .... 1.2592 to 1.2598 in (31.984 to 32.000 mm)
    C .... 0.9835 to 0.9840 in (24.980 to 24.993 mm)
  Wear limit:
    A .... 1.022 in (25.95 mm)
    B .... 1.257 in (31.93 mm)
    C .... 0.98 in (24.93 mm)
Mainshaft run-out:
  New .... 0.0016 in (0.04 mm)
  Wear limit .... 0.004 in (0.10 mm)

Gear wheels
Thickness:
  2nd gear:
    New .... 1.198 to 1.200 in (30.42 to 30.47 mm)
    Wear limit .... 1.192 in (30.3 mm)
  3rd gear:
    New .... 1.158 to 1.160 in (29.42 to 29.47 mm)
    Wear limit .... 1.15 in (29.3 mm)
  4th gear:
    New .... 1.158 to 1.160 in (29.42 to 29.47 mm)
    Wear limit .... 1.15 in (29.3 mm)
  5th gear:
    New .... 1.060 to 1.062 in (26.92 to 26.97 mm)
    Wear limit .... 1.06 in (26.8 mm)

Countershaft
Outside diameter:
  New:
    A .... 1.2992 to 1.2998 in (33.000 to 33.015 mm)
    B .... 1.3380 to 1.3386 in (33.984 to 34.000 mm)
    C .... 0.9835 to 0.9840 in (24.980 to 24.993 mm)
  Wear limit:
    A .... 1.297 in (32.95 mm)
    B .... 1.336 in (33.93 mm)
    C .... 0.981 in (24.93 mm)

Torque wrench settings

	lbf ft	Nm
Oil filler plug	33	45
Drain plug	29	39
Gearchange shift rod fork bolt	16	22
Torque rod front mounting bolt	7	9
Mainshaft locknut	65	88
Countershaft locknut	65	88
Shift rod guide bolt	22	30
Shift rod holder bolts	13	18
Mainshaft bearing retainer bolts	21	28
Reversing light switch	18	24
Transmission-to-clutch housing bolts	21	28
Detent plugs	16	22
Transmission end cover bolts	9	12
Differential ring gear bolts *(left-hand thread)*	76	103

1 General description

The transmission unit consists of the clutch, gearbox and differential combined in an aluminium housing, bolted to the engine.

The clutch housing and rear cover are detachable; the rear cover enclosing the fifth gear and fifth gear shift forks.

Drive from the engine is transmitted, via the clutch, through the gearbox and differential unit to the driveshafts and front roadwheels.

Gear selection is by a central, floor-mounted gearshift lever, which is connected to the gearbox by a shift rod.

Gearbox lubrication is separate from the engine, having its own filler and drain plugs.

2 Routine maintenance

At the intervals given in Routine Maintenance (at the front of the book) undertake the following service tasks.
1 Check the transmission oil level and top up as necessary.
2 Drain the transmission and refill with new oil.
3 With the vehicle standing on level ground and the engine at normal operating temperature (ie warm), undo the drain plug in the transmission unit, and allow the oil to drain into a suitable container (photo).
4 Refit the drain plug, using a new sealing washer.
5 Remove the filler plug and, using a funnel to avoid spillage, refill

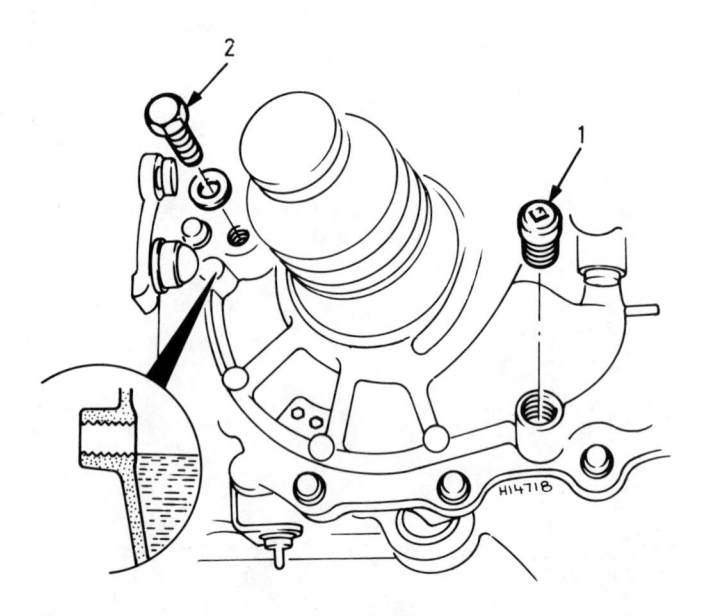

Fig. 6.1 Gearbox drain plug (1) and filler plug (2) (Sec 2)

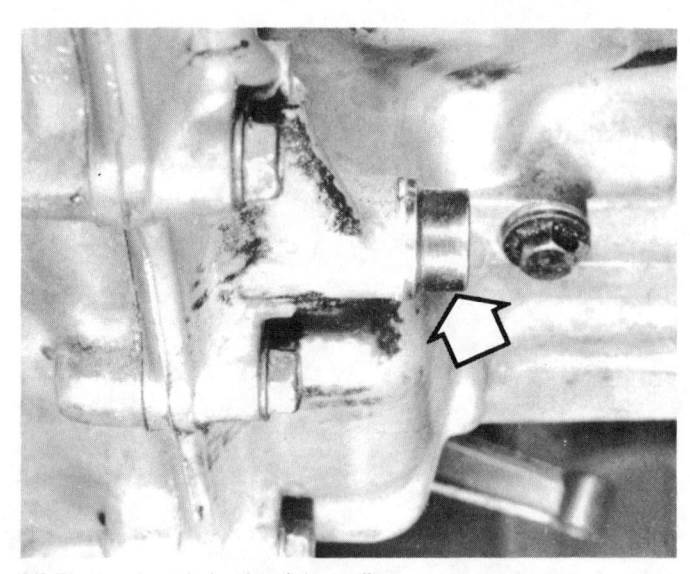

2.3 The gearbox drain plug (arrowed)

the transmission unit using fresh oil until oil just begins to flow from the filler plug hole.
6 Refit the filler plug.
7 Start the engine and check for leaks.

3 Transmission – removal and refitting

1 The transmission unit can be removed with the engine (refer to Chapter 1) or it can be removed as a separate unit as described here.
2 Release the steering lock and select neutral gear.
3 Disconnect the battery negative (earth) cable, then the positive cable, and remove the battery from the vehicle.
4 Disconnect the battery positive cable from the starter motor and the black/white cable from the starter solenoid.
5 Disconnect the reversing light switch.
6 Undo the clip holding the wiring harness to the clutch housing, and pull the harness out of the way.
7 Refer to Chapter 5 and disconnect the clutch cable at the release arm.
8 Remove the two upper transmission mounting bolts.
9 Drain the transmission oil.
10 Raise the front end of the vehicle on to axle stands.
11 Remove both driveshafts, as described in Chapter 8.
12 Position a suitable jack under the transmission unit, and extend it to just take the weight of the unit (after disconnection, the transmission unit is lowered to the ground, not lifted out).
13 Remove the bolt securing the speedometer housing to the transmission casing, and remove the whole unit. Do not pull the cable from the housing first as the drivegear may fall into the transmission unit.
14 On vehicles equipped with power steering, remove the speed sensor as a complete unit and tie the hoses and cables back out of the way, without kinking any of them unduly. If the hoses are disconnected, power steering fluid will be lost.
15 Disconnect the shift lever torque rod from the clutch housing (see Section 12).
16 Remove the bolt from the gearchange shift rod fork end, and swing the rod free (see Section 12).
17 Remove the engine rear upper torque arm bolts, remove the torque arm, and remove the brackets from the clutch housing (refer to Chapter 1 for details of engine mounting brackets).
18 Remove the bolts from the engine damper bracket and remove the bracket.
19 Remove the two bolts from the front transmission mounting which enter the clutch housing.
20 Remove the bolts from the rear transmission mounting bracket which enter the clutch housing.
21 Remove the clutch cover.
22 Remove the bolts securing the starter motor and remove the starter motor.
23 Assistance will be needed to steady the transmission unit whilst it is pulled away from the engine to clear the locating dowels and input shaft, before lowering the transmission unit and removing it from underneath the vehicle.
24 Refitting the assembly is a reversal of removal, with the following points:
25 Apply a smear of molybdenum disulphide grease to the splines of the input shaft.
26 If the clutch has been dismantled, ensure the driven plate is centralised (Chapter 5).
27 Tighten all bolts and nuts to the specified torque and reconnect the driveshafts as described in Chapter 8.
28 Check clutch cable adjustment and refill the transmission assembly with oil.

4 Transmission – dismantling into components

Note: *the detection of wear in the various components by measurement may be called for on disassembly or reassembly. A lot of building and rebuilding may be avoided if measurements are taken at each step of dismantling, so that worn parts may be identified and renewed as necessary. The reader is therefore advised to study the whole of this Chapter before dismantling begins, in order to establish which measurements are best taken when. Where tolerances are regularly exceeded, especially on high mileage vehicles, consideration should be given to obtaining a reconditioned exchange unit.*

1 Clean the grime from the outside of the transmission casing.
2 Place the unit on a suitable work bench.
3 Remove the end cover.
4 Measure the endfloat between the spacer collar and the shoulder on fifth gear on both the main and countershafts, as a reference on reassembly.
5 Bend back the locking tabs on the locknuts of both shafts.
6 Before the nuts can be undone, the gears will have to be prevented from turning. One method is by removing the spring roll pin from the fifth gear selector forks and shaft, selecting a gear other than fifth, then selecting fifth gear. This will lock the geartrains.
7 Undo the locknuts, noting that the mainshaft locknut has a **left-hand thread**.

151

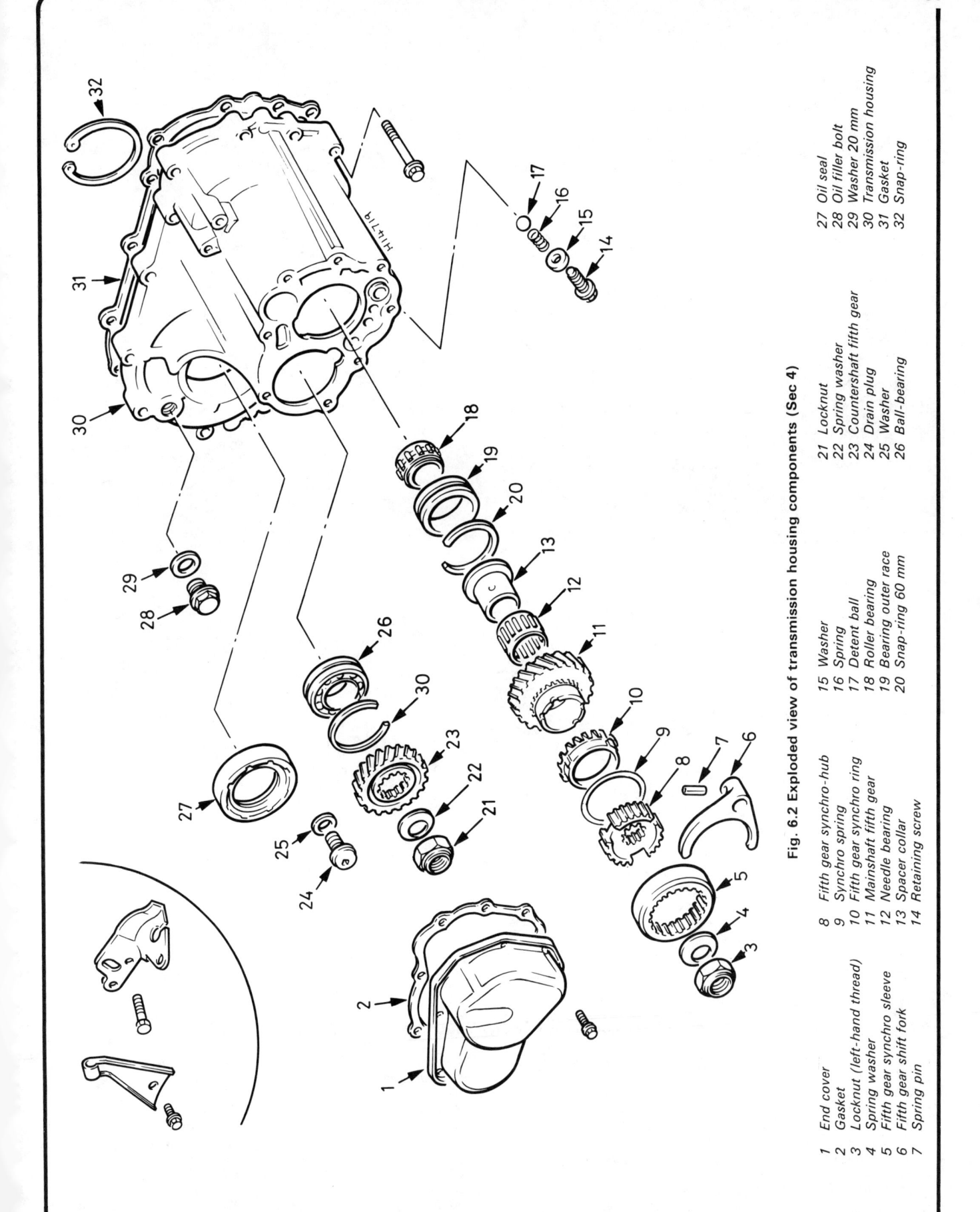

Fig. 6.2 Exploded view of transmission housing components (Sec 4)

1 End cover
2 Gasket
3 Locknut (left-hand thread)
4 Spring washer
5 Fifth gear synchro sleeve
6 Fifth gear shift fork
7 Spring pin
8 Fifth gear synchro-hub
9 Synchro spring
10 Fifth gear synchro ring
11 Mainshaft fifth gear
12 Needle bearing
13 Spacer collar
14 Retaining screw
15 Washer
16 Spring
17 Detent ball
18 Roller bearing
19 Bearing outer race
20 Snap-ring 60 mm
21 Locknut
22 Spring washer
23 Countershaft fifth gear
24 Drain plug
25 Washer
26 Ball-bearing
27 Oil seal
28 Oil filler bolt
29 Washer 20 mm
30 Transmission housing
31 Gasket
32 Snap-ring

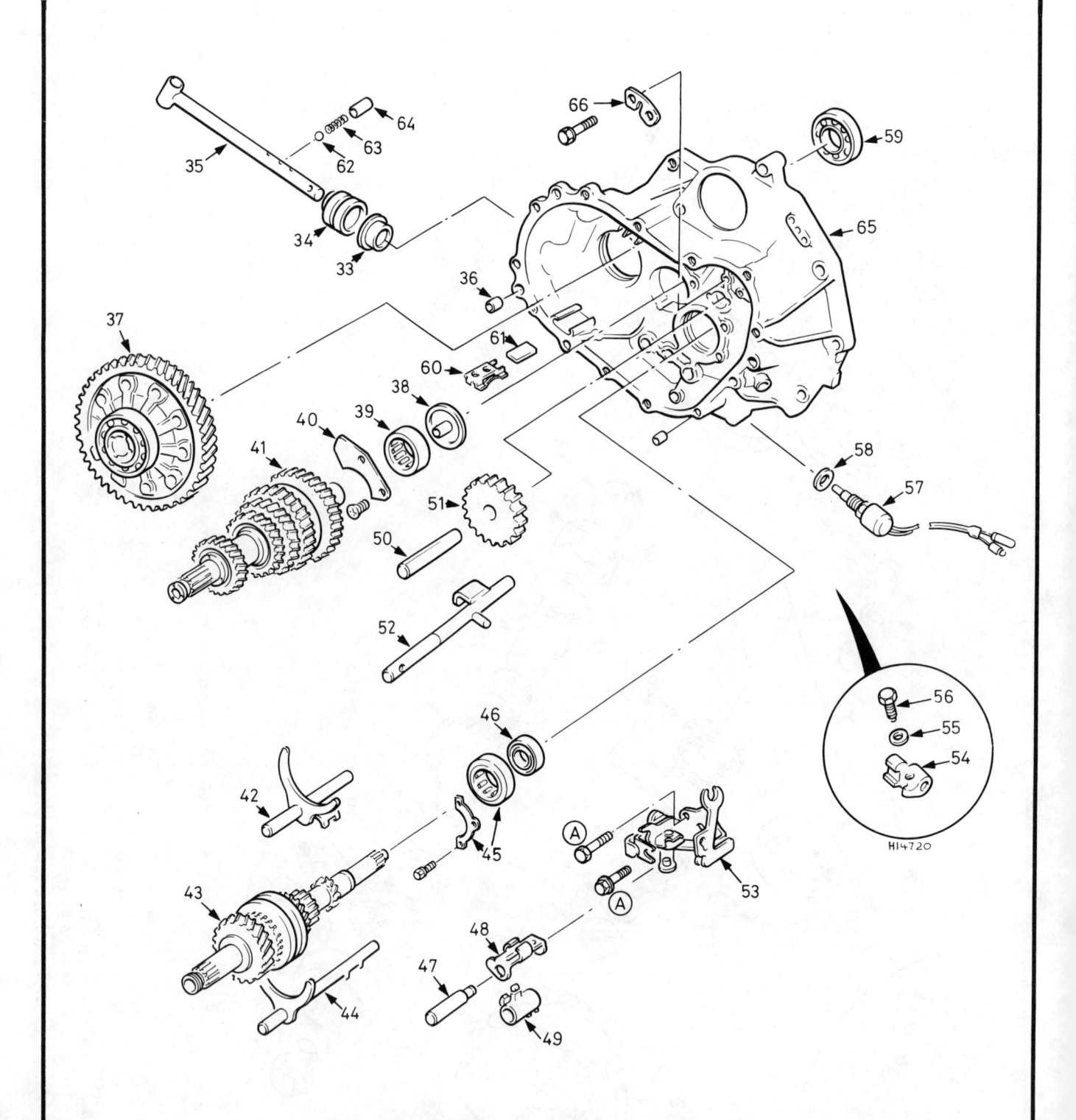

Fig. 6.3 Exploded view of clutch housing components (Sec 4)

33 Oil seal
34 Boot
35 Shift rod
36 Dowel pin
37 Differential assembly
38 Oil barrier plate
39 Countershaft bearing
40 Bearing retainer plate
41 Countershaft assembly
42 First & second gearshift shaft
43 Mainshaft
44 Third & fourth gearshift shaft
45 Mainshaft bearing and bearing retainer plate
46 Oil seal
47 Shift guide shaft
48 Interlock
49 Shift guide
50 Reverse idler gear shaft
51 Reverse idler gear
52 Fifth & reverse gearshift shaft
53 Shift arm holder
54 Shift rod guide
55 Washer
56 Bolt
57 Reversing light switch
58 Washer
59 Oil seal
60 Set plate
61 Magnet
62 Detent ball
63 Spring
64 Spring collar
65 Clutch housing
66 Breather chamber plate

Chapter 6 Manual transmission

8 If not all ready done, drive out the spring roll pin securing the fifth gear shift fork to the shift shaft.
9 Remove the fifth gear, shift fork, synchroniser sleeve, hub, ring and spring as one unit.
10 Remove the countershaft fifth gear.
11 Remove the three plugs, springs and detent balls from the side of the transmission casing.
12 Remove the reversing light switch.
13 Remove the bolts securing the transmission housing to the clutch housing.
14 A special housing puller is used by Honda to separate the two casings. If this is not available, gently tap the transmission casing using a soft-faced mallet, until the housings separate. Do not attempt to lever them apart, as the facings of the aluminium casings are easily damaged, resulting in oil leaks.
15 Lift off the transmission casing from the clutch bellhousing. The mainshaft bearing will remain on the shaft and the countershaft bearing will remain in the casing.
16 Using feeler gauges, measure the clearance between fifth/reverse shift shaft pin and the reverse shift fork, comparing the results with the tolerances in the Specifications.
17 If the given tolerances are exceeded, measure the width of the slot in the reverse shift fork.
18 Measure the clearance between the reverse idler gear and the reverse idler gear shift fork.
19 If the given tolerances are exceeded, pull out the reverse collar and its shaft, and measure the distance between the fingers on the reverse idler gear shift fork.
20 Renew all worn parts as necessary, to allow tolerances to be met.
21 Return the transmission to neutral.
22 Remove the mainshaft bearing retaining plate. Pull out the shift guide shaft.
23 If still in place, pull out the reverse idler gear and shaft.
24 Pull the 3rd/4th and 1st/2nd shift shafts up, to shift into fourth and second.
25 Remove the 5th/reverse shift shaft by pulling it up, at the same time lifting the reverse shift fork.
26 Tilt the interlock and shift guide to the side, and lift them out.
27 The mainshaft and countershaft, together with 1st/2nd and 3rd/4th shift shafts, may now be lifted out as an assembly.

Mainshaft – dismantling
28 Before dismantling the mainshaft, measure the endfloat of the gears, as described in Section 6.
29 Worn components should be renewed as necessary.
30 Refer to Fig. 6.6 and slide each component off the shaft, removing the snap-rings as necessary.
31 Keep each component in its relative position, and inspect each component, as described in Section 5.

Countershaft – dismantling
32 Dismantle the countershaft in the same way as for the mainshaft, referring to Fig. 6.7.
33 Again, measure the endfloat tolerances as described in Section 6. Do not mix any components from the mainshaft and countershaft assemblies.

5 Mainshaft, countershaft and components – inspection

Note: *before inspecting any components, they should be washed in solvent and dried off, using an air line or non-fluffy rag. Keep all parts in their relative order, and on no account mix the mainshaft and countershaft components.*

Mainshaft
1 With all the components removed from the shaft, and cleaned as above, inspect the shaft splines and gearteeth for wear, cracks, pitting and chipping etc.
2 Ensure the oil passageways are clean and free from obstruction.
3 The outside diameter of the shaft should be measured in three places; points A, B and C in Fig. 6.8 to detect any radial wear.
4 The shaft should also be checked for run out, as shown in Fig. 6.9.
5 Shafts which exceed the wear limits in the Specifications should be renewed.

Countershaft
6 Inspect the countershaft in the same manner as the mainshaft, but as shown in Figs. 6.10 and 6.11.
7 Again, any shaft which exceeds the limits in the Specifications should be renewed.

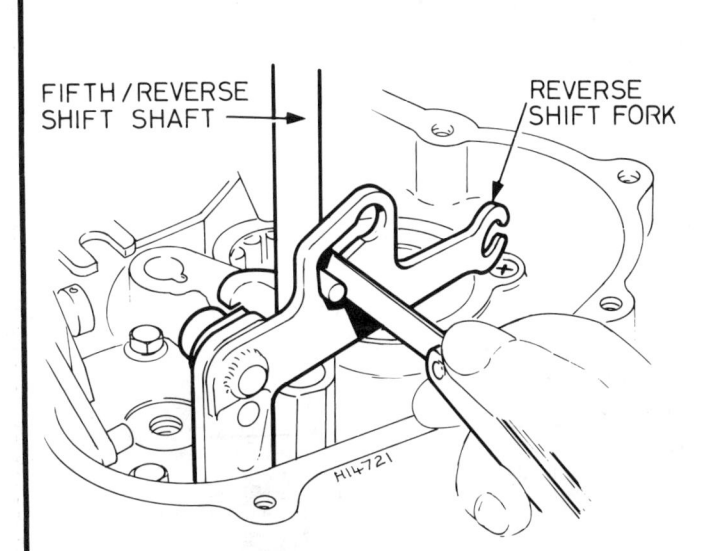

Fig. 6.4 Measuring the clearance between fifth/reverse shift shaft pin and the reverse shift fork (Sec 4)

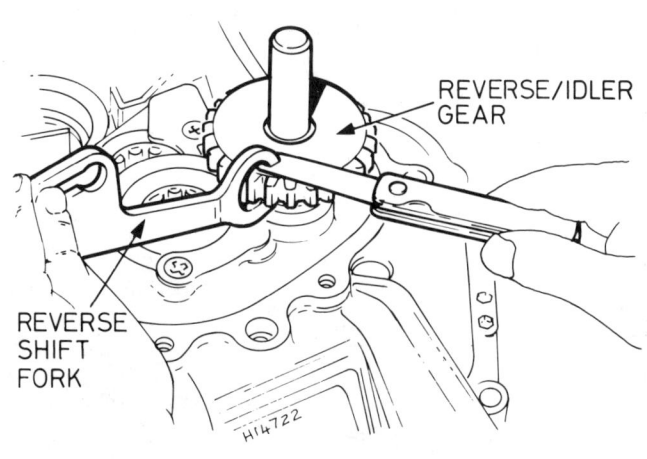

Fig. 6.5 Measuring the clearance between the reverse idler gear and reverse idler gear shift fork (Sec 4)

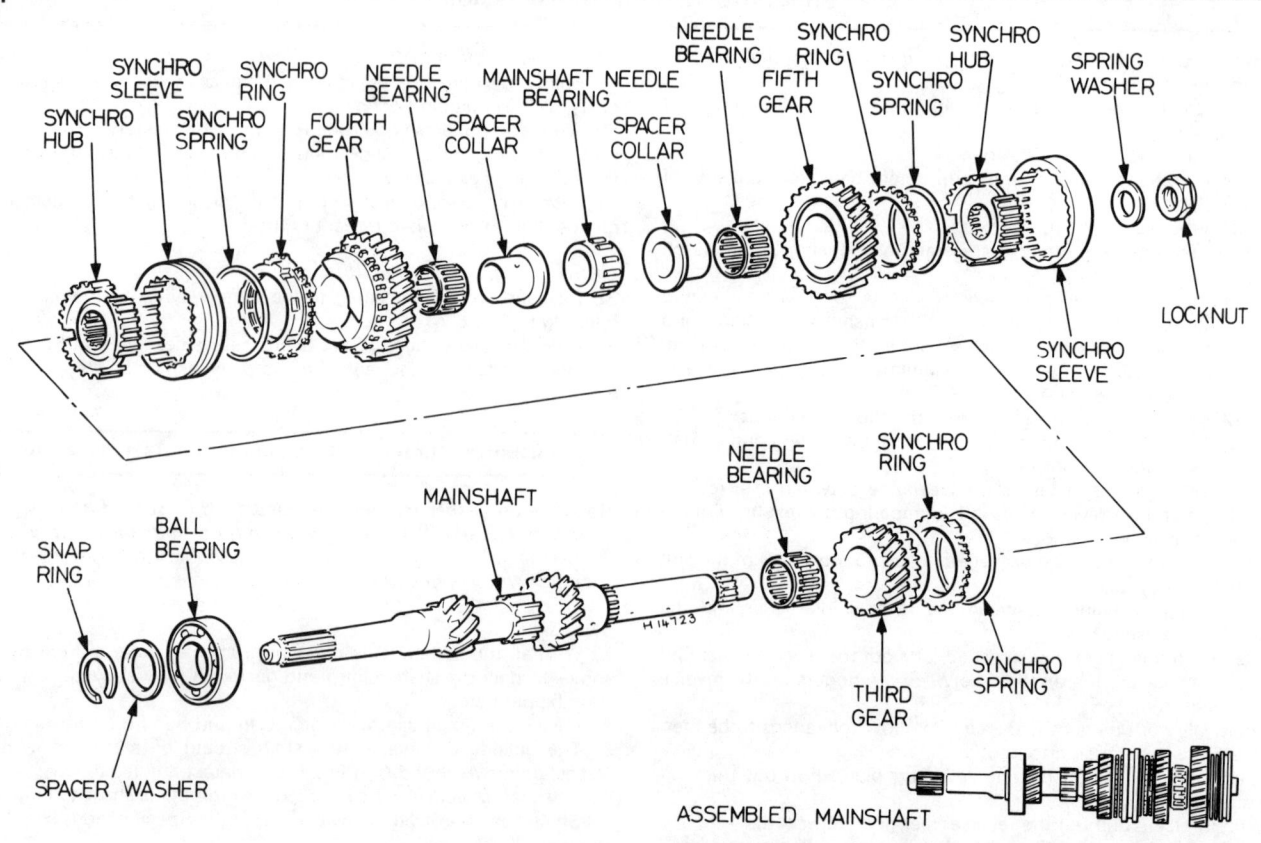

Fig. 6.6 Mainshaft components (Sec 4)

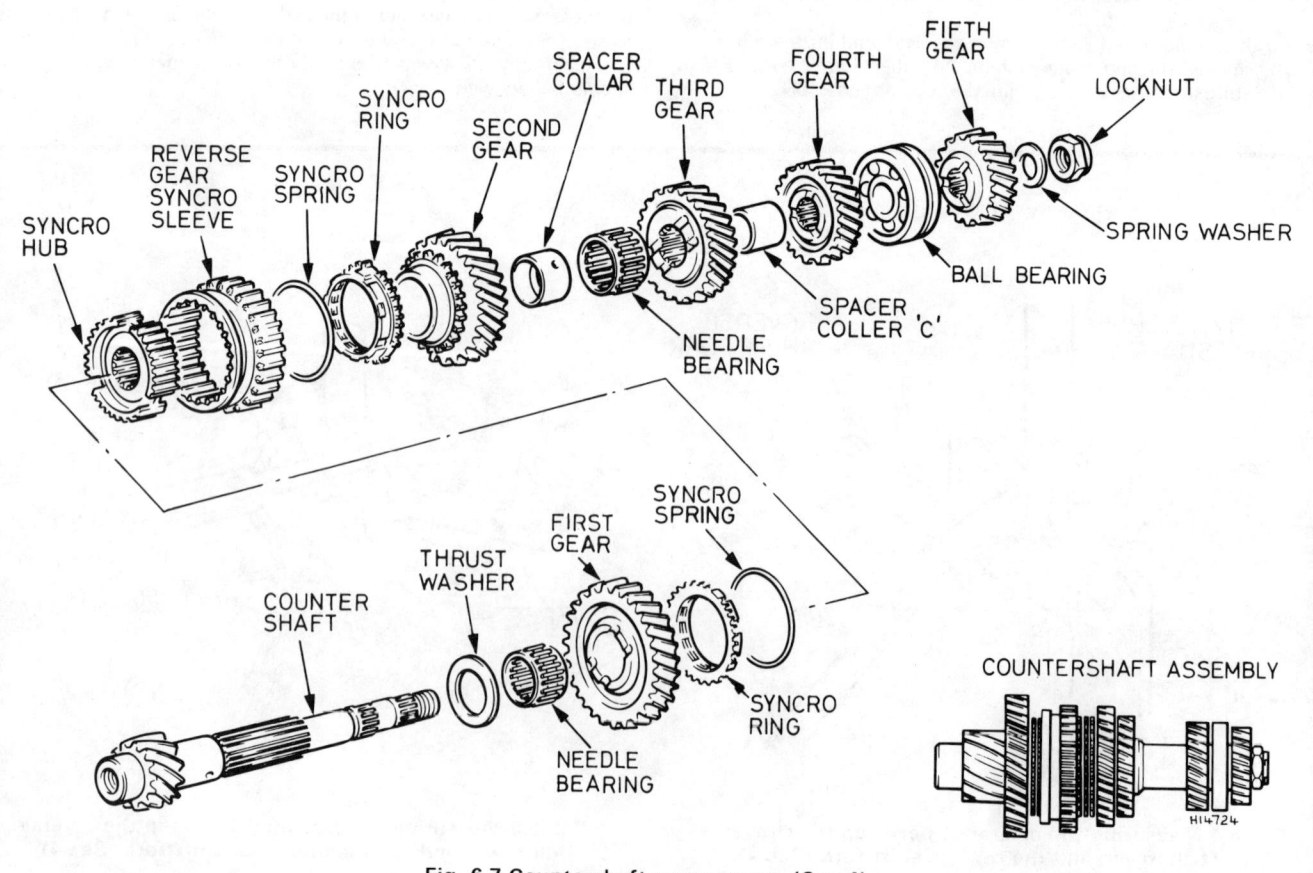

Fig. 6.7 Countershaft components (Sec 4)

Chapter 6 Manual transmission

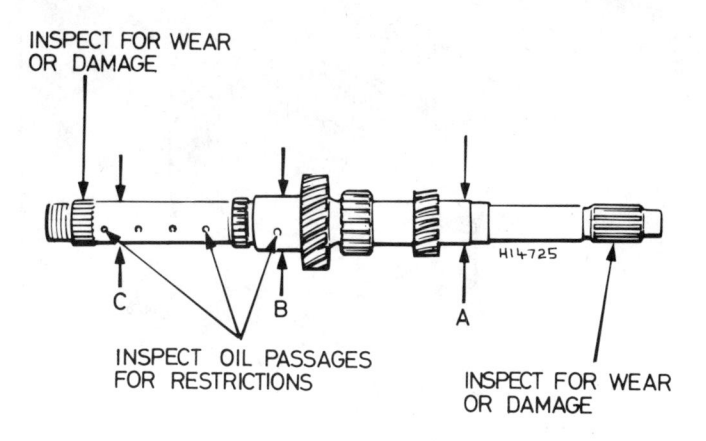

Fig. 6.8 Mainshaft wear inspection points A, B and C (Sec 5)

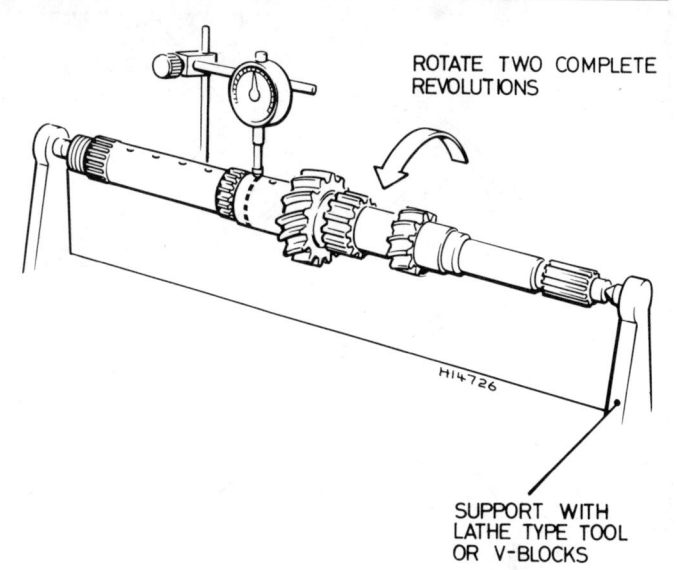

Fig. 6.9 Checking the mainshaft for run-out (Sec 5)

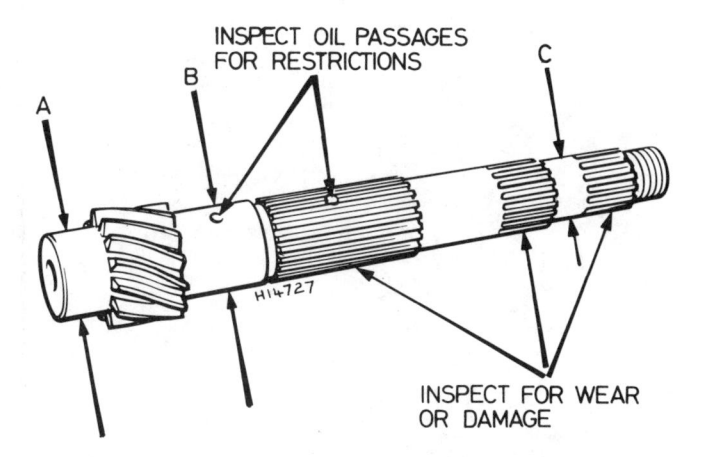

Fig. 6.10 Countershaft wear inspection points A, B and C (Sec 5)

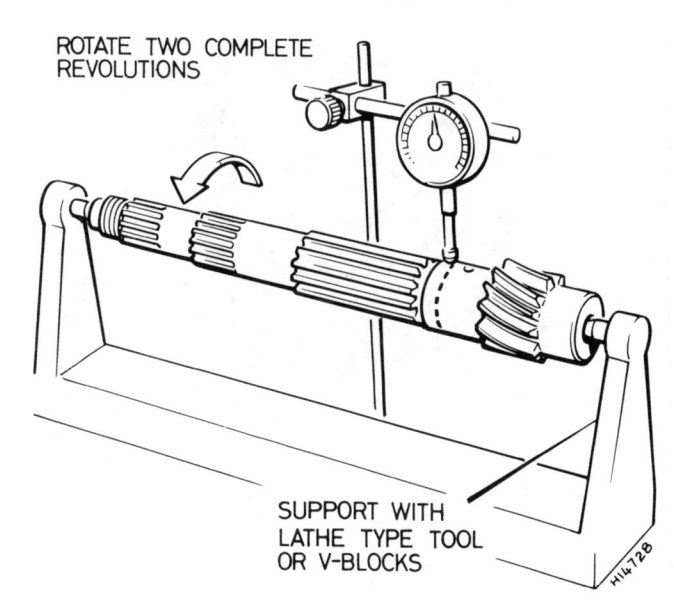

Fig. 6.11 Countershaft run-out inspection (Sec 5)

Gear and synchroniser ring
8 Take each gear and synchroniser ring assembly in turn and inspect them as follows.
9 Inspect the inside of the synchroniser ring for wear.
10 Inspect the teeth on both the synchroniser ring and gear for wear.
11 Inspect the thrust surface on the gear hub for wear.
12 Inspect the cone surface on 1st and 2nd countershaft gears and 3rd, 4th and 5th mainshaft gears for wear.
13 Inspect teeth on all gears for uneven wear, scoring, cracks, chipping etc.
14 Fit each synchroniser ring on its matching gear cone and rotate it until it stops (approximately 10 to 20°), then measure the clearance between the ring and gearteeth with feeler gauges.
15 Renew worn rings.
16 Remove the ring from the gear, coat all parts in oil and reassemble, using a new synchroniser spring.

Shift fork-to-synchroniser sleeve clearance
17 The clearance between the fingers of each shift fork and its synchroniser sleeve should be measured using feeler gauges, and compared with the tolerances in the Specifications.
18 If out of limits, renew the synchroniser sleeve.
19 The tolerances are the same for all three forks.

Synchroniser sleeve and hub
20 Inspect each hub and sleeve for 'rounding off' of gearteeth, indicating wear, and renew as necessary.

21 Install the synchroniser sleeve onto its hub and check for freedom of movement.

Needle and roller bearings
22 Inspect each bearing for signs of overheating, shown by blueing or discoloration.
23 Excessive wear in the bearings is indicated by a 'rattle' noise if the bearing is shaken.
24 The three needle roller bearings are identical, but should be refitted in their original positions.

Gearwheels
25 Before reassembly commences, measure the thickness of each gearwheel, as shown in Fig. 6.15.
26 Renew any gearwheels which are worn beyond limits.

156

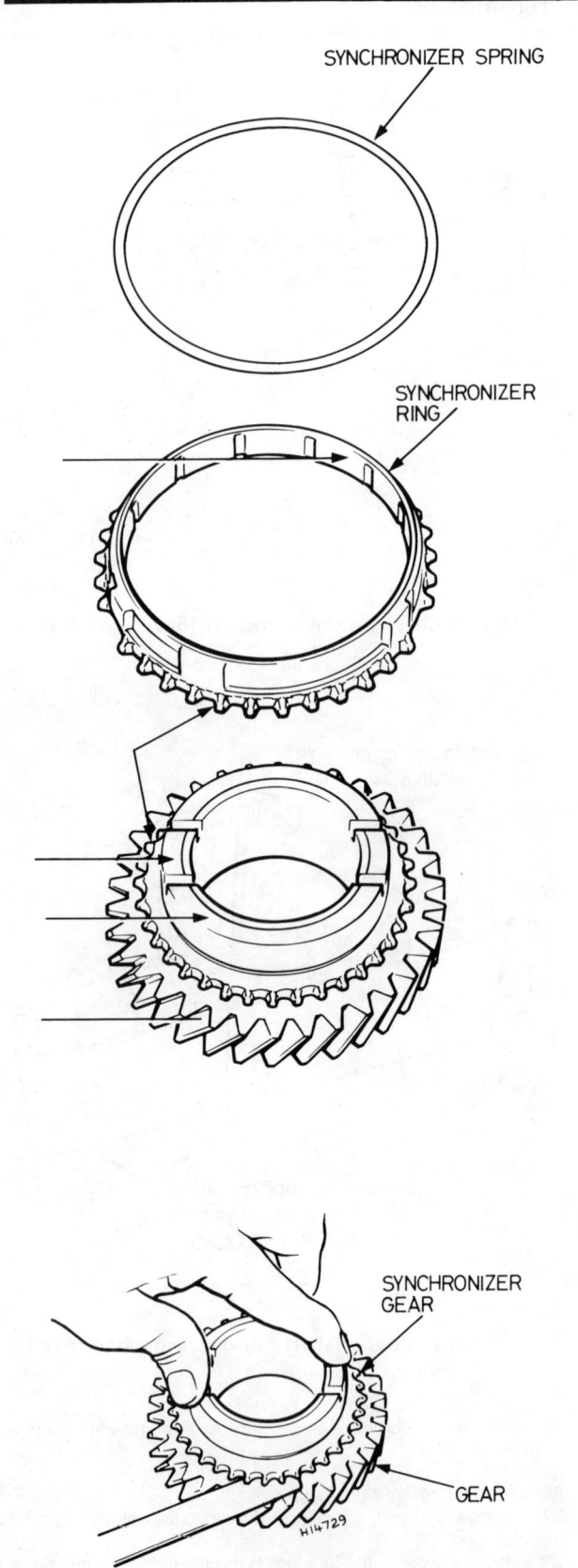

Fig. 6.12 Gear and synchroniser ring inspection (Sec 5)

Arrows indicate areas to inspect for wear

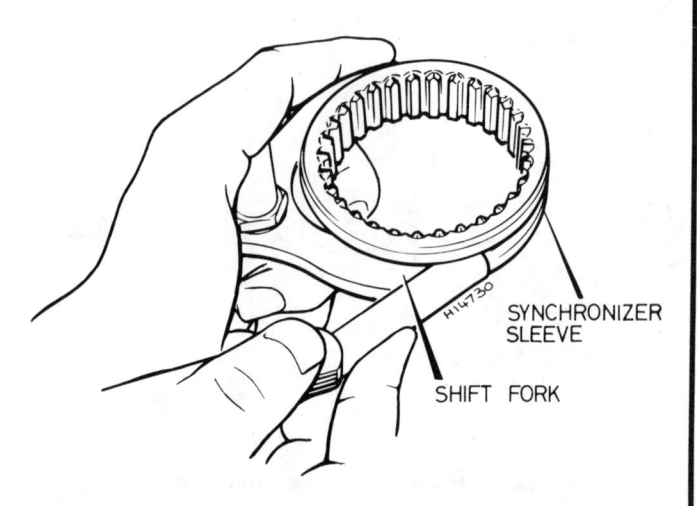

Fig. 6.13 Checking the shift fork-to-synchroniser sleeve clearance (Sec 5)

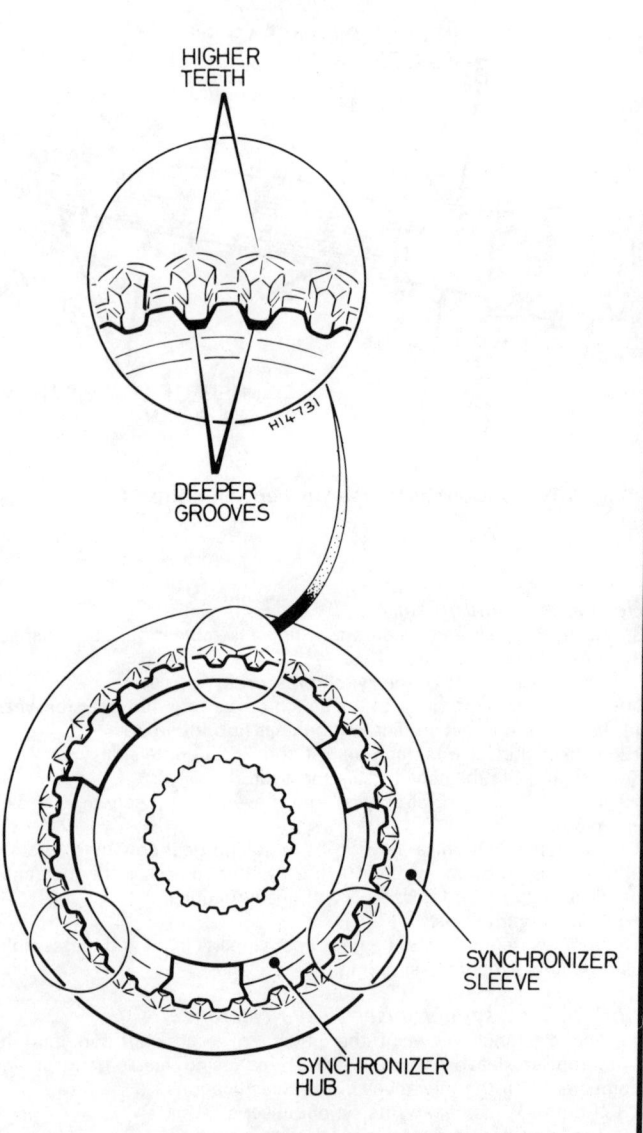

Fig. 6.14 Correct assembly of synchroniser sleeve and hub (Sec 5)

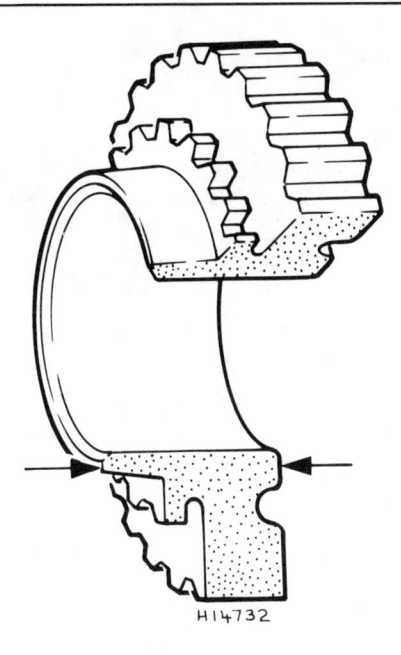

Chapter 6 Manual transmission

Fig. 6.15 Measuring gearwheel thickness (Sec 5)

6.2 Fit the ball-bearing, spacer washer and snap-ring

6 Mainshaft and countershaft – reassembly

Mainshaft
1 Refer to Figs. 6.6 and 6.7 during reassembly to ensure all components are replaced in their correct order.
2 Fit the mainshaft ball-bearing, spacer washer and snap-ring (photo).
3 Slide on the needle roller bearing (photo).
4 Fit the third gear, synchro ring and spring (photos).
5 Next, fit the synchroniser hub and sleeve (photos).
6 Fit the remaining synchroniser ring and spring, then slide on fourth gear, the needle roller bearing, and the spacer collar (photo).
7 Fit the mainshaft roller bearing (photo).
8 The remaining components are fitted after the main and countershaft assemblies are fitted into the transmission housing.
9 However, in order to establish correct clearances, the remaining components should be fitted to the mainshaft, and the locknut tightened to the specified torque. Do not over torque, or incorrect clearances may result.

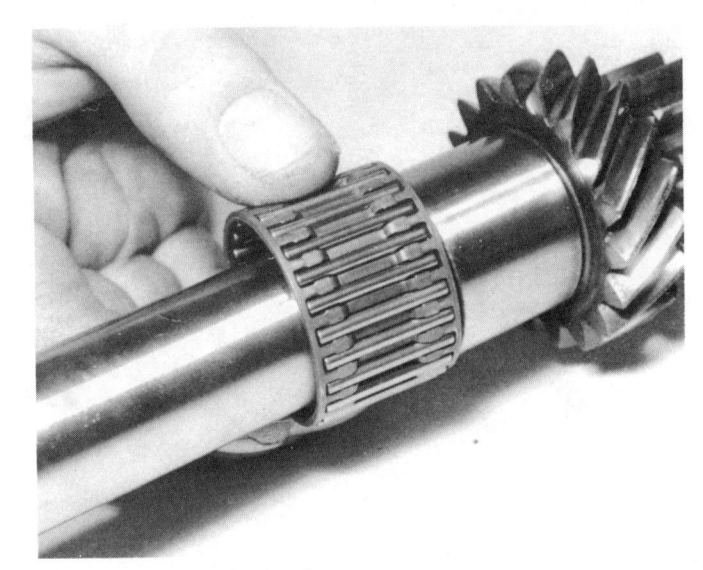

6.3 Sliding on the roller bearing

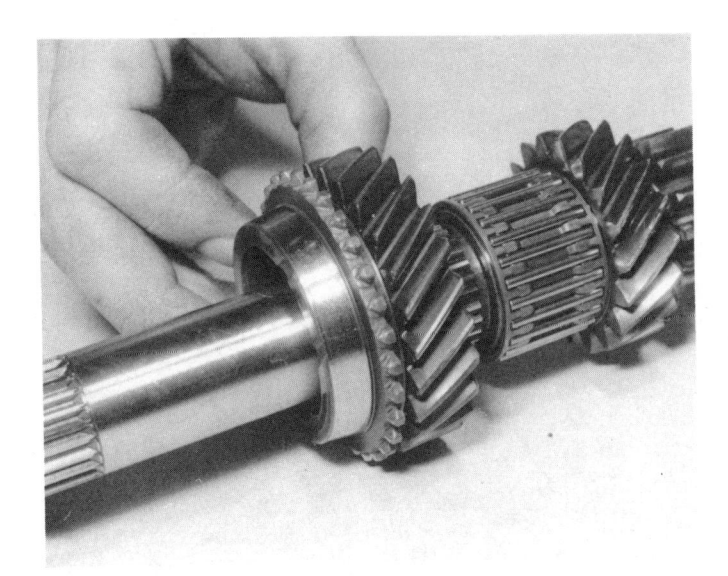

6.4A Fitting third gear ...

6.4B ... and the synchro ring and spring

158   **Chapter 6 Manual transmission**

6.5A Fitting the synchro-hub ...

6.5B ... and sleeve

6.6 Slide on fourth gear, the needle roller bearing and spacer collar

6.7 Fitting the roller bearing

10 Now carry out the following clearance checks, in conjunction with the tolerances in the Specifications.
11 Measure the clearance between the shoulder on third gear and the shoulder on second gear.
12 If out of limits, measure thickness of third gear, and renew as necessary.
13 If third gear is within limits, then fit a new synchroniser hub.
14 Measure the clearance between spacer collar and shoulder on fourth gear.
15 If out of limits, measure thickness of fourth gear, and renew if necessary.
16 If fourth gear is within limits, then fit a new synchroniser hub.
17 Measure the clearance between spacer collar and shoulder on fifth gear.
18 If out of limits, measure thickness of fifth gear.
19 If fifth gear is worn, renew the fifth gear.
20 Measure the clearance between the spacer washer and mainshaft ball-bearing. If out of tolerance, fit a new spacer washer of suitable thickness. Refer to the Specifications for available sizes.

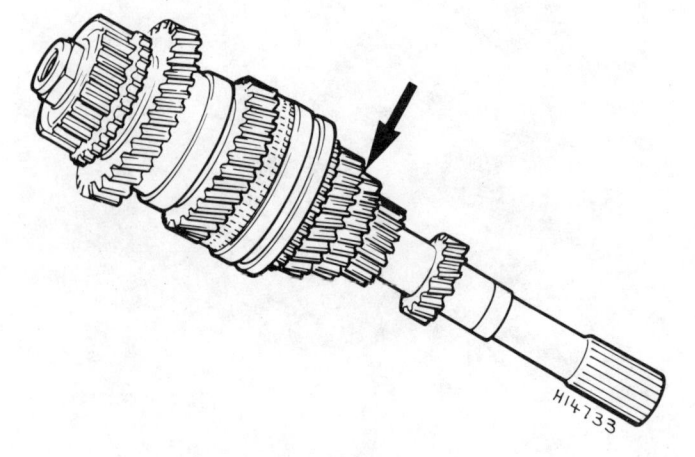

Fig. 6.16 Measuring 3rd gear clearance (Sec 6)

Chapter 6 Manual transmission 159

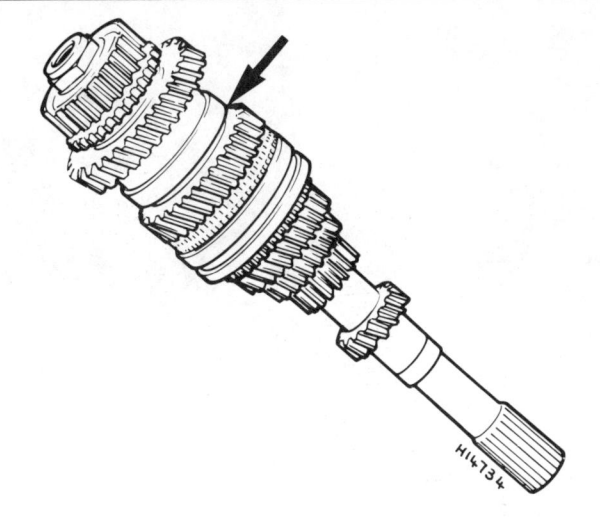

Fig. 6.17 Measuring 4th gear clearance (Sec 6)

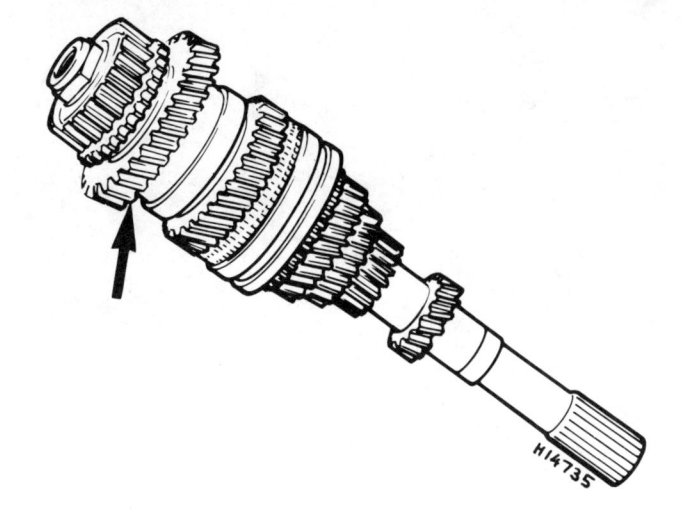

Fig. 6.18 Measuring 5th gear clearance (Sec 6)

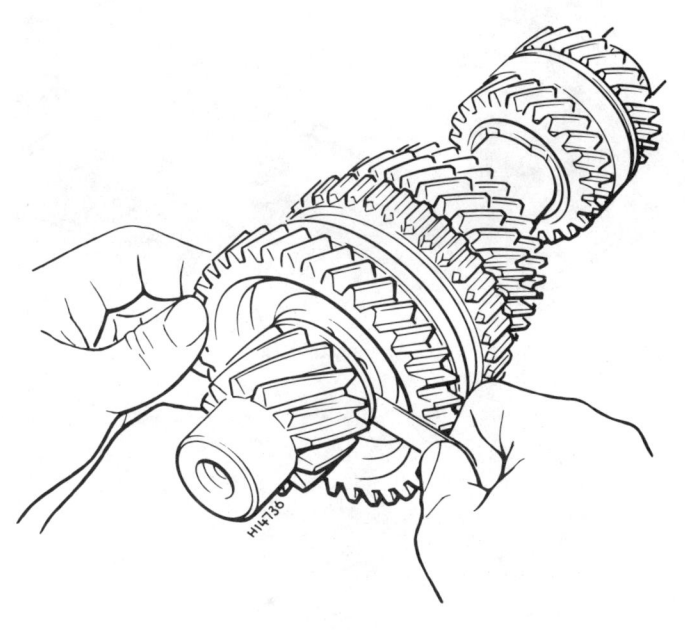

Fig. 6.19 Measuring mainshaft ball-bearing clearance (Sec 6)

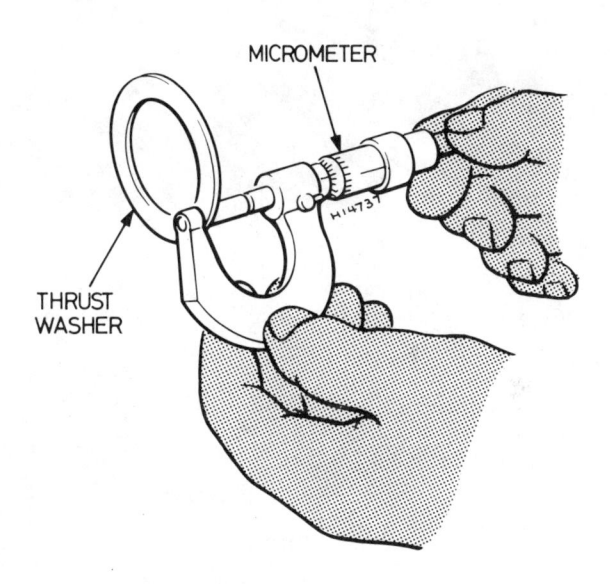

Fig. 6.20 Measuring the thickness of a thrust washer (Sec 6)

Countershaft
21 Commence reassembly by fitting the thrust washer to the countershaft (photos).
22 Slide on the needle roller bearing (photo).
23 Fit first gear, followed by synchro ring and spring (photos).
24 Fit the synchro-hub and reverse gear synchro sleeve (photo).
25 Slide on the needle roller bearing (photo).
26 Fit the synchro spring, synchro ring and second gear (photo).
27 Fit third gear (photo).
28 Slide on the spacer collar and fourth gear (photos).

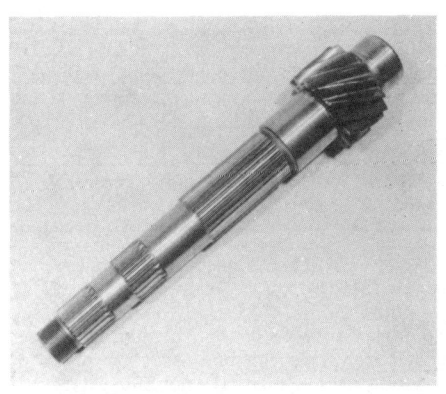

6.21A The countershaft before reassembly

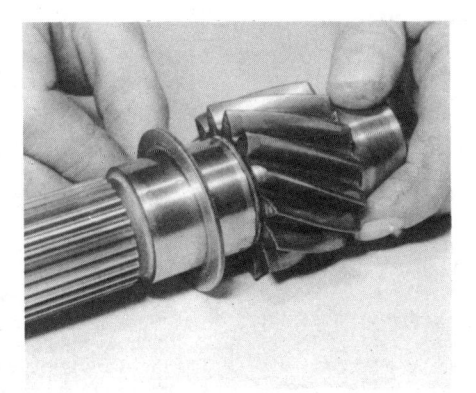

6.21B Fitting the thrust washer

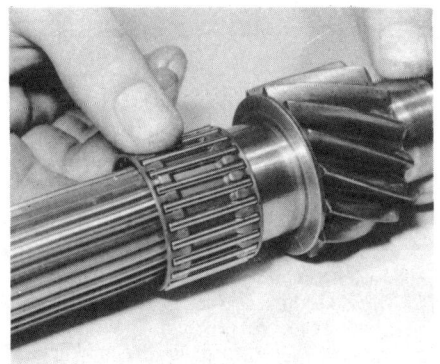

6.22 Slide on the roller bearing

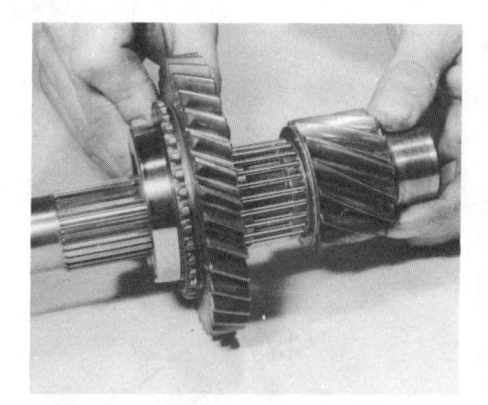

Chapter 6 Manual transmission

6.23A Fitting first gear ...

6.23B ... and the synchro ring and spring

6.24 Fitting the synchro-hub and reverse gear

6.25 Slide on the needle roller bearing

6.26 Fitting the synchro ring and second gear

6.27 Fitting third gear

6.28A Slide on the spacer collar ...

6.28B ... and fourth gear

29 The remaining components are fitted after the mainshaft and countershaft are fitted in the transmission housing.
30 However, as with the mainshaft, these components must be assembled onto the shaft and the locknut fitted and torque loaded in order to carry out the following measurements. Do not over torque the locknut, or incorrect tolerances may result.
31 Refer to the Specifications for tolerances.
32 Measure the clearance between the first gear thrust washer and the shoulder on first gear, changing the thickness of the thrust washer as necessary if the tolerances cannot be met.
33 Measure the clearance between the shoulder on third gear and the shoulder on second gear.
34 If out of limits, measure the thickness of second gear, renewing second gear if the limits are exceeded.
35 After all clearances have been checked, and brought into limits, reassemble both main and countershafts and recheck all clearances.
36 Once they are correct, remove the fifth gear components and reinstall the bearings in the transmission housing.

7 Countershaft bearing and mainshaft oil seal – removal, inspection and refitting

Clutch housing
1 The countershaft bearing and mainshaft oil seal in the clutch housing are shown in the photo.
2 To remove the bearing, first remove the bearing retainer plate. Note that the two securing screws are centre punched to lock them, so an impact driver may be needed to remove them. Do not forget to punch lock the screws on refitting.

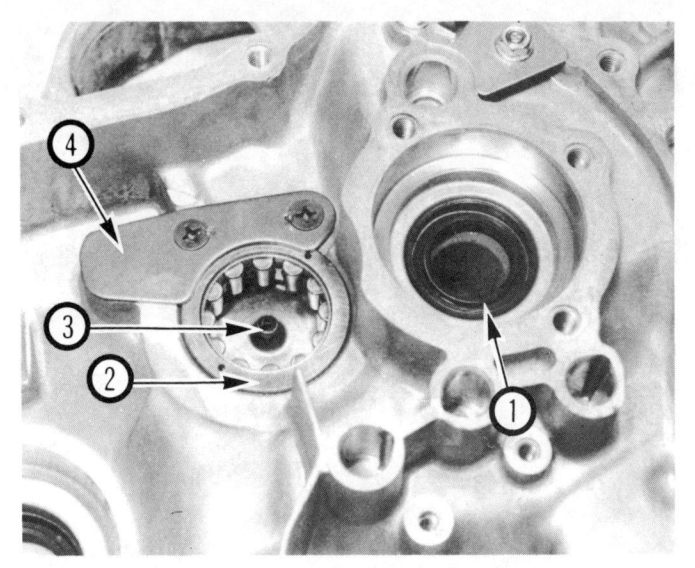

Chapter 6 Manual transmission

7.1 The countershaft bearing and mainshaft oil seal
1 Oil seal
2 Bearing
3 Oil guide plate
4 Bearing retainer

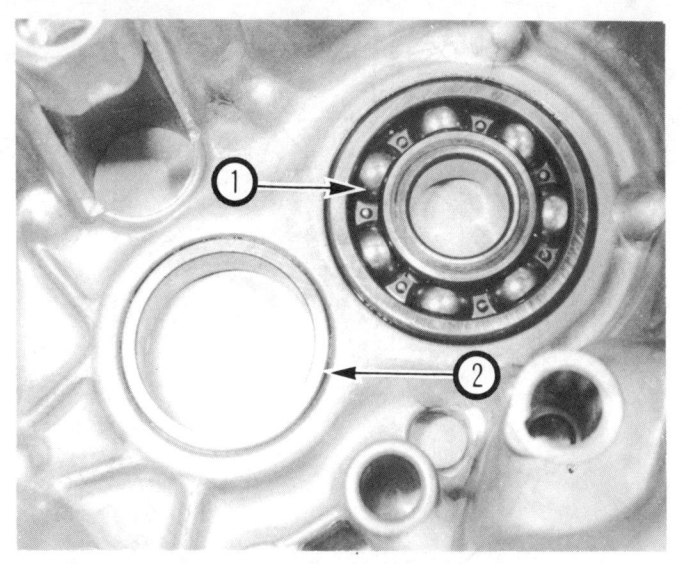

7.12 The countershaft ball-bearing (1) and mainshaft bearing outer race (2)

3 Pull the bearing from the housing, then lift out the plastic oil guide plate.
4 Clean out the bearing housing and the oil guide plate, and wash and dry the bearing.
5 Inspect the bearing for signs of wear, discolouration due to overheating and cracking or scoring of the rollers.
6 Refit the plastic oil guide.
7 To refit the bearing, first liberally coat it in clean engine oil then, with a reaction block positioned under the casing, and using a mandrel of equal diameter to the bearing, drive the bearing fully home into the housing.
8 Refit the bearing retainer plate, and punch lock the screws.
9 Pry out the mainshaft oil seal using a screwdriver, being careful not to damage the housing.
10 Coat a new oil seal with engine oil and, again using a mandrel equal in diameter to the oil seal, gently tap the oil seal into the housing.
11 The sealing lips on the seal face toward the bearing, and be careful on fitting not to tilt the seal.

Transmission housing

12 The photo shows the countershaft ball-bearing and the mainshaft bearing outer race in the transmission casing.
13 Both items are held in place by snap-rings, and are removed by expanding the snap-rings using suitable pliers and lifting out the bearing and/or the outer race.
14 Do not expand the snap-rings any more than is necessary to remove the bearings, and avoid damaging the casing during this operation.
15 On refitting, place the bearings in position, expand the snap-rings so that the bearings can enter the housings, then push the bearings down by hand, feeling for the 'click' that will tell when the snap-ring has snapped into the groove on the bearings. Note that the bearing is fitted with the part number facing out.
16 As a check that the snap-ring is positioned correctly, measure dimension 'A' in Fig. 6.23 and compare it against the tolerances given in the Specifications.
17 If the dimension 'A' is out of tolerance, either reseat the bearing and snap-ring or renew the snap-ring.

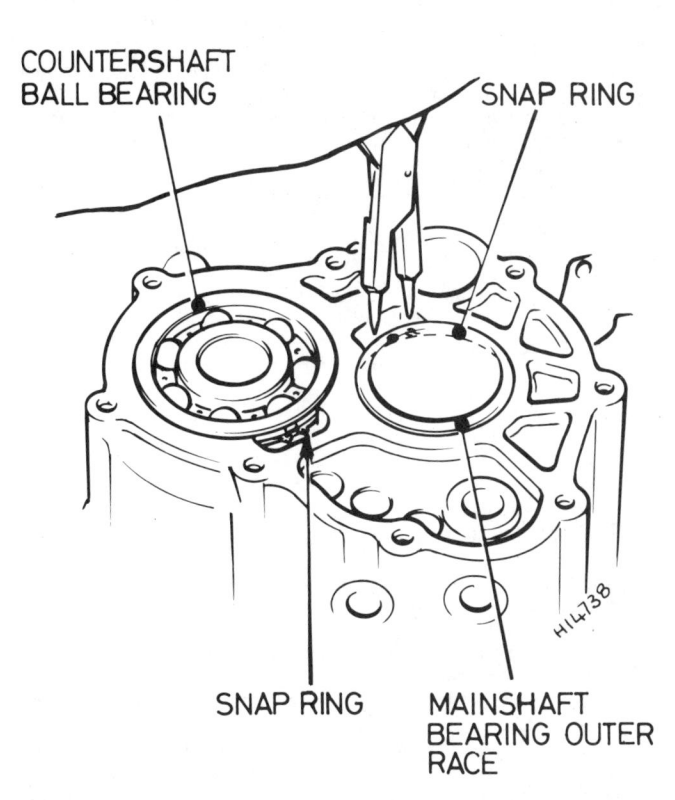

Fig. 6.21 Removing the countershaft ball-bearing and mainshaft bearing outer race snap-rings (Sec 7)

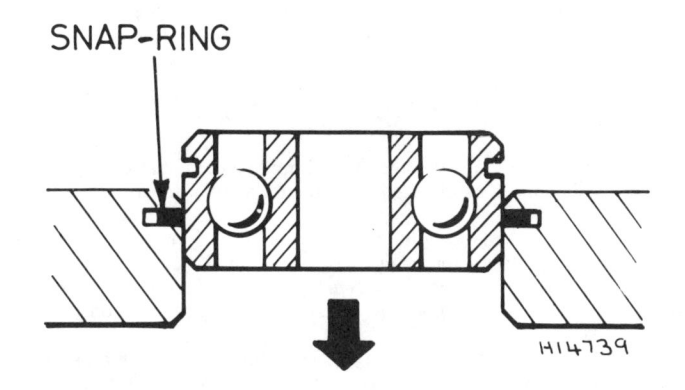

Fig. 6.22 Reinstalling the bearing (Sec 7)

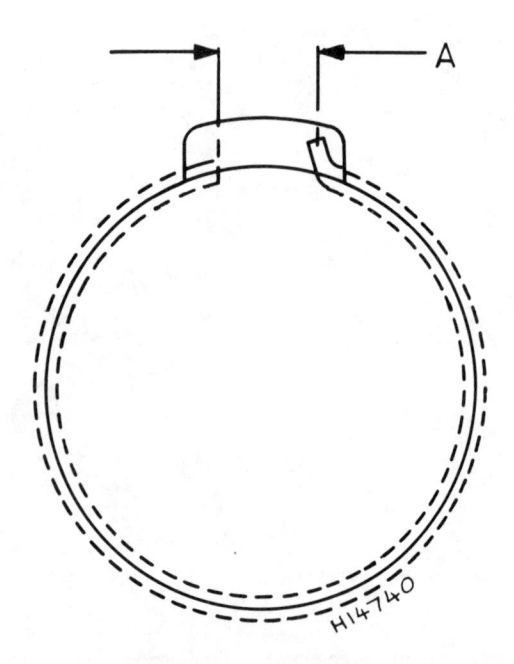

Fig. 6.23 Measure dimension 'A' to ensure the snap-ring is seated correctly (Sec 7)

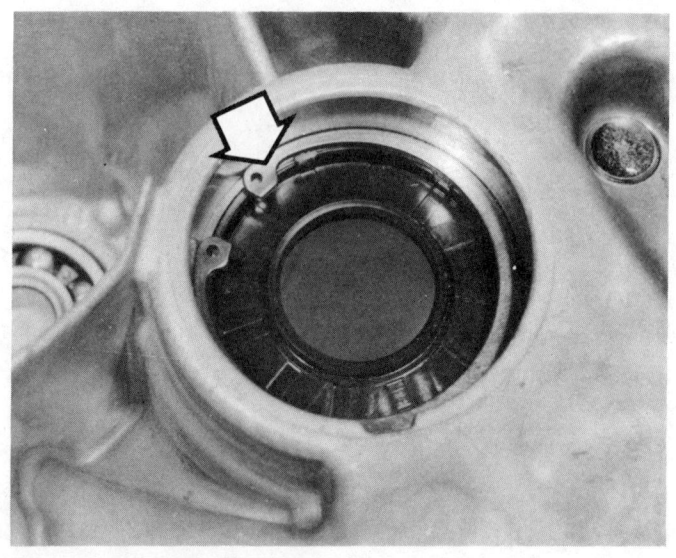

8.4 Transmission housing differential oil seal snap-ring (arrowed)

8 Differential oil seals – removing and refitting

1 To renew the differential oil seals, the transmission components and differential assembly must first be removed from the transmission housings.
2 Both seals should be renewed whenever the transmission assembly is dismantled.
3 To remove the seal in the clutch housing, use a mandrel of suitable diameter and drive the seal from the housing.
4 Similarly, drive out the seal from the transmission housing, but first remove the snap-ring (photo).
5 Refitting is a reversal of this procedure, noting the following.
6 The transmission housing oil seal is fitted with its part number side facing away from the snap-ring.
7 The clutch housing seal is fitted with its part number side facing away from the bearing.
8 If the differential bearings or carrier were renewed, then a snap-ring of suitable thickness to give the correct clearance between the snap-ring and the bearing outer race will have to be selected and fitted (see Section 9).

9 Differential unit – removal, inspection and reassembly

Removal
1 The differential unit will have already been lifted out of the transmission housing, together with the geartrains and gearshift mechanism.
2 Before proceeding any further with dismantling, the backlash in the pinion gears should be checked, with the differential set up as in Fig. 6.26.
3 Backlash can be brought within limits by fitting thicker thrust washers behind the pinion gears.
4 The thrust washers on each side should be of equal thickness. Refer to the Specifications for thrust washer sizes and backlash tolerances.
5 If the backlash cannot be brought within tolerance by the fitting of thicker assemblies, then the unit will have to be completely dismantled and inspected, as described in the following paragraphs. Consideration as to obtaining a reconditioned unit should be given at this point.

6 Inspect the bearings for wear, discolouration due to overheating and roughness during rotation.
7 If the bearings are in reasonable condition and backlash not excessive, then the unit may be refitted. If further dismantling is necessary, proceed as follows.
8 Use a bearing puller to remove the two bearings.
9 The ring gear bolts have **left-handed** threads. Remove these and the ring gear, and inspect the teeth for wear and damage.
10 On vehicles fitted with automatic transmission, pry the snap-ring off the carrier, then remove the speedometer drive gear and dowel pin. **Caution:** the speedometer drivegear has very sharp edges, take care when handling it.
11 Remove the pinion shaft spring pin with a pin punch and hammer, and remove the pinion shaft, pinion gears and thrust washers.
12 Inspect all parts for wear, scoring, overheating, burrs and damage and renew any that are defective.

Reassembly
13 Coat all gears with molybdenum disulphide grease before reassembly.
14 Place the side gears in the differential carrier.
15 Position the pinion gears in place, exactly opposite each other and in mesh with the side gears, then install a thrust washer of selected thickness behind each gear. Remember that these thrust washers must be of equal thickness.
16 Rotate the gears until the shaft holes in the pinion gears line up with those on the carrier.
17 Insert the pinion shaft, lining up the spring pin hole with the hole in the carrier.
18 Fit a new spring pin, using a punch and hammer.
19 Check the backlash of both pinion gears, as described at the beginning of this Section.
20 If the tolerances cannot be met using thicker washers, renew the pinion gears, and if this fails, then renew the side gears.
21 If backlash is still excessive, renew the whole carrier assembly.
22 On automatic models, fit the speedometer drivegear, its chamfer on the inside diameter facing the carrier, and secure it with the snap-ring.
23 The snap-ring should be fitted so that it covers the roll pin securing the pinion shaft, as shown in Fig. 6.28.
24 Fit the ring gear, noting the differences between manual and automatics, as shown in Fig. 6.29, and tighten the bolts, which have **left-handed threads,** to the specified torque.
25 The procedure for refitting the differential will be found in Section 11 of this Chapter for manual transmission and Section 15 of Chapter 7 for automatics.

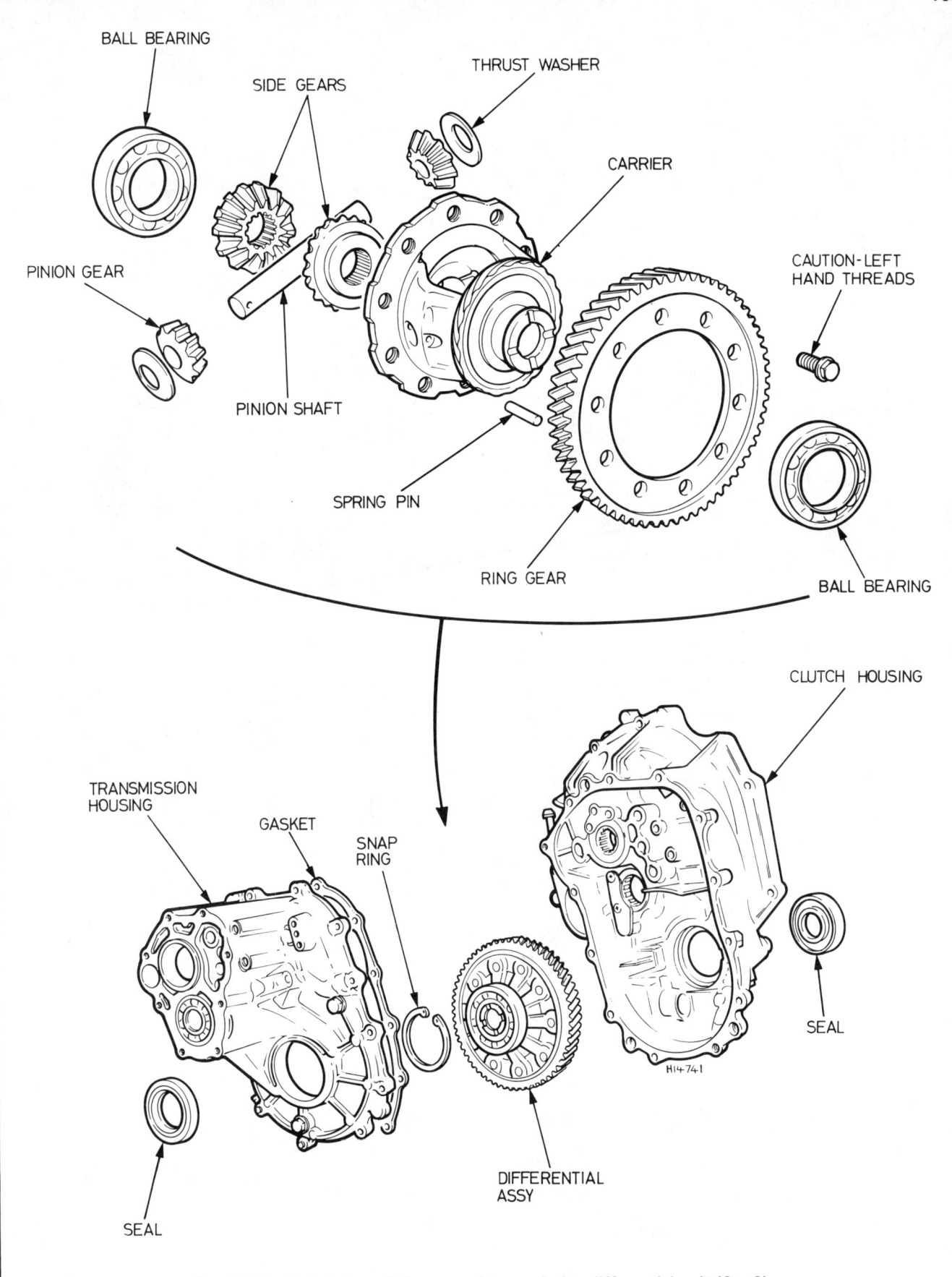

Fig. 6.24 Exploded view of the manual transmission differential unit (Sec 9)

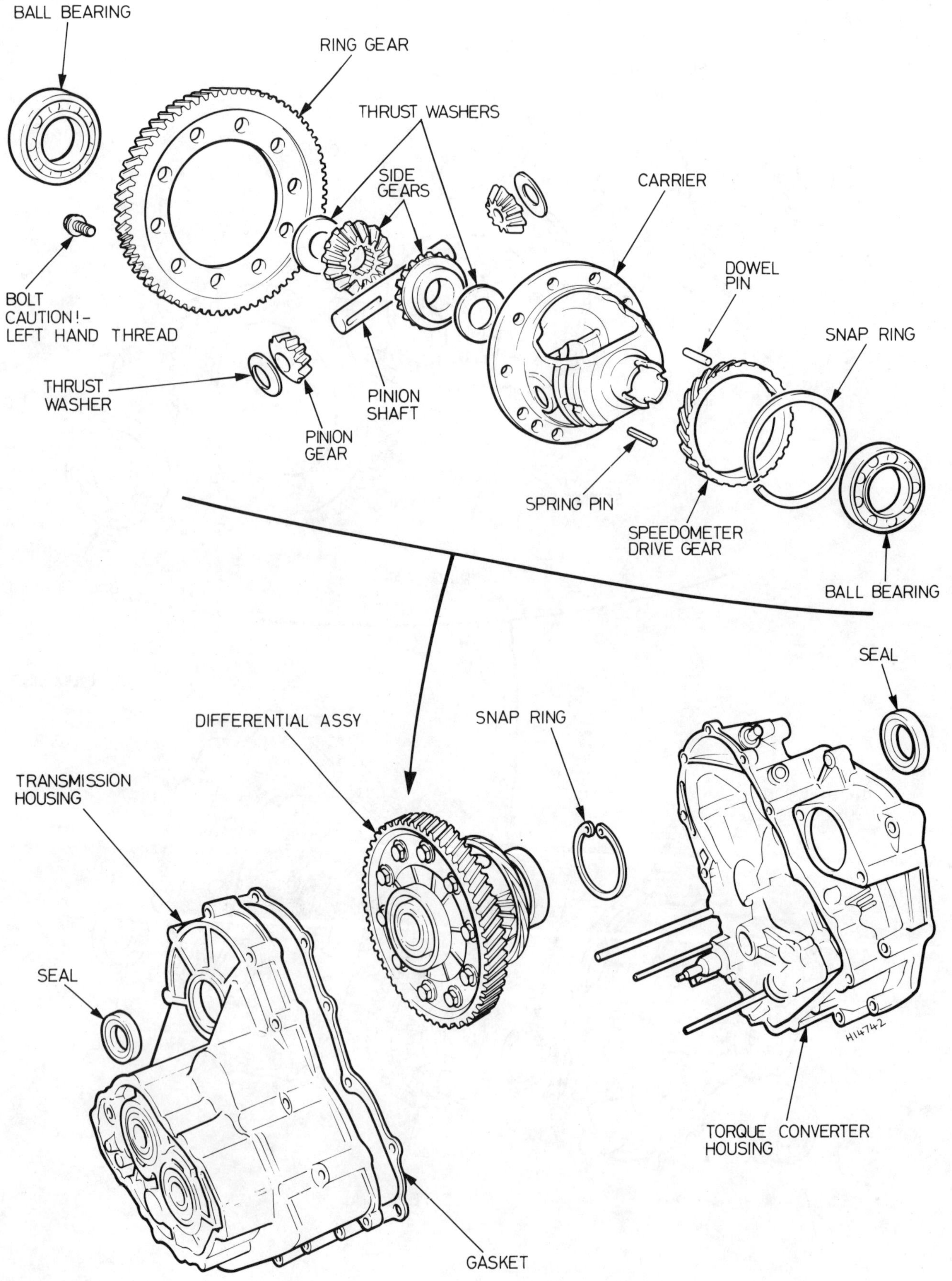

Fig. 6.25 Exploded view of the automatic transmission differential unit (Sec 9)

Chapter 6 Manual transmission

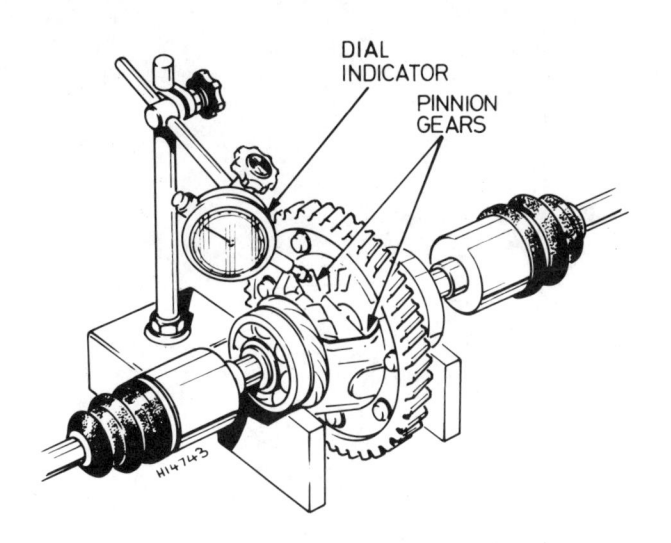

165

Fig. 6.26 Setting up the differential for backlash measurement (Sec 9)

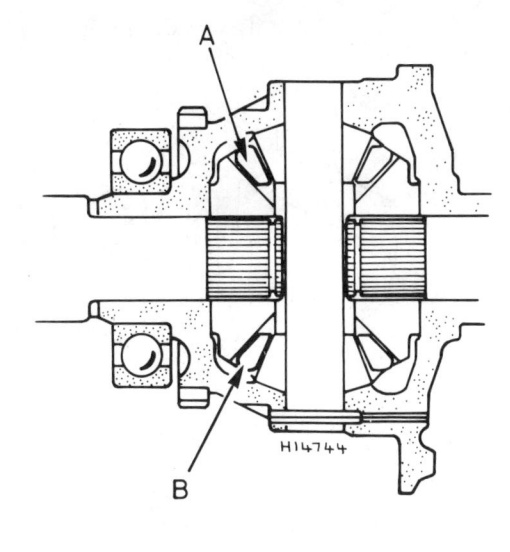

Fig. 6.27 The differential backlash should be checked at both A and B (Sec 9)

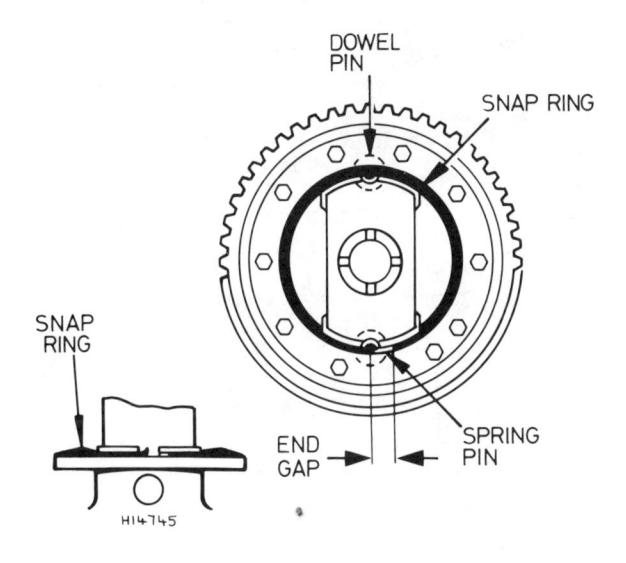

Fig. 6.28 Correct alignment of snap-ring and roll pin (Sec 9)

10 Gearshift selector mechanism – removal, inspection and refitting

1 The shift arm holder index will still be bolted in place in the transmission housing.
2 Before removing it, measure the clearance between the end collar and the shim. If it is out of tolerance, fit a thicker shim. Refer to the Specifications for details.
3 To remove the selector arm from the holder for shimming, use a pin punch to drive out the spring pin.
4 Use a new spring pin on reassembly.
5 The shift arm-to-shift guide clearance should be measured, and if found to be out of limits, measure the width of the slot in the shift guide.
6 If the slot is wider than the standard measurement given, renew the shift guide.
7 Check the selector arm-to-interlock clearance and if out of limits, measure the gap between the selector arm fingers.
8 If tolerances cannot be met, renew the selector arm.
9 If further dismantling is required, this can be done after the main and countershaft assemblies have been removed.
10 The shift arm holder may be removed after undoing the three bolts which secure it in place.

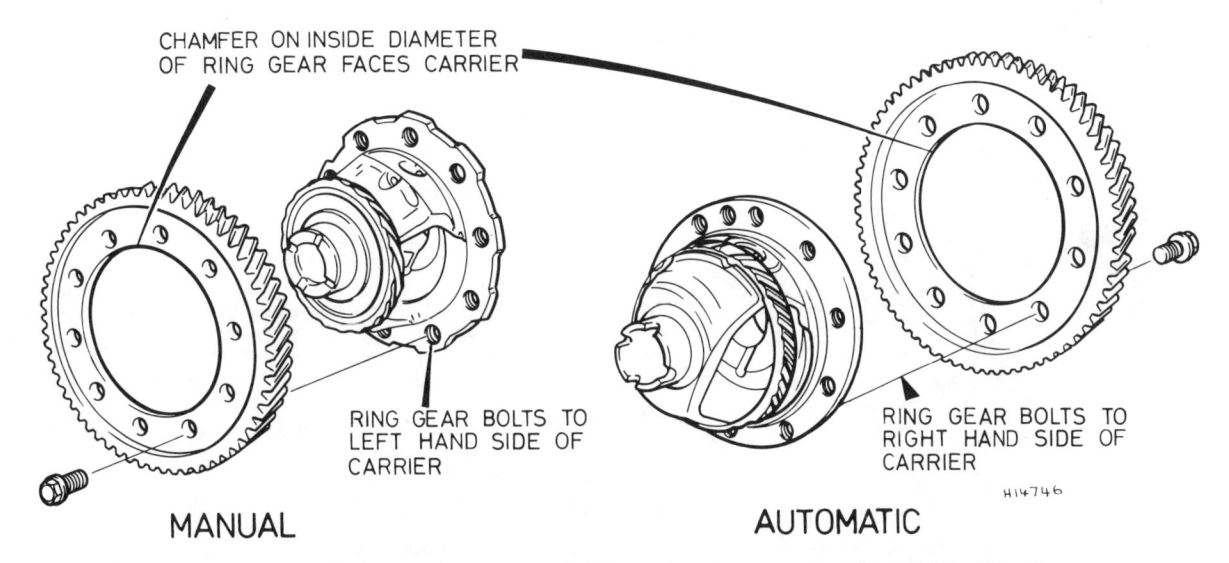

Fig. 6.29 The differences between the manual and automatic differentials (Sec 9)

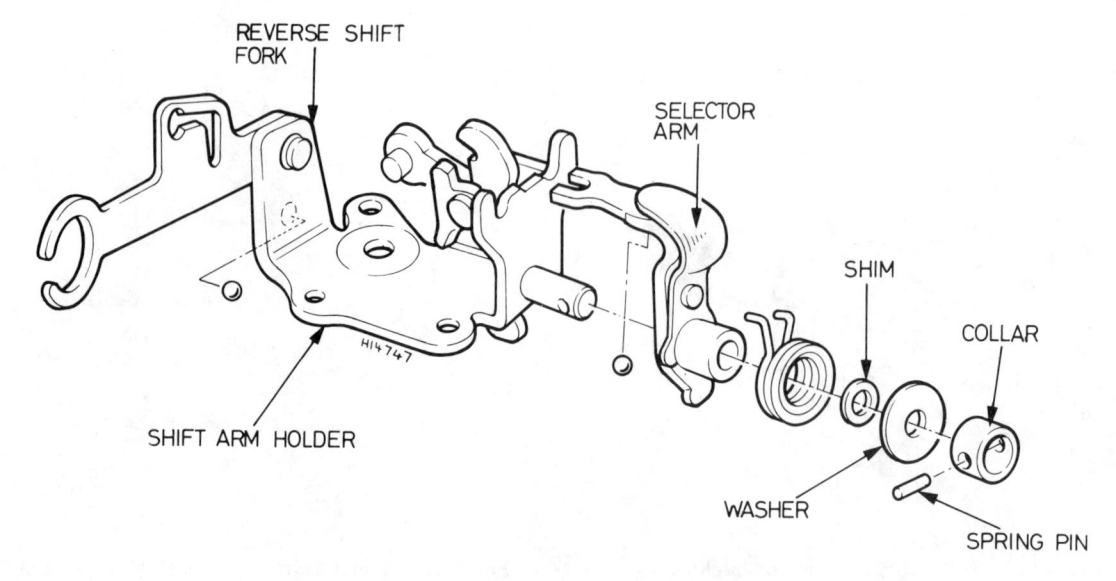

Fig. 6.30 Exploded view of the shift selector mechanism (Sec 10)

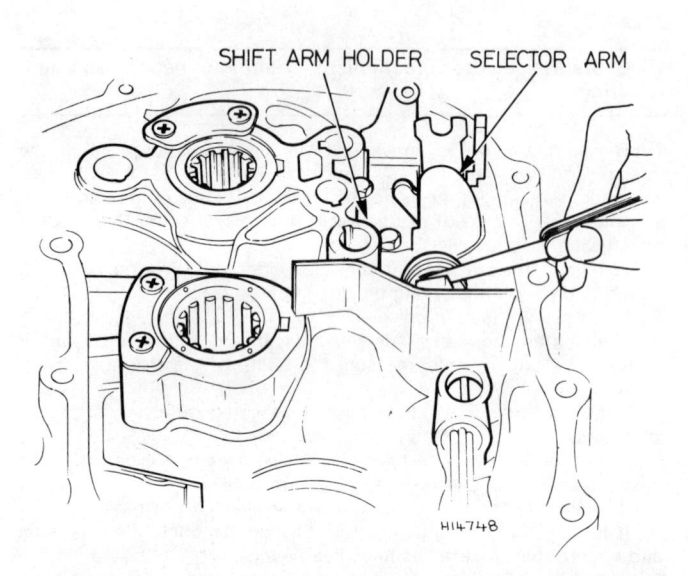

Fig. 6.31 Measuring shift arm holder-to-shim clearance (Sec 10)

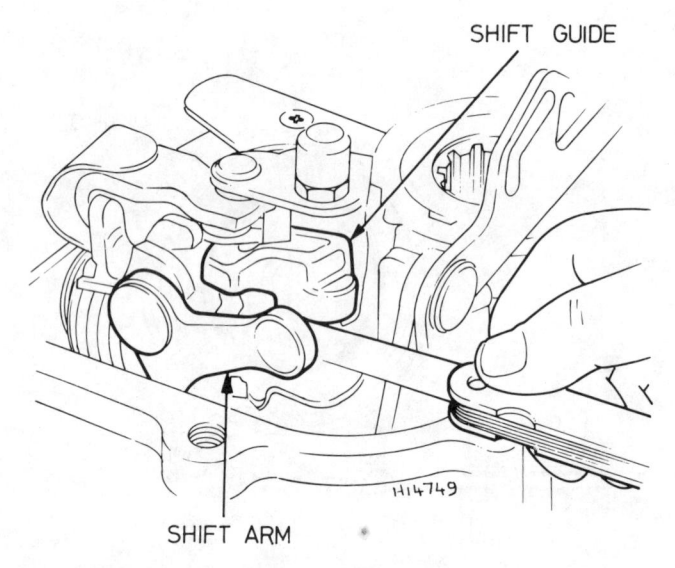

Fig. 6.32 Measuring shift arm-to-guide clearance (Sec 10)

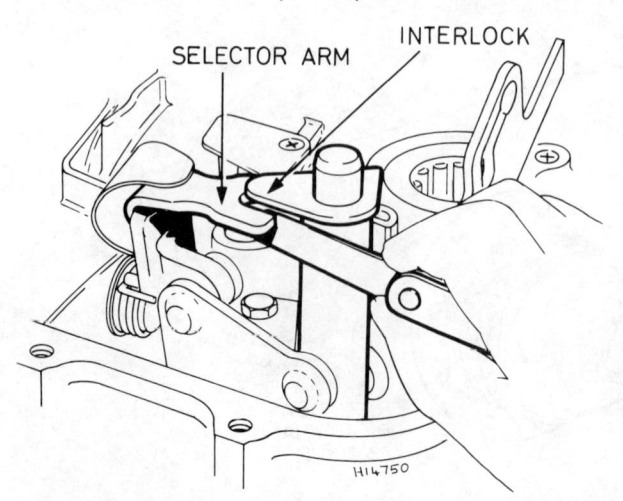

Fig. 6.33 Measuring selector arm-to-interlock clearance (Sec 10)

11 Inspect the shift rod for damage.
12 To remove the shift rod for further inspection or to renew the shift rod oil seal and rubber bellows, undo the shoulder bolt.
13 This will release the shift rod guide, and the rod may be pulled from the housing.
14 Watch for the detent ball and spring as the shift rod clears the detent hole.
15 Prise out the shift rod oil seal and press in a new oil seal.
16 Remove the old rubber bellows from the shift rod, lubricate the inside of the new bellows with grease, and fit it to the shift rod.
17 Remove and clean the magnetic filter and its holder, refitting them on completion (photos).
18 Refit the shift rod into the housing (photo).
19 Replace the detent spring into its housing, followed by the steel ball (photos).
20 Depress the ball and spring and push the shift rod right through (photos).
21 Fit the shift rod guide and shoulder bolt and snap the rubber bellows over the oil seal (photos).
22 Check the clearance between the shift arm and shift rod guide, if out of limits, measure the width of the slot in the shift rod guide.
23 If this is worn beyond limits, renew the shift rod guide.

10.17A Refitting the magnetic filter ...

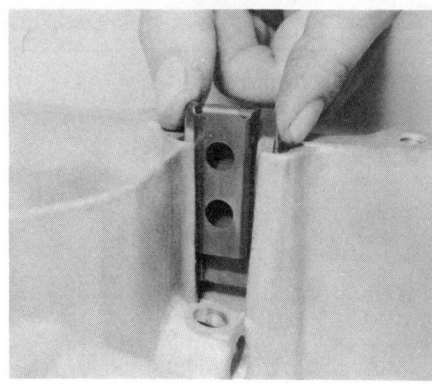

10.17B ... and its holder

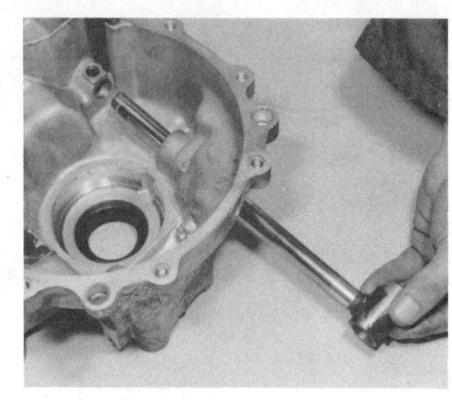

10.18 Fitting the shift rod

10.19A Fit the spring ...

10.19B ... and steel ball

10.20A Depressing the ball and spring ...

10.20B ... to push the rod through

10.21A Fit the shift rod guide and bolt ...

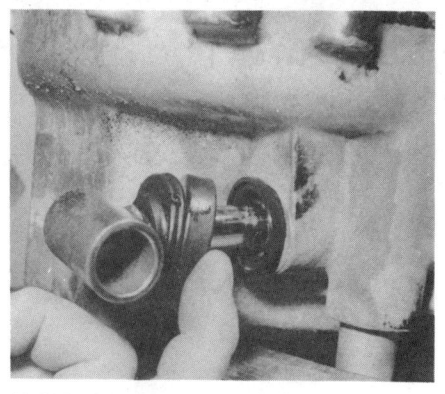

10.21B ... and snap the bellows ...

10.21C ... over the oil seal

10.26A Fitting the shift arm holder assembly ...

10.26B ... and the three bolts (arrowed)

Chapter 6 Manual transmission

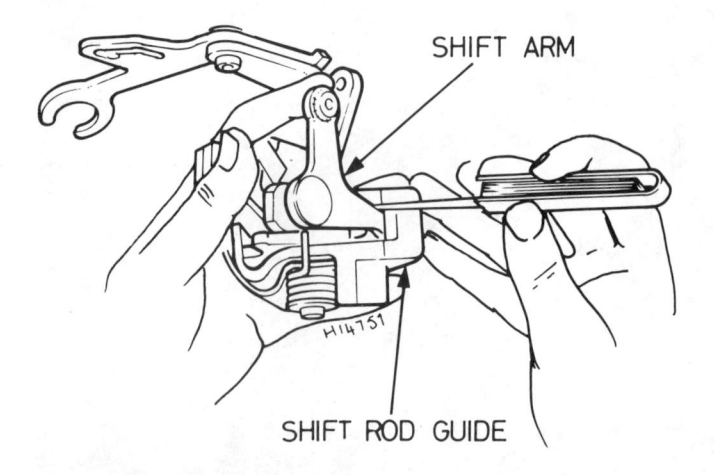

Fig. 6.34 Measuring the clearance between the shift arm and shift rod guide (Sec 10)

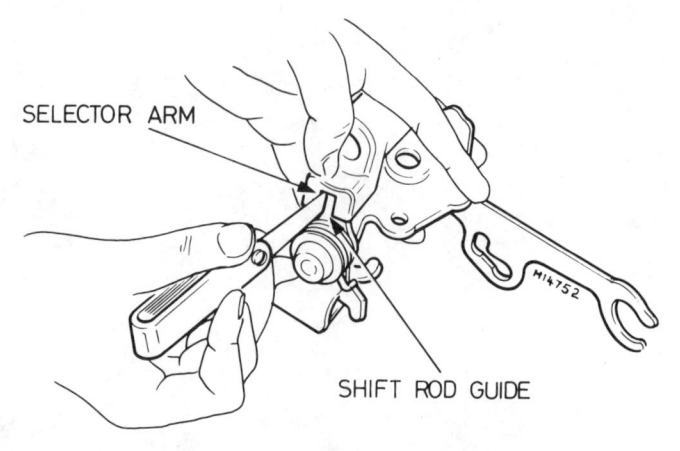

Fig. 6.35 Measuring selector arm-to-shift rod guide clearance (Sec 10)

24 Check the selector arm-to-shift rod guide clearance, and if outside the limit, measure the width of the tab on the selector arm.
25 If this is worn beyond limits, replace the arm.
26 Refit the shift arm holder assembly and tighten the three bolts (photos).
27 The remainder of the selector mechanism is fitted during reassembly of the transmission unit.

11 Transmission – reassembly

1 Ensure that the countershaft ball-bearing and mainshaft bearing outer race are fitted correctly to the transmission housing.
2 Ensure the countershaft roller bearing and mainshaft oil seal are fitted correctly to the clutch housing.
3 Check that the shift rod is fitted correctly, the magnetic oil filter is fitted in place and that the oil smeared pipe screen is in position.
4 If not all ready done, fit the shift arm holder.
5 Finally check the differential oil seals are fitted correctly. Reassembly may now commence.
6 Fit the differential assembly into the clutch housing (photos).
7 Fit the main and countershaft assemblies, meshing them together as they go into position (photos).
8 Fit the first and second gearshift shaft, and the third and fourth gearshift shaft (photo).

9 The main and countershaft assemblies may have to be lifted slightly to allow the shift shafts to be positioned.
10 It will also be found helpful if the shift forks are placed in second and fourth gear.
11 Lift the mainshaft and install the interlock into the selector arm (photo).
12 Return the shift rod to neutral then hook the interlock into the selected arm, first and second gearshift shaft and third and fourth gearshift shaft.
13 Hook the shift guide into the shift arm, then install fifth and reverse shift shaft, making sure its pin locates in the reverse shift fork slot (photo).
14 Finally, install the shift guide shaft so that it bottoms securely in its housing in the clutch housing (photo).
15 The end of the shift guide shaft, if it is fitted correctly, should not extend more than 0.5 in (12 mm) above the interlock (photo).
16 If it does, check the complete installation for correct assembly.
17 Fit the mainshaft bearing retainer plate (photo).
18 Fit the reverse idler gear and shaft (photo).
19 Fit the reverse light switch, using a new washer (photo).
20 Place a new gasket in position on the clutch housing and ensure the dowel pins are fitted.
21 Place the transmission in third gear to position the shift guide shaft correctly for reassembly, then fit the transmission housing, making sure the main and countershafts line up with the bearings, and that the housing locates over the dowel pins (photo).

11.6A The differential assembly ...

11.6B ... being fitted into the transmission housing

11.7A Fitting the main and countershaft assemblies ...

11.7B ... into the transmission housing

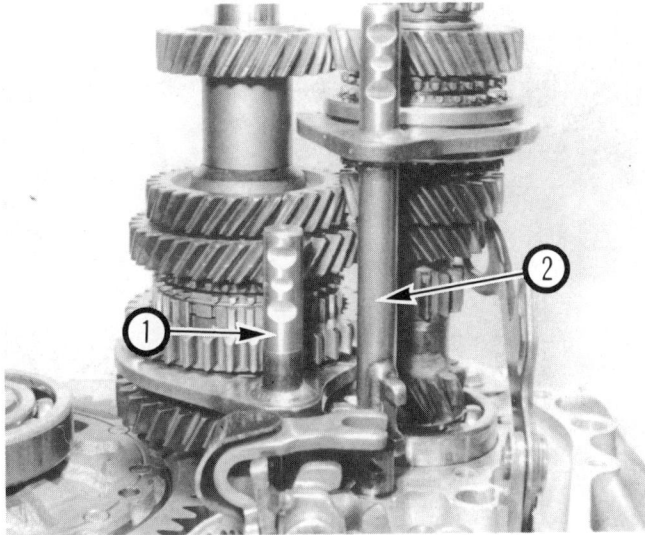

11.8 Fitting the first and second gearshift shaft (1) and the third and fourth gearshift shaft (2)

11.11 Installing the selector mechanism
1 Interlock
2 Selector arm
3 Shift guide
4 Shift arm
5 Fifth/reverse shift shaft

11.13 Fifth/reverse shift shaft pin location in the reverse shift fork slot (arrowed)

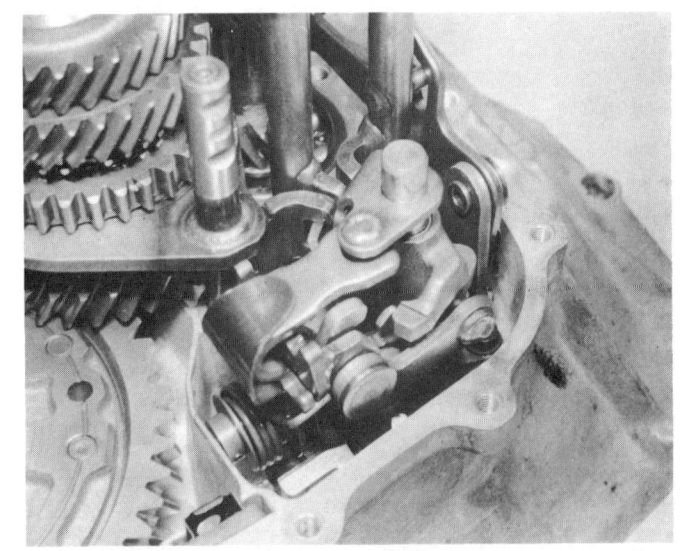

11.14 General view of the selector mechanism in position

11.15 Measuring the extension of the shift guide shaft

11.17 Fitting the mainshaft bearing retainer plate (arrowed)

11.18 The reverse idler gear and shaft in position

11.19 Fitting the reverse light switch

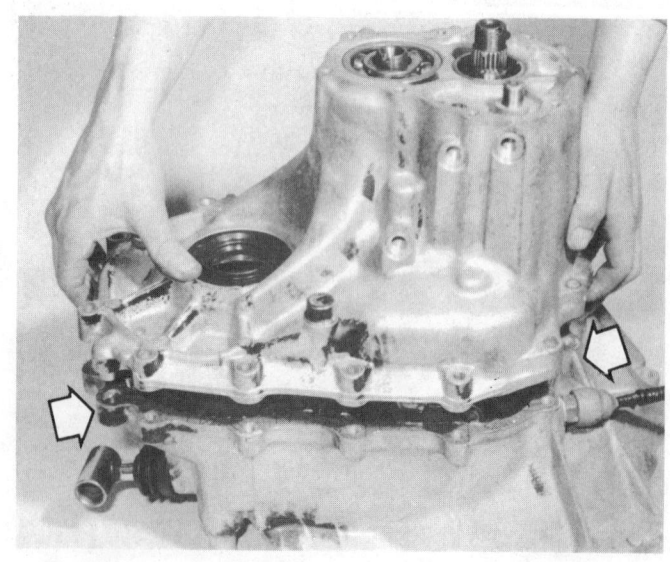

11.21 Fitting the transmission housing (arrows show location of dowel pins)

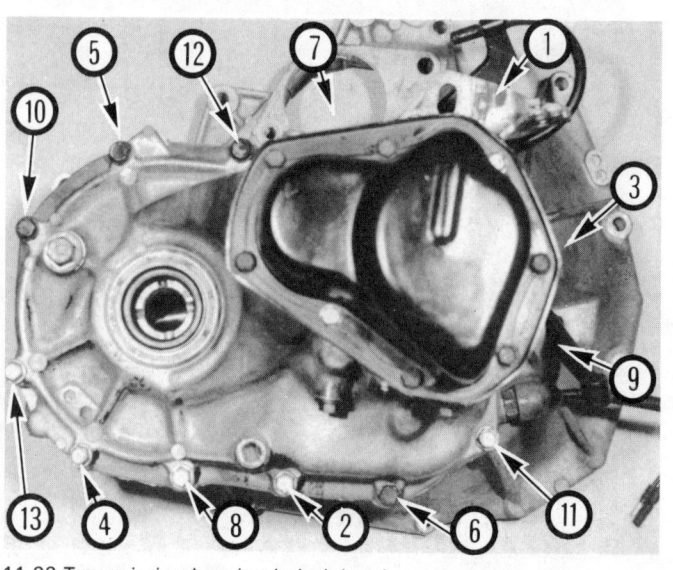

11.22 Transmission housing bolt tightening sequence

Chapter 6 Manual transmission

22 Tighten the retaining bolts to the specified torque, in the sequence shown (photo).
23 Install the three detent balls, springs, washers and plugs, and load them to the specified torque (photos).
24 Fit the countershaft fifth gear high side facing down (photo).
25 Fit the spacer collar and needle roller bearing over the mainshaft (photos).
26 Install the mainshaft fifth gear, synchro ring, spring, hub and sleeve onto the mainshaft, at the same time fitting the shift fork to the sleeve and over the shaft (photos).
27 Fit spring washers to both countershaft and mainshaft, concave side facing down, and fit the retaining nuts (photo).
28 Fit a new spring pin to the fifth gear shift fork, and drive it home (photo).
29 Lock the transmission, then tighten the locknuts on the counter and mainshafts to the specified torques, slacken the nuts off, and then retighten to the same torque (photo).
30 Stake the locking collar of the nuts into the slots on the shafts (photo).
31 Fit a new end cover gasket, fit the end cover and tighten the retaining nuts to the specified torque (photos).
32 Install the transmission assembly, as described in Section 3.

12 Gearchange lever, shift rod and torque rod – removal and refitting

1 Remove the bolts from the front and rear gearshift rod forks and remove the rod (photos).
2 Remove the nylon or rubber bushes and inspect them for hardening or cracking, renewing as necessary (photo).
3 Remove the bolt from the torque arm front mounting (photos).

11.23A Fitting a steel ball to the detent ...

11.23B ... and the spring and plug

11.24 Fitting the countershaft fifth gear

11.25A Spacer collar ...

11.25B ... and needle roller bearing being fitted to the mainshaft

11.26A Fitting the fifth gear synchro components

11.26B ... and the shift fork

11.27 Main and countershaft retaining nuts in position

11.28 Fitting a new spring pin to the fifth gear shift fork

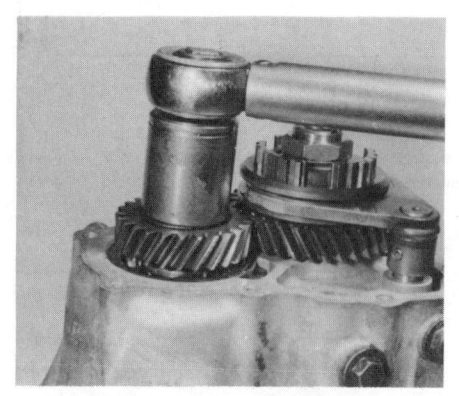

11.29 Tightening the main and countershaft locknuts

11.30 Staking the locking collar

11.31A The end cover gasket ...

11.31B ... and end cover ...

11.31C ... bolted in position

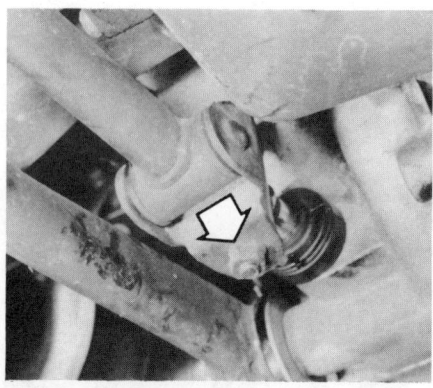

12.1A Gearshift lever front fork bolt ...

12.1B ... and the rear fork bolt

12.2 Inspect the bushes

12.3A Torque arm front mounting bolt ...

12.3B ... and washer removed ...

12.3C ... together with the torque arm

Chapter 6 Manual transmission 173

4 From inside the car, remove the gear lever shift knob and the rubber boot.
5 Remove the four screws from the ball housing retaining plate, and lift out the gearshift lever.
6 Inspect the components of the gearshift lever balljoint assembly, and renew any which are worn or show signs of deterioration.
7 The torque rod can be removed by removing the bolts from the rear support housing.
8 Reassembly is a reversal of removal, but lubricate all bushes and joints with general purpose grease.

13 Reversing light switch – testing

1 Test the switch by selecting reverse gear and switching on the ignition.
2 If the lights do not come on, check the fuse.
3 If the fuse is working, then remove the reversing light switch and check its continuity when the plunger is depressed.
4 If there is no continuity, renew the reversing light switch.

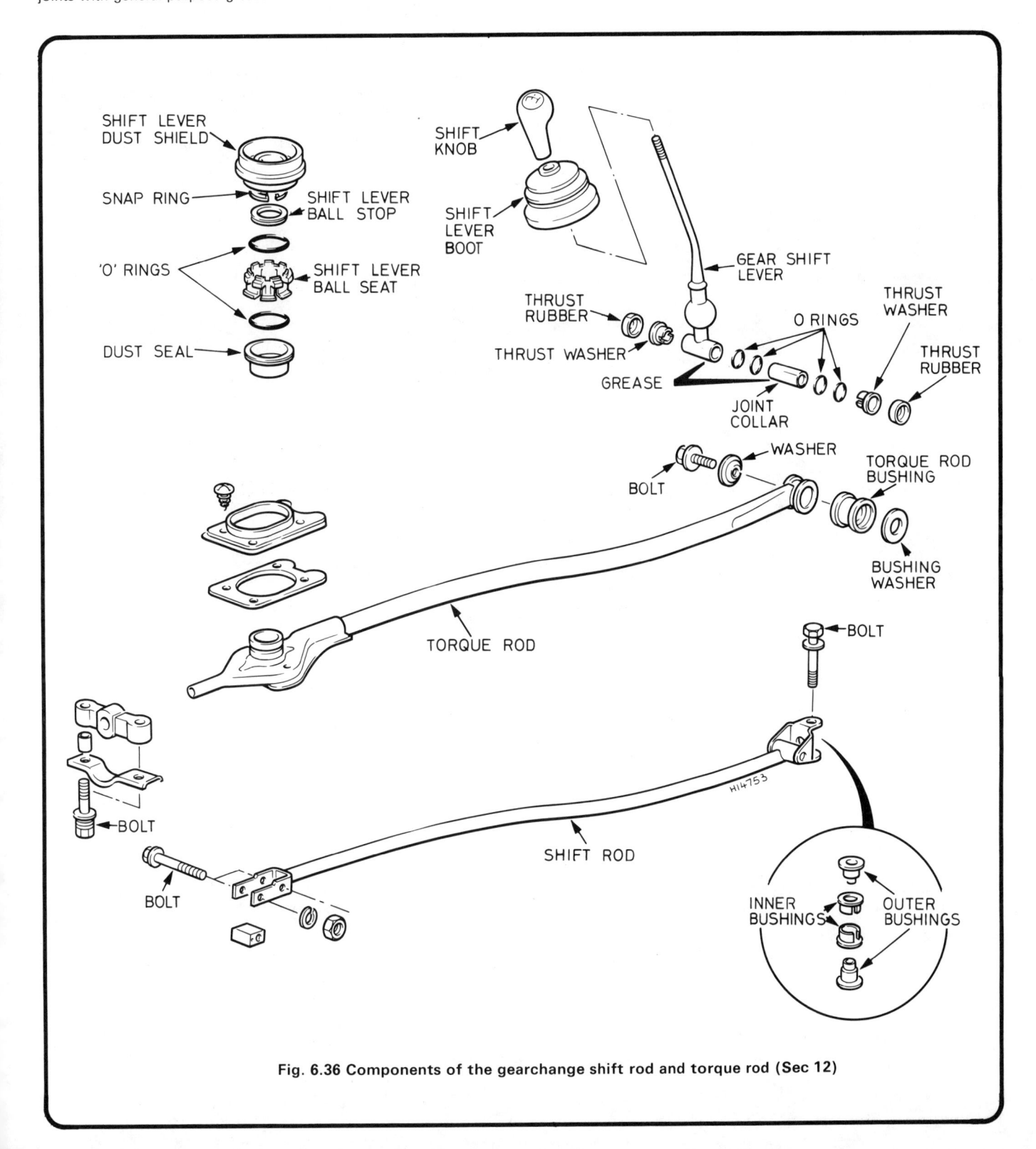

Fig. 6.36 Components of the gearchange shift rod and torque rod (Sec 12)

14 Fault diagnosis – manual transmission

Symptom	Reason(s)
Ineffective synchromesh	Worn baulk rings or synchro-hubs
Jumps out of one or more gears (on drive or over-run)	Weak detent springs or worn selector forks or worn gears
Noisy, rough, whining and vibration	Worn bearing and/or thrust washers (initially) resulting in extended wear generally due to play and backlash
Noisy and difficult engagement of gears	Clutch fault (see Chapter 5)

Note: *It is sometimes difficult to decide whether it is worthwhile removing and dismantling the gearbox for a fault which may be nothing more than a minor irritant. Gearboxes which howl, or where the synchromesh can be 'beaten' by a quick gearchange, may continue to perform for a long time in this state. A worn gearbox usually needs a complete rebuild to eliminate noise because the various gears, if re-aligned on new bearings will continue to howl when different wearing surfaces are presented to each other.*

The decision to overhaul, therefore, must be considered with regard to time and money available, relative to the degree of noise or malfunction that the driver has to suffer.

Chapter 7 Automatic transmission

Contents

Automatic transmission – removal and refitting	4
Differential oil seals – renewal	14
Fault diagnosis – automatic transmission	19
Fluid level – checking, topping-up and renewing	3
Gearshift indicator – adjustment and checks	17
Gearshift selector – removal and refitting	16
General description	1
Governor – removal and refitting	7
Mainshaft and countershaft – gear clearance measurements	12
Mainshaft and countershaft – removal	6
Mainshaft and countershaft bearings and seals – removal and refitting	10
Main valve body – removal	8
Major assemblies – inspection and overhaul	9
Reverse idler gear – renewal	11
Routine maintenance	2
Shift cable – adjustment	18
Torque converter – removal and refitting	13
Transmission unit – dismantling	5
Transmission unit – reassembly	15

Specifications

Type .... 3-element torque converter, dual shaft, fully automatic with 4 forward and one reverse gear. Lock-up overdrive on 4th gear

Ratios
UK models:
  1600 cc:
    1st .... 2.380 : 1
    2nd .... 1.560 : 1
    3rd .... 1.032 : 1
    4th .... 0.729 : 1
    Reverse .... 1.954 : 1
  1800 cc:
    1st .... 2.380 : 1
    2nd .... 1.560 : 1
    3rd .... 0.969 : 1
    4th .... 0.729 : 1
    Reverse .... 1.954 : 1
North American models:
  1st .... 2.380 : 1
  2nd .... 1.560 : 1
  3rd .... 1.032 : 1
  4th .... 0.667 : 1*
  Reverse .... 1.954 : 1
*California and Canadian 4th gear ratio = 0.729 : 1
Final drive ratio .... 3.875 : 1

Fluid capacity
On reassembly .... 9.8 Imp pint (5.9 US qt, 5.6 litre)
At fluid change .... 5.0 Imp pint (3.0 US qt, 2.8 litre)

Accumulator springs

Outside diameter:
- 1st .... 0.79 in (20.0 mm)
- 2nd .... 0.79 in (20.0 mm)
- 3rd .... 0.81 in (20.6 mm)
- 4th .... 0.70 mm (17.9 mm)

Spring free length:
- 1st:
  - Standard .... 1.96 in (49.8 mm)
- 2nd:
  - Standard .... 3.28 in (83.4 mm)
  - Limit .... 3.20 in (81.4 mm)
- 3rd:
  - Standard .... 4.21 in (107.0 mm)
  - Limit .... 4.13 in (105.0 mm)
- 4th:
  - Standard .... 3.72 in (94.6 mm)
  - Limit .... 3.65 in (92.6 mm)

Main valve springs

Free length:
- Orifice control spring .... 1.51 (38.3 mm)
- Torque converter check valve spring .... 1.43 in (36.4 mm)
- Relief valve spring .... 1.88 in (47.7 mm)
- 1st-to-2nd shift spring .... 1.50 in (38.1 mm)
- 2nd-to-3rd shift spring .... 2.56 in (65.0 mm)
- 3rd-to-4th shift spring .... 1.30 in (33.1 mm)

Oil pump

- Drive and driven gear axial clearance (unit) .... 0.003 in (0.08 mm)
- Drive gear side clearance .... 0.008 to 0.010 in (0.21 to 0.27 mm)
- Driven gear side clearance .... 0.002 to 0.004 in (0.05 to 0.09 mm)

Regulator valve

- Outer spring free length .... 3.30 in (83.9 mm)
- Inner spring free length .... 1.73 in (44.0 mm)
- Stator reaction spring free length .... 1.19 in (30.3 mm)

Servo valve

- Return spring free length (limit) .... 1.44 in (36.7 mm)

Countershaft clearances

- 4th gear-to-selector hub .... 0.003 to 0.006 in (0.07 to 0.15 mm)
- Replacement spacer collars .... Available in increments from 1.534 in (38.97 mm) to 1.547 in (39.30 mm)
- 2nd gear clearance .... 0.003 to 0.006 in (0.07 to 0.15 mm)
- Replacement splined thrust washers .... Available in increments from 0.117 in (2.97 mm) to 0.134 in (3.40 mm)

Mainshaft clearances

- 2nd gear clearance .... 0.003 to 0.006 in (0.07 to 0.15 mm)
- Replacement thrust washers .... Available in increments from 0.117 in (2.97 mm) to 0.134 in (3.40 mm)

Clutch endplate-to-top disc clearance

- Low .... 0.016 to 0.028 in (0.40 to 0.70 mm)
- 2nd .... 0.026 to 0.031 in (0.65 to 0.80 mm)
- 3rd .... 0.016 to 0.023 in (0.40 to 0.60 mm)
- 4th .... 0.016 to 0.023 in (0.40 to 0.60 mm)
- Replacement clutch end plates .... Available in increments from 0.094 in (2.4 mm) to 0.123 in (3.3 mm)

Torque wrench settings

	lbf ft	Nm
Main valve body shift valve cover bolts	6	8
Pressure control valve lockbolt	12	16
Servo valve throttle control valve 'B' lockbolt	9	12
Governor mounting bolts	9	12
1st accumulator mounting bolts	9	12
1st accumulator stop bolt	12	16
Control lever lockbolt	9	12
Regulator valve mounting bolts	9	12
Main valve body mounting bolts	9	12
Servo valve mounting bolts	9	12
Clutch pressure control valve mounting bolts	9	12
2nd/3rd accumulator cover bolts	9	12
4th accumulator bolts	9	12

Chapter 7 Automatic transmission

Torque wrench settings (continued)

	lbf ft	Nm
Reverse shift fork lockbolt	10	14
Transmission housing-to-torque converter housing bolts (in several stages)	20	27
Parking lever lock bolt	10	14
Throttle control lever bolt	6	8
Countershaft locknut	70	95
Mainshaft locknut *(left-hand thread)*	70	95
End cover bolts	9	12
Banjo bolt	19	26
Cooler end fitting	19	26
Driveplate-to-torque converter bolts	9	12
Torque converter to crankshaft	54	73
Fluid drain plug	29	39

1 General description

General

The Honda automatic transmission unit is a combination of a 3-element torque converter and a dual shaft automatic transmission which provides 4 forward speeds and 1 reverse, with overdrive on 4th gear (lock-up 4th).

The torque converter is bolted to the driveplate, which in turn is splined to the engine crankshaft, turning together as one unit as the engine turns. The starter ring gear is located around the periphery of the torque converter unit, into which the starter motor pinion meshes, while the engine is being started. The whole assembly serves as the flywheel, while transmitting power to the transmission mainshaft.

The transmission assembly consists of two parallel shafts called the mainshaft and countershaft, the mainshaft being in line with the engine crankshaft.

The gears on the mainshaft are in constant mesh with those on the countershaft, and under certain combinations of gear selection, power is transmitted to the countershaft.

The hydraulic control system consists of an oil pump, regulator valve and manual valve (selector valve).

The regulator valve controls the fluid pressure within the system. Fluid from the regulator passes to the various clutches via oil passageways, and to the servo valve which operates the shift forks.

The manual valve (or shift selector) has six positions – Park, Reverse, Neutral, D4, D3, and 2. Starting is only possible in Park or Neutral.

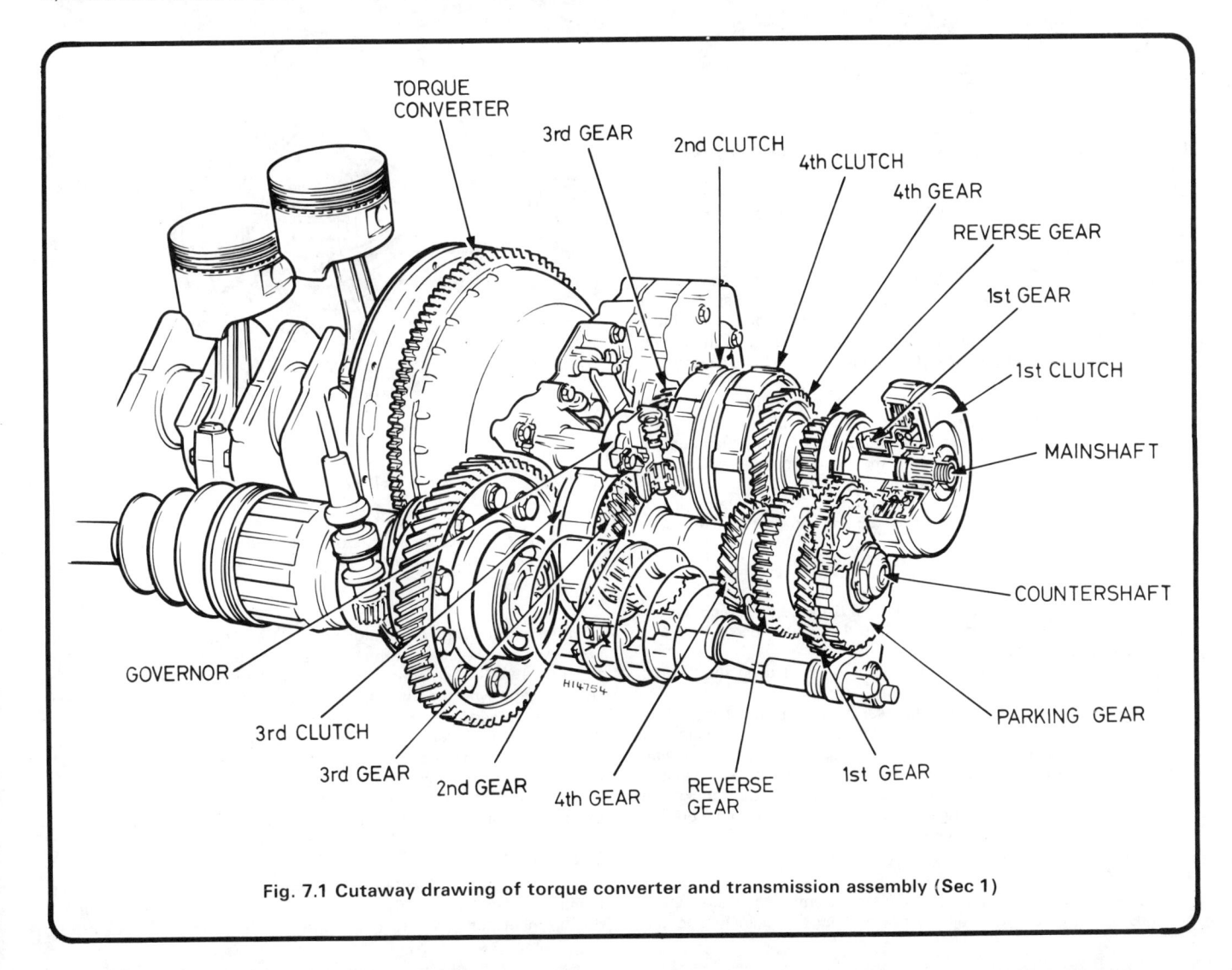

Fig. 7.1 Cutaway drawing of torque converter and transmission assembly (Sec 1)

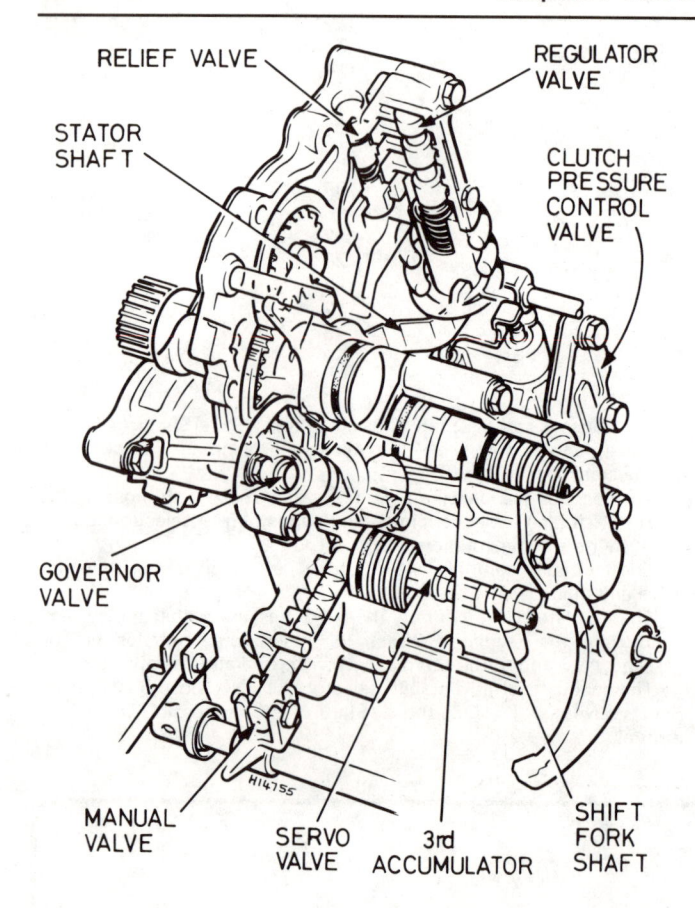

Fig. 7.2 Cutaway view of hydraulic control assembly (Sec 1)

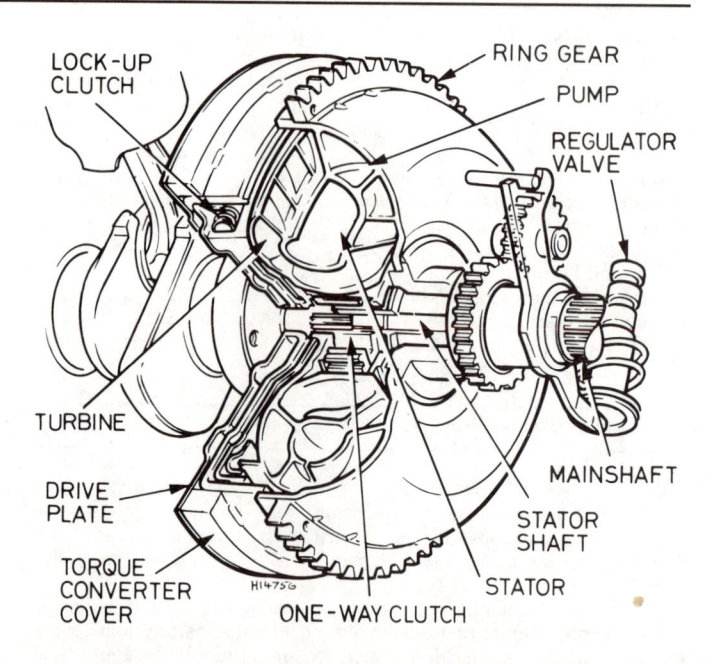

Fig. 7.3 Cutaway view of torque converter (Sec 1)

A position indicator situated in the instrument panel indicates which gear position has been selected.

Torque converter

The design characteristics of the torque converter are such that it can increase the torque produced by the engine. The converter is filled with oil and, as the engine turns, the oil strikes the vanes of the driven member (or turbine), which produces the 'push' which causes the driven member to rotate.

On the driving member (or stator), there is a series of stationary curved vanes which reverse the direction of the oil as it leaves the driven member, causing the oil to pass through the driving member in a 'helping' direction before re-entering the driven member, so adding more torque. The stator rides on a one-way clutch, and as the driven member approaches pump speed, the clutch moves the stator out of the way of the oil. The convertor then acts as a straightforward fluid coupling.

Transmission

The transmission assembly consists of a 3-phase torque converter, as previously described, and a dual shaft, 4 forward speeds and one reverse speed automatic transmission.

It changes gears, up or down, depending on roadspeed and throttle position.

In the D4 drive range, automatic gear changing occurs from first gear through to fourth, as conditions dictate. On D3, gear changing will occur only up to third gear, and in 2 only up to second gear. Reverse gear speaks for itself. Whilst if P (Parking) is selected, the parking gear on the countershaft is locked, which in turn, locks the front wheels. N (Neutral) disengages the transmission from the engine and allows the vehicle to be pushed or towed.

1st clutch

The 1st clutch is mounted on the right-hand end of the mainshaft. It engages in D3 or D4 transmitting power from the mainshaft to the

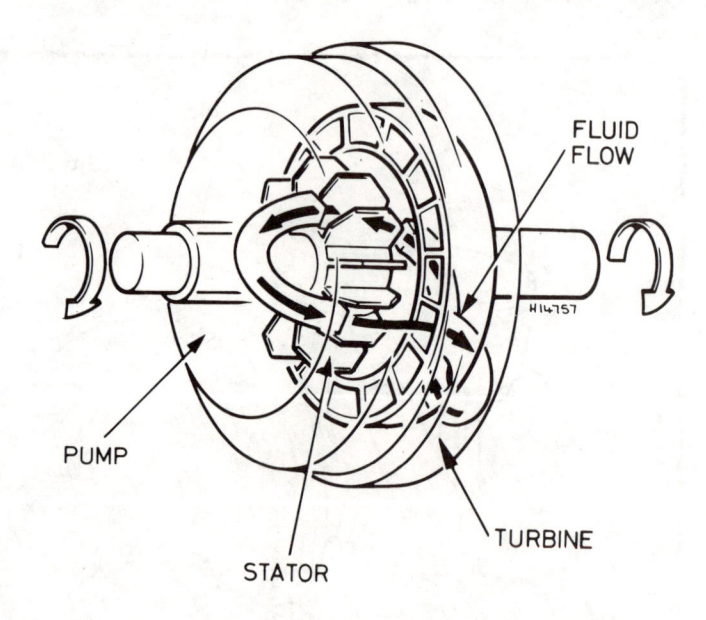

Fig. 7.4 Fluid flow through torque convertor (Sec 1)

countershaft low gear, giving extra torque for better pulling and acceleration. The teeth on the outside of the clutch plate engage with the inner teeth of the clutch drum, while the inner teeth of the clutch plate engage with the outer teeth of the mainshaft low gear. The low gears are carried by a needle bearing on the main and countershaft. With the manual selector lever in the 2, D3 or D4 drive range, pressure is applied through the right side lever and mainshaft to the clutch piston, engaging the clutch. The power path is as follows:

Countershaft low gear – one-way clutch – parking gear – countershaft. The one-way clutch locks only in first gear in D3 or D4.

2nd clutch

The 2nd clutch is located at the centre of the mainshaft, as shown in Fig. 7.5, and is similar in construction to the 1st clutch. It is mounted on the mainshaft, back to back with the 4th clutch and rides on a

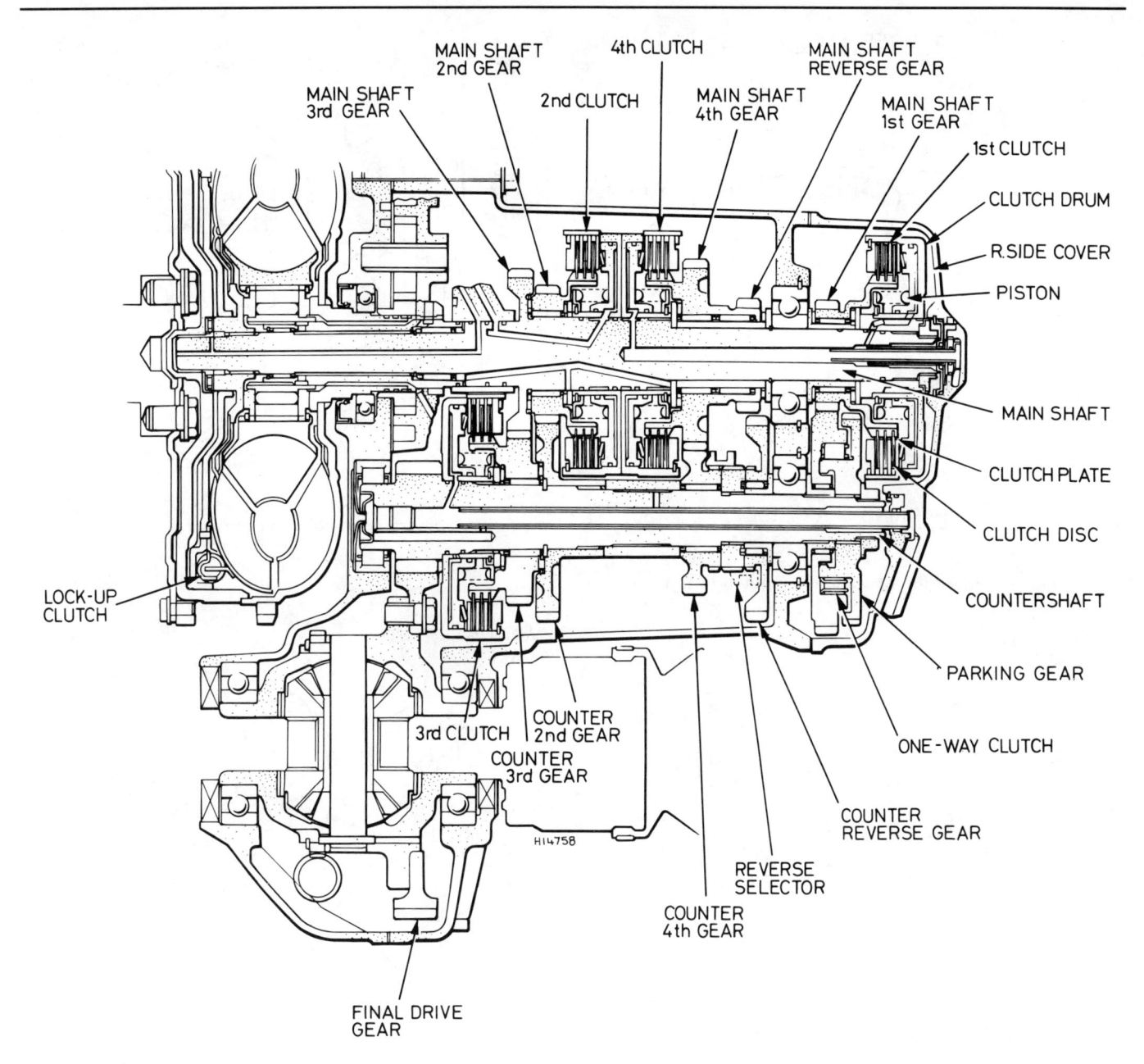

Fig. 7.5 Sectional view of transmission assembly (Sec 1)

needle bearing. The counter 2nd gear is splined to the countershaft, and in 2, D3 or D4 drive range, the clutch is engaged; transmitting power from the main 2nd gear to the counter 2nd gear.

3rd clutch

The 3rd clutch is mounted on the left-hand end of the countershaft. Similar again in construction to the other clutches, the 3rd clutch hub is integral with the counter 3rd gear, and carried on a needle roller bearing. The clutch locks when D3 or D4 is selected.

4th clutch

The 4th clutch hub is integral with main 4th gear and main reverse gear and is carried on the mainshaft by two needle roller bearings. It is engaged only when D4 or Reverse are selected.

One-way clutch

A sprag type one-way clutch is used between the counter low gear and parking gear, mounted on the countershaft. When the gears are turning in the direction of the black arrows in Fig. 7.6, the sprags incline and jam between the gears, locking them together. When the transmission changes up from first to 2nd, the countershaft will turn faster than the 1st gear, by means of 2nd gear, resulting in the parking gear being turned in the direction of the white arrow (Fig. 7.6), and freewheels. The same action takes place in 2nd and 3rd gears when D3 is selected, and to 2nd, 3rd and 4th gears when D4 is selected.

Reverse

When Reverse is selected, the reverse selector engages with the counter reverse gear. Power is transmitted by way of the main reverse gear, reverse idler gear and counter reverse gear. The reverse selector is actuated by a shift fork through a hydraulic servo. Return selection is assisted by spring.

Parking brake system

The parking brake system consists of the shift shaft, locking pawl and parking gear, which is splined to the countershaft. When the

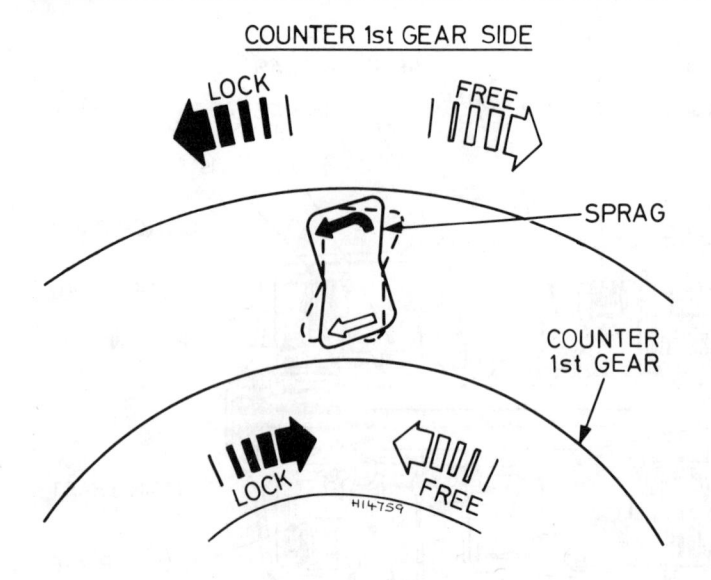

Fig. 7.6 Operation of sprag type clutch (Sec 1)

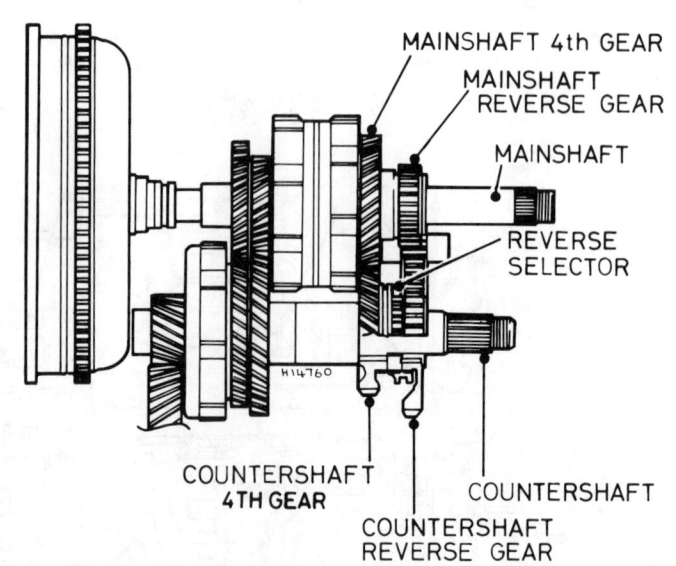

Fig. 7.7 Diagrammatical view of reverse gear (Sec 1)

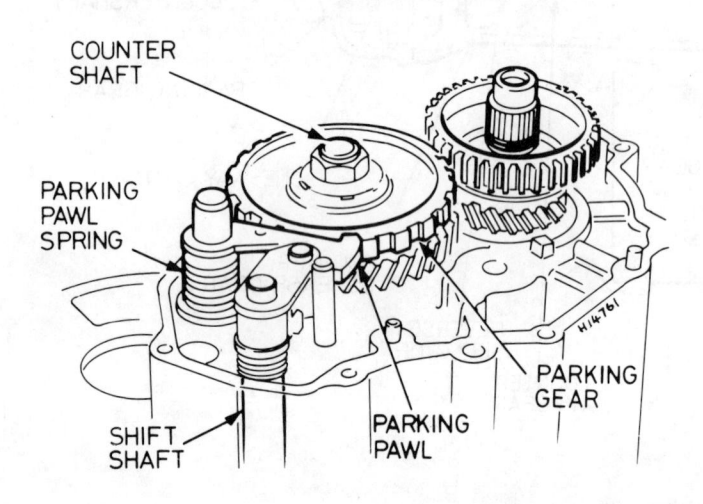

Fig. 7.8 The parking mechanism (Sec 1)

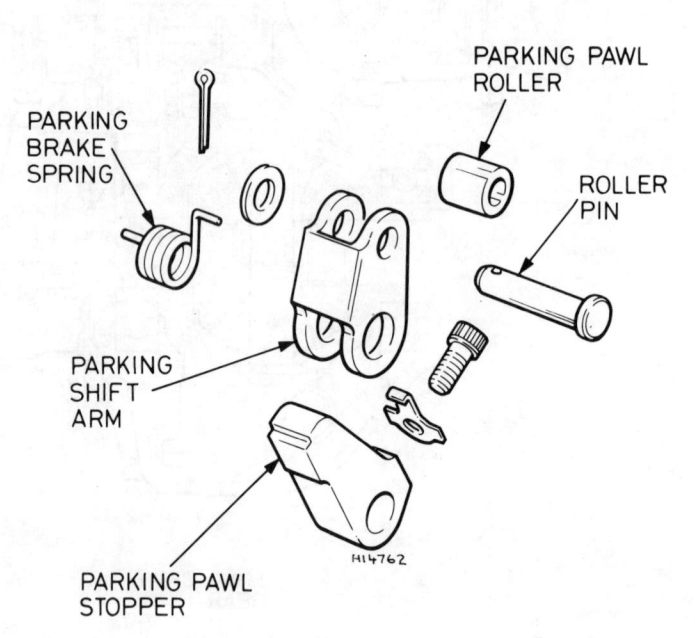

Fig. 7.9 Exploded view of parking pawl assembly (Sec 1)

selector lever is placed in the P (Park) position, the movement is transmitted through the shift shaft to the locking pawl, which engages with the parking gearteeth. When any other selection is made, the spring forces the locking pawl out of engagement with the parking gear.

Hydraulic control

The transmission shifts up or down automatically depending on roadspeed and throttle opening. The hydraulic system directs oil pressure to, or from, the various clutches and servo unit. Oil pressure from both the throttle valve and governor are used in opposition to one another, so determining which shift valves operate. In drive range 2 or Reverse, the clutches are controlled through the servos by the selector lever. Also in Reverse, a servo operates the shift fork, engaging reverse gear.

2 Routine maintenance

At the intervals specified in Routine Maintenance (at the front of the book) undertake the following service tasks.

1 Check the transmission fluid level and top up if necessary. Check the fluid for contamination, indicating internal wear.
2 Renew the transmission fluid.

3 Fluid level – checking, topping-up and renewing

Checking and topping-up

1 With the vehicle on level ground, and the engine at normal operating temperature (having been driven for at least 5 miles), and within 1 minute of shutting off the engine, unscrew and remove the transmission fluid level dipstick.
2 Dry the end of the dipstick with non-fluffy rag, then replace the dipstick, but do not screw it down.
3 Withdraw the dipstick again and check the fluid level is between the FULL and LOW marks.
4 If it is low add the recommended fluid, using a funnel, via the dipstick hole.

Chapter 7 Automatic transmission

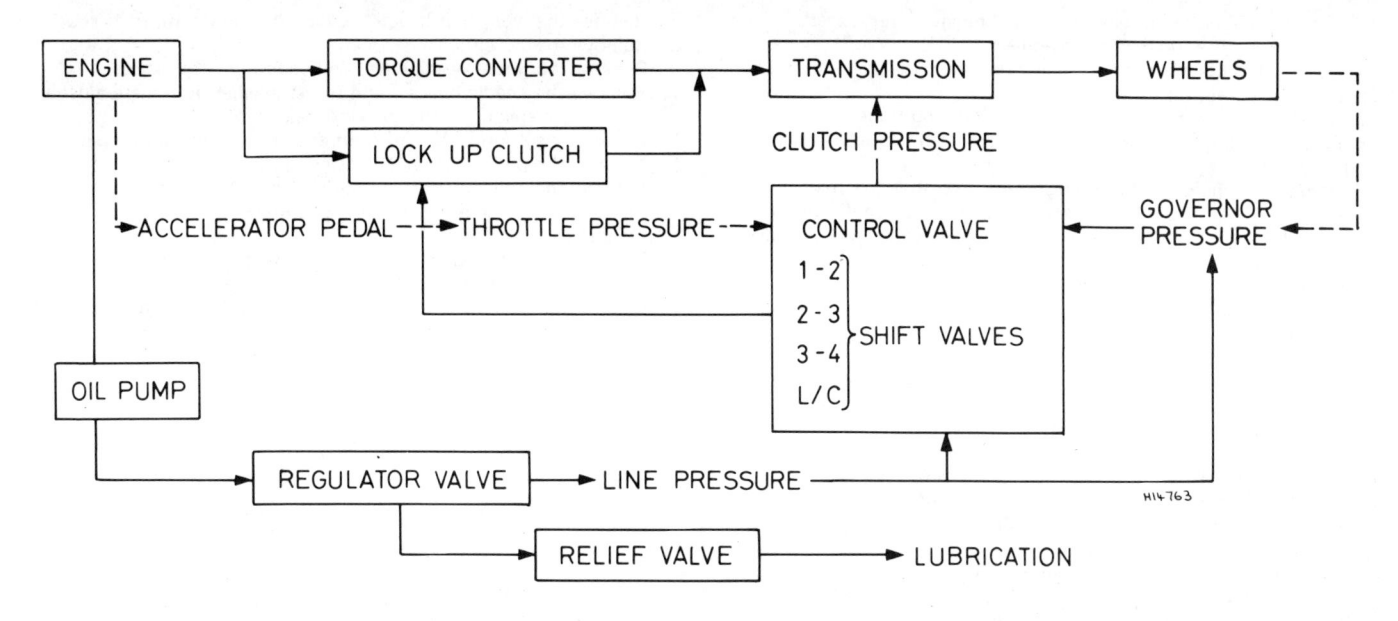

Fig. 7.10 Hydraulic control circuit diagram (Sec 1)

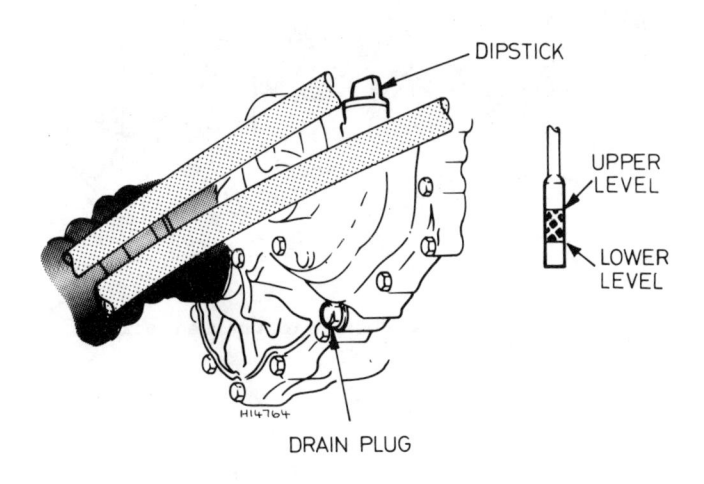

Fig. 7.11 Automatic transmission unit drain plug (Sec 3)

Renewing

5 The transmission fluid should be at normal operating temperature (see paragraph 1), and the vehicle on level ground.
6 Unscrew and remove the drain plug, and catch the fluid in a suitable container. Removing the dipstick will ease fluid flow.
7 Refit the drain plug using a new washer.
8 Using a funnel in the dipstick hole, refill the transmission unit to the FULL mark with the recommended fluid.
9 Remove the funnel, refit the dipstick, run the vehicle on the road for about 5 miles and then recheck the fluid level and also check for leaks.

4 Automatic transmission – removal and refitting

1 Disconnect the battery (refer to Chapter 12).
2 Release the steering lock, and select Neutral.
3 Disconnect the battery positive cable from the starter.
4 Disconnect the electrical lead from the starter solenoid.
5 Disconnect the oil cooler hoses, and wire them up next to the radiator to prevent transmission fluid from leaking out.
6 Remove the top mounting bolt from the starter.
7 Loosen the front wheel nuts.

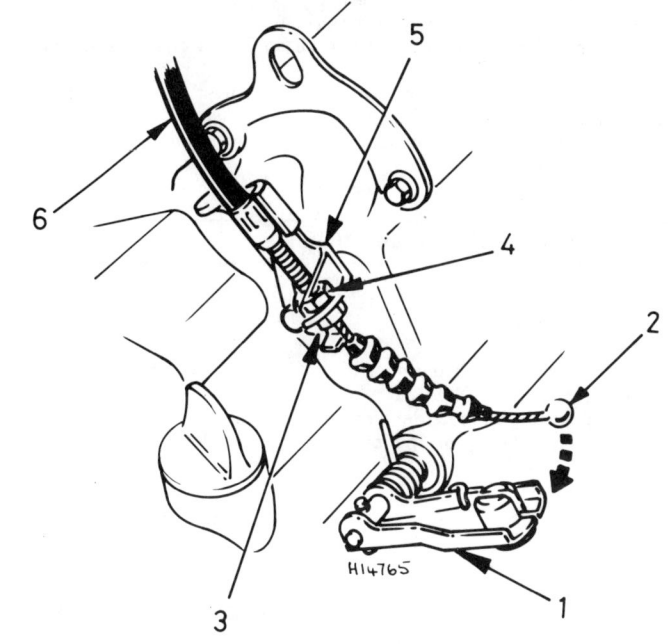

Fig. 7.12 Throttle control cable end fitting (Sec 4)

1 Throttle control lever
2 Cable end
3 Locknut B
4 Locknut A
5 Bracket
6 Throttle control cable

8 Apply the handbrake, chock the rear wheels, then raise the front end of the vehicle onto axle stands.
9 Remove the front wheels.
10 Remove the transmission drain plug and drain the transmission fluid. Refit the drain plug, using a new washer.
11 Remove the throttle control cable by removing the end from the throttle lever, loosening locknut B, and removing the cable from bracket (Fig. 7.12).
12 For models without power steering, remove the cable clip and pull the speedometer cable from its holder.

13 For models with power steering, remove the speed sensor complete with speedometer cable and hoses.
14 Remove the bolt from the transmission side of the starter motor and remove the starter motor.
15 Remove the two upper transmission mounting bolts.
16 Place a hydraulic trolley jack beneath the transmission assembly and raise it to just take the weight.
17 Assemble a hoist to the lifting bracket so that the transmission assembly may be steadied from above, to prevent it from falling off the trolley jack (the assembly is lowered from the vehicle, not lifted out).
18 Remove the subframe centre beam.
19 Remove the balljoint pinch-bolt from the right side lower suspension control arm, then, using a soft-faced mallet, tap the control arm free from the knuckle.
20 Remove the right side radius rod.
21 Disconnect the stabiliser bar at the right side lower control arm.
22 Remove the stabiliser bar from the radius rods.
23 Remove the front self-locking nuts from both left and right radius rods.
24 Remove the lower arm bolt from both sides of the subframe.
25 Refer to Chapter 8 and remove the right-hand driveshaft.
26 Remove the engine front stiffener bracket located in front of the torque converter cover plate.
27 Remove the torque converter cover plate.
28 Refer to Section 9 for control shaft removal and disconnect the shift cable; tying the cable out of the way on completion.
29 Remove the eight bolts, securing the torque converter assembly to the drive plate (rotate the crankshaft to reach all eight bolts, **do not** undo the recessed bolts holding the ring gear to the torque converter).
30 Remove the rear engine mounting.
31 Remove the transmission lower mounting bolts.
32 Pull the transmission away from the engine to clear the locating dowels.
33 Disconnect the left-hand drive shaft. Lower the transmission assembly and remove it from beneath the vehicle.
34 Refitting is a reversal of removal. For throttle cable adjustment, refer to Chapter 3.

5 Transmission unit – dismantling

1 Remove the dipstick.
2 Remove the end cover.
3 Set transmission to P (Park).
4 Lock the mainshaft, for which a special tool is needed (Honda tool number 07923 – 6890201).
5 Remove the gasket dowel pins and O-rings from the end cover.
6 Relieve the staking on the locknut of the 1st clutch.
7 Remove the mainshaft locknut, which has a **left-hand thread**.
8 Remove the 1st clutch.
9 Remove the thrust washer, needle bearing and 1st gear.
10 Remove the needle bearing and thrust washer from the mainshaft.
11 Relieve the staking on the parking gear locknut.
12 Remove the countershaft locknut and parking pawl stop pin.
13 Remove the parking pawl, shaft and spring.
14 Remove the parking gear and countershaft 1st gear as a unit.
15 Remove the needle roller bearing and 1st gear collar from the countershaft.
16 Remove the 0-ring and 1st gear collar from the mainshaft.
17 Remove the reverse collar bearing holder.
18 Relieve the tab on the locking plate of the bolt holding the parking shift arm and spring, remove the bolt and the arm and spring.

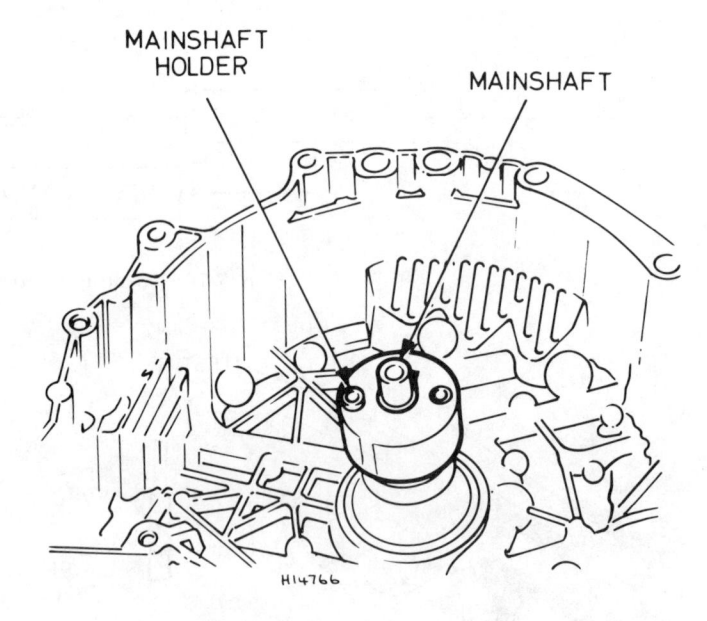

Fig. 7.13 Mainshaft locking tool in use (Sec 5)

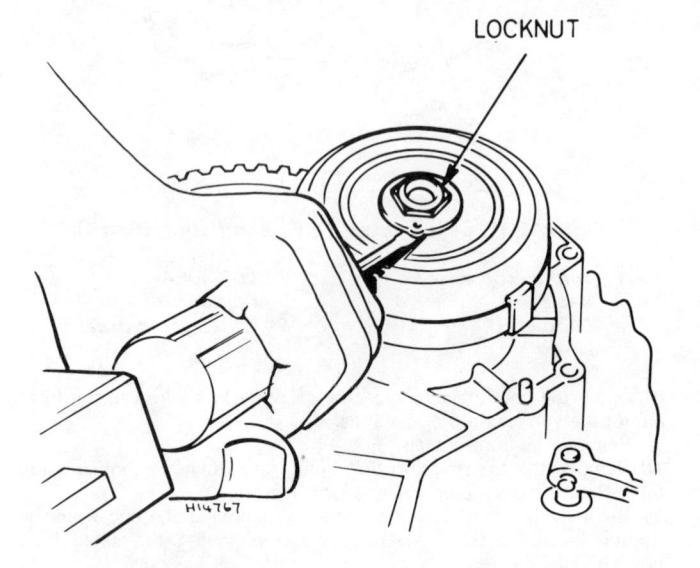

Fig. 7.14 Relieving the locknut staking (Sec 5)

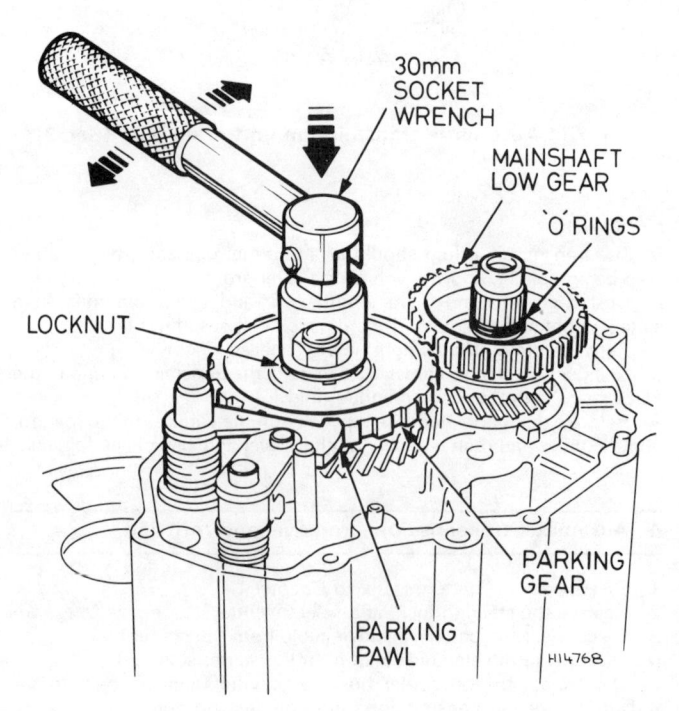

Fig. 7.15 Removing a shaft locknut (Sec 5)

Chapter 7 Automatic transmission

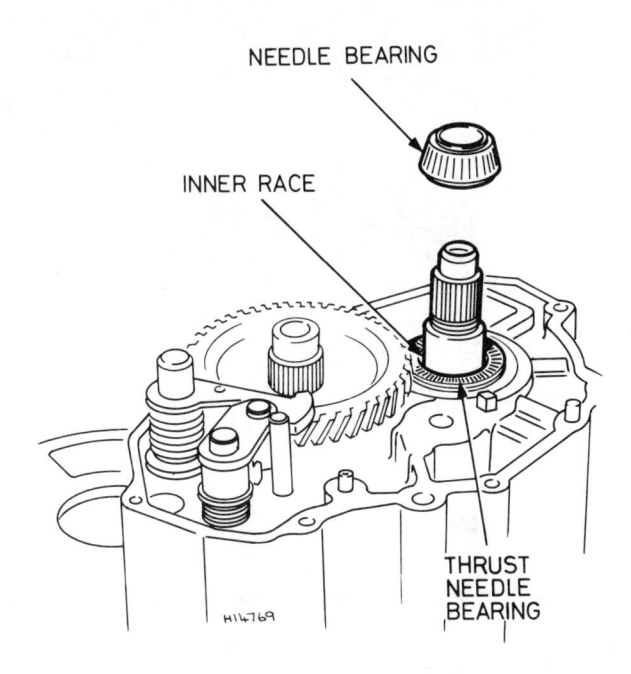

Fig. 7.16 Removing the bearings (Sec 5)

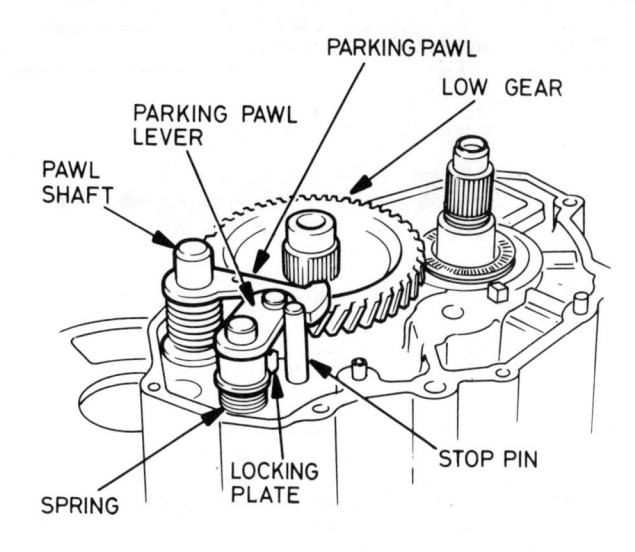

Fig. 7.17 Components of the parking pawl (Sec 5)

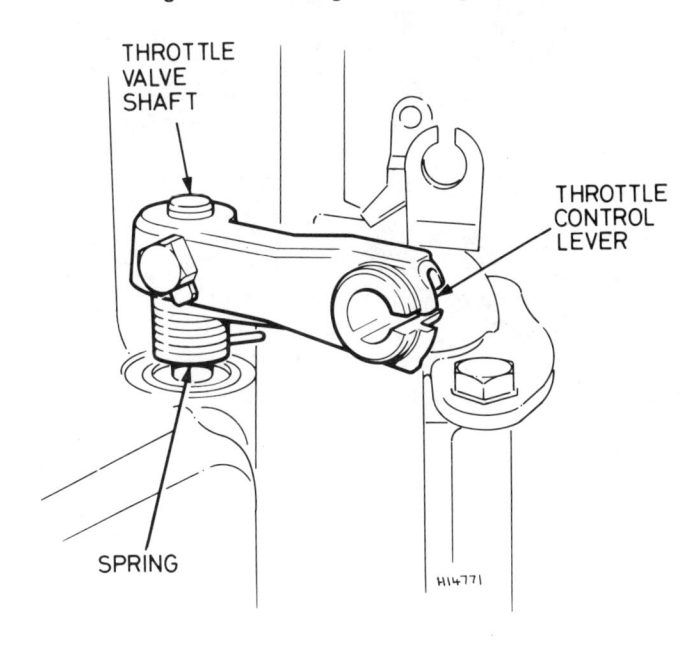

Fig. 7.18 Throttle control lever assembly (Sec 5)

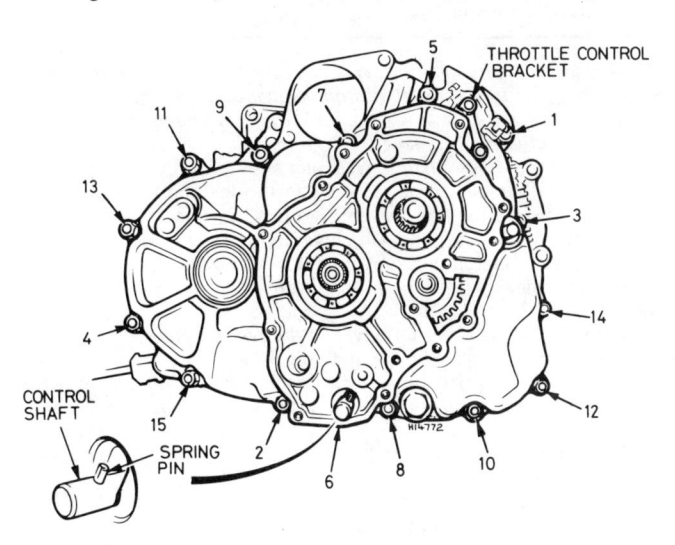

Fig. 7.19 Transmission housing bolt removal sequence (Sec 5)

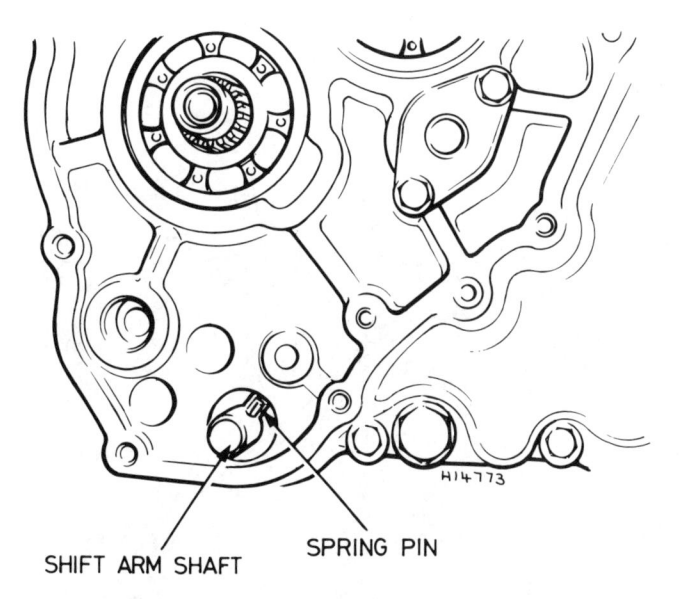

Fig. 7.20 The control shaft spring pin and cut-out (Sec 5)

19 Similarly, remove the bolt from the throttle control lever and remove the lever and spring from the throttle valve shaft.
20 Remove the transmission housing bolts in the sequence shown in Fig. 7.19. **Note:** Bolt number 1 will not come right out, because the throttle control cable bracket is in the way, but it can be withdrawn sufficiently to allow the transmission housing to be removed. If the throttle control bracket is removed, it must be adjusted correctly on reassembly – see Section 15.
21 Line up the control shaft spring pin with the cut-out in the transmission housing.
22 Using the special puller – Honda tool number 07933-6890200 or 6890201, loosen the transmission housing from the torque converter housing.
23 Remove the puller and separate the two housings. If the puller is not available, the housings may be separated by gentle tapping with a soft-faced mallet, but over zealous use of this method will lead to damage of the casings.

6 Mainshaft and countershaft – removal

1 Remove the gasket and dowel pins.
2 Lift off the reverse gear collar and needle roller bearing from the countershaft.
3 Relieve the locking tab on the lockplate and remove the bolt from the reverse shift fork. Lift off the shift fork and reverse selector as a unit.
4 Remove the selector hub, 4th gear and needle bearing from the countershaft.
5 Remove the mainshaft and countershaft as one unit, lifting them both out together, tilting them at an angle to clear the governor.

7 Governor – removal and refitting

1 Bend down the locking tabs and remove the three bolts.
2 Lift out the governor from the torque converter housing.
3 Little can be done by way of repair to the governor and, if it is malfunctioning, it should be renewed as a complete unit.
4 Refitting is a reversal of removal.

8 Main valve body – removal

Warning: *the accumulator covers are spring-loaded and, to prevent stripping of the threads in the torque converter housing, unscrew the bolts diagonally, a little at a time, while pressing down on the covers until spring pressure is relieved.*
1 Remove the accumulator covers and springs.
2 Undo and remove the three bolts in the lock-up valve body and move the valve to the right to remove the oil transfer pipes.
3 Remove the 1st, 3rd and 4th clutch oil pipes.
4 Remove the clutch pressure control valve body.
5 Remove the E-clip securing the throttle control shaft and remove the shaft.
6 Remove the servo valve body, and the two oil pipes.
7 Remove the separator plate and dowel pins.
8 Remove the four steel balls from the valve body oil passageways, but do not use a magnet to do this or the balls may become magnetized.
9 Remove the regulator valve body held by three bolts, being careful not to lose the steel ball.
10 Remove the stator shaft arm, dowel pins, stop pin and four bolts holding the valve body to the converter housing.

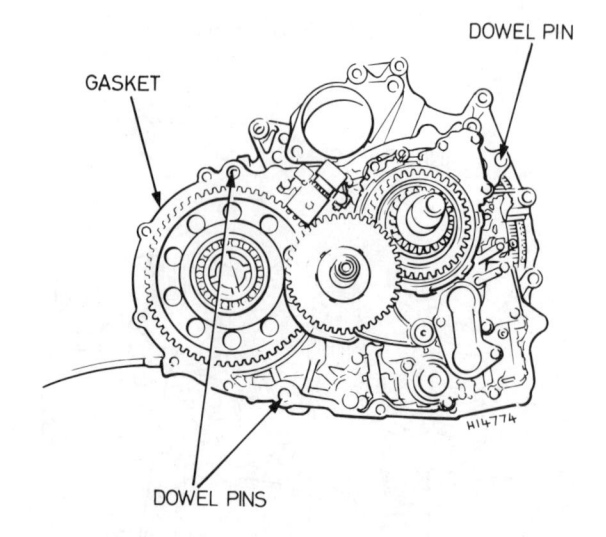

Fig. 7.21 Positions of the dowel pins (Sec 6)

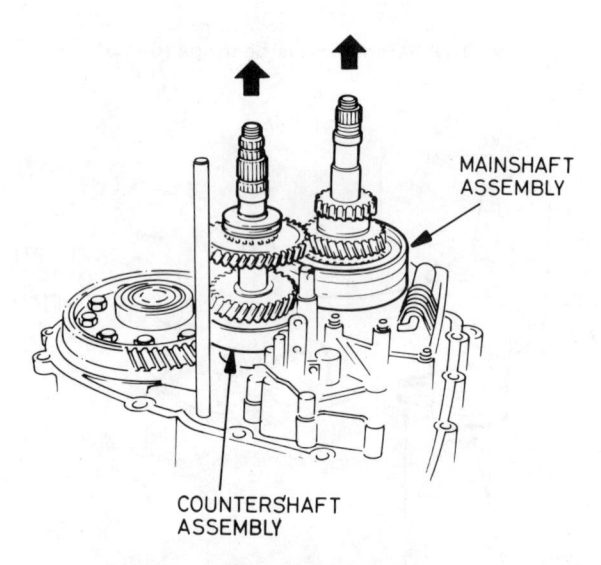

Fig. 7.22 Lifting out the main and countershafts as one (Sec 6)

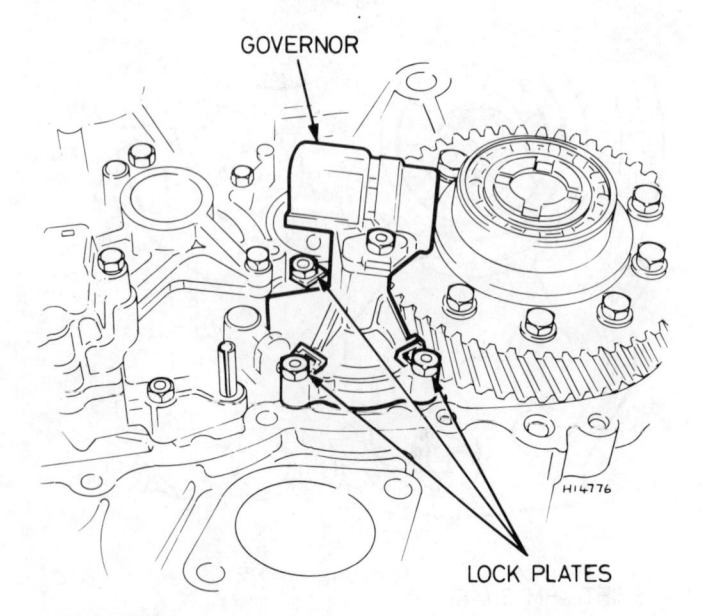

Fig. 7.23 Governor assembly and retaining bolts (Sec 7)

Fig. 7.24 The accumulator covers (Sec 8)

185

Fig. 7.25 The accumulator springs (Sec 8)

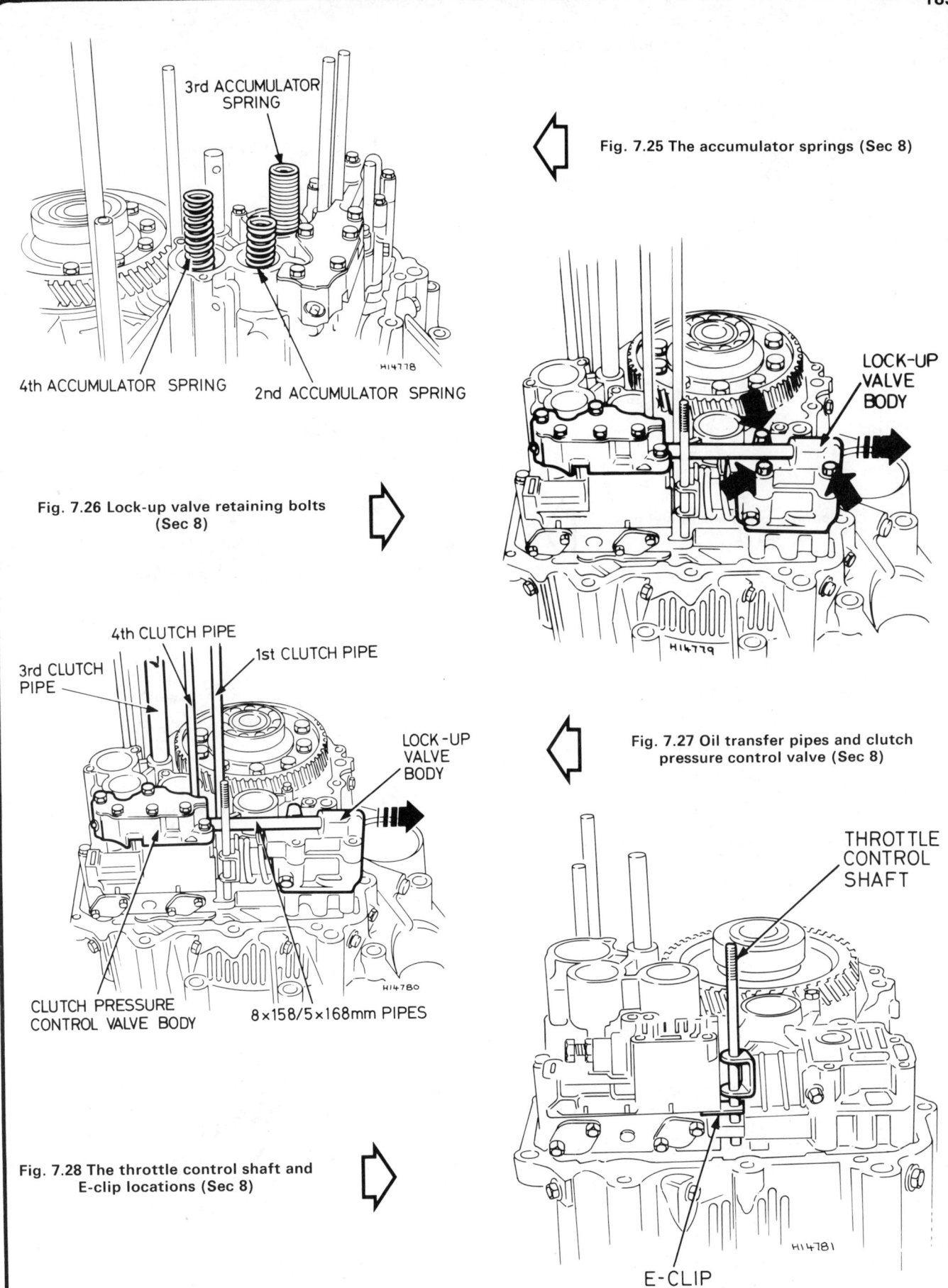

3rd ACCUMULATOR SPRING
4th ACCUMULATOR SPRING
2nd ACCUMULATOR SPRING

LOCK-UP VALVE BODY

Fig. 7.26 Lock-up valve retaining bolts (Sec 8)

Fig. 7.27 Oil transfer pipes and clutch pressure control valve (Sec 8)

4th CLUTCH PIPE
3rd CLUTCH PIPE
1st CLUTCH PIPE
LOCK-UP VALVE BODY
CLUTCH PRESSURE CONTROL VALVE BODY
8 x 158/5 x 168mm PIPES

THROTTLE CONTROL SHAFT

Fig. 7.28 The throttle control shaft and E-clip locations (Sec 8)

E-CLIP

Fig. 7.29 The servo valve body and oil pipes (Sec 8)

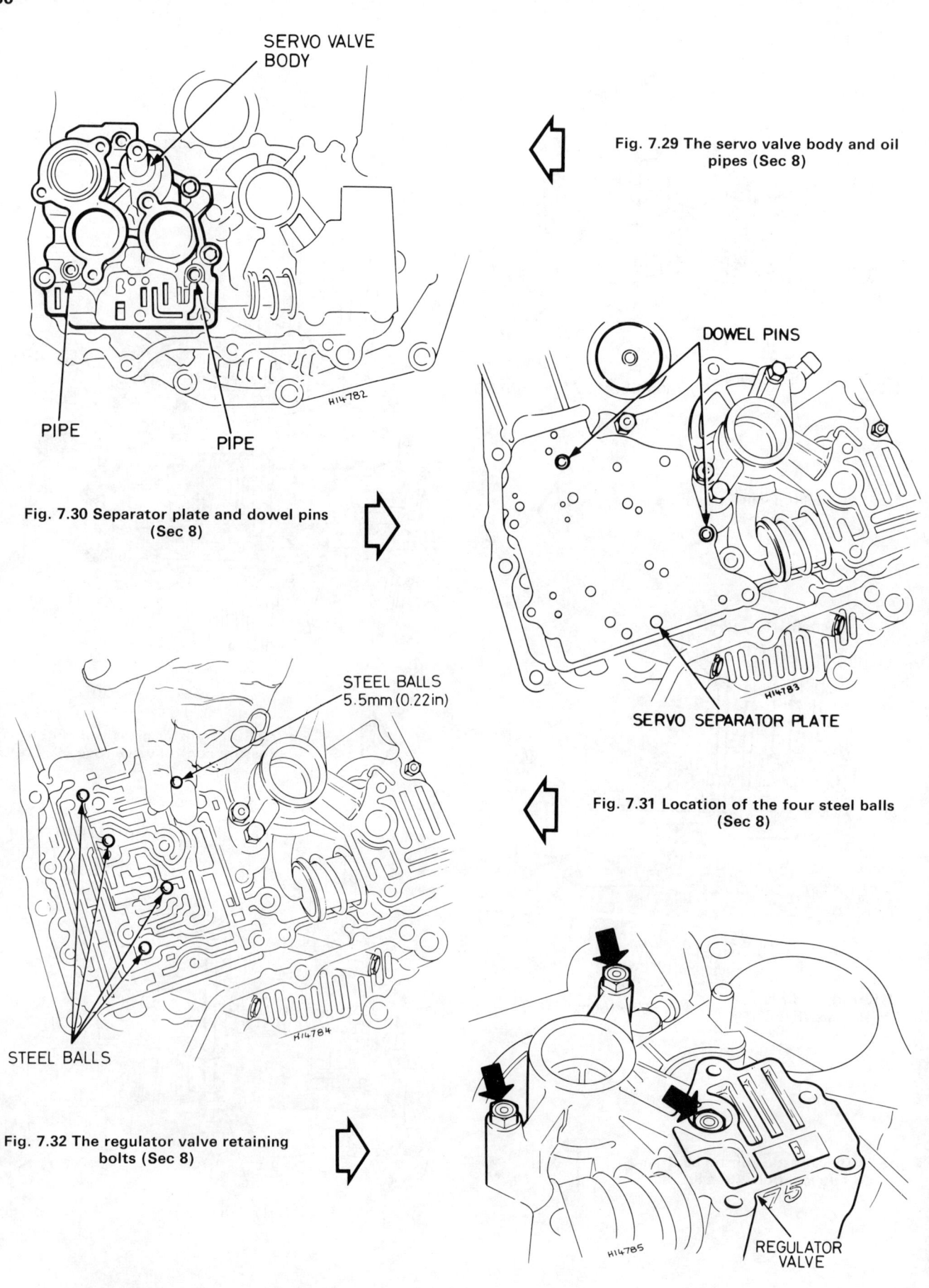

Fig. 7.30 Separator plate and dowel pins (Sec 8)

Fig. 7.31 Location of the four steel balls (Sec 8)

Fig. 7.32 The regulator valve retaining bolts (Sec 8)

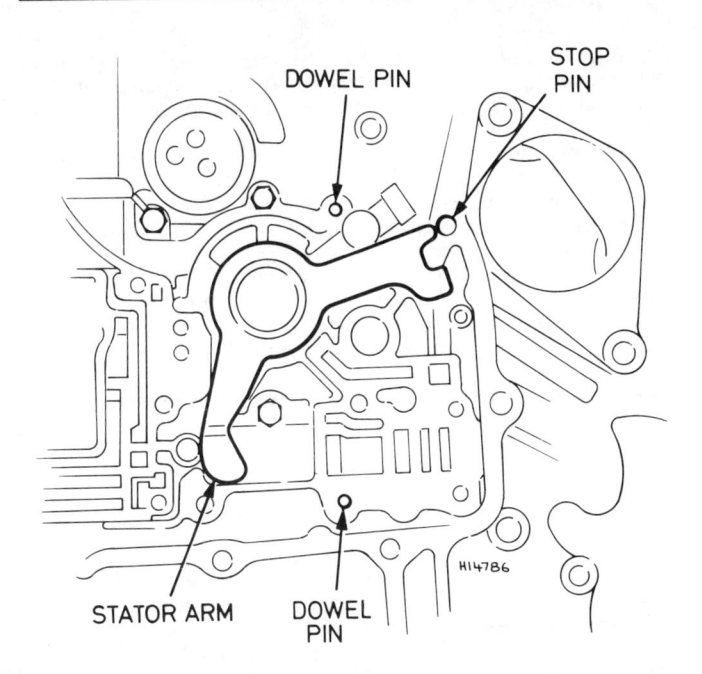

Fig. 7.33 The stator arm assembly (Sec 8)

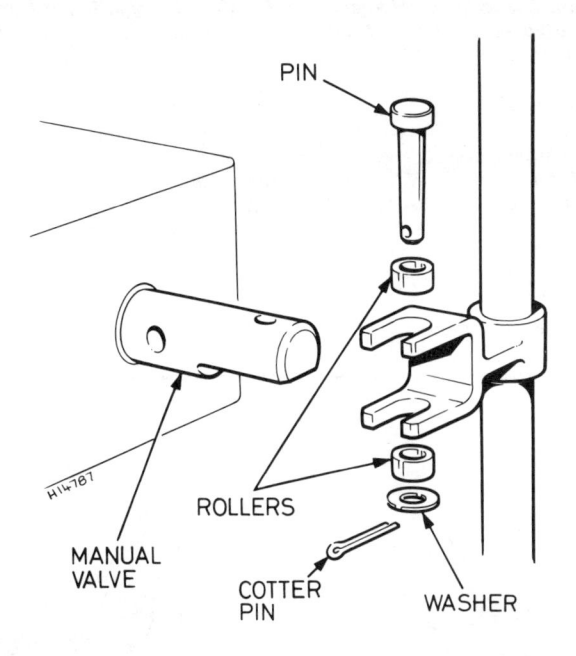

Fig. 7.34 The manual valve attachment fork (Sec 8)

11 Remove the split pin, washer rollers and shackle pin from the manual valve attachment fork.
12 Remove the main valve body, being prepared to catch the torque converter check valve and spring.
13 Lift out the oil pump gears and shaft.
14 Remove the separator plate, dowel pins and check valve and spring.
15 Lift out the filter screen and oil suction pipe. Once removed the filter screen should be discarded and replaced with a new item.

16 The transmission is now dismantled and the individual components should be inspected as described in the following Sections.
17 For removal and inspection of the differential unit, refer to Chapter 6.

9 Major assemblies – inspection and overhaul

Note: *The importance of cleanliness during overhaul of hydraulic control assemblies cannot be over emphasised. Even the minute hairs from cleaning cloths can cause malfunction. Do not use wire to probe out channels or drillings, nor use abrasive paper to clean valve surfaces. Use a solvent for cleaning and non-fluffy rag (nylon for*

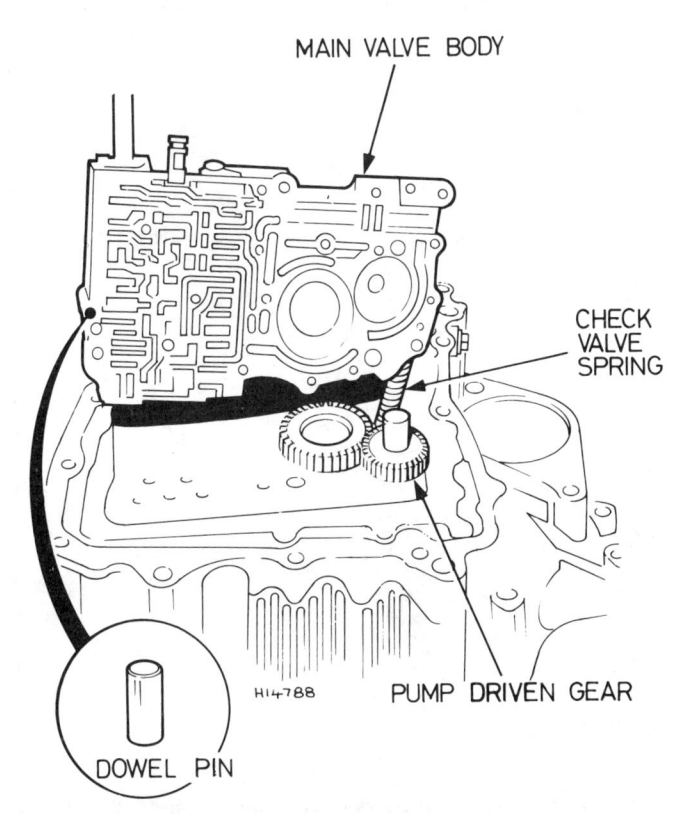

Fig. 7.35 Main valve body being removed to show oil pump and check valve (Sec 8)

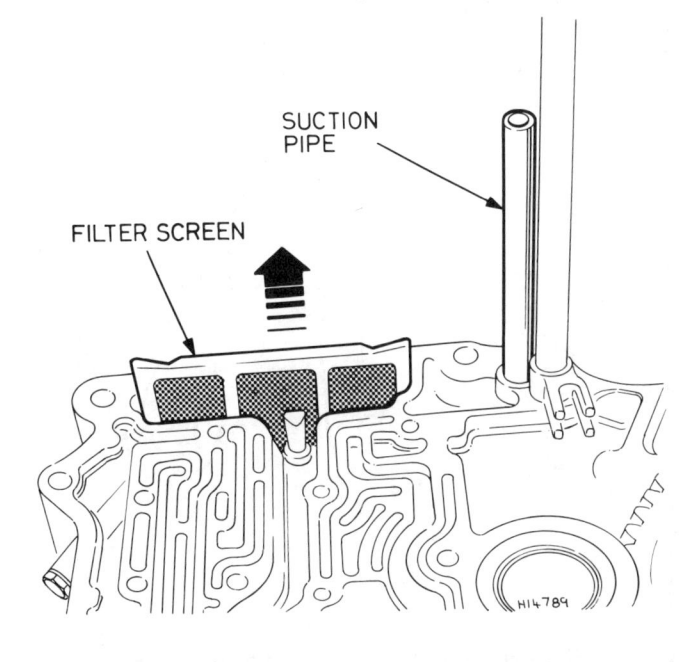

Fig. 7.36 Removing the filter screen and suction pipe (Sec 8)

188 Chapter 7 Automatic transmission

instance), and blow dry all components. After cleaning and inspection, renew all seals, and coat all assemblies in clean transmission fluid before reassembly.

Control shaft

1 The control shaft can be removed from the converter housing after removing the control lever.
2 Refer to Fig. 7.39 for the component parts, which makes dismantling and reassembly self evident.
3 Renew any parts which are worn and use new spring pins on reassembly.

Governor

4 Refer to Section 7.

Servo valve

Warning: *Do not remove or adjust the throttle pressure adjustment bolt. It is factory adjusted to give the correct gear change points.*

5 Dismantling of the valve may be undertaken with reference to Fig. 7.40.
6 Clean all parts as described previously and inspect the components for signs of wear.
7 Renew all O-ring seals.
8 Inspect the valve springs and check their free lengths against the tolerances in the Specifications.
9 Ensure all valves move freely in their bores.
10 Any signs of scratching or scoring and burnt or discoloured valves or bores should be viewed with suspicion and components renewed.
11 Lubricate all parts with clean transmission fluid on reassembly.

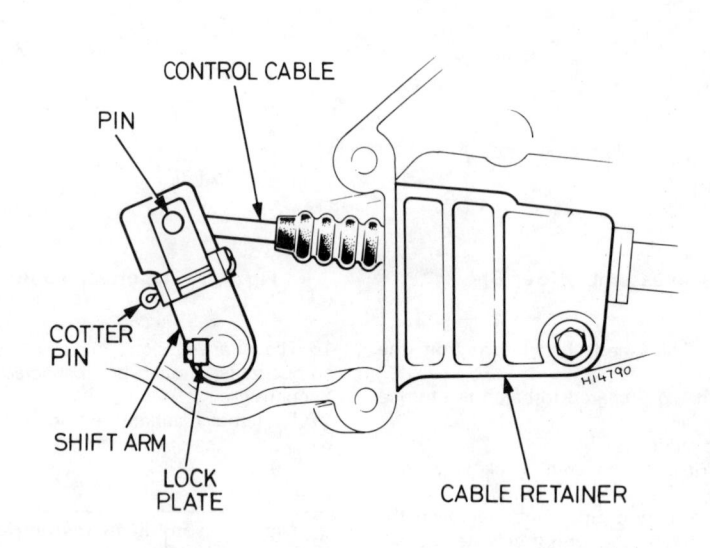

Fig. 7.37 Control shaft and cable end (Sec 9)

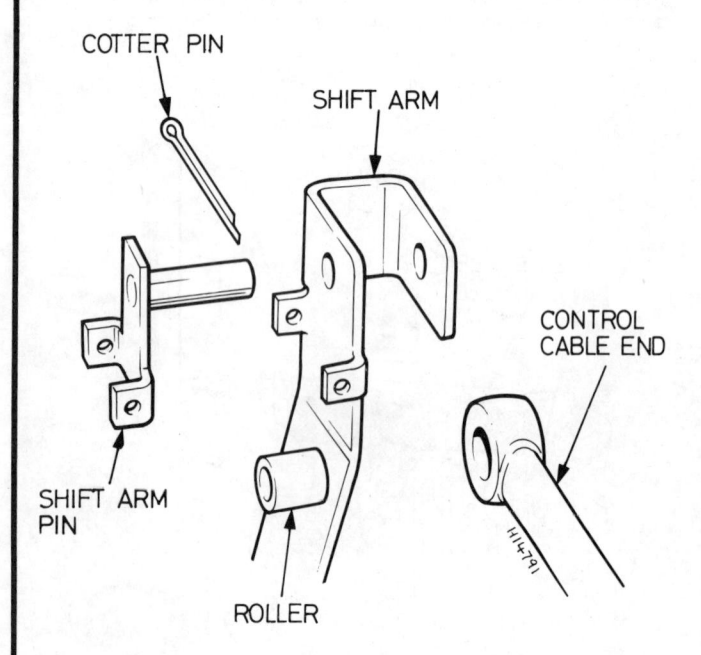

Fig. 7.38 Cable-to-control arm fittings (Sec 9)

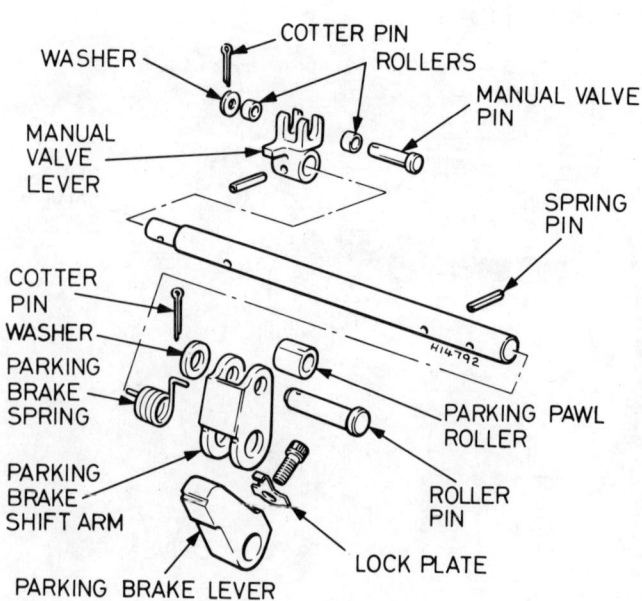

Fig. 7.39 Component parts of the control shaft (Sec 9)

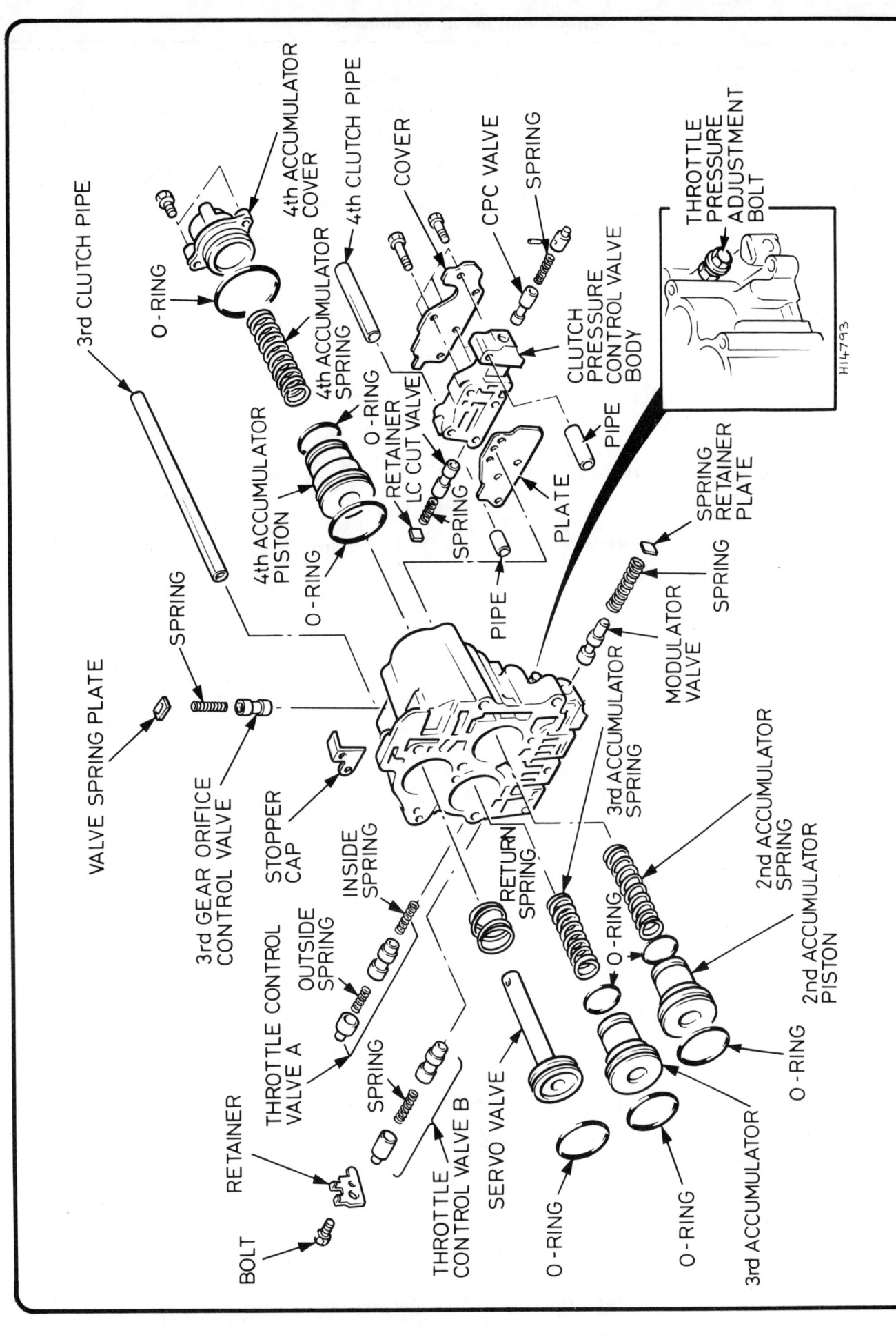

Fig. 7.40 Exploded view of the servo valve (Sec 9)

Chapter 7 Automatic transmission

Pressure regulator valve

12 The valve can be dismantled after removing the lockbolt; being careful to retain the springs by releasing the retainer slowly.
13 Clean the valve components and inspect for signs of wear, renewing any which are worn.
14 Lubricate all parts with clean transmission fluid on reassembly.
15 Align the hole in the valve retainer with the hole in the valve body to install the lockbolt.

Main valve body

16 An exploded view of the main valve body appears in Fig. 7.42.
17 Remove all valves carefully, retaining all springs and detent steel balls.
18 Remove the oil pump gears and inspect them for wear.
19 Check all valves for freedom of movement, signs of scoring or scratching, and burning or discolouration.
20 Wear limits and tolerances are given in the Specifications and if any parts are worn beyond limits, then the complete valve must be removed.
21 Again, lubricate all parts with transmission fluid on reassembly, which should be undertaken with reference to Fig. 7.43 to 7.46.

Pressure control valve

22 Clean and inspect the component parts as described for the other valves, renewing the complete valve if there are signs of excessive wear.

Mainshaft

23 The mainshaft can be dismantled by extracting the snap-rings and sliding off each part of the shaft.

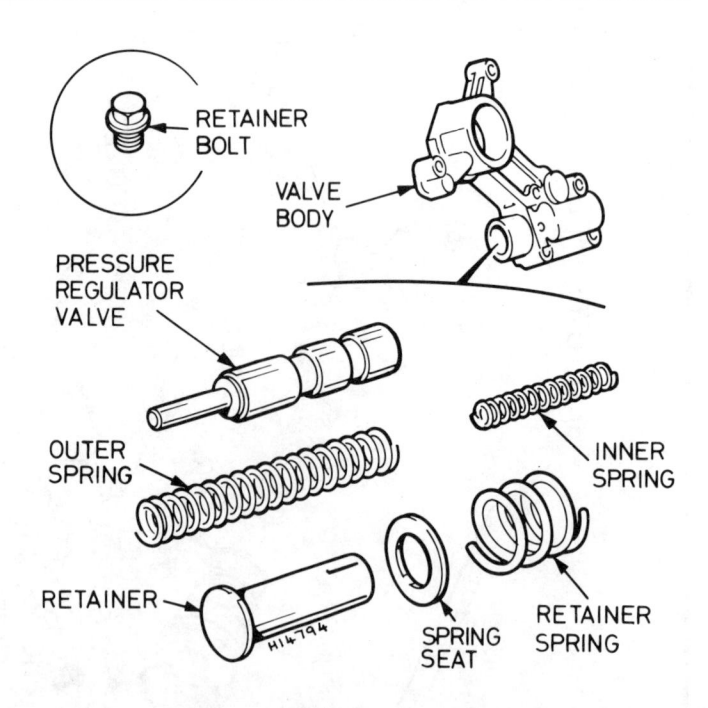

Fig. 7.41 Components of the pressure regulating valve (Sec 9)

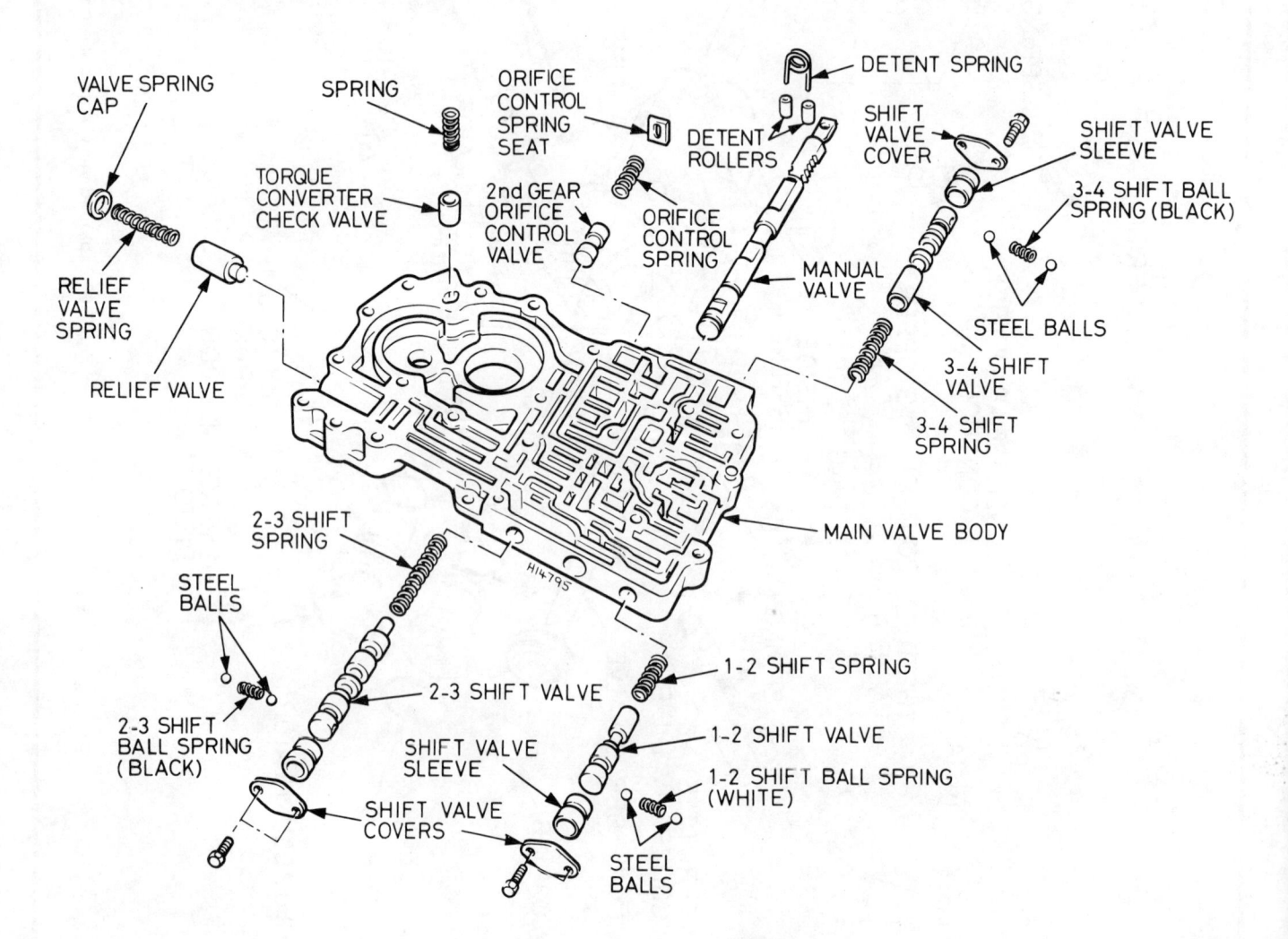

Fig. 7.42 Exploded view of the main valve (Sec 9)

191

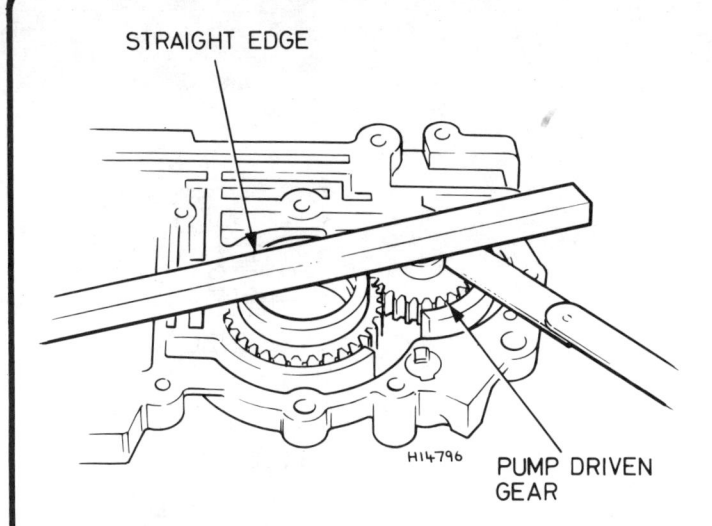

Fig. 7.43 Measuring axial clearance of the oil pump gears (Sec 9)

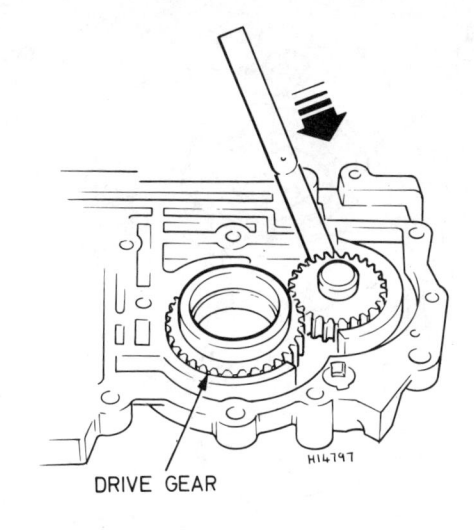

Fig. 7.44 Measuring oil pump side clearance (Sec 9)

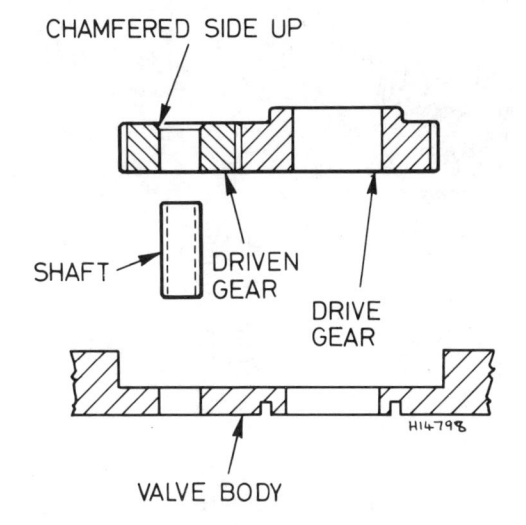

Fig. 7.45 Correct orientation of oil pump gears (Sec 9)

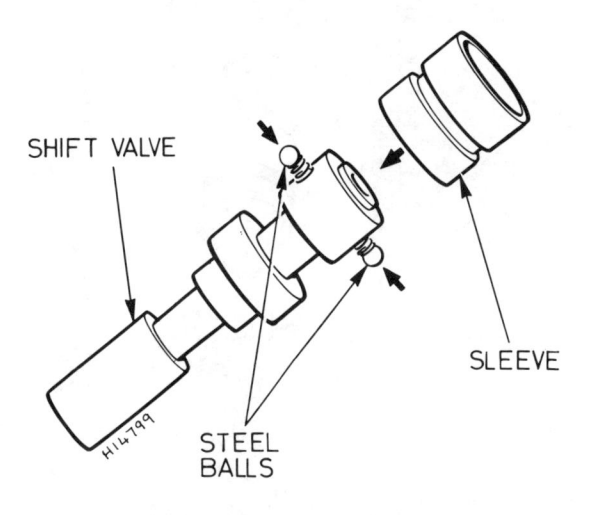

Fig. 7.46 Refit the springs and steel balls into the valve and slide the sleeve over them (Sec 9)

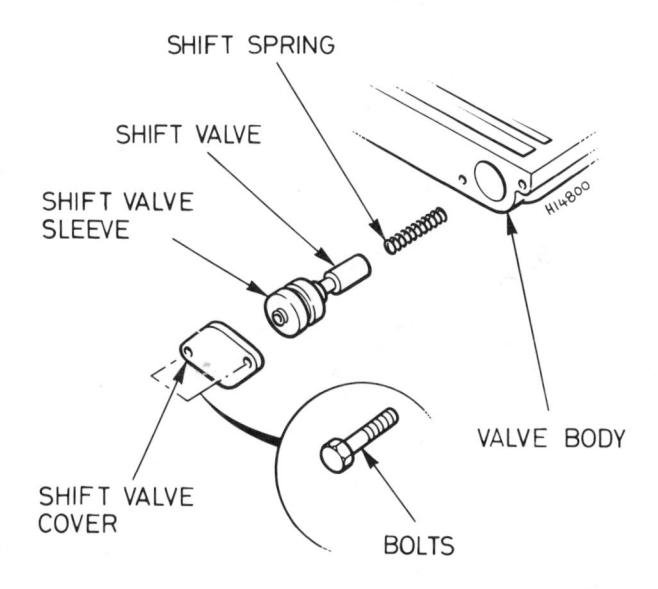

Fig. 7.47 Control valve assembly (Sec 9)

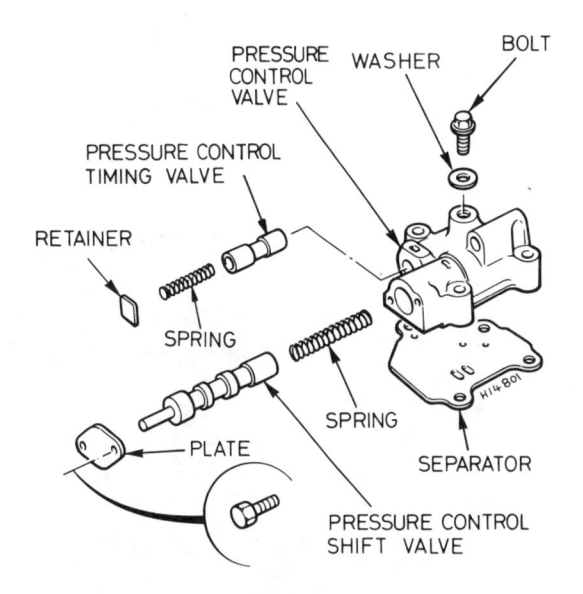

Fig. 7.48 Exploded view of pressure control valve (Sec 9)

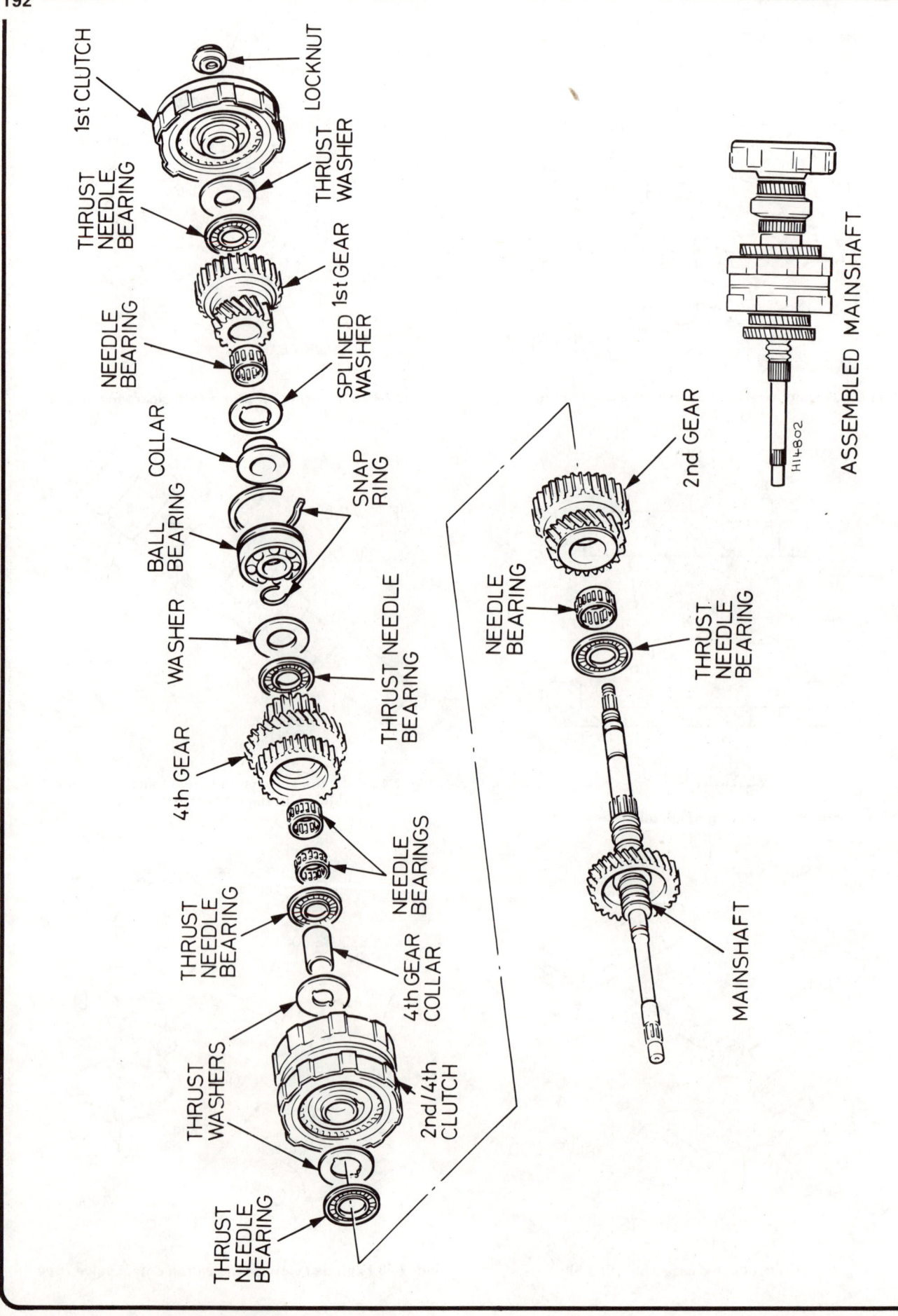

Fig. 7.49 Component parts of the mainshaft (Sec 9)

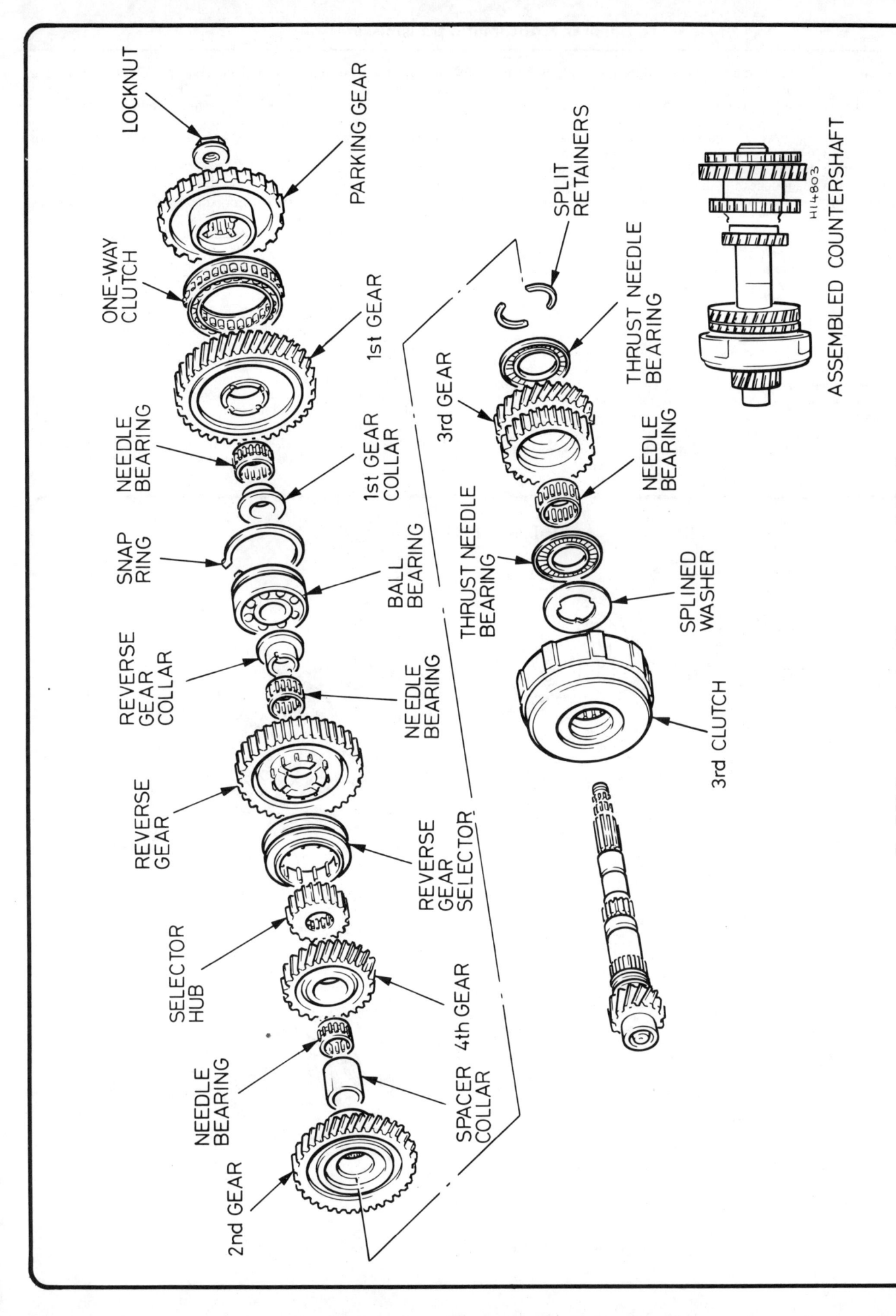

Fig. 7.50 Component parts of the countershaft (Sec 9)

24 Clean and inspect the shaft and splines and the gearteeth for wear, damage, cracks, chipping etc.
25 Inspect the needle roller bearings for roughness, excessive 'rattle' indicating wear and signs of burning.
26 On reassembly, use new O-rings and metal sealing rings.
27 Inspection of the clutch assemblies is given in later paragraphs.
28 When refitting the thrust needle bearings, note that the unrolled edge of the bearing cage faces the thrust washer.
29 Gear clearances are measured on reassembly, see Section 12.

Countershaft

30 Inspect the countershaft and its components in the same way as described for the mainshaft.

Clutch disc assemblies

Note: *to dismantle the clutch assemblies a special spring compressor is required, or a suitable long bolt and spacers can be adapted, but care will be needed in order not to damage the clutch assemblies. If it is known that the clutch assemblies are worn, then consideration should be given to renewing the assemblies. The 1st and 3rd clutch assemblies are not interchangeable, so do not mix them.*
31 Remove the snap-ring.
32 Remove the end plate, clutch discs and plates.
33 Use the special tool or adaptor as described, and compress the spring sufficiently to allow the snap-ring to be removed.
34 Remove the snap-ring and then the special tool, and lift out the spring retainer and spring.
35 Apply low air pressure to the oil passageway and eject the piston by holding a finger tip over the other end.
36 Inspect the piston for wear and check that the check valve is not loose.
37 Measure the return spring free length against the tolerances in the Specifications.
38 Inspect the clutch plates for wear or damage, and renew all parts as necessary.
39 Before reassembly, the clutch discs should be soaked in transmission fluid for a minimum of 30 minutes.
40 Renew all O-ring seals and snap-rings on reassembly.
41 Build up the clutch assembly in the reverse order to dismantling, noting the following points.
42 Make sure the spring washer is properly positioned (Fig. 7.53).
43 Avoid pinching the O-rings when installing the piston.
44 When fitting the return spring snap-ring, the sharp edge of the snap-ring should be facing up (away from the piston).
45 The clutch end plate is installed with its flat side toward the disc.
46 After reassembly, the clearance between the clutch end plate and top disc should be measured with feeler gauges and compared with the tolerances in the Specifications.
47 There are ten different sizes of end plate, and one should be selected to bring the clearance within the laid down tolerances.

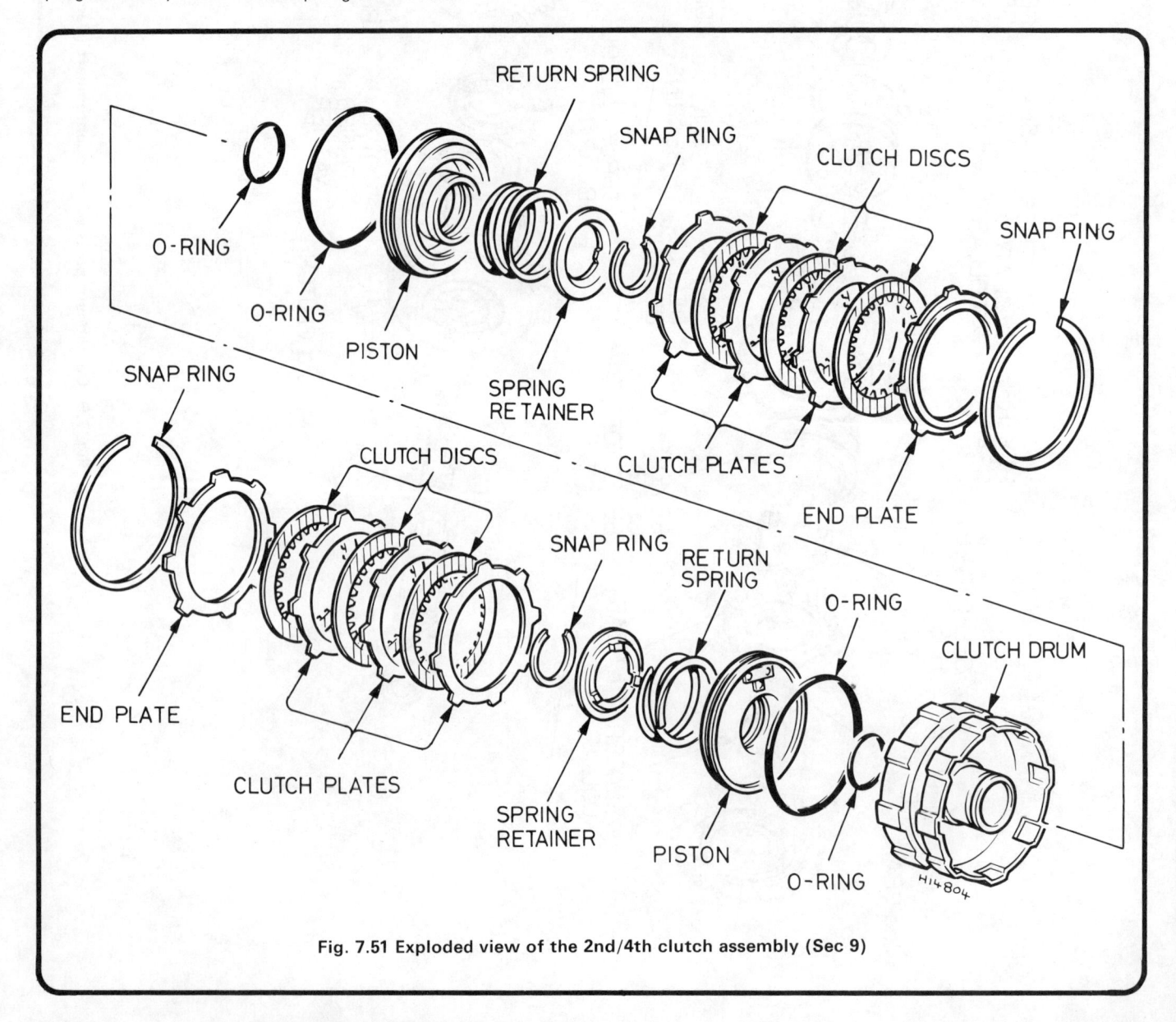

Fig. 7.51 Exploded view of the 2nd/4th clutch assembly (Sec 9)

Chapter 7 Automatic transmission

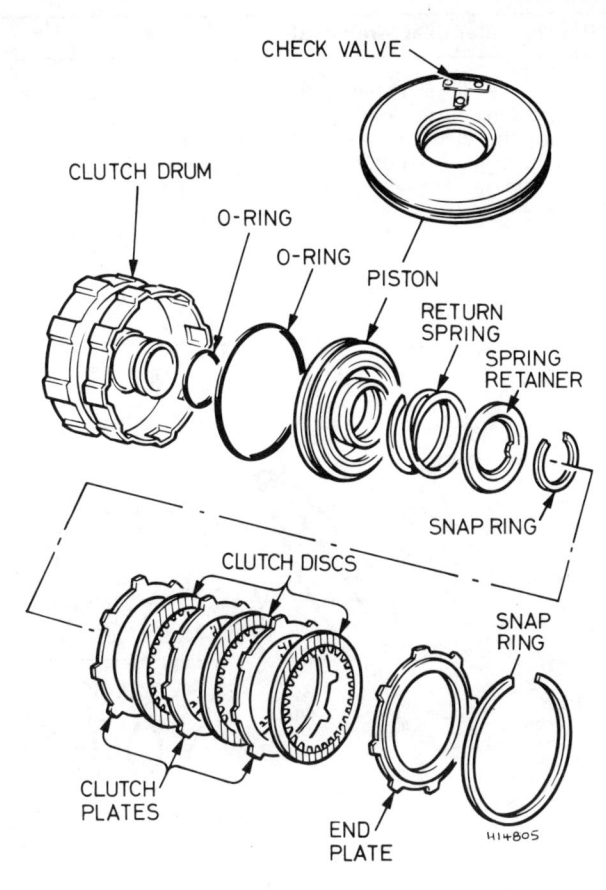

Fig. 7.52 Exploded view of the 1st/3rd clutch assembly (Sec 9)

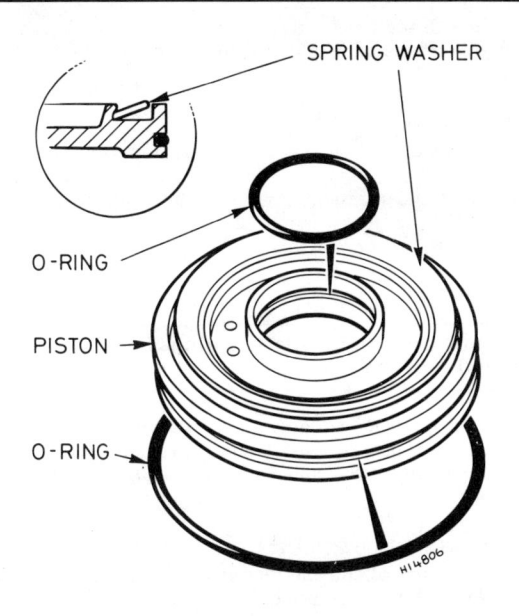

Fig. 7.53 Correct positioning of the spring washer in the piston (Sec 9)

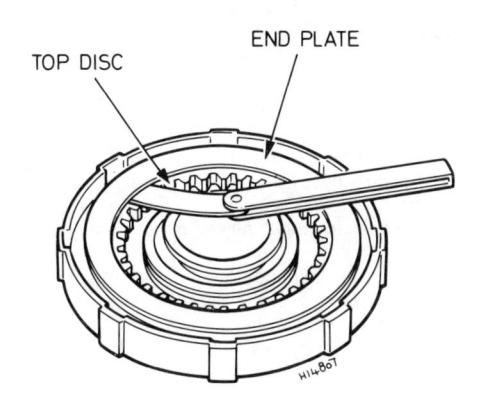

Fig. 7.54 Checking clutch end plate-to-disc clearance (Sec 9)

48 The clutch operation can be checked by applying air pressure to the oil passageway, when the clutch should engage, and disengage when air pressure is removed.

End cover

49 The components of the end cover may be removed by extracting the snap-rings.

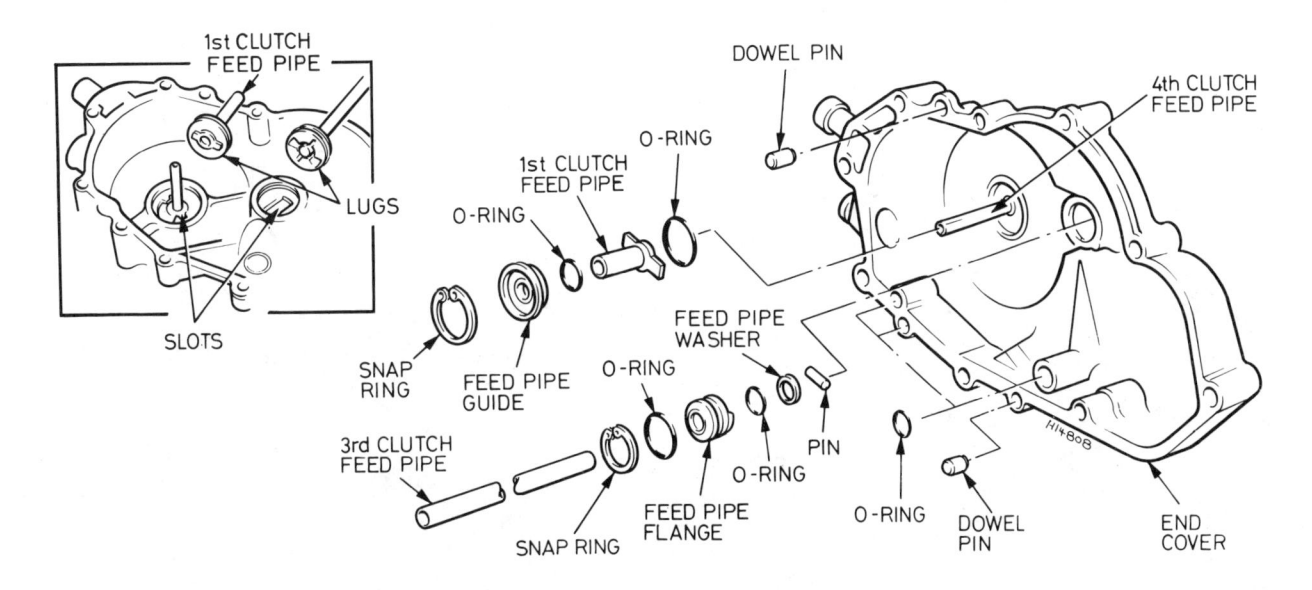

Fig. 7.55 Components of the end cover (Sec 9)

Chapter 7 Automatic transmission

50 Renew the O-rings.
51 When refitting the oil feed pipes, line up the lugs with the slots in the end cover.

1st Accumulator

52 The accumulator can be removed without removing the end cover if so desired.
53 Dismantling and reassembly is straightforward, referring to Fig. 7.56.
54 Renew all O-rings on reassembly.

10 Mainshaft and countershaft bearings and seals – removal and refitting

Note: *In order to establish correct gear clearances on the main and countershafts, the bearings in the transmission housing must be removed and fitted to the main and countershaft, as described in Section 12.*

Converter housing

1 Using a mandrel of suitable diameter, drive the bearing and seal from the torque converter housing.
2 Drive in a new bearing until it bottoms in the housing.
3 Install a new mainshaft seal, so that it is flush with the housing.
4 Remove the countershaft bearing, using a bearing puller.
5 Clean out the bearing housing and oil guide plate.
6 Refit the guide plate.
7 Drive in a new bearing until its face is flush with the housing surface.

Transmission housing

8 The bearings in the transmission housing are held in by snap-rings.
9 To remove the bearings, expand the snap-rings and push the bearings out by hand.
10 Do not remove the snap-rings unless it is necessary to clean out the grooves.
11 Install the bearings by expanding the snap-rings pushing in the bearing by hand, and listening for the 'click' as the snap-ring enters the groove in the bearing.
12 Bearings are installed with their part number facing the snap-ring.

11 Reverse idler gear – removal

1 Remove the idler shaft holder, then remove the idler gear shaft and bearing from the outside of the transmission housing.
2 Remove the idler gear and needle roller bearing.
3 Refitting is a reversal of removal, but install the idler gear with the larger chamfer facing towards the torque converter housing.

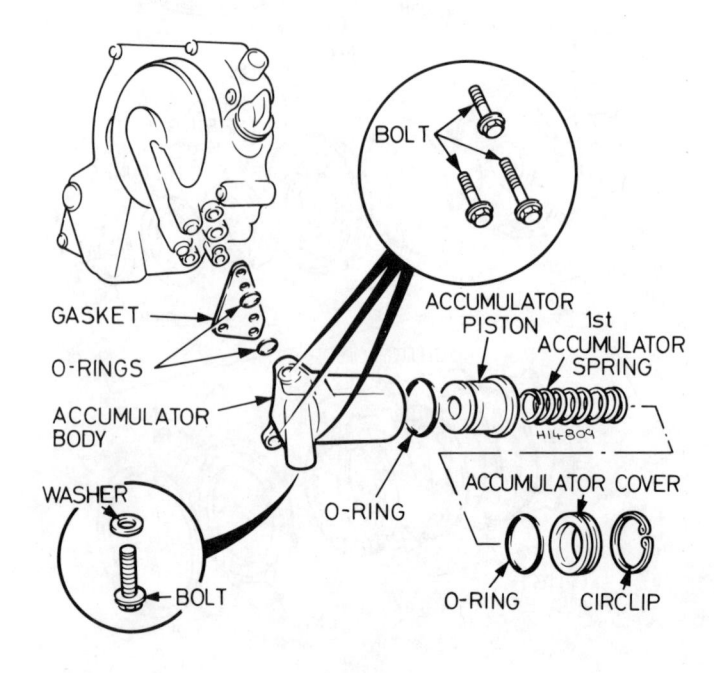

Fig. 7.56 1st accumulator components (Sec 9)

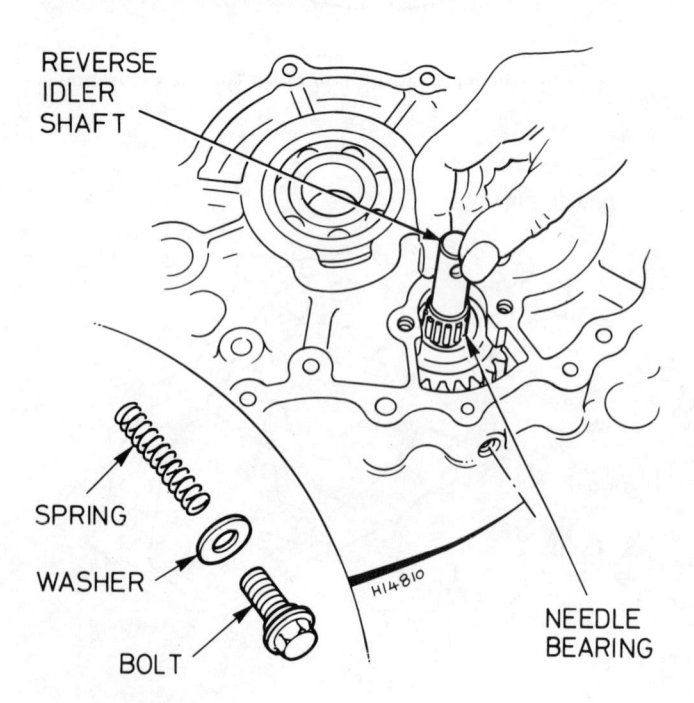

Fig. 7.57 Removing the reverse idler shaft (Sec 11)

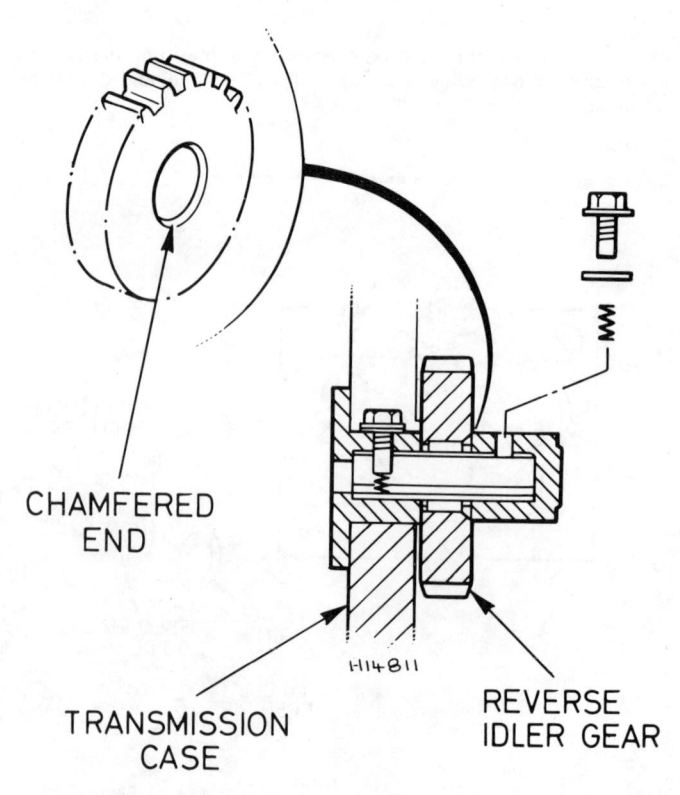

Fig. 7.58 Correct orientation of chamfer on idler gear (Sec 11)

12 Mainshaft and countershaft – gear clearance measurement

1 Before reassembling the transmission unit, the mainshaft and countershaft bearings have to be removed from the transmission housing and fitted to the shafts, so that the following measurements may be taken, as the shafts are built-up.
2 Reassemble the main and countershafts with reference to Figs. 7.49 and 7.50.
3 On all thrust needle bearings, the unrolled edge of the bearing cage faces the thrust washer.
4 As previously described, the mainshaft and countershaft bearings must be removed from the transmission casing and fitted to the shafts.
5 Install the built-up shafts into place in the converter housing.
6 Prevent the shafts from turning and tighten the locknuts to 25 lbf ft (34 Nm).
(**Note**: this torque figure is for measurement purposes only – final torque figures are given in the Specifications).
7 Measure the following clearances, using feeler gauges.

Countershaft

8 Measure the clearance between the shoulder on the selector hub and the shoulder on 4th gear.
9 If the clearance exceeds the service limit given in the Specifications, measure the thickness of the spacer collar, and select a spacer collar which will bring the clearance within limits.
10 Leave the feeler gauge in position between 4th gear and the selector hub, and measure the clearance between 2nd and 3rd gears, firstly with 3rd gear pushed fully out, and again with 3rd gear pushed fully in.
11 Calculate the difference between the two readings to give the actual clearance between the two gears.
12 If the service limit is exceeded, select a new splined thrust washer of suitable thickness to bring the clearance within limits.

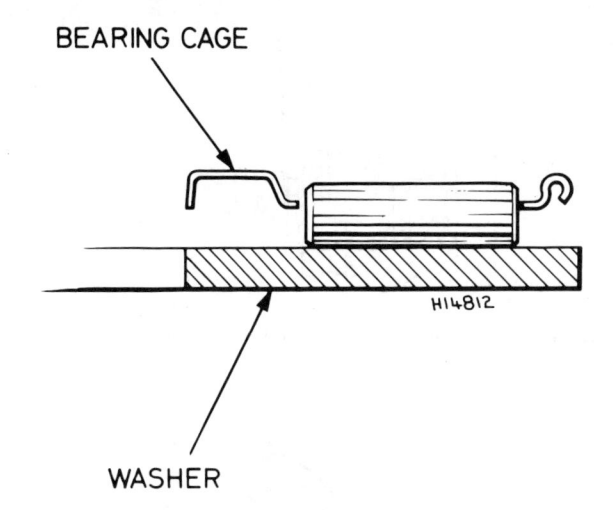

Fig. 7.59 Correct positioning of needle roller bearings (Sec 12)

Mainshaft

13 Measure the clearance between the shoulder of 2nd gear and 3rd gear.
14 If the clearance exceeds the service limit, measure the thickness of 2nd clutch thrust washer and select one to give the correct clearance.

All shafts

15 Both the main and countershaft should now be dismantled again, ready for assembly of the transmission unit.

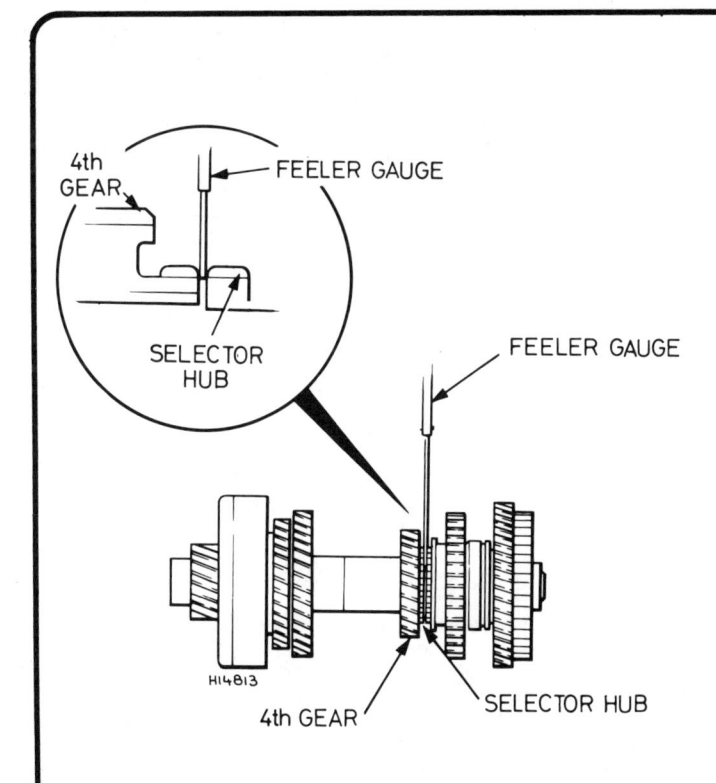

Fig. 7.60 Measure 4th gear clearance on the countershaft (Sec 12)

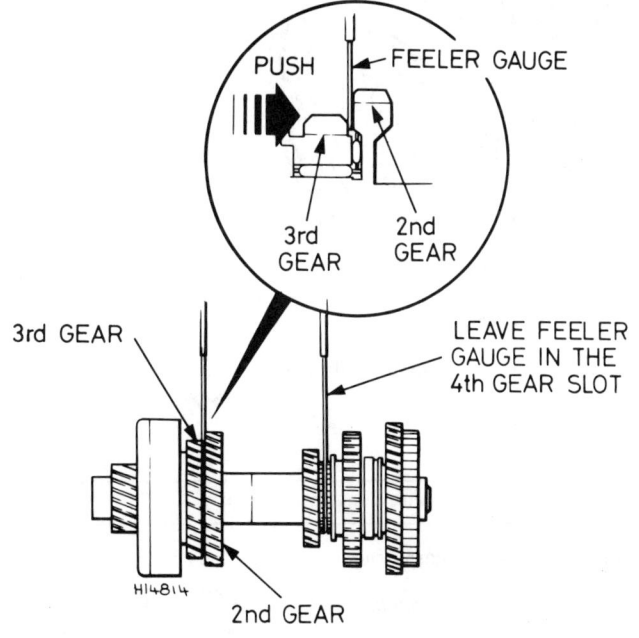

Fig. 7.61 Measure 3rd gear clearance on the countershaft (Sec 12)

198　　　　　　　　　　　Chapter 7 Automatic transmission

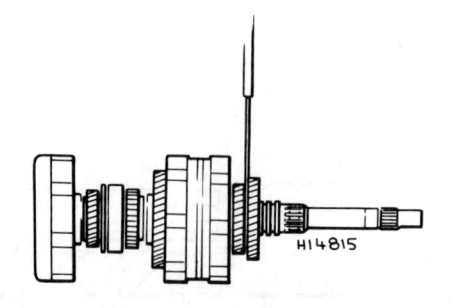

Fig. 7.62 Mainshaft 2nd gear clearance on the mainshaft (Sec 12)

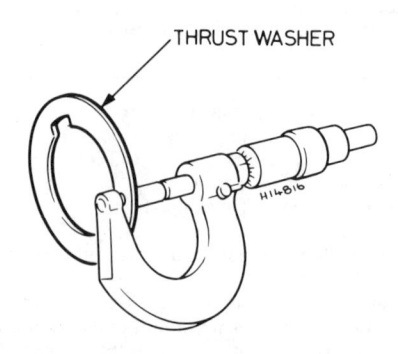

Fig. 7.63 Measuring a thrust washer using a micrometer (Sec 12)

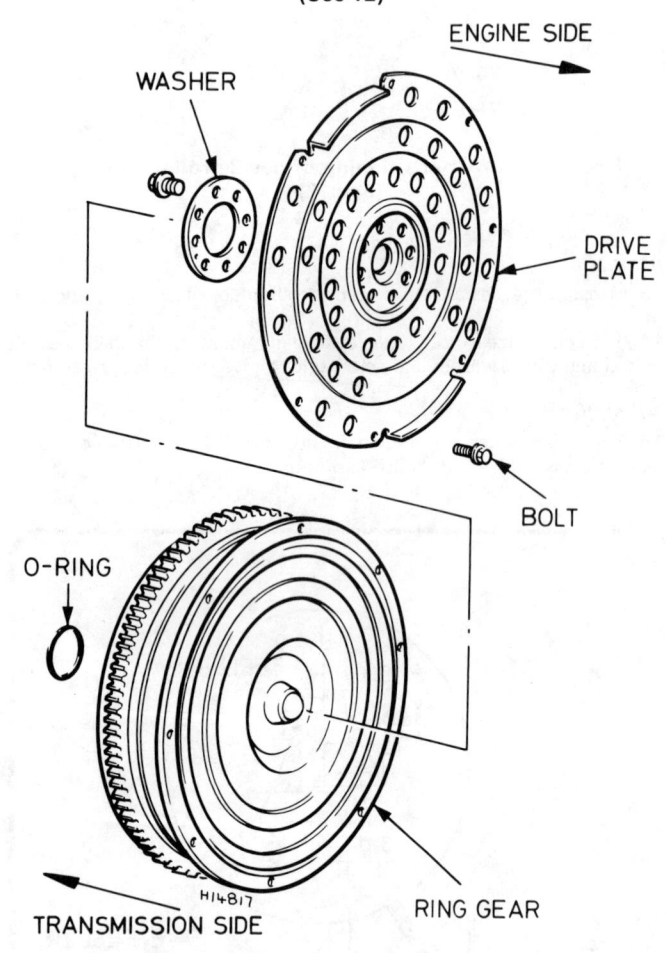

Fig. 7.64 Torque converter and driveplate assembly (Sec 13)

13 Torque converter – removal and refitting

1　On removal of the transmission unit, the torque converter will have been put to one side.
2　The driveplate will have remained bolted to the crankshaft.
3　To remove the driveplate, undo the eight bolts and remove it and the large washer.
4　Inspect the driveplate for cracks and distortion.
5　On refitting the driveplate, torque load the eight bolts to the correct loading shown in the Specifications in a criss-cross pattern.
6　The torque converter is a sealed assembly and cannot be dismantled for repair if it is faulty. It should be renewed as a complete unit.
7　When refitting the transmission unit, always use a new O-ring seal.

14 Differential oil seals – renewal

1　The procedure given in Chapter 6 for differential oil seal removal will suffice, substituting torque converter housing for clutch housing.
2　Clearances and tolerances are the same for manual and automatic differentials.

15 Transmission unit – reassembly

1　Lubricate all components liberally with transmission fluid during reassembly.
2　Install the differential assembly, referring to Chapter 6. If either the torque converter or transmission housing have been renewed, or the differential side bearings renewed, the differential side clearance must be checked.
3　Fit the manual valve lever to the control shaft and install it in the torque converter housing.
4　Turn the converter housing over and fit the control lever, tighten the bolt to the correct torque and bend over the locking tab.
5　Turn the housing over again, and fit the suction pipe and a new oil filter screen, with the support webs facing inwards.
6　Fit the separator plate, dowel pin, pump gears and shaft.
7　Fit the check valve and spring and install the main valve body onto the torque converter housing.
8　Tighten the four valve body bolts in the sequence shown (Fig. 7.66), to the torque figure given in the Specifications, frequently checking that the pump drivegear and shaft operate freely without binding.
9　Fit the stator shaft arm, stop pin and dowels.
10　Fit the regulator valve, torque load the bolts to their specified torque.
11　Fit the steel ball to its location in the regulator valve.
12　Fit the four steel balls to the main valve body oil passageways as shown in Fig. 7.67.
13　Fit the separator plate and dowel pins.
14　Install the servo valve, ensuring the correct length bolt is used in each thread, and tighten to the specified torque.
15　Fit the throttle control shaft and use a new E-clip to retain it in place.
16　Fit the rollers to each side of the manual valve stem, attach the valve to the lever with the shackle pin, and lock with the washer and split pin.
17　Install the clutch pressure control valve body, cover and separator plate and tighten the bolts to the torque setting.
18　Fit the 1st, 3rd and 4th clutch feed pipes.
19　Fit the separator plate.
20　Position the oil transfer pipes between the pressure control valve body and clutch pressure control valve body, and slide the pressure control valve into position, fit and tighten the bolts to torque.
21　Fit the accumulator springs in position, then fit the covers and depress the springs and covers when tightening the bolts to avoid stripping the threads in the housing. Tighten the bolts to torque.
22　Fit the governor valve, tighten the bolts to torque and bend over the locking tabs.
23　Lift the mainshaft and countershaft, mesh them together and fit to

Chapter 7 Automatic transmission

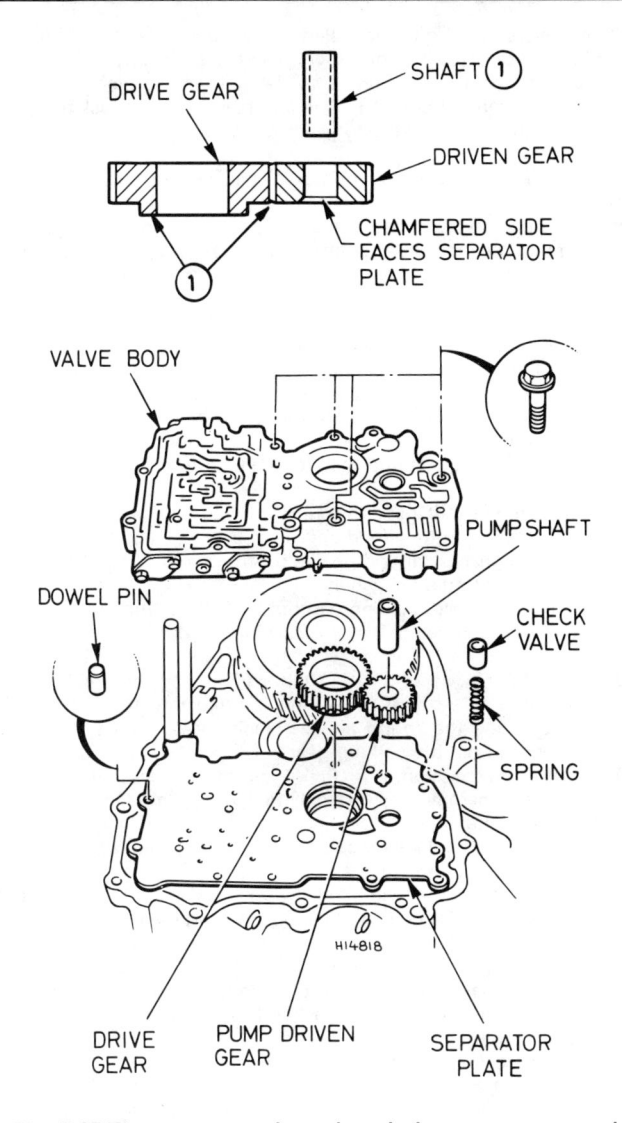

Fig. 7.65 The separator plate, dowel pin, pump gears and shaft (Sec 15)

1 Lubricate with ATF before reassembly

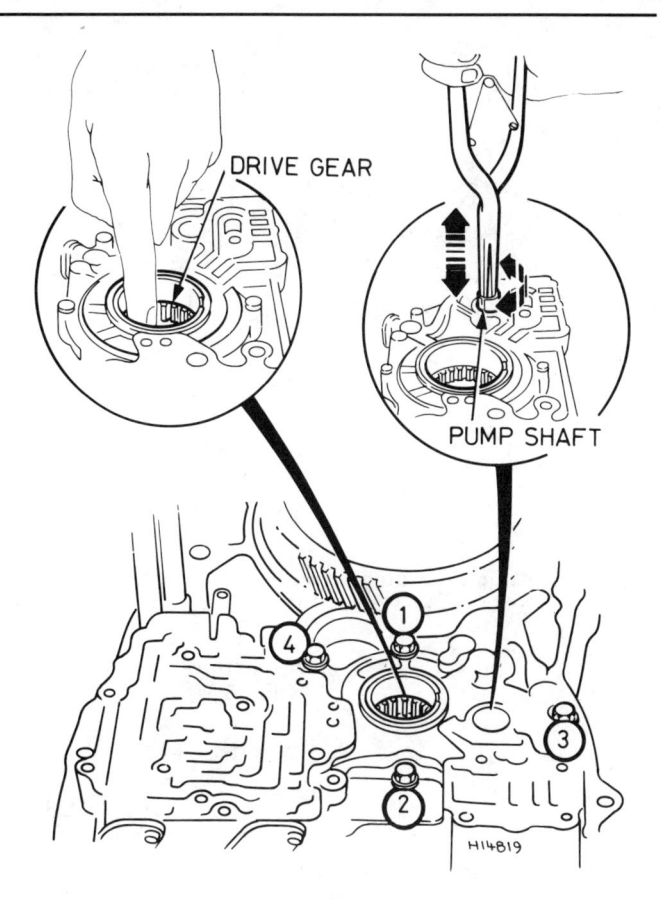

Fig. 7.66 Check valve retaining bolt tightening sequence (Sec 15)

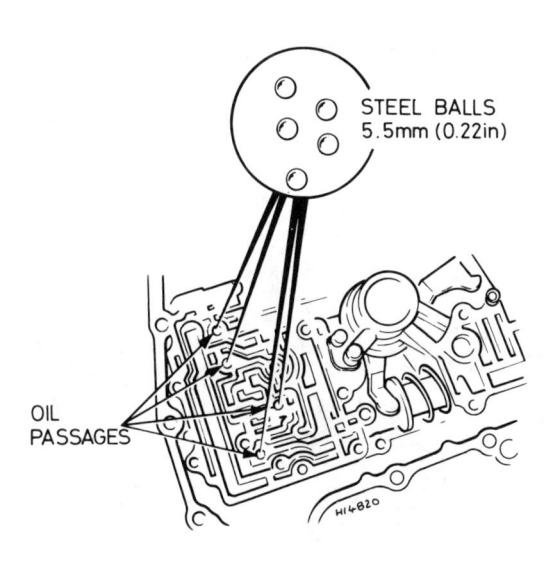

Fig. 7.67 Location of the 4 steel balls in the main valve (Sec 15)

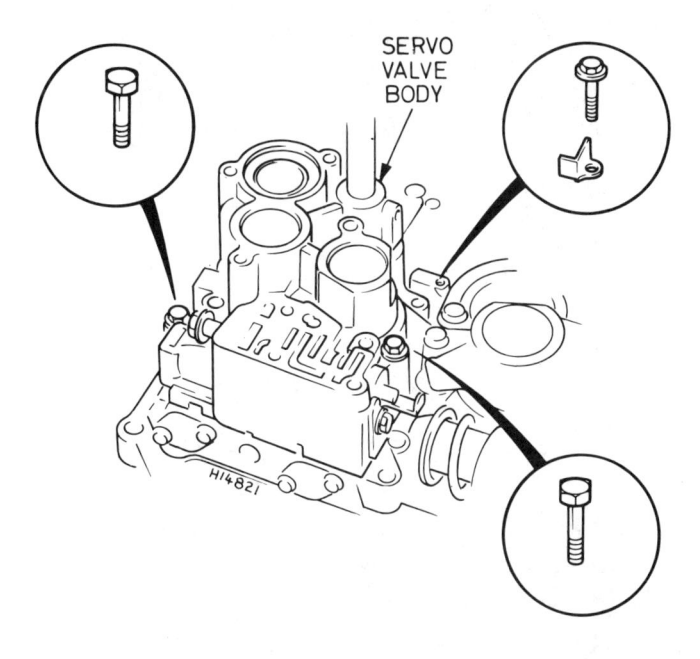

Fig. 7.68 Servo valve retaining bolt lengths (Sec 15)

the transmission housing as an assembly. (Do not be tempted to tap them home with a hammer).

24 Install the 4th gear and its needle roller bearing and the countershaft 4th gear and its selector hub.

25 Assemble the reverse shift fork and selector sleeve, and install them as an assembly on the countershaft, the flat face of the sleeve facing upward.

26 The reverse shift fork is installed with its unmarked side facing up.
27 Slide the reverse shift fork down over the servo valve stem.
28 Align the hole in the fork with the hole in the stem, fit a new lockplate, tighten the bolt to torque and bend over the locking tab.
29 Fit the countershaft reverse gear, needle bearing and reverse gear collar.
30 Fit a new gasket and three dowel pins to the torque converter housing.
31 Position the transmission housing onto the converter housing; ensuring that the main valve control shaft lines up with the hole in the housing, and that the reverse idler gear meshes with the mainshaft and countershaft or the housing will not go on.
32 Also make sure the housing is sited correctly on the dowels, before fitting and torque loading the retaining bolts in the sequence shown (Fig. 7.71) in two or more stages. Take care when tightening the housing bolts that the throttle control bracket is not damaged or distorted, or the transmission shift points will be altered.
33 Fit the throttle control lever and spring onto the throttle control shaft, fit a new lockplate and bolt, torque load the bolt and bend over the locking tab.
34 Fit the parking shift arm and spring onto the shift shaft with the bolt and new lockplate, torque load and bend over the tab.
35 Fit the 1st gear collar and needle bearing on the countershaft.
36 Fit the collar to the mainshaft.
37 Fit the reverse idler bearing and holder bolt, spring and washer.
38 Install the countershaft 1st gear and parking gear onto the countershaft.
39 Fit the stop pin, parking pawl shaft, parking pawl and pawl release spring, so that the end of the spring is fitted into the hole on the parking pawl and the spring applies clockwise tension to the pawl, forcing it away from the parking gear.
40 Shift the transmission to P (Park), and fit the mainshaft holder, or devise some other method of locking the transmission, while the locknuts are tightened.
41 Fit and torque load a new countershaft locknut and then stake the locking collar to the groove in the gear.
42 Fit the thrust washer and needle roller bearing to the mainshaft.
43 Install 1st gear, the needle thrust bearing and the thrust washer on the mainshaft.
44 Fit the 1st clutch
45 Again, the shafts must be locked while the locknut is tightened.
46 Fit and torque load a new mainshaft locknut, remembering it has a **left-hand thread**.
47 Stake the locknut to the groove in 1st clutch.
48 Fit a new gasket, dowel pins and O-rings to the transmission housing, fit the end cover and torque load the retaining bolts.
49 Fit the transmission cooler pipe banjo union, but do not tighten it until the oil cooler pipe is positioned correctly after the transmission is installed.

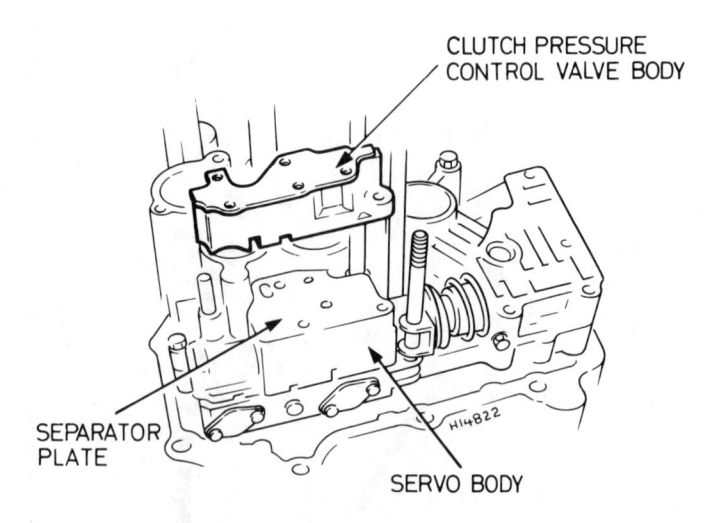

Fig. 7.69 Fitting the clutch pressure control valve assembly (Sec 15)

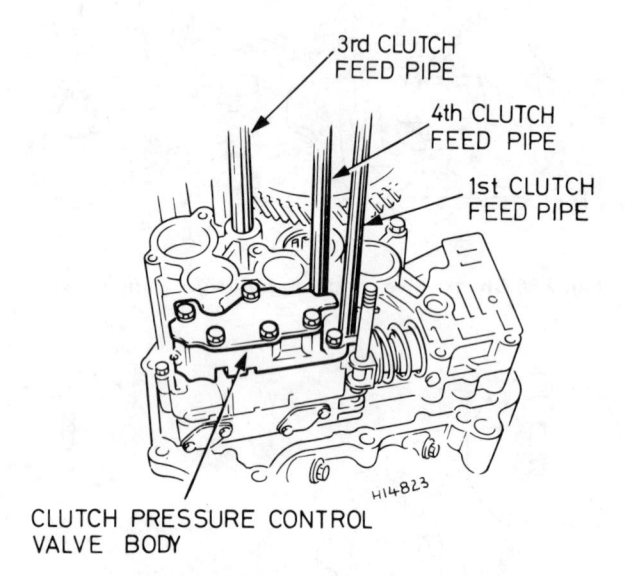

Fig. 7.70 Oil feed pipe locations (Sec 15)

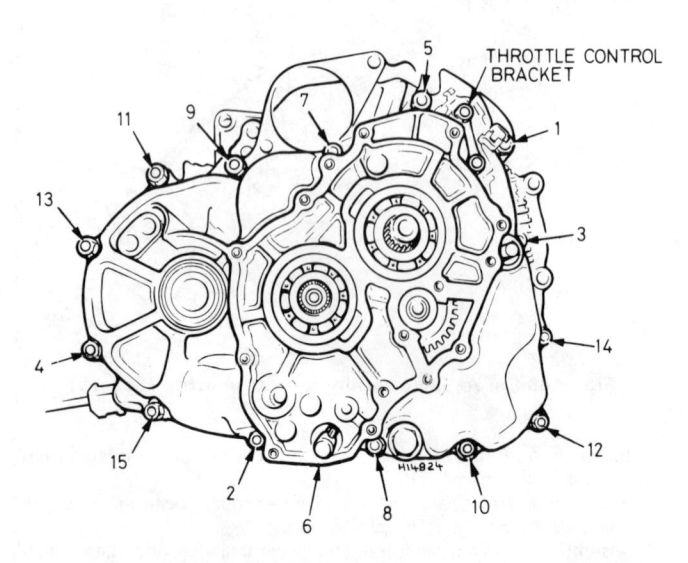

Fig. 7.71 Transmission housing retaining bolt tightening sequence (Sec 15)

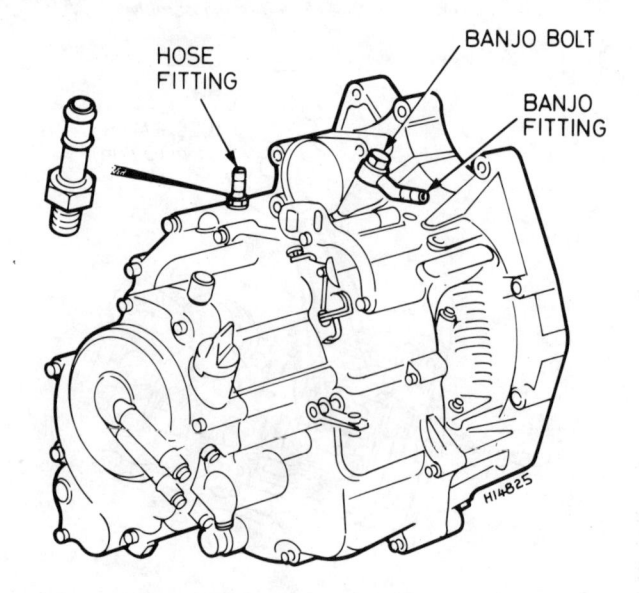

Fig. 7.72 Oil cooler fittings on transmission housing (Sec 15)

Chapter 7 Automatic transmission

50 Fit and tighten the oil cooler hose fitting.
51 Fit the dipstick.
52 The transmission unit is now ready for installation.

16 Gearshift selector – removal and refitting

1 An exploded view of the gear shift selector mechanism appears in Fig. 7.73.
2 The selector mechanism may be removed after removing the centre console and disconnecting the shift cable at its attachment to the selector lever.
3 Adjust the cable as described in Section 18.

17 Gearshift indicator – adjustment and checks

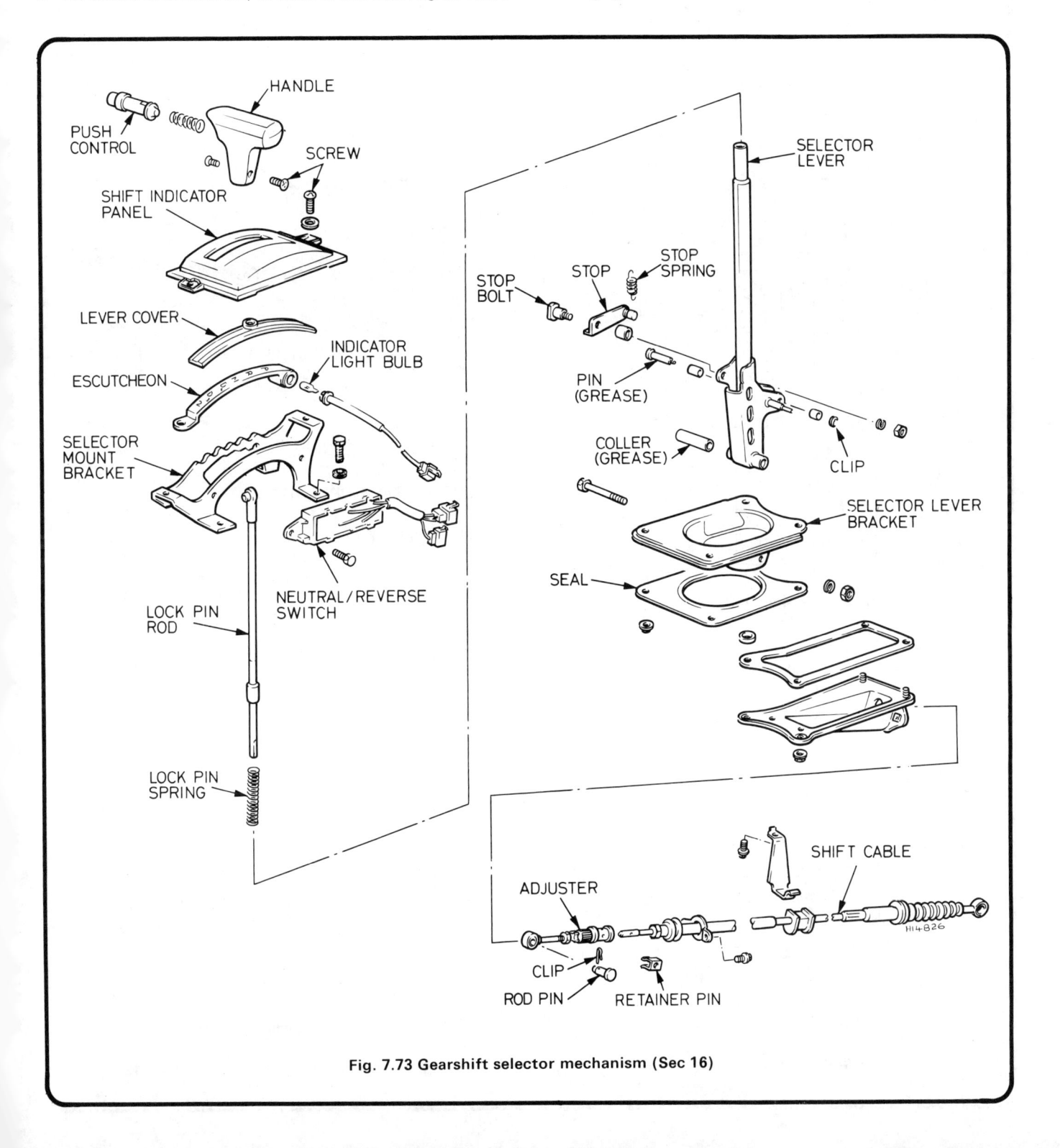

Neutral/reversing light switch test

1 Move the selector lever to Park, Reverse and Neutral to check continuity of the combined neutral safety (inhibitor) switch and the reversing light switch.

Fig. 7.73 Gearshift selector mechanism (Sec 16)

INHIBITER SWITCH

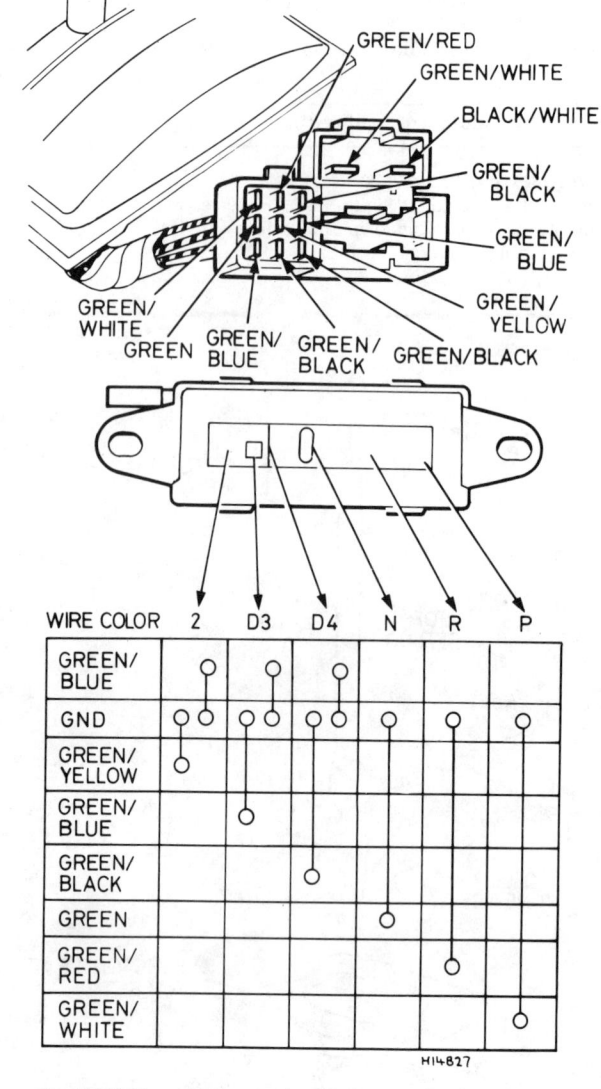

WIRE COLOR	N	R	P
BLACK/WHITE	O——————————O		O——————O
GREEN/BLACK	O——————O	O——————O	
BLACK/WHITE	O	O	O
GREEN/BLACK		O	O

Wire labels on connector: GREEN/RED, GREEN/WHITE, BLACK/WHITE, GREEN/BLACK, GREEN/BLUE, GREEN/YELLOW, GREEN/BLACK, GREEN/BLUE, GREEN, GREEN/WHITE

WIRE COLOR	2	D3	D4	N	R	P
GREEN/BLUE		O	O			
GND	O O	O O	O O	O	O	O
GREEN/YELLOW	O					
GREEN/BLUE		O				
GREEN/BLACK			O			
GREEN				O		
GREEN/RED					O	
GREEN/WHITE						O

Fig. 7.74 Neutral/reversing light continuity check (Sec 17)

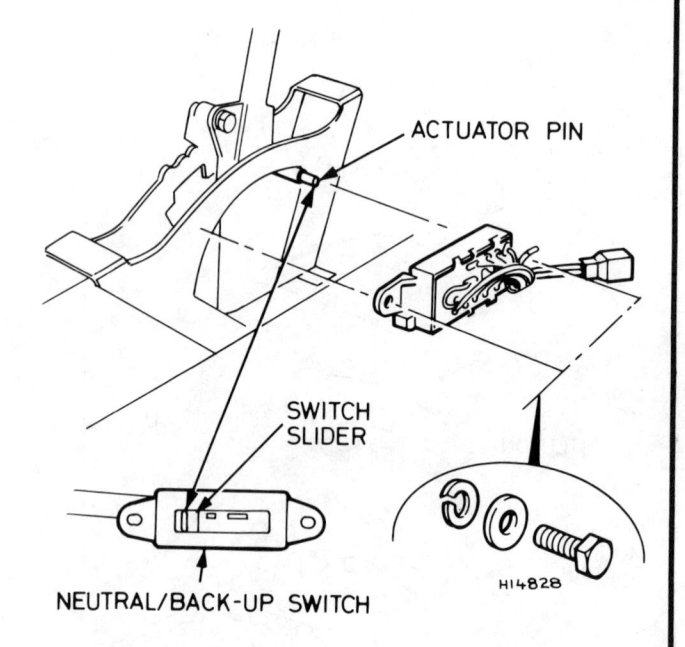

Fig. 7.75 Inhibitor switch continuity check (Sec 17)

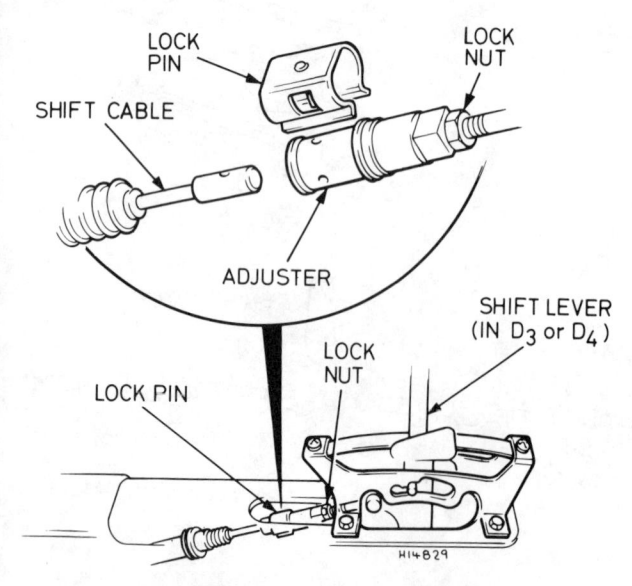

Fig. 7.76 Shift cable adjuster (Sec 18)

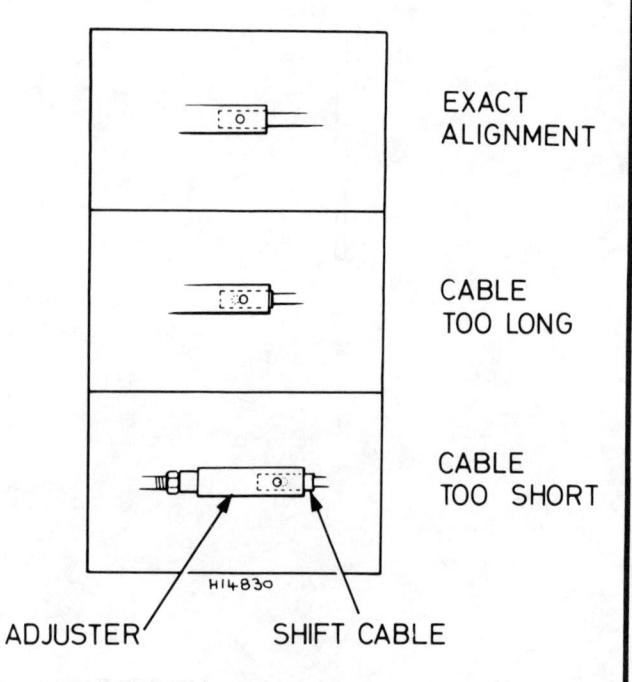

Fig. 7.77 The cable alignment hole (Sec 18)

2 Replace the switch if there is no continuity between the terminals shown in Fig. 7.74.
3 The switch is bolted within the gearshift mounting bracket.
4 The shift indicator light is a push-fit in a holder on the underside of the indicator escutcheon.
5 Whenever the escutcheon is removed, realign it with the index mark on Neutral before tightening the retaining screws.

18 Shift cable – adjustment

1 Remove the centre console.
2 Shift to Drive and remove the lock pin from the cable adjuster.
3 Check that the hole in the adjuster is perfectly aligned with the hole in the shift cable.
4 There are two holes in the end of the shift cable, positioned 90° apart to allow cable adjustment in $1/4$ turn increments.
5 If not perfectly aligned, loosen the locknut on the shift cable and adjust as necessary.

19 Fault diagnosis – automatic transmission

The tracing of faults in automatic transmission units requires the use of specialist equipment which only your local dealer will have. Most minor problems will be due to low fluid level and badly adjusted control cables. It is recommended that the tracing of major faults be left to your dealer.

Chapter 8 Driveshafts

Contents

Driveshaft inboard joint – dismantling, inspection and reassembly	4	Fault diagnosis – driveshafts	6
Driveshaft outboard joint – inspection	5	General description	1
Driveshafts – removing and refitting	3	Routine maintenance	2

Specifications

Type .... Open shafts of unequal length with constant velocity joints at each end driving the front roadwheels

Lubrication
Type .... Molybdenum disulphide grease
Quantity .... 3.50 to 3.88 oz (100 to 110g)

Length setting
Right .... 20.2 to 20.4 in (514.0 to 518.5 mm)
Left .... 31.9 to 32.0 in (809.0 to 813.5 mm)

Torque wrench settings
	lbf ft	Nm
Spindle nut	137	186
Balljoint pinch-bolt	40	54
Stabiliser bar bolt	16	22
Tie-rod balljoint nut	32	43

1 General description

The driveshafts are of open type, transmitting engine power from the final drive unit in the gearbox to the front wheels.

The driveshafts are of unequal length simply because the transmission unit is not on the centre line of the vehicle.

There are constant velocity joints at each end of each driveshaft covered by a protective rubber boot. Only the two inboard joints may be dismantled. If the outboard joints become worn then the whole driveshaft must be renewed.

Chapter 8 Driveshafts

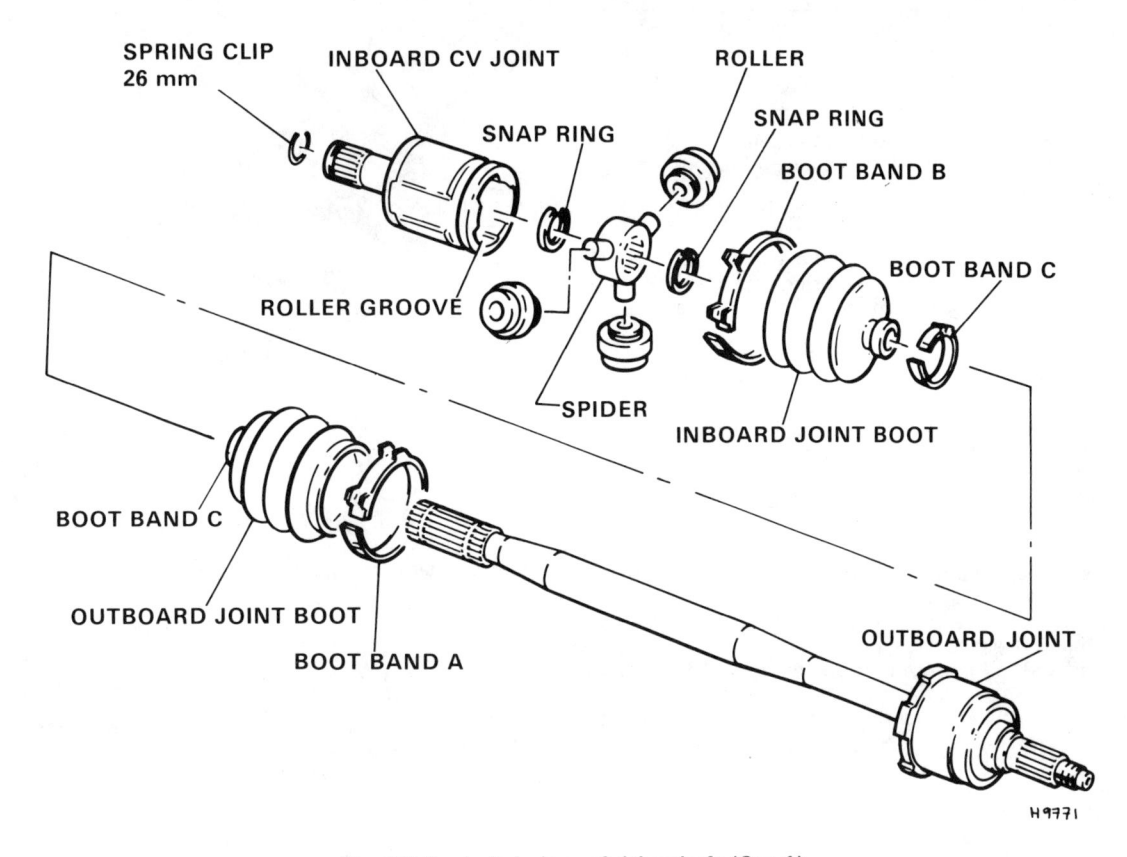

Fig. 8.1 Exploded view of driveshaft (Sec 1)

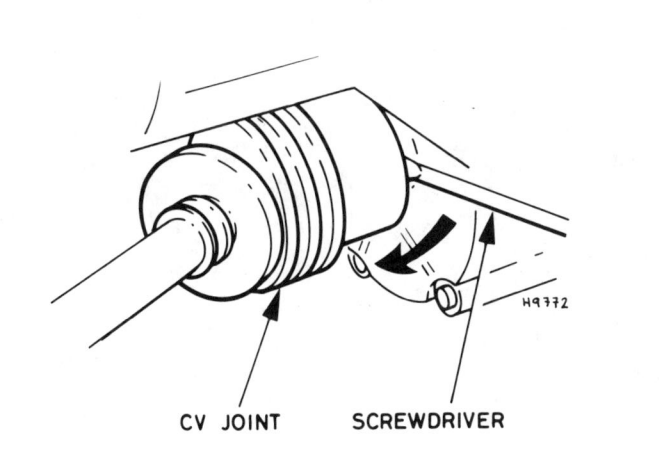

Fig. 8.2 Prising out the inboard joint (Sec 3)

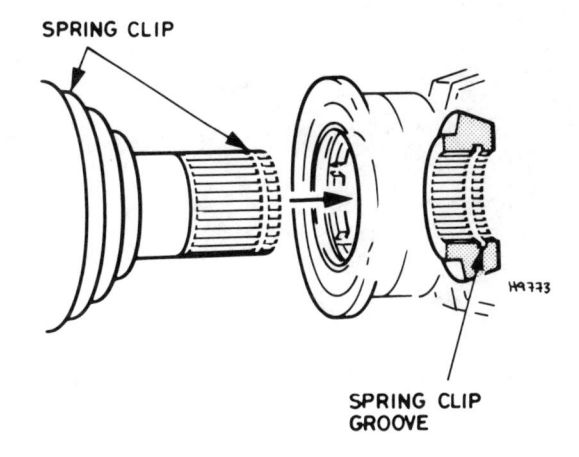

Fig. 8.3 The spring clip and locating groove in the side gear (Sec 3)

2 Routine maintenance

At regular intervals (see Routine Maintenance at the front of the book).

Check the driveshaft constant velocity joints protective rubber bellows for splits, cracking, leakage and security. Where the vehicle is operated under severe conditions, reduce the service interval.

3 Driveshafts – removal and refitting

1 Refer to Chapter 10 and separate the lower arm from the knuckle.
2 On vehicles equipped with anti-lock brakes, remove the bolts securing the sensor lead to the suspension strut.
3 Drain the transmission oil, or transmission fluid on automatic models.
4 Push the front hub outwards off the driveshaft, being careful not to apply any pulling force on the inboard joint, or it may come apart.
5 Pry the inboard joint out from its housing in the differential unit by about 1/2 in (12 mm), using a large screwdriver or flat bar, being careful not to damage the oil seal or housing. **Note**: do not attempt to pull the joint out, it may come apart.
6 Once the circlip type spring has been dislodged from the groove in the differential side gear, the driveshaft can be lifted clear.
7 Refitting is a reversal of removal, with the following addition.
8 Always use a new spring circlip and ensure that the inboard joint bottoms correctly in the differential, so that the circlip locks in the groove.

Chapter 8 Driveshafts

4 Driveshaft inboard joint – dismantling, inspection and reassembly

Dismantling

1 Place the driveshaft in a vice, inboard joint uppermost.
2 Prise up the locking tabs on the bellows retaining clips and remove the clips (photo).
3 Pull the bellows down over the joint, towards the shaft, exposing the joint.
4 Clean off any excess grease from around the joint.
5 Mark each roller in relation to the joint, using a centre punch. so that each roller is refitted into the same groove on the joint that it came out of.
6 Remove the joint, being careful not to dislodge the rollers.
7 Remove each roller from the spider, and remove the inner snap-ring.
8 Remove the outer snap-ring from the driveshaft, pull off the spider and remove the inner snap-ring. (Note the thicknesses of the snap-rings and replace in same position).
9 Pull off the rubber bellows.

Inspection

10 Wash all parts in petrol and dry off using a non-fluffy rag. Inspect each part for signs of wear, damage and cracks. If excessive wear is evident, then replace the whole unit.
11 Inspect the rubber bellows for signs of perishing, cracking or splitting. It would be sound practice to renew these anyway, having dismantled the joint.
12 Do not forget to inspect the rubber bellows on the outboard joint at this stage, as the inboard joint must be removed to fit the outboard bellows.
13 Lubricate all parts with molybdenum disulphide grease before commencing reassembly.

Reassembly

14 Lubricate the inside of the bellows with the recommended grease and slide it on to the driveshaft (photo).

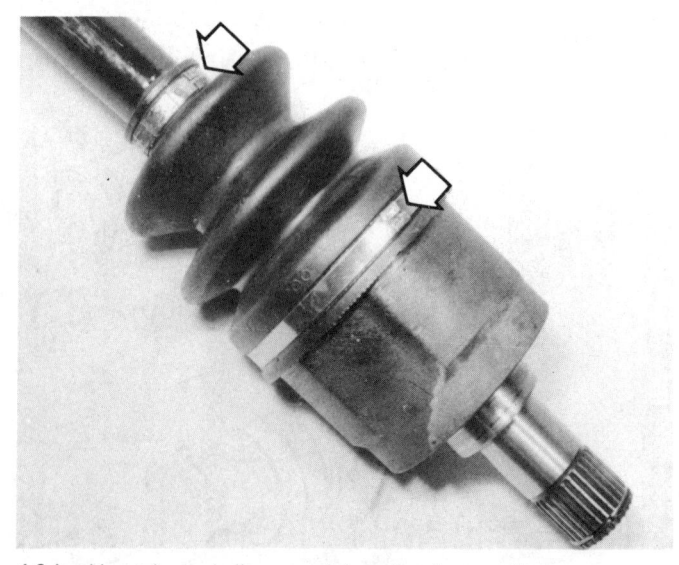

4.2 Locking tabs on bellows retaining clips (arrowed)

15 Fit the inner snap ring (photo).
16 Grease the shaft splines, and slide on the spider (photo).
17 Fit the outer snap-ring (photo).
18 Refit the rollers to the spider, ensuring they are in their original positions (photo).
19 Refer to the Specifications and pack the inside of the joint housing with the correct quantity of recommended grease, before fitting it over the rollers (photo).
20 Pull the rubber bellows up over the joint (photo).
21 Set the driveshaft to the correct length (see Specifications and Fig.

4.14 Slide the bellows onto the driveshaft

4.15 Fit the inner snap-ring

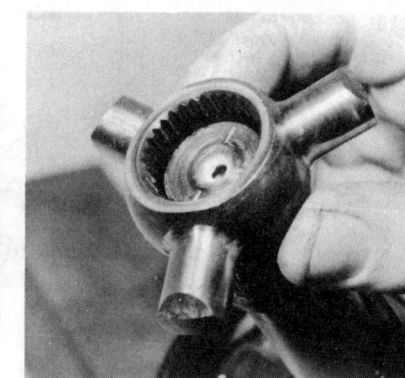

4.16 Slide on the spider

4.17 Fit the outer snap-ring

4.18 Refit the rollers on the spider

4.19 Fitting the joint housing

Chapter 8 Driveshafts

8.4) and adjust the rubber bellows so that they are halfway between full compression and full extension.

22 Fit new bellows retaining clips and lock them in position by bending down the tabs, so that the clips are positioned midway between the locating ridges on the driveshaft (photo).

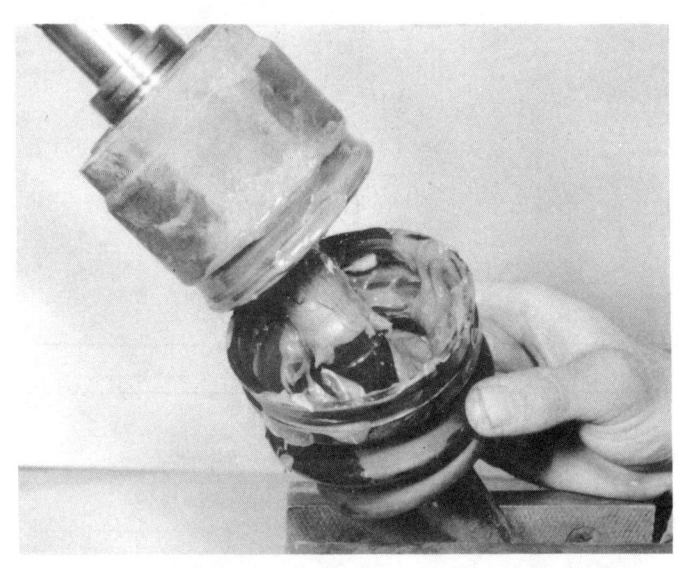

4.20 Fitting the rubber bellows

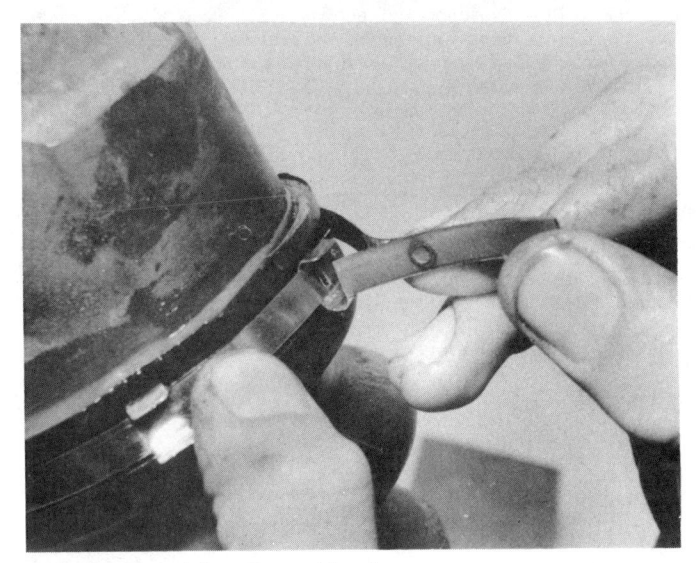

4.22 Fit new retaining clips and bend over the locking tabs

5 Driveshaft outboard joint – inspection

1 As mentioned previously the outboard joint cannot be dismantled, and should be replaced as a unit if the joint becomes worn.
2 Whenever the driveshafts are removed and the inboard joint dismantled, the opportunity should be taken to renew the outboard rubber bellows.
3 Remove the old bellows in the same manner as the inboard bellows. Clean old grease from the joint and pack the joint with fresh grease.
4 Lubricate the inside of the new bellows with grease and fit it over the driveshaft and down onto the joint.
5 Fit the retaining clips.

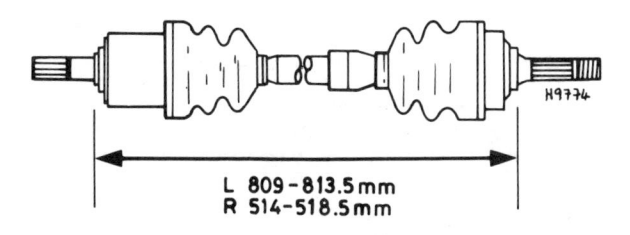

Fig. 8.4 Driveshaft length setting diagram (Sec 4)

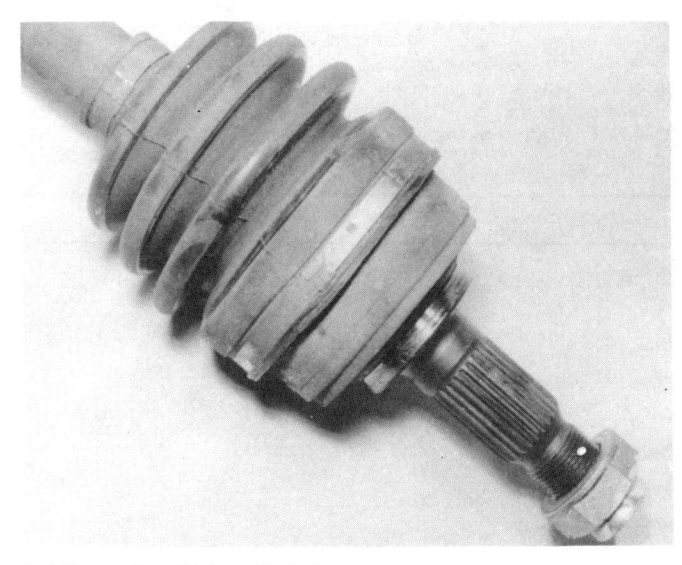

L 809 – 813.5 mm
R 514 – 518.5 mm

5.3 The outboard joint with bellows retaining clips correctly locked

6 Fault diagnosis – driveshafts

Symptom	Reason(s)
Vibration	Worn joints
	Worn wheel bearings
	Worn differential bearings
Noise on taking up drive	Worn driveshaft splines
	Worn joints
	Loose spindle nut

Chapter 9 Braking system

Contents

Anti-lock braking system (ALB)	26
Brake booster servo unit – description and maintenance	18
Brake booster servo unit – removal and refitting	19
Brake fluid level switch – testing	23
Brake light switch – removal, testing and refitting	25
Brake pedal – removal, refitting and adjusting	24
Brake system – bleeding	17
Duel proportioning valve – general description	14
Fault diagnosis – braking system	27
Front brake disc – inspection and renovation	7
Front disc brake caliper – removal, overhaul and refitting	6
Front disc pads – inspection and renewal	3
General description	1
Handbrake – adjustment	20
Handbrake cable – removal and refitting	21
Handbrake warning light switch – adjusting and testing	22
Hydraulic pipes and hoses – general	16
Load sensing valve – removal, refitting and adjusting	15
Master cylinder – removal, overhaul and refitting	13
Rear brake disc – inspection and renewal	8
Rear brake drum – inspection and renewal	12
Rear disc brake caliper – removal and refitting	10
Rear disc pads – inspection and renewal	4
Rear drum brake lining – inspection and renewal	5
Rear wheel cylinder – removal, overhaul and refitting	11
Routine maintenance	2
Wheels hubs, bearings and discs – general	9

Specifications

Type .... Hydraulic, servo-assisted, ventilated front disc, rear drum or disc, diagonally split system, self-adjusting with cable-operated handbrake

Front disc
Diameter:
   UK models .... 7.5 in (190 mm)
   North American models .... 7.6 in (194 mm)
Disc run-out (maximum):
   UK models .... 0.006 in (0.15 mm)
   North American models .... 0.004 in (0.10 mm)
Minimum thickness:
   UK models .... 0.59 in (15.0 mm)
   North American models .... 0.67 in (17.0 mm)
Pad thickness – lining only:
   Standard .... 0.39 in (10.0 mm)
   Minimum .... 0.06 in (1.6 mm)

Rear disc
Diameter .... 9.40 in (239 mm)
Run-out (maximum) .... 0.006 in (0.15 mm)
Minimum thickness .... 0.31 in (8.0 mm)
Pad thickness – lining only:
   Standard .... 0.31 in (8.0 mm)
   Minimum .... 0.06 in (1.6 mm)

Rear drum brake
Maximum inside diameter .... 7.91 in (201.0 mm)
Shoe thickness – lining only:
   Standard .... 0.177 in (4.5 mm)
   Minimum .... 0.079 in (2.0 mm)

Master cylinder
Piston-to-pushrod clearance .... 0 to 0.016 in (0 to 0.4 mm)

Chapter 9 Braking system

Brake pedal
Pedal height	7.4 in (187.0 mm)
Free play	0.04 to 0.20 in (1.0 to 5.0 mm)

Torque wrench settings

	lbf ft	Nm
Bleed screw:		
Disc brake	7	9
Drum brake	5	7
Banjo bolt	25	34
Front caliper guide bolts	20	27
Rear caliper guide bolts	22	30
Front caliper mounting bolts	56	76
Rear caliper mounting bolts	28	38
Rear caliper protector bolts	7	9
Brake booster servo unit-to-bulkhead nuts	9	12
Master cylinder-to-brake booster servo unit nuts	5	7
Master cylinder stop-bolt	6	8
Brake booster servo unit pushrod locknut	7	9
Drum brake backplate bolts	22	30
Brake pipe unions:		
Brake pipe to disc caliper	25	34
Brake pipe to rear wheel cylinder	10	14
All other unions	10	14
Handbrake cable mounting bolts	7	9

1 General description

The brake system is hydraulic, servo-assisted acting upon all four wheels. Depending on model, either front disc brakes with rear drum, or front and rear disc brakes are fitted. In either case, the brakes are self-adjusting.

The brake piping, apart from being rust treated, is as far as possible routed inside the vehicle to give greater protection.

Incorporated in the system, again depending on model, are a brake load sensing valve, and a dual proportioning valve, fitted to give greater breaking efficiency.

The ratchet type, centrally-mounted handbrake is cable-operated, acting upon the rear wheels.

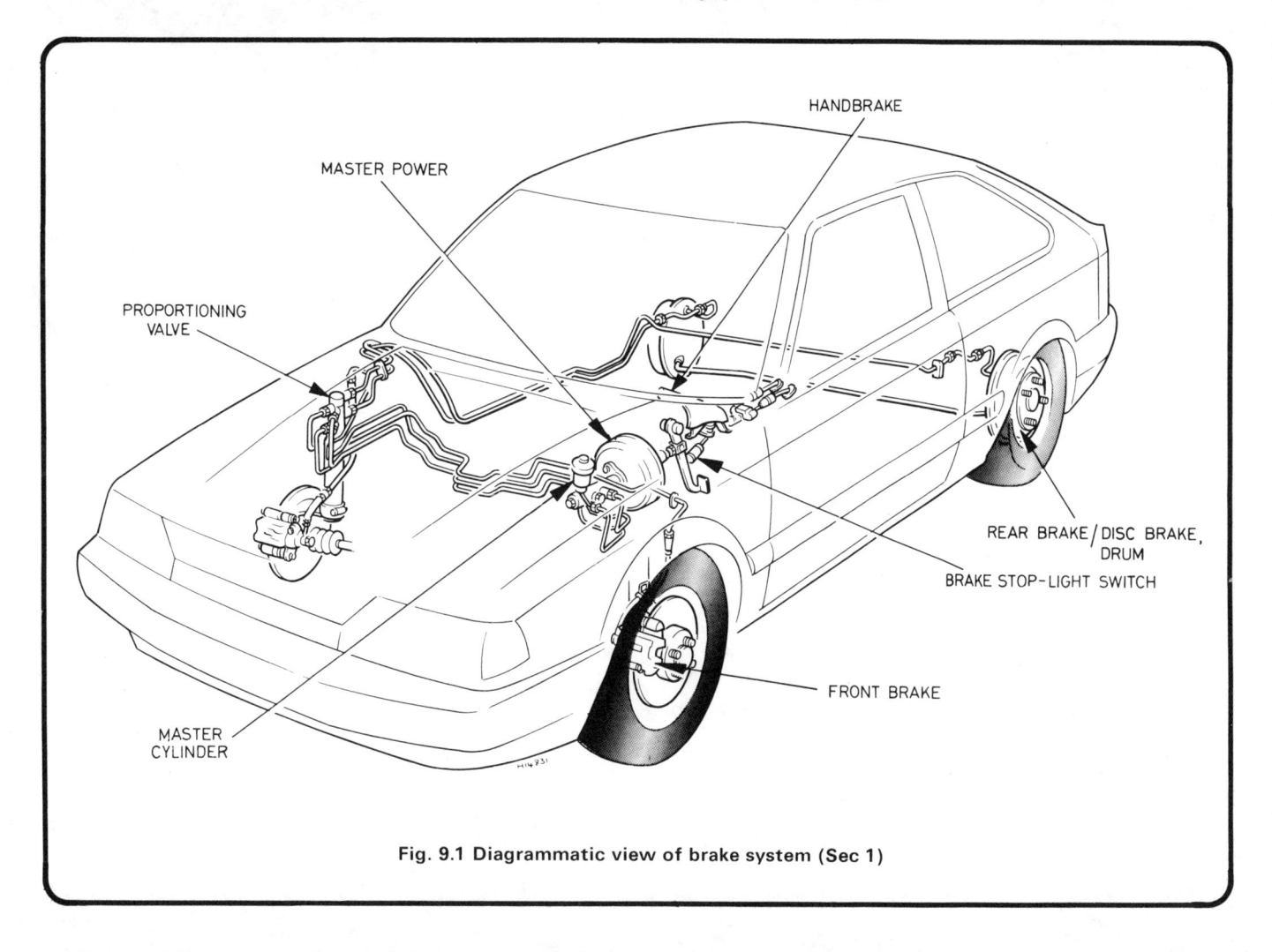

Fig. 9.1 Diagrammatic view of brake system (Sec 1)

2 Routine maintenance

At the intervals specified in Routine Maintenance (at the front of the book) undertake the following service tasks.
1 Check the level of hydraulic fluid in the reservoir on the brake master cylinder. The frequent need to top up would indicate a leak at some point in the circuit.
2 Inspect the front brake pads for wear.
3 Inspect all brake hoses and pipe lines (including ALB, if fitted). Inspect the front brake discs and calipers. Inspect the handbrake for correct operation, and check the operation of the ALB if fitted. Check the load sensing valve for correct operation.
4 Renew the brake system hydraulic fluid (including ALB). Inspect the rear brakes, and, on ALB systems, renew the high pressure hoses.

3 Front disc pads – inspection and renewal

1 Raise the front of the vehicle, support it safely on stands, and remove the road wheels. **Note:** *brake pads should always be renewed in pairs to avoid uneven braking and should never be swopped around to compensate for wear.*
2 Remove the lower guide bolt and pivot the brake caliper up out of the way (photos).

3.2A Remove the lower guide bolt ...

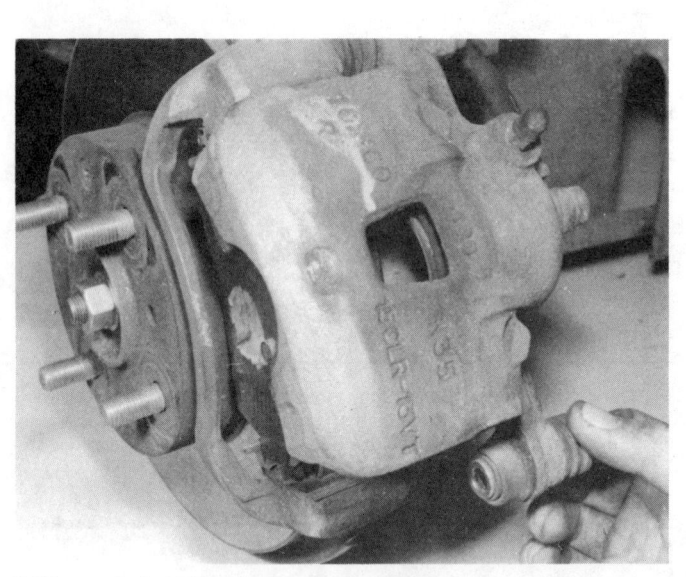

3.2B ... and pivot the caliper up out of the way

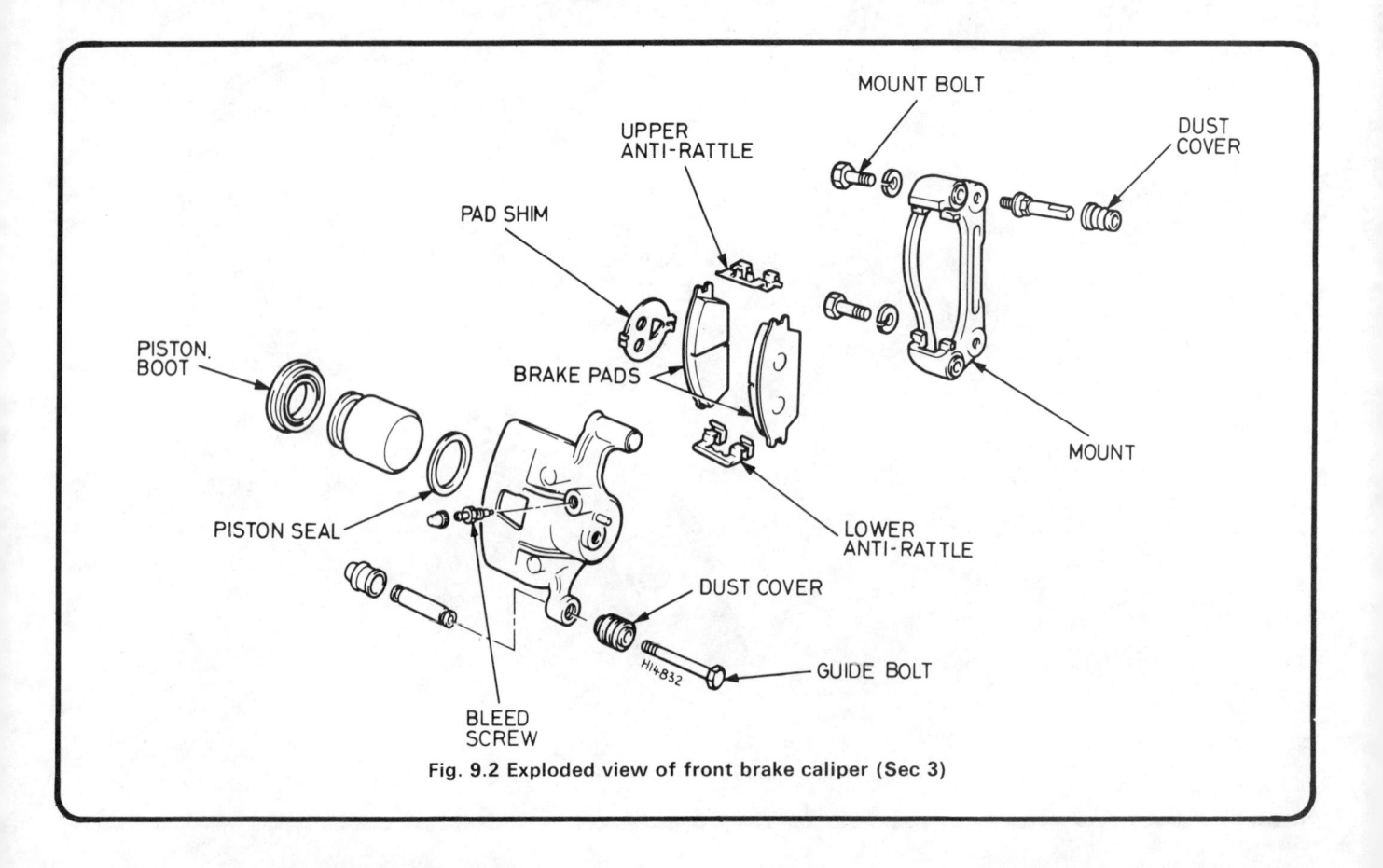

Fig. 9.2 Exploded view of front brake caliper (Sec 3)

Chapter 9 Braking system

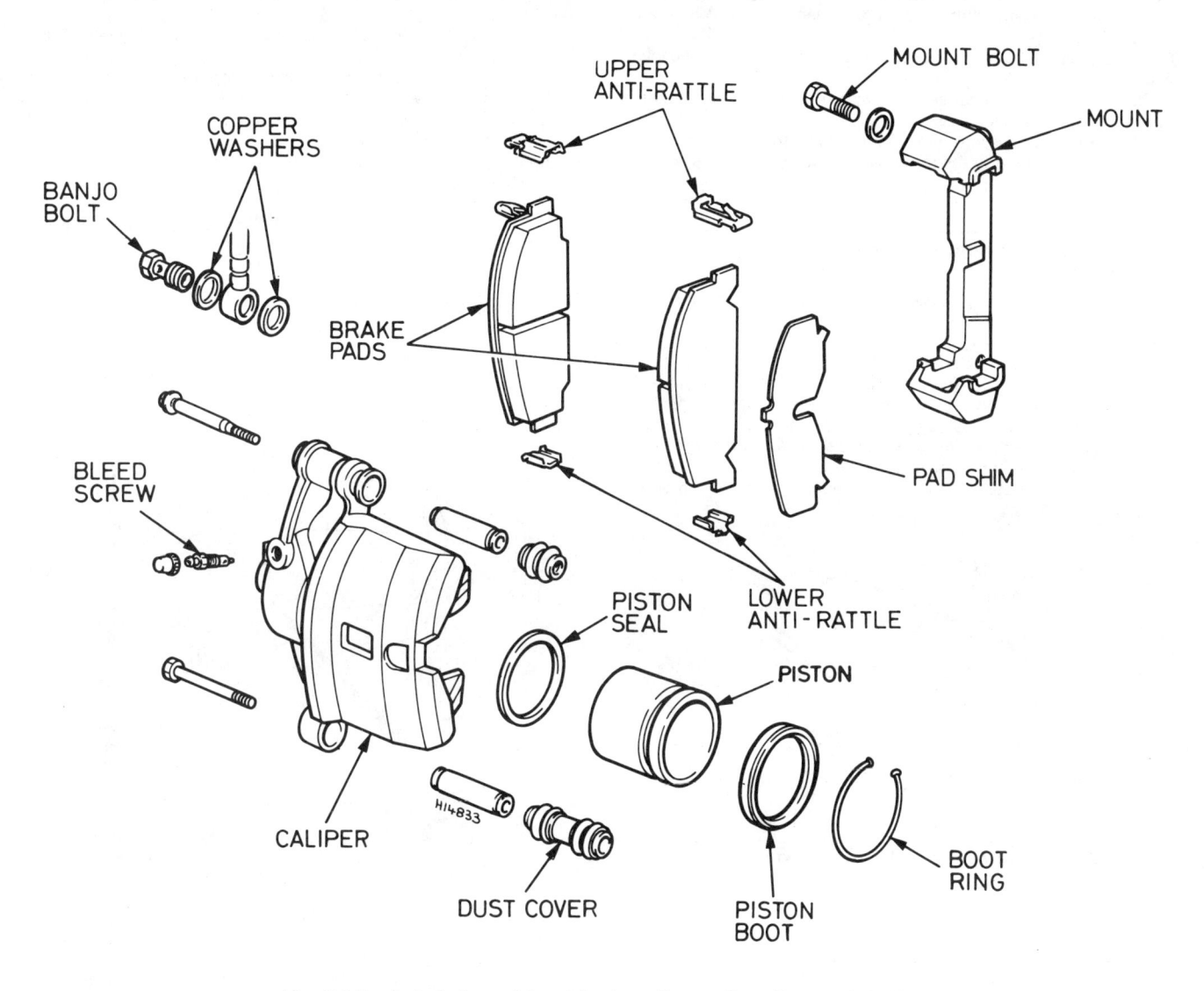

Fig. 9.3 Exploded view of front brake caliper – Canadian models (Sec 3)

3 Remove the brake pads, shims and anti-rattle springs (photos).
4 Measure the thickness of each pad and renew them if they are worn beyond limits (see Specifications).
5 Use a wire brush to clean away all dust from the caliper and caliper mount, and give the disc a brush down also. *Wear a mask as the dust formed from the brake pads may be injurious to health.*

6 Apply a silicone-based grease to all metal-to-metal contact areas, but do not allow any on to the disc or brake pad lining.
7 Refit the pads in the reverse order.
8 On brakes fitted with a wear indicator device, the pad with the wear indicator goes on the inside.
9 If there is insufficient clearance between the brake caliper and

3.3A Remove the brake pads ...

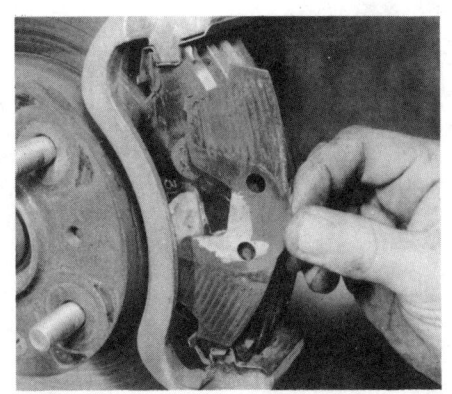

3.3B ... the pad shims

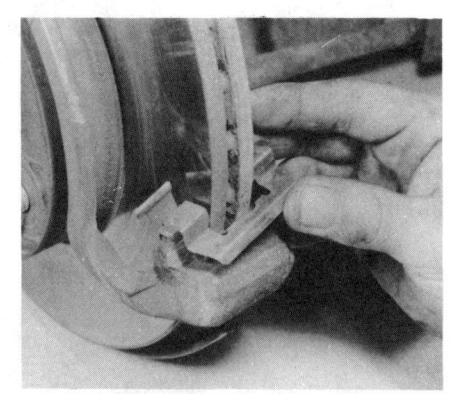

3.3C ... and anti-rattle springs

piston to insert the pads, loosen the bleed screw 1/4 turn and push the piston in. Mop up any spilt fluid, and tighten the bleed screw on completion.
10 Pivot the caliper back down over the pads, fit and tighten the lower guide bolt.
11 Depress the brake pedal several times to position the pads and ensure the caliper works.
12 Refit the roadwheels and remove the vehicle from jacks/stands.
13 Check the fluid level in the reservoir, topping-up as necessary, and test drive the vehicle.

4 Rear disc pads – inspection and renewal

1 Chock the front wheels, release the handbrake and raise the rear of the car on to stands.
2 Remove the roadwheels.
3 Remove the caliper protection shield.
4 Remove the shackle pin from the handbrake cable operating lever.
5 Remove both guide bolts from the caliper and swing the caliper out of the way. Do not put undue strain on the flexible hydraulic hose.
6 The procedure from now on is much the same as for the front pads, except that, when refitting the caliper, the piston should be rotated until one of the cut-outs in the piston will engage with the tab on the inner brake pad. Refer to Fig. 9.13.

7 Refit the caliper guide bolts, handbrake cable and protection shield.
8 Depress the brake pedal several times and operate the handbrake to ensure all parts are working.
9 If necessary, adjust the handbrake as described in Section 20.
10 Refit the roadwheels, remove the vehicle from stands/jacks, remove the chocks from the front wheels.
11 Check the fluid level in the reservoir, topping-up as necessary, before test driving the vehicle.

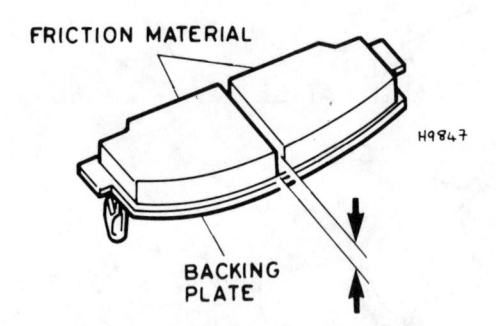

Fig. 9.4 Measuring the brake pad thickness (Sec 4)

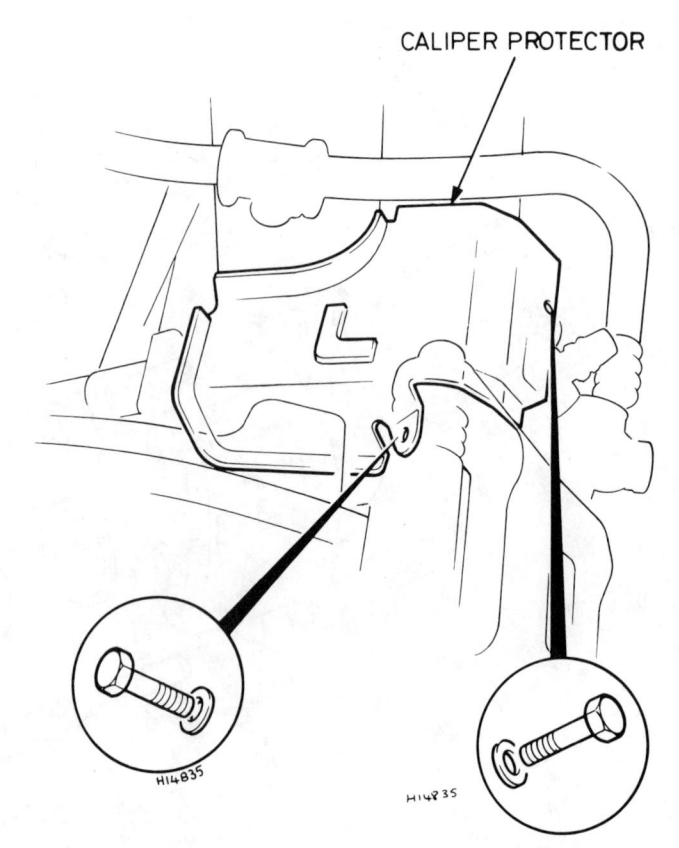

Fig. 9.5 The caliper protection shield and securing bolts (Sec 4)

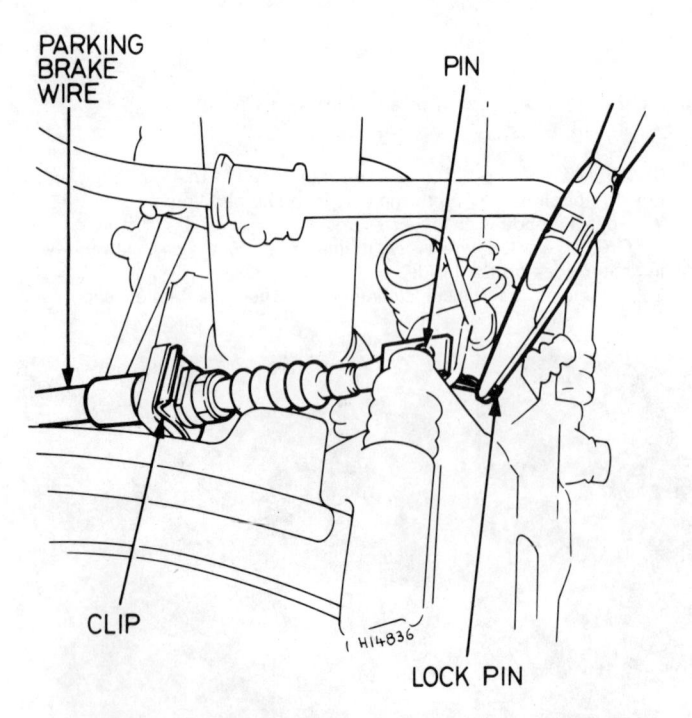

Fig. 9.6 The handbrake cable operating lever shackle pin (Sec 4)

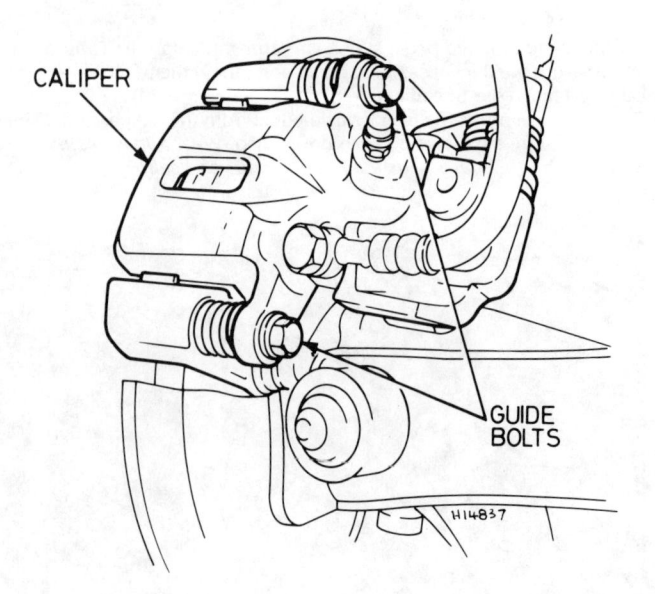

Fig. 9.7 Caliper guide bolts (Sec 4)

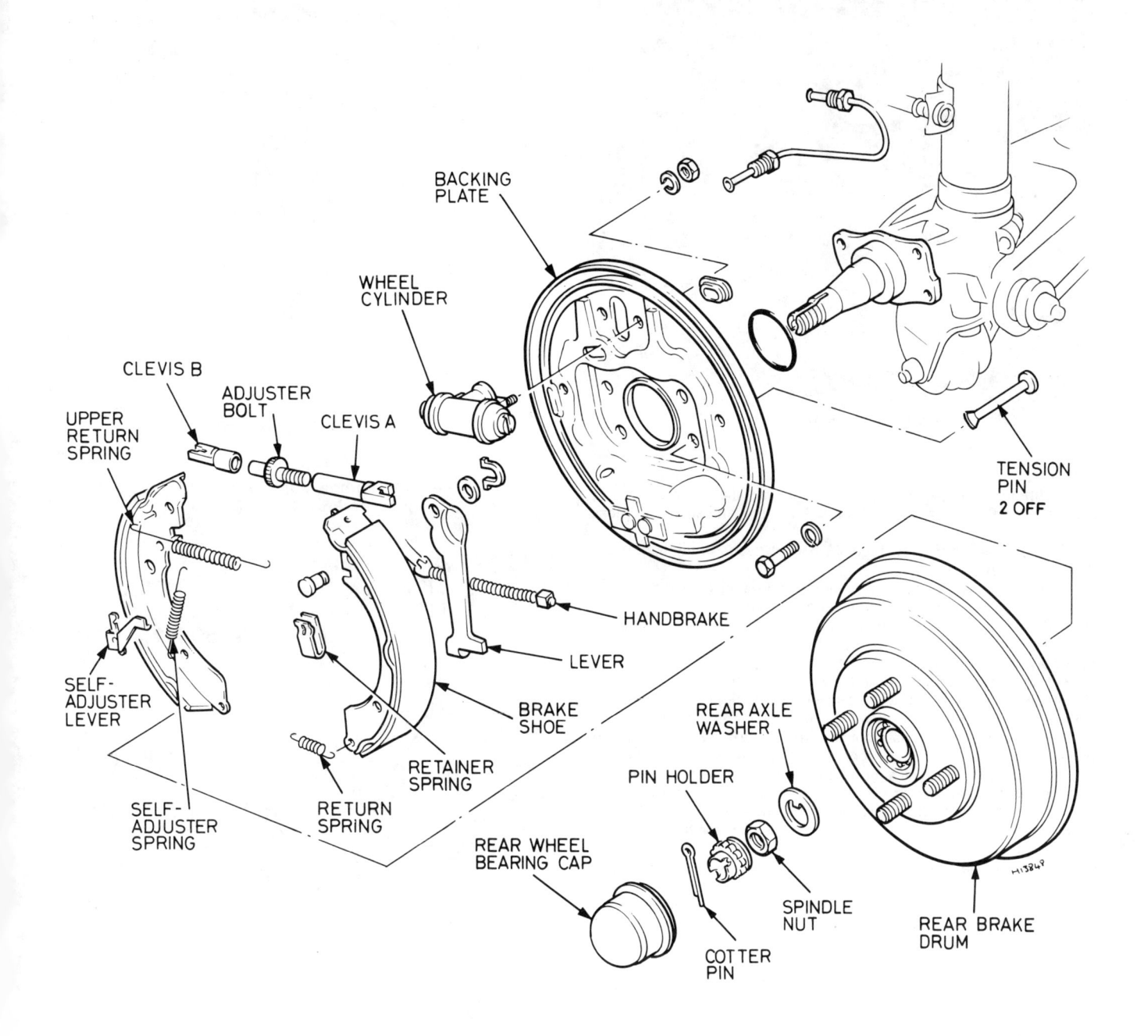

Fig. 9.8 Exploded view of rear drum brake (Sec 5)

5 Rear drum brake lining – inspection and renewal

1 Chock the front wheels. Raise the rear of the vehicle, support it safely, remove the roadwheels and release the handbrake.
2 Tap off the hub cap, and remove the split pin and pin holder from the spindle nut.
3 Remove the spindle nut and pull off the brake drum (photo).
4 The brake linings may be inspected and measured *in situ*. If they are worn and need replacing proceed as follows:
5 Remove the retaining springs by depressing them and turning them through 90°, to release them from the tension pin.
6 Pull the brake linings forward slightly and unhook the upper and lower return springs.
7 Release the handbrake cable from its lever.
8 The linings may now be removed from the brake backplate.
9 Clean all dust from the backplate, using a brush and wearing a face mask (brake dust is injurious to health).
10 Remove the handbrake lever from the worn brake lining by removing the circlip, washer and pivot pin, and fit it to the new brake lining.
11 Similarly, remove the adjuster bolt and lever from the remaining brake lining and fit them to the other replacement lining.
12 The brake lining assembly may now be built up as shown in Fig.

5.3 View of rear brake with drum removed

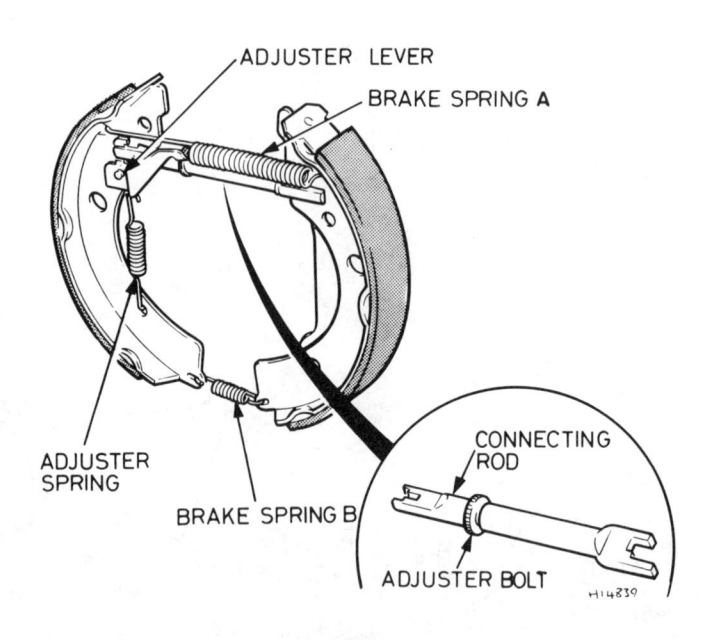

Fig. 9.9 The brake linings, adjuster and springs built-up prior to installation (Sec 5)

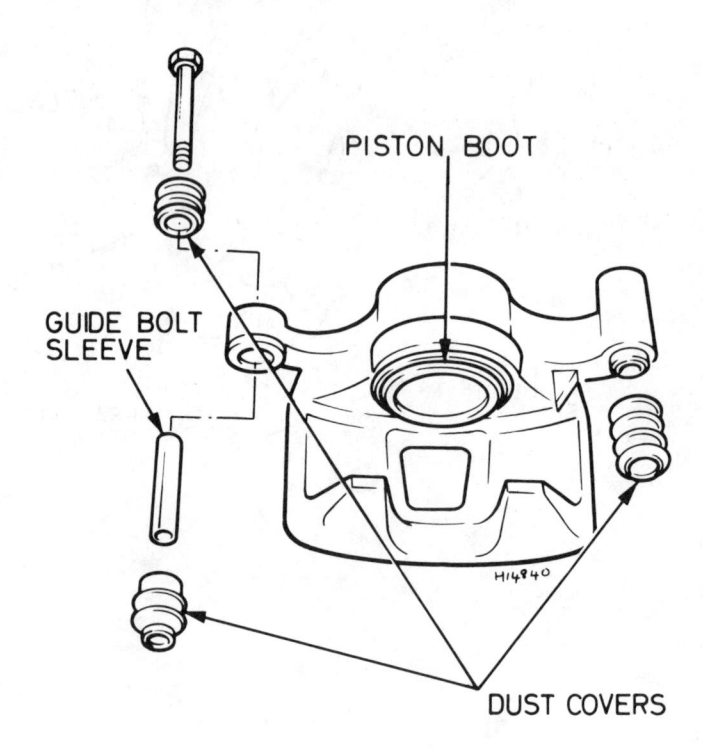

Fig. 9.10 Caliper guide bolts and dust covers (Sec 6)

9.9 by fitting the upper and lower springs, and the adjuster springs. Note that the longest connecting rod on the adjuster bolt faces forwards.

13 Offer the assembly up to the brake backplate and fit the handbrake cable and spring to the lever.

14 Now fit the lower ends of the linings into the slots behind the lower retaining plate, ensuring the return spring is routed behind the plate.

15 Pull the top ends of the linings apart and fit them into the slots on the brake cylinder.

16 Apply silicone grease to all metal-to-metal contact areas, but do not get any on the linings or the brake drum surface.

17 Fit the retaining springs to the tension pins, fitting the pin ends into the slot on the springs, and then rotating the springs through 90°.

18 Centralise the whole assembly on the brake backing plate, and ensure all springs are correctly located and routed.

19 Also ensure that the self-adjuster bolt is at its shortest position, and that the self adjuster lever is located correctly in the adjuster bolt.

20 Refit the rear brake drum, washer and spindle nut.

21 Refer to Chapter 10 and tighten the spindle nut, refit the pin holder and split pin.

22 Coat the inside of the bearing cap with a generous layer of grease and tap it back in position.

23 Operate the footbrake and handbrake lever several times to position the linings and adjust the handbrake, then check that the drum will turn freely by hand.

24 Refit the roadwheel, remove the vehicle from the jack/stands and road test.

6 Front disc brake caliper – removal, overhaul and refitting

1 Raise the front of the vehicle, support it safely and remove the roadwheels.

2 To prevent loss of hydraulic fluid when disconnecting the banjo bolt, either apply a clamp to the flexible part of the hydraulic hose or create an air lock in the system by screwing the brake fluid reservoir cap down onto a piece of polythene sheet.

3 Remove the banjo bolt from the unit on the brake caliper. Note the copper washers either side of the hydraulic union, these should be renewed on reassembly.

4 Remove the upper and lower guide bolts and remove the brake pads (see Section 3) and the brake caliper.

5 Brush away loose dirt and dust, *taking precautions against inhalation of the dust, which can be harmful.*

6 Remove the dust covers from the guide bolts, and the guide bolt sleeves from the guide bolts.

7 Clean them in hydraulic fluid, and inspect the boots for cracking, splits and general deterioration. Renew any which are defective.

8 Similarly, clean the guide bolts and sleeves and inspect them for corrosion.

9 Remove the piston boot.

10 Place a piece of wood in the caliper, and then apply low pressure air (as from a foot pump) to the fluid entry hole in the caliper to eject the piston.

11 Clean the piston and the piston bore with clean hydraulic fluid and inspect all parts for corrosion, scoring and general wear.

12 Carefully remove the piston seal from the caliper; prying it from its housing.

13 Renew all parts which show signs of deterioration.

Chapter 9 Braking system 215

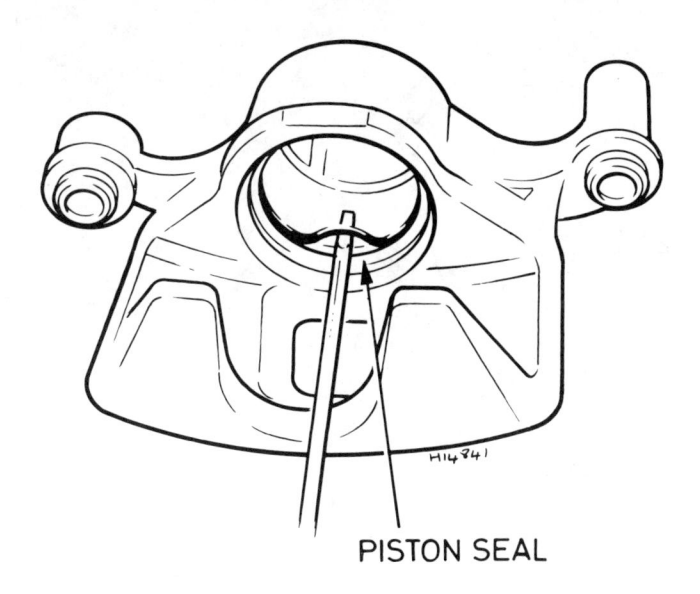

Fig. 9.11 Prise out the piston seal (Sec 6)

14 Repair kits containing all the necessary seals are available, and it would be false economy not to renew the seals having dismantled the caliper thus far.
15 Lubricate all parts and seals in clean hydraulic fluid before reassembly.
16 Fit the piston seal in its groove in the piston bore in the caliper.
17 Fit the piston boot to the caliper, ensuring it seats in its groove.
18 Gently push the piston into the piston bore, until the lip on the piston boot engages in the groove on the piston.
19 Apply silicone grease to the guide bolt boots and fit them to the caliper.
20 Fit the caliper to the caliper bracket, grease the guide bolts and sleeves, and fit and tighten the bolts.
21 Check the caliper is a free sliding fit on the guide bolts (the principle of operation is that. on application of hydraulic pressure, the piston will push against one side of the disc, and an equal and opposite reaction will cause the whole caliper to slide on the guide bolts, applying pressure to the other side of the disc, in a 'pincer' type fashion).
22 Reconnect the banjo union, using new copper sealing washers, and remove the clamp from the hydraulic line, or polythene from the fluid reservoir.
23 Refit the brake pads, as described in Section 3.
24 Bleed the hydraulic system, as described in Section 17.
25 Refit the roadwheels, remove the vehicle from jack/stands, and carry out a road test.

7 Front brake disc – inspection and renovation

1 Whenever the brake calipers are removed for brake pad renewal, inspect the disc for cracks, scoring or grooving, and corrosion.
2 The disc should be inspected for run-out using a dial test indicator or feeler blades inserted between a fixed point and the disc, while the disc is rotated. Tolerances will be found in the Specifications.
3 The thickness of the disc should also be checked using a micrometer (photo).
4 Measure the disc at eight different points, approximately 45° apart and 0.39 in (10 mm) in from the outer edge.
5 The disc should be replaced if the tolerances given in the Specifications are exceeded, and if the difference between any thickness measurement exceeds 0.0006 in (0.015 mm).
6 Provided that the refinishing limit given in the Specifications is not exceeded, the disc may be resurfaced by specialist engineers. **Note:** Some discs may have a refinishing tolerance stamped on them. This tolerance should take precedence over that given in the Specifications.
7 The procedure for removing the disc will be found in Chapter 10.

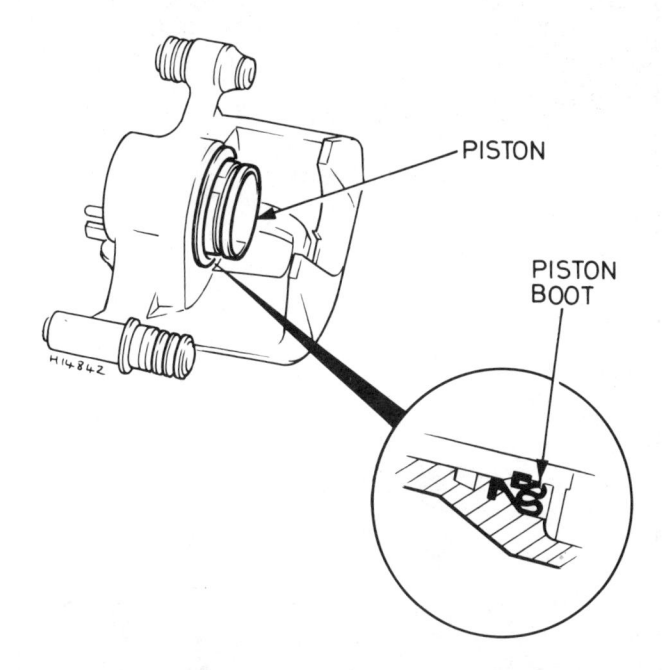

Fig. 9.12 Correct positioning of piston boot (Sec 6)

7.3 Measuring a brake disc using a micrometer

8 Rear brake disc – inspection and renewal

1 The rear brake disc is inspected in the same way as described for a front disc, but complying with the tolerances laid down for rear discs in the Specifications.
2 The procedure for removal and refitting of the disc are given in Chapter 10.

9 Wheel hubs, bearings and discs – general

1 Although the bearings form part of the disc/drum assembly, the bearings are actually part of the wheel hub, and instructions for removing the disc/drum for bearing replacement and adjustment will be found in Chapter 10.

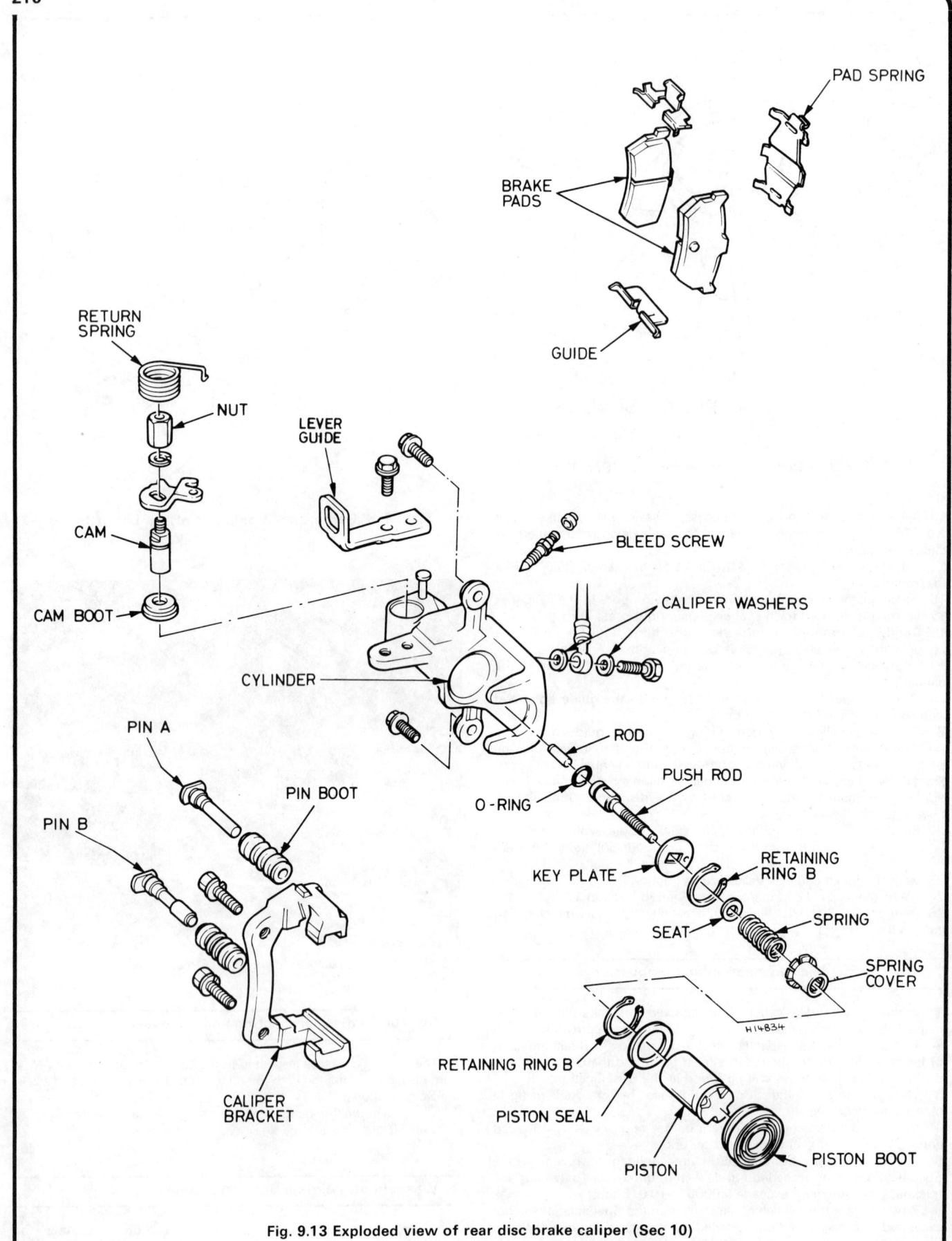

Fig. 9.13 Exploded view of rear disc brake caliper (Sec 10)

Chapter 9 Braking system

10 Rear disc brake caliper – removal, overhaul and refitting

Note: *Special tools are required to remove and refit the self-adjusting mechanism and, as these are only likely to be available to Honda dealers, any problems in this area should be left to your dealer, and home servicing limited to seal replacement, as described here.*

1 Remove the caliper protector, handbrake cable and caliper guide bolts, along with the brake pads, as described in Section 4.
2 Disconnect the hydraulic pipeline at the banjo union on the brake caliper. **Note:** *To avoid loss of hydraulic fluid, comply with the instructions as for disconnecting the front pipe lines in Section 6.*
3 Cover the end of the hydraulic pipeline with a small polythene bag to prevent ingress of dirt.
4 Brush off any dust and dirt from the caliper, *avoiding breathing in of the dust.*
5 Twist the piston anti-clockwise and, at the same time, pull it and the piston boot out of the caliper, being careful to avoid damaging them.
6 The internal components of the piston may be removed after removal of the spring clip (see Fig. 9.14). Ensure each part goes back in its original position on reassembly.
7 Use a blunt instrument to prise out the piston seal, being extremely careful not to damage the piston bore in the caliper, or leakage will result.
8 Clean all parts in clean hydraulic fluid and inspect them for signs of wear.
9 Light corrosion may be removed with fine emery cloth, but do not be too drastic in its application.
10 Lubricate a new seal and piston boot with rubber grease, and apply a film of suitable grease to the piston bore in the caliper.
11 Coat the outside of the piston with rubber grease before refitting it to the pushrod in the caliper.
12 Twist the piston clockwise back into position, being careful not to twist the piston boot at the same time.
13 Refit the brake pads, as described in Section 4.
14 Connect the hydraulic pipe line to the brake caliper; using new sealing washers either side of the banjo bolt. Tighten the bolt to the specified torque.
15 The handbrake operating cam may be removed for inspection and renewal of the rubber boot after removing the return spring, nut, washer and operating lever.
16 Clean all parts in clean hydraulic fluid before reassembly, and coat the cam boot and needle roller bearing with rubber grease.
17 After refitting, connect the handbrake cable shackle pin.
18 Bleed the brakes, as described in Section 17.
19 Operate the footbrake and handbrake several times to bed in the components and set the adjuster.
20 Refit the roadwheel, remove the vehicle from jacks and road test.

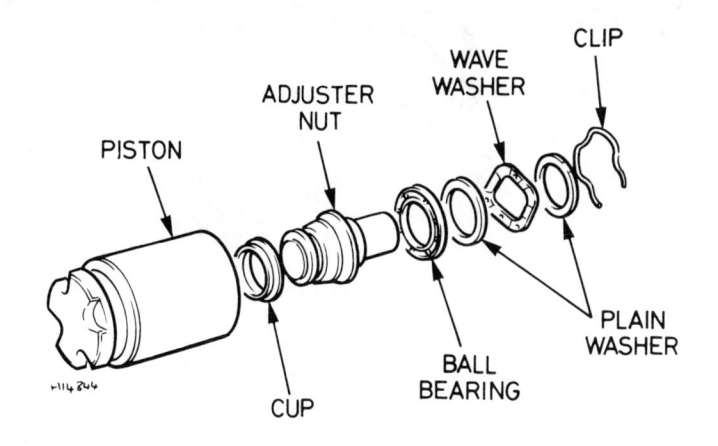

Fig. 9.14 The components of the rear disc brake piston (Sec 10)

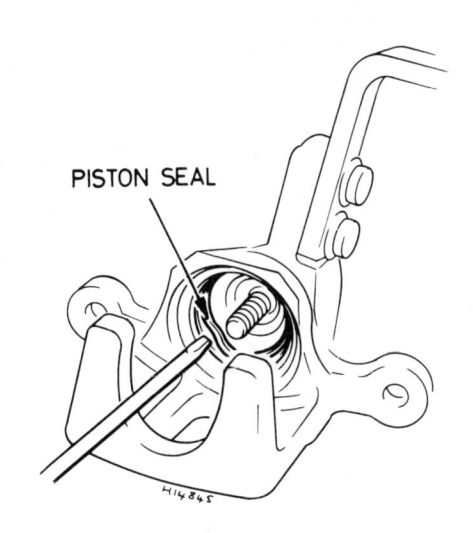

Fig. 9.15 Levering out the piston seal (Sec 10)

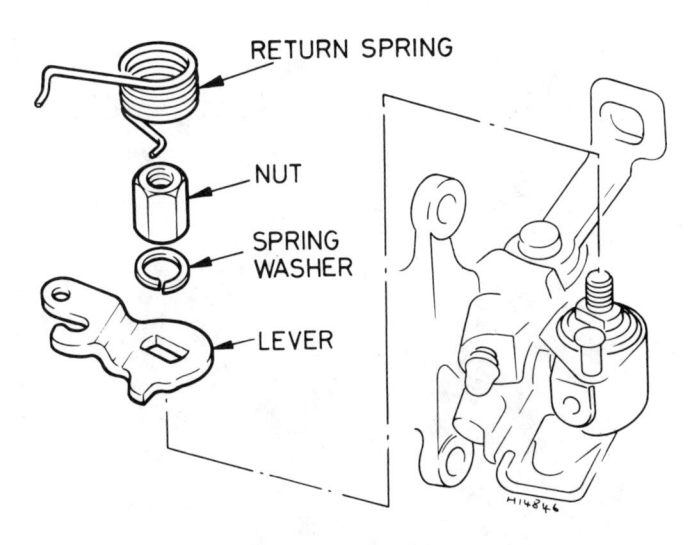

Fig. 9.16 Components of the handbrake operating lever (Sec 10)

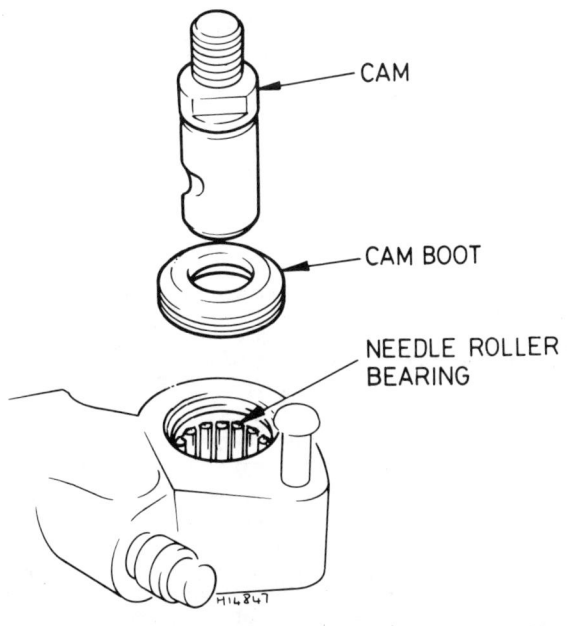

Fig. 9.17 The cam and needle roller bearing assembly (Sec 10)

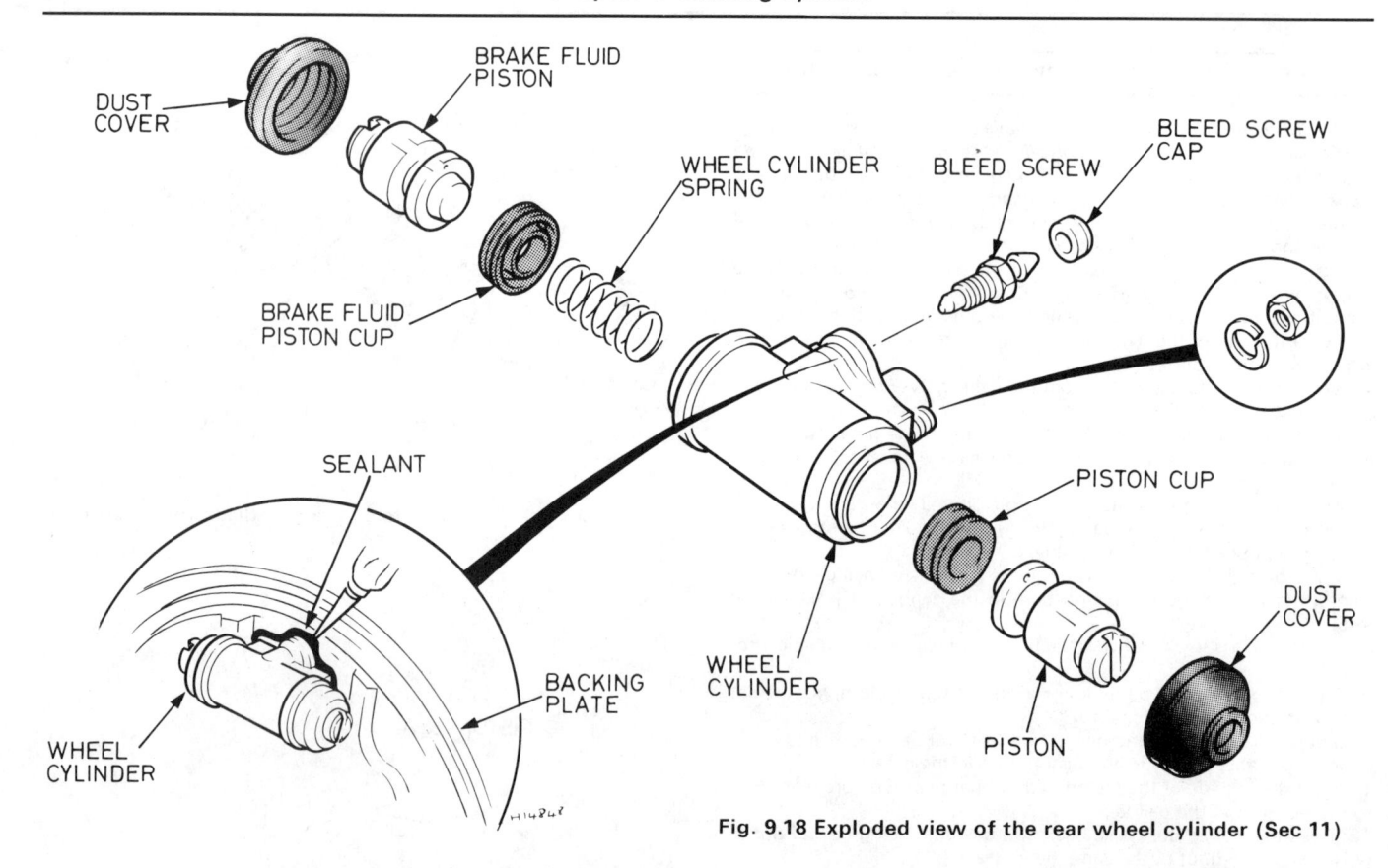

Fig. 9.18 Exploded view of the rear wheel cylinder (Sec 11)

11 Rear wheel cylinder – removal, overhaul and refitting

1 Remove the brake shoes, as described in Section 5.
2 Disconnect the hydraulic pipe line from the wheel cylinder, and cover the pipe line end with a polythene bag to prevent ingress of dirt.
3 Undo the nut on the cylinder which holds it to the backplate.
4 Clean away all external dirt before pulling off the dust covers.
5 Remove the internal components by applying air pressure from a foot pump to the fluid entry point.
6 Note the sequence in which components come out, and the orientation of seals.
7 Wash all parts in clean hydraulic fluid before inspecting the rubber covers and seals for hardening, splits and cracking.
8 Inspect the cylinder and piston for corrosion, scoring, pitting and wear.
9 Renew any parts which show signs of deterioration.
10 Depending upon the severity of this, the complete wheel cylinder may be renewed, or an overhaul kit obtained and the seals and dust covers only changed.
11 Lubricate all parts in clean hydraulic fluid before reassembly.
12 Fit the internal components in the same order as they were removed, being particularly careful to fit the cups the right way round.
13 Refit the wheel cylinder to the backplate; applying sealant to its mating surface.
14 Reconnect the hydraulic pipe line, refit the brake shoes, as described in Section 5.

12 Rear brake drum – inspection and renovation

1 Whenever the rear drum is removed for shoe inspection, examine the interior of the drum, especially the friction surface.
2 Check for cracks, scoring and pitting.
3 It is possible to have the friction surface reworked provided the maximum internal diameter of the drum is not exceeded. Refer to the Specifications section for tolerances. **Note:** if a refinishing tolerance is stamped on the drum, use that figure in preference to that shown in the Specifications.

4 Inspect the bearings in the hub for smooth operation. Refer to Chapter 10 for bearing inspection and renewal.

13 Master cylinder – removal, overhaul and refitting

1 The brake master cylinder is bolted to the brake booster unit in the engine compartment (photo).
2 To remove the master cylinder, disconnect the fluid lines, allowing the fluid to drain into a suitable container.
3 Unbolt the master cylinder from the brake booster unit, disconnect the low fluid warning terminals, and withdraw the unit from the booster.

13.1 The brake master cylinder

Chapter 9 Braking system

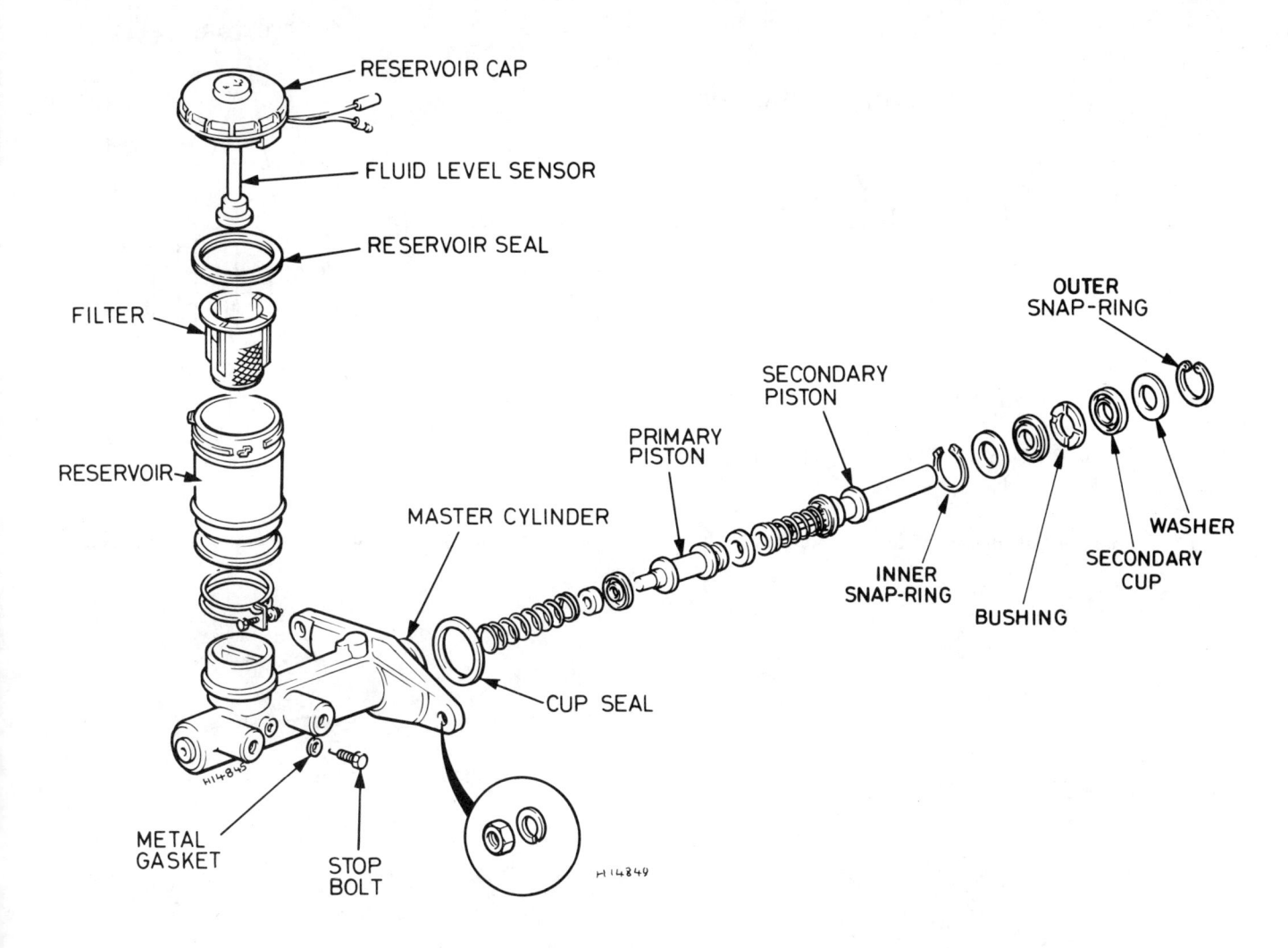

Fig. 9.19 The components of the master cylinder (Sec 13)

4 Clean away external dirt.
5 Remove the reservoir cap and tip out any remaining fluid before undoing the reservoir retaining clip, and removing the reservoir.
6 Clean the reservoir, using non-fluffy rag and check the air hole in the cap is clear.
7 Clean the filter, check the operation of the float switch and the condition of the reservoir cap seal, renewing as necessary.
8 Remove the outer snap-ring from the master cylinder body, and extract the secondary cup and bushing. Note the order and orientation of each component as it comes out.
9 Remove the stop bolt and then remove the inner snap-ring with circlip pliers, whilst pushing inwards on the secondary piston assembly.
10 Remove the secondary and primary piston assembly.
11 Remove the screw from the secondary piston assembly, which will allow the secondary spring to be removed.
12 Clean all parts in clean hydraulic fluid before inspecting them.
13 Inspect the inside bore of the master cylinder for wear, scoring, cracks, pitting and corrosion. Renew the cylinder assembly if necessary.
14 An overhaul kit, containing new seals, washers etc should be available from your dealer, in which case discard the old seals.
15 Lubricate all components in clean hydraulic fluid before reassembly.
16 Refit the new seals to the primary and secondary piston assemblies, using only your fingers to manipulate them into position.
17 Refit the springs to the piston assemblies and insert the assemblies into the bore of the master cylinder, in the order shown in Fig. 9.21.

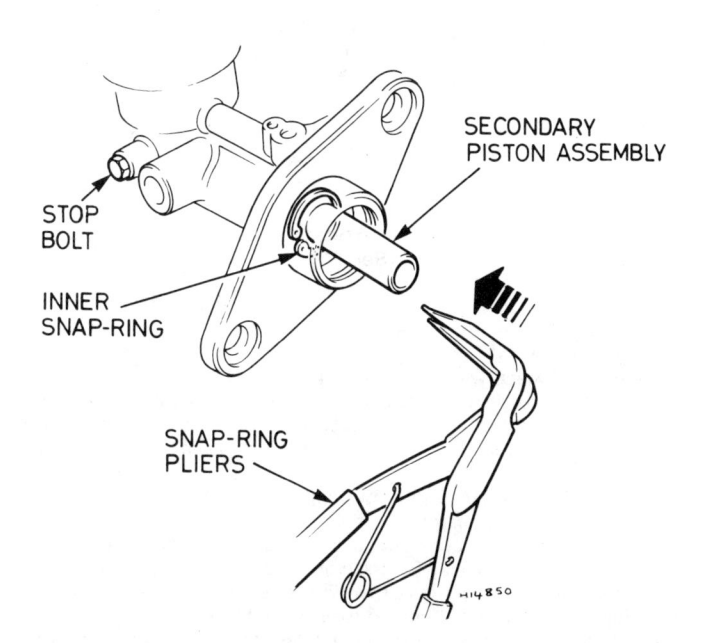

Fig. 9.20 Removing the inner snap-ring (Sec 13)

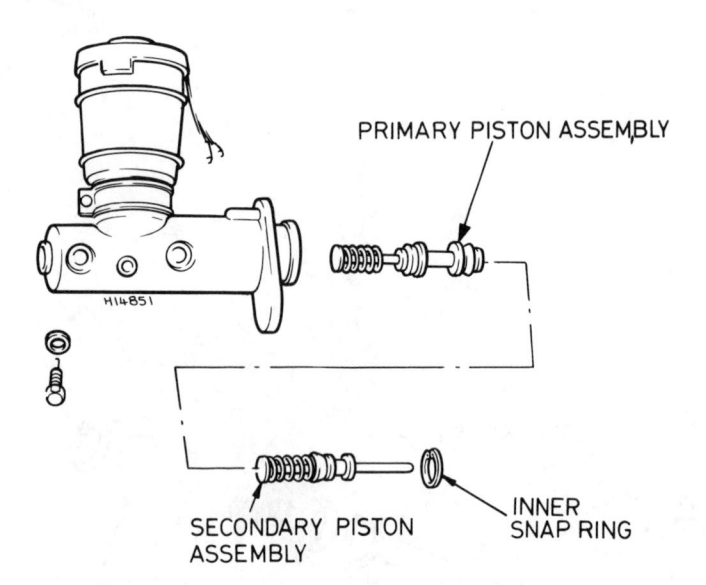

Fig. 9.21 Primary and secondary piston assemblies (Sec 13)

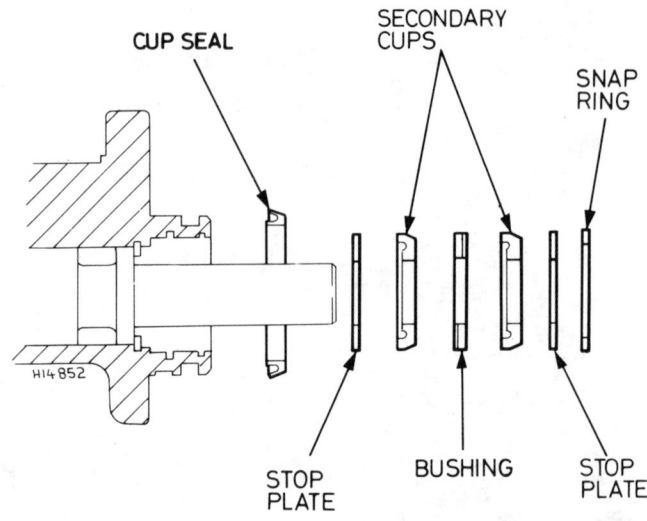

Fig. 9.22 Order of assembly of secondary cups (Sec 13)

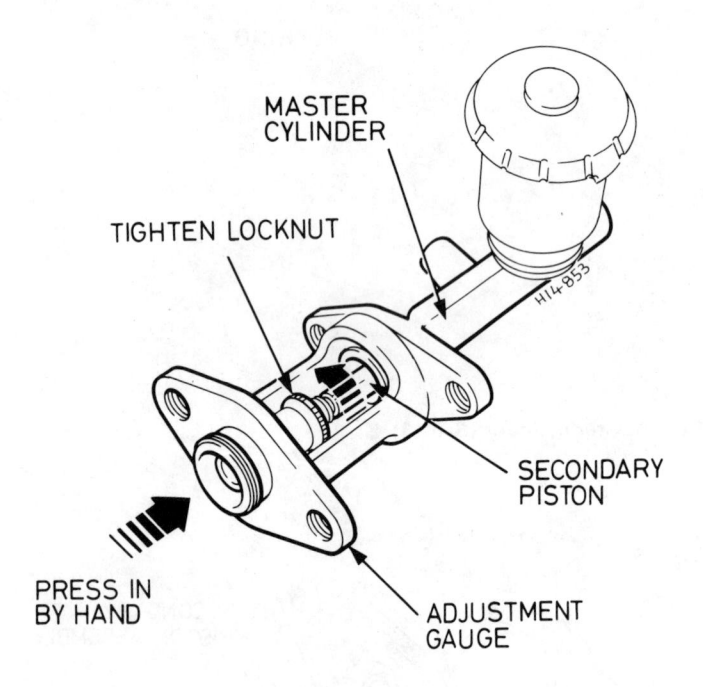

Fig. 9.23 Special gauge fitted to the master cylinder (Sec 13)

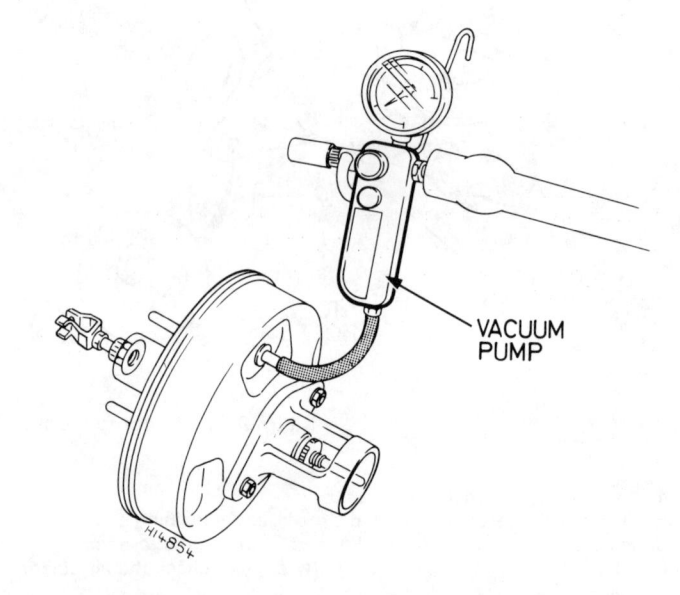

Fig. 9.24 The vacuum pump fitted to the servo unit (Sec 13)

18 It will be found advantageous if the piston assemblies are rotated during insertion.
19 Fit a new metal washer to the piston stop bolt and then turn the master cylinder so that it is standing on the piston end, push down on the master cylinder to compress the springs, and insert the stop bolt.
20 Tighten the stop bolt to the torque figure given in the Specifications.
21 Install the inner snap-ring, pushing in on the secondary piston assembly so that the snap-ring can enter its groove.
22 Fit the secondary cups and bushing, using silicon grease as a lubricant. Refer to Fig. 9.22 for assembly order and orientation of seals.
23 Install the outer snap-ring, then fit the seal to the master cylinder mounting flange, using silicone grease as a lubricant.
24 Before installing the master cylinder, the bush rod-to-piston clearance must be checked and adjusted as necessary.

25 This requires a special gauge and a vacuum pump, which will have to be obtained from your dealer. If you cannot borrow them, it may be possible to take both the master cylinder and the booster to your dealer for him to set up.
26 The procedure is as follows:
27 Fit the adjustment bolt gauge to the master cylinder, press in on the end and adjust the bolt so that it is flush with the end of the master cylinder piston.
28 Tighten the adjuster bolt locknut.
29 Transfer the gauge to the brake booster, apply a vacuum of 20 in (500 mm) Hg using the vacuum pump, and measure the distance between the output rod and adjusting bolt.
30 The clearance should be as given in the Specifications, and adjustment is made on the adjusting bolt at the rear of the booster.
31 Having set the pushrod-to-piston clearance, refit the master cylinder to the front face of the booster unit.
32 Refit the plastic reservoir and tighten its retaining clip, and connect up the low level warning leads.
33 Reconnect the brake fluid lines, and fill the master cylinder reservoir with fresh hydraulic fluid.

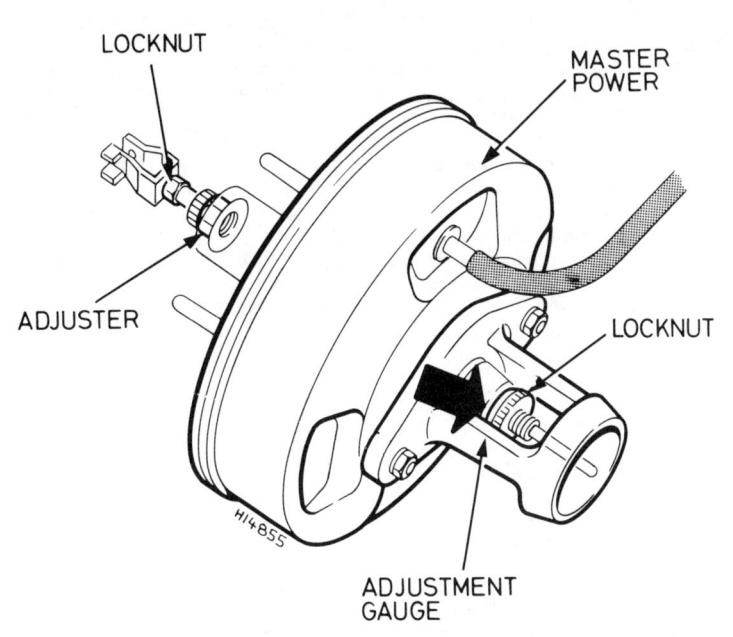

Chapter 9 Braking system

Fig. 9.25 The special gauge transferred to the servo unit (Sec 13)

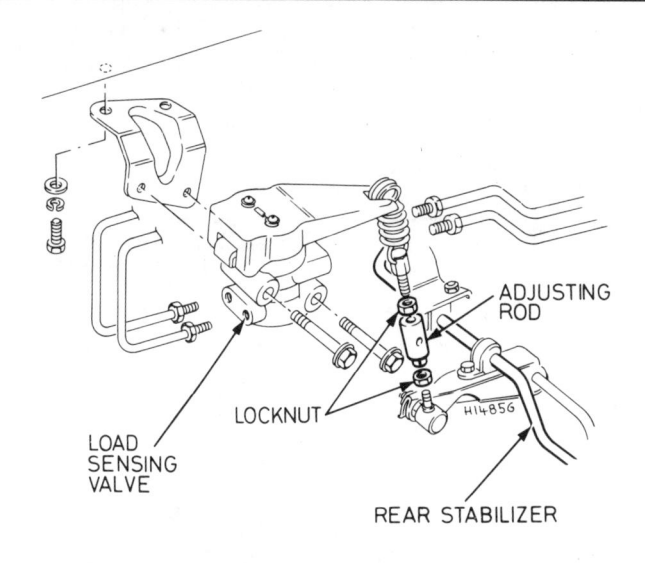

Fig. 9.26 Exploded view of the load sensing valve (Sec 15)

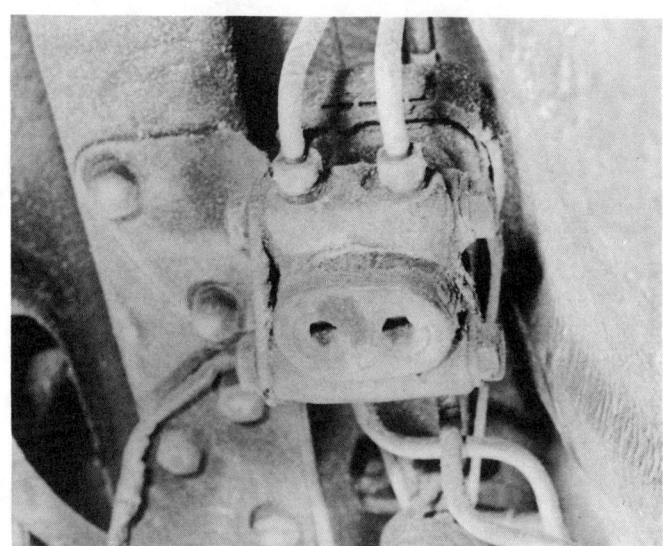

34 The complete system should now be bled free of air, as described in Section 17.

14 Dual proportioning valve – general description

1 The dual proportioning valve is mounted in the engine compartment, by the brake master cylinder.
2 Its purpose is to prevent the rear wheels locking under certain adverse braking conditions, and does this by aportioning brake fluid pressure equally in an X pattern between the front right and rear left wheel, and the left front and rear right wheel.
3 Little can be done by way of maintenance to the valve, and testing of the valve to evaluate its performance should be left to your dealer.
4 If it is known that the valve is malfunctioning it can be renewed by removing the brake lines, and unbolting it from the engine compartment wall.
5 Fit a new valve, reconnect the brake lines, and bleed the complete circuit, as described in Section 17.

15 Load sensing valve – removal, refitting and adjusting

1 The load sensing valve fitted to European models is located underneath the car at the left rear side, bolted to the suspension stabiliser bar.
2 It provides stable, lock-free braking of both the front and rear wheels under most braking conditions.
3 If it is suspected of malfunction, renew the complete valve.
4 Jack up the vehicle, support it on stands, and drain the brake fluid from the system.
5 Disconnect the brake lines to the load sensing valve (photo).
6 Loosen the locknuts and remove the linkage between the valve and rear stabiliser (photo).
7 Remove the bolts securing the valve and lift it out.
8 Transfer the spring and adjuster rod to the new valve.
9 Refit the new valve by reversing the above procedure.
10 Bleed the brake system on completion.
11 Adjust the valve as follows:
12 Remove the vehicle from jacks/stands, and, with it standing empty, refer to Fig. 9.27 and measure distance 'A'.
13 Adjust the turnbuckle to correct dimension 'A', if necessary.
14 To check the valve for correct operation, stand the vehicle on level

15.5 Hydraulic lines to the load sensing valve

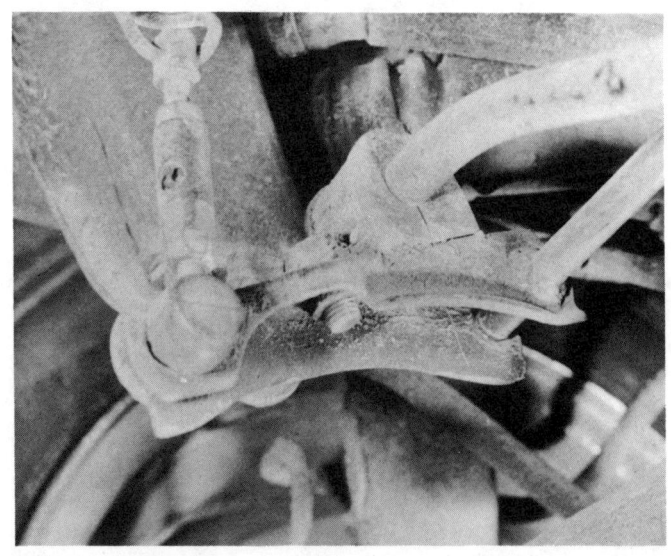

15.6 The load sensing valve adjuster linkage and spring

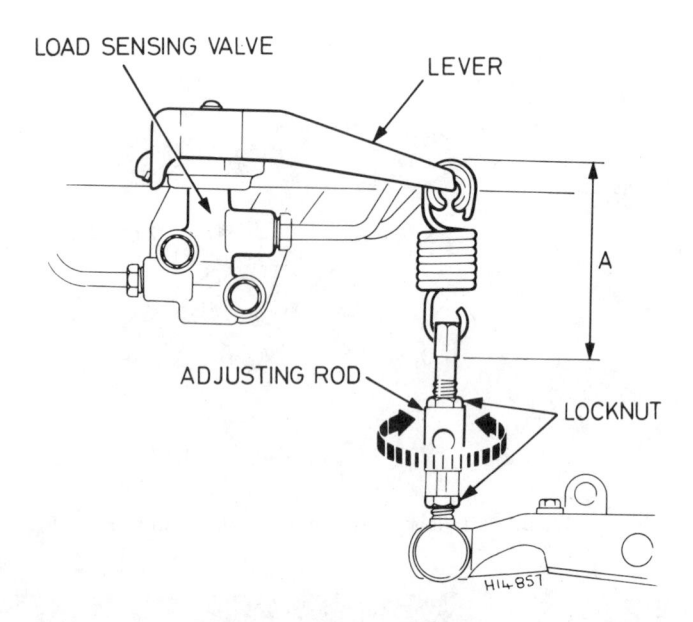

Fig. 9.27 Load sensing valve setting dimension (Sec 15)

A = 3.61 in (91.7 mm)

ground, with a driver, spare tyre, service tools and jack in position, start the engine and push down hard on the brake pedal.
15 Remove foot from the brake pedal, and the lever on the load sensing valve should operate.
16 If it doesn't, repeat the test operation; ensuring that the brake pedal is pressed down with a force of at least 44 lb (20 kg), before deciding to renew the valve.

16 Hydraulic pipes and hoses – general

1 Periodically inspect the condition of the flexible brake hoses. If they appear swollen, chafed or when bent with the fingers tiny cracks appear, they must be renewed.
2 Always uncouple the rigid pipe from the flexible hose first, then release the end of the flexible hose from the support bracket (photo). Do this by pulling out the lockplate wth pliers.
3 Now unscrew the flexible hose from the caliper or connector. On calipers a banjo type connector is used. When refitting a banjo union, always use new sealing washers.
4 After installation, check the routing of all pipe lines to ensure they will not chafe against adjacent components. Flexible hose attitude may be altered by pulling out the lockplate from the support bracket and twisting the hose in the desired direction, by not more than a quarter turn, and refitting the lockplate.
5 At regular intervals, wipe the steel brake pipes clean and examine them for signs of corrosion or damage.
6 Examine the fit of the pipes in their insulated securing clips, bending the clips as necessary to ensure a positive fit (photo).
7 Check rigid pipes are not rubbing against other components, bending them gently to achieve this.
8 Any section of pipe which is rusty or chafed should be renewed. Brake specialists will make up pipe lines to pattern.
9 When installing new pipes, do not bend them any more than is necessary, or they may collapse internally.
10 Any disturbance of components or pipelines will necessitate bleeding the system.

17 Brake system – bleeding

Note: *Refer to Section 26 if an anti-lock system is fitted.*
1 The reservoir on the brake master cylinder must be full at the start of the bleeding operation, and checked after each cylinder has been bled. The level must not be allowed to drop below the minimum mark during any stage of bleeding or air may be introduced into the system. Fill as required, using only the recommended brake fluid.
2 Have an assistant slowly pump the brake pedal several times, and then depress and hold the brake pedal down.
3 Using the sequence shown in Fig. 9.28, attach a length of plastic tube firmly over the brake cylinder bleed nipple (photo) and insert the other end into a jar which is half full of brake fluid.
4 Loosen the bleed nipple sufficiently to allow air to escape, tightening the nipple once an air free flow of fluid emerges from the nipple.
5 If the brake pedal bottoms before all the air is ejected, tighten the nipple, release the brake pedal and then repeat the above procedure.
6 Bleed each brake in sequence.
7 On completion, the brake pedal should offer firm resistance when depressed. If it feels 'spongy' or is slowly depressed under pressure, then there is either still air in the system or there is a leak.

16.2 Typical rigid-to-flexible pipe line connection

16.6 An insulated pipe securing clip

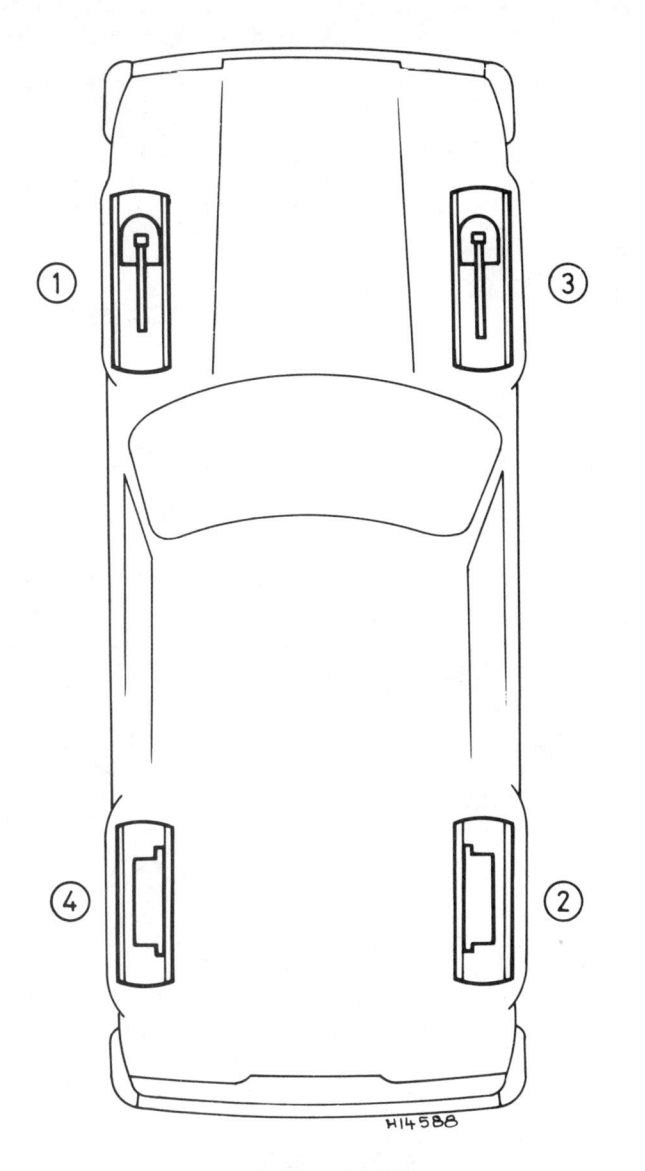

Chapter 9 Braking system

Fig. 9.28 Brake bleeding sequence (Sec 17)

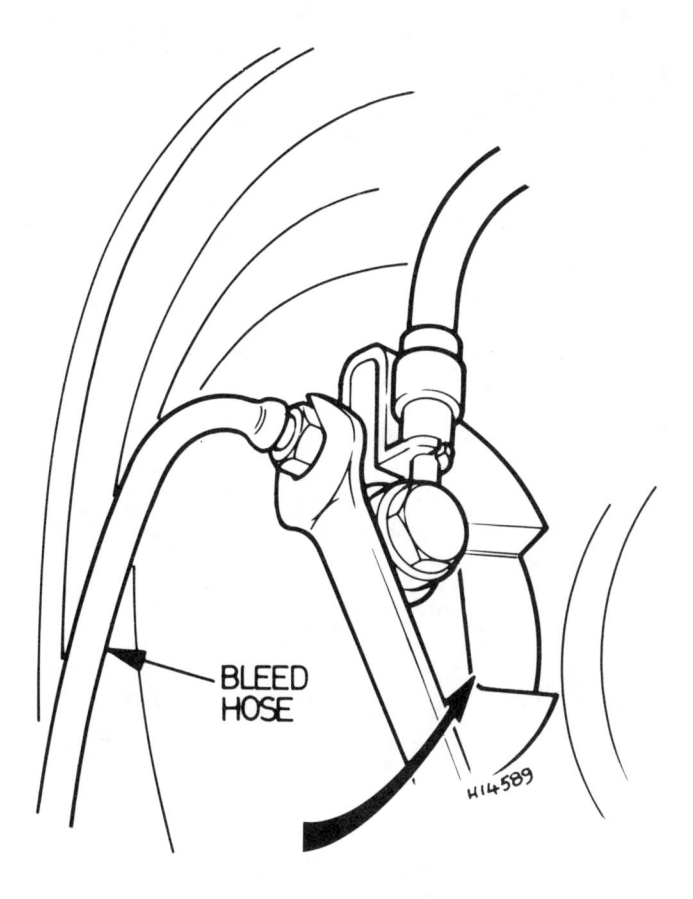

Fig. 9.29 A hose fitted to the bleed nipple (Sec 17)

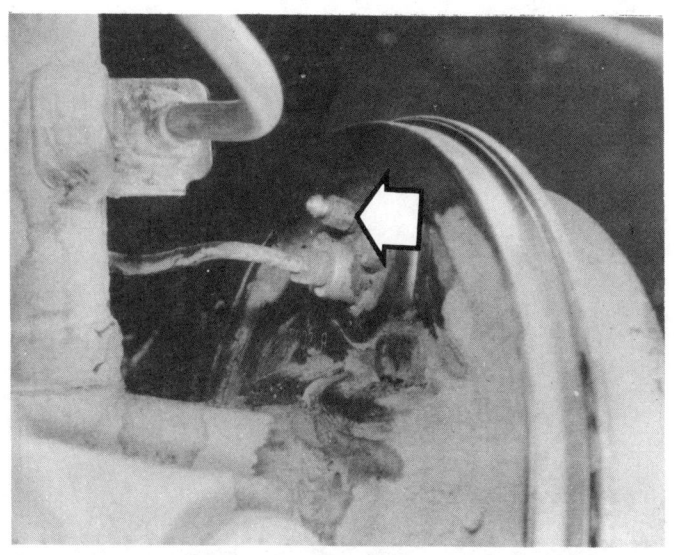

8 Check the brake unions and cylinders for signs of seepage and repeat the bleeding operation.
9 If the brake pedal still continues to depress under pressure, suspect an internal leak in the master cylinder.
10 Remove the bleeding equipment and test drive the vehicle.
11 **Note:** There are several brake bleeding kits offered on the market. If it is intended to use one of these, follow the manufacturer's instructions, but use the brake bleeding sequence given here.

18 Brake booster servo unit – description and maintenance

1 The brake booster servo unit, located on the rear bulkhead of the engine compartment, and to which the brake master cylinder is bolted, is fitted to provide braking assistance to the driver.
2 It reduces the effort required to depress the brake pedal, and, should the unit fail, it does not mean that braking efficiency will be lost, but rather that more effort will be required to apply the brakes.
3 The unit operates by vacuum obtained by hose from the inlet manifold, and comprises a large diaphragm and a check valve. The servo unit and hydraulic brake master cylinder are connected together, so that the pushrod in the servo unit acts upon the pushrod in the master cylinder.
4 When the brake pedal is depressed, a pushrod at the top of the pedal arm pushes against the servo unit pushrod, which in turn acts on the pushrod in the master cylinder.
5 With the engine running, both sides of the diaphragm in the servo unit are open to the vacuum in the inlet manifold.
6 When the brake pedal is depressed, the rear chamber in the servo is opened to the atmosphere, and the diaphragm moves forward, pushing on its pushrod, thus assisting the braking action.
7 Under normal operating conditions the brake servo unit is extremely reliable and requires little maintenance, and if it does fail, it is far better to obtain a replacement unit than to repair the original.

17.3 Brake cylinder bleed nipple (arrowed)

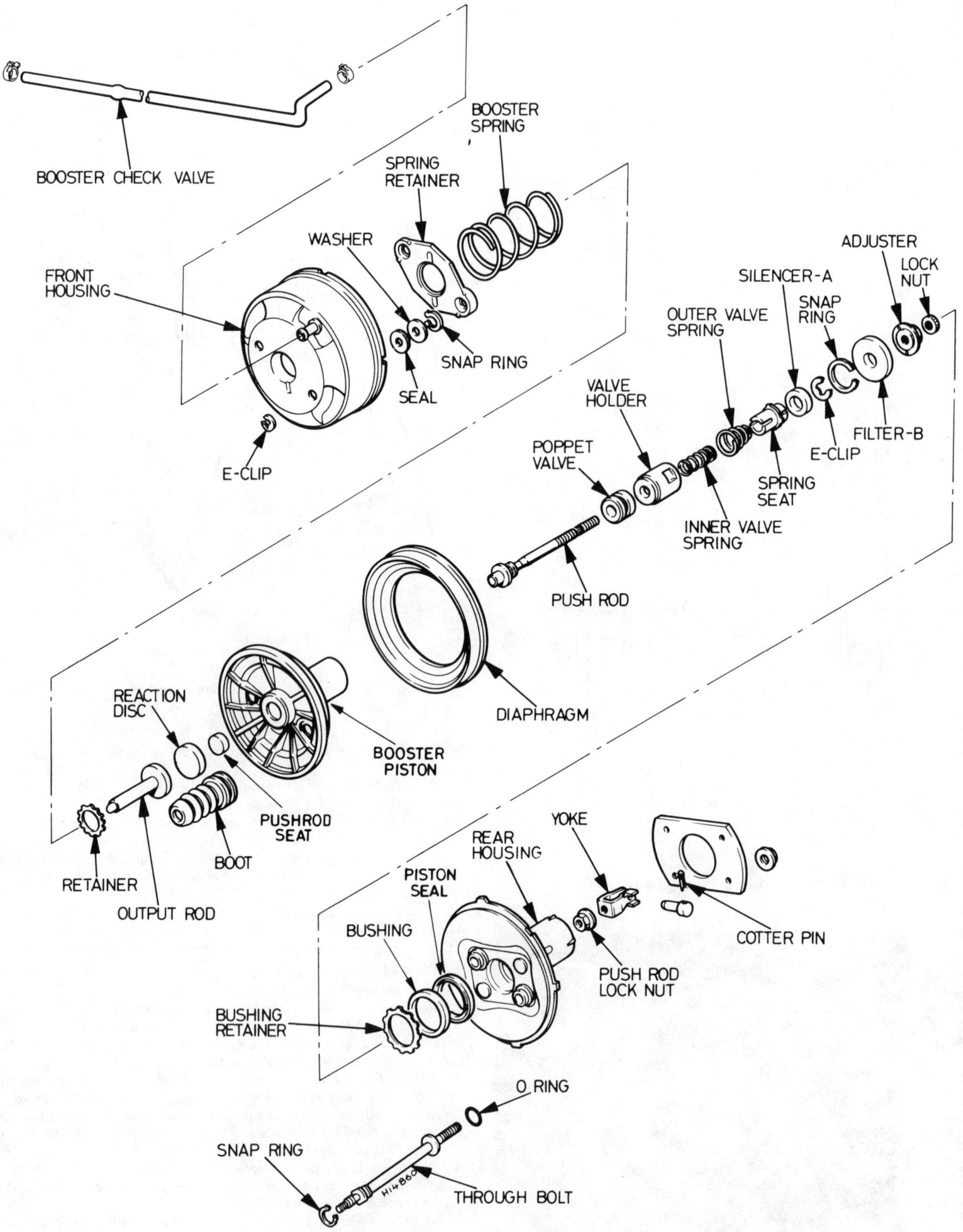

Fig. 9.30 Exploded view of the brake booster servo unit (Sec 18)

Chapter 9 Braking system

Functional test

8 The following functional test will establish the serviceability of the brake servo unit.
9 With the engine stopped, depress the brake pedal several times, so evacuating the vacuum in the servo unit, then hold the brake pedal down under pressure for 15 seconds.
10 If the pedal sinks, the master cylinder, brake lines or hoses, or wheel cylinders are leaking. Inspect and repair them as described in other Sections of this Chapter.
11 Start the engine, with the brake pedal depressed.
12 The brake pedal should sink slightly. If it does not, the brake servo unit is faulty, and the vehicle should be taken to your dealer for further tests.
13 Periodically inspect the hose between the servo unit and the inlet manifold for deterioration and cracking, and that the two connections are sound.
14 A check valve is fitted in this vacuum hose and may be tested as follows:
15 Mark each end of the hose so that it is refitted correctly, then undo the clips and remove it from the inlet manifold and the servo unit.
16 Being careful not to inhale any air through the hose, blow through the hose from the brake servo end. No resistance should be felt.
17 Now blow through from the inlet manifold end, this should offer resistance.
18 Renew the complete hose if the check valve is defective, as the valve cannot be renewed.

19 Brake booster servo unit – removal and refitting

1 Remove the brake master cylinder, as described in Section 13.
2 Disconnect the vacuum hose from the inlet manifold.
3 From inside the car, remove the split pin and shackle pin attaching the brake pedal to the servo unit pushrod.
4 Remove the two nuts holding the servo unit to the bulkhead, and then remove the unit from the engine compartment.
5 Refitting is a reversal of removal, but the pushrod clearances will have to be adjusted, as described in Section 13.

Note: On cars fitted with fuel injection, remove the windscreen wiper motor for better access to the brake booster, then remove the brake master cylinder from the booster before removing the booster.

20 Handbrake – adjustment

1 The handbrake cable is adjusted automatically on both rear drum and disc brake models by the action of the automatic adjusters in the rear brake mechanism.
2 The cable should only need adjustment after high mileage, when the cable may have stretched, or on fitting a new cable.
3 The handbrake cable is correctly adjusted if the rear brakes are fully applied after the handbrake lever has been pulled up 4 to 8 clicks. Adjust as follows:
4 Chock the front wheels and raise the rear wheels clear of the ground.
5 Refer to Chapter 11 and gain access to the handbrake cable equaliser adjuster beneath the centre console, and remove the cover plate (photo).
6 Release the handbrake and undo the adjuster until all tension on the handbrake cables is relieved (photo).
7 Pull the handbrake up one notch on the ratchet and tighten the adjuster until the rear wheels just begin to bind on the brake shoes as they are turned.

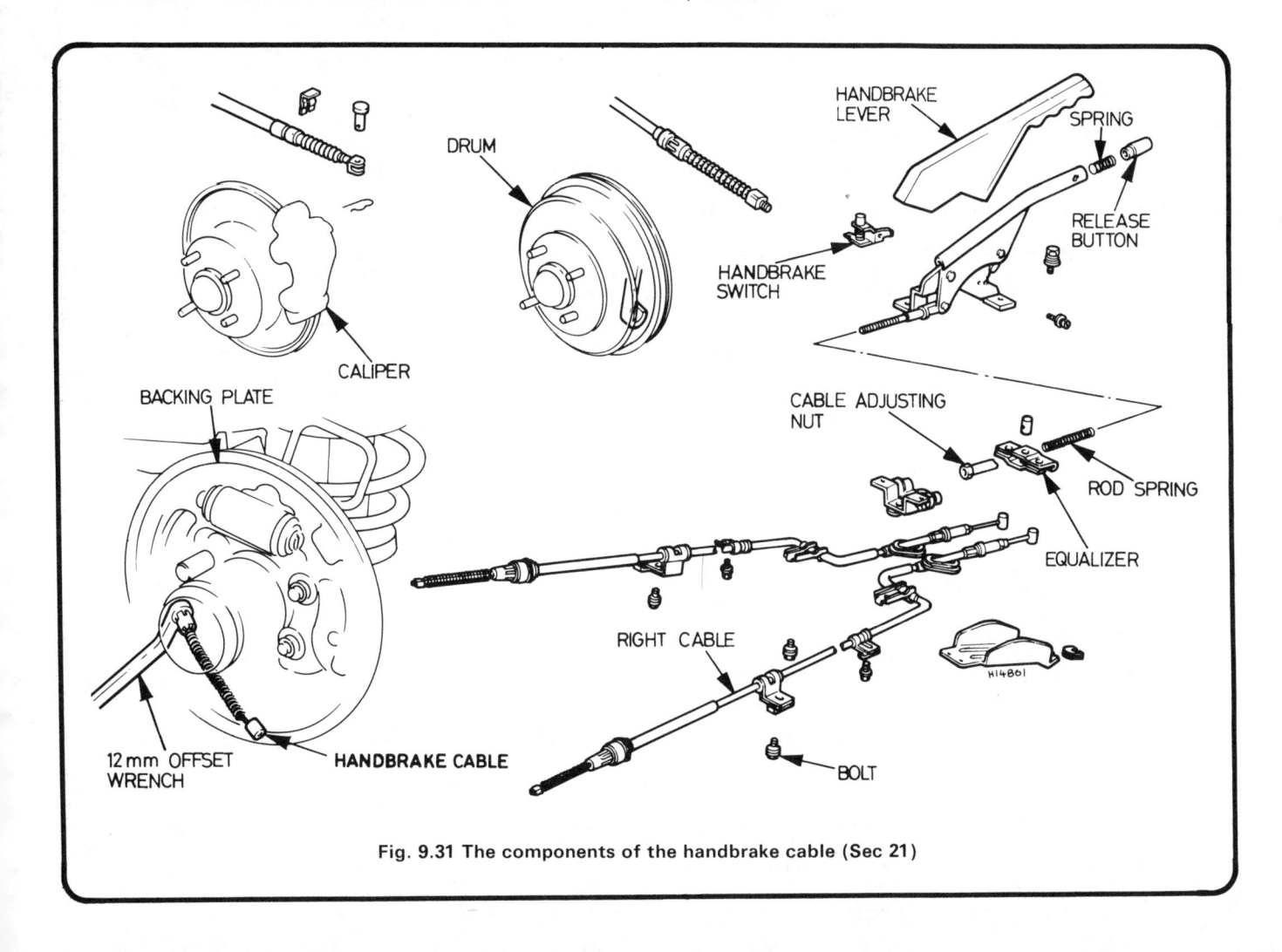

Fig. 9.31 The components of the handbrake cable (Sec 21)

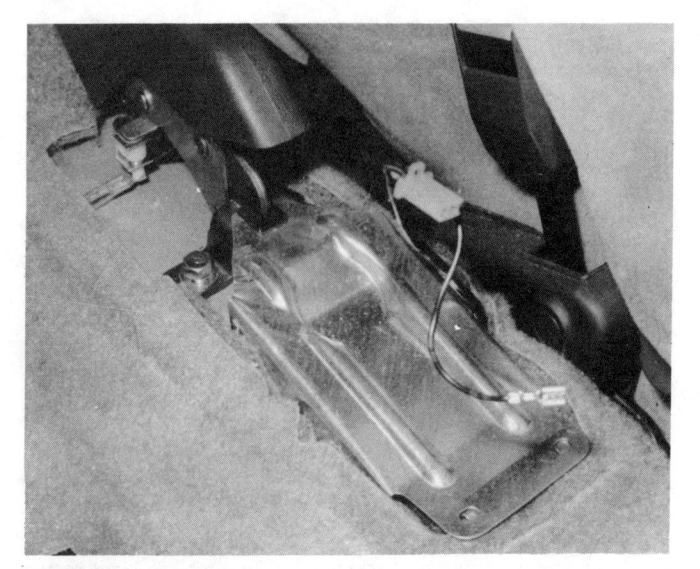

20.5 Handbrake equaliser assembly coverplate

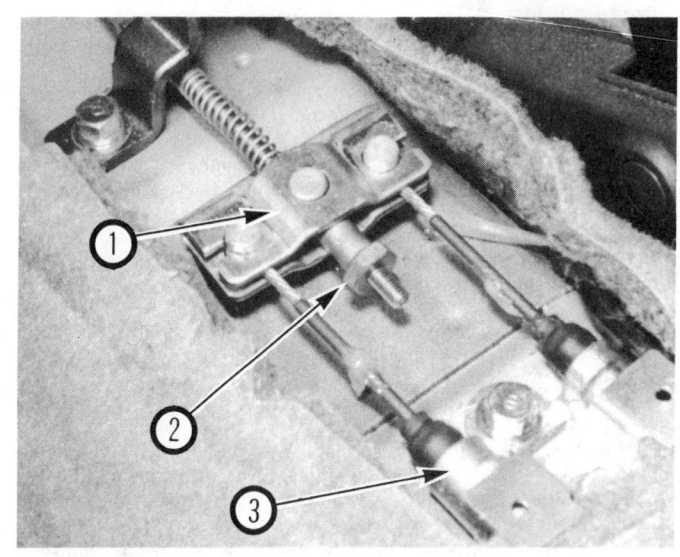

20.6 Handbrake equaliser and adjuster
1 Equaliser
2 Adjuster locknut
3 Cable clamp

8 Release the handbrake and check that the wheels turn freely without binding.
9 With the handbrake fully applied, check that the wheels cannot be turned. Readjust as necessary.
10 Remove the vehicle from jacks and refit the centre console.

21 Handbrake cable – removal and refitting

1 Chock the front wheels and raise the rear wheels clear of the ground.
2 Gain access to the equalised adjuster under the centre console (refer to Chapter 11).
3 Slacken off the adjuster until the cables can be unhooked from the equaliser.
4 Undo and remove the three support brackets on each cable on the underside of the car.
5 Refer to the relevant Sections and gain access to the cable ends at the rear wheels.
6 On vehicles with rear calipers, remove the shackle pin and spring plate to disconnect the cable.
7 On vehicles with rear drums, unhook the cable from the operating arm and unscrew the cable from the backplate.
8 Feed the cables down out of the vehicle, through the two holes in the floor and from around any other obstacles on the underside of the vehicle.
9 Refitting is a reversal of the above.
10 On completion, adjust the cable, as described in Section 20, and remove the vehicle from the jacks.

22 Handbrake warning light switch – adjusting and testing

1 The handbrake warning light switch is situated under the centre console beneath the handbrake lever (photo).
2 To adjust the switch, bend its support bracket so that the microswitch begins to operate when the handbrake lever is raised one notch, at which point the warning light should come on.
3 The switch can be tested by connecting a test lamp or ohmmeter between its terminal and a good earth.
4 With the handbrake off, the test lamp should be off, or no reading shown on the ohmmeter.
5 Application of the handbrake should result in the test lamp illuminating, or a reading showing on the ohmmeter.
6 If the handbrake warning light fails to come on during normal use, check the warning light bulb before changing the switch.

23 Brake fluid level switch – testing

1 The brake fluid level switch, in the master cylinder reservoir, is a simple bi-metal switch actuated by a float in which is a magnet.
2 With the correct level in the reservoir, the float and magnet are at the top of the switch, and the switch will remain open.
3 With a low fluid level, the float and magnet are at the bottom of the switch, and will close the switch, thus completing the circuit and switching on the warning light.
4 The switch can be tested simply by lifting the reservoir cap and switch out of the reservoir with the ignition switched on, and the low level warning light should come on.
5 Check the warning light bulb before replacing the switch.

24 Brake pedal – removal, refitting and adjustment

1 The brake pedal is of the pendant type being hung from a cross shaft to which it is attached at its upper end.
2 Connected to the brake pedal is a pushrod which operates the brake servo unit and master cylinder when the brakes are applied (photo).
3 The pedal may be removed from the cross shaft by removing either the E-clip or nut which holds it in place (depending upon model and market), disconnecting the servo unit pushrod, and unhooking the pedal return spring.
4 The pedal may now be slipped off the cross shaft.
5 Refit in the reverse order, and then adjust as follows.
6 Loosen the brake light switch locknut and back it off, so it no longer makes contact with the pedal.
7 Loosen the brake servo pushrod and screw the pushrod in or out until the pedal is at the prescribed height from the floor measured without the floor mat in position (see Fig. 9.32).
8 Tighten the pushrod locknut on completion.
9 Screw the brake light switch in until the plunger which contacts the brake pedal arm is fully depressed.
10 Back the switch off a half turn and tighten the locknut.
11 Check that the brake lights go off when the pedal is released.
12 With the engine stopped and all vacuum in the brake servo released (operate the pedal several times) check the free play which should be as shown in Fig. 9.33.
13 Apply only light finger pressure when checking free play and do not apply pressure to the servo unit pushrod when adjusting free play.
14 If the correct free play cannot be achieved, check all adjustments are correct.

Chapter 9 Braking system

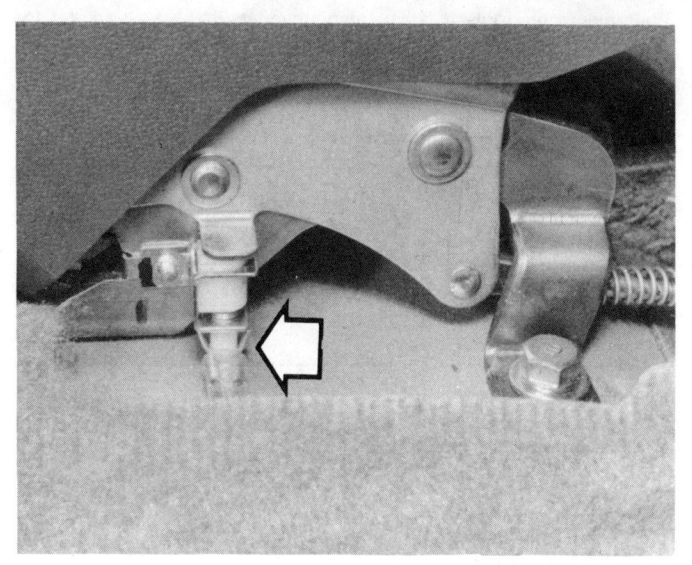

22.1 Handbrake warning light microswitch (arrowed)

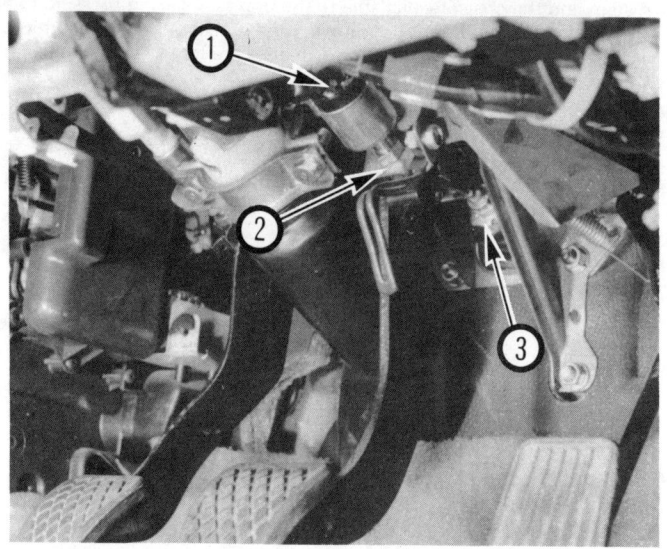

24.2 Brake pedal assembly
1 Stop-light switch
2 Adjusting nut
3 Brake servo pushrod

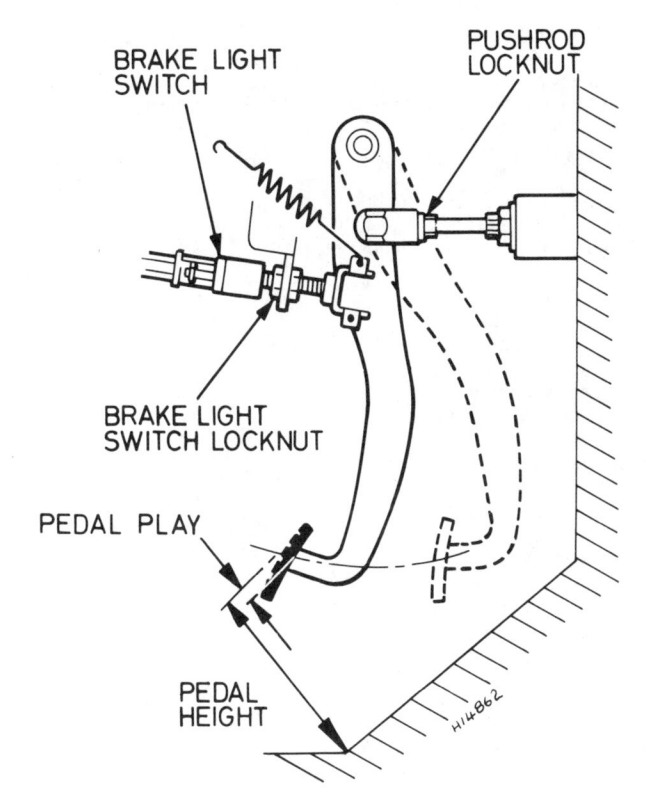

Fig. 9.32 Brake pedal adjustment setting (Sec 24)

Pedal height = 7.4 in (187 mm) measured without floormats

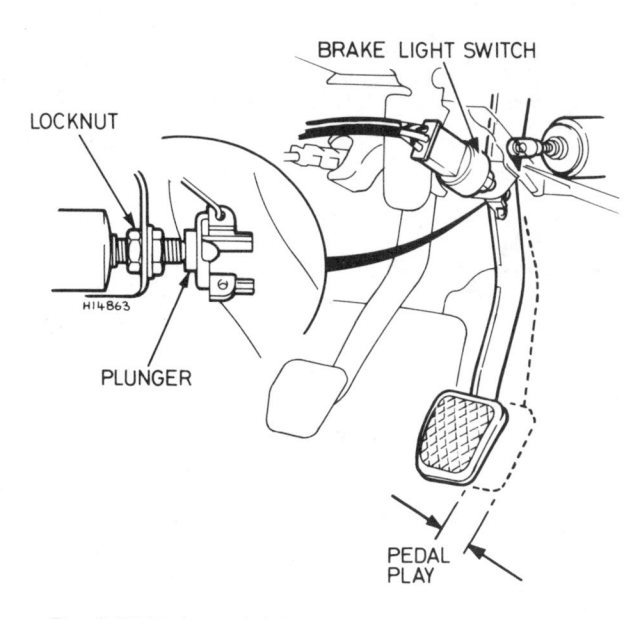

Fig. 9.33 Brake pedal free play adjustment (Sec 24)

Pedal play = 0.04 to 0.20 in (1 to 5 mm)

15 On completion, check again that the brake lights go off when the brake pedal is released.

25 Brake light switch – removing, testing and refitting

1 The brake light switch is screwed into a bracket just above the brake pedal.
2 To remove it, undo the locknuts and the electrical leads, then screw the switch out of the bracket.
3 The switch can be tested by connecting an ohmmeter between the terminals and operating the microswitch plunger by hand.
4 If there is no reading, renew the switch.
5 Refitting is a reversal of removing, and adjust as described in Section 24 on completion.

26 Anti-lock braking system (ALB)

General description

1 The ALB is an electro-hydraulic system, designed to prevent the wheels from locking under all braking conditions.
2 Gear pulsers, attached to the driveshafts in the case of the front wheels, and to the brake disc in the rear, rotate with those components.
3 Wheel sensor pick-ups detect this rotation and relay an electrical signal back to the control unit.
4 The control unit constantly monitors the operation of the hydraulic

system and the pressure available to the brakes, relieving pressure to a wheel which is about to lock, allowing the wheel to just keep turning.
5 The whole system has a fail safe ability, reverting to normal braking should a malfunction occur. The ALB warning light will come on when the fail safe system is in operation.

Servicing

6 Because of the complexity of the system, and need for special test equipment and tools it is not recommended that any work on the system be undertaken by the DIY mechanic.
7 Even bleeding of air from the system, which would be present after a component change is a complicated business, involving first the rear of the vehicle and then the front. As it is unlikely that this operation can be safely carried out at home, bearing in mind that one needs to be under the car to bleed the brakes, there is no recourse but to entrust any work on the ALB system to your dealer.

27 Fault diagnosis – braking system

Symptom	Reason(s)
Brake grab	Out-of-round drums Excessive run-out of discs Rust on drum or disc Oil stained linings or pads
Brake drag	Faulty master cylinder Foot pedal return impeded Reservoir breather blocked Seized caliper or wheel cylinder piston Incorrect adjustment of handbrake Weak or broken shoe return springs Crushed, blocked or swollen pipe lines
Excessive pedal effort required	Linings or pads not yet bedded-in Drum, disc or linings contaminated with oil or grease Scored drums or discs Faulty booster servo unit
Brake pedal feels hard	Glazed surfaces of friction material Rust on disc surfaces Seized caliper or wheel cylinder piston
Excessive pedal travel	Low reservoir fluid level Disc run-out excessive Worn front wheel bearings Air in system Worn pads or linings Rear brakes require adjustment
Pedal creep during sustained application	Fluid leak Internal fault in master cylinder Faulty booster unit check valve
Pedal "spongy"	Air in system Perished flexible hose Loose master cylinder mounting nuts Cracked brake drum Faulty master cylinder Reservoir breather blocked Linings not bedded-in
Fall in reservoir fluid level	Normal, due to pad or lining wear Leak in hydraulic system

Chapter 10 Suspension and steering

Contents

Fault diagnosis – suspension and steering	35
Front and rear spring height – inspection	15
Front and rear wheels – bearing adjustment	8
Front and rear wheel toe-in – adjustment	9
Front radius rod – removal and refitting	4
Front stabiliser bar – removal and refitting	3
Front suspension lower arm – removal and refitting	6
Front suspension steering knuckle and wheel hub bearings – removal and refitting	7
Front suspension strut – removal, overhaul and refitting	5
General description	1
Manual steering column – removal, overhaul and refitting	20
Manual steering rack and gearbox – overhaul	22
Manual steering rack and gearbox – refitting	23
Manual steering rack and gearbox – removal	21
Manual steering rack guide – adjustment	18
Power steering assist – inspection	29
Power steering column – removal and refitting	26
Power steering fluid – removal and topping-up	27
Power steering pump – overhaul	33
Power steering pump – removal and refitting	32
Power steering pump belt – adjustment	28
Power steering rack – removal and refitting	34
Power steering rack guide – adjustment	30
Power steering speed sensor – removal and refitting	31
Rear shock absorber – removal and refitting	10
Rear suspension lower arm – removal and refitting	12
Rear suspension radius rod – removal and refitting	11
Rear suspension stabiliser bar – removal and refitting	13
Rear wheel hub carrier and wheel bearings – removal and refitting	14
Routine maintenance	2
Steering lock – removal and refitting	24
Steering wheel – removal and refitting	19
Steering wheel rotational play – inspection	17
Tie-rod end balljoints – removal	25
Wheels and tyres – general care and maintenance	16

Specifications

Front suspension
Type .... Independent MacPherson strut, with coil spring and telescopic damper. Stabiliser bar

Spring height:
Without air conditioning .... 26.6 in (675 mm)
With air conditioning .... 26.4 in (670 mm)

Rear suspension
Type .... Independent MacPherson strut, with coil spring and telescopic damper. Stabiliser bar

Spring height .... 25.7 in (652 mm)

Steering
Type .... Rack and pinion. Manual or power-assisted
Ratios:
Manual .... 18.7 to 1
Power-assisted .... 14.7 to 1
Turns (lock-to-lock):
Manual 3.5
Power-assisted 2.76

Fluid capacity (power-assisted)	3.0 Imp pint (1.8 US qt, 1.7 litre)	
Steering wheel rotational play (measured at rim)	0.4 in (10 mm) maximum	
Steering effort:		
Manual	3.3 lb (1.5 kg)	
Power-assisted:		
Saloon	4 lb (1.8 kg)	
Hatchback	5 lb (2.3 kg)	

Roadwheels

Type	Pressed steel or light alloy
Size	4½ J x 13 or 5J x 13

Tyres

Size	165 SR 13
	185/70 R 13
	185/70 SR 13
	P 185/70 R 13
	T 105/80 D 13 (spare)
Pressures	Refer to handbook or tyre placard on vehicle

Wheel settings

Steering angles (unladen):
  UK models:
    Camber (front) .... 0 ± 1°
    Castor .... 1° 25′ ± 1°
    King pin inclination (front) .... 12° 30′ ± 1° 30′
    Toe-in:
      Front .... 0 ± 0.118 in (0 ± 3 mm)
      Rear .... 0 + 0.157 in (0 + 4 mm
                     − 0 in   − 0 mm)

  North American models:
    Camber (front) .... 0 ± 1°
    Castor .... 1° 30′ ± 1°
    Kingpin inclination (front) .... 12° 30′ ± 1° 30′
    Toe-in:
      Front .... 0 ± 0.118 in (0 ± 3 mm)
      Rear:
        All models except SEi .... 0.079 ± 0.079 in (2 ± 2 mm)
        SEi models .... 0.079 ± 0.118 in (2 ± 3 mm)
Wheel endplay:
  Front .... 0 to 0.002 in (0 to 0.05 mm)
  Rear .... Zero

Torque wrench settings

	lbf ft	Nm
Front suspension		
Shock absorber upper mounting nuts	28	38
Shock absorber centre rod locknut	33	45
Shock absorber-to-knuckle/hub pinch-bolt	47	64
Lower control arm inner bolt	36	49
Lower control arm balljoint bolt	40	54
Radius rod front mounting nut	32	43
Radius rod-to-lower arm bolts	40	54
Stabiliser bar mountings	16	22
Splash guard screws	4	5
Caliper mounting bolts	56	78
Front wheel spindle nut	137	186
Front wheel lug nuts	80	108
Rear suspension		
Shock absorber top mounting nuts	16	22
Shock absorber centre rod locknut	40	54
Shock absorber-to-hub carrier bolt	40	54
Radius rod front adjusting bolt	47	64
Radius rod rear attachment bolt	51	69
Lower arm inner attachment bolt	40	54
Lower arm-to-hub carrier bolt	60	81
Stabiliser bar mounting bolts	16	22
Brake back plate/caliper mount-to-hub carrier bolts	22	30
Rear wheel bearing initial torque	18	24
Rear wheel bearing final torque	4	5
Steering (manual)		
Steering wheel shaft nut	36	49
Steering column upper bracket nuts	9	12
Steering column lower bracket bolts	16	22
Universal joint bolts	22	30
Steering rack mounting bracket bolts	16	22
Rack guide screw locknut	18	24

Chapter 10 Suspension and steering

Torque wrench settings (continued)

	lbf ft	Nm
Steering (manual)		
Tie-rod-to-rack locknut	54	73
Tie-rod-to-balljoint locknut	33	45
Balljoint castle nut	32	43
Power steering (where different to manual)		
Steering column upper mounting nuts	10	14
Steering column lower mounting bolts	10	14
Pump pivot bolt	32	43
Pump adjusting bolt	20	27
Pump pulley nut	33	45
Gearbox fluid pipelines (Fig. 10.29):		
pipe D	9	12
pipe P	27	37
pipe S	9	12
pipe T	22	30

1 General description

Both front and rear suspension is independent by MacPherson strut, coil spring and telescopic damper, with front and rear stabiliser bars.

Steering is by rack and pinion, manual or power-assisted, and the steering column is collapsible on impact.

Although several tasks may be undertaken by the home mechanic, many of these require the use of special tools or test equipment, and the reader is advised to study the relevant Section to determine what he can do, and what is best left to a garage.

2 Routine maintenance

At the intervals given in Routine Maintenance (at the front of the book) undertake the following service tasks.

1 Check the tyre pressures (cold) including the spare. Check the tyres for excessive wear, cuts, bulges and tread depth. Check the

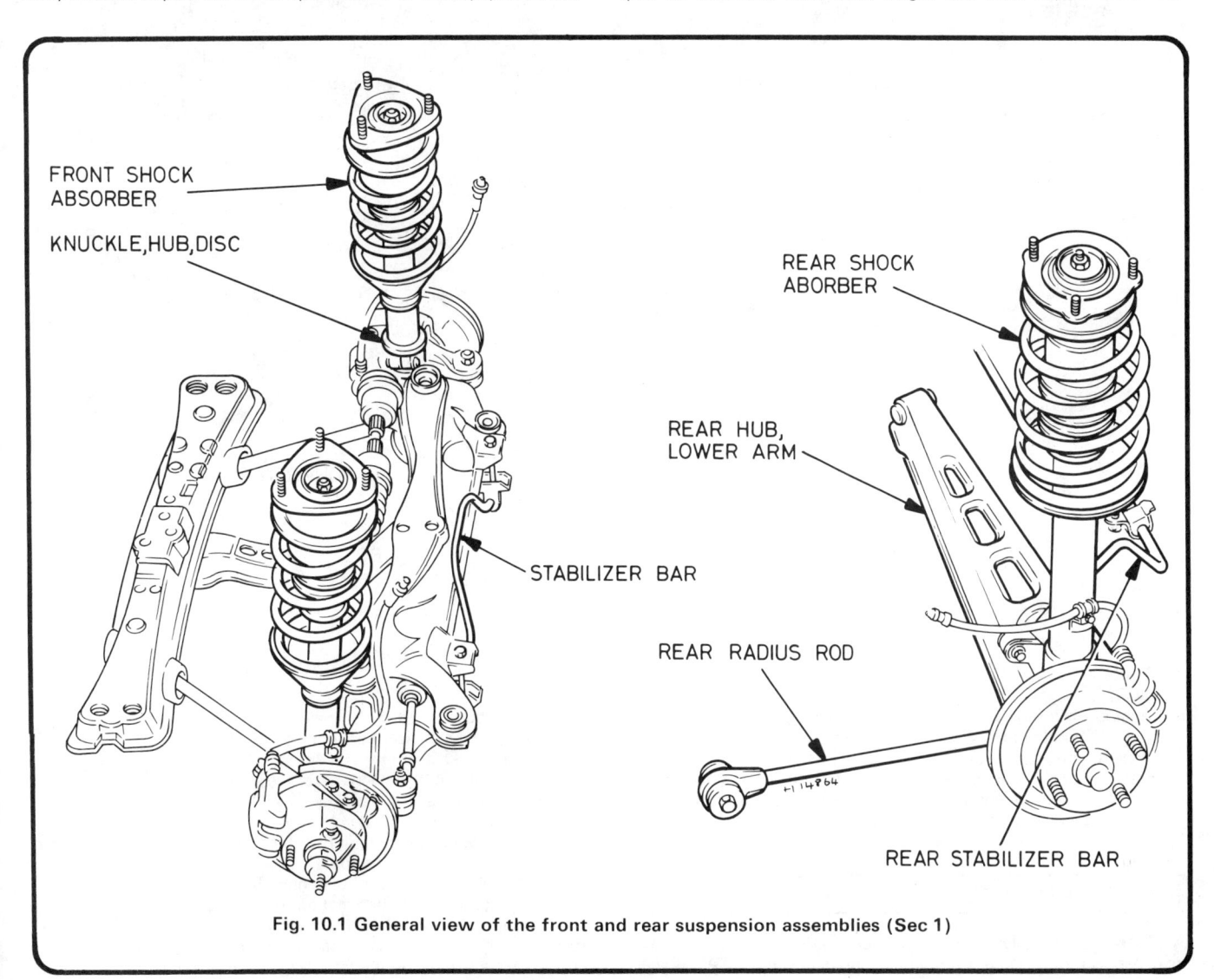

Fig. 10.1 General view of the front and rear suspension assemblies (Sec 1)

232 Chapter 10 Suspension and steering

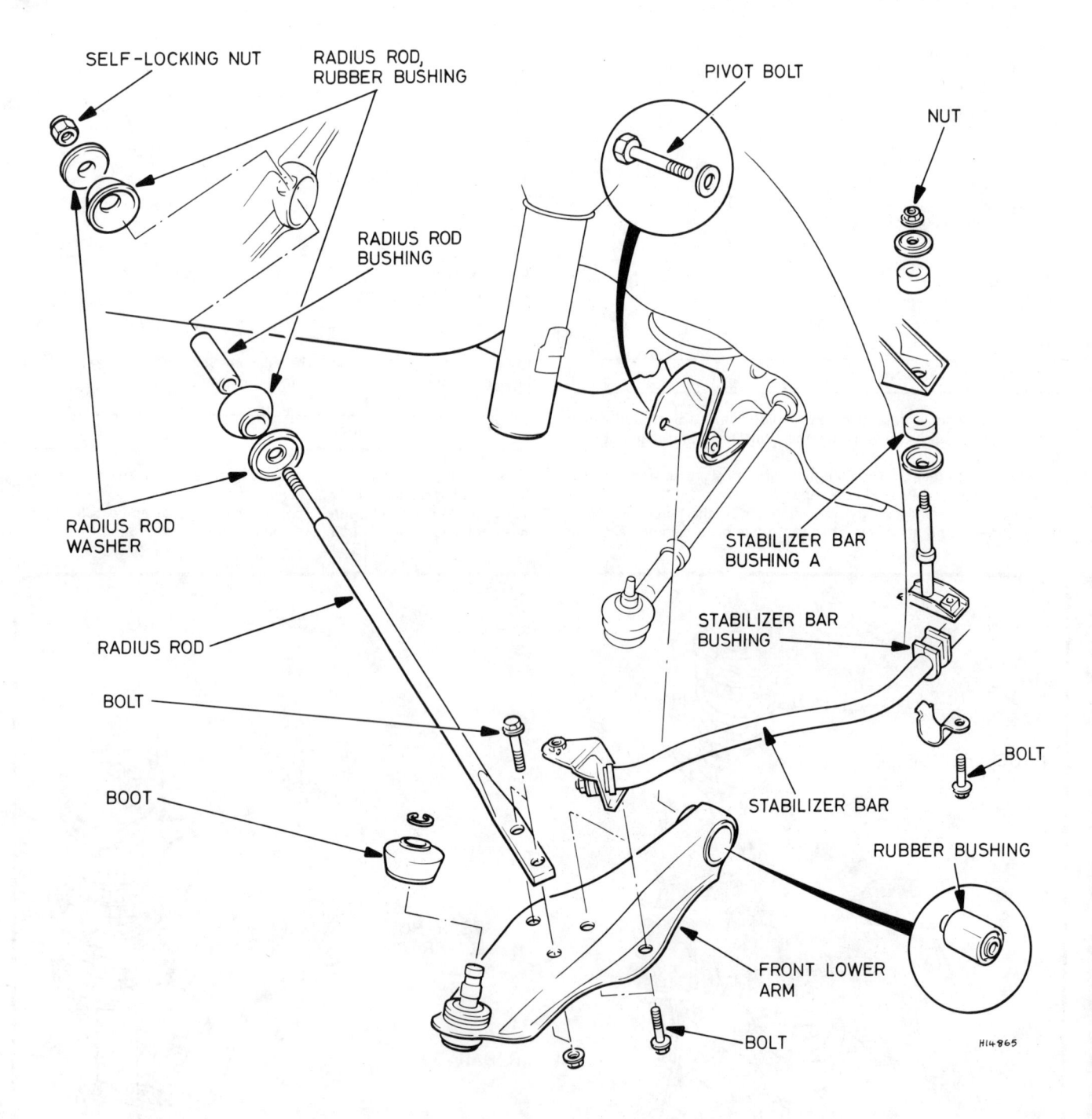

Fig. 10.2 Exploded view of the front suspension (Sec 1)

power steering reservoir fluid level and fill as necessary, and check the tension and condition of the pump drivebelt.
2 Inspect the suspension mounting bolts, check front wheel alignment, and the operation of steering, tie-rod ends, steering rack and gearbox, and the condition of all protective rubber boots. Check the power steering system where fitted.

3 Front stabiliser bar – removal and refitting

1 Refer to the relevant Chapters and disconnect the exhaust downpipe, and the gearbox torque arm and gear shift rod. Allow them to hang, supported, out of the way.
2 Unscrew the stabiliser bar support bracket bolts on both sides and remove the brackets (photo).
3 Remove the nut from each end of the stabiliser bar and unscrew the bolts from the end support brackets (photo).
4 Withdraw the stabiliser bar.
5 If necessary, remove the nut from the top end of each inner support bracket rod, and remove the rods.
6 Inspect all bushes and rubber components for cracking, deterioration and oil or grease contamination. Renew as necessary.
7 Refitting is a reversal of this procedure, but if the vehicle is on stands, do not fully tighten the stabiliser bar bolts/nuts until the weight of the vehicle is back on the wheels.

Chapter 10 Suspension and steering

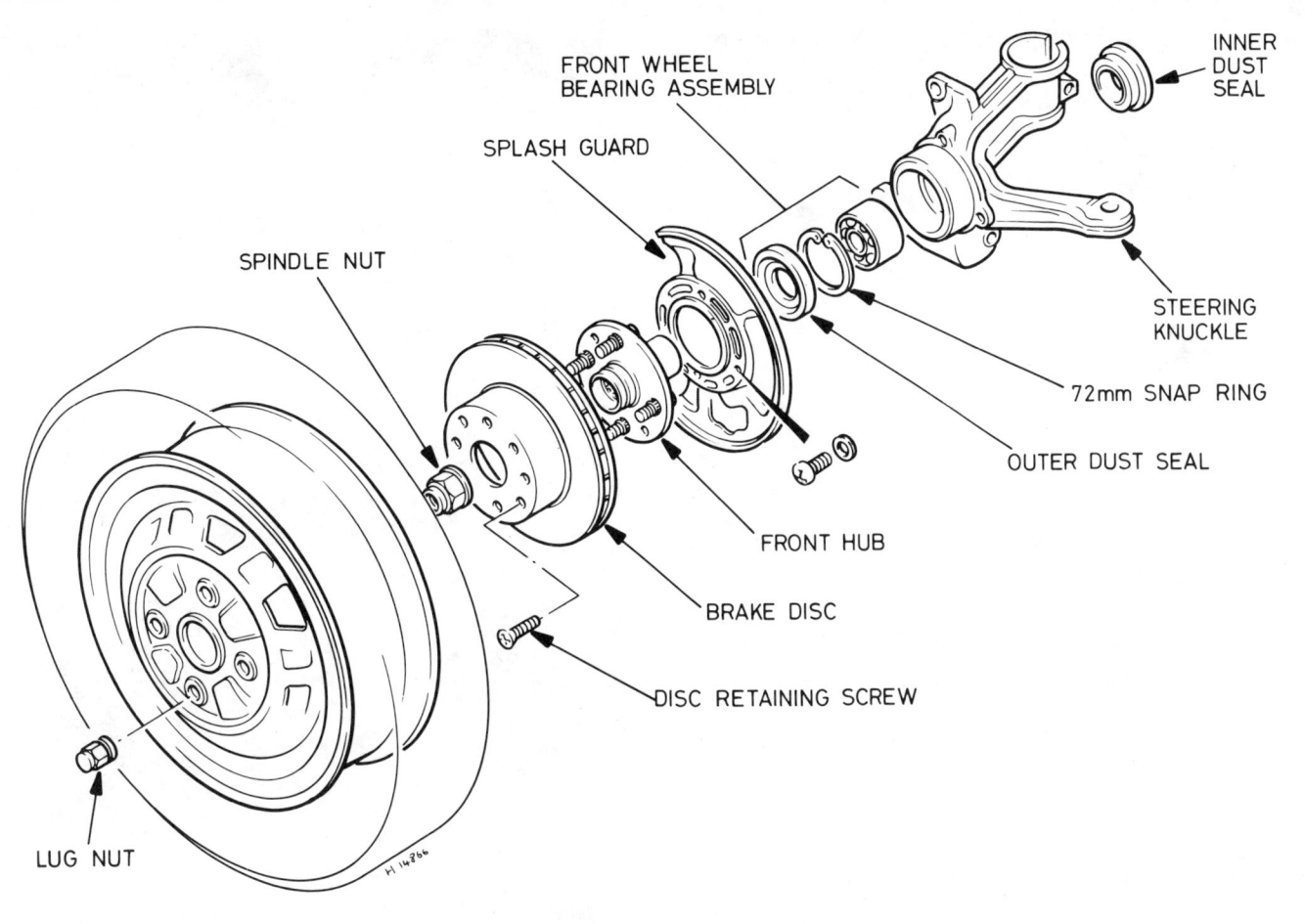

Fig. 10.3 Exploded view of the front hub (Sec 1)

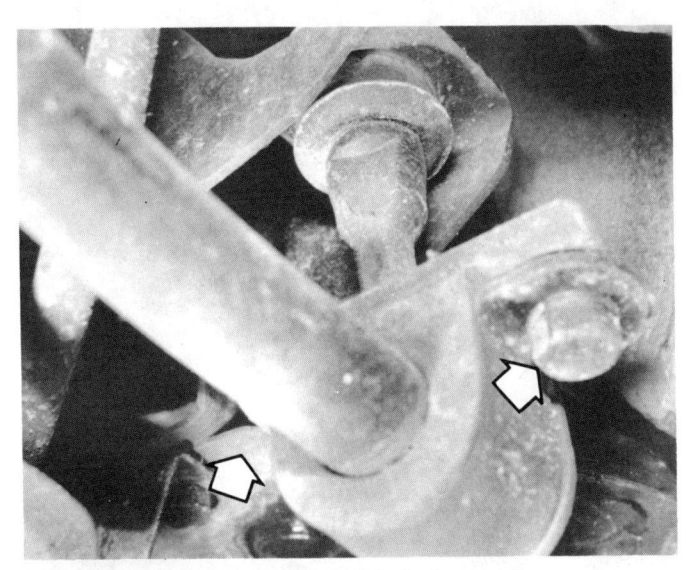

3.2 The stabiliser bar support bracket bolts (arrowed)

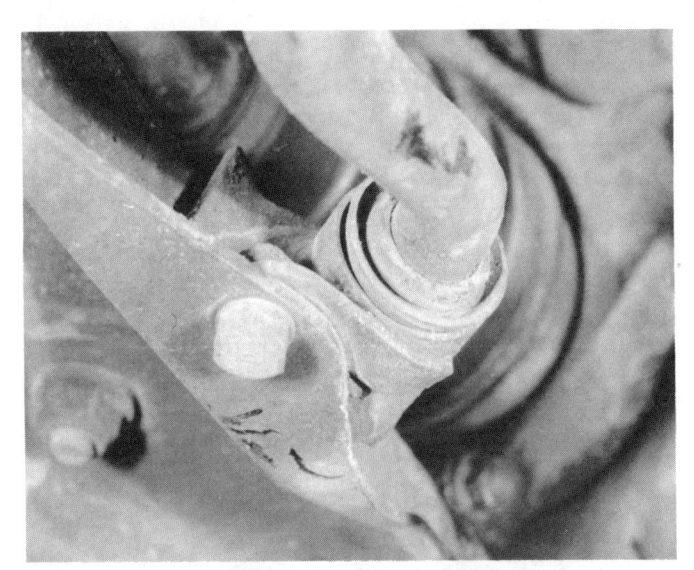

3.3 The stabiliser bar end support bracket

4 Front radius rod – removal and refitting

1 Remove the two bolts from the front lower arm (photo).
2 Remove the nut from the bush housing at the front end of the radius rod (photo).
3 Withdraw the rod.
4 Extract the bushing from the housing and inspect the rubber parts for deterioration, cracking and contamination by oil or grease. Renew as necessary.
5 Refitting is a reversal of this procedure, but use a new self-locking nut on the bush housing and leave final tightening until the weight of the vehicle is on the wheels.

4.1 Radius rod-to-lower arm bolts

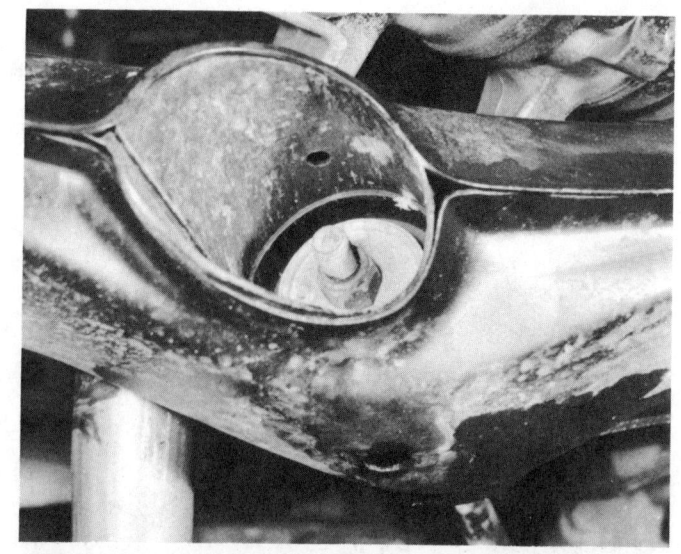

4.2 Self-locking nut on the radius arm front bush housing

5 Front suspension strut – removal, overhaul and refitting

1 Raise the front end of the vehicle and support it safely on stands.
2 Remove the front wheels.
3 Disconnect the brake hydraulic hose from the suspension strut (photo).
4 Position a suitable jack under the front lower suspension arm, just beginning to take the weight.
5 Refer to Chapter 9 and remove the brake caliper mounting bolts and support the caliper with wire so that it is out of the way.
6 Remove the stabiliser bar from the lower arm (Section 3).
7 Remove the self-locking bolt from the steering knuckle/hub assembly (photo).
8 Using a lead or hide-faced hammer, tap the knuckle down off the shock absorber, lowering the supporting jack as you do so, allowing the arm to fall away (photo).
9 Remove the rubber cap from the shock absorber upper mounting in the engine compartment, and remove the three mounting bolts (photo).

5.3 Disconnect the brake hydraulic hose from the suspension strut mounting (arrowed)

5.7 The self-locking bolt in the steering knuckle (arrowed)

5.8 The balljoint removed from the knuckle

Chapter 10 Suspension and steering

Fig. 10.4 Removing the steering knuckle from the MacPherson strut (Sec 5)

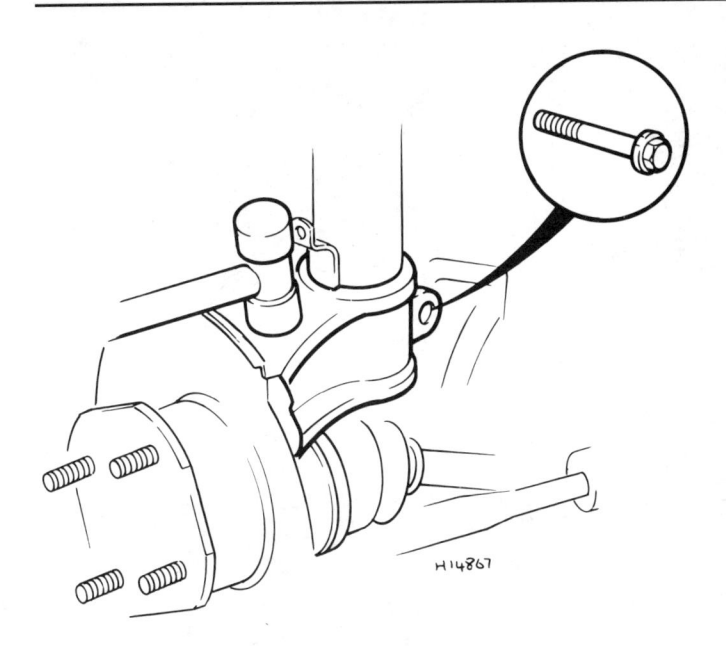

5.9 Remove the three mounting bolts (arrowed)

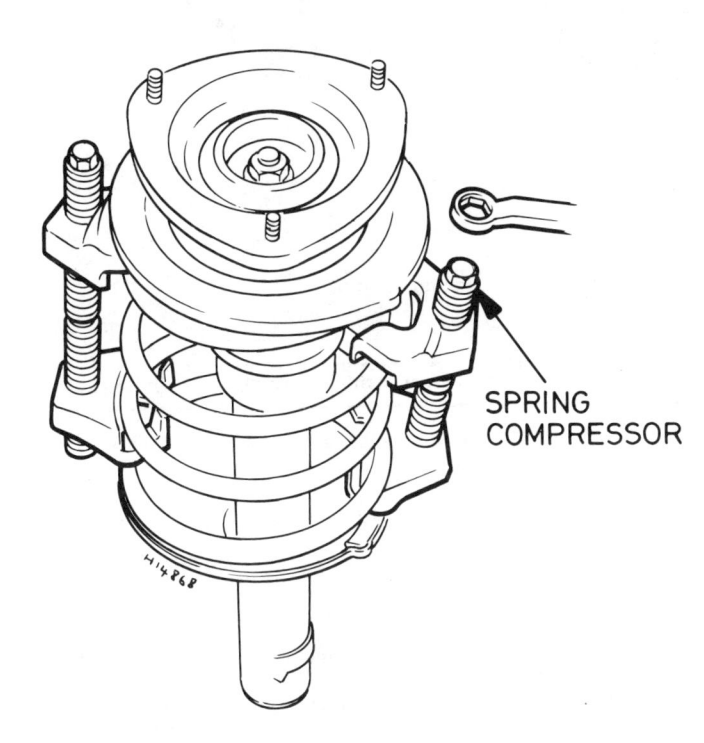

Fig. 10.5 Using spring compressors to compress the front suspension spring (Sec 5)

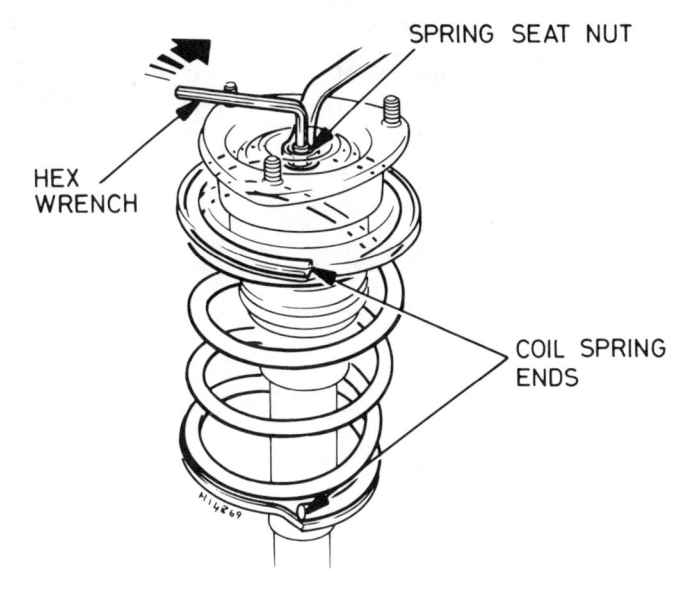

Fig. 10.6 Removing the seat nut using a ring spanner and Allen key (Sec 5)

10 Lower the shock absorber out from under the wing.
11 For further dismantling, coil spring compressors are required, readily available from motor accessory shops.
12 Fit the coil spring compressors, following the manufacturer's instructions.
13 Compress the coil spring; compressing it no further than is necessary to gain access to the seat nut.
14 Remove the seat nut using a ring spanner, with an Allen key inserted in the spindle end to prevent it from turning.
15 Remove the spring compressor to relieve the tension on the spring, at the same time removing the components from the top mounting.
16 Operate the damper unit several times using long, slow strokes and then short sharp strokes of 2 to 3 in (50 to 75 mm).
17 Check that operation is smooth and there is no sign of jerkiness, no binding or abnormal noises, and that the unit is free from oil leaks. Renew faulty shock absorbers. It is advised that this be done in pairs.
18 Inspect the components of the upper bearing housing for wear or damage and the bearing for smooth, free running, renewing any defective parts.
19 Inspect the dust cover and rubber bump stop for general condition, renewing as necessary.
20 Inspect the coil spring for damage and corrosion, and check for weakened compression, which will have been noticeable under normal driving. Your Honda dealer can check springs for compression.
21 Reassembly is a reversal of this procedure, but coat both sides of the needle roller bearing with grease and ensure that the coil spring ends are located correctly in the steps in the upper spring seat and shock absorber. Partially tighten all nuts and bolts.

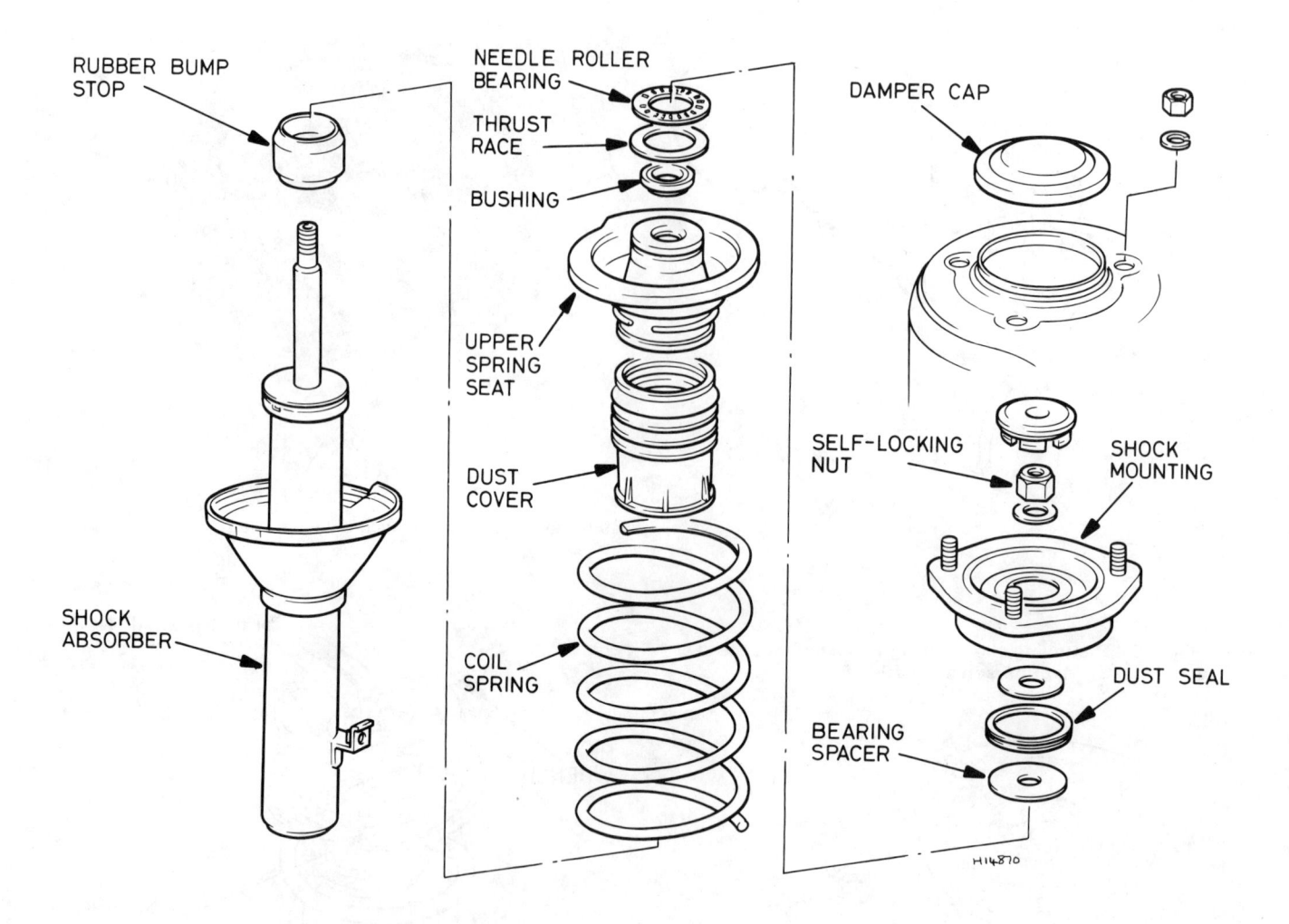

Fig. 10.7 Exploded view of the shock absorber upper mounting (Sec 5)

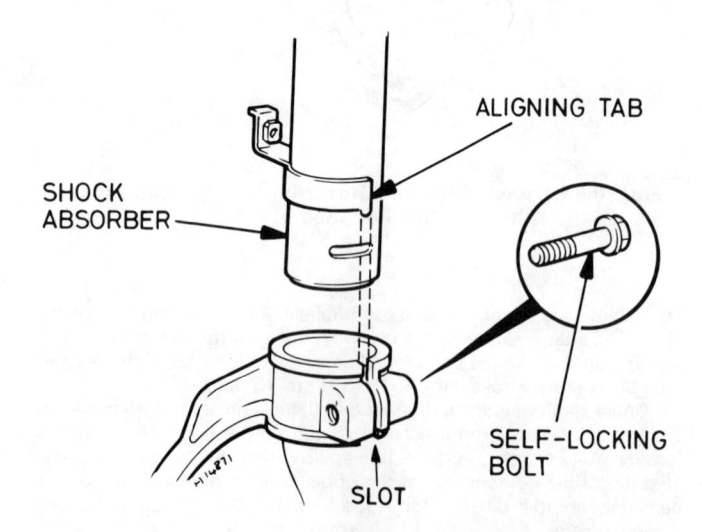

Fig. 10.8 Ensure the aligning tap enters the slot in the knuckle (Sec 5)

22 Ensure that the aligning tab, which is part of the brake hose support bracket welded to the shock absorber, lines up with the slot in the knuckle.
23 Carry out final tightening of the suspension bolts when the weight of the vehicle is on the wheels.

6 Front suspension lower arm – removal and refitting

1 Raise the front of the vehicle and support on stands placed under the side-members.
2 Remove the roadwheel.
3 Remove the stabiliser bar bolts at the mounting on the lever control arm (see Section 3).
4 Remove the bolt from the lower arm balljoint and the bolts from the radius rod-to-control arm (photo).
5 Remove the bolt from the lower arm inner attachment point (photo) and lower the arm away.
6 Check the rubber bushes for deterioration, damage or contamination.
7 Check the condition of the rubber boot on the balljoint.
8 A special tool is required to remove/refit the boot, so if it is in need of renewal, take the control arm to your dealer.
9 Similarly, if the balljoint is worn, consult your dealer.
10 Renew all other parts as necessary.
11 Refitting is a reversal of this procedure, but do not fully tighten nuts and bolts to their specified torque until the weight of the vehicle is on the wheels.

7 Front suspension steering knuckle and wheel hub bearings – removal and refitting

1 Relieve the staking on the wheel hub spindle nut and loosen the spindle nut a half turn.

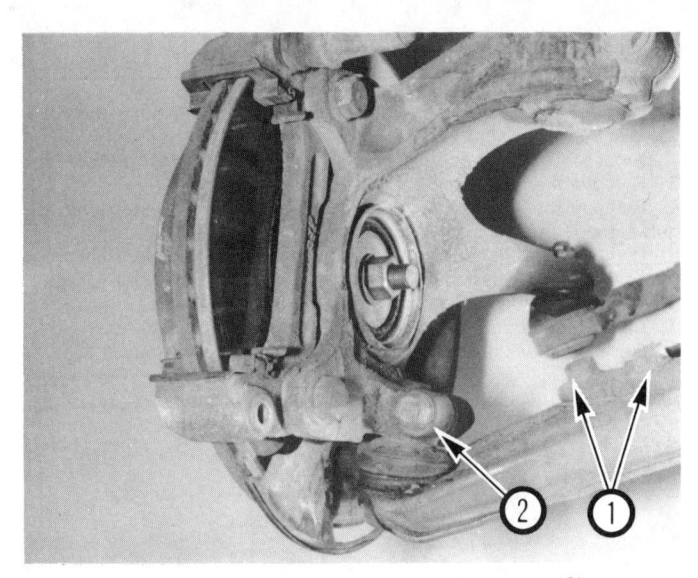

Chapter 10 Suspension and steering

6.4 The radius rod bolts (1) and the lower balljoint bolt (2) Photo shows dummy driveshaft fitted

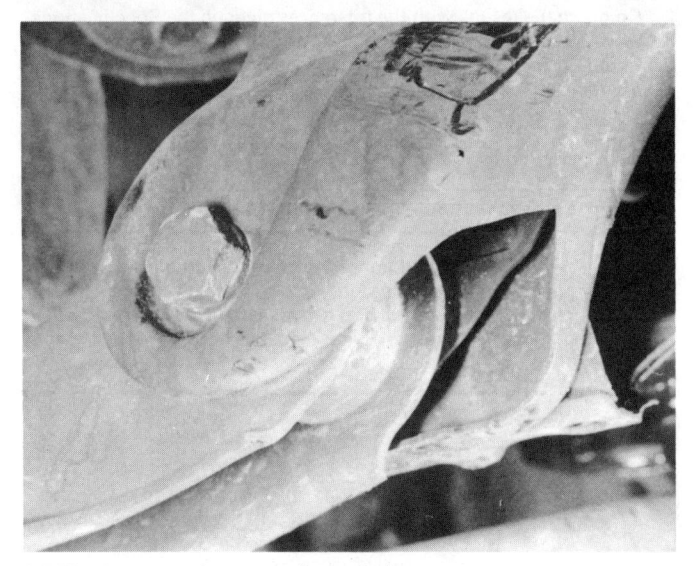

6.5 The lower arm inner attachment bolt

2 Loosen the wheel lug nuts and then raise the front end of the vehicle onto axle stands.
3 Remove the lug nuts, wheel and spindle nut.
4 Refer to Chapter 9 and unbolt the brake caliper; supporting it to one side.
5 Remove the brake disc by removing the two disc retaining screws and then screwing suitable screws into the other threaded holes in the disc. this will push the disc away from the hub. Turn each screw in a little at a time to prevent tilting the disc excessively.
6 Refer to Section 21 and disconnect the steering tie-rod balljoint.
7 Refer to Section 6 and remove the lower arm balljoint pinch-bolt, and tap the lower arm free.
8 Remove the shock absorber pinch-bolt and remove the knuckle/hub assembly by gently tapping it downward using a soft-faced mallet.
9 Pull the knuckle and hub assembly off the driveshaft, ensuring that the inboard joint is not pulled out, or it may come apart (refer to Chapter 8).
10 For any further dismantling special tools and presses are required; firstly to separate the hub from the knuckle, and then to remove the bearings. It is advisable to take the knuckle/hub assembly to your dealer and have him separate the unit and remove and install the bearings.
11 With new bearings and seals fitted, the knuckle and hub assembly may be refitted, using the reverse sequence to that given here, and paying attention to the following points:
12 If the nuts on self-locking bolts can be easily threaded over the nylon insert, renew the bolts.
13 Use a new spindle nut on reassembly, tighten the nut to the specified torque, and stake the locking collar with a punch on completion (photo).
14 Tighten all suspension nuts and bolts to the specified torque after the weight of the vehicle is back on the wheels.

8 Front and rear wheels – bearing adjustment

1 The front and rear wheel bearing endplay may be measured using a dial test indicator set up as shown in Fig. 10.9.

7.13 Staking the spindle nut locking collar

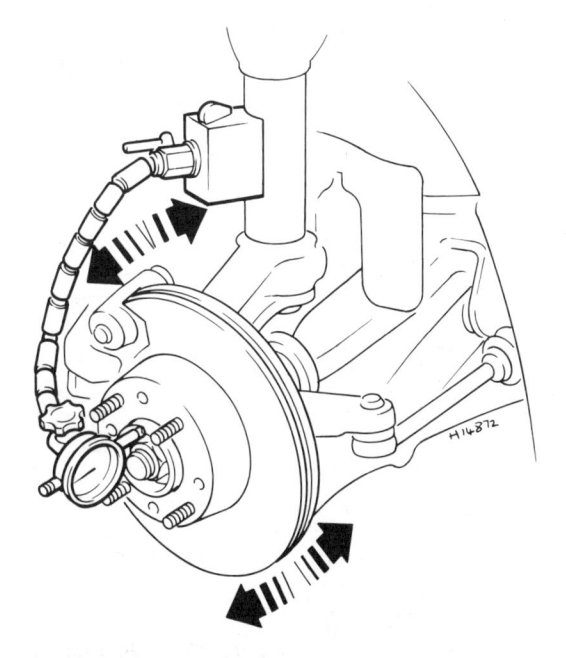

Fig. 10.9 Dial test indicator set up to measure wheel bearing endplay (Sec 8)

Chapter 10 Suspension and steering

2 Compare the results with the tolerances given in the Specifications.
3 Where tolerances are exceeded, retorque the spindle nut and check the endplay again, before deciding on renewing the bearings.
4 Spindle nut tightening procedures are given in Section 7 for the front wheel and Section 14 for the rear.

9 Front and rear wheel toe-in – adjustment

1 Whenever any suspension components are disturbed the front and rear wheel toe-in should be measured and adjusted, as necessary.
2 As this requires special checking gauges it is best left to specialists (local garage or tyre fitting firm) who have the necessary equipment.

10 Rear shock absorber – removal and refitting

1 Raise the rear of the vehicle onto stands and remove the road wheels.
2 Refer to Chapter 9 and disconnect and remove the rear brake drum.
3 Remove the stabiliser bar from the lower arm.
4 Remove the self-locking bolt securing the shock absorber to the hub carrier (photo).
5 Remove the rubber cap and the three nuts from the shock absorber upper mounting. These can be reached from inside the boot (photo).
6 Lower the shock absorber from under the wheel arch.
7 Refer to Section 5 and carry out the same dismantling procedure as for the front shock absorber, referring to Fig. 10.11 for details.

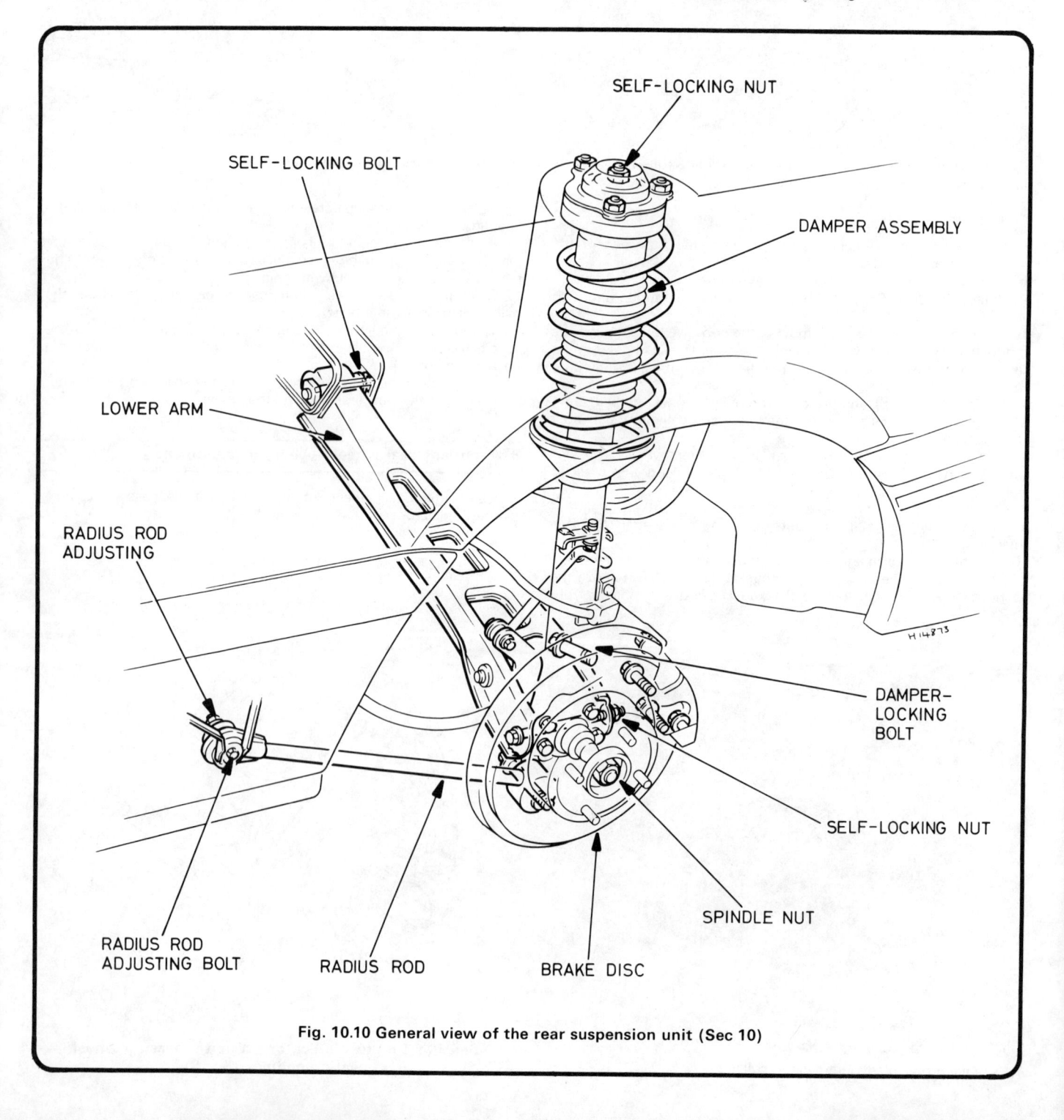

Fig. 10.10 General view of the rear suspension unit (Sec 10)

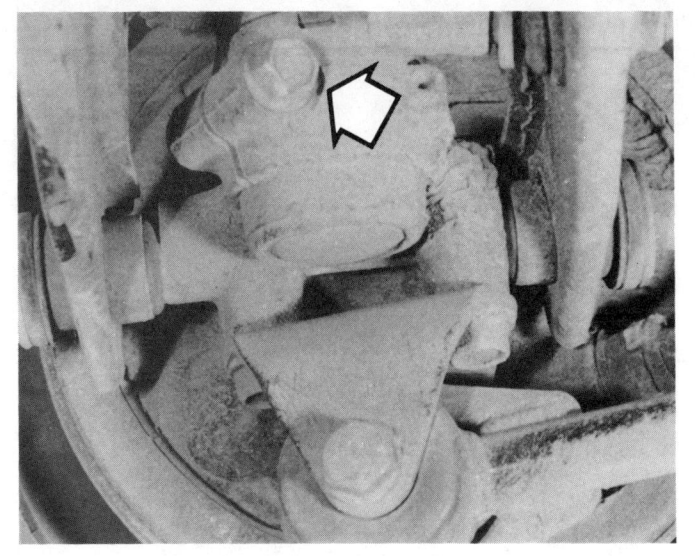

10.4 Remove the self-locking bolt (arrowed)

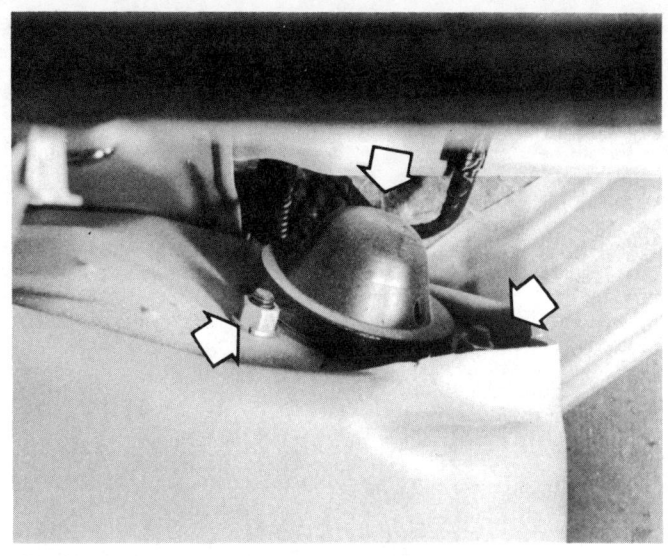

10.5 Shock absorber upper mounting nuts (arrowed)

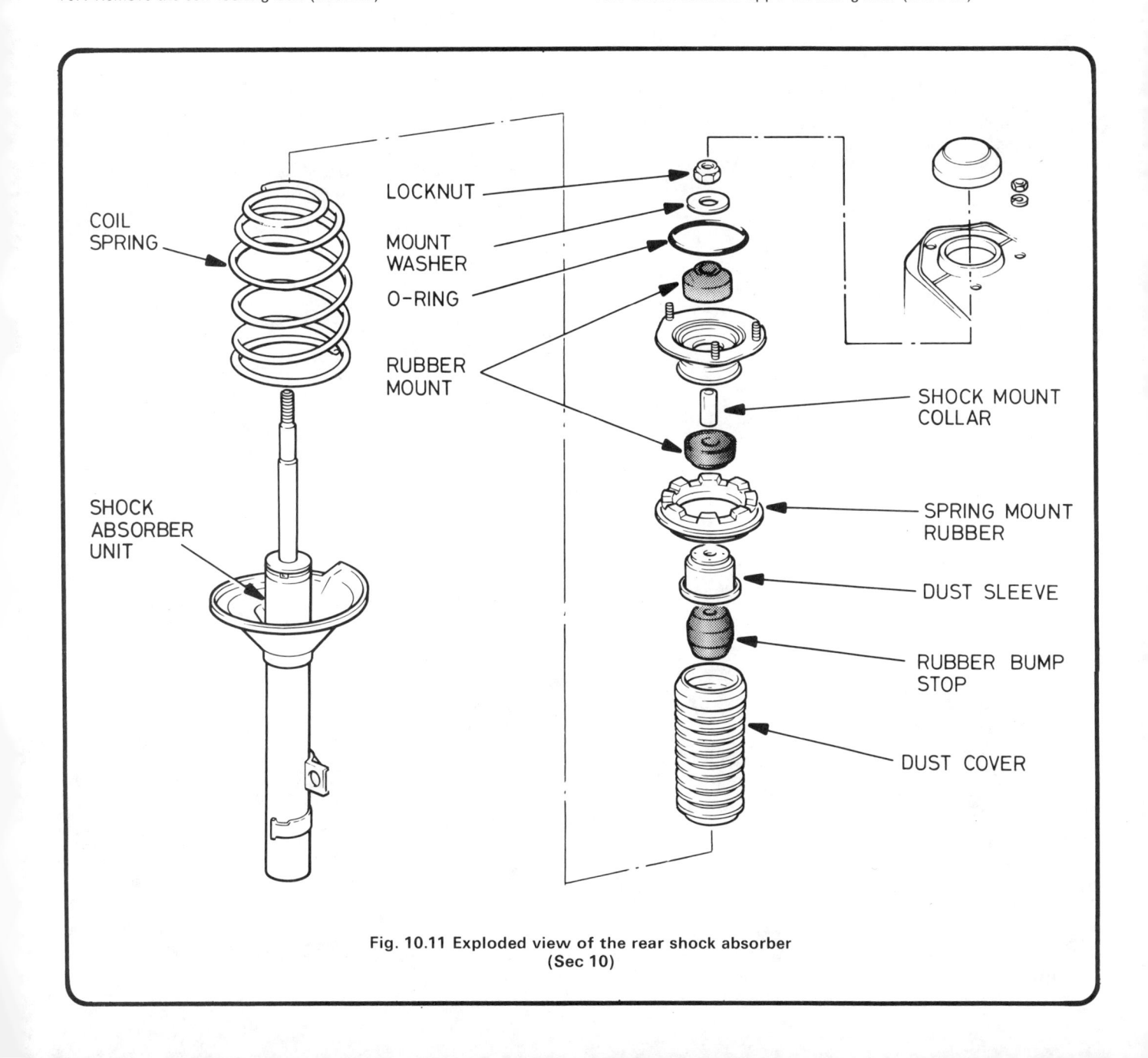

Fig. 10.11 Exploded view of the rear shock absorber
(Sec 10)

Chapter 10 Suspension and steering

8 The same test and inspections apply as for the front shock absorber.
9 Reassemble the shock absorber by compressing the spring with spring compressors, and fitting it to the shock absorber.
10 Install the rubber bump stop, dust sleeve, dust cover and spring mount rubber.
11 Install the shock absorber damper mount base so that the OUT mark is opposite the index mark on the damper strut (see Fig. 10.12).
12 Install the remaining components of the upper mounting, and loosely tighten the locknut, using an Allen key to prevent the centre shaft turning.
13 Refit the hub carrier, aligning the slot in the carrier with the tab on the shock absorber.
14 Tighten the self-locking bolt to the specified torque.
15 Refit the stabiliser bar.
16 Refit the brake drum, connect the hydraulic line and handbrake cable, and bleed the brake system (refer to Chapter 9).
17 Refit the road wheel, lower the vehicle to the ground, and torque load all disturbed bolts and nuts to their specified torque.

11 Rear suspension radius rod – removal, refitting

1 The rear radius rods, one on either side of the vehicle, provide positive location for each lower arm.
2 One end is attached to a bracket bolted to the bottom of the rear wheel hub carrier (photo), and the other to a bracket bolted to the chassis forward of the rear suspension (photo).
3 To remove the radius rod, remove the bolts from both attachment points and pull the rod down.
4 Inspect the rubber bushes for deterioration and cracking, and the attachment brackets for corrosion and cracking.
5 Renew any defective parts.
6 Refitting is a reversal of removing.
7 Having refitted the radius rod, the toe-in of the rear wheels must be checked and set correctly.
8 As this requires special gauges, the vehicle should be taken to your dealer as soon as is practicable. Setting is done by adjusting the forward housing bolt, which is actually a cam plate, and effectively moves the rear suspension forward or back as it is turned.
9 If follows that the front mounting should never be disturbed unless necessary.

12 Rear suspension lower arm – removal and refitting

1 Raise the rear of the vehicle on to axle stands.
2 Remove the bolt securing the stabiliser bar to the lower arm.

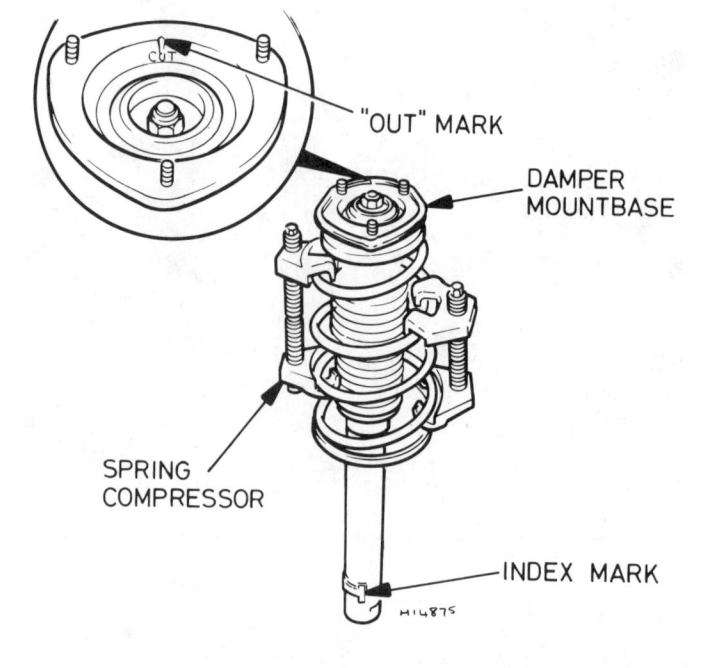

Fig. 10.12 Align the 'OUT' mark with the index mark (Sec 10)

3 Remove the bolt securing the lower arm to the rear wheel hub carrier.
4 Remove the bolt from the inner end of the lower arm securing the arm to the chassis (photo).
5 The lower arm may now be removed.
6 Inspect all rubber bushes for damage and deterioration, renewing any as found necessary.
7 Examine the lower arm for damage, cracks and corrosion.
8 New bushes should be lubricated with petroleum jelly to ease re-assembly. Note that the inner bush has smaller inserts fitted into it. These should be pressed in from the front.
9 Refitting is a reversal of removal, but tighten all bolts to their specified torque after the weight of the car is on the wheels.
10 The rear wheel toe-in should be checked on completion (see Section 9).

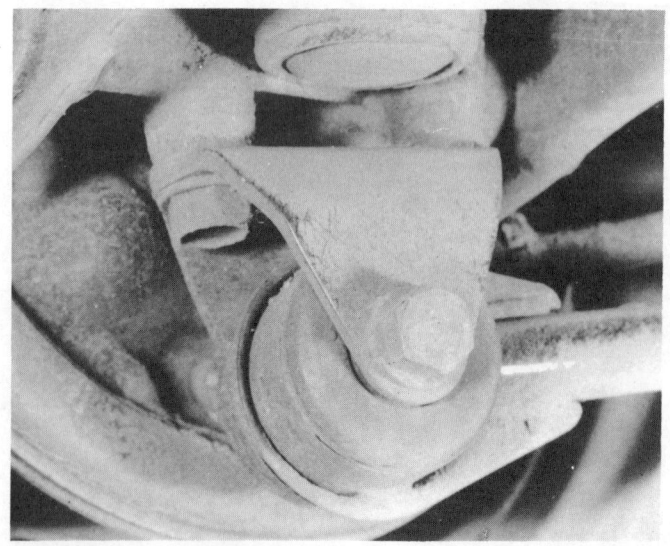

11.2A Rear suspension radius rod attachment point to the rear wheel hub carrier ...

11.2B ... and to the chassis

Chapter 10 Suspension and steering

12.4 The lower arm inner mounting bolt

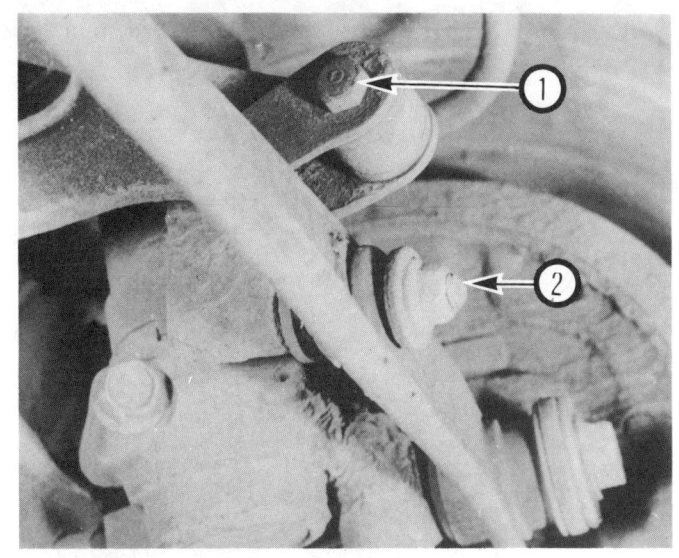

13.1 View of the rear stabiliser bar attachment point
1 Stabiliser bar-to-bracket bolt
2 Mounting bracket-to-lower arm bolt

13 Rear suspension stabiliser bar – removal and refitting

1 Remove the bolt securing the stabiliser bar to the mounting brackets on the lower arms (photo).
2 Undo and remove the bolts from the two support brackets holding the bar to each chassis side-member.
3 Refer to Chapter 9 and remove the bracket from the load sensing valve where the stabiliser bar is attached.
4 The stabiliser bar may now be manoeuvred out from under the vehicle.
5 Inspect the bar for damage, distortion and corrosion, renewing if necessary.
6 Inspect all the mounting rubber bushes for deterioration, cracking and contamination 'by oil' and grease. Do not forget the mounting bracket bushes on the lower arm.
7 Refitting is a reversal of removal, but leave final tightening of bolts to their specified torque until the weight of the vehicle is back on the wheels.
8 Refer to Chapter 9 and check and adjust the load sensing valve if necessary.

14 Rear wheel hub carrier and wheel bearings – removal and refitting

1 Loosen the rear wheel lug nuts and then raise the rear of the vehicle onto safety stands.
2 Remove the rear wheels.
3 On vehicles fitted with drum brakes, remove the hub cap by gently tapping with a mallet to dislodge it, remove the split pin from the locking collar and remove the spindle nut.
4 Ensure the hand brake is off (chock the front wheels) and remove the brake drum.
5 Refer to Chapter 9 and remove the components of the rear drum brake.
6 Remove the four bolts securing the brake backplate to the hub carrier and remove the backplate.
7 On vehicles equipped with disc brakes, refer to chapter 9 and disconnect the brake caliper and swing it clear.
8 Remove the four bolts securing the caliper bracket and the backplate to the hub carrier and remove them.
9 The wheel bearings will have come away with either the disc or drum, these will be dealt with later.
10 Remove the bolt securing the radius rod to the hub carrier and allow the radius rod to hang free.
11 Remove the bolt securing the hub carrier to the lower arm, and allow the lower arm to hang down free from the hub carrier. It may be necessary to disconnect the stabiliser bar from the lower arm to give enough play.
12 Remove the bolt securing the hub carrier to the shock absorber, then tap the hub carrier free from the shock absorber, using a mallet.
13 If necessary, remove the radius rod bracket by removing the two bolts.
14 Clean mud and road dirt from the hub carrier, and inspect it for cracks, damage and corrosion, renewing or treating corrosion as necessary.
15 Clean the stub axle and apply a thin film of high melting-point grease to it.
16 Inspect the bearings and their seals for damage, corrosion, blueing due to burning, and check the bearings for smooth free running. Renew the bearings as follows:
17 Where drum brakes are fitted, the bearings are a loose fit and can be removed by hand. For discs proceed as follows:
18 Drive the bearings out of the disc by supporting the disc on blocks, and then using a hammer and drift to drive out the bearing using a criss-cross pattern to avoid locking the bearing.
19 Turn the disc over and drive out the remaining bearing. The seals will come out with the bearings.
20 Clean the bearing seats thoroughly and then apply a film of grease to the bearing surfaces.
21 Pack the inner recess of the disc with grease.
22 Using a socket or tube of the same diameter as the outer periphery of the bearing, drive in the inboard bearing. Turn the disc over and repeat for the outer bearing.
23 Check that the bearings are properly seated against the inner flanges.
24 Pack the bearings with high melting-point grease.
25 Where drum brakes are fitted, place the inboard bearing in position, then grease a new seal and, using a mallet, tap it into place.
26 Similarly, fit a new seal in place on the disc.
27 Refitting of the various components is a reversal of removal, with the following points.
28 Use a new O-ring seal between the caliper mount/splash guard and the hub carrier.
29 Leave final tightening of the suspension components until the weight of the vehicle is on the wheels.
30 With the brake units refitted, and the disc/drum in position, adjust the wheel bearings as follows.
31 Where drum brakes are fitted, fit the drum over the spindle and brake unit.
32 On both discs and drums, fit the hub washer and spindle nut.
33 Tighten the spindle nut to the initial torque figure, then rotate the disc/drum by hand several turns in each direction.

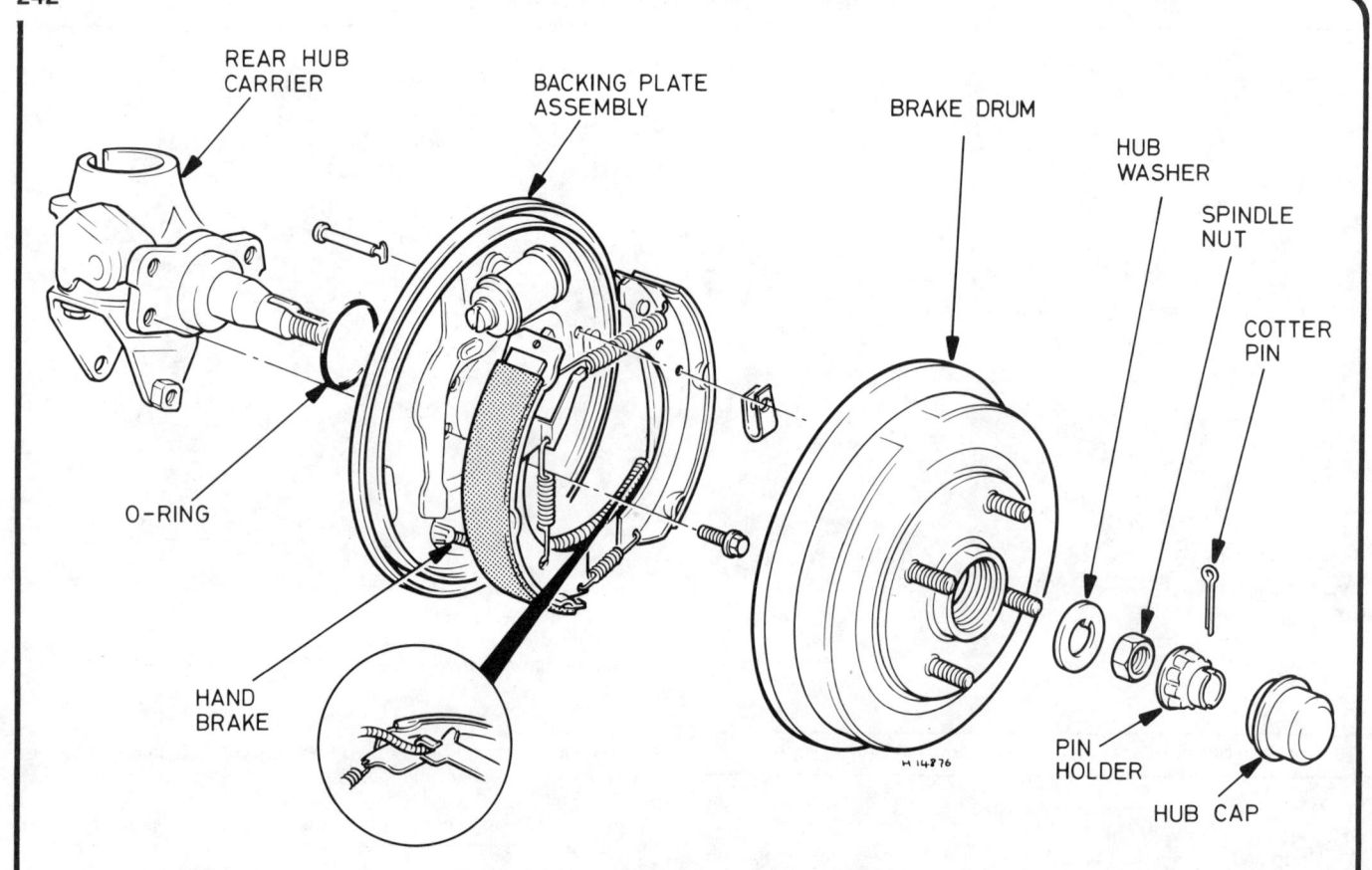

Fig. 10.13 Exploded view of rear hub assembly (drum brakes) (Sec 14)

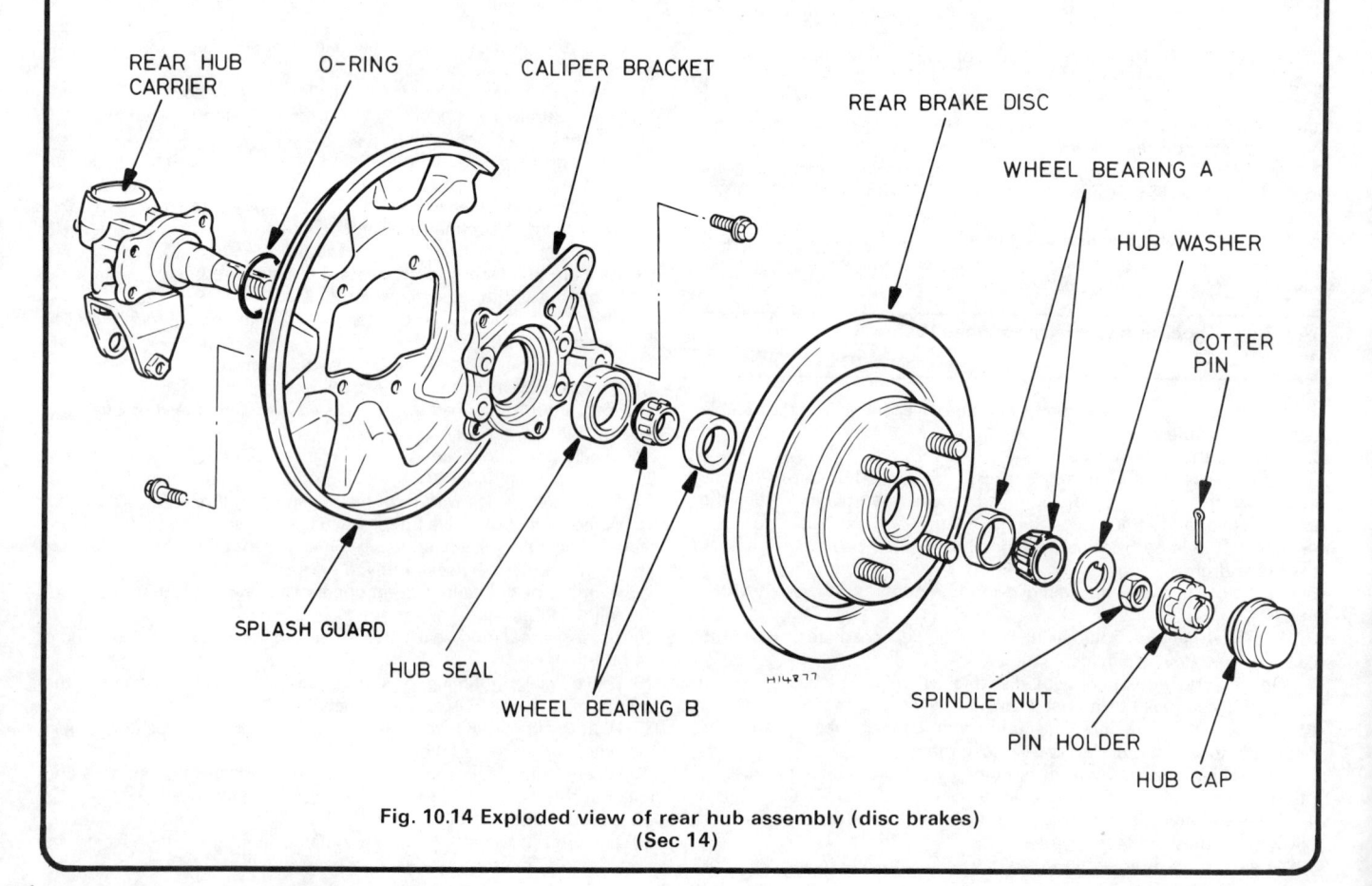

Fig. 10.14 Exploded view of rear hub assembly (disc brakes) (Sec 14)

Chapter 10 Suspension and steering

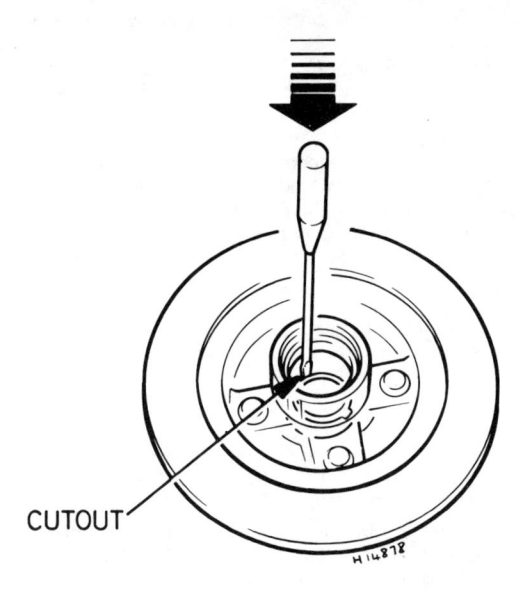

Fig. 10.15 Driving out the bearings from the disc hub (Sec 14)

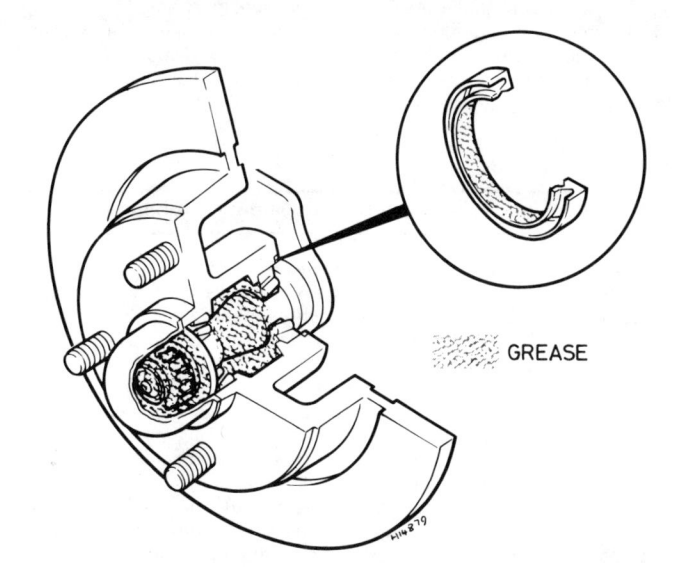

Fig. 10.16 Areas to pack with grease on reassembly (Sec 14)

34 Loosen the spindle nut.
35 Now tighten the spindle nut to the final torque figure.
36 Fit the split pin locking collar so that its holes are as close as possible to lining up with the hole in the spindle, then tighten the spindle nut just enough to allow the split pin to be fitted.
37 Check the rotational drag of the bearing by turning the brake drum with a spring balance as shown in Fig. 10.17.
38 If the reading exceeds the specified figure, recheck the bearing torque.
39 Bleed the brakes as described in Chapter 9, fit the rear road wheels and check and adjust the load sensing valve and handbrake.
40 Remove the vehicle from jacks/stands and finally torque all suspension bolts when the weight of the vehicle is on the wheels.

15 Front and rear spring height – inspection

1 The vehicle should be empty, parked on level ground and the tyres inflated to their specified pressure.
2 Bounce the car up and down several times before measuring.
3 Refer to Fig. 10.18 and measure the distance X.

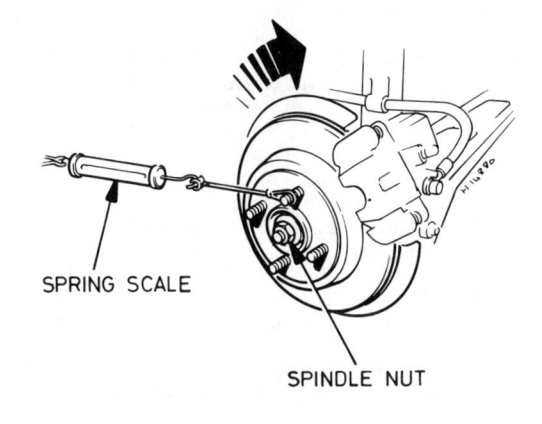

Fig. 10.17 Using a spring balance to measure bearing rotational drag (Sec 14)

Scale reading = 0.9 to 4.0 lb (0.4 to 1.8 kg)

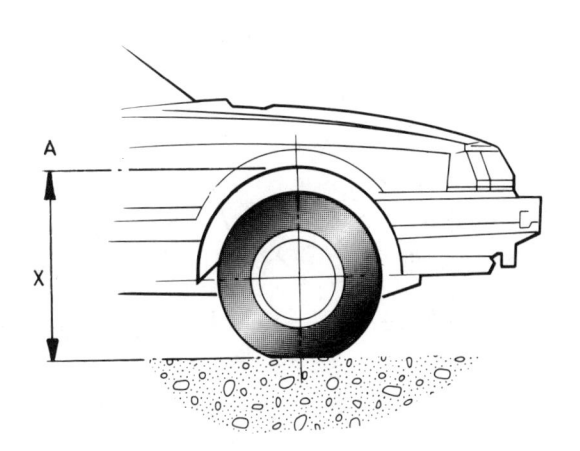

Fig. 10.18 Suspension spring height measurement (Sec 15)

A Front suspension   *B Rear suspension*

4 Compare with the figures given in the specifications.
5 If there is any significant deviation from the specified figure, check the suspension components for signs of failure or damage.
6 If there is no obvious sign of failure or damage then suspect the coil springs of losing their tension.

16 Wheels and tyres – general care and maintenance

Wheels and tyres should give no real problems in use provided that a close eye is kept on them with regard to excessive wear or damage. To this end, the following points should be noted.

Ensure that tyre pressures are checked regularly and maintained correctly. Checking should be carried out with the tyres cold and not immediately after the vehicle has been in use. If the pressures are checked with the tyres hot, an apparently high reading will be obtained owing to heat expansion. Under no circumstances should an attempt be made to reduce the pressures to the quoted cold reading in this instance, or effective underinflation will result.

Underinflation will cause overheating of the tyre owing to excessive flexing of the casing, and the tread will not sit correctly on the road surface. This will cause a consequent loss of adhesion and excessive wear, not to mention the danger of sudden tyre failure due to heat build-up.

Overinflation will cause rapid wear of the centre part of the tyre tread coupled with reduced adhesion, harsher ride, and the danger of shock damage occurring in the tyre casing.

Regularly check the tyres for damage in the form of cuts or bulges, especially in the sidewalls. Remove any nails or stones embedded in the tread before they penetrate the tyre to cause deflation. If removal of a nail *does* reveal that the tyre has been punctured, refit the nail so that its point of penetration is marked. Then immediately change the wheel and have the tyre repaired by a tyre dealer. Do *not* drive on a tyre in such a condition. In many cases a puncture can be simply repaired by the use of an inner tube of the correct size and type. If in any doubt as to the possible consequences of any damage found, consult your local tyre dealer for advice.

Where a compact spare tyre is fitted it should be noted that the tyre pressure for this differs to the other roadwheel tyres. Although the specified pressure is higher, care must be taken when adding pressure to the compact spare as it gains pressure very quickly due to its size. The compact spare tyre must not be used on any other wheel. For other precautionary notes regarding the compact spare refer to the manufacturer's handbook supplied with the vehicle.

Periodically remove the wheels and clean any dirt or mud from the inside and outside surfaces. Examine the wheel rims for signs of rusting, corrosion or other damage. Light alloy wheels are easily damaged by 'kerbing' whilst parking, and similarly steel wheels may become dented or buckled. Renewal of the wheel is very often the only course of remedial action possible.

The balance of each wheel and tyre assembly should be maintained to avoid excessive wear, not only to the tyres but also to the steering and suspension components. Wheel imbalance is normally signified by vibration through the vehicle's bodyshell, although in many cases it is particularly noticeable through the steering wheel. Conversely, it should be noted that wear or damage in suspension or steering components may cause excessive tyre wear. Out-of-round or out-of-true tyres, damaged wheels and wheel bearing wear/maladjustment also fall into this category. Balancing will not usually cure vibration caused by such wear.

Wheel balancing may be carried out with the wheel either on or off the vehicle. If balanced on the vehicle, ensure that the wheel-to-hub relationship is marked in some way prior to subsequent wheel removal so that it may be refitted in its original position.

General tyre wear is influenced to a large degree by driving style – harsh braking and acceleration or fast cornering will all produce more rapid tyre wear. Interchanging of tyres may result in more even wear, but this should only be carried out where there is no mix of tyre types on the vehicle. However, it is worth bearing in mind that if this is completely effective, the added expense of replacing a complete set of tyres simultaneously is incurred, which may prove financially restrictive for many owners.

Front tyres (and rear tyres on IRS models) may wear unevenly as a result of wheel misalignment. The wheels should always be correctly aligned according to the settings specified by the vehicle manufacturer.

Legal restrictions apply to the mixing of tyre types on a vehicle. Basically this means that a vehicle must not have tyres of differing construction on the same axle. Although it is not recommended to mix tyre types between front axle and rear axle, the only legally permissible combination is crossply at the front and radial at the rear. When mixing radial ply tyres, textile braced radials must always go on the front axle, with steel braced radials at the rear. An obvious disadvantage of such mixing is the necessity to carry two spare tyres to avoid contravening the law in the event of a puncture.

In the UK, the Motor Vehicles Construction and Use Regulations apply to many aspects of tyre fitting and usage. It is suggested that a copy of these regulations is obtained from your local police if in doubt as to the current legal requirements with regard to tyre condition, minimum tread depth, etc.

17 Steering wheel rotational play – inspection

1 Place the steering in the straight-ahead position and measure the distance the steering wheel can be turned in either direction before the front wheels start to turn.
2 If the movement exceeds the limit shown in the Specifications, examine all steering components for signs of wear or damage.

18 Manual steering rack guide – adjustment

1 Raise the front wheels off the ground.
2 Using a spring balance, check the effort required to turn the steering wheel in either direction (Fig. 10.19).
3 If outside the limits shown in the Specifications adjust as follows.
4 Loosen the steering rack screw locknut using the special tool or a modified spanner.
5 Tighten the steering rack guide adjusting screw until it just bottoms.
6 Back the screw off 45° from the bottomed position and tighten the locknut.
7 Check the steering for tightness or looseness of operation, and then recheck steering wheel effort as described above.
8 If the effort still does not fall within the specified limits, overhaul the steering rack and gearbox, as described in Sections 21 and 22 (manual) or Section 34 (power-assisted).

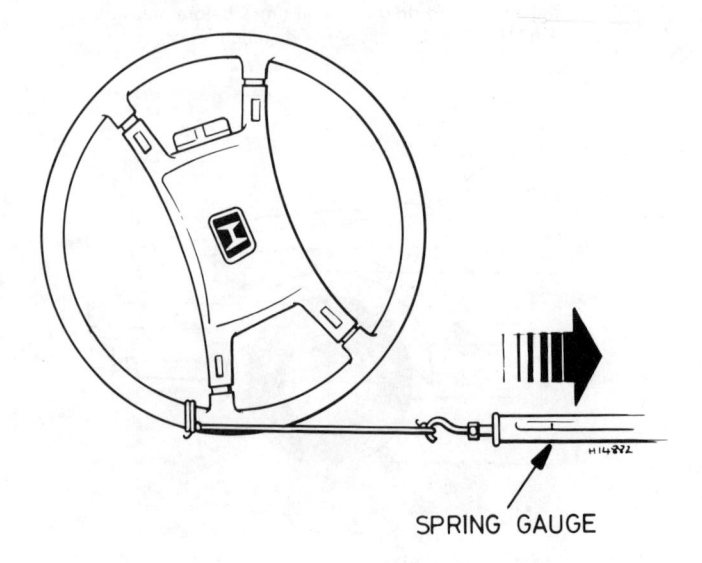

Fig. 10.19 Checking steering effort using a spring balance (Sec 18)

Chapter 10 Suspension and steering

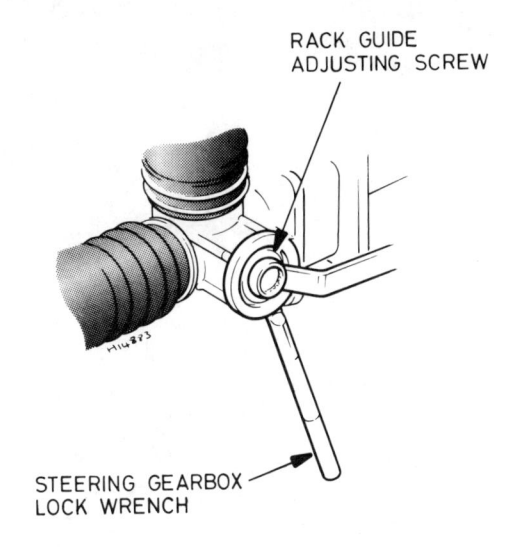

Fig. 10.20 Adjusting the steering rack guide on the steering gearbox (Sec 18)

19 Steering wheel – removal and refitting

1 Prise the centre pad out of the horn cover to reveal the steering shaft nut.
2 Set the wheels to the straight-ahead position.
3 Remove the steering shaft nut.
4 Rock the steering wheel from side to side and at the same time pull it up and off the steering shaft.
5 Disconnect the electrical leads from the horn and cruise control switches.
6 For dismantling of the steering wheel for access to the horn switches and cruise control switch, refer to Chapter 12.
7 Refitting is a reversal of this procedure, but tighten the steering shaft nut to its specified torque. On completion check the horn and cruise control operation.

20 Manual steering column – removal, overhaul and refitting

1 The steering column is of an impact absorbing kind. The upper bracket will bend on impact, and the lower bracket is mounted over a plastic collar on the column, allowing the column to slide forward on

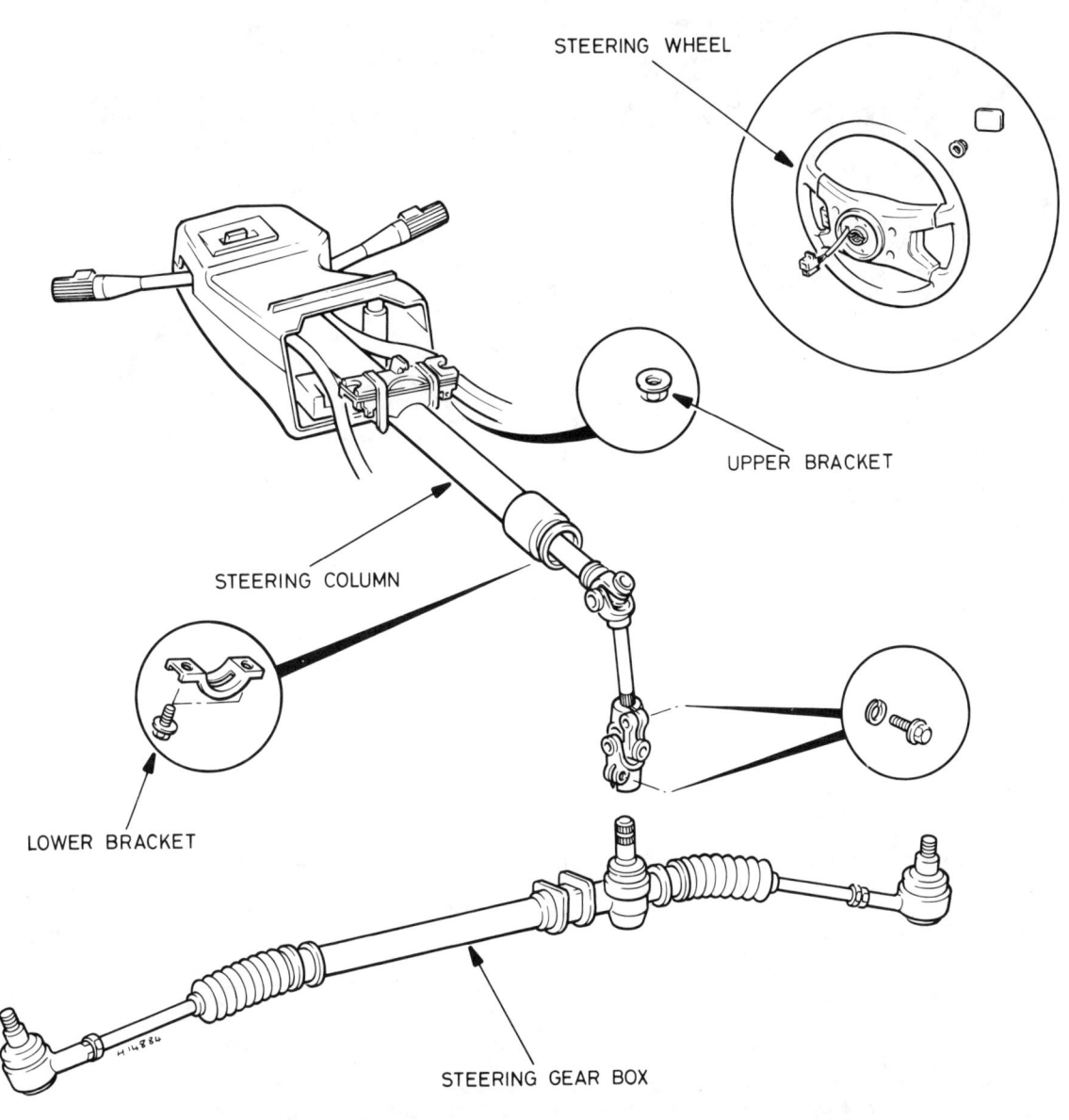

Fig. 10.21 Exploded view of the manual steering components (Sec 20)

Chapter 10 Suspension and steering

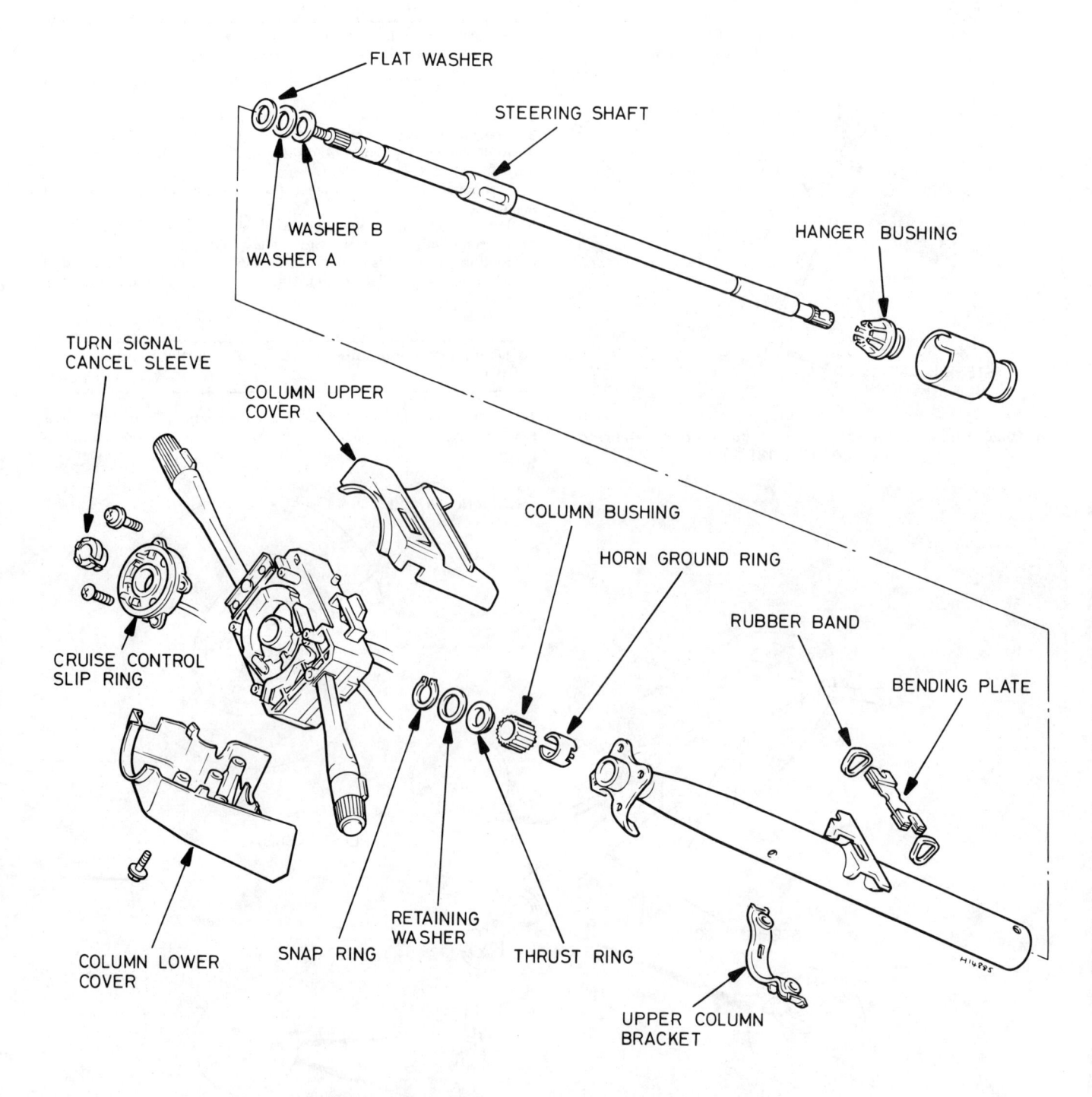

Fig. 10.22 Exploded view of the manual steering column (Sec 20)

impact. The lower universal joints, where the column is attached to the steering rack, are mounted at an angle, allowing them to 'fold' on impact, so that the column can slide forward.
2 To remove the column, proceed as follows:
3 Remove the steering wheel (Section 19).
4 Undo and remove the nuts securing the upper column bracket, which also forms part of the bending plate.
5 Remove the bottom bolt of the universal joint where it joins the steering rack gearbox.
6 Remove the bolts from the lower bracket (photo).
7 Supporting the column, prise it off the steering rack gearbox shaft, disconnect the electrical leads, and remove the assembly from the vehicle.

8 Remove the top bolt from the universal joint and remove the joint from the column.
9 Remove the upper and lower column covers by removing the recessed screws in the lower cover (photo) and lifting the covers away.
10 Remove the cruise control slip ring by undoing the four screws (photo).
11 Remove the turn signal cancelling sleeve, then remove the screws holding the turn signal switch and remove the switch.
12 Take off the rubber bands and remove the bending plate from the column upper bracket.
13 Remove the snap ring from the steering shaft.
14 Turn the ignition switch to the 'I' position.
15 Remove the plastic collar from the bottom end of the shaft and

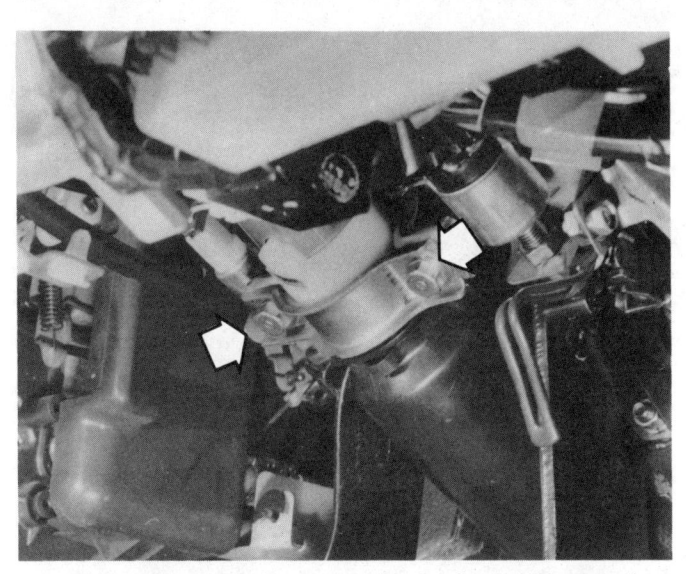

Chapter 10 Suspension and steering

247

column, and then withdraw the shaft from the column, pulling it out from the bottom end of the column.

16 Inspect the components as follows:

> Thrust ring and column bushing – wear or damage
> Bending plate – distortion
> Steering shaft – bending
> Steering shaft splines – wear or damage
> Hanger bushing – wear or damage (if this is evident the steering shaft must be renewed)

17 Renew all parts as necessary.
18 Reassembly is a reversal of disassembly, but coat the steering shaft with oil before inserting it in the column, and, when inserting it, be careful not to bend the horn earth ring.
19 Before pushing the hanger bushing into the lower end of the column, apply grease both to the bushing and the steering shaft and inside of column where the bushing sits.
20 Details of the electrical components will be found in Chapter 12.
21 Repair procedure for the ignition lock switch will be found in Section 24 of this Chapter.
22 With the steering column assembled it may be refitted to the vehicle.
23 Support the column roughly in position and loosely fit the upper mounting bracket nuts.
24 Similarly, install the lower bracket.
25 Ensure the bending plate is seated properly and then tighten the upper and lower bracket nuts and bolts to their specified torque.
26 Connect the electrical harness connectors.
27 Slip the longer (upper) end of the universal joint onto the steering shaft, so that its bolt will line up with the cut-out flat portion of the shaft.
28 Slide the lower end of the universal joint onto the pinion driveshaft of the steering rack, so that its bolt may be inserted through the groove in the pinion shaft.
29 Fit and tighten the lower bolt to its correct torque.
30 While pulling down on the steering shaft, fit and tighten the upper bolt to its torque figure.
31 Refit the steering wheel, and test all circuits.

21 Manual steering rack and gearbox – removal

1 Raise the front of the vehicle onto stands and remove the front roadwheels.
2 Remove the lower bolt from the steering column universal joint, and pull the joint up off the pinion shaft.
3 Remove the split pin and nut from both tie-rod end balljoints and use a balljoint separater to remove the balljoints from the steering knuckles (photos).

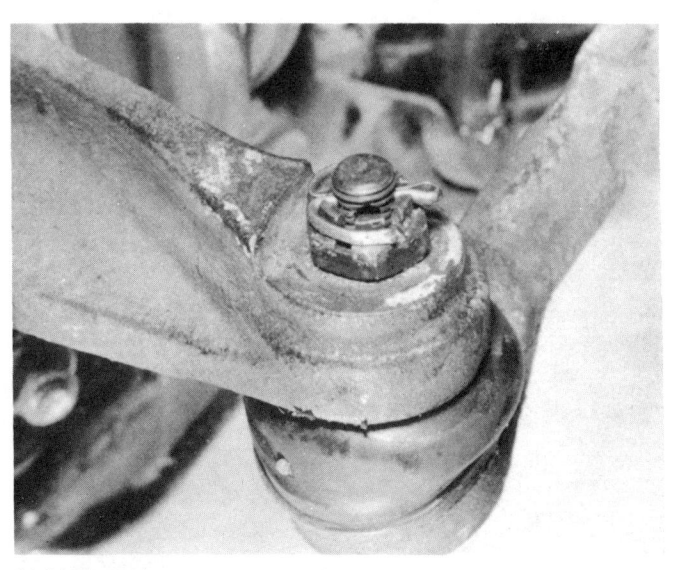

20.6 The steering column lower mounting bracket bolts (arrowed)

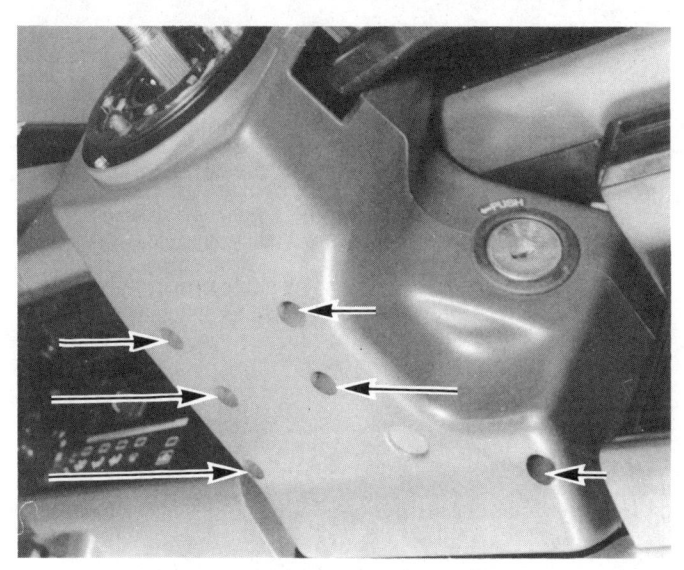

20.9 Recessed screws hold the lower and upper covers

20.10 The cruise control slip-ring securing screws (arrowed)

21.3A Removing ...

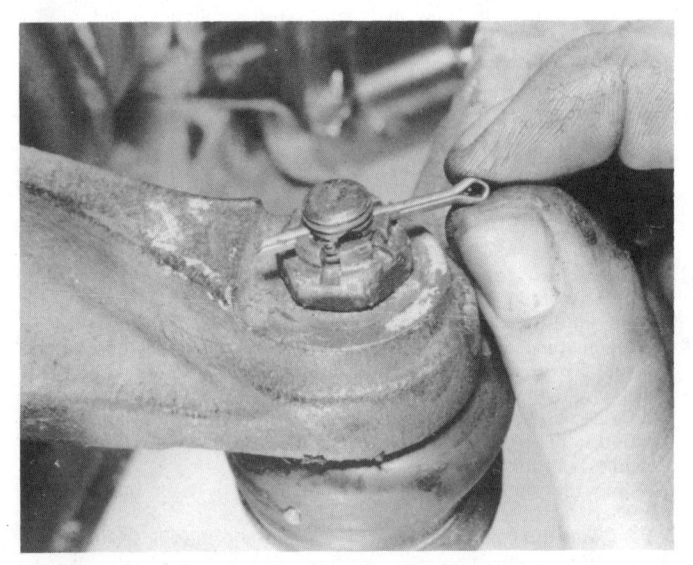

248 Chapter 10 Suspension and steering

21.3B ... the split pin ...

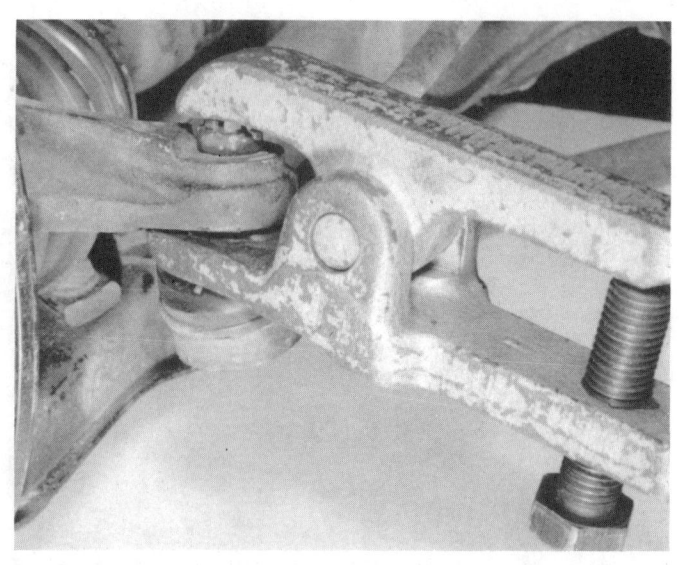

21.3C ... and using a balljoint separator ...

21.3D ... to remove the balljoint

4 Remove the engine centre support beam.
5 On manual transmission vehicles disconnect the gear change shift rod and torque arm.
6 On automatic transmission vehicles, remove the shift cable guide from the floor, and pull the shift cable down out of the way.
9 Remove the nuts from the exhaust downpipe-to-manifold connection, and the nuts from the exhaust pipe-to-engine bracket.
8 Push the steering rack all the way to the right (simulating a left turn).
9 Remove the bolts from the rack mounting brackets (photo).
10 Manoeuvre the steering rack and gearbox out from under the vehicle.

22 Manual steering rack and gearbox – overhaul

1 Gently clamp the rack in a vice, using protection against damage by the vice jaws.
2 Loosen the boot bands and pull the boots back down along the tie-rods.

21.9A Steering rack mounting bolts ...

21.9B ... at each end

249

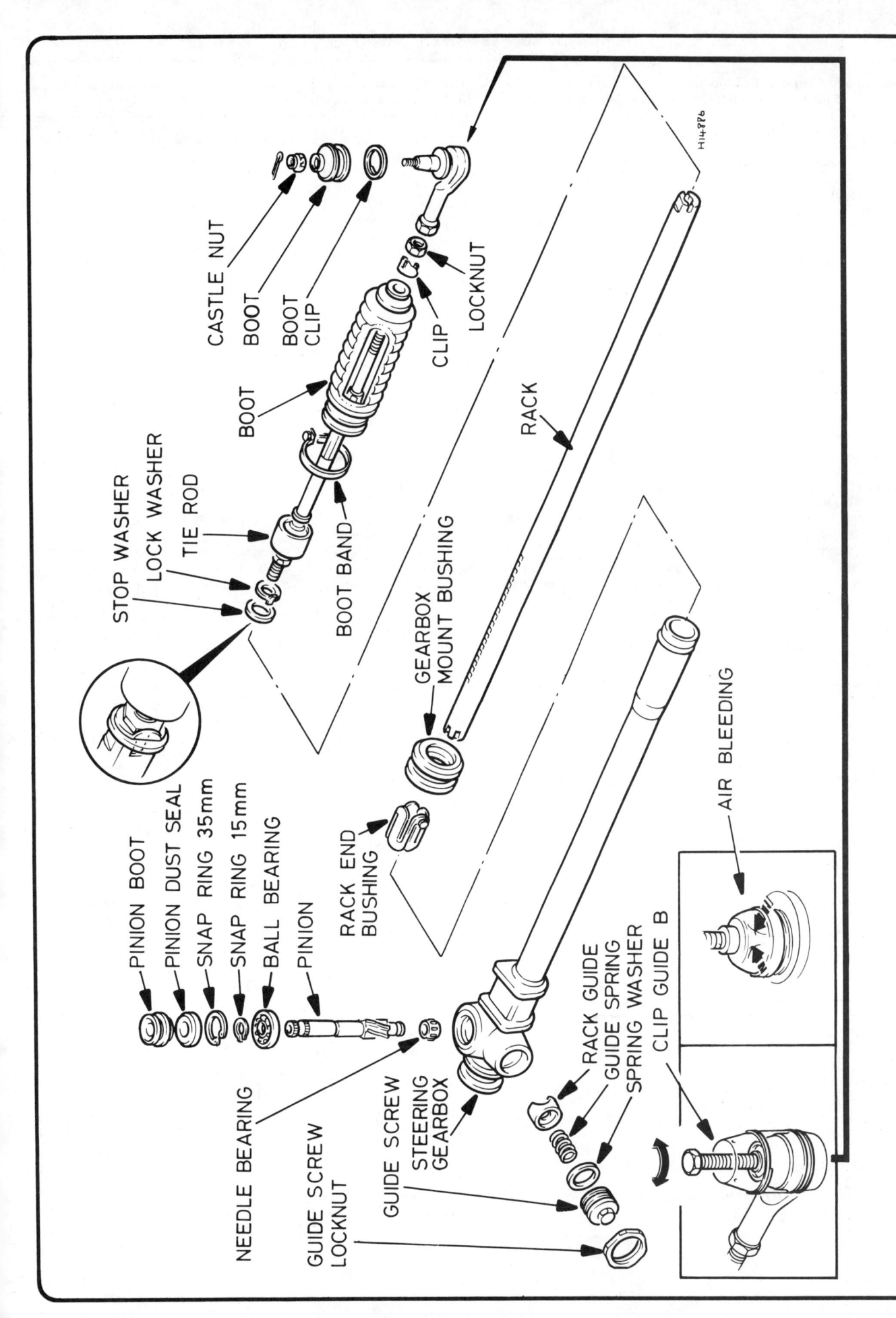

Fig. 10.23 Exploded view of the manual steering rack and gearbox (Sec 22)

3 Remove the locking effect of the lockwashers on the locknuts on the inner ends of the tie-rods, where they join the rack, and unscrew the tie-rods.
4 Remove the guide screw locknut, and carefully remove the rack guide components.
5 Remove the boot, seal and snap-ring from the gearbox pinion, and then pull the pinion out of the gearbox.
6 Finally, slide the rack out of the gearbox housing.
7 Clean all parts thoroughly before examining them as follows, and renewing worn parts as necessary.
8 Inspct the pinion shaft for wear, scoring or damaged teeth, and its needle roller and ball bearing for wear and general condition.
9 Inspect the rack and rack guide for wear, especially where the guide contacts the rack, and the rack teeth.
10 All seals and bushing should be renewed as a matter of course.
11 Commence reassembly by coating all parts in multi-purpose grease.
12 Fit the rack end bushing into the rack so that their projections fit into the holes on the rack.
13 Ensure that the slots in the bushing are kept free of grease, as they serve as air breather passages.
14 Slide the rack mount bushes onto the rack ends, so that the rack projection is as shown in Fig. 10.24.
15 Slide the rack into the gearbox housing, pack the gearbox with 0.9 to 1.2 oz (25 to 35g) of grease, then insert the pinion shaft, with needle bearing into the gearbox.
16 Fit the ball bearing, snap-rings and dust seal and boot.
17 Fit the rack guide components and adjust the rack guide as described in Section 18.
18 Tie-rod end balljoint overhaul procedure is given in Section 25.
19 Slide new rubber boots onto the tie-rods and fit the tie-rods to the rack ends, tighten to the correct torque and bend over the locking tabs on the lockwashers.
20 Fill the rubber boots with grease, slide them over the rack ends and fit and tighten the securing clips.
21 The steering rack is now ready for installation.

23 Manual steering rack and gearbox – refitting

1 Refitting the steering rack is largely a reversal of the removal sequence, with the following points.
2 Tighten all bolts/nuts to their specified torque.
3 When refitting the engine centre beam, ensure the central engine insulator rubber bush fits comfortably, adjusting as necessary.
4 On automatic transmission, check the shift cable adjustment after fitting.
5 Set the tie-rod ends to their original lengths, and have the front wheel track adjusted as soon as practicable after completing the job.
6 Carry out tests on the steering gear to comply with Sections 17 and 18.

24 Steering lock – removal and refitting

1 The steering lock consists of two parts – the electrical ignition switch and the mechanical lock.
2 The ignition switch is easily removed from the lock by removing the two retaining screws.
3 To remove the mechanical lock, the screws which hold it to the steering column must first be drilled out. Centre punch the screws to give the drill bit initial purchase, drill off the heads, and remove the bracket and lock.
4 No repair is possible to an unserviceable lock and a replacement lock will have to be obtained.
5 Fit the new lock without the ignition key inserted, tightening the new shear screws by hand only.
6 Refit the ignition switch, and check that the lock operates satisfactorily.
7 Note that the lockbolt should go through the hole in the column and into the steering shaft.
8 Finally, tighten the sheer-bolts until the heads twist off.

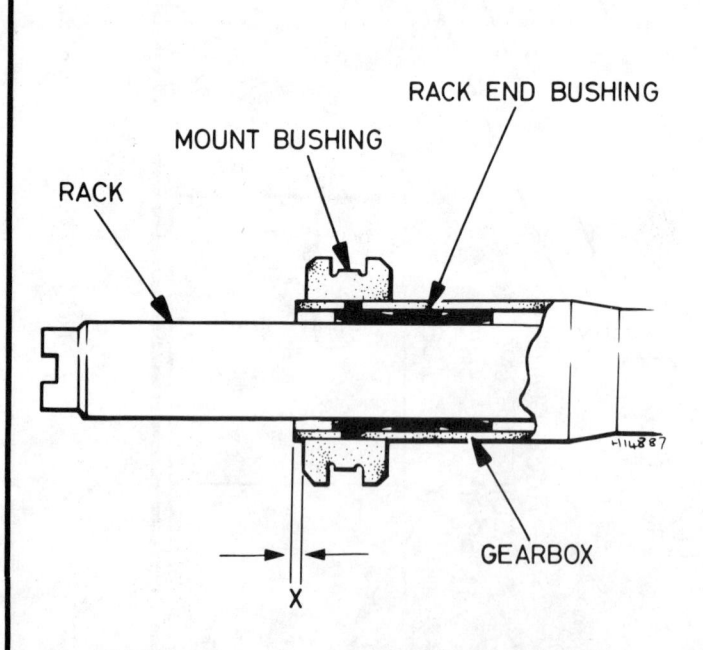

Fig. 10.24 Cutaway view of rack end bushing (Sec 22)

Dimension X = 0.08 in ± 0.04 in (2.0 ± 1.0 mm)

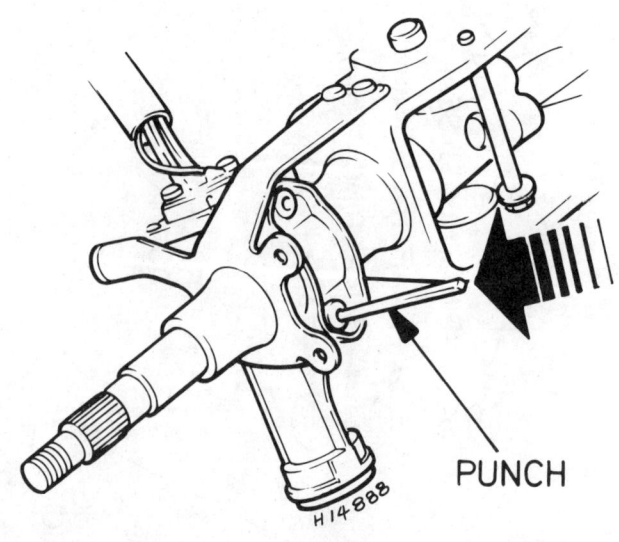

Fig. 10.25 The steering lock shear screws which have to be drilled out (Sec 24)

Chapter 10 Suspension and steering

25 Tie-rod end balljoint – renewal

1 The removal of the complete tie-rod and balljoint is covered in Section 21.
2 To remove a balljoint for renewal or to renew the rubber boots on the steering rack, without removal of the complete steering rack, proceed as follows:
3 Disconnect the balljoint from the steering knuckle as described in Section 21.
4 Undo the balljoint locknut and unscrew the balljoint from the tie-rod, counting the number of turns it takes to remove it, so that the balljoint end can be refitted in the same position, so keeping the track setting reasonably accurate.
5 The balljoint cannot be separated from the tie-rod and so a new unit must be obtained. **Note:** A balljoint should be considered worn if, with the vehicle on jacks and using a good strong lever to move the joint, any up and down play is evident.
6 Special tool guides are also required to fit the rubber boots to the balljoints, and there are two different typres of balljoint boot – a clip type and a press-on type, which is another good reason for buying a replacement tie-rod end.
7 In the case of simply renewing the rubber boot, consult your dealer.
8 Refit the tie-rod end to the tie-rod, screwing it on the same number of turns it took to remove it, and then tighten the locknut.
9 Refit the balljoint to the steering knuckle, as described in Section 23.
10 On completion, have the steering toe-in checked by your dealer as soon as possible.

26 Power steering column – removal and refitting

1 The steering column used on power-assisted systems is basically the same as the manual column, and the removal and refitting instructions given in Section 20 will suffice with the following notes being adhered to:
2 Some of the torque figures are different – refer to the Specifications.
3 A special tool is required to set the adjustment of the steering column after installation (Fig. 10.26).
4 This tool will have to be obtained and used as follows:
5 Before tightening any nuts or bolts and with the column loosely fitted, place the adjustment guide on to the steering column and push it on as far as it will go, so that it abuts on the turn signal switch.
6 Carry out the remaining installation operations, as described in Section 20, but using the correct torque figures from the Specifications.
7 On completion check that the legs of the adjustment guide are still in contact with the turn signal switch.
8 When properly adjusted, the steering wheel will move 0.04 in (1 mm) out when turned to the left and the same distance in when turned to the right.
9 Remove the adjustment guide, fit the steering wheel and test for correct movement, as described in paragraph 8:

27 Power steering fluid – renewal and topping-up

1 The power steering fluid reservoir is a plastic header tank mounted in the front left-hand corner of the engine compartment.
2 The fluid level should be maintained at the upper mark, and is checked with the fluid cold (photo).
3 To renew the complete fluid in the system, which should only be necessary if the fluid is contaminated or for repair work, proceed as follows:
4 Disconnect the return hose at the reservoir, and plug the hole in the reservoir (see photo 27.2).
5 Place the disconnected hose into a suitable container to catch ejected fluid.
6 Start the engine, allowing it to idle.
7 Turn the steering wheel from lock to lock several times.
8 Once fluid has stopped running from the hose, switch off the engine and discard the old fluid.
9 Refit the return hose to the reservoir.
10 Fill the reservoir with the recommended fluid and then start the engine and run it at a fast idle.

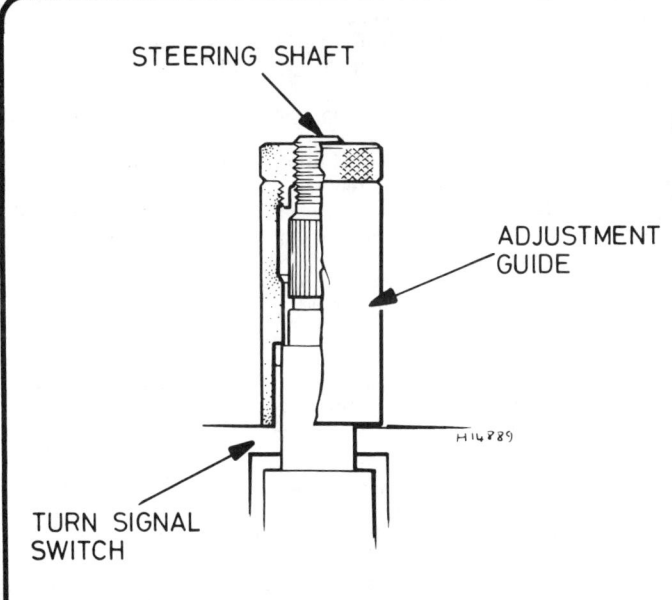

Fig. 10.26 Power steering column adjustment tool in position (Sec 26)

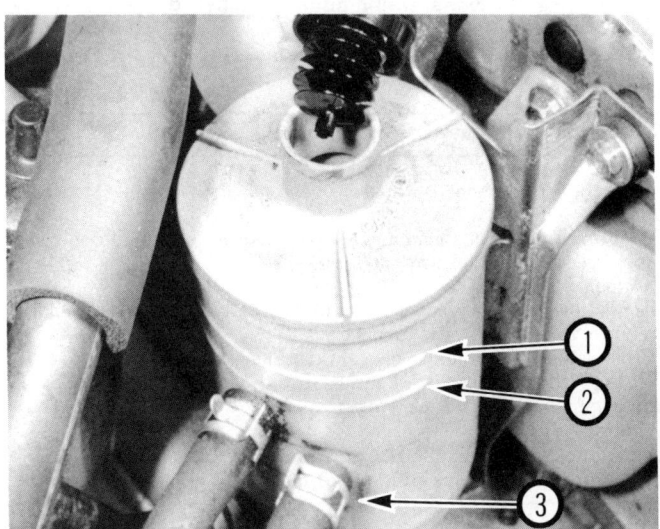

27.2 Power steering fluid reservoir
1 Upper level mark
2 Lower level mark
3 Return hose

11 Turn the steering from lock to lock several times to bleed the system free from air.
12 Switch off the engine on completion, and recheck the fluid level. Do not overfill.

28 Power steering pump belt – adjustment

1 The correct deflection of the power steering pump drivebelt is shown in Fig. 10.27.
2 To adjust the belt, loosen the pump pivot bolt and adjusting nut and prise the pump outward to give the correct deflection on the belt before tightening the pivot and adjuster nuts and bolts.

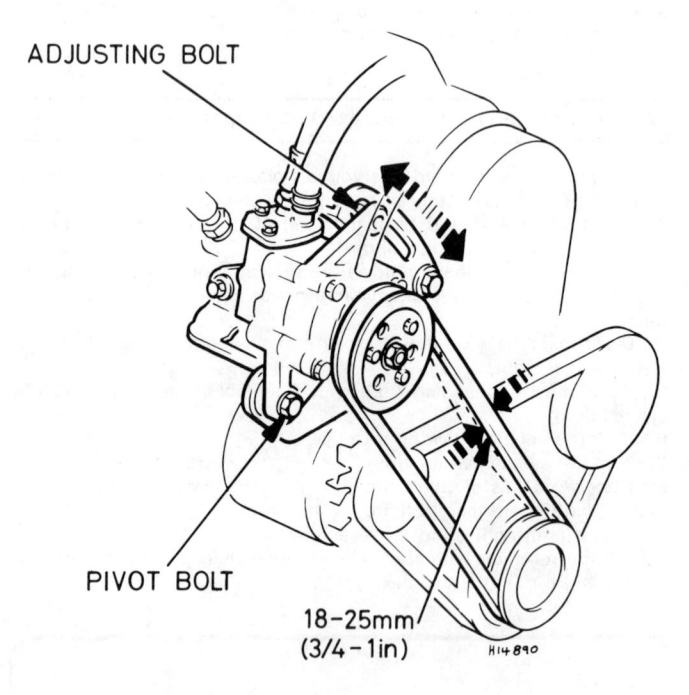

Fig. 10.27 Power steering pump drivebelt deflection (Sec 28)

29 Power steering assist – inspection

1 Check the power steering fluid level and pump drivebelt tension are correct.
2 Start the engine, allowing it to idle, and turn the steering from lock to lock several times, to warm up the fluid.
3 Attach a spring balance to the steering wheel and pull on the steering wheel with the balance, noting the effort required at the point when the front wheels start to move.
4 The reading should be as given in the Specifications.
5 If the figure is exceeded, consult your dealer who will carry out more comprehensive fault diagnosis.

30 Power steering rack guide – adjustment

1 The procedure is as given for manual systems, except that the guide screw should be screwed in and then backed off 35° (about one tenth of a turn), and the locknut tightened to the specified torque.

31 Power steering speed sensor – removal and refitting

1 The speed sensor unit, mounted on the transmission unit, controls the amount of power assist according to road speed (photo).
2 At low speed it allows maximum assist, and then progressively gives less as roadspeed increases.
3 To remove the unit, proceed as follows:
4 Pull up the rubber boot on the speedometer cable, remove the clip, and pull the cable from the speed sensor.
5 Disconnect and plug the hydraulic hoses to the speed sensor unit.
Note: When removing the transmission unit, do not disconnect the hoses, but remove the speed sensor, complete with speedometer cable, and tie it out of the way.
6 Remove the set bolt securing the speed sensor to the transmission housing and lift the sensor out.
7 Refitting is a reversal of removal.
8 On completion, turn the steering from lock to lock with the engine idling to bleed any air from the system.

32 Power steering pump – removal and refitting

1 Drain the fluid, as described in Section 27.
2 Remove the drivebelt, as described in Section 28.
3 Disconnect the inlet, outlet and lubrication return hoses at the pump.
4 Remove the adjusting bolt and pivot bolt, loosened in paragraph 2.
5 Remove the pump.
6 Refitting is a reversal of removal, adjust the drivebelt and refill and bleed the system as described in Sections 27 and 28.

33 Power steering pump – overhaul

1 Due to the complex nature of the pump, the need for special tools and for absolute cleanliness during overhaul, it is not recommended that the power steering pump be overhauled by the home mechanic.
2 Should the pump fail in service, it should be replaced by a service exhange unit.
3 Should a leak occur from the driveshaft seal, this may be renewed as follows:
4 Remove the pump, as described in Section 32.
5 Hold the pump in a vice, prevent the pulley wheel from turning by using a strap wrench, and remove the pulley nut.
6 Remove the pulley wheel from the shaft using a three or four-legged puller, do not attempt to lever it off with a screwdriver or it may become distorted.
7 Lever out the shaft seal, being careful not to damage the front cover.
8 Do not remove the front cover, as a special tool is required to line up the cover with the shaft so that the seal is fully effective.
9 Lubricate the new seal with grease, push it squarely into its housing by hand as far as it will go, and then drive it fully home using a socket of suitable size.
10 Refit the remaining components in a reversal of removal, refit the pump and bleed the system free of air.

34 Power steering rack – removal and refitting

1 Due to the complex nature of overhaul work, and the need for special tools, it is not recommended that the home mechanic attempts to overhaul the steering rack or gearbox.
2 If it malfunctions or becomes worn during service, it should be replaced by a new or reconditioned unit.
3 The removal and refitting procedure is very similar to that for the manual steering rack with the following additions.
4 Drain the fluid from the system, as described in Section 27.
5 Remove the gearbox shield.
6 Disconnect the fluid lines to the gearbox, maintaining absolute cleanliness, and blank off the ends of the pipes to prevent ingress of dirt.
7 Remove the steering rack as described for manual versions.
8 Refitting is a reversal of removal, but carry out bleeding of the system and the power assist checks with reference to the relevant Sections.

Chapter 10 Suspension and steering

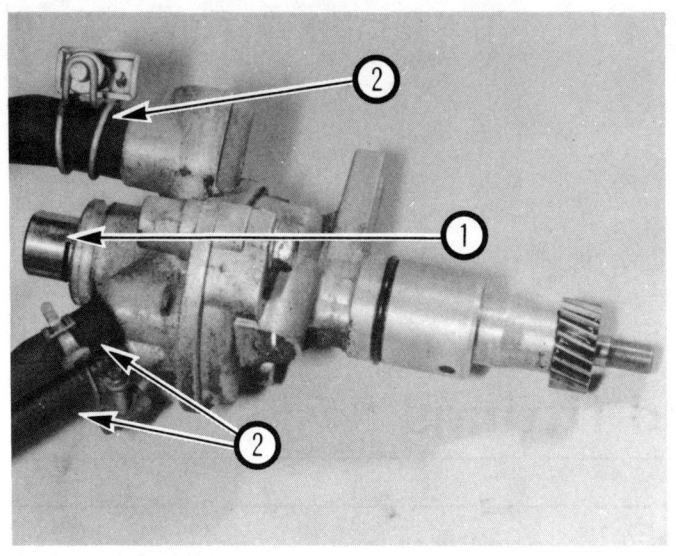

31.1 The power steering speed sensor unit removed from the transmission housing
1 Speedometer drive housing
2 Fluid hoses

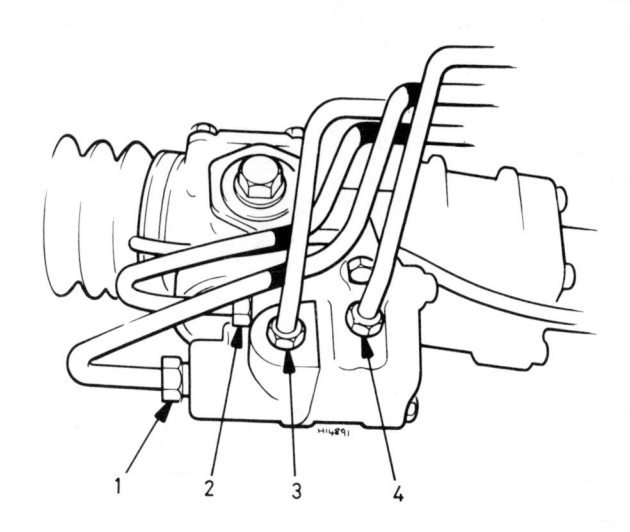

Fig. 10.28 Power steering gearbox pipeline connections (Sec 34)

1 To reservoir
2 To reservoir through speed sensor
3 To speed sensor
4 From pump

35 Fault diagnosis – suspension and steering

Symptom	Reason(s)
Steering feels vague, car wanders and floats at speed	Tyre pressures uneven Shock absorbers worn Spring broken Steering gear balljoints badly worn Suspension geometry incorrect Steering mechanism free play excessive Front and rear suspension out of alignment
Stiff and heavy steering	Tyre pressure too low Corroded swivel joints Corroded steering and suspension balljoints Front wheel toe setting incorrect Suspension geometry incorrect Steering gear incorrectly adjusted too tightly Steering column badly misaligned
Wheel wobble and vibration	Wheel nuts loose Front wheels and tyres out of balance Steering balljoints badly worn Hub bearings badly worn Steering gear free play excessive Front springs weak or broken
Power-assisted steering (additional to above) Stiff action or no return action	Slipping pump drivebelt Air in fluid Steering column misaligned Castor angles incorrect
Unequal steering effort (lock-to-lock)	Leaking seal in steering rack Clogged fluid passage in gearbox
Noisy pump	Loose pulley Kinked hose Clogged filter in reservoir Low fluid level

Chapter 11 Bodywork and fittings

Contents

Air conditioner – description and maintenance	48
Air conditioner evaporator – removal and refitting	51
Air conditioner compressor – removal and refitting	49
Air conditioner condenser – removal and refitting	50
Bonnet – removal and refitting	6
Bonnet lock and release assembly – removal, refitting and adjustment	7
Boot lid – removal and refitting	25
Centre console – removal and refitting	34
Dashboard – removal and refitting	36
Door – removal and refitting	21
Door interior handle – removal and refitting	11
Door latch – removal and refitting	14
Door lock cylinder – removal and refitting	13
Door side moulding – removal and refitting	19
Door striker – removal, refitting and adjusting	22
Door trim panel – removal and refitting	10
Door window regulator scissors assembly – removal and refitting	18
Exterior rear view mirror – removal and refitting	23
Front and rear bumpers – removal and refitting	9
Front console – removal and refitting	35
Front door window – removal and refitting	16
Front seats – removal and refitting	37
General description	1
Heater and ventilation – general description	42
Heater blower – removal and refitting	44
Heater controls and cables – removal, refitting and adjusting	46
Heater unit – removal and refitting	43
Interior rear view mirror – removal and refitting	41
Interior trim – general	33
Maintenance – bodywork and underframe	2
Maintenance – upholstery and carpets	3
Major body damage – repair	5
'Mild flow' assembly – removal and refitting	45
Minor body damage – repair	4
Outside door handle – removal and refitting	12
Power door lock – removal and refitting	15
Radiator grille – removal and refitting	8
Rear door window – removal and refitting	17
Rear hatch – removal and refitting	24
Rear hatch/boot and fuel filler lid remote release lever and latches – general	26
Rear quarter light – removal and refitting	28
Rear seat (Hatchback) – removal and refitting	39
Rear seat (Saloon) – removal and refitting	38
Rear quarter light – removing and refitting	28
Seat belts – general	40
Sunroof – adjustment	30
Sunroof – removal and refitting	29
Sunroof motor, drain tube and frame – removal and refitting	31
Vacuum-operated components – general	47
Wind deflector – removal and refitting	32
Window weatherstrip – removal and refitting	20
Windscreen and rear window – removal and refitting	27

Specifications

Torque wrench settings

	lbf ft	Nm
Rear seat mounting bolts	7	10
Seat belt anchorage bolts	23	32
Bumper retaining bolts	16	22
Bonnet hinge bolts	16	22
Heater mounting nut	12	17
Heater mounting bolts	7	10
Heater blower mounting bolts	7	10

1 General description

The bodywork is of all-steel, welded, unitary construction. The front wings are bolted on for easy repair, and impact absorbing bumpers are fitted to most models.

Plastic, clip-on type panels are used extensively for interior trim.

All vehicles are protected against corrosion and, according to model, have various body mouldings, fitted against day-to-day knocks. Some versions also have wheel arch shields bolted to the wings for added protection.

All models in the range are well-equipped, and several factory-fitted optional extras are available.

2 Maintenance – bodywork and underframe

1 The general condition of a vehicle's bodywork is the one thing that significantly affects its value. Maintenance is easy but needs to be regular. Neglect, particularly after minor damage, can lead quickly to further deterioration and costly repair bills. It is important also to keep

Chapter 11 Bodywork and fittings

watch on those parts of the vehicle not immediately visible, for instance the underside, inside all the wheel arches and the lower part of the engine compartment.

2 The basic maintenance routine for the bodywork is washing – preferably with a lot of water, from a hose. This will remove all the loose solids which may have stuck to the vehicle. It is important to flush these off in such a way as to prevent grit from scratching the finish. The wheel arches and underframe need washing in the same way to remove any accumulated mud which will retain moisture and tend to encourage rust. Paradoxically enough, the best time to clean the underframe and wheel arches is in wet weather when the mud is thoroughly wet and soft. In very wet weather the underframe is usually cleaned of large accumulations automatically and this is a good time for inspection.

3 Periodically, except on vehicles with a wax-based underbody protective coating, it is a good idea to have the whole of the underframe of the vehicle steam cleaned, engine compartment included, so that a thorough inspection can be carried out to see what minor repairs and renovations are necessary. Steam cleaning is available at many garages and is necessary for removal of the accumulation of oily grime which sometimes is allowed to become thick in certain areas. If steam cleaning facilities are not available, there are one or two excellent grease solvents available which can be brush applied. The dirt can then be simply hosed off. Note that these methods should not be used on vehicles with wax-based underbody protective coating or the coating will be removed. Such vehicles should be inspected annually, preferably just prior to winter, when the underbody should be washed down and any damage to the wax coating repaired. Ideally, a completely fresh coat should be applied. It would also be worth considering the use of such wax-based protection for injection into door panels, sills, box sections, etc, as an additional safeguard against rust damage where such protection is not provided by the vehicle manufacturer.

4 After washing paintwork, wipe off with a chamois leather to give an unspotted clear finish. A coat of clear protective wax polish will give added protection against chemical pollutants in the air. If the paintwork sheen has dulled or oxidised, use a cleaner/polisher combination to restore the brilliance of the shine. This requires a little effort, but such dulling is usually caused because regular washing has been neglected. Care needs to be taken with metallic paintwork, as special non-abrasive cleaner/polisher is required to avoid damage to the finish. Always check that the door and ventilator opening drain holes and pipes are completely clear so that water can be drained out (photo). Bright work should be treated in the same way as paint work. Windscreens and windows can be kept clear of the smeary film which often appears by the use of a proprietary glass cleaner. Never use any form of wax or other body or chromium polish on glass.

3 Maintenance – upholstery and carpets

Mats and carpets should be brushed or vacuum cleaned regularly to keep them free of grit. If they are badly stained remove them from the vehicle for scrubbing or sponging and make quite sure they are dry before refitting. Seats and interior trim panels can be kept clean by wiping with a damp cloth. If they do become stained (which can be more apparent on light coloured upholstery) use a little liquid detergent and a soft nail brush to scour the grime out of the grain of the material. Do not forget to keep the headlining clean in the same way as the upholstery. When using liquid cleaners inside the vehicle do not over-wet the surfaces being cleaned. Excessive damp could get into the seams and padded interior causing stains, offensive odours or even rot. If the inside of the vehicle gets wet accidentally it is worthwhile taking some trouble to dry it out properly, particularly where carpets are involved. *Do not leave oil or electric heaters inside the vehicle for this purpose.*

4 Minor body damage – repair

The photographic sequences on pages 262 and 263 illustrate the operations detailed in the following sub-sections.
Note: *For more detailed information about bodywork repair, the Haynes Publishing Group publish a book by Lindsay Porter called The Car Bodywork Repair Manual. This incorporates information on such aspects as rust treatment, painting and glass fibre repairs, as well as details on more ambitious repairs involving welding and panel beating.*

Repair of minor scratches in bodywork

If the scratch is very superficial, and does not penetrate to the metal of the bodywork, repair is very simple. Lightly rub the area of the scratch with a paintwork renovator, or a very fine cutting paste, to remove loose paint from the scratch and to clear the surrounding bodywork of wax polish. Rinse the area with clean water.

Apply touch-up paint to the scratch using a fine paint brush; continue to apply fine layers of paint until the surface of the paint in the scratch is level with the surrounding paintwork. Allow the new paint at least two weeks to harden; then blend it into the surrounding paintwork by rubbing the scratch area with a paintwork renovator or a very fine cutting paste. Finally, apply wax polish.

Where the scratch has penetrated right through to the metal of the bodywork, causing the metal to rust, a different repair technique is required. Remove any loose rust from the bottom of the scratch with a penknife, then apply rust inhibiting paint to prevent the formation of rust in the future. Using a rubber or nylon applicator fill the scratch with bodystopper paste. If required, this paste can be mixed with cellulose thinners to provide a very thin paste which is ideal for filling narrow scratches. Before the stopper-paste in the scratch hardens, wrap a piece of smooth cotton rag around the top of a finger. Dip the finger in cellulose thinners and then quickly sweep it across the surface of the stopper-paste in the scratch; this will ensure that the surface of the stopper-paste is slightly hollowed. The scratch can now be painted over as described earlier in this Section.

Repair of dents in bodywork

When deep denting of the vehicle's bodywork has taken place, the first task is to pull the dent out, until the affected bodywork almost attains its original shape. There is little point in trying to restore the original shape completely, as the metal in the damaged area will have stretched on impact and cannot be reshaped fully to its original contour. It is better to bring the level of the dent up to a point which is about ⅛ in (3 mm) below the level of the surrounding bodywork. In cases where the dent is very shallow anyway, it is not worth trying to pull it out at all. If the underside of the dent is accessible, it can be hammered out gently from behind, using a mallet with a wooden or plastic head. Whilst doing this, hold a suitable block of wood firmly against the outside of the panel to absorb the impact from the hammer blows and thus prevent a large area of the bodywork from being 'belled-out'.

Should the dent be in a section of the bodywork which has a double skin or some other factor making it inaccessible from behind, a different technique is called for. Drill several small holes through the

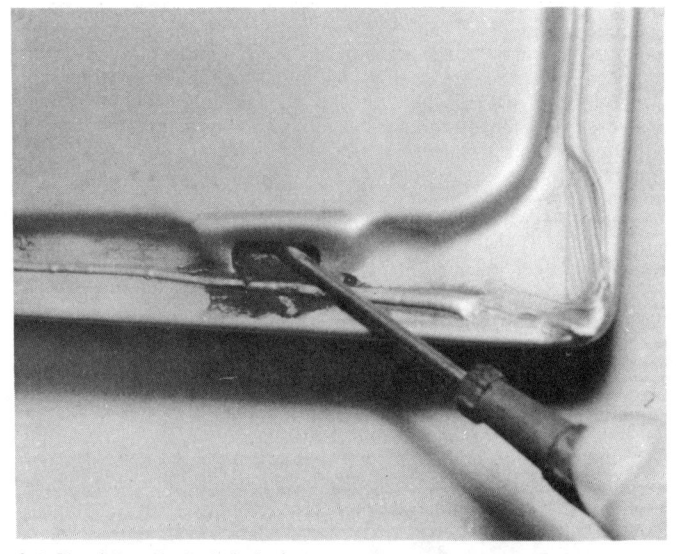

2.4 Clearing a door drain hole

metal inside the area – particularly in the deeper section. Then screw long self-tapping screws into the holes just sufficiently for them to gain a good purchase in the metal. Now the dent can be pulled out by pulling on the protruding heads of the screws with a pair of pliers.

The next stage of the repair is the removal of the paint from the damaged area, and from an inch or so of the surrounding 'sound' bodywork. This is accomplished most easily by using a wire brush or abrasive pad on a power drill, although it can be done just as effectively by hand using sheets of abrasive paper. To complete the preparation for filling, score the surface of the bare metal with a screwdriver or the tang of a file, or alternatively, drill small holes in the affected area. This will provide a really good 'key' for the filler paste.

To complete the repair see the Section on filling and re-spraying.

Repair of rust holes or gashes in bodywork

Remove all paint from the affected area and from an inch or so of the surrounding 'sound' bodywork, using an abrasive pad or a wire brush on a power drill. If these are not available a few sheets of abrasive paper will do the job just as effectively. With the paint removed you will be able to gauge the severity of the corrosion and therefore decide whether to renew the whole panel (if this is possible) or to repair the affected area. New body panels are not as expensive as most people think and it is often quicker and more satisfactory to fit a new panel than to attempt to repair large areas of corrosion.

Remove all fittings from the affected area except those which will act as a guide to the original shape of the damaged bodywork (eg headlamp shells etc). Then, using tin snips or a hacksaw blade, remove all loose metal and any other metal badly affected by corrosion. Hammer the edges of the hole inwards in order to create a slight depression for the filler paste.

Wire brush the affected area to remove the powdery rust from the surface of the remaining metal. Paint the affected area with rust inhibiting paint; if the back of the rusted area is accessible treat this also.

Before filling can take place it will be necessary to block the hole in some way. This can be achieved by the use of aluminium or plastic mesh, or aluminium tape.

Aluminium or plastic mesh is probably the best material to use for a large hole. Cut a piece to the approximate size and shape of the hole to be filled, then position it in the hole so that its edges are below the level of the surrounding bodywork. It can be retained in position by several blobs of filler paste around its periphery.

Aluminium tape should be used for small or very narrow holes. Pull a piece off the roll and trim it to the approximate size and shape required, then pull off the backing paper (if used) and stick the tape over the hole; it can be overlapped if the thickness of one piece is insufficient. Burnish down the edges of the tape with the handle of a screwdriver or similar, to ensure that the tape is securely attached to the metal underneath.

Bodywork repairs – filling and re-spraying

Before using this Section, see the Sections on dent, deep scratch, rust holes and gash repairs.

Many types of bodyfiller are available, but generally speaking those proprietary kits which contain a tin of filler paste and a tube of resin hardener are best for this type of repair. A wide, flexible plastic or nylon applicator will be found invaluable for imparting a smooth and well contoured finish to the surface of the filler.

Mix up a little filler on a clean piece of card or board – measure the hardener carefully (follow the maker's instructions on the pack) otherwise the filler will set too rapidly or too slowly. Using the applicator apply the filler paste to the prepared area; draw the applicator across the surface of the filler to achieve the correct contour and to level the filler surface. As soon as a contour that approximates to the correct one is achieved, stop working the paste – if you carry on too long the paste will become sticky and begin to 'pick up' on the applicator. Continue to add thin layers of filler paste at twenty-minute intervals until the level of the filler is just proud of the surrounding bodywork.

Once the filler has hardened, excess can be removed using a metal plane or file. From then on, progressively finer grades of abrasive paper should be used, starting with a 40 grade production paper and finishing with 400 grade wet-and-dry paper. Always wrap the abrasive paper around a flat rubber, cork, or wooden block – otherwise the surface of the filler will not be completely flat. During the smoothing of the filler surface the wet-and-dry paper should be periodically rinsed in water. This will ensure that a very smooth finish is imparted to the filler at the final stage.

At this stage the 'dent' should be surrounded by a ring of bare metal, which in turn should be encircled by the finely 'feathered' edge of the good paintwork. Rinse the repair area with clean water, until all of the dust produced by the rubbing-down operation has gone.

Spray the whole repair area with a light coat of primer – this will show up any imperfections in the surface of the filler. Repair these imperfections with fresh filler paste or bodystopper, and once more smooth the surface with abrasive paper. If bodystopper is used, it can be mixed with cellulose thinners to form a really thin paste which is ideal for filling small holes. Repeat this spray and repair procedure until you are satisfied that the surface of the filler, and the feathered edge of the paintwork are perfect. Clean the repair area with clean water and allow to dry fully.

The repair area is now ready for final spraying. Paint spraying must be carried out in a warm, dry, windless and dust free atmosphere. This condition can be created artificially if you have access to a large indoor working area, but if you are forced to work in the open, you will have to pick your day very carefully. If you are working indoors, dousing the floor in the work area with water will help to settle the dust which would otherwise be in the atmosphere. If the repair area is confined to one body panel, mask off the surrounding panels; this will help to minimise the effects of a slight mis-match in paint colours. Bodywork fittings (eg chrome strips, door handles etc) will also need to be masked off. Use genuine masking tape and several thicknesses of newspaper for the masking operations.

Before commencing to spray, agitate the aerosol can thoroughly, then spray a test area (an old tin, or similar) until the technique is mastered. Cover the repair area with a thick coat of primer; the thickness should be built up using several thin layers of paint rather than one thick one. Using 400 grade wet-and-dry paper, rub down the surface of the primer until it is really smooth. While doing this, the work area should be thoroughly doused with water, and the wet-and-dry paper periodically rinsed in water. Allow to dry before spraying on more paint.

Spray on the top coat, again building up the thickness by using several thin layers of paint. Start spraying in the centre of the repair area and then, using a circular motion, work outwards until the whole repair area and about 2 inches of the surrounding original paintwork is covered. Remove all masking material 10 to 15 minutes after spraying on the final coat of paint.

Allow the new paint at least two weeks to harden, then, using a paintwork renovator or a very fine cutting paste, blend the edges of the paint into the existing paintwork. Finally, apply wax polish.

Plastic components

With the use of more and more plastic body components by the vehicle manufacturers (eg bumpers, spoilers, and in some cases major body panels), rectification of damage to such items has become a matter of either entrusting repair work to a specialist in this field, or renewing complete components. Repair by the DIY owner is not really feasible owing to the cost of the equipment and materials required for effecting such repairs. The basic technique involves making a groove along the line of the crack in the plastic using a rotary burr in a power drill. The damaged part is then welded back together by using a hot air gun to heat up and fuse a plastic filler rod into the groove. Any excess plastic is then removed and the area rubbed down to a smooth finish. It is important that a filler rod of the correct plastic is used, as body components can be made of a variety of different types (eg polycarbonate, ABS, polypropylene).

If the owner is renewing a complete component himself, he will be left with the problem of finding a suitable paint for finishing which is compatible with the type of plastic used. At one time the use of a universal paint was not possible owing to the complex range of plastics encountered in body component applications. Standard paints, generally speaking, will not bond to plastic or rubber satisfactorily. However, it is now possible to obtain a plastic body parts finishing kit which consists of a pre-primer treatment, a primer and coloured top coat. Full instructions are normally supplied with a kit, but basically the method of use is to first apply the pre-primer to the component concerned and allow it to dry for up to 30 minutes. Then the primer is applied and left to dry for about an hour before finally applying the special coloured top coat. The result is a correctly coloured component where the paint will flex with the plastic or rubber, a property that standard paint does not normally possess.

Chapter 11 Bodywork and fittings

5 Major body damage – repair

This should be left to your dealer or specialist body repairer who has the necessary body jigs and alignment gauges needed to check for body and frame distortion. This must be corrected if the vehicle is to retain its roadholding and handling characteristics.

6 Bonnet – removal and refitting

1 The bonnet is hinged at its forward end, and the release latch is mounted on the engine compartment rear bulkhead.
2 To remove the bonnet, open it by operating the interior release handle and supporting it on its strut.
3 Mark the position of the hinges in soft pencil before undoing and removing the two bolts from each hinge (photo).
4 Have an assistant standing by to help lift the bonnet clear, and place it safely on rags to protect the corner paintwork.
5 Refitting is a reversal of this procedure, lining up the hinges with the previously made pencil marks.
6 Do not fully tighten the hinge bolts until its alignment has been checked, and the bonnet lock operates smoothly and positively.
7 Fore and aft adjustment of the bonnet is achieved by placing shims under the hinge mountings on the front valance.

6.3 Bonnet hinge retaining bolts (arrowed)

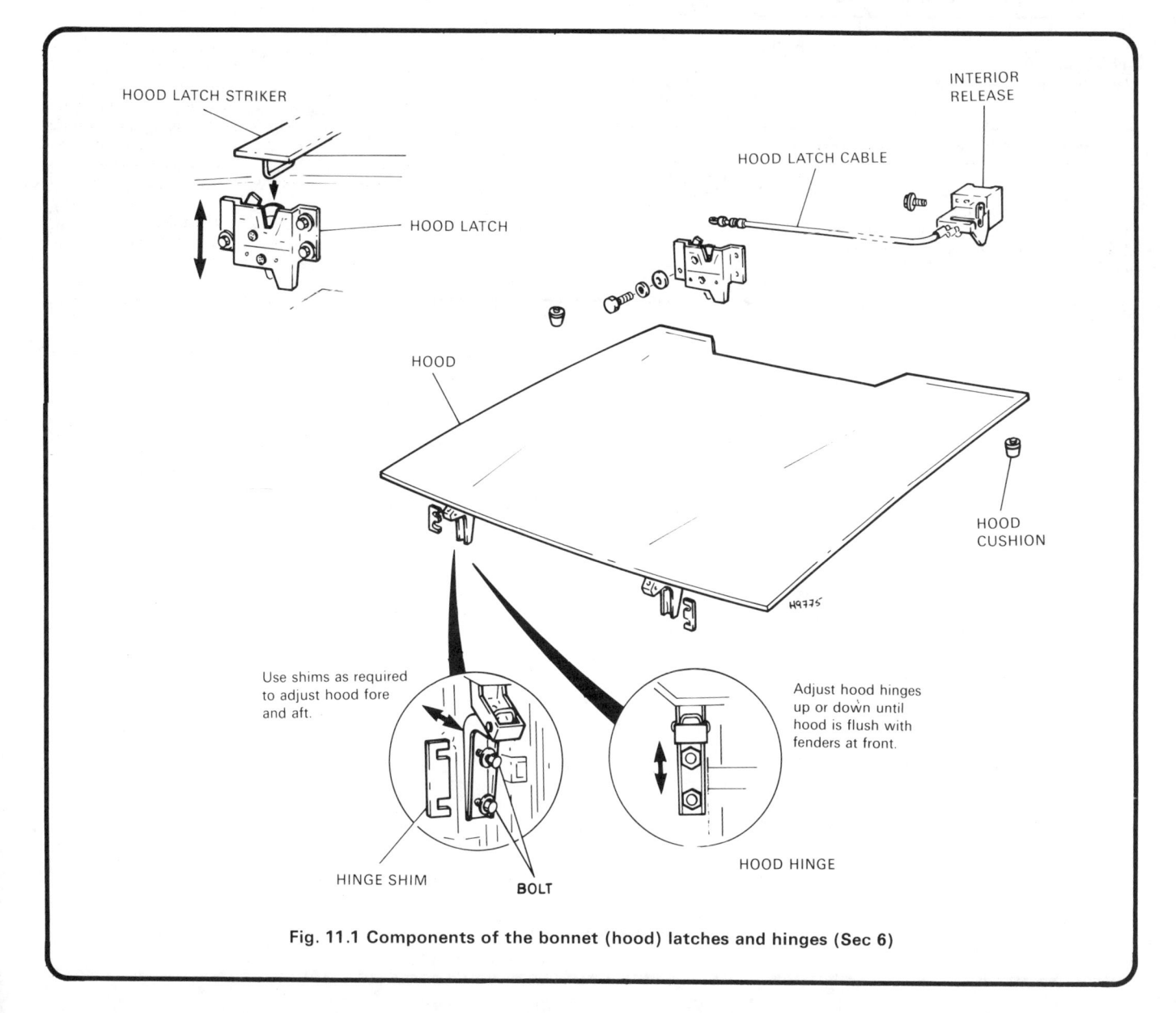

Fig. 11.1 Components of the bonnet (hood) latches and hinges (Sec 6)

7 Bonnet lock and release assembly – removal, refitting and adjustment

1 Open the bonnet and support it on its strut.
2 Remove the bolts securing the lock assembly to the bulkhead (photo).
3 Loosen the locknut on the cable adjustment point and unhook the cable.
4 If the cable needs renewing, release it from the release handle in the car and feed it through the grommet in the bulkhead.
5 The release handle assembly may be removed by undoing its retaining bolts.
6 Refitting is a reversal of this procedure, but do not tighten the lock assembly bolts until the bonnet has been gently closed, and the lock assembly centralised so that the bonnet lock striker enters the lock assembly.
7 Minor up and down adjustments can be made to the lock assembly by loosening the retaining bolts.
8 Adjust the cable at the adjuster to give a smooth positive operation of the release handle and lock.
9 Finally tighten all bolts.

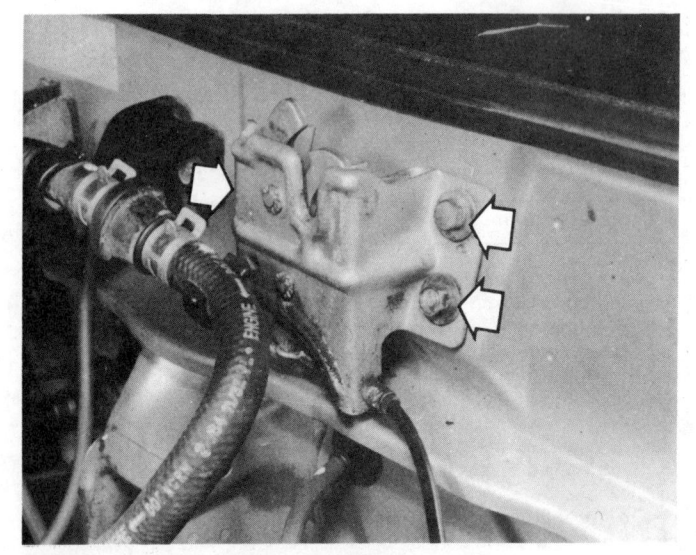

7.2 Bonnet lock assembly (retaining bolts arrowed)

8 Radiator grille – removal and refitting

1 The radiator grille is easily removed by undoing the retaining screws and lifting it out.
2 Refit in the reverse order.

9 Front and rear bumpers – removal and refitting

1 The front and rear bumpers are of moulded plastic compound material and are bolted either directly to the bodyframe or to hydraulic dampers, if they are of the impact absorbing type.
2 To remove the bumpers, refer to the relevant diagram, according to models (Figs. 11.3, 11.4 and 11.5).
3 Disconnect the electrical leads to any light units contained within the bumper. (The light units can be removed when the bumper is off).
4 Unscrew the bumper mounting bolts and lift the bumper away.
5 Refit in the reverse order.
6 The efficiency of the bumper damper units may be checked by placing the vehicle square and close to a wall and using a padded jack placed between the wall and bumper, compress the damper unit by about 0.5 in (0.12 in). Retract the jack and check that the bumper returns to its original position. If it does not, or hydraulic fluid leaks from the damper, renew the damper unit.

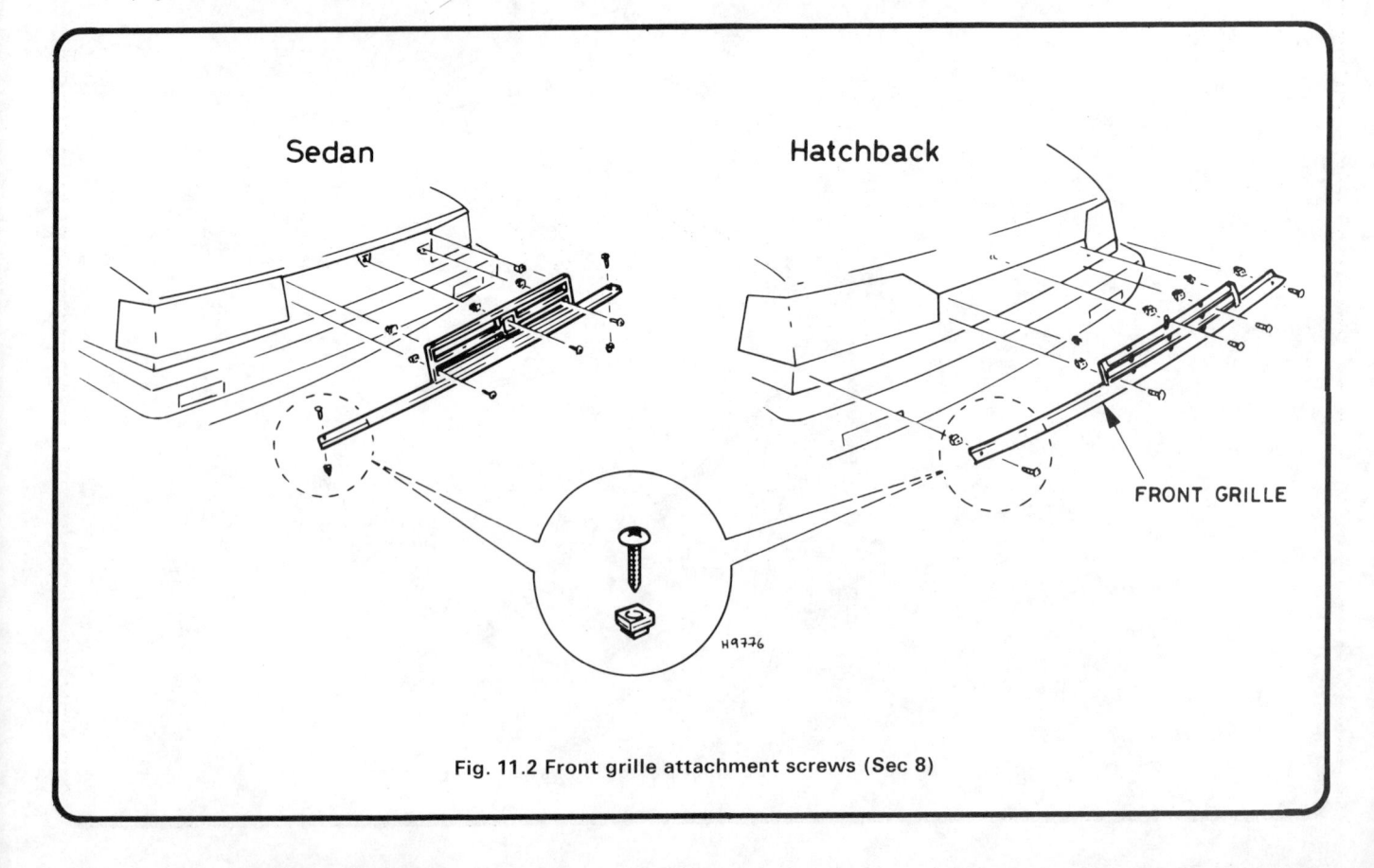

Fig. 11.2 Front grille attachment screws (Sec 8)

259

FRONT BUMPER LEFT STAY

FRONT BUMPER STAY BRACKET

BOLT

3

2

1

CANADIAN MODEL

FRONT BUMPER UPPER BEAM

FRONT BUMPER LOWER BEAM

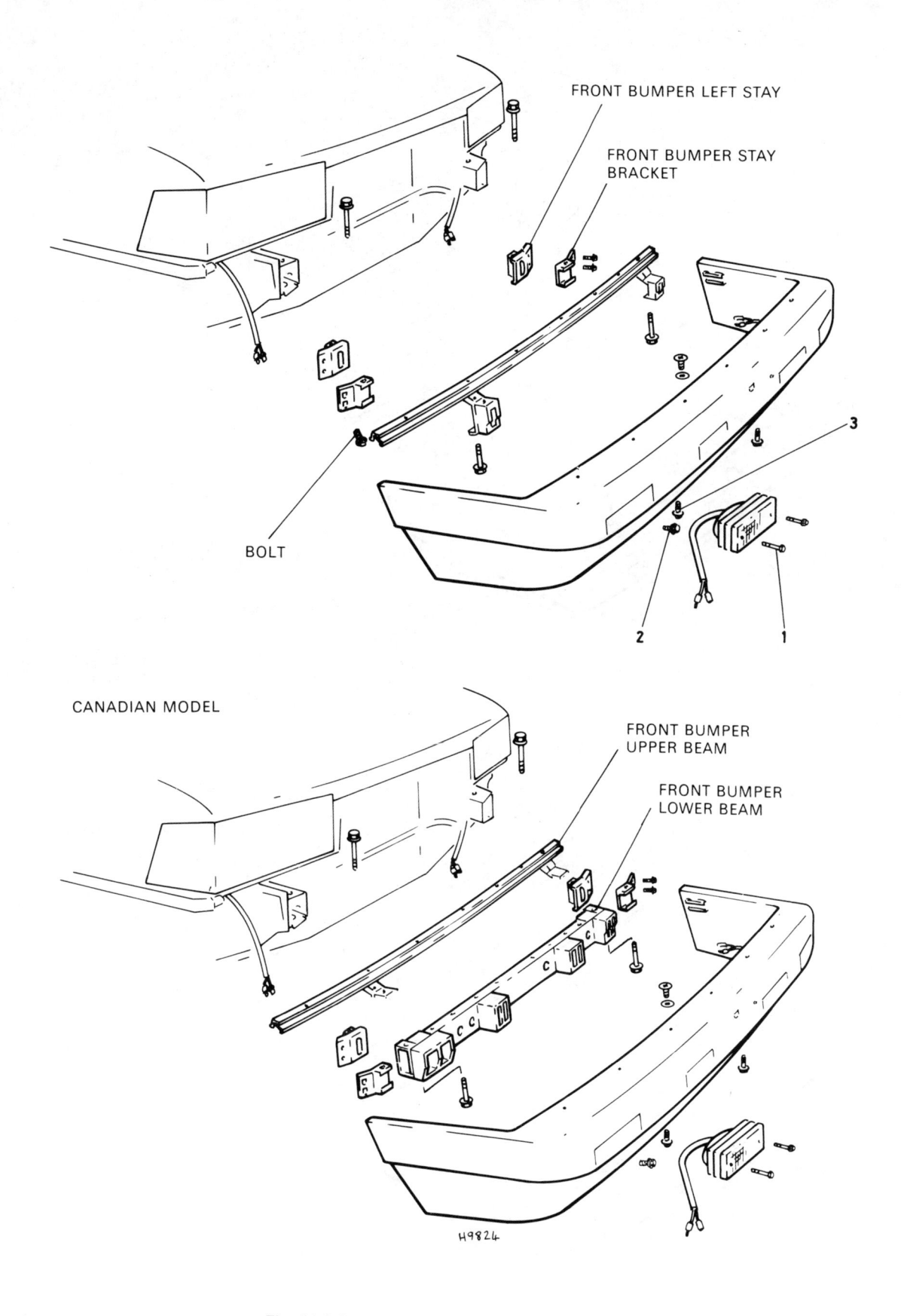

H9824

Fig. 11.3 Components of the front bumper (Sec 9)

Fig. 11.4 Components of the rear bumper (Sec 9)

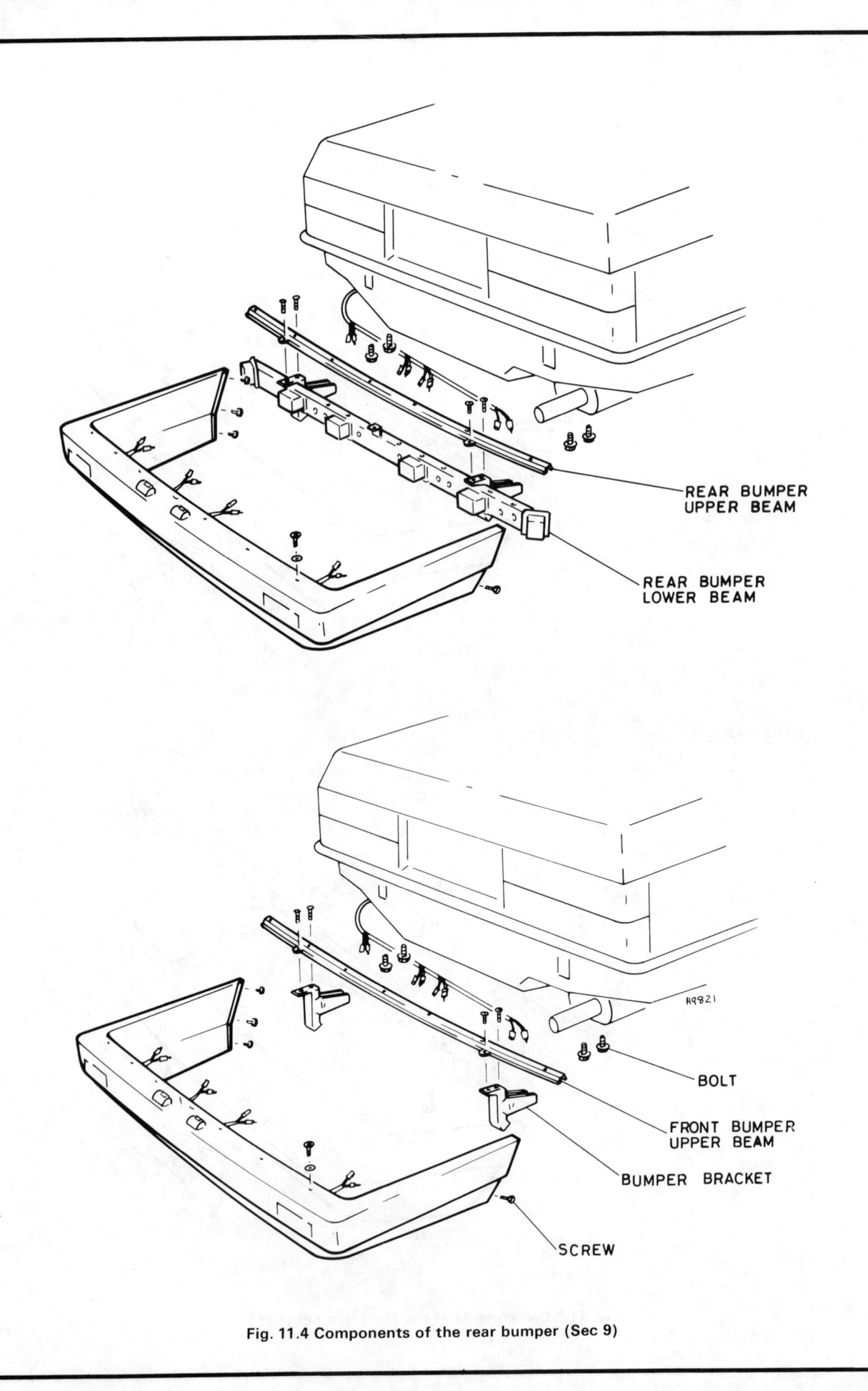

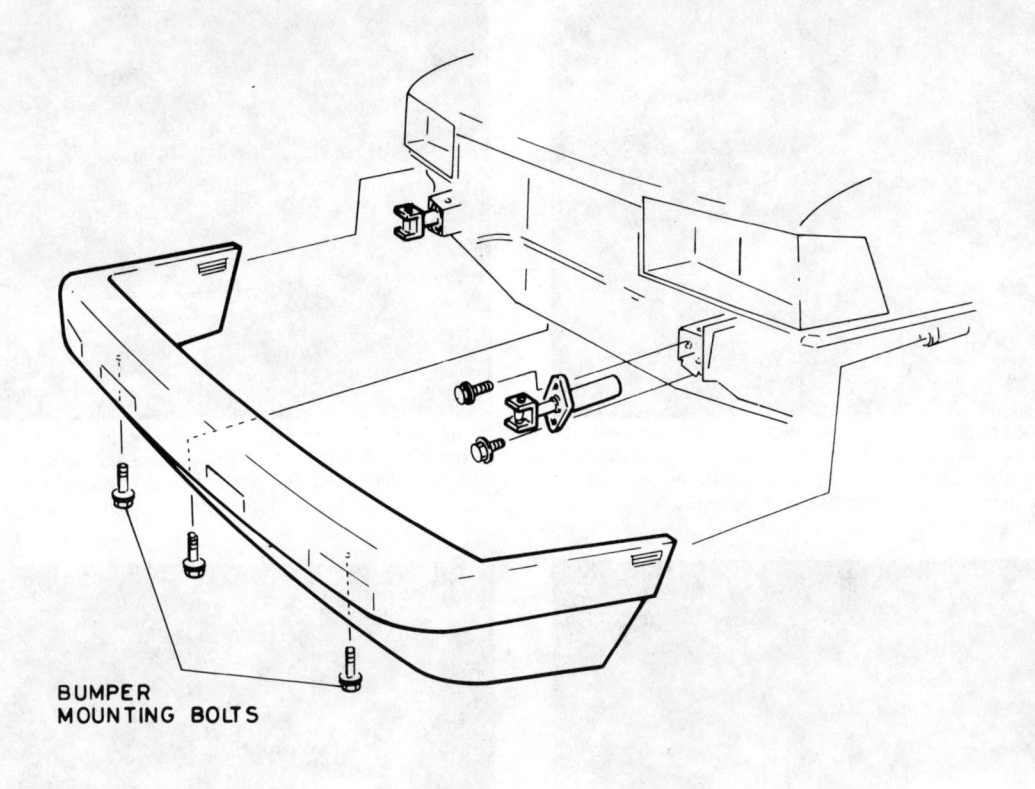

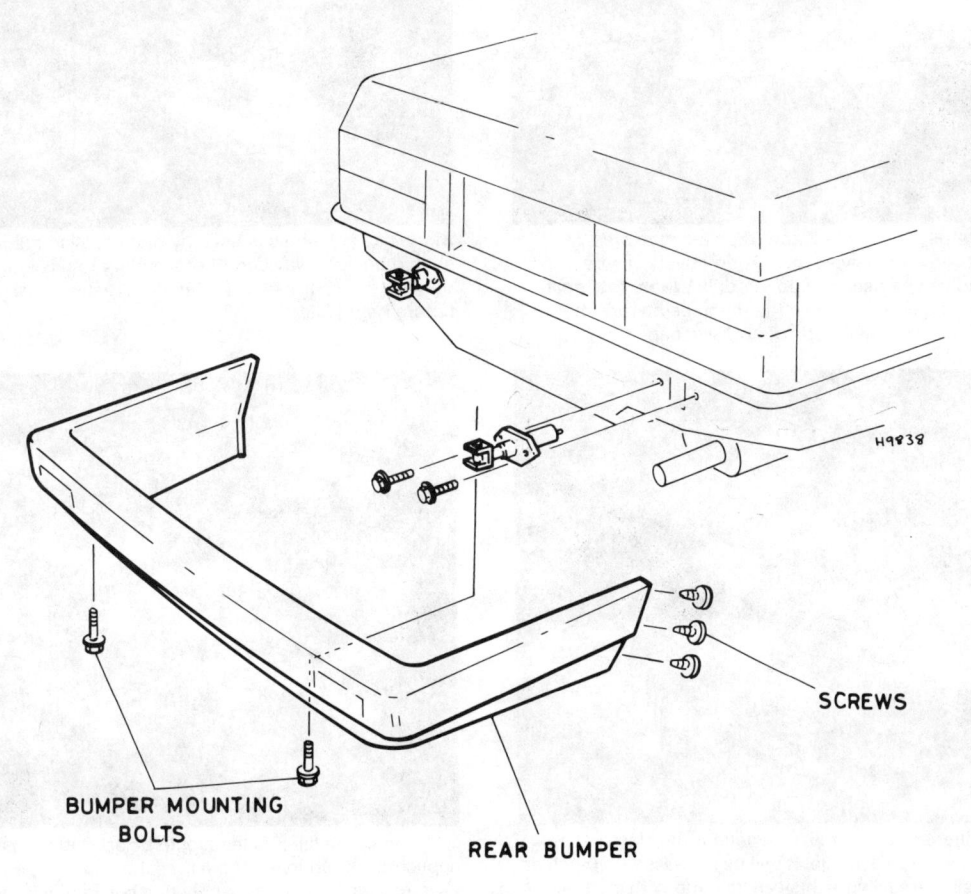

Fig. 11.5 Impact absorbing bumper (Sec 9)

This sequence of photographs deals with the repair of the dent and paintwork damage shown in this photo. The procedure will be similar for the repair of a hole. It should be noted that the procedures given here are simplified – more explicit instructions will be found in the text

In the case of a dent the first job – after removing surrounding trim – is to hammer out the dent where access is possible. This will minimise filling. Here, the large dent having been hammered out, the damaged area is being made slightly concave

Now all paint must be removed from the damaged area, by rubbing with coarse abrasive paper. Alternatively, a wire brush or abrasive pad can be used in a power drill. Where the repair area meets good paintwork, the edge of the paintwork should be 'feathered', using a finer grade of abrasive paper

In the case of a hole caused by rusting, all damaged sheet-metal should be cut away before proceeding to this stage. Here, the damaged area is being treated with rust remover and inhibitor before being filled

Mix the body filler according to its manufacturer's instructions. In the case of corrosion damage, it will be necessary to block off any large holes before filling – this can be done with aluminium or plastic mesh, or aluminium tape. Make sure the area is absolutely clean before ...

... applying the filler. Filler should be applied with a flexible applicator, as shown, for best results; the wooden spatula being used for confined areas. Apply thin layers of filler at 20-minute intervals, until the surface of the filler is slightly proud of the surrounding bodywork

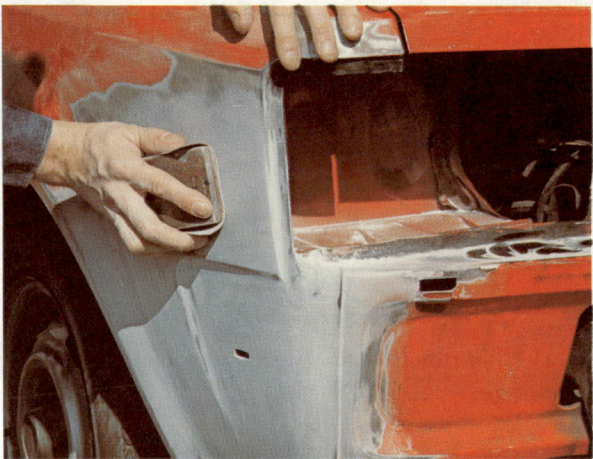

Initial shaping can be done with a Surform plane or Dreadnought file. Then, using progressively finer grades of wet-and-dry paper, wrapped around a sanding block, and copious amounts of clean water, rub down the filler until really smooth and flat. Again, feather the edges of adjoining paintwork

The whole repair area can now be sprayed or brush-painted with primer. If spraying, ensure adjoining areas are protected from over-spray. Note that at least one inch of the surrounding sound paintwork should be coated with primer. Primer has a 'thick' consistency, so will find small imperfections

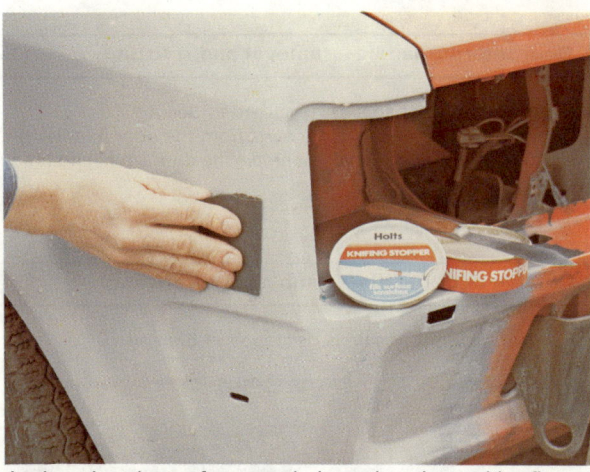

Again, using plenty of water, rub down the primer with a fine grade wet-and-dry paper (400 grade is probably best) until it is really smooth and well blended into the surrounding paintwork. Any remaining imperfections can now be filled by carefully applied knifing stopper paste

When the stopper has hardened, rub down the repair area again before applying the final coat of primer. Before rubbing down this last coat of primer, ensure the repair area is blemish-free – use more stopper if necessary. To ensure that the surface of the primer is really smooth use some finishing compound

The top coat can now be applied. When working out of doors, pick a dry, warm and wind-free day. Ensure surrounding areas are protected from over-spray. Agitate the aerosol thoroughly, then spray the centre of the repair area, working outwards with a circular motion. Apply the paint as several thin coats

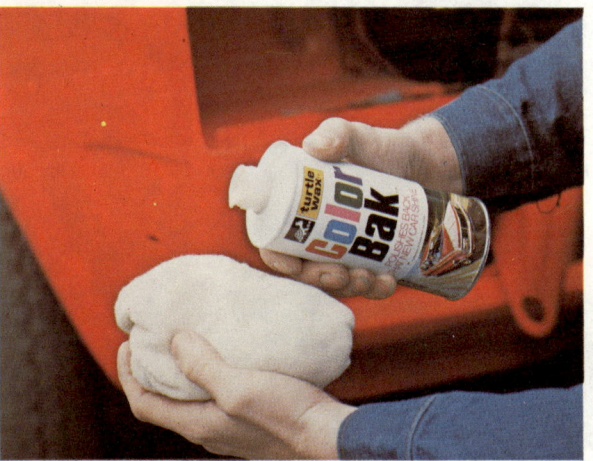

After a period of about two weeks, which the paint needs to harden fully, the surface of the repaired area can be 'cut' with a mild cutting compound prior to wax polishing. When carrying out bodywork repairs, remember that the quality of the finished job is proportional to the time and effort expended

264 Chapter 11 Bodywork and fittings

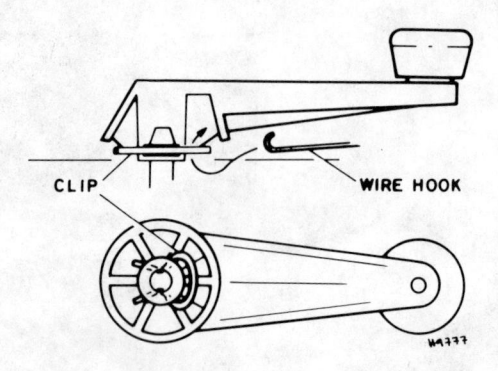

Fig. 11.6 Removing the door window regulator handle (Sec 10)

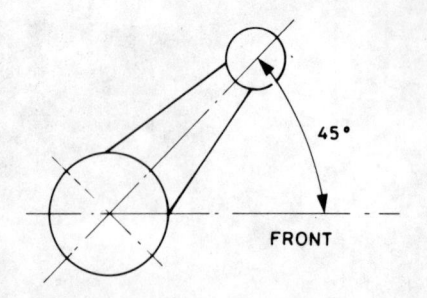

Fig. 11.7 Refitting the door window regulator handle at the correct angle (Sec 10)

10 Door trim panel – removal and refitting

1 Using a wire hook, remove the retaining clip from the door window regulator, and remove the handle (manual windows only) – see Fig. 11.6.
2 Again on manual window versions only, remove the self-tapping screws from the armrest, and remove the armrest.
3 On power operated windows, remove the screw from the door pull recess and lift out the door pull (photos).
4 Remove the plug from the inside handle trim plate, remove the screw, and lift off the trim plate (photos).
5 Using a wide-bladed instrument, prise the door panel from the door.
6 Once all the clips are prised from their locations lift the door panel to remove it.
7 Disconnect all electrical leads at their connectors (eg power windows, door locks and courtesy light).
8 Remove the two screws from the support bracket and remove the bracket (photo).
9 Carefully remove the plastic sheet covering the door. It is important this sheet remains intact and is refitted to prevent ingress of moisture.
10 Refit the door panel in the reverse order, with the window regulator handle pointing forwards and upwards at 45° with the window closed.

11 Door interior handle – removal and refitting

1 Remove the door panel as described in Section 10.
2 Remove the three retaining screws from the door handle (photo).
3 Unhook the wire rod which operates the door latch, and lift out the handle.
4 Refit in the reverse order.

12 Outside door handle – removal and refitting

1 Remove the door panel, as described in Section 10.

10.3A Remove the screw ...

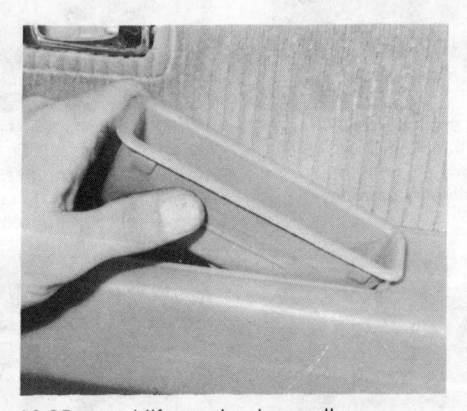

10.3B ... and lift out the door pull recess

10.4A Remove the screw ...

10.4B ... lift the handle and remove the plate

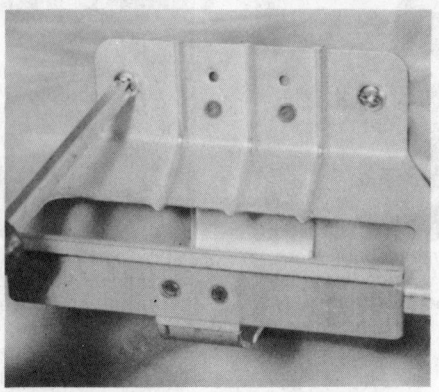

10.8 Remove the screws from the support bracket

11.2 Unscrew the three retaining screws

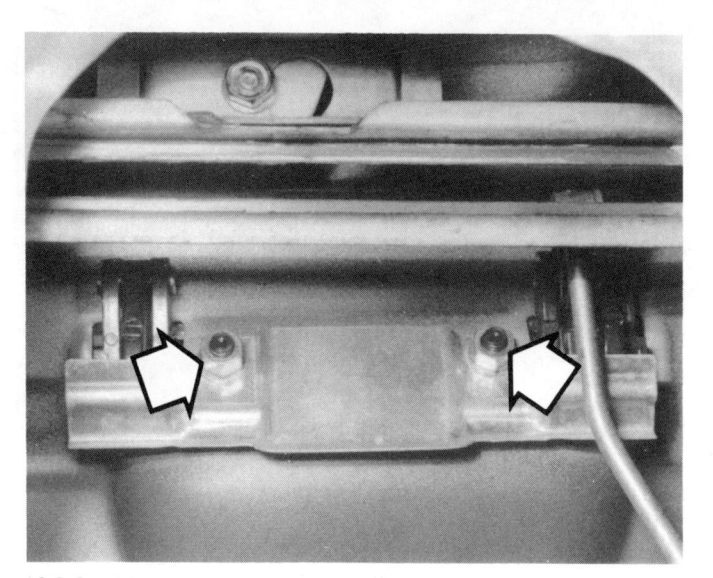

Chapter 11 Bodywork and fittings

12.3 Outside door handle retaining nuts (arrowed)

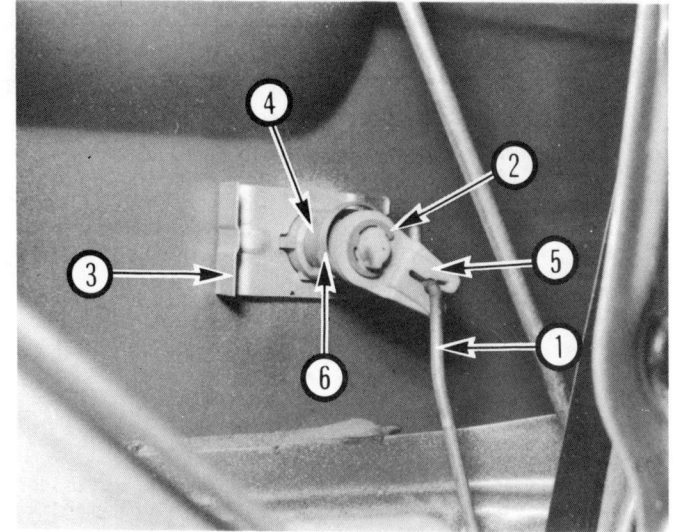

13.2 A door lock cylinder
1 Lock operating rod
2 Circlip
3 Lock retainer
4 Lock cylinder
5 Lock arm
6 Spring

2 Disconnect the latch and lock operating rods.
3 Remove the two retaining nuts and remove the door handle (photo).
4 Refit in the reverse order.

13 Door lock cylinder – removal and refitting

1 Remove the door trim panel, as described in Section 10.
2 Unhook the lock operating rod (photo).
3 Remove the circlip and take off the lock arm and spring.
4 Pull the lock retainer out of the recess on the lock.
5 Lift out the lock from outside the vehicle.
6 Refit in the reverse order.

14 Door latch – removal and refitting

1 With the door trim panel removed, as described in Section 10, remove the three latch retaining screws (photos).
2 Remove the inside door handles, as described in Section 11.
3 Remove the latch and handle as a complete assembly.
4 Little can be done to repair the latch and if it is faulty, it should be renewed.
5 Refitting is a reversal of removal.

15 Power door lock – removal and refitting

1 Remove the door trim panel, as described in Section 10.
2 Disconnect the lead to the door lock actuator.
3 Remove the bolts securing the door lock actuator in place.
4 Unhook the actuator assembly from the door lock rod and remove the actuator.
5 Little can be done to repair the actuator if it is faulty, and it is best renewed.
6 Refitting is a reversal of removal.

16 Front door window – removal and refitting

1 Remove the door trim panel, as described in Section 10.
2 Lower the window to bring the glass mounting bolts into view (photo). Remove the bolts.
3 Standing on the inside of the door, draw the window glass upward and inward, tilting it as you do so, to remove it from the door.
4 Refitting is a reversal of this procedure.
5 Refer to Section 18 for fitting of the scissors assembly and window adjustment.

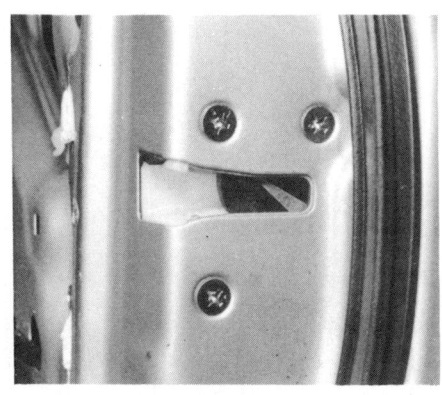

14.1A Front door latch retaining screws

14.1B Rear door latch retaining screws

16.2 The window glass mounting bolt on one of the two mountings (arrowed)

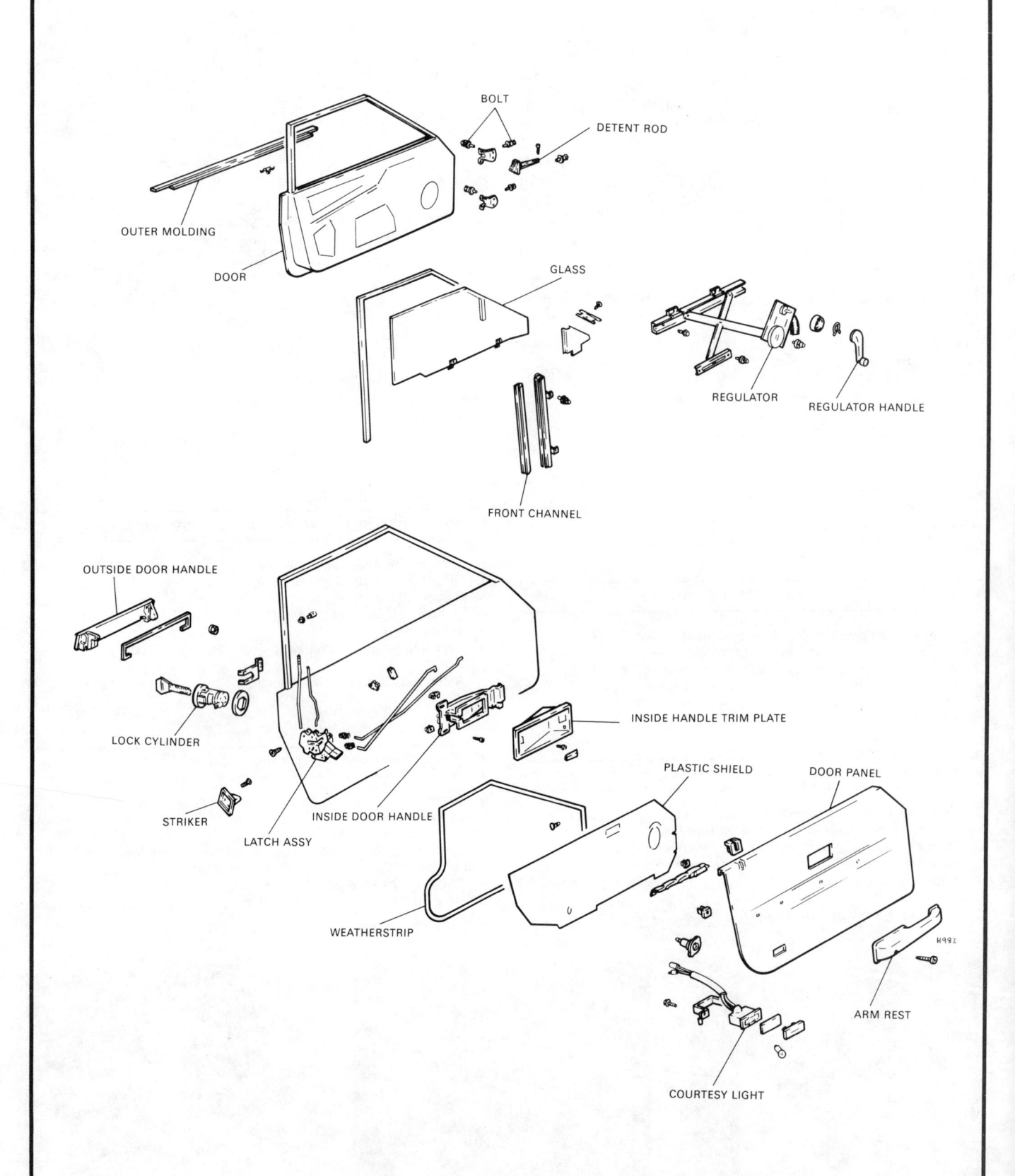

Fig. 11.8 Components of the front door – hatchback (Sec 16)

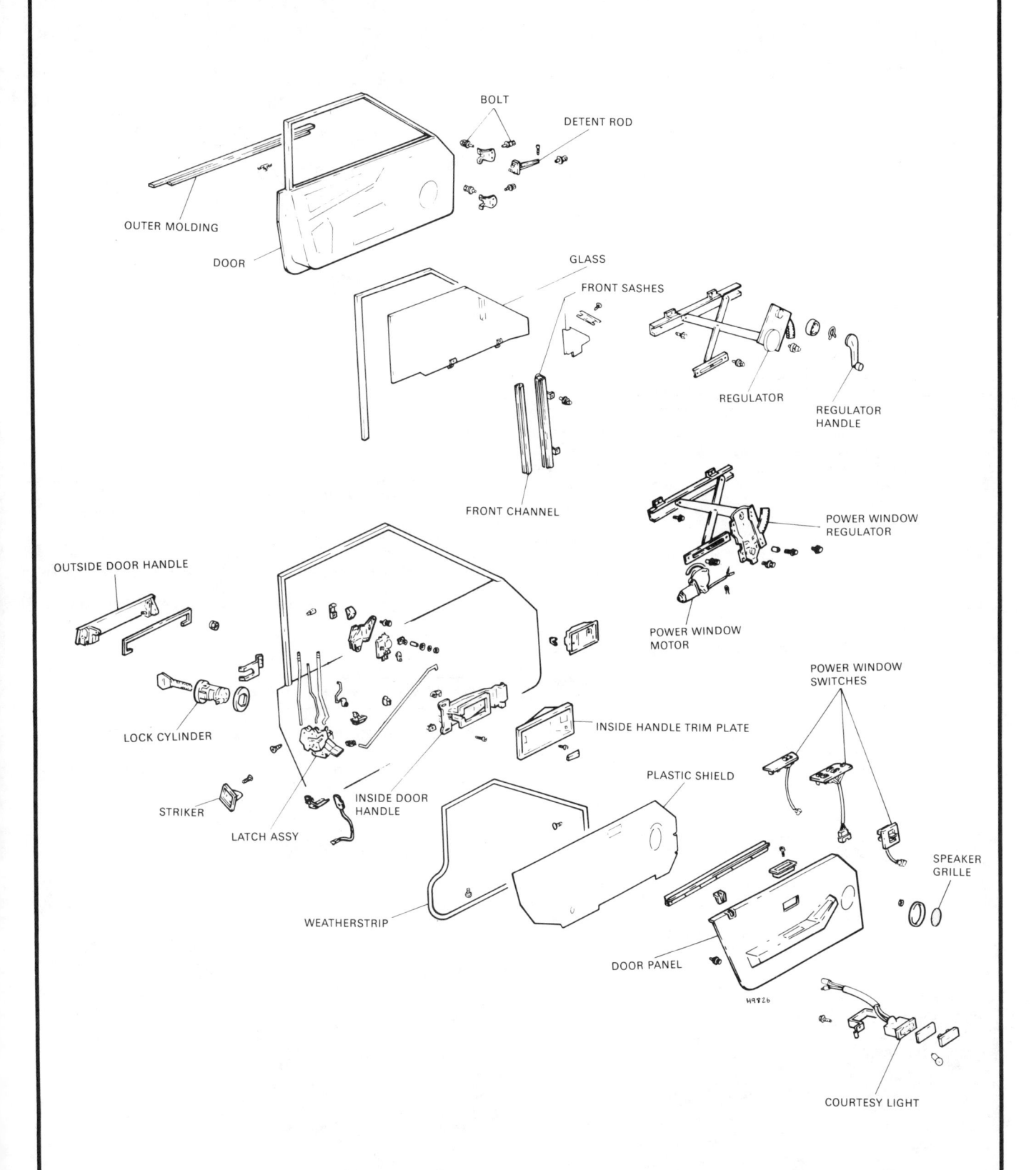

Fig. 11.9 Components of the front door – Saloon (Sec 16)

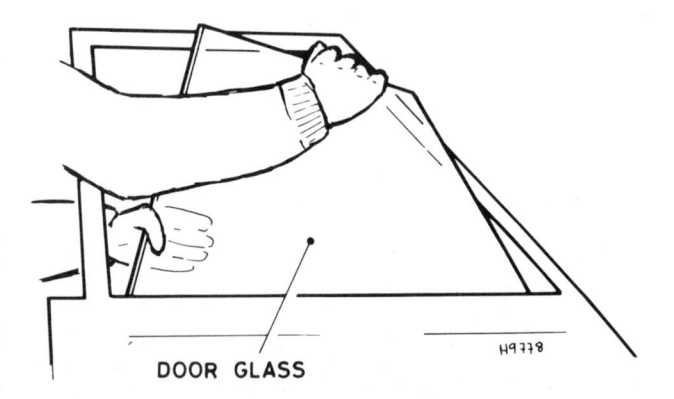

Fig. 11.10 Removing a window (Sec 16)

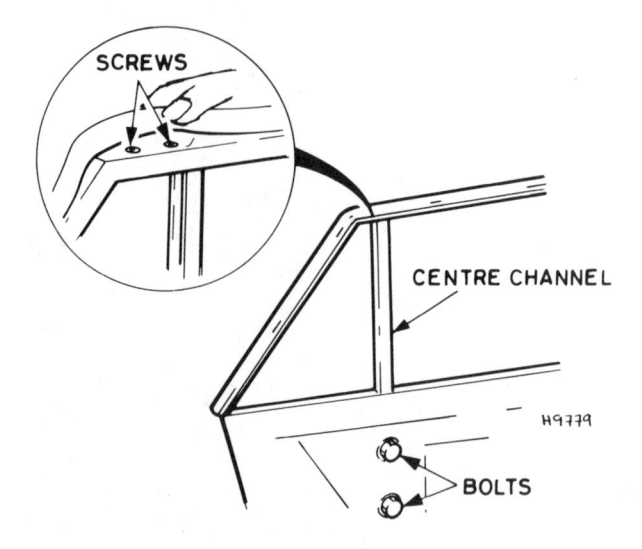

Fig. 11.11 Rear quarter light retaining screws and bolts (Sec 17)

17 Rear door window – removal and refitting

1 The procedure is identical to that for the front window, except that the rear quarter light must also be removed.
2 Remove the two bolts securing the centre channel and the two screws securing the quarter light at the top of the door.
3 Remove the quarter light.

18 Door window regulator scissors assembly – removal and refitting

1 On the front door, remove the two bolts which secure the glass guide channel and remove it.
2 Remove the bolts holding the scissors assembly to the door and remove the scissors (photos).
3 Inspect the gearteeth for wear, and the spring and linkage for wear or damage.
4 Lubricate the gears and pivot points with general purpose grease.
5 Refit in the reverse order of removal.
6 If, on installation of the window, it is found that the window is not central, loosen the regulator roller bolts and slide the window backwards or forwards until it fits centrally.

19 Door side moulding – removal and refitting

1 Remove the door trim panel, as described in Section 10.
2 Remove the nut from the moulding at the door outside edge.
3 Pull the moulding from outside the vehicle to release the remaining clips.
4 If they are tight, use pliers to squeeze them together from inside the vehicle to get them started.
5 Refitting is a reversal of removal.

20 Window weathersrtip – removal and refitting

1 Remove the door trim panel, as described in Section 10.
2 Pull the weather strip from the door (Fig. 11.13).
3 In all probability, this will break the plastic fixings and a new weatherstrip will have to be fitted.

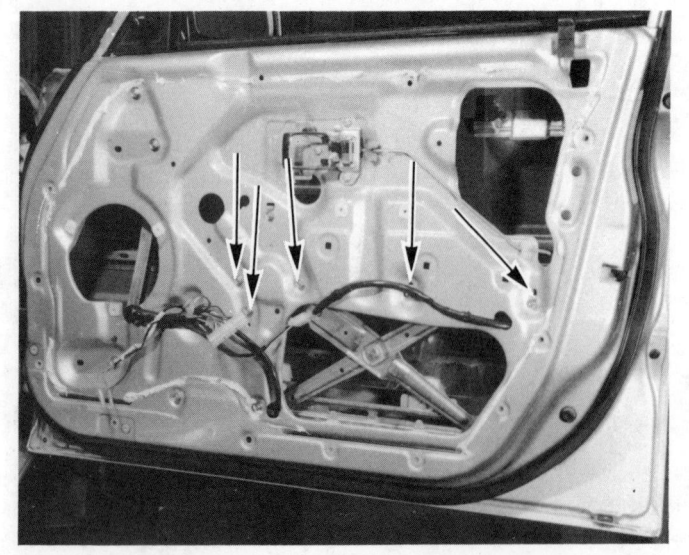

18.2A Window regulator scissors mounting bolts (arrowed) ...

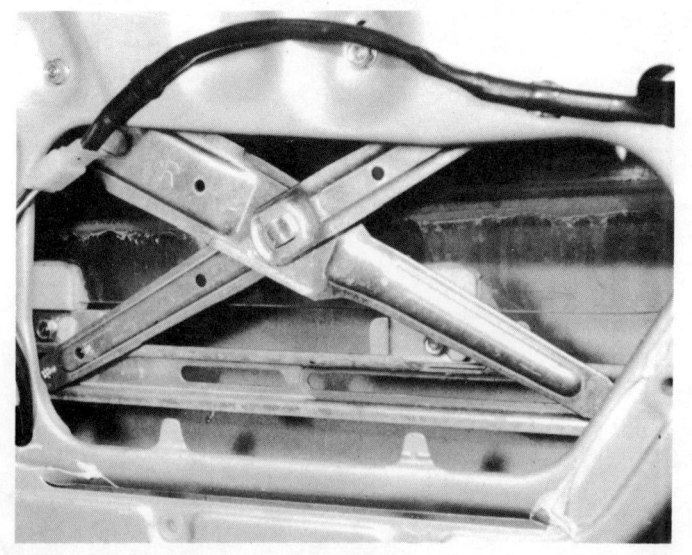

18.2B ... and a close-up of the scissors

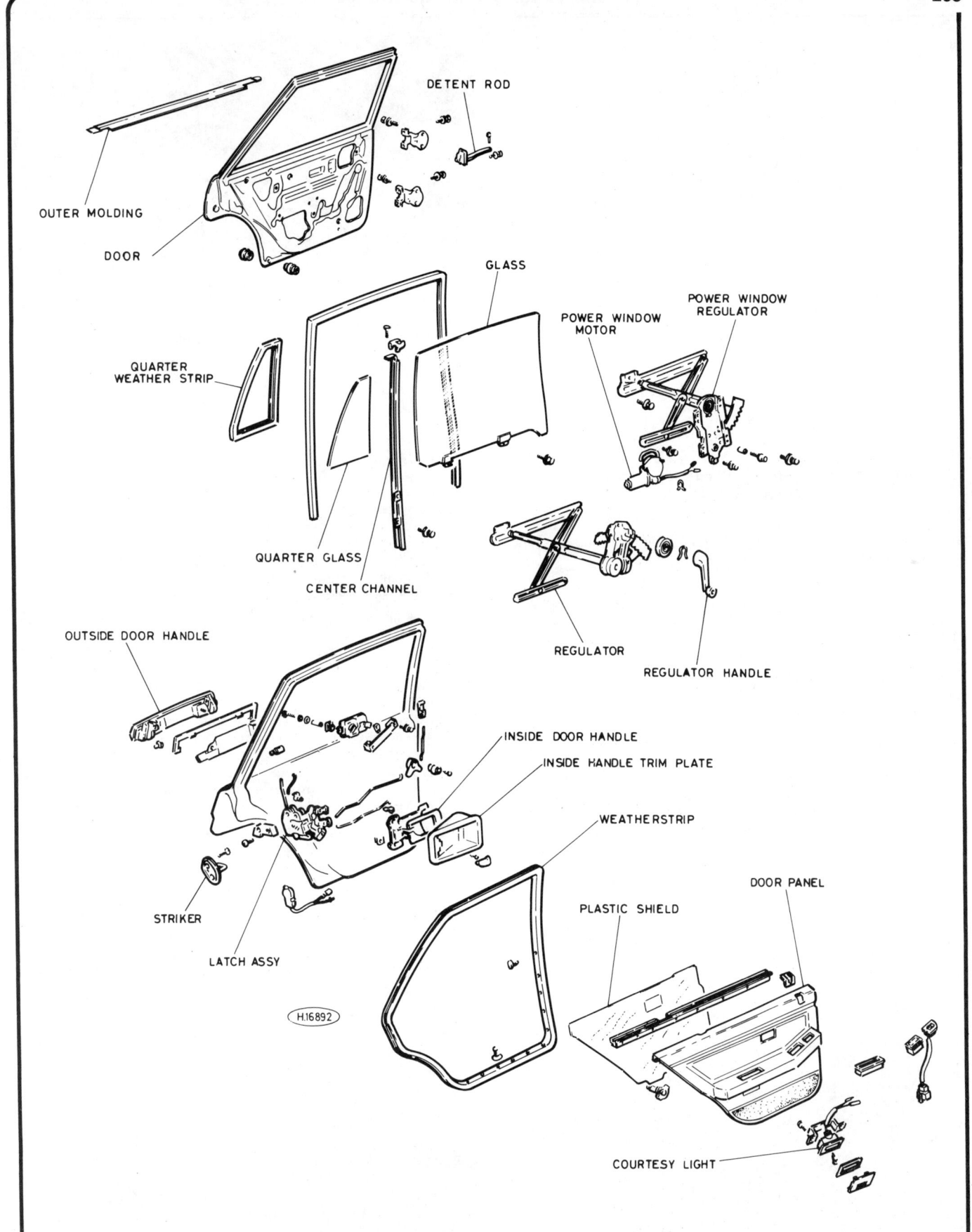

Fig. 11.12 Components of the rear door – Saloon (Sec 17)

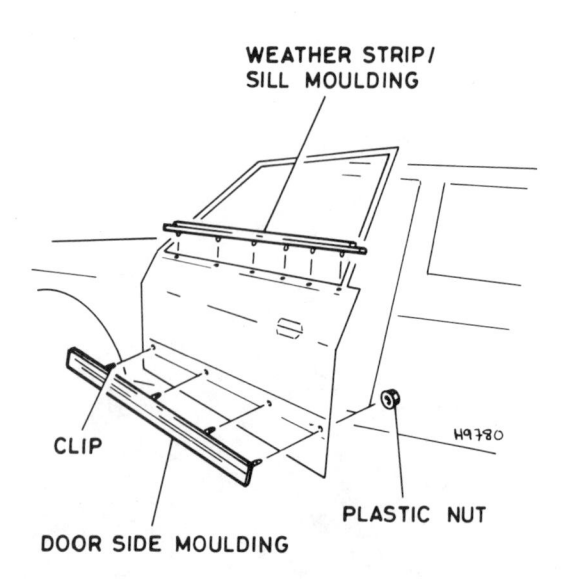

Fig. 11.13 Door side moulding and weatherstrip (Secs 19 & 20)

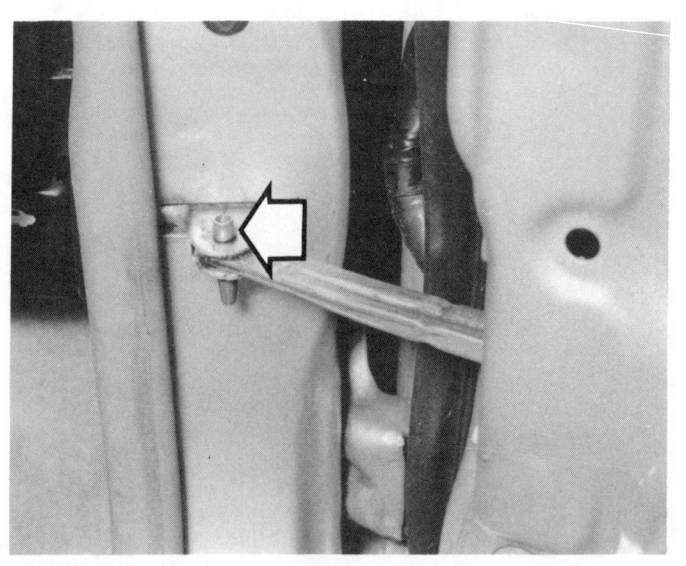

21.2 Door check strap pin (arrowed)

4 Position the new strip over the fixing location holes and press the strip into position.
5 Refit the door trim panel.

21 Door – removal and refitting

1 Support the door in the open position on a jack or blocks of wood, suitably protected.
2 Remove the pin from the door check strap (photo).
3 This may require the use of a hammer.
4 Remove the door trim panel and disconnect the electrical leads at their connectors.
5 With a helper steadying the door, remove the bolts from the door hinges (photos).
6 If necessary, remove the bolts from the hinge mountings (On the front doors this will entail removing the wheel arch shield).
7 Remove the door.
8 Refit in the reverse order.
9 Before finally tightening the hinge bolts on the door, or the hinge mounting bolts on the vehicle, ensure the door is in alignment with the surrounding bodywork.
10 If necessary, shims can be fitted under the hinges to achieve this.

22 Door striker – removal, refitting and adjusting

1 Draw a line around the striker plate with a soft lead pencil as a reference mark.
2 Remove the securing screws and lift the striker plate off.
3 Refit the plate and line it up with the previously made pencil mark.
4 Ensure that the door closes neatly without any excessive slamming, and remains tightly closed.
5 If necessary, loosen the screws and move the striker plate up and down or in and out to achieve a good door fit (photo).

23 Exterior rear view mirror – removal and refitting

1 The mirror glass may be renewed without removing the complete assembly.
2 Locate the screw head which secures the mirror glass through the hole in the bottom edge of the unit.

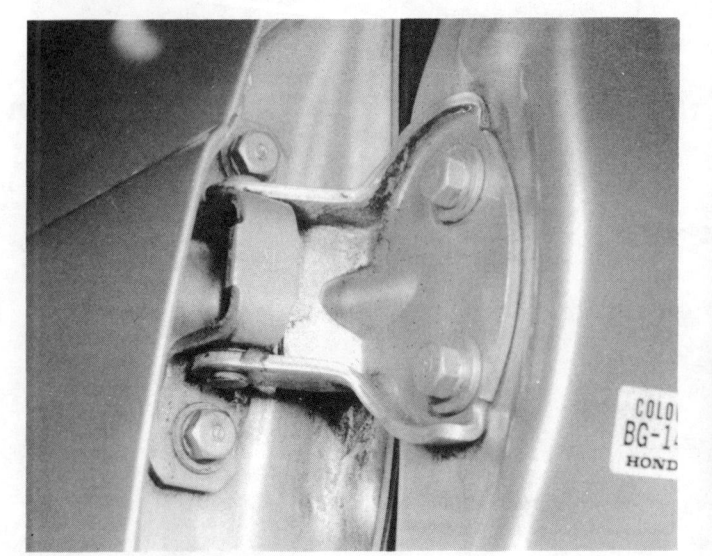

21.5A The door upper hinge ...

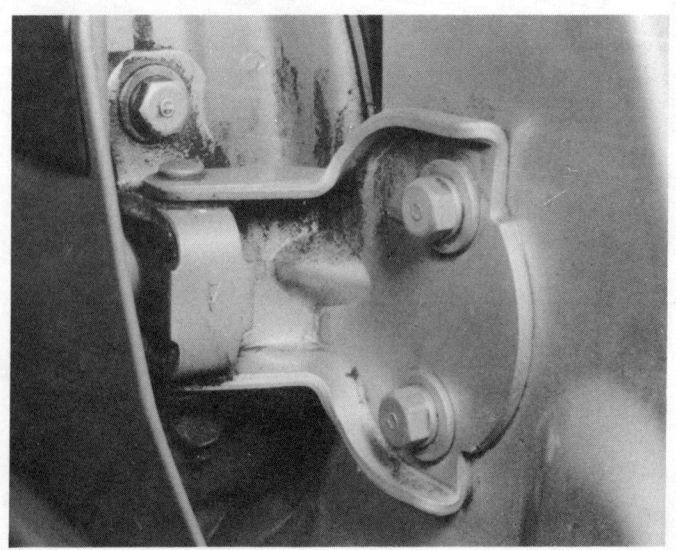

21.5B ... and the lower hinge

Chapter 11 Bodywork and fittings

3 Loosen the screw completely.
4 Carefully pry the mirror glass off the balljoint. Use padding to prevent damage to the mirror and rear view unit.
5 Apply general purpose grease to the balljoint and fork end before refitting the mirror.
6 To remove the complete unit, unscrew the screw from the front edge of the interior control knob, and remove the knob.
7 Carefully pry off the cover panel.
8 Support the mirror assembly and remove the three retaining screws, and remove the mirror.
9 Refit in the reverse order.

24 Rear hatch – removal and refitting

1 Remove the rear trim panel and plastic shield and disconnect the electric loom at the multi-block connector.
2 Disconnect the washer hose.
3 Tie a piece of stout string to the harness and then pull the harness up through the rear pillar.
4 Once it is pulled through, untie the string and leave it routed through the pillar, so that, on refitting, the string can be used to pull the harness back through the pillar.
5 Support the hatch in the open position, and remove the nuts from both support struts.

22.5 Door striker adjustment

Fig. 11.14 Removing the glass from the exterior rear view mirror (Sec 23)

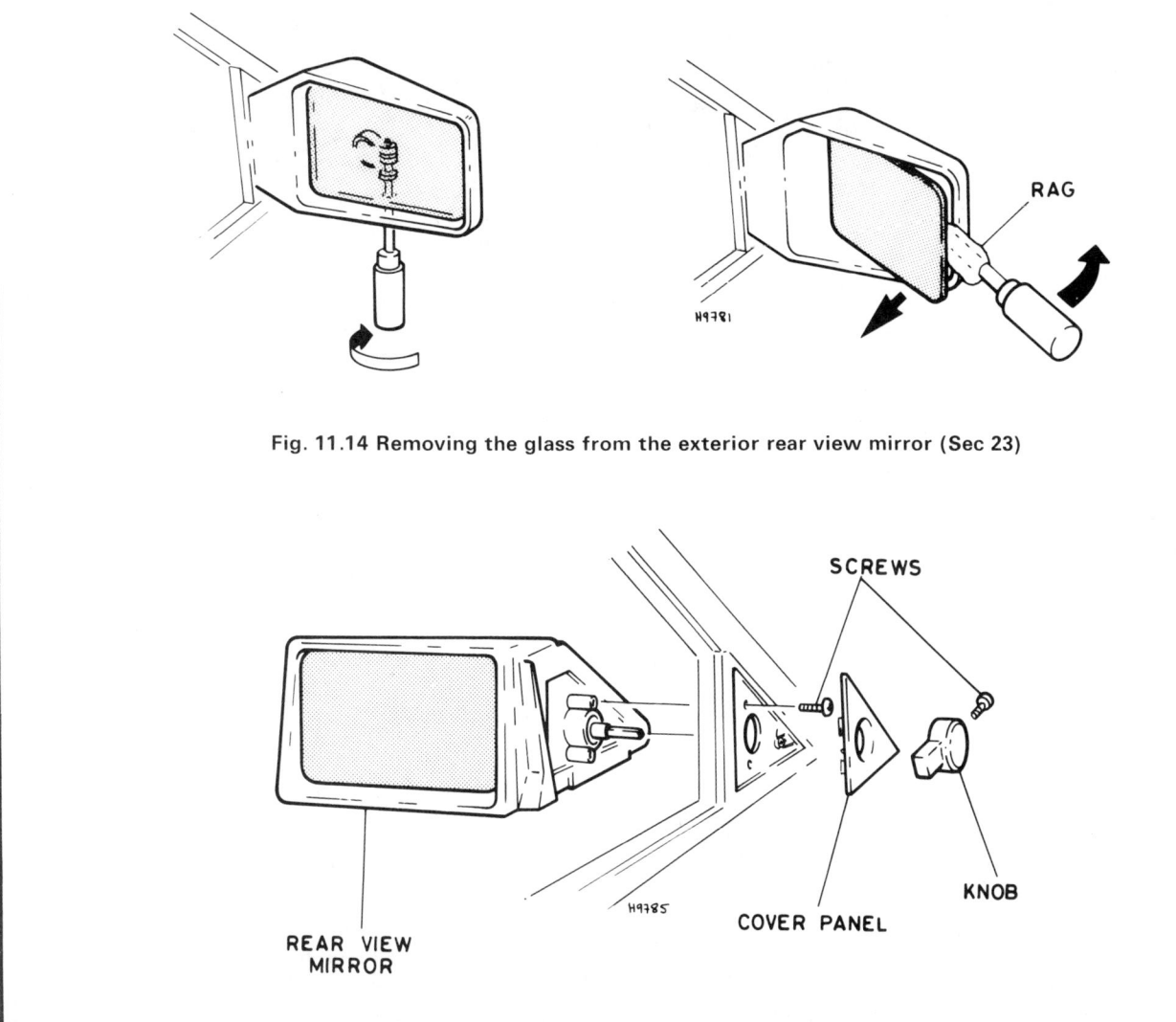

Fig. 11.15 The exterior rear view mirror assembly (Sec 23)

272 Chapter 11 Bodywork and fittings

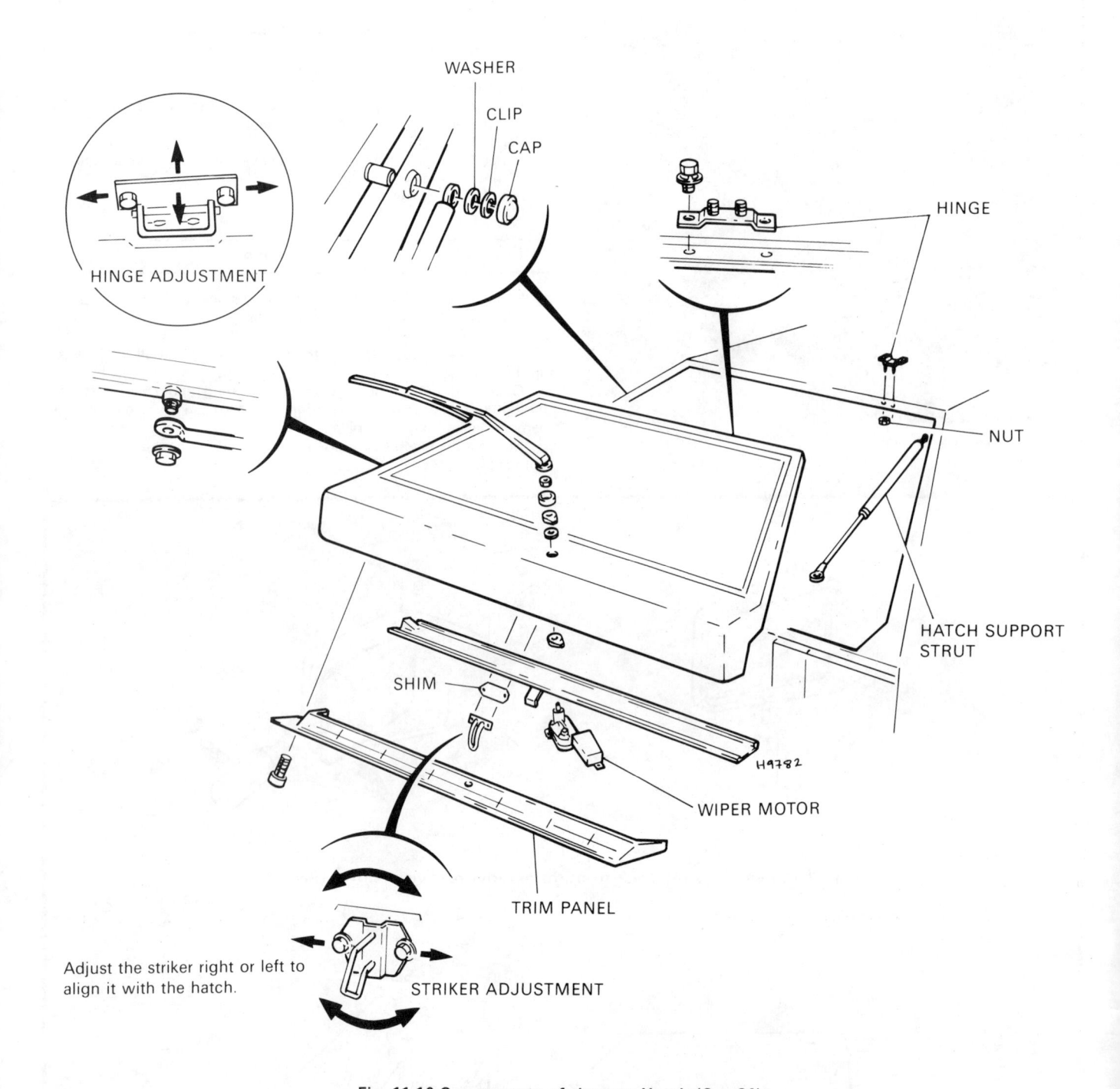

Fig. 11.16 Components of the rear Hatch (Sec 24)

6 Remove the headlining from around the hinges.
7 Remove the hinge bolts and remove the hatch.
8 Refitting is a reversal of removal, but do not tighten the hinge or striker bolts fully until the hatch has been centralised and closes properly.

25 Boot lid – removal and refitting

1 Remove the hinge bolts and lift off the boot lid (photo).
2 Remove the torsion rods by hand.
3 From inside the vehicle, remove the rear shelf.
4 Remove the hinge bracket nuts and then remove the hinges.

5 Refitting is a reversal of removal, but note that the left-hand torsion rod is marked with white paint.
6 Do not fully tighten the hinge bracket mounting nuts until the boot lid is central and closes properly, and adjust the striker as required (photo).

26 Rear hatch/boot and fuel filler lid remote release lever and latches – general

1 The rear hatch/boot lid remote control is cable-operated from a lever handle positioned by the driver's seat.
2 A second lever and cable control the fuel filler lid.

273

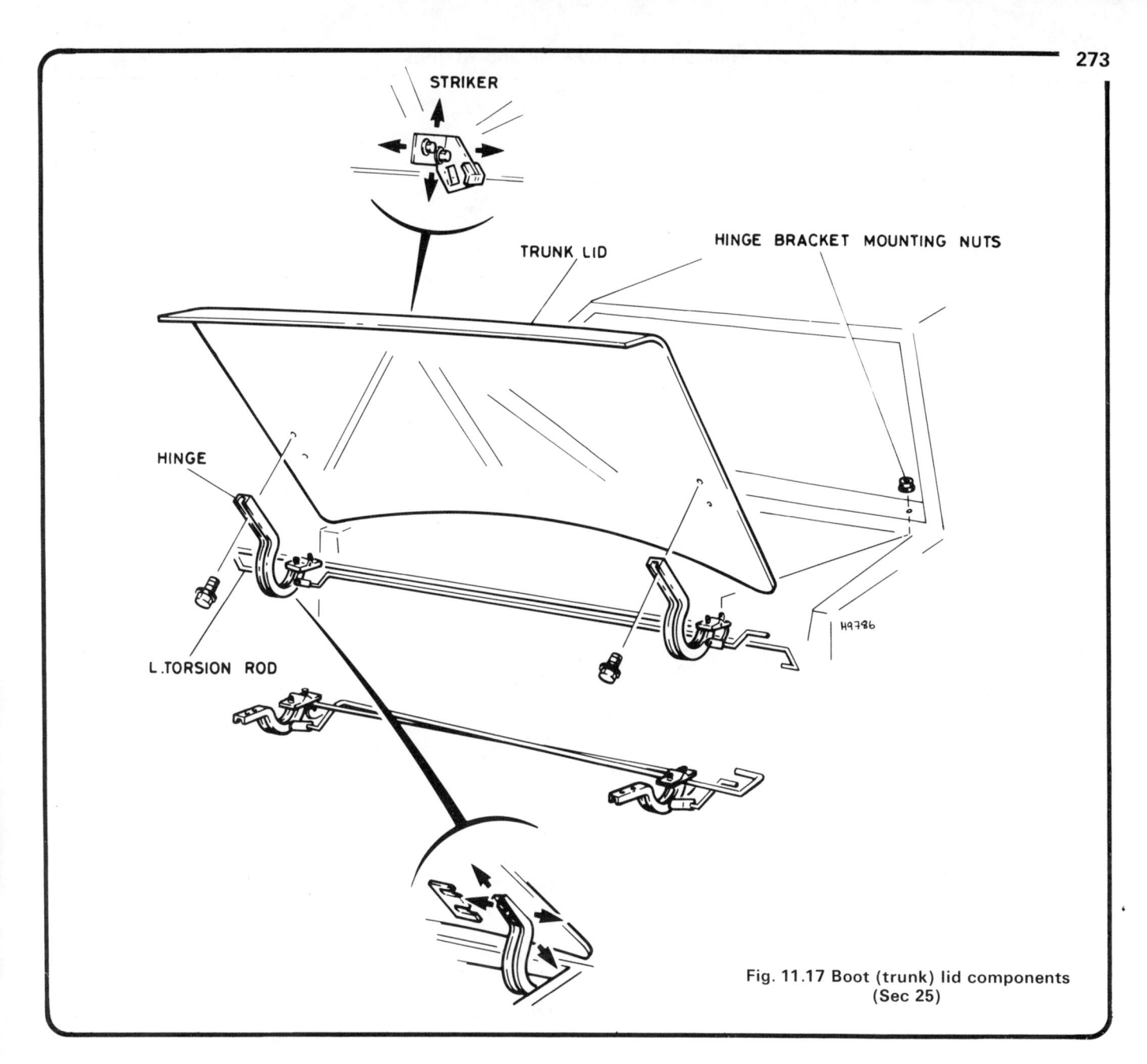

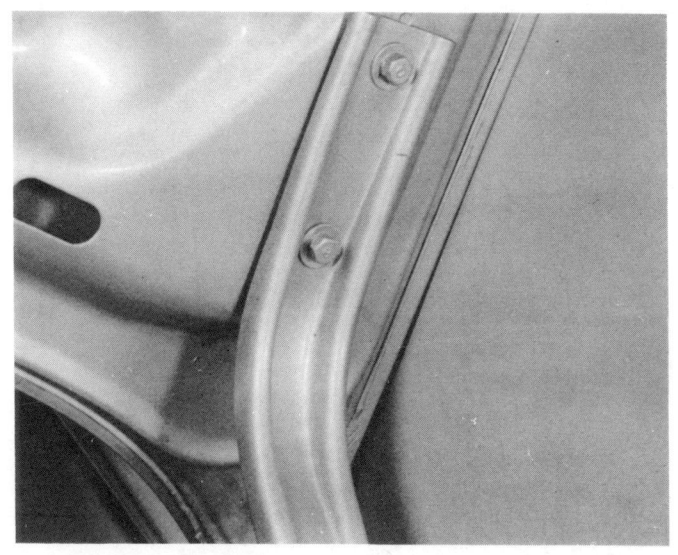

Fig. 11.17 Boot (trunk) lid components (Sec 25)

25.1 The boot lid hinge bolts

25.6 Boot striker

274   Chapter 11 Bodywork and fittings

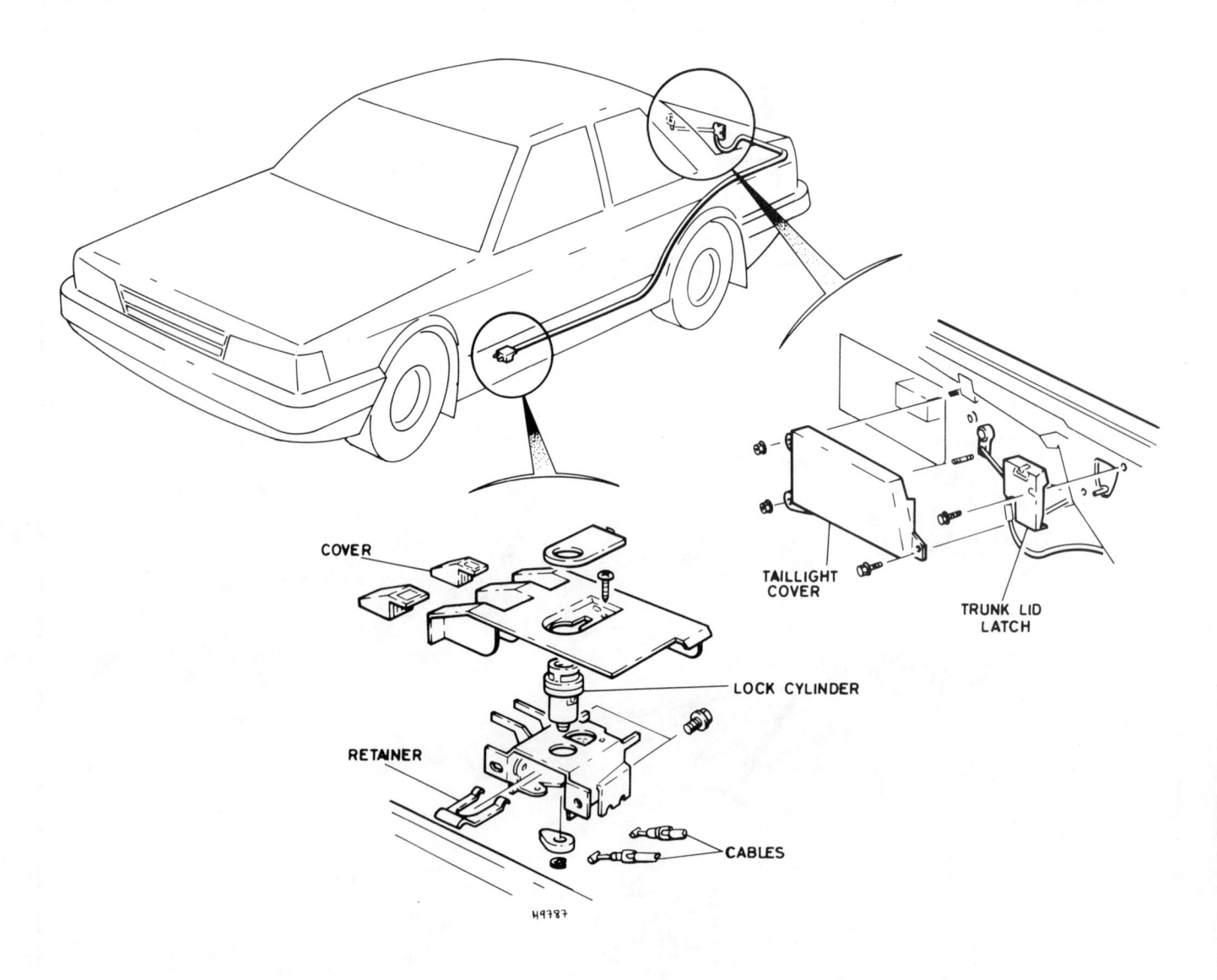

Fig. 11.18 Components of the remote release lever assembly (Sec 26)

3 To remove the release handle assembly, refer to Fig. 11.18 and remove the cover from the levers.
4 Pry out the cap on the top of the cover and remove the retaining screw, then the cover.
5 Remove the two bolts securing the unit to the door sill.
6 Unhook the cables and remove the unit.
7 To remove the fuel filler lid cable at the fuel tank end, turn the complete cable end unit a quarter turn to release it from the frame.
8 The lock can be removed from the door by removing the retainer plate.
9 The rear hatch/boot lid latch is accessible after removing the rear trim panel on the Hatchback version, or the plastic cover on Saloons (photo).
10 To remove the cable, undo the locknut and screw the adjuster right in to allow enough slack to unhook the cable eye end.
11 The latch assembly is removed by undoing its retaining screws or bolts, and disconnecting the door lock solenoid lead and earth wire, where fitted.
12 To renew the cables, remove as much trim as is necessary, to feed the old cables out of the vehicle, and new ones in.
13 Avoid excessive bending or kinking during this operation.
14 Refitting the release levers and latch assemblies is a reversal of removal.
15 On completion, adjust the cables so that the latches operate correctly, and tighten the locknuts.

26.9 Saloon boot lid latch

27 Windscreen and rear window – removal and refitting

1 The windscreen and rear window on both the Saloon and Hatchback are bonded to the frame using special adhesive and sealants.
2 It is recommended therefore that the replacement of these windows is left to your Honda dealer or local windscreen fitting specialists.

28 Rear quarter light – removal and refitting

1 Prise the plastic cover off the latch mounting and remove the screws.
2 Remove the upper and lower trim panels and the quarter light trim.
3 Remove the quarter light hinge bolts and remove the quarter light.
4 To remove the latch from the glass, prise off the circlip and remove the securing stud and latch.
5 Remove the nuts and bolts from the pillar trim to remove the trim from the glass.
6 Refitting is a reversal of removal.

29 Sunroof – removal and refitting

1 Move the sunshade to the fully open position.
2 Prise the plugs from the bracket covers, extract the screw and remove the cover.
3 Close the glass and remove the mounting nuts.
4 Remove the glass by raising and pulling it forward.
5 Remove the guide rail mounting nuts, raise and spread the ends of the rails and slide the sunshade out.
6 Refitting is a reversal of removal, but adjust as described in the following Section.

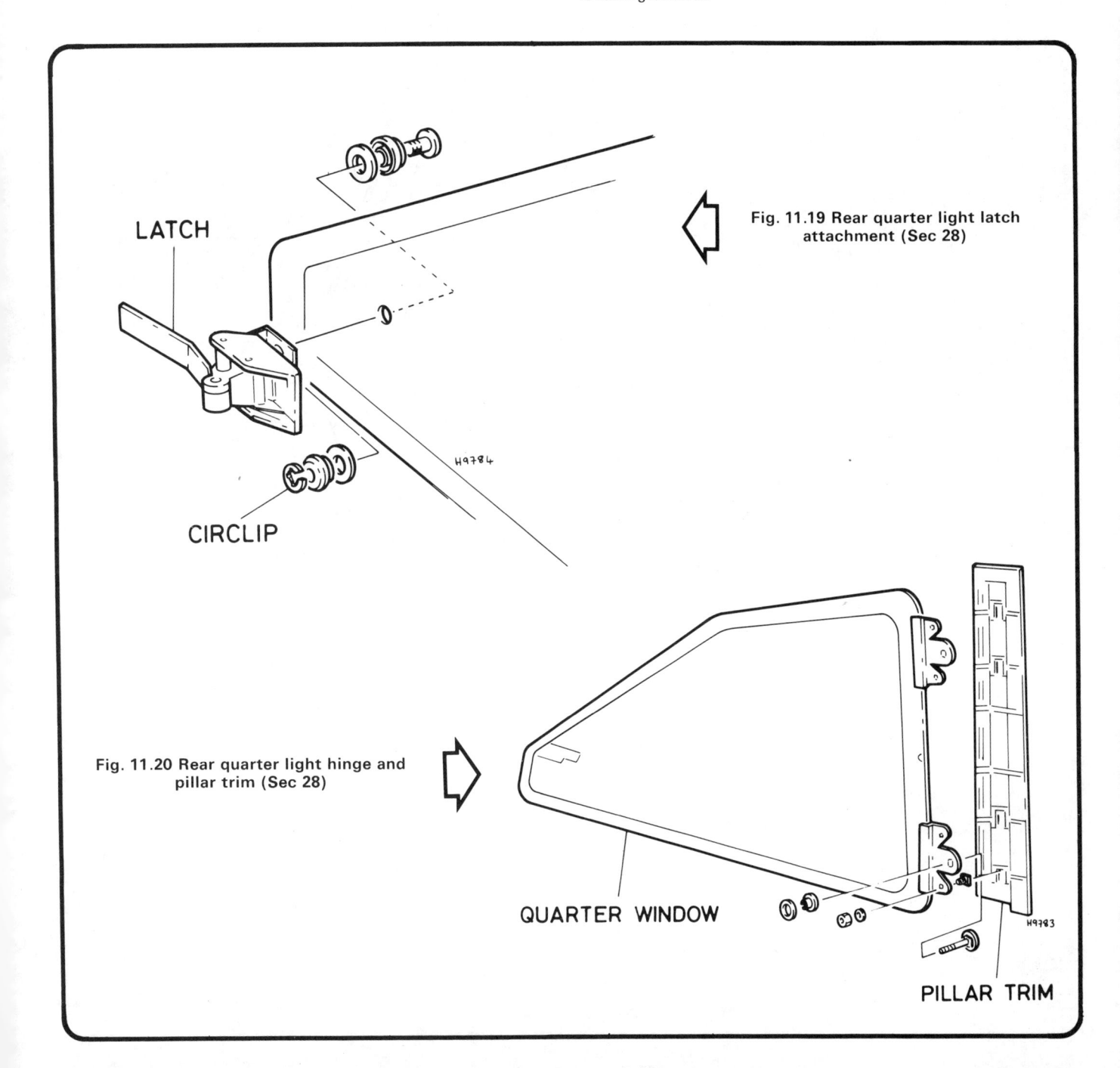

Fig. 11.19 Rear quarter light latch attachment (Sec 28)

Fig. 11.20 Rear quarter light hinge and pillar trim (Sec 28)

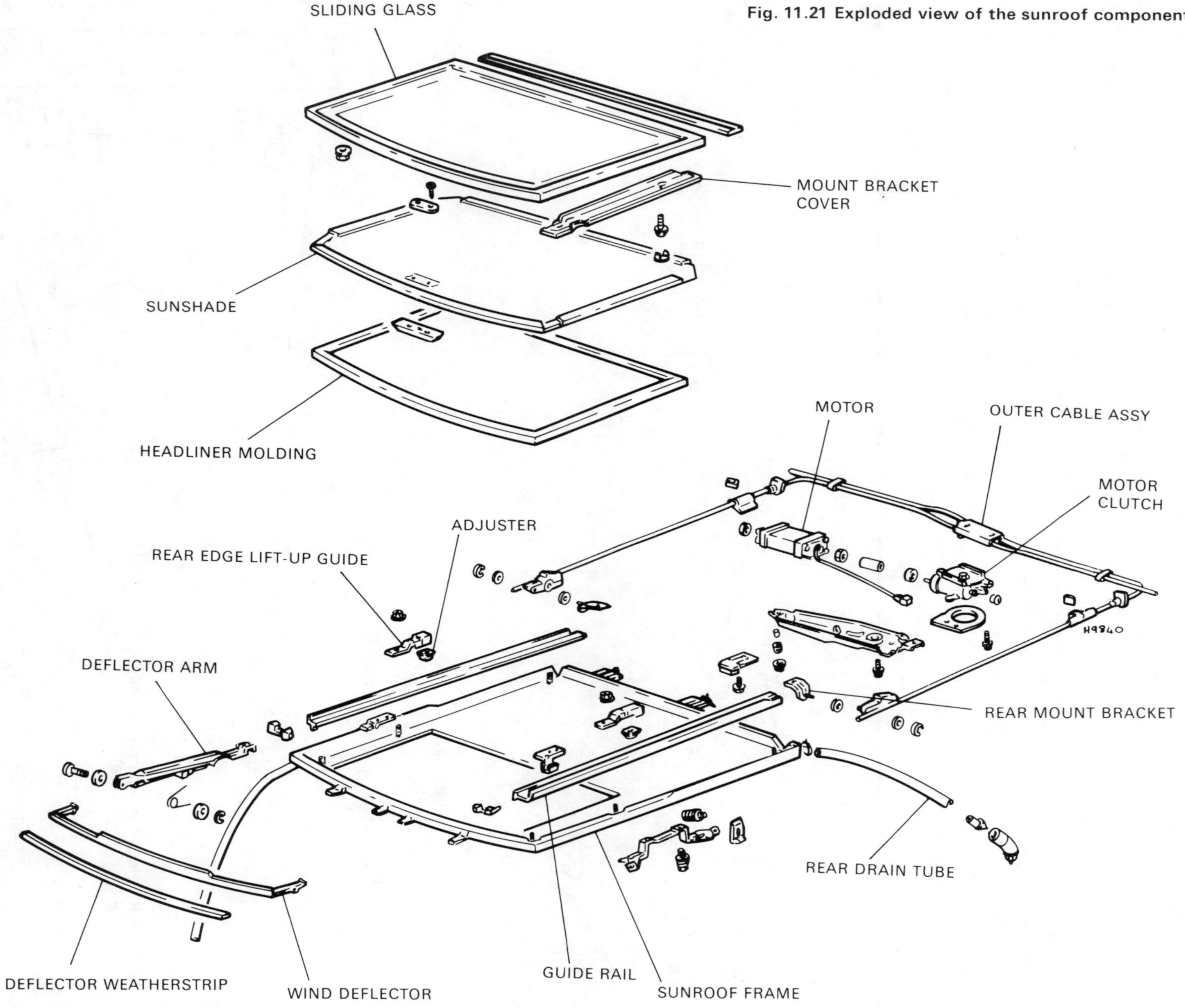

Fig. 11.21 Exploded view of the sunroof components (Sec 29)

Chapter 11 Bodywork and fittings

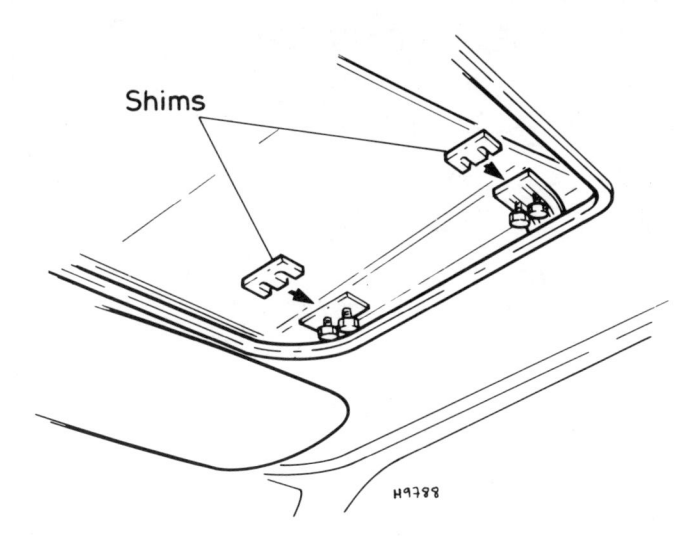

Fig. 11.22 Sunroof glass height adjustment shims (Sec 30)

Height adjustment = 0.04 ± 0.06 in (1 ± 1.5 mm) all the way round

30 Sunroof – adjustment

Glass side clearance
1 Binding on one side or the other may be corrected by loosening the mounting bracket nuts and moving the glass as necessary.

Glass height
2 The sunroof should be level and within 0.04 ± 0.06 in (1.0 ± 1.5 mm) of the weather strip all the way round.
3 The glass height may be adjusted by placing shims under the brackets.

Rear edge closing adjustment
4 Open the sunroof about 12.0 in (300 mm) and close it, noting the point at which the rear edge begins to rise.
5 If it rises too soon or too late to give a tight seal when it is closed adjust as follows:
6 Remove the rail covers from both sides, and loosen the lift-up guide nuts.
7 Move the guide forwards or back on the toothed plate, tighten the nuts and recheck. (Note: the pitch of the teeth on the toothed plate is 0.06 in (1.5 mm), so the distance to move the guides can be estimated from this).

Wind deflector adjustment
8 The wind deflector should be adjusted so that the edge of its seal contacts the roof evenly, or wind noise will be excessive.
9 Only the fore and aft plane of the deflector can be adjusted (as shown in Fig. 11.24), its opening height being fixed.

31 Sunroof motor, drain tube and frame – removal and refitting

1 Remove the headliner.
2 Unbolt the sunroof motor from the mounting at the rear of the sunroof, disconnect the wiring harness and remove the motor.
3 Disconnect the front and rear drain tubes.
4 Remove the frame mounting bolts and remove the frame.
5 In this condition, the cables can be removed by removing the guide rails and lifting out the cables with the rear mount brackets attached.
6 The rear mount brackets are shown in Fig. 11.26.

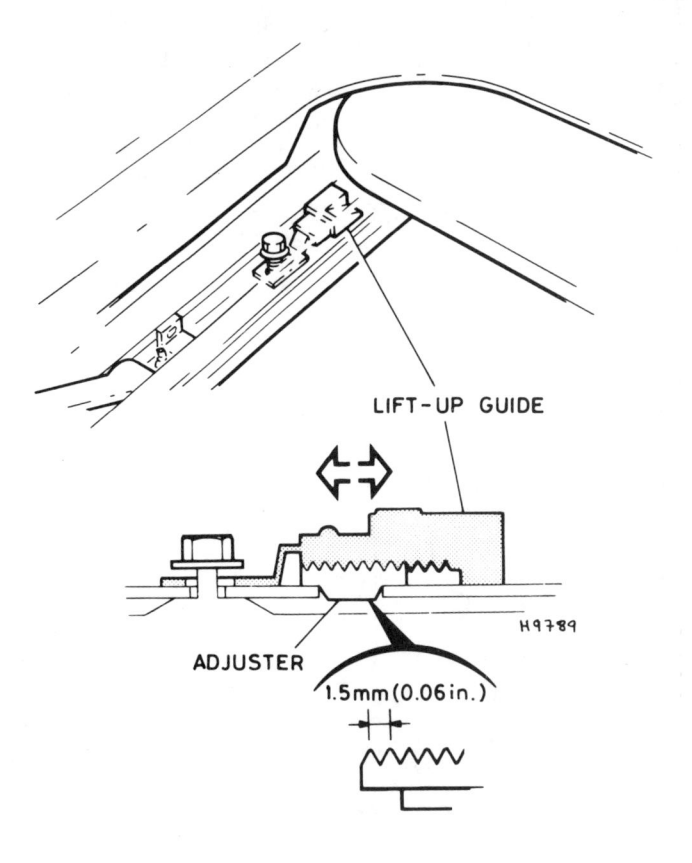

Fig. 11.23 Lift-up guide and adjuster for rear edge closing (Sec 30)

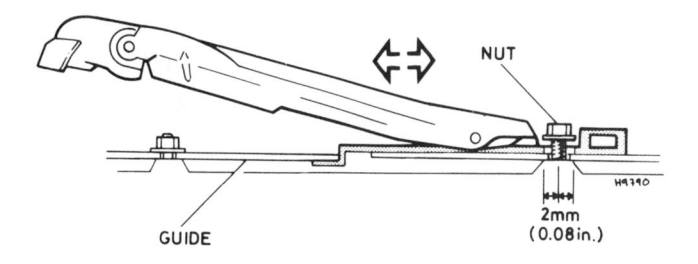

Fig. 11.24 Wind deflector adjustment (Sec 30)

7 Refitting is a reversal of removal, but apply sealant to the groove in each cable grommet when fitting the cable.
8 The effort required to open the sunroof should be checked by hand without the motor installed, as shown in Fig. 11.27.
9 With the motor installed, measure the force required to stop the motor closing the sunroof, again using a spring balance.
10 Adjust the sunroof motor clutch by turning the adjuster in to increase and out to decrease the force.

32 Wind deflector – removal and refitting

1 Remove the nuts securing the deflector to the guide rails.
2 Slide the deflector rearwards to unhook it and lift it clear.
3 Refit in the reverse order.

278

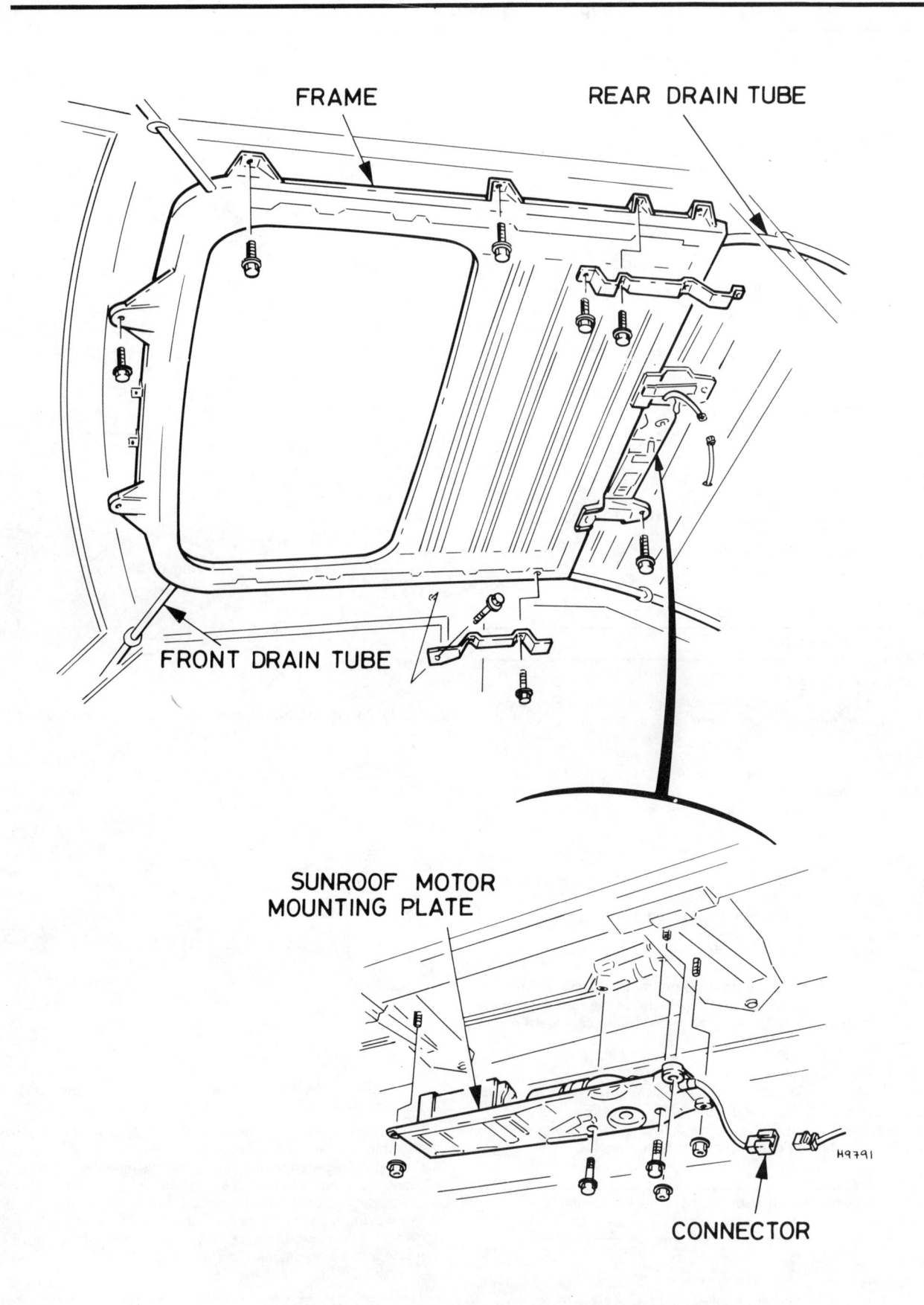

Fig. 11.25 Sunroof frame, motor mounting and drain tubes (Sec 31)

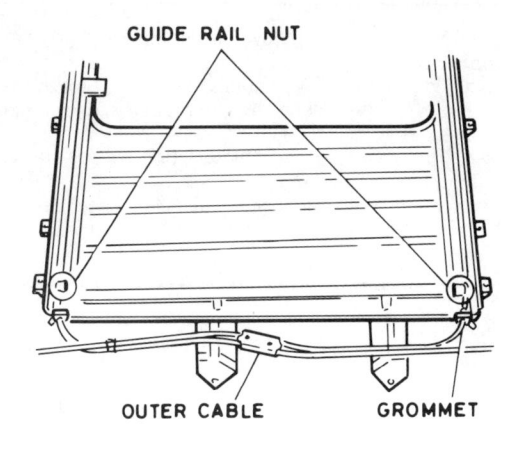

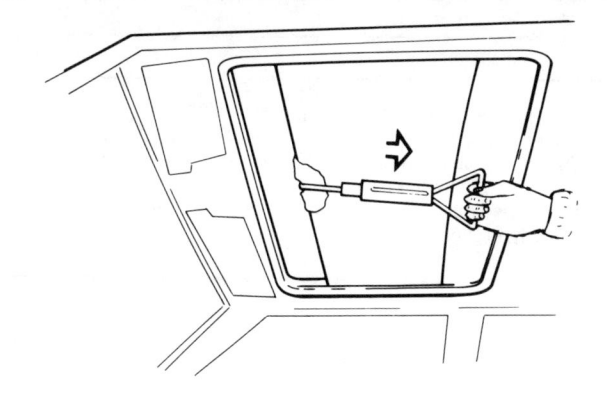

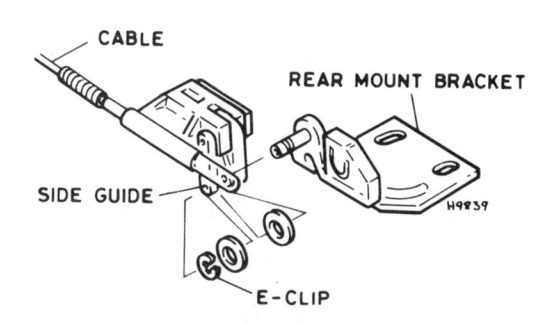

Fig. 11.26 Sunroof cable mounting and rear mount assembly (Sec 31)

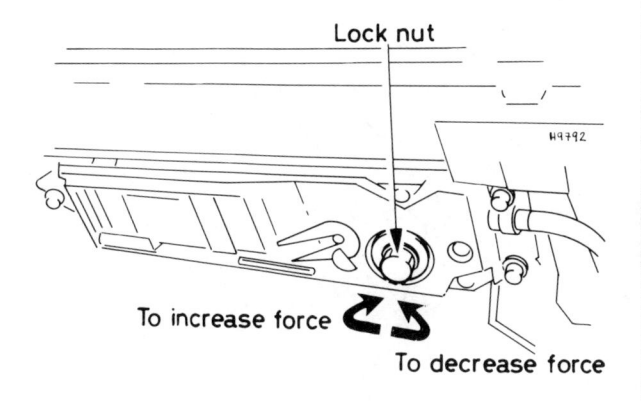

Fig. 11.27 Checking sunroof closing force (Sec 31)

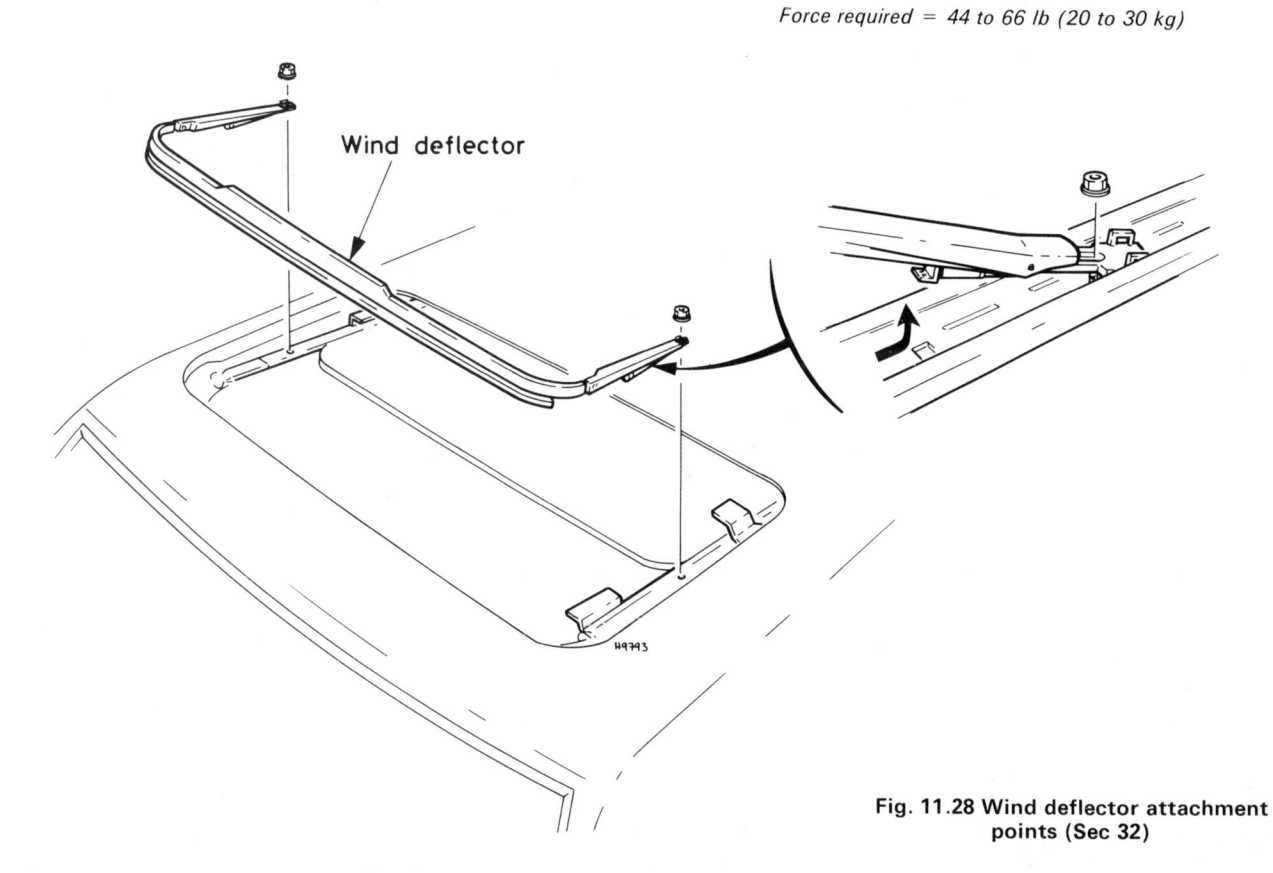

Force required = 44 to 66 lb (20 to 30 kg)

Fig. 11.28 Wind deflector attachment points (Sec 32)

33 Interior trim – general

1 The interior trim sections are shown in Figs. 11.29 and 11.30.
2 They are held in place by a variety of different fixings and it is a good idea to build up a stock of spare fixings, as these are easily broken when removing trim panels.
3 Care should be exercised when removing or fitting trim panels, as these too are easily broken.
4 The carpeting too, is held in by plastic fittings, and the same comments apply here.
5 The headliner is held in place by concealed spring bars in the roof, and it can be removed after taking out the rear view mirror, courtesy light and all necessary trim.
6 The edges of the headliner are adhesive backed and fit over the weld edges all around the interior.
7 It is recommended that headliner replacement be undertaken by your local dealer, as a satisfactory fit will be very difficult to obtain.

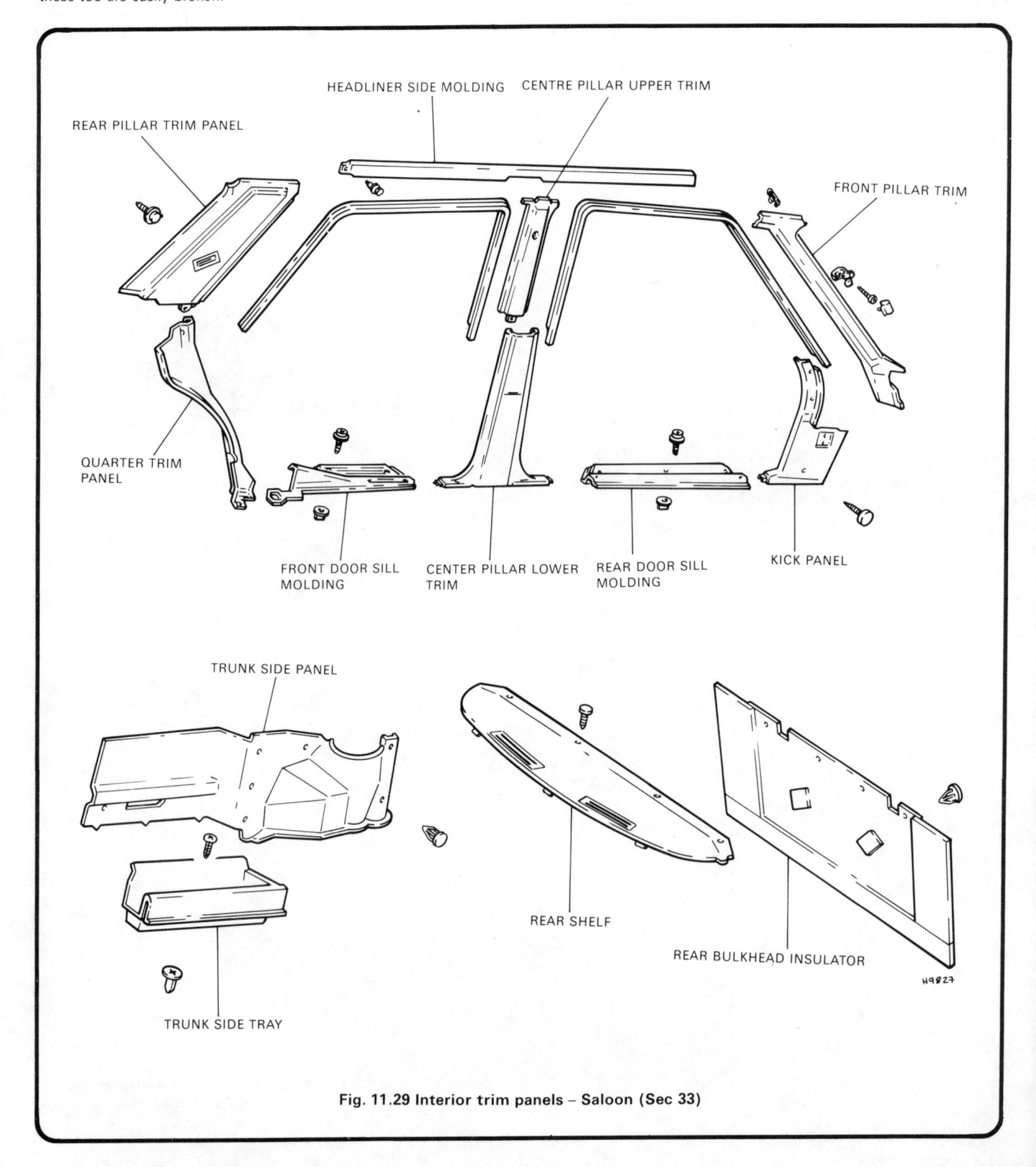

Fig. 11.29 Interior trim panels – Saloon (Sec 33)

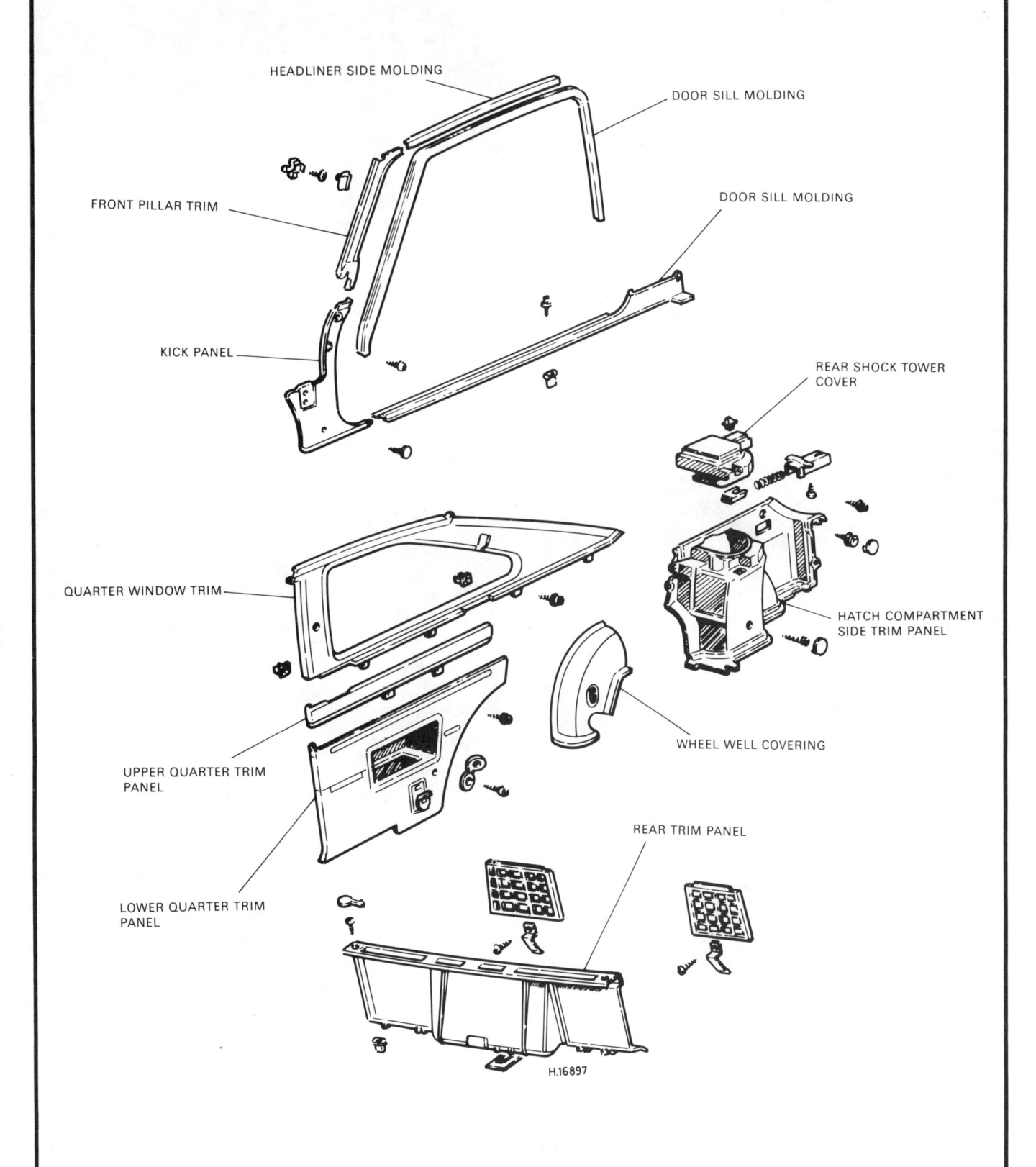

Fig. 11.30 Interior trim panels – Hatchback (Sec 33)

34 Centre console – removal and refitting

1 Remove the screw from the rear cigar lighter panel, remove the panel and disconnect the cigar lighter (photos).
2 Remove the two screws from the rear edge of the console and the screws at either side of the front end, and remove the console (photos).
3 Refit in the reverse order.

35 Front console – removal and refitting

1 Remove the centre console, as described in Section 34.
2 Remove the gear lever knob or automatic selector lever handle.
3 On models equipped with remote headlight beam adjusters, remove the adjuster knob, plate and retaining screws.
4 Remove the two retaining screws or each side of the console (photo).
5 Lift the console clear of the gear lever, and disconnect the electrical leads, before lifting the console out.

34.1A Cigar lighter panel retaining screw

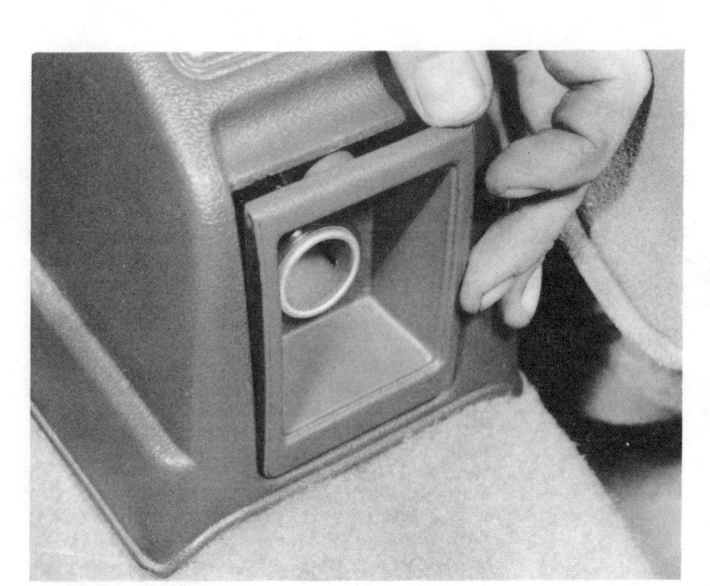

34.1B Lift out the panel ...

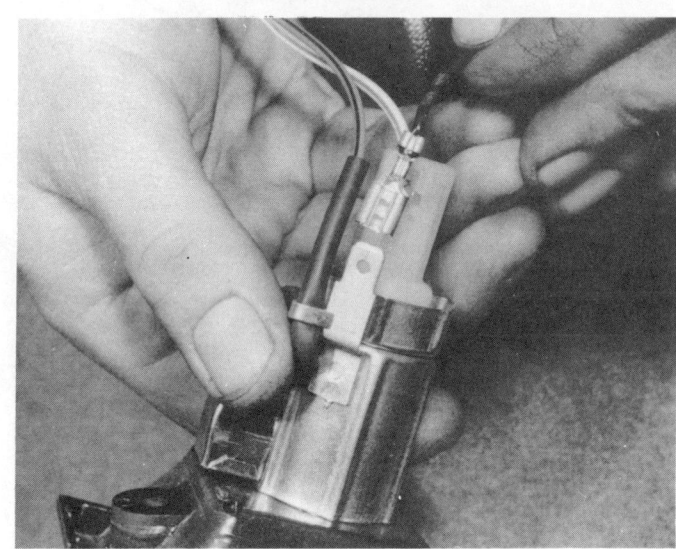

34.1C ... and disconnect the leads

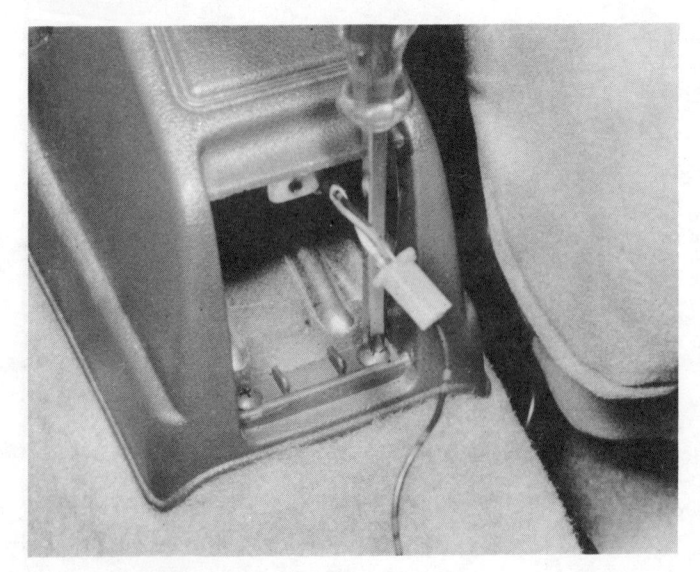

34.2A Remove the screws from the rear edge ...

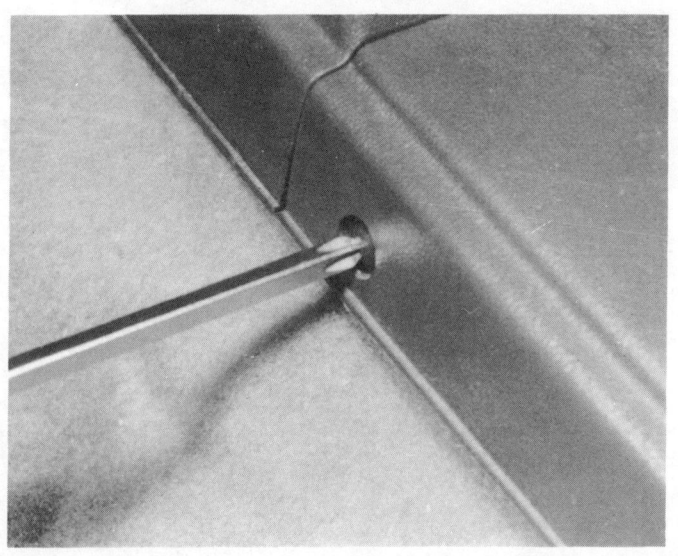

34.2B ... and either side of the front edge

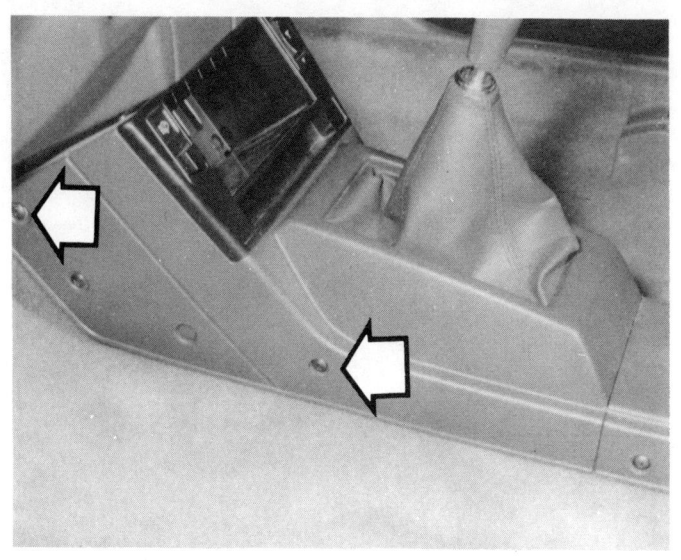

Chapter 11 Bodywork and fittings

35.4 Front console retaining screws (arrowed)

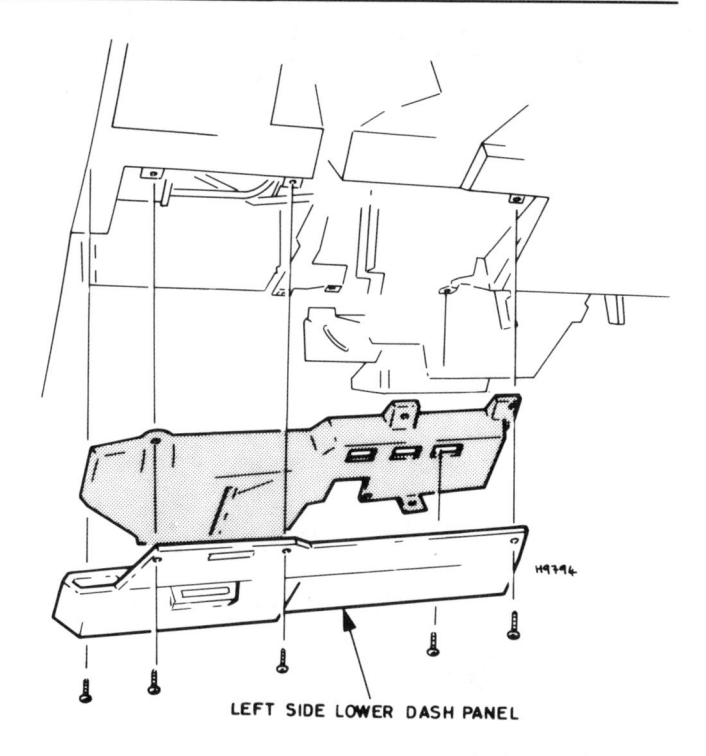

Fig. 11.31 Lower dash panel fixing screws (Sec 36)

36 Dashboard – removal and refitting

Note: *refer to Chapter 12 for removal of the instrument panel and to Sections 43 and 49 of this Chapter for heater and air conditioner. The instructions given here are for left-hand drive models and should be reversed for right-hand drive, where necessary.*

1 Remove the left-hand lower dash panel and fusebox cover.
2 Refer to Chapter 10 and lower the steering column.
3 Disconnect all instrument cables at their connectors.
4 Remove the heater control panel and bracket.
5 Remove the centre access panel and clock.
6 Remove the ashtray.
7 Remove the dashboard mounting bolts, and, supporting the dashboard, lift and pull it from the central guide pin, and manoeuvre it out of the vehicle.
8 Refitting is a reversal of removal.

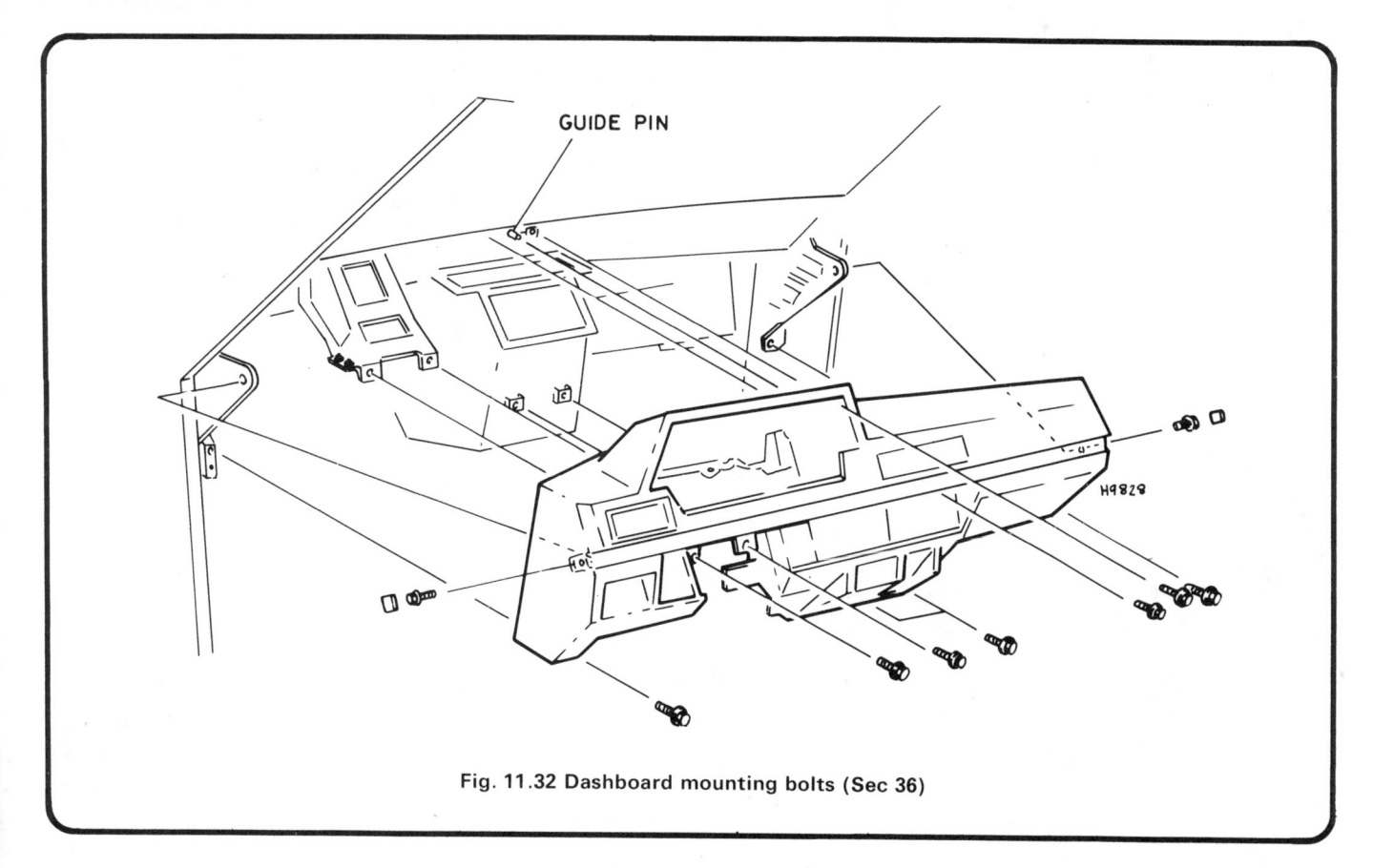

Fig. 11.32 Dashboard mounting bolts (Sec 36)

Chapter 11 Bodywork and fittings

37 Front seats – removal and refitting

1 The front seats are mounted on rails which are bolted to the floorpan.
2 To remove the seats, undo the four bolts (photo).
3 The sliding rails, which allow for fore and aft adjustment, can be removed by removing their attachment bolts (photo).
4 Access to the seat back adjuster is by removing the cover panel (photos).
5 Apply general purpose grease to all sliding/moving parts before refitting in the reverse order.

38 Rear seat (Saloon) – removal and refitting

1 On Saloon versions, pull the upholstery away from the middle of the seat back bottom edge to reveal the centre mounting bolt, and remove the bolt.
2 The seat squab can now be raised, back edge first, unhooked from the front fastenings, and removed (photo).
3 Remove the remaining two bolts from the seat back.
4 Push on the middle of the top edge of the seat back, pulling upward firmly with the other hand to dislodge the seat back from the centre bracket (photo).

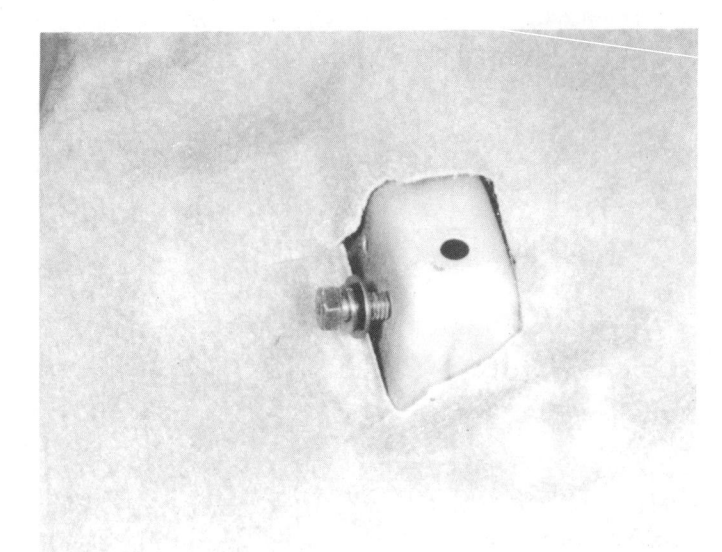

37.2 A front seat mounting point in the floorpan

37.3 Sliding rail-to-seat attachment bolt

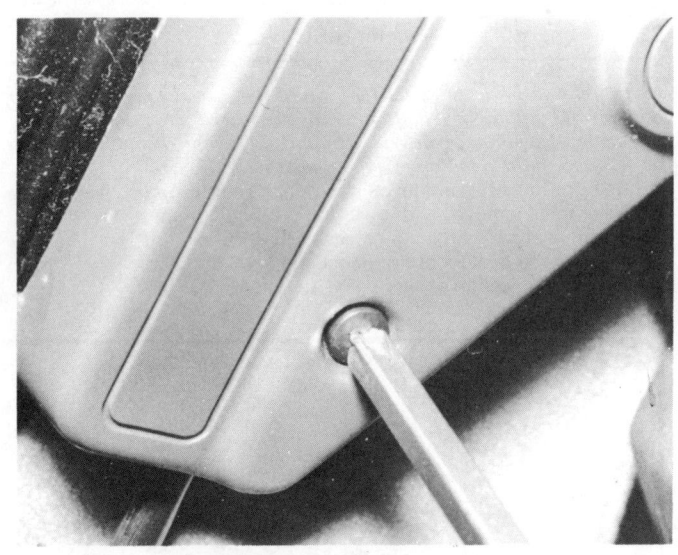

37.4A Removing the seat back adjuster cover panel ...

37.4B ... to reveal the adjuster mechanism

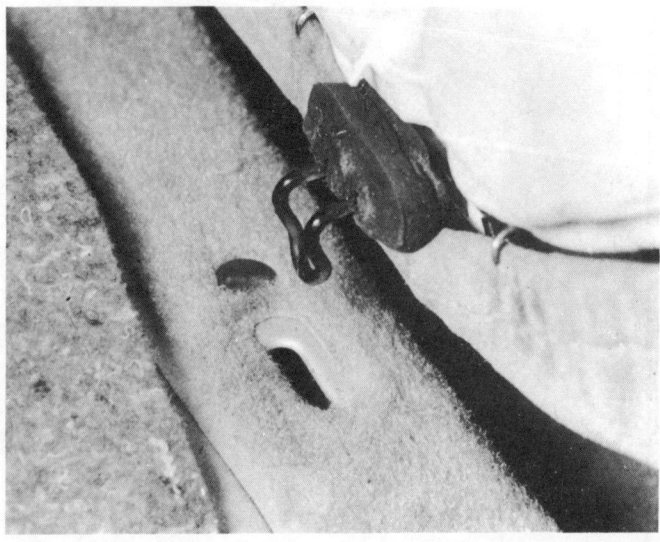

38.2 The seat squab front hooks

Chapter 11 Bodywork and fittings

5 Lift the seat belt out of the vehicle.
6 Refit in the reverse order.

39 Rear seat (Hatchback) – removal and refitting

1 Pull the seat back forward.
2 Remove the carpet from the rear seat by removing the retaining clips.
3 Remove the plastic collar from the seat pivot.
4 Slide each seat toward the centre to release the spigot from the pivot bracket and lift the seat backs out.
5 The seat squab is removed in the same way as for the Saloon, after having threaded the seat belt buckles through the seat cushion.
6 Refit in the reverse order.

40 Seat belts – general

1 Check the condition of all seat belts regularly. If they become frayed or cut, or are contaminated by oil or grease, they should be renewed.

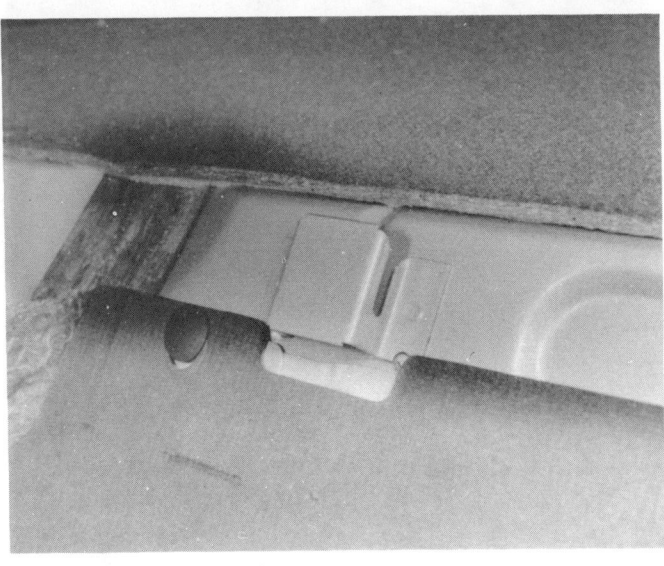

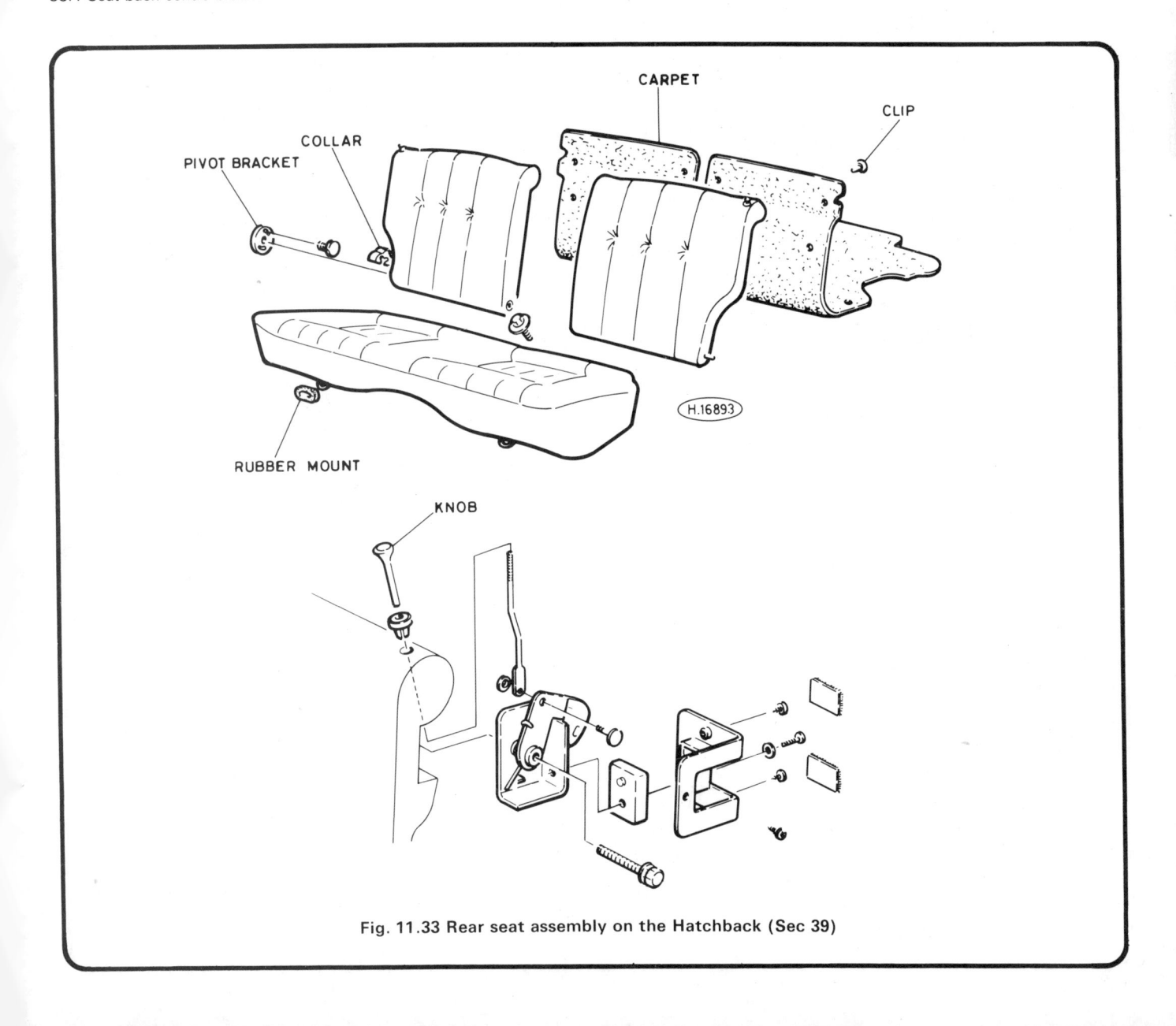

38.4 Seat back centre bracket

Fig. 11.33 Rear seat assembly on the Hatchback (Sec 39)

286

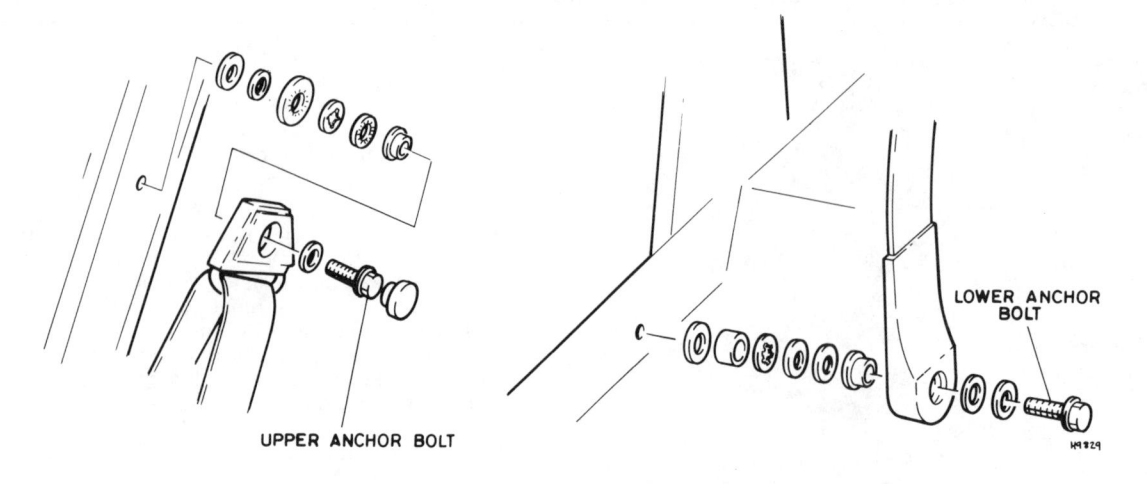

Fig. 11.34 Front seat belt upper and lower mounting points (Sec 40)

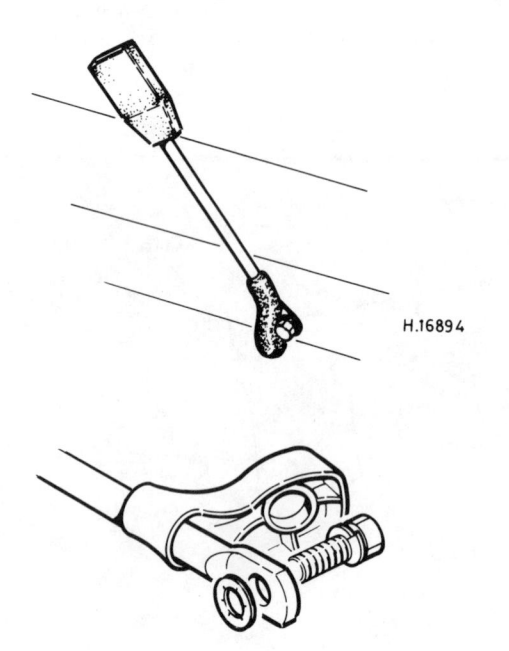

Fig. 11.35 Front seat belt seat anchor strap (Sec 40)

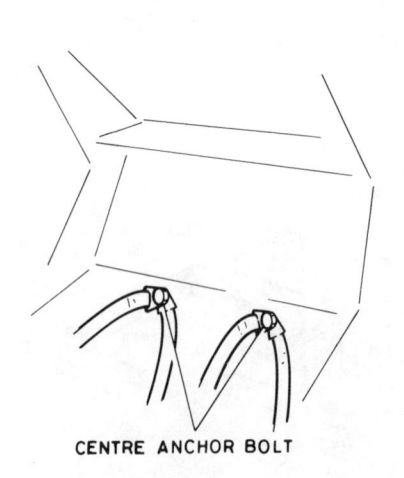

Fig. 11.36 Rear seat inertia reel mounting (Sec 40)

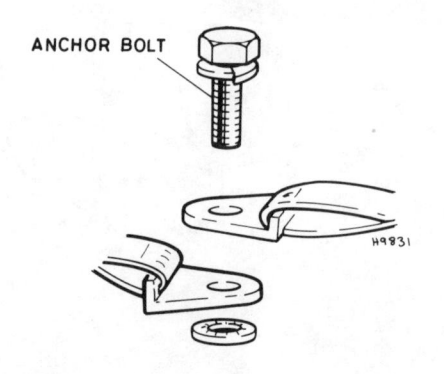

Fig. 11.37 Rear seat belt centre mounting (Sec 40)

Chapter 11 Bodywork and fittings 287

2 The belts may be cleaned by using warm water and a mild detergent. Allow the belts to dry before letting them retract.
3 The belts are anchored to strong points in the vehicle frame by single belts, and if they are removed make sure all washers and spacers go back in their original positions.
4 Lightly oil the anchor bolt assemblies occasionally, but do not allow oil onto the webbing.
5 The inertia reel assemblies are bolted to frame strong points and can be removed by uncovering the trim panels or seats as necessary and removing the mounting bolts.

41 Interior rear view mirror – removing and refitting

1 The interior rear view mirror is clipped into its base support, which in turn is screwed to the roof lining.
2 To remove the mirror, first remove the rubber damper between the mirror and windscreen.
3 Slide the mirror backwards to detach it from the base.
4 Remove the two screws and remove the mirror base.
5 Refitting is a reversal of removal, but note that the cut-out in the base faces toward the windscreen.

42 Heater and ventilation – general description

1 The heater has at its core a matrix which is, in effect a heat exchanger, which is fed hot coolant from the engine cooling system.
2 Air is drawn through the matrix by an electric fan, where it is heated in the process, and is then distributed to the interior of the car via

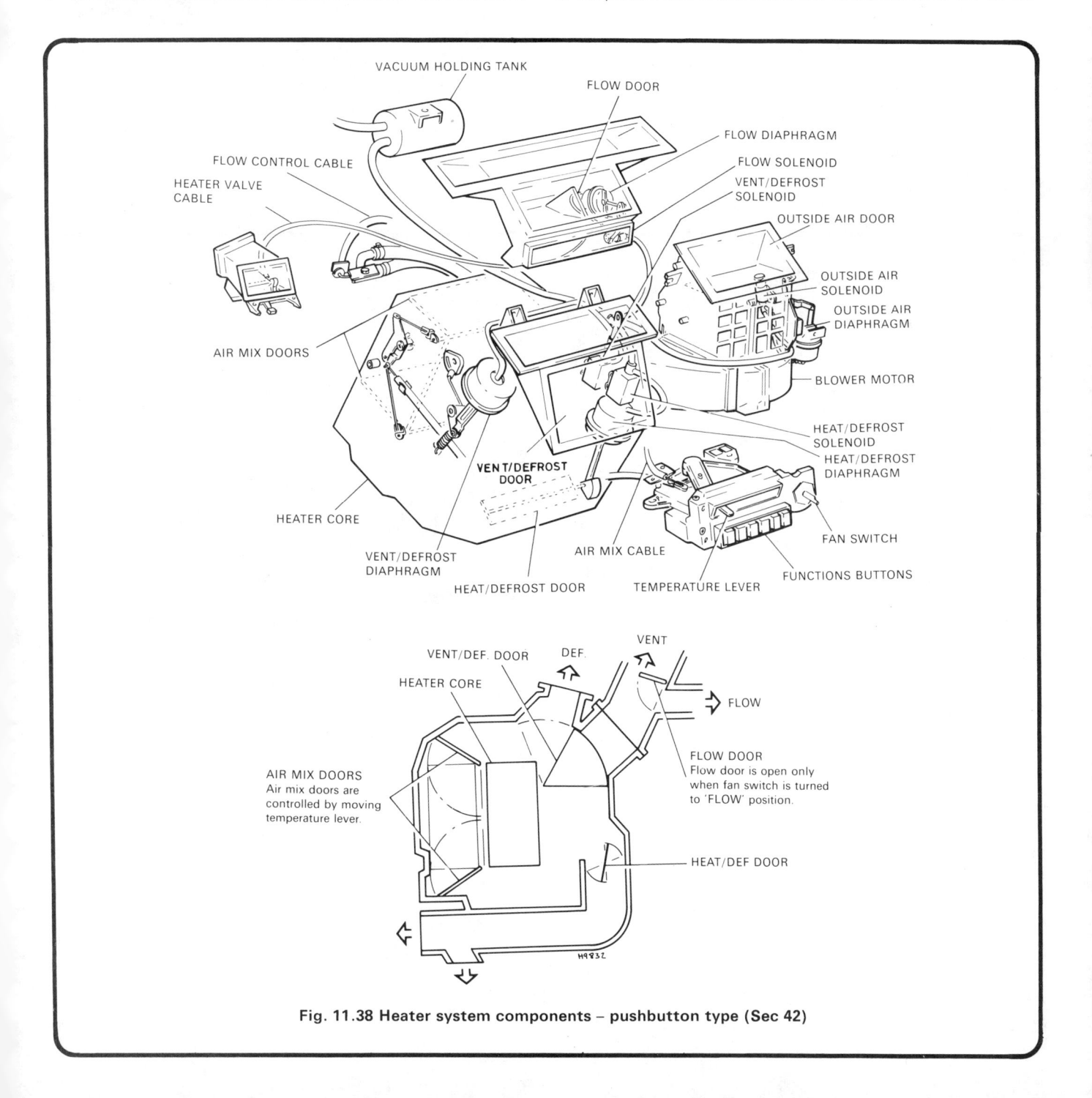

Fig. 11.38 Heater system components – pushbutton type (Sec 42)

ducting, and directed to the feet or windscreen as the controls are set.
3 Fresh air can also be drawn in by the fan through the grille at the base of the windscreen, and is again directed through the ducting by the controls.
4 Thus any degree of hot/cold air can be mixed and delivered to the interior of the car, according to how the controls are set.
5 Stale air is exhausted automatically through grilles in the rear support pillars.
6 There are two types of control panel, the lever type and the pushbutton.

43 Heater unit – removal and refitting

1 Drain the cooling system, as described in Chapter 2. Although this will save a lot of coolant, there will still be coolant in the heater unit, so be prepared for spillage when the heater unit is removed.
2 Disconnect the heater hoses, first the connections in the engine bay and, when all coolant has drained, those inside the vehicle.
3 Disconnect the control cable to the heater valve at the valve in the engine bay (photo).
4 Remove the heater unit lower mounting nut in the engine bay.
5 Remove the heater control panel as described in Section 46.
6 Remove the dashboard, as described in Section 36.
7 Remove those sections of the heater ducting necessary to remove the heater unit.

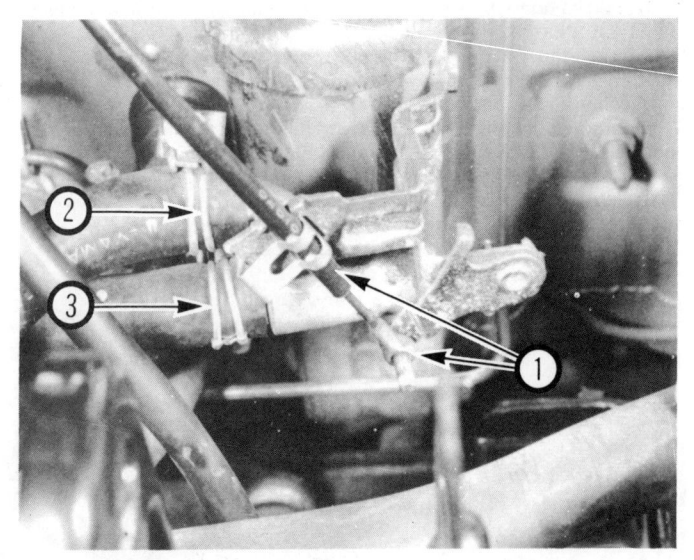

43.3 The heater water control valve
1 Cable end and clip
2 Inlet pipe to heater
3 Outlet pipe from heater

Fig. 11.39 External components of the heater unit (Sec 43)

Chapter 11 Bodywork and fittings

8 Disconnect the heater function cable and the air mix cable, as applicable.
9 Prise out the retaining clips and remove the floor ducts.
10 Remove the heater mounting nuts and pull the heater away from the bulkhead.
11 Disconnect the vacuum tube, and lift the heater unit from the vehicle.
12 Further dismantling of the heater is by removing the casing clips and splitting the unit in two.
13 Refitting is a reversal of removal, adjust the cables as described in Section 46, and fill and bleed the cooling system, as described in Chapter 2.

44 Heater blower – removal and refitting

1 Remove the glovebox.
2 Remove the floor ducting and moulded insulator pad.
3 Remove the blower duct.
4 Remove the three blower motor mounting bolts and lower the unit.
5 Disconnect the wiring connector and vacuum tube from the rear of the blower and remove the unit.
6 Refitting is a reversal of removal.

45 'Mild flow' assembly – removing and refitting

1 The 'mild flow' vent allows a trickle flow of fresh air to flow to the horizontal vent across the dashboard.
2 To remove the assembly, which is hidden away behind the dashboard, the dashboard must first be removed.
3 Remove the components of the 'mild flow' ducting in the order shown in Fig. 11.42.
4 Refit in the reverse order and adjust the cable as described in Section 46.
5 The 'mild flow' control cable should be adjusted so that there is no play in the cable when the air flap is fully closed.

46 Heater controls and cables – removal, refitting and adjusting

Air mix cable
1 Slide the temperature lever to HOT. Release the cable conduit spring clamp.
2 Pull the arm of the air mix door (which is located in front of the heater matrix) upwards.
3 Slide the cable conduit away from the end of the inner cable to eliminate any slack, but not enough to move the control lever. Secure the conduit with the clamp.

Heater control valve cable
4 Slide the temperature control lever to COLD.
5 Close the heater valve with the fingers.
6 Release the cable conduit spring clamp.
7 Clamp the cable conduit with the spring clamp.

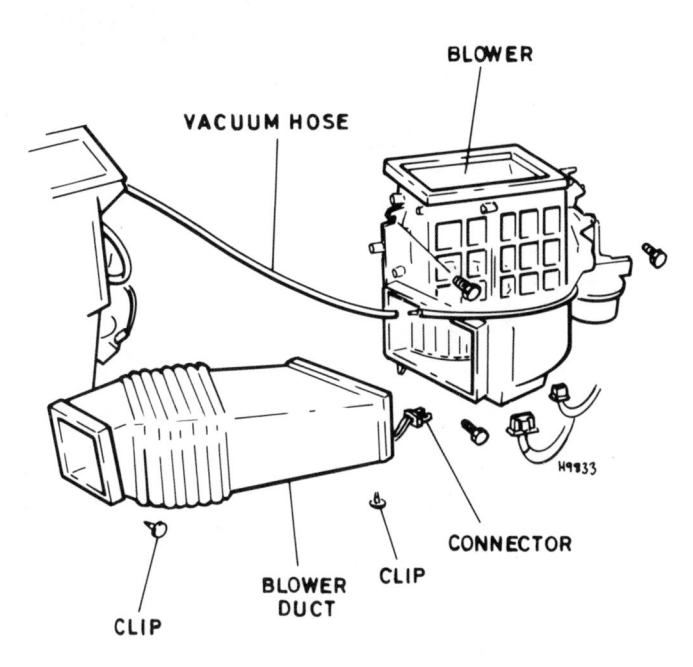

Fig. 11.40 Heater blower assembly (Sec 44)

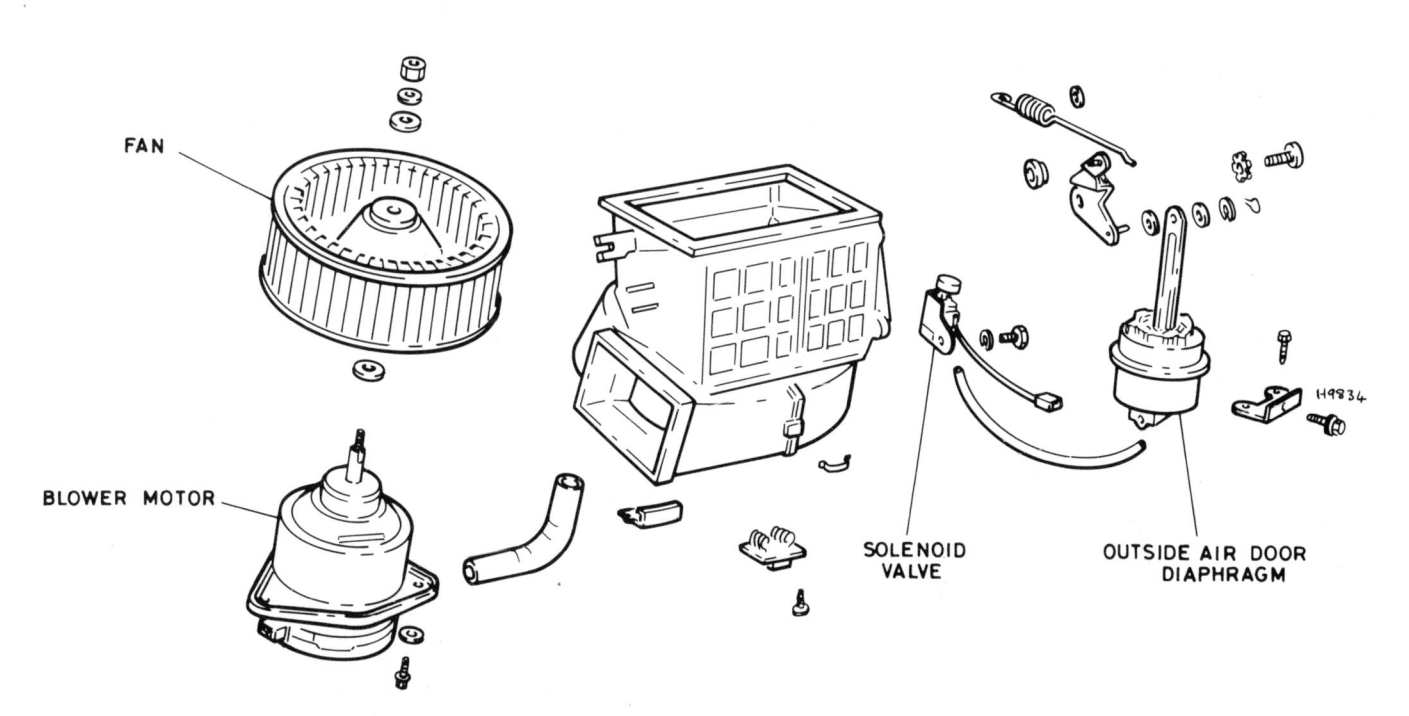

Fig. 11.41 Heater blower motor dismantled (Sec 44)

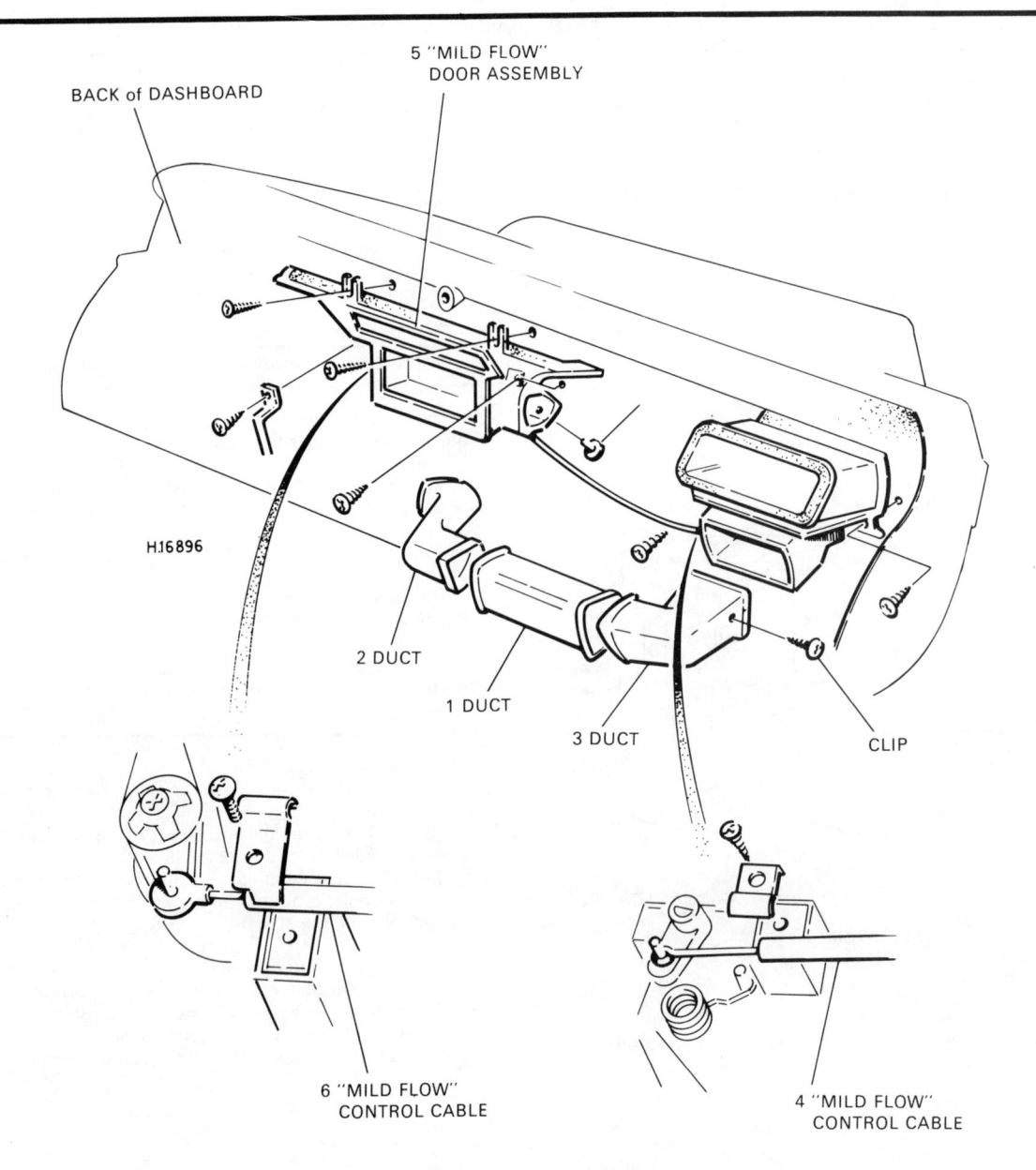

Fig. 11.42 'Mild flow' ducting (Sec 45)

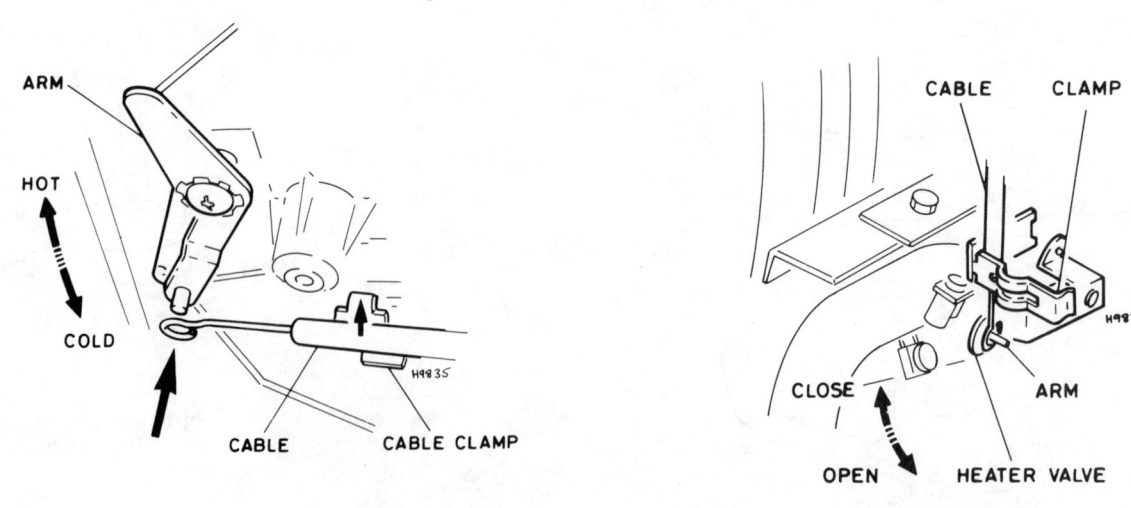

Fig. 11.43 Air mix cable (Sec 46)

Fig. 11.44 Heater control valve cable (Sec 46)

291

Fig. 11.45 Heater function cable (Sec 46)

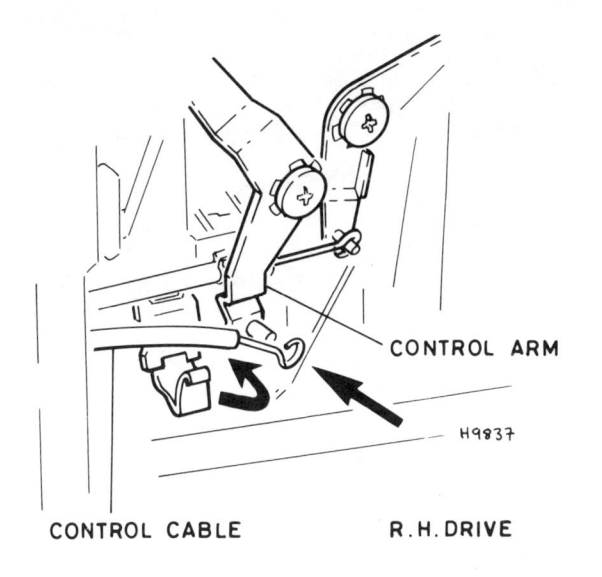

Fig. 11.46 Lever type control panel (Sec 46)

Fig. 11.47 Pushbutton type control panel (Sec 46)

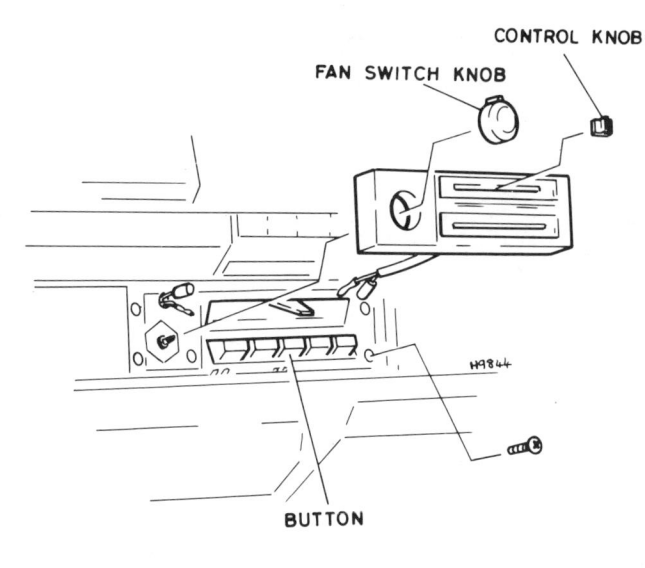

Fig. 11.48 Heater lever type controls (Sec 46)

1 Blower switch
2 Air direction (function) lever
3 Temperature lever

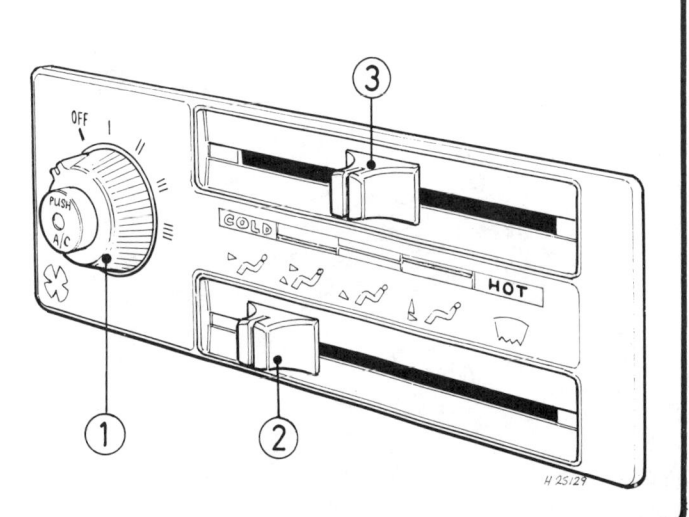

Chapter 11 Bodywork and fittings

'Mild flow' cable
8 Close the air flap fully, release the cable clamp screw.
9 Slide the cable conduit away from the end of the inner cable to eliminate any slack then tighten the cable clamp screw.

Function (air distribution) cable
10 Slide the function lever to VENT.
11 Release the cable clamp.
12 Set the control arm on the heater to VENT.
13 Clamp the cable.

Lever type heater control panel – removal and refitting
14 Pull off the control knobs.
15 Prise the heater control panel escutcheon out with the screwdriver. Release the cable clips.
16 Disconnect the wiring plugs and remove the panel fixing screws. Remove the panel.
17 Refitting is a reversal of removal.

Pushbutton type heater control panel – removal and refitting
18 Pull off the control knobs.
19 Prise out the control panel escutcheon with a screwdriver.
20 Extract the fixing screws, disconnect the wiring plugs and the cable clamps and withdraw the control panel.
21 Refitting is a reversal of removal.

47 Vacuum-operated components – general

1 The outside air door, 'mild flow' door and vent/defrost doors are all additionally controlled by vacuum and electrical solenoid circuits (see Fig. 11.39).
2 The vacuum is obtained from the inlet manifold and 'stored' in a vacuum holding tank located under the scuttle in front of the windscreen.
3 The system requires specialist equipment to test and adjust, and, if any of these components malfunction, it is best left to your dealer.

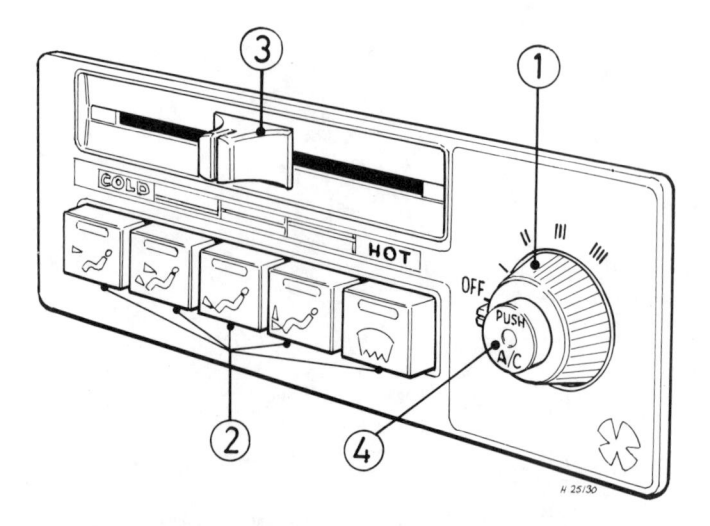

Fig. 11.49 Heater pushbutton type controls (Sec 46)

1 Blower switch
2 Air direction (function) button
3 Temperature lever
4 Air conditioner switch (optional)

48 Air conditioner – description and maintenance

1 The air conditioner is available as a factory-fitted option on certain models. The main components consist of a compressor, a condenser, an evaporator and a receiver/dryer.
2 It is not recommended that the system is discharged by the home mechanic.
3 Once the system has been discharged professionally, however,

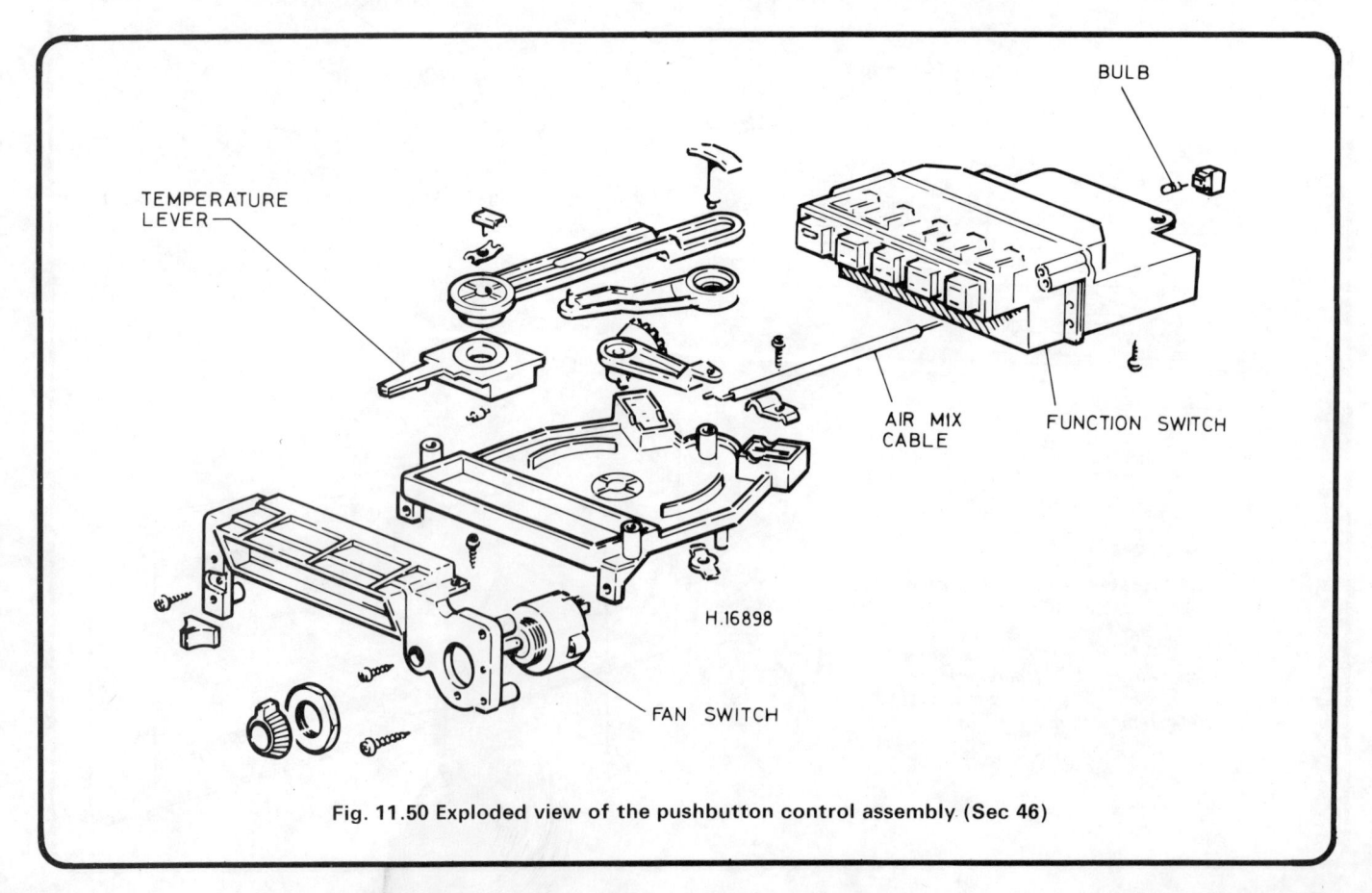

Fig. 11.50 Exploded view of the pushbutton control assembly (Sec 46)

Chapter 11 Bodywork and fittings 293

BLOWER
Forces air thru the evaporator

EVAPORATOR
As refrigerant circulates, heat is absorbed from the surrounding passenger compartment air

Sight glass.

COMPRESSOR
Compresses the refrigerant and then forces it thru the condenser

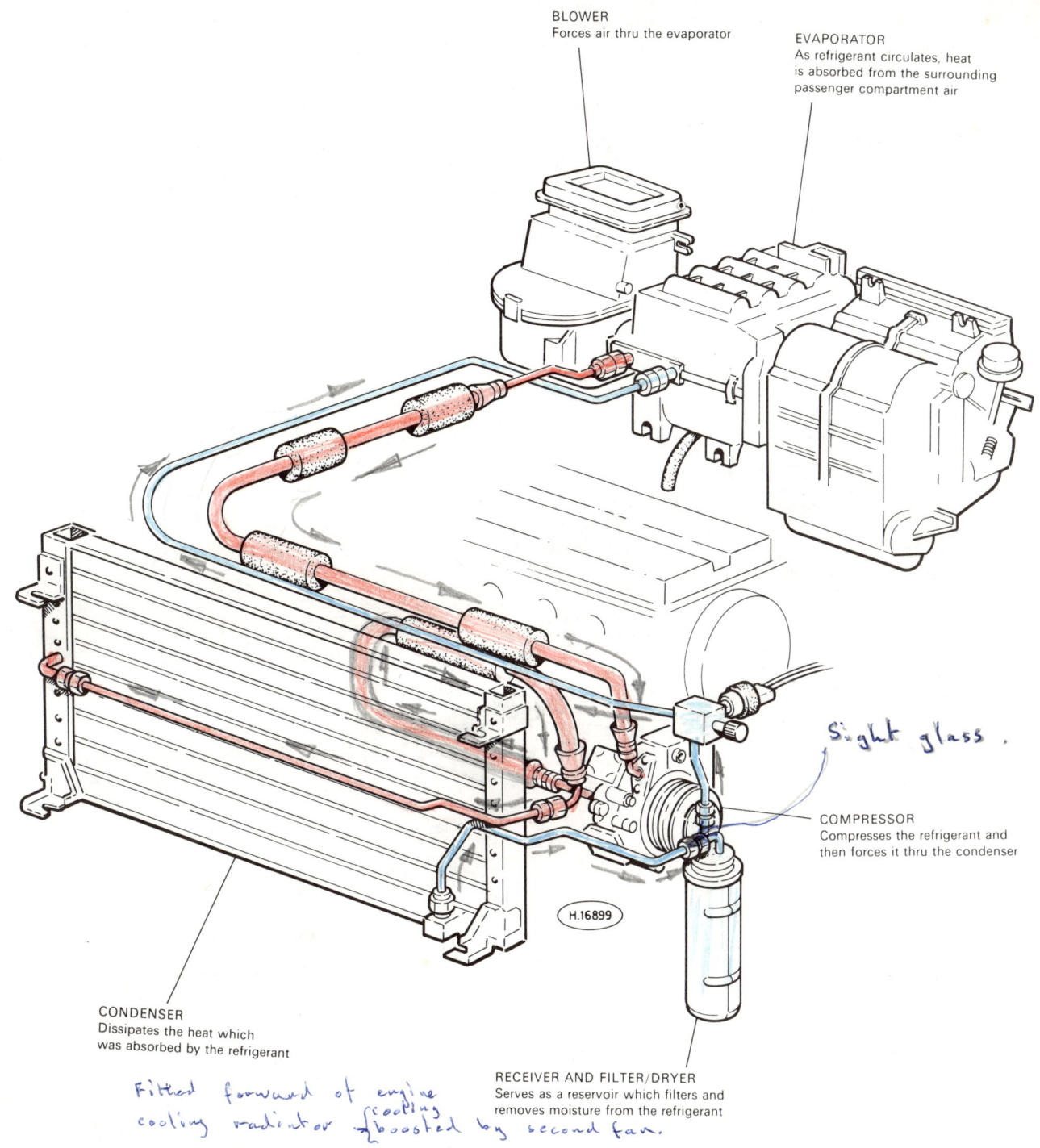

CONDENSER
Dissipates the heat which was absorbed by the refrigerant

Fitted forward of engine cooling radiator cooling boosted by second fan.

RECEIVER AND FILTER/DRYER
Serves as a reservoir which filters and removes moisture from the refrigerant

Fig. 11.51 Components of the air conditioner system – North American type shown (Sec 48)

there is no reason why the individual components cannot be removed and new ones fitted provided the open lines or ports are sealed immediately after disconnection. The seals on new units should not be removed until just before they are to be connected to the system. These precautions are to prevent the admission of moisture.
4 To keep the system in perfect order, carry out the following maintenance items.
5 Regularly brush the fins of the condenser free from dirt and flies.
6 Keep the compressor drivebelt in good condition and correctly tensioned with a deflection of 0.4 to 0.5 in (10 to 12 mm) at the mid-point of the belt run.
7 Occasionally, start the engine and allow it to run at fast idle for a few minutes while looking through the sight glass on the receiver dryer. The appearance of continuous bubbles or foam indicates a low refrigerant level which should be rectified by your dealer or refrigeration engineer.
8 During the winter months, operate the air conditioner for ten minutes once a week to lubricate the seals and the interior of the compressor.

49 Air conditioner compressor – removal and refitting

1 Start the engine, allow it to idle and switch on the air conditioner for a few minutes.

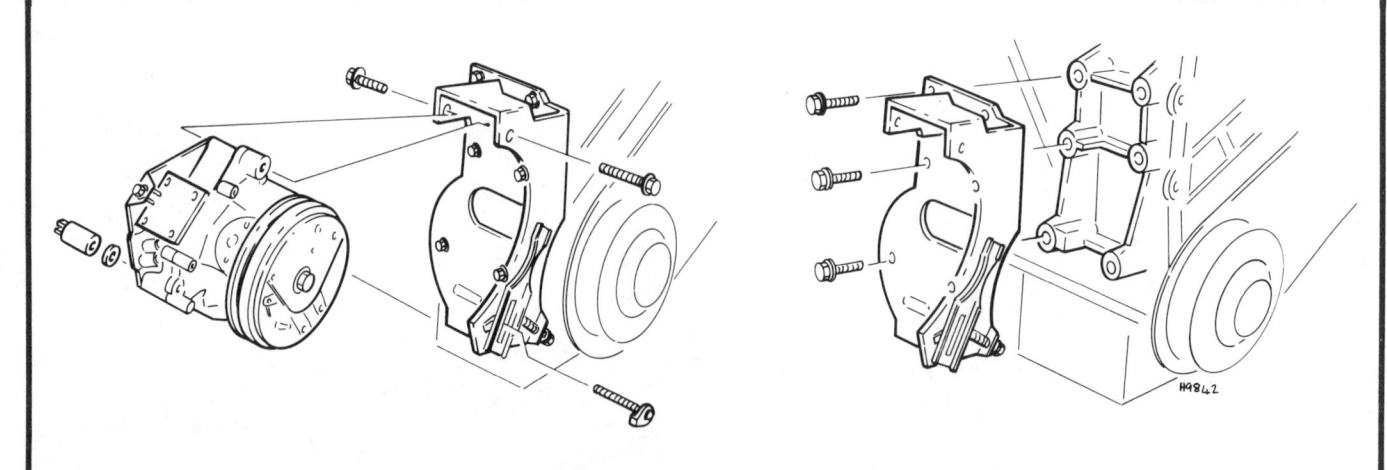

Fig. 11.52 The air conditioner compressor and mounting bracket (Sec 49)

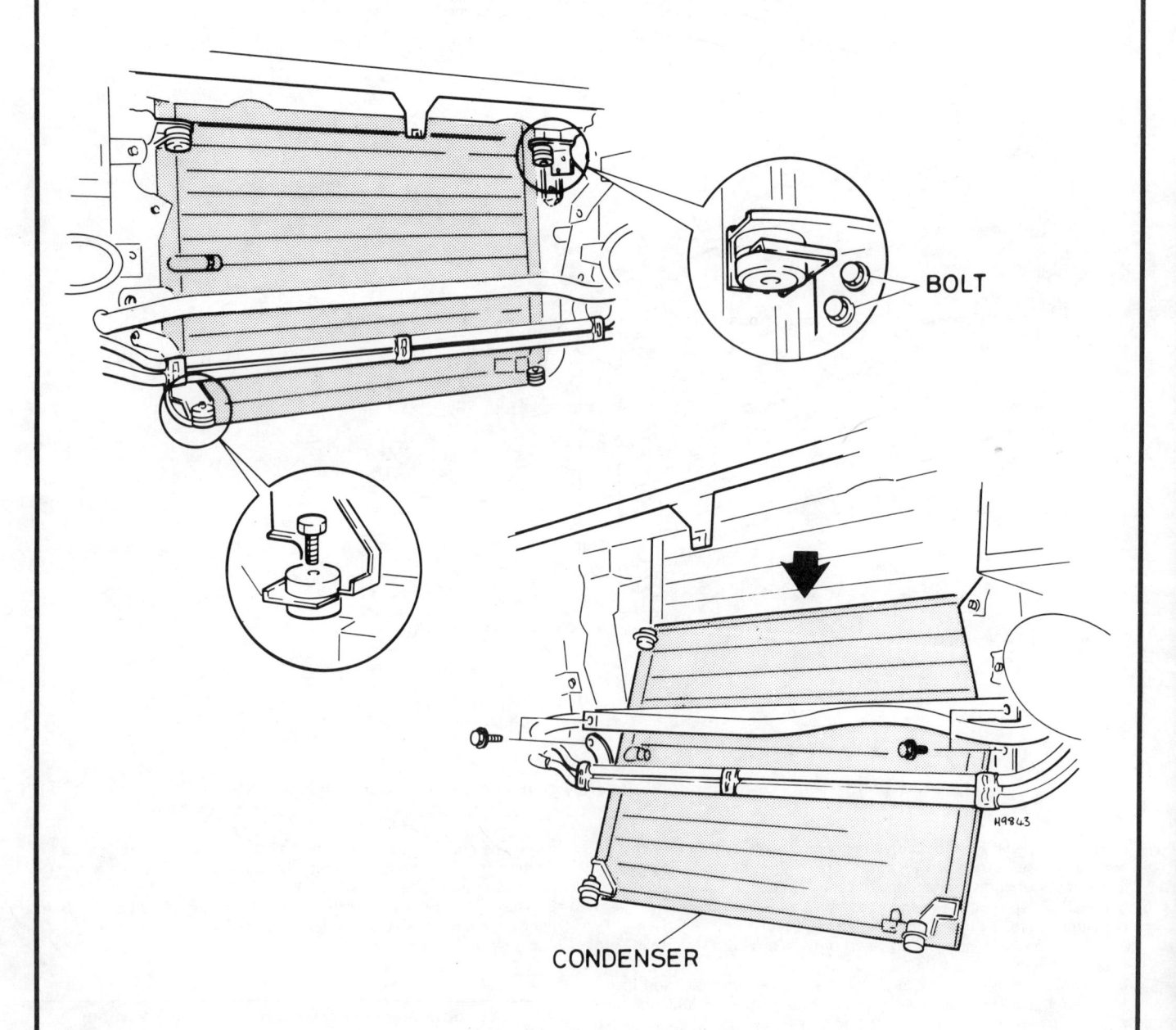

Fig. 11.53 Removing the air conditioner condenser (Sec 50)

Chapter 11 Bodywork and fittings

2 Switch off the engine and air conditioner and disconnect the lead from the battery negative terminal.
3 Disconnect the compressor clutch lead.
4 Have the system discharged by your dealer or a competent refrigeration engineer.
5 Disconnect the suction and discharge hoses from the compressor. Cap all openings immediately.
6 Loosen the compressor adjusting and mounting bolts and remove the belt cover.
7 Remove the compressor drivebelt.
8 Remove the mounting bolts and lift the compressor from its mountings.
9 Refit by reversing the removal operations. If a new compressor is being fitted, pour 0.95 fl oz (30.0 cc) through the suction port on the compressor. Tighten all fixings.
10 Adjust the drivebelt and have the system recharged. Adjustment of the drivebelt will be found in Chapter 2.

50 Air conditioner condenser – removal and refitting

1 Disconnect the negative lead from the battery.
2 Have the system discharged professionally.
3 Disconnect the pipeline and hose from the condenser. Cap the openings immediately.
4 Extract the fixing screws and remove the front skirt.
5 Remove the condenser mounting bolts and withdraw the condenser downwards.
6 Refitting is a reversal of removal, but if a new condenser is being fitted, pour 0.3 fl oz (10.0 cc) of refrigerant oil into it.
7 Tighten all fixings. Have the system recharged.

51 Air conditioner evaporator – removal and refitting

1 Disconnect the lead from the battery negative terminal.
2 Have the system discharged professionally.
3 Disconnect the pipe line and hose from the evaporator and immediately cap all openings.
4 Remove the grommets.
5 Remove the glove box and its frame.
6 Unscrew the three evaporator mounting bolts.
7 Refitting is a reversal of removal. Make sure the sealing bands are securely fitted, tighten all fixings.
8 Have the system recharged.

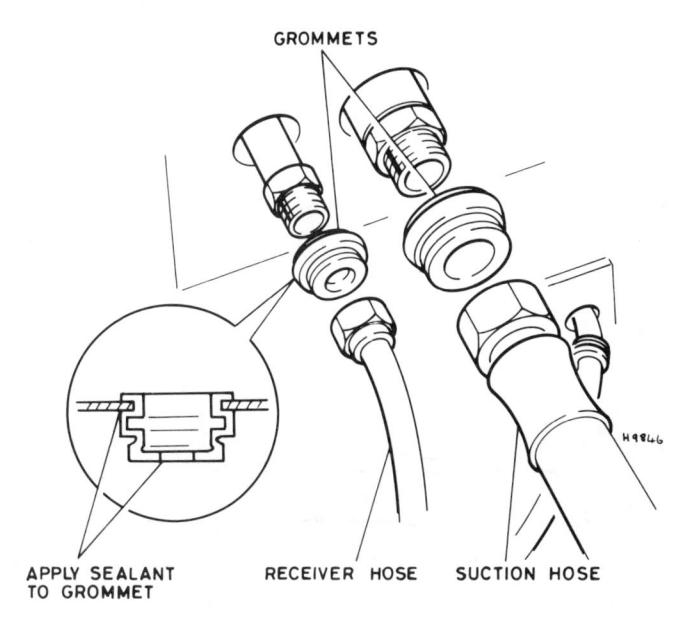

Fig. 11.54 Evaporator hose connections (Sec 51)

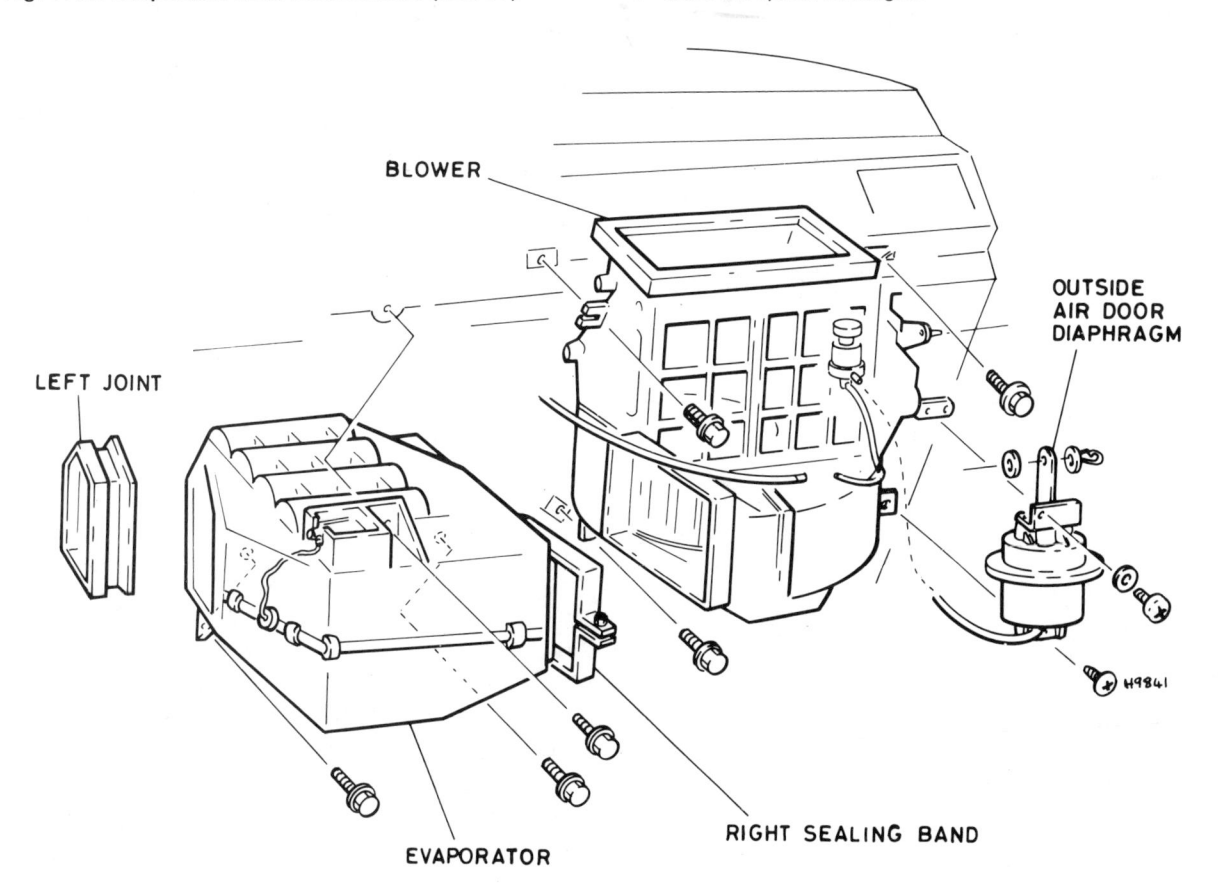

Fig. 11.55 Evaporator, blower and seals (Sec 51)

Chapter 12 Electrical system

Contents

Alternator – description, maintenance and precautions	6
Alternator – overhaul	9
Alternator – removal and refitting	8
Alternator – testing *in situ*	7
Battery – charging	5
Battery – inspection	4
Battery – removal and refitting	3
Cigar lighter – removal and refitting	22
Cruise control system – general	30
Fault diagnosis	34
Fuses, relays and control units – general	12
General description	1
Headlamp beam – adjusting	17
Headlamp bulbs – removal	16
Headlamps – removal and refitting	15
Headlamp wiper unit – removal and refitting	28
Instrument panel – removal and refitting	13
Interior lamps – removal, refitting and bulb renewal	21
Mobile radio equipment – interference-free installation	25
Optional systems – general	33
Parking light and side indicator light – removal and refitting	18
Radio – removal and refitting	23
Radio aerial – removal and refitting	24
Rear lamp cluster – removal, refitting and bulb renewal	19
Routine maintenance	2
Speedometer cable – removal and refitting	31
Starter motor – description	10
Starter motor – testing *in situ*	11
Steering wheel and column switches – removal and refitting	14
Supplementary exterior lights – removal, refitting and bulb renewal	20
Warning lights – general	32
Windscreen wiper and headlamp wash system – general	29
Windscreen wiper blades and arms – removal and refitting	26
Windscreen wiper motor – removal and refitting	27
Wiring diagrams – general	35

Specifications

System type .... 12 volt, negative earth

Battery
Capacity:
- UK models .... 50 amp hr
- North American models .... 47 amp hr

Alternator
- Rating .... 65 amp
- Output at 5500 rev/min .... 14 volts
- Minimum brush length .... 0.20 in (5.0 mm)

Starter motor
- Type .... Pre-engaged
- Output rating (UK):
  - Nippondenso .... 0.8, 1.0 or 1.4 kW
  - Mitsuba .... 1.0 kW or 1.4 kW
  - Hitachi .... 0.8 kW
- Output rating (North America):
  - Nippondenso .... 1.4 kW
  - Mitsuba .... 1.4 kW

Chapter 12 Electrical system

Fuses

UK models:	
Saloon	6 x 20 amp, 10 x 15 amp, 6 x 10 amp
Hatchback	1 x 20 amp, 7 x 15 amp, 6 x 10 amp
North American models:	
Sedan:	
STD	4 x 20 amp, 8 x 15 amp, 9 x 10 amp
LX and SEi	5 x 20 amp, 12 x 15 amp, 9 x 10 amp
Hatchback	4 x 20 amp, 8 x 15 amp, 9 x 10 amp
Canadian models:	
Sedan	6 x 20 amp, 10 x 15 amp, 6 x 10 amp
Hatchback	1 x 20 amp, 7 x 15 amp, 6 x 10 amp
Main fuse:	
UK models	65 amp
North American models	65 amp
ALB (anti-lock brake) fuse	35 amp
Sunroof	45 amp
Air conditioning	45 amp

Torque wrench settings

	lbf ft	Nm
Alternator top bracket bolt	16	22
Alternator adjusting bolt	16	22
Alternator mounting-to-bracket bolt	33	45
Alternator lower bracket to cylinder block	33	45
Alternator pulley bolt	53	72

1 General description

The major components of the 12 volt negative earth system consist of a 12 volt battery, an alternator (driven from the crankshaft pulley), and a starter motor.

The battery supplies a steady amount of current for the ignition, lighting and other electrical circuits and provides a reserve of power when the current consumed by the electrical equipment exceeds that being produced by the alternator.

The alternator has its own regulator which ensures a high output if the battery is in a low state of charge and the demand from the electrical equipment is high, and a low output if the battery is fully charged and there is little demand from the electrical equipment.

When fitting electrical accessories to cars with a negative earth system it is important, if they contain silicon diodes or transistors, that they are connected correctly, otherwise serious damage may result to the components concerned. Items sich as radios, tape players, electronic ignition systems, electronic tachometer, automatic dipping etc. should all be checked for correct polarity.

2 Routine maintenance

At the intervals given in Routine Maintenance (at the front of the Manual) the following tasks should be undertaken.
1 Check the battery electrolyte level, or the battery condition indicator. Top up as necessary. Check that the horn and all lights are working. Renew the bulbs as necessary. Check the condition and security of all wiring looms, connectors and terminals.
2 Check the condition and tension of the alternator drivebelt.

3 Battery – removal and refitting

1 The battery is mounted in the front right-hand corner of the engine bay.
2 It is held to the battery tray by a clamp (photo).
3 Whenever the vehicle is being worked on, and the task in hand does not require the battery to be in use, disconnect the battery.
4 Disconnect the battery earth strap first (negative terminal), then the positive terminal.
5 When working in the engine compartment, it is advisable to remove the battery altogether, which precludes the possibility of tools being laid across the two terminals.
6 To remove the battery, disconnect the two terminals; negative (earth) first.
7 Remove the clamp.
8 Lift the battery out of the engine bay.
9 Refit in the reverse order, connecting the negative (earth) lead last.

4 Battery – inspection

1 Check the battery case for cracks and integrity.
2 Ensure the clamp is tightened down and the cell plugs are fully tightened and not leaking.
3 Check the electrolyte level in each cell and, if the level is low, remove the cell caps and add distilled water to each cell to bring the level to the upper mark. Do not overfill. In some countries a fully sealed battery is fitted which cannot be refilled.
4 **Warning:** *The battery contains sulphuiric acid. Do not allow it to contact the skin or clothing. Wash off any spillage immediately. Batteries also discharge flammable gases, so observe fire precautions when working on the battery, especially when charging.*
5 Check the battery terminals for corrosion and clean them off with a wire brush.
6 Coat the terminals in petroleum jelly to prevent the build-up of corrosion.

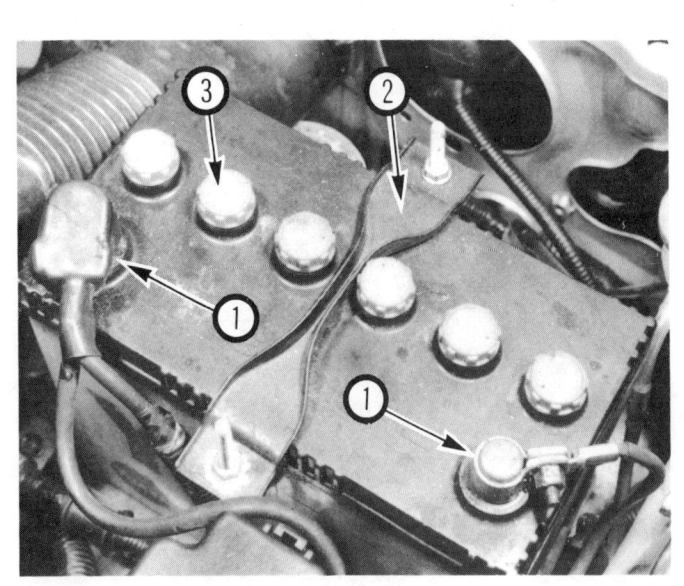

3.2 View of the battery (UK type)
1 Terminals
2 Clamp
3 Cell plugs

5 Battery – charging

1 If the battery has become discharged, or the indicator, if fitted, shows white, the battery should be charged.
2 Remove the battery from the engine bay and connect up a battery charger, in accordance with the manufacturer's instructions.
3 Remove the cell caps, if fitted, during the charging process and ensure the electrolyte level does not fall below the low mark.
4 Charge the battery, at no more than 10% of the ampere hour rating, for 24 hours.
5 Sealed batteries are charged in the same way.
6 If the charge warning light in the instrument panel comes on and remains on, or stays on at idle and goes off with an increase in engine speed, have the charging system checked by your dealer.

6 Alternator – description, maintenance and precautions

1 The alternator is mounted on the crankcase at the timing belt end of the engine.
2 The unit is driven by a belt from the crankshaft pulley. A voltage regulator is integral with the brush holder plate.
3 Keep the drivebelt correctly tensioned (see Chapter 2) and the electrical connections tight.
4 Keep the outside of the alternator free from grease and dirt.
5 It is important that the battery leads are always disconnected if the battery is to be charged. Also, if body repairs are to be carried out using electrical welding equipment, the alternator must be disconnected otherwise serious damage can be caused.
6 Do not stop the engine by pulling a lead from the battery.

7 Alternator – testing in situ

1 Turn the ignition off and disconnect the white wire from the terminal on the alternator.
2 Connect the ammeter between the white wire and the terminal (Fig. 12.1).
3 Start the engine.

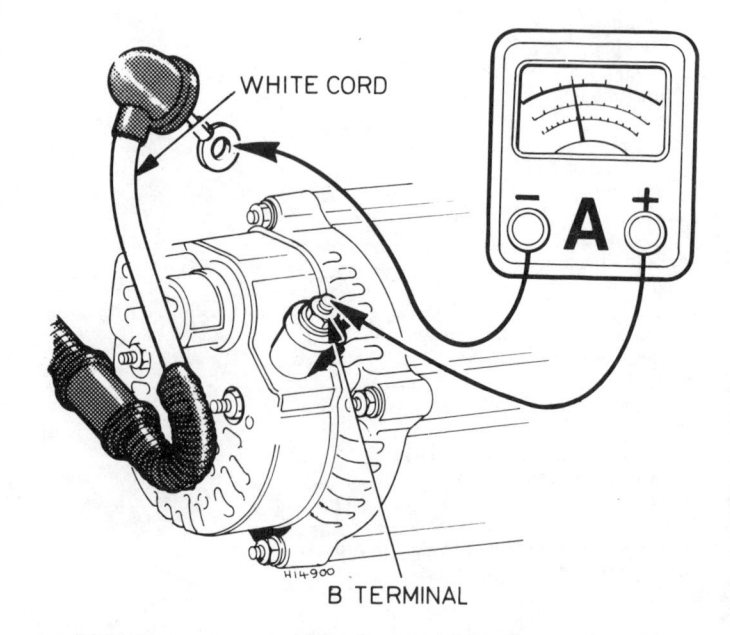

Fig. 12.1 Testing the alternator (Sec 7)

4 Turn the headlights on to full beam, the rear window demister on, and the heater fan on to full.
5 Check alternator output against the curves shown in Figs. 12.2 and 12.3.

8 Alternator – removal and refitting

1 Disconnect the battery earth (negative) terminal.
2 Disconnect the multi-plug and terminal first on the alternator.
3 Loosen the alternator adjuster bolt and mounting bolt.

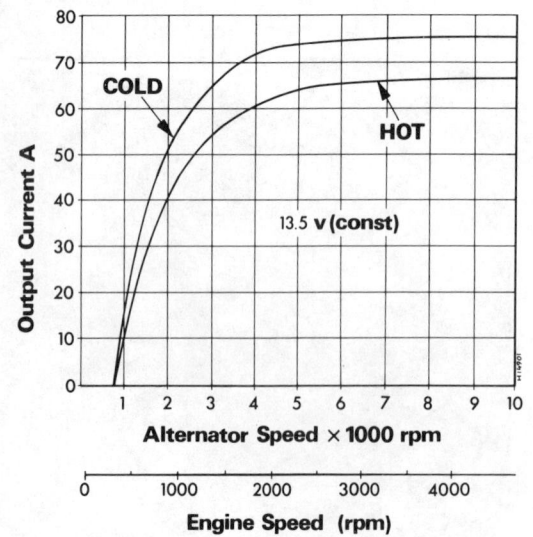

Fig. 12.2 Output graph for carburettor engine (Sec 7)

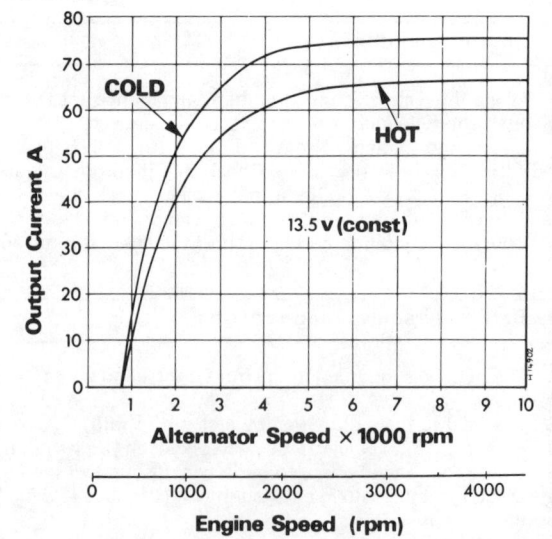

Fig. 12.3 Output graph for fuel injection engine (Sec 7)

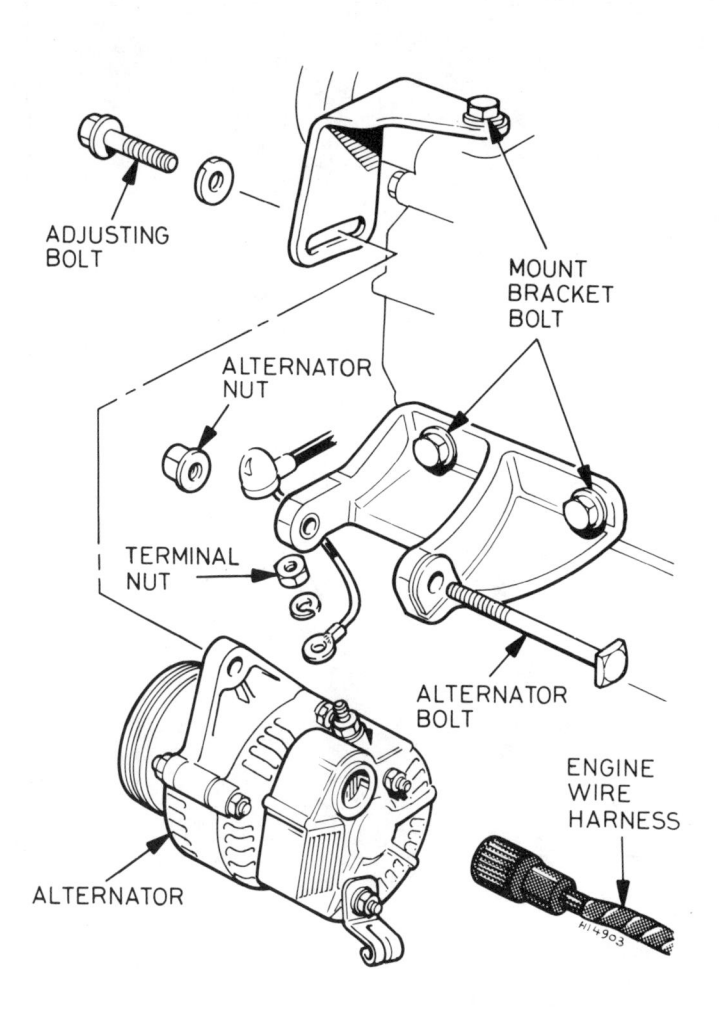

Chapter 12 Electrical system

Fig. 12.4 Alternator mounting brackets and bolts (Sec 8)

9.4 Removing the alternator rear cover

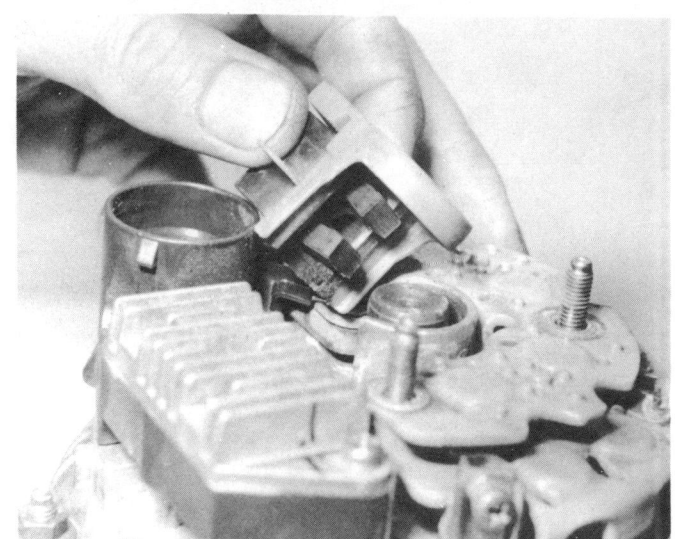

9.5 Removing the brush holder

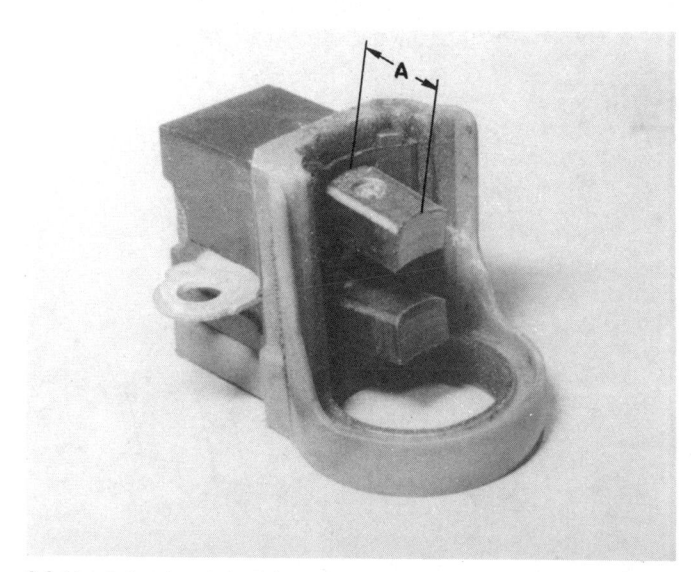

9.6 Measuring brush length

A = *minimum length*

4 Remove the alternator belt.
5 Remove the adjuster bolt and the mounting bolt and lift out the alternator.
6 Refitting is a reversal of removal, but adjust the belt tension as described in Chapter 2.

9 Alternator – overhaul

1 If the charge (ignition) warning light stays on after the engine has been started, or the battery requires frequent topping-up, indicating overcharging, the brushes and regulator may be changed as described here.
2 Further overhaul of the alternator should be left to your dealer, or an exchange alternator fitted.
3 Remove the alternator.
4 Remove the rear housing cover (photo).
5 Remove the brush holder (photo).
6 Measure the length of the brushes (photo).
7 If they are shorter than the minimum length given in the Specifications, they should be renewed.
8 Do this by unsoldering the brush wires from the back of the holder, removing them and fitting new brushes, soldering the wires in place. Use only resin core solder, to prevent corrosion.
9 Refit the brush holder, depressing the brushes with a screwdriver to fit them over the slip ring (photo).
10 To renew the regulator, undo its retaining screws and lift it off. On some alternators the brush holder must be removed first.

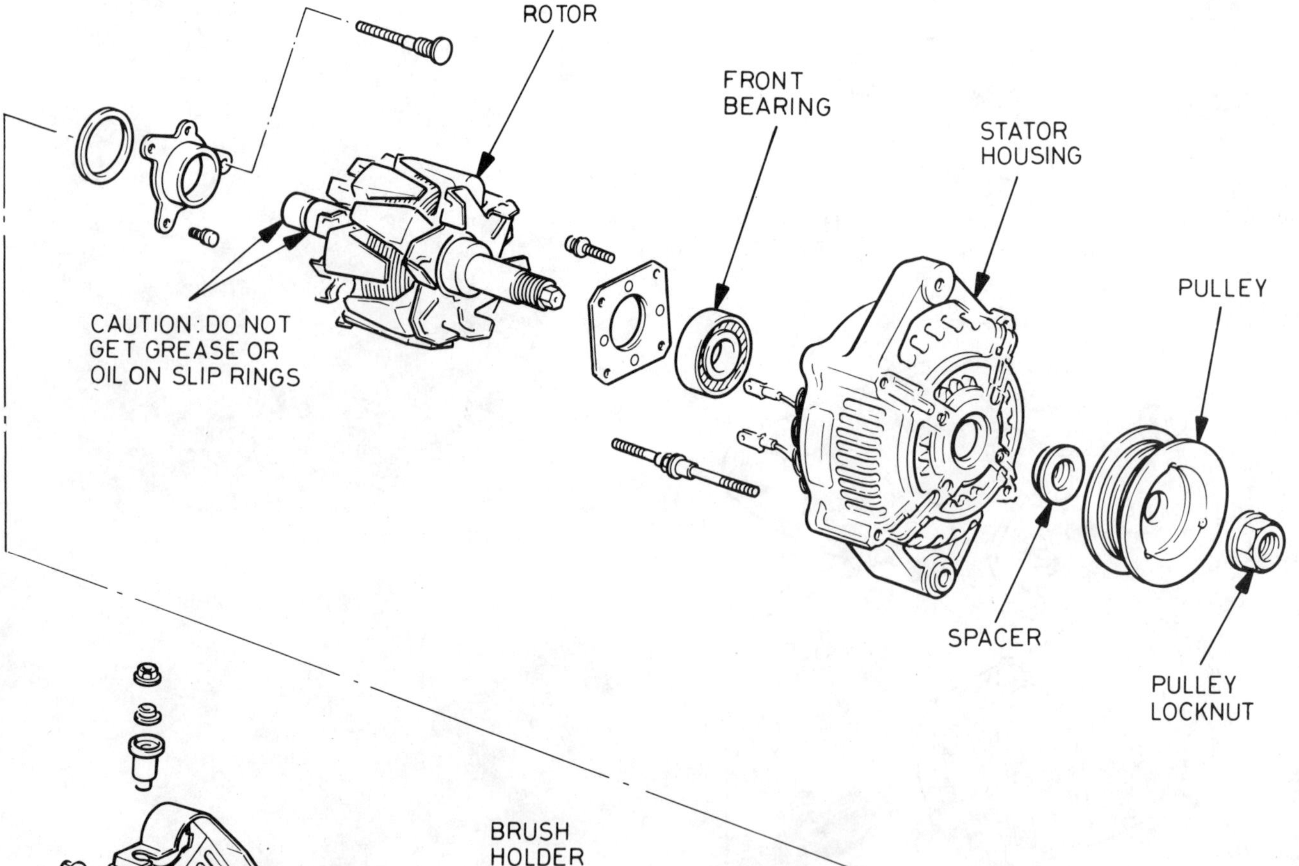

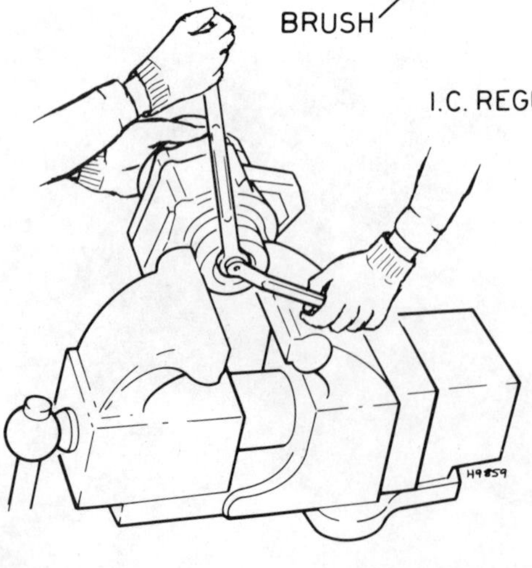

Fig. 12.5 Exploded view of the alternator (Sec 9)

Chapter 12 Electrical system   301

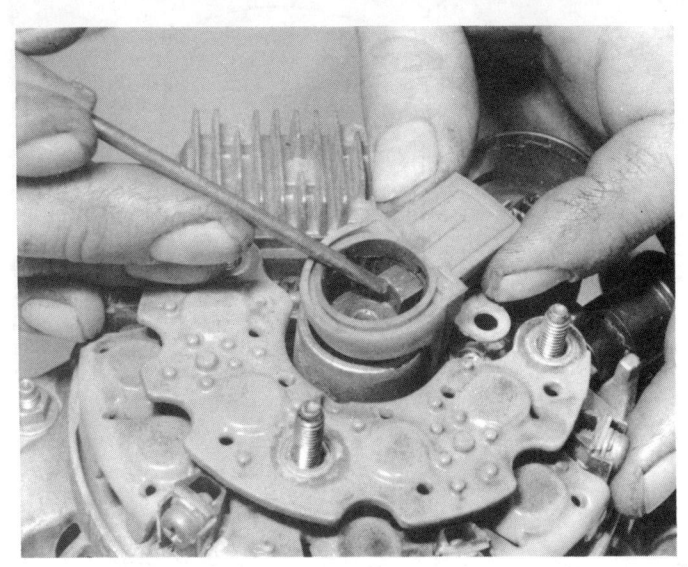

9.9 Refitting the brushes

11 Fit a new regulator, and brush holder if it was removed, and fit the rear cover.
12 Refit the alternator.

10 Starter motor – description

1 The starter motor is of pre-engaged type.
2 When the starter switch is operated, current flows from the battery to the solenoid switch which is mounted on the starter body. The plunger in the solenoid moves inwards, so causing a centrally pivoted lever to push the drive pinion into mesh with the starter ring gear. When the solenoid plunger reaches the end of its travel, it closes an internal contact and full starting current flows to the starter field coils. The armature is then able to rotate the crankshaft, so starting the engine.
3 A special freewheel clutch is fitted to the starter drive pinion so that as soon as the engine fires and starts to operate on its own it does not drive the starter motor.
4 When the starter switch is released, the solenoid is de-energised and a spring moves the plunger back to its rest position. This operates the pivoted lever to withdraw the drive pinion from engagement with the starter ring.
5 On automatic transmission models, an idler gear is incorporated at the drive end of the starter motor.

11 Starter motor – testing *in situ*

1 If the starter motor fails to turn the engine when the switch is operated there are five possible causes:
   (a) The battery is faulty
   (b) The electrical connections between the switch, solenoid battery and starter motor are somewhere failing to pass the necessary current from the battery through the starter to earth
   (c) The solenoid switch is faulty
   (d) The starter motor is mechanically or electrically defective
   (e) The starter motor pinion and/or flywheel ring gear is badly worn and in need of replacement

2 To check the battery, switch on the headlights. If they dim after a few seconds the battery is in a discharged state. If the lights glow brightly, operate the starter switch and see what happens to the lights. If they dim then you know that power is reaching the starter motor but failing to turn it. If the starter turns slowly when switched on, proceed to the next check.

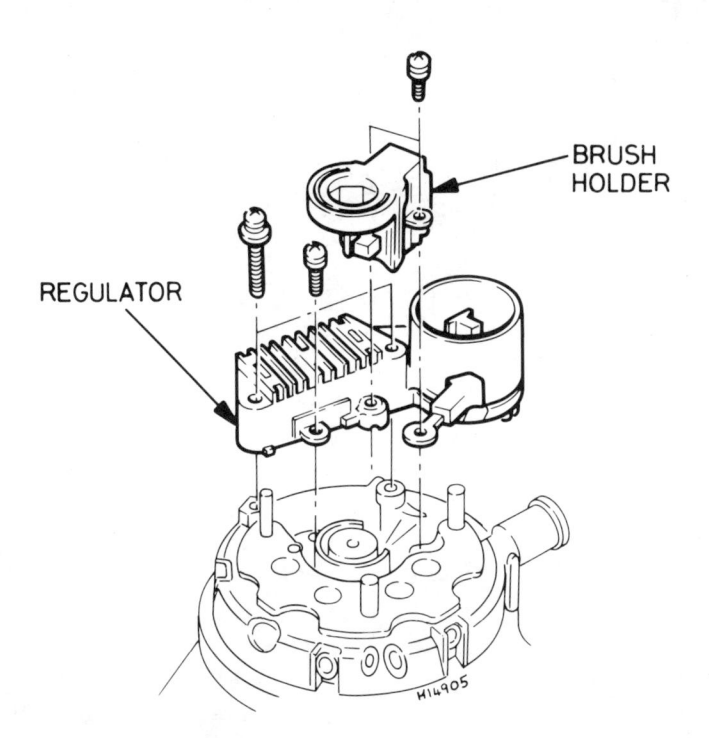

Fig. 12.6 Alternator brush holder and regulator mounting screws (Sec 9)

3 If, when the starter switch is operated, the lights stay bright, then insufficient power is reaching the motor. Remove the battery connections, starter/solenoid power connections and the engine earth strap and thoroughly clean them and refit them. Smear petroleum jelly around the battery connections to prevent corrosion. Corroded connections are the most frequent cause of electric system malfunctions.
4 When the above checks and cleaning tasks have been carried out, but without success, you will probably have heard a clicking noise each time the starter switch was operated. This was the solenoid switch operating, but it does not necessarily follow that the main contacts were closing properly (if no clicking has been heard from the solenoid, it is certainly defective). The solenoid contact can be checked by putting a voltmeter or bulb across the main cable connection on the starter side of the solenoid and earth. When the switch is operated, there should be a reading or lighted bulb. If there is no reading or lighted bulb, the solenoid unit is faulty and should be renewed.
5 If the starter motor operates but doesn't turn the engine over then it is most probable that the starter pinion and/or flywheel ring gear are badly worn, in which case the starter motor will normally be noisy in operation.
6 Finally, if it is established that the solenoid is not faulty and 12 volts are getting to the starter, then the motor is faulty and should be removed for inspection.
7 It is not recommended that the starter be overhauled by the home mechanic.
8 Even brush renewal requires the renewal of the armature housing, so it is better to obtain a reconditioned or new replacement unit.
9 Remove the starter by disconnecting the battery, disconnecting the starter leads at the starter, and removing the transmission housing bolts which hold the starter to it.
10 Refit in the reverse order.

12 Fuses, relays and control units – general

1 The location of the various fuses, relays and control units are shown in Fig. 12.7 and 12.8.
2 Not all the units are fitted to all models.

302

Fig. 12.7 Relay and control unit location – vehicle (Sec 12)

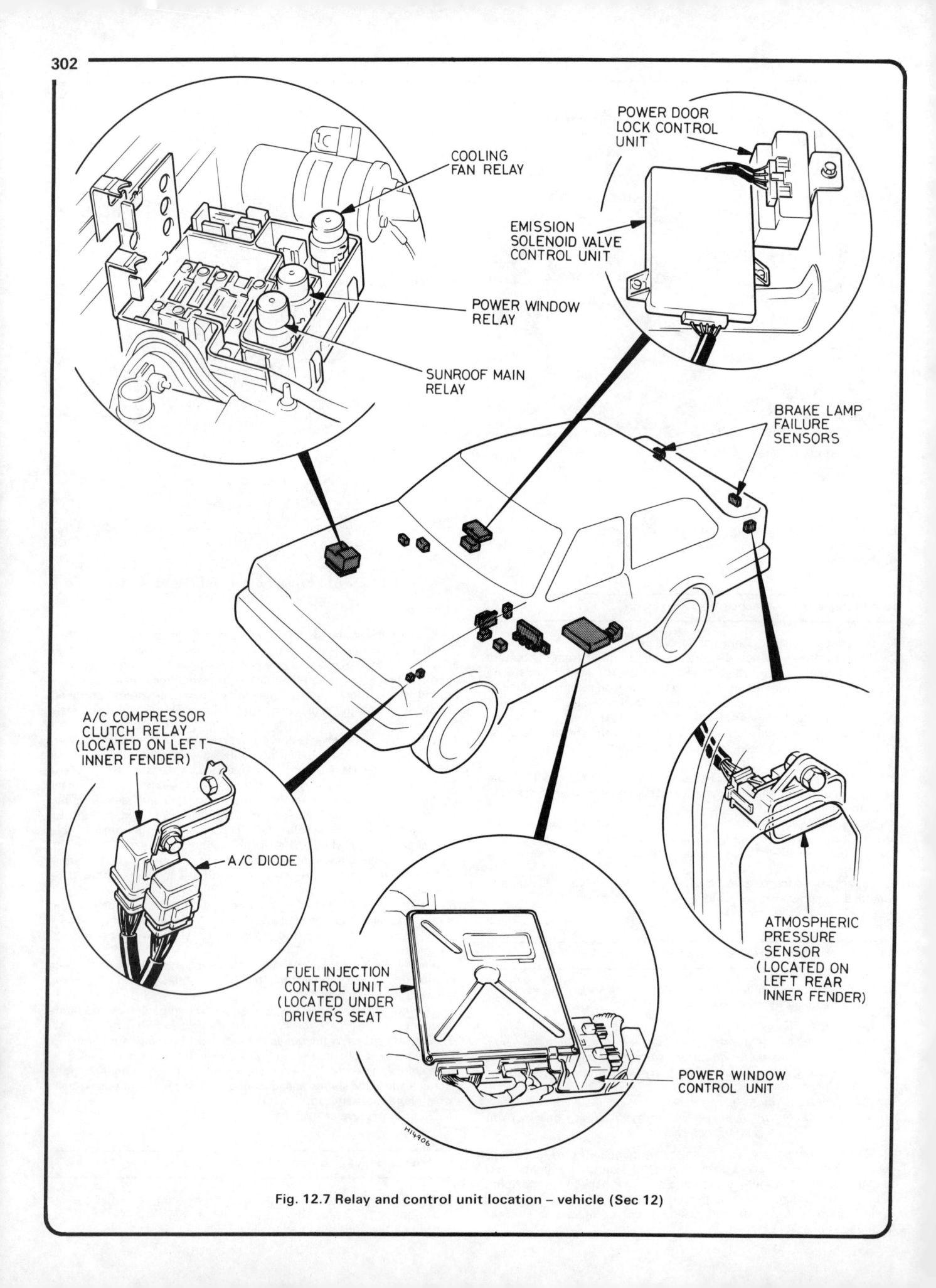

Fig. 12.8 Relay and control unit location – instruments (Sec 12)

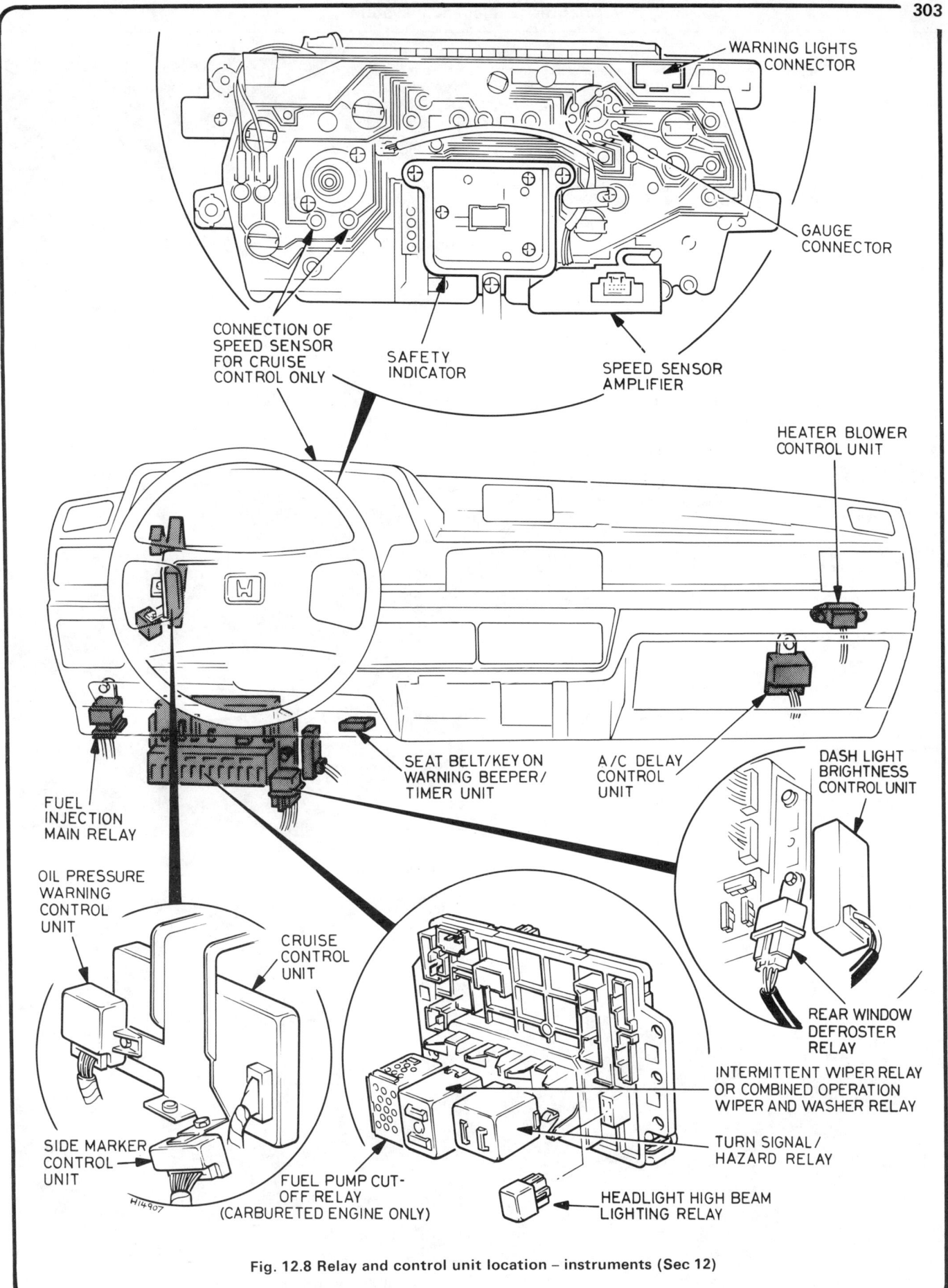

Fuses

3 The fusebox (photo) is located under the facia panel either to the left or right-hand side, depending on whether the vehicle is LHD or RHD.
4 The fusebox is hinged and simply pulls down.
5 Further fuses are located in the control box by the battery.
6 If any lights, accessories or controls fail, check for blown fuses and renew as necessary.
7 If the main fuse in the engine compartment control box blows, have the electrical system checked by your dealer.

Relays

8 A relay isolates heavy current used by system components from the system control switches, allowing the switches to operate on a much lower current. They are, in effect, remote control switches.
9 If a relay is suspected of malfunction, test it by substituting a known serviceable relay in its place.
10 They are relatively cheap to buy and are easily removed from their locations.

Control units

11 There are several control units/boxes in the various systems used on these vehicles, especially those with the higher accessory specification.
12 On some, like the main control box in the engine bay, individual components can be removed and tested/renewed.
13 Others, like the fuel injection control unit, are a sealed unit.
14 Where control boxes are suspected of malfunction, have your dealer check the system involved.

Multi-block connectors

15 There are many multi-block connectors around the electrical system.
16 Keep the terminals clean by the periodic use of a water repellant spray and pack the rear of connectors, where the wires enter, with silicone grease (photo).

Switches

17 Several different kinds of switches are used for control of the various systems.
18 Some simply clip into their housings and are levered out.
19 Others, like the power window switch block, mounted in the door trim panel, are held by screws (photo).
20 A third type, like the instrument panel dimmer switch (photo), are held by a locknut which is screwed onto its threaded body. Remove the control knob (lever it off) and the locknut can be undone with a box spanner or deep socket.

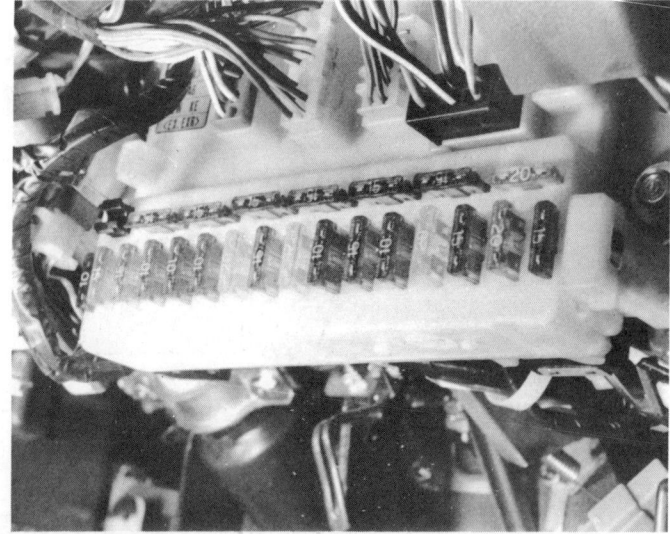

12.3 View of the fusebox

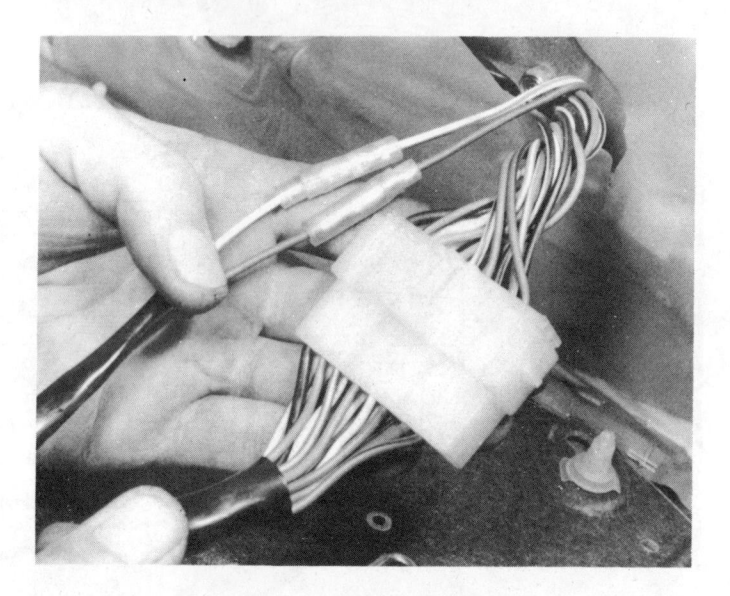

12.16 Multi-block connector

12.19 Power window switch

12.20 Instrument panel dimmer switch

Chapter 12 Electrical system

21 Disconnect their electrical cables, and remove the switches from their housings.
22 Switches rarely fail in service, but can become worn after long use, when it is best to renew them.

13 Instrument panel – removal and refitting

1 Lower the steering column, as desribed in Chapter 10.
2 Remove the five instrument panel bezel retaining screws.
3 There are 3 along its top edge (photo), and two at the lower edge (photo).
4 Pull the instrument panel bezel forward and disconnect the multi-blocks from its switches, and remove the bezel.
5 Four screws hold the instrument gauge assembly (photo). Remove these screws, and lift the assembly to enable the wire connectors to be reached.
6 Separate the connectors. Do not pull on the cables.
7 Disconnect the speedometer cable, and remove the unit from the dashboard.
8 The various components are easily removed by undoing their retaining screws (photos).
9 The instrument bulbs and their holders are bayonet fixings (photo).
10 Refitting is a reversal of removal.

13.3A Instrument panel bezel retaining screws at the top edge ...

13.3B ... and two at the bottom (arrowed)

13.5 Instrument gauge assembly retaining brackets (arrowed)

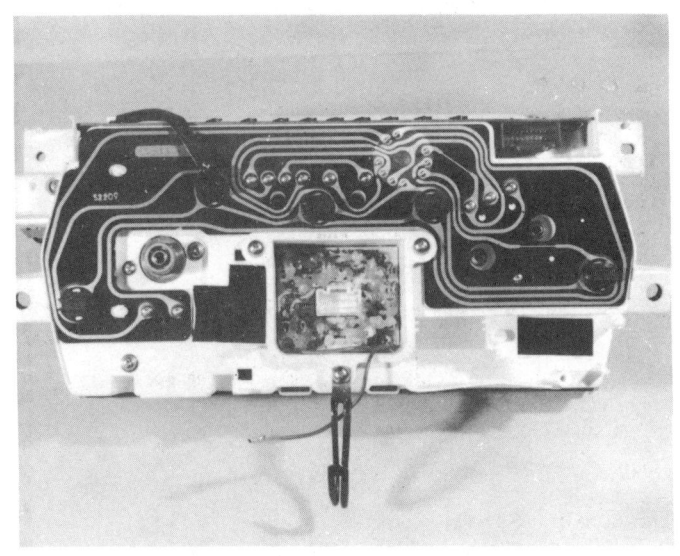

13.8A Two views of the rear ...

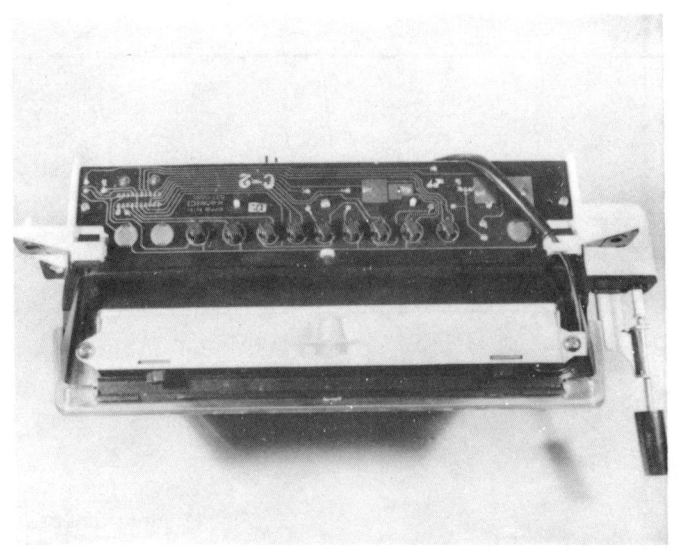

13.8B ... of the instrument panel

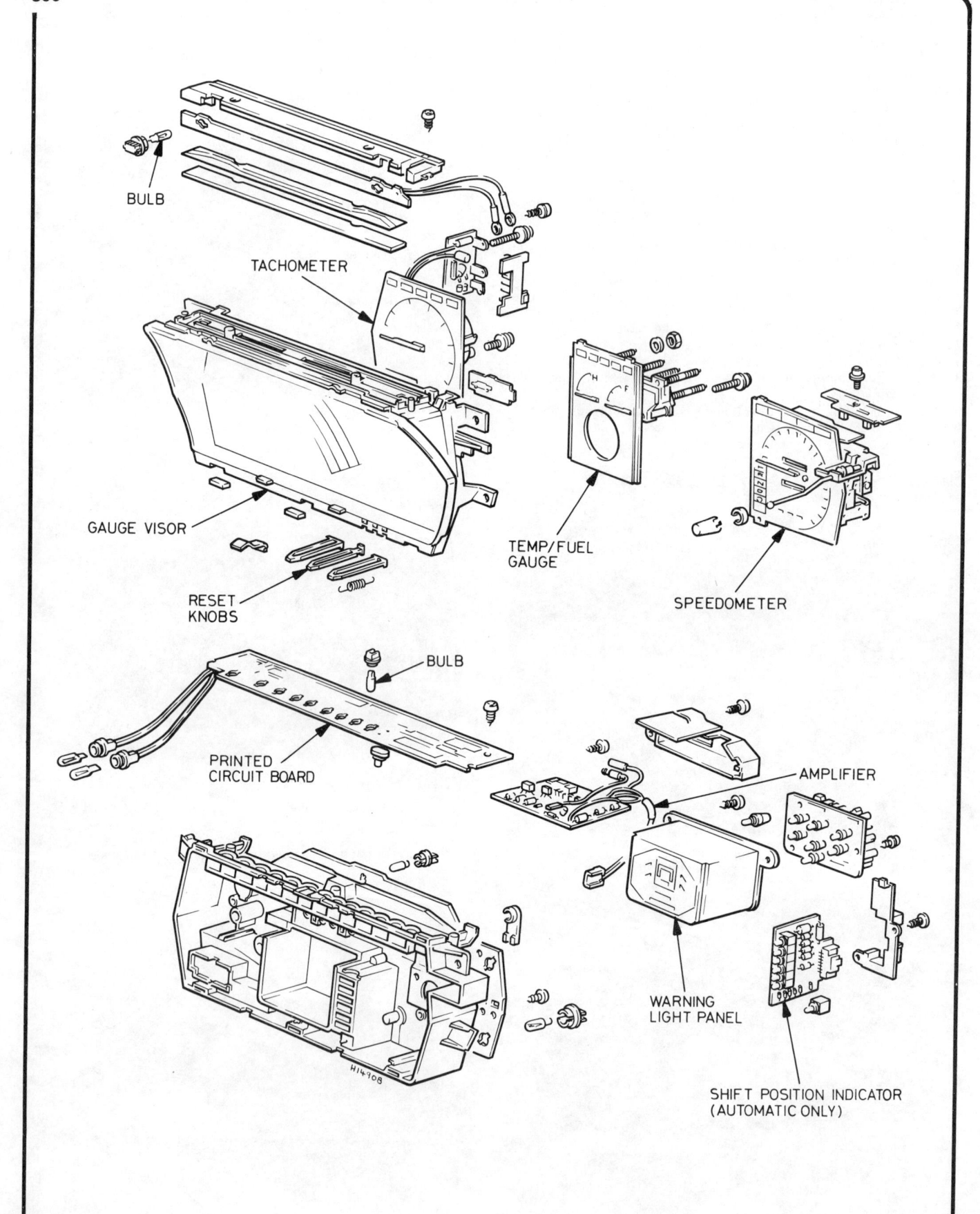

Fig. 12.9 Instrument panel components (Sec 13)

Chapter 12 Electrical system

14 Steering wheel and column switches – removal and refitting

Horn switch
1 Undo the four screws which hold the steering wheel centre trim panel in place and are reached from behind the steering wheel. Remove the panel (photo).
2 The horn push switches and terminals are screwed to the rear of the panel (photo).
3 If the switches become faulty, they should be renewed.

Cruise control switches
4 The cruise control switches are mounted on the steering wheel. Remove the wheel centre trim panel.
5 If the switches become defective, they should be renewed (photo).

Combination switch
6 Refer to Chapter 10 for details of removal of the combination switch, which houses the cruise control slip-ring, indicators, headlight, wiper and hazard warning controls.
7 Renew the switch as a unit if it becomes defective.

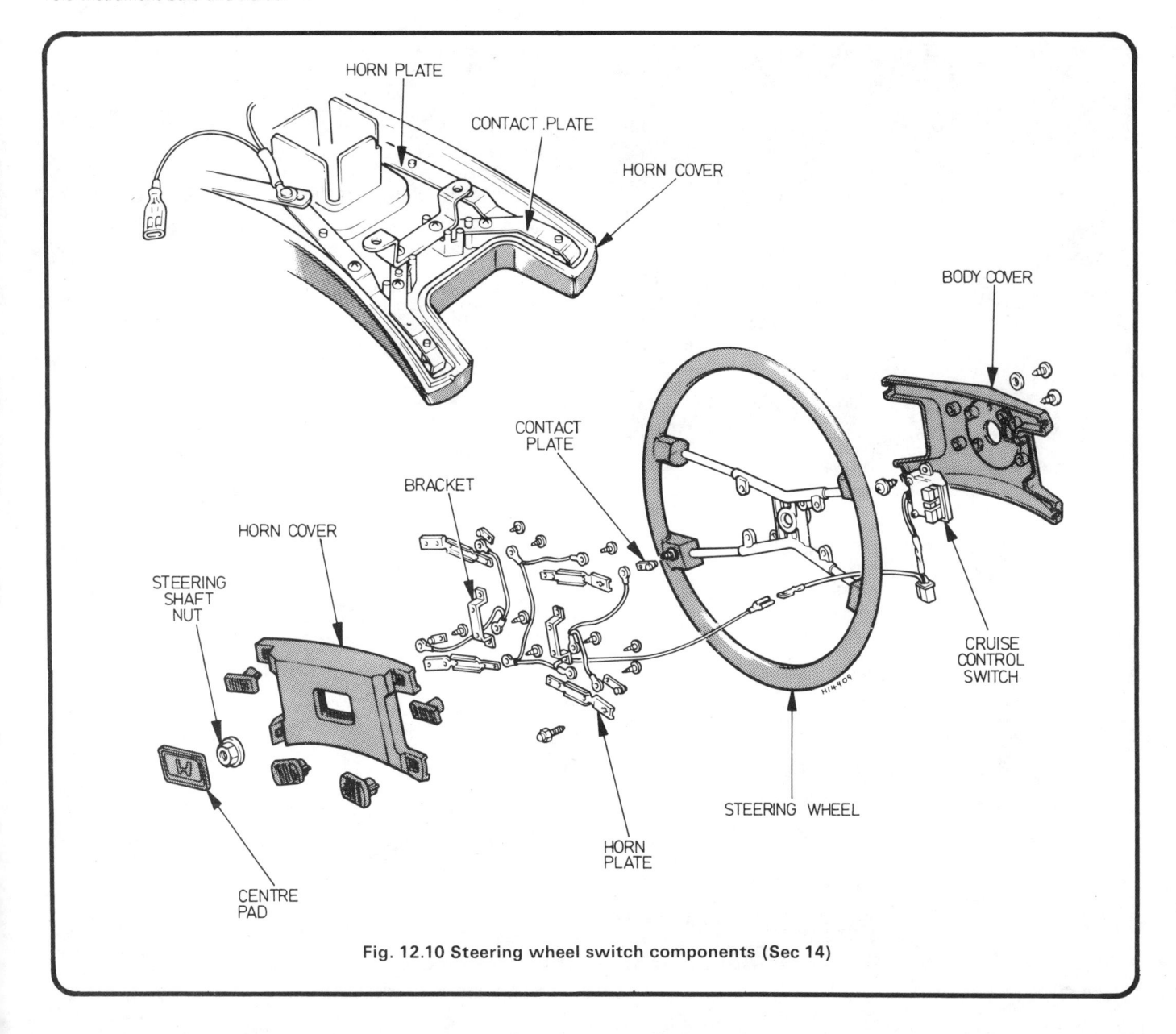

13.9 Instrument bulb and holder

Fig. 12.10 Steering wheel switch components (Sec 14)

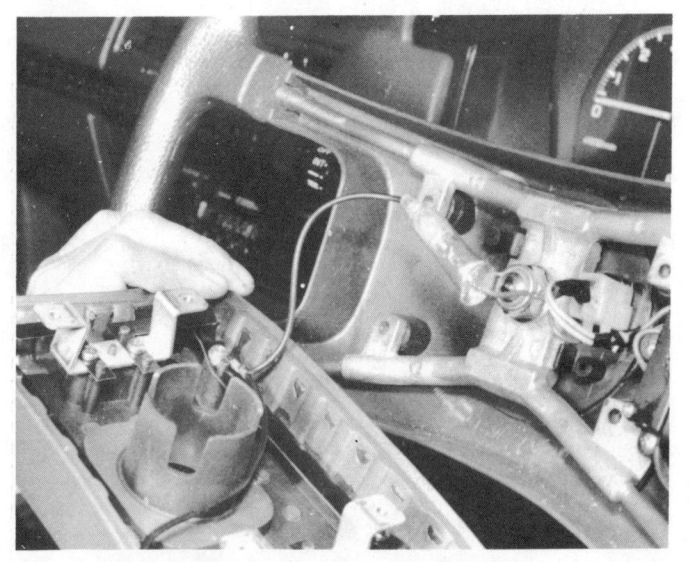

308　Chapter 12 Electrical system

14.1 Removing the steering wheel centre panel

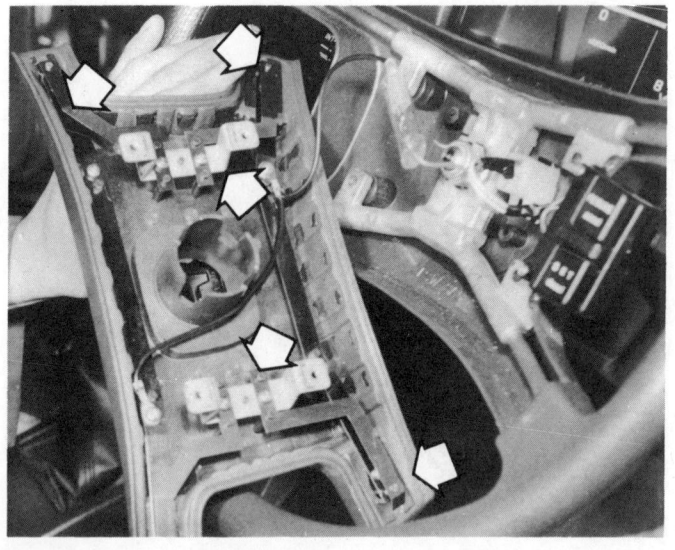

14.2 The horn switches at the rear of the panel (arrowed)

14.5 Cruise control switch screws (arrowed)

15 Headlamps – removal and refitting

1　Both single and twin headlamp clusters are used, according to model.
2　Remove the grille (Chapter 11).
3　Disconnect the multi-block connectors at the rear of the headlamp.
4　On single lamp units, remove the side turn indicator light (Section 18).
5　Remove the two bolts from the side mounting bracket (photo).
6　There is another bolt at the inboard edge and one screw at the rear, reached from inside the engine bay.
7　Remove the headlamp unit (photo).
8　On the twin headlamp types, simply remove the four retaining screws.
9　Refitting is a reversal of removal.

16 Headlamp bulbs – renewal

1　The headlamp bulbs can be renewed without removing the complete unit.
2　Pull off the connector block from the rear of the bulb, reached from inside the engine bay (photo).

15.5 Headlamp mounting bolts (arrowed)

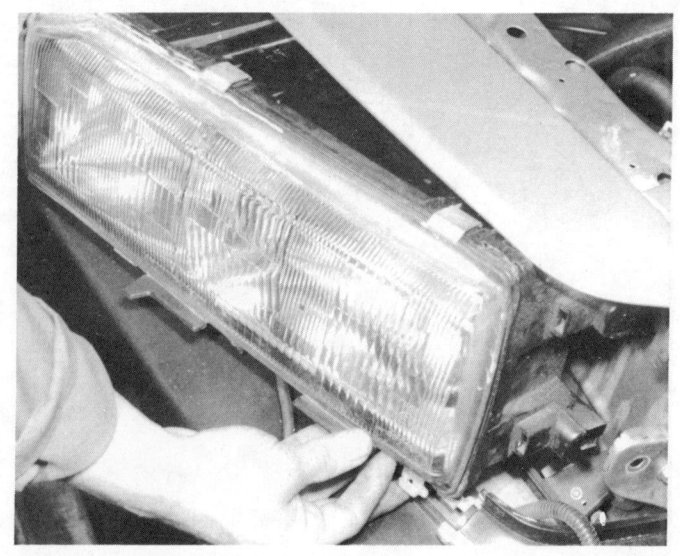

15.7 Removing the headlamp unit

SIDE TURN SIGNAL LIGHT

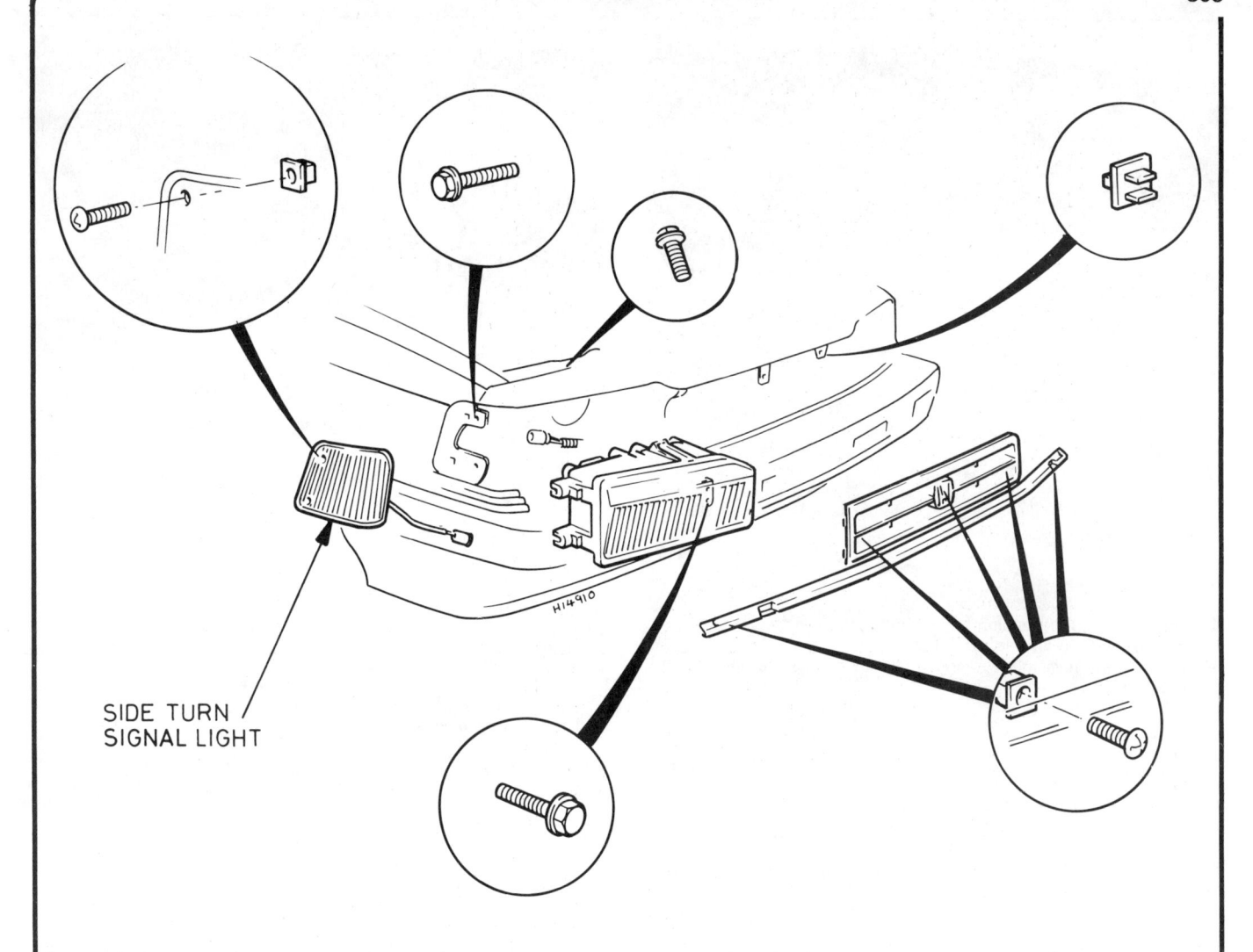

Fig. 12.11 Single unit headlamp retaining screws (Sec 15)

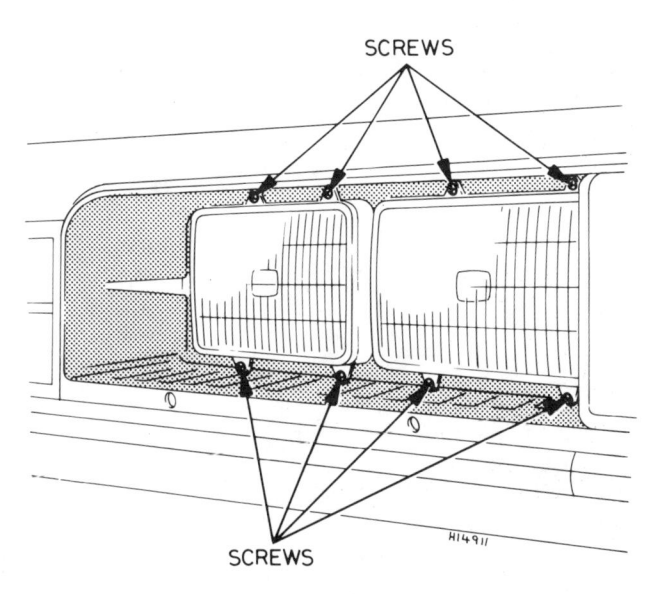

SCREWS

SCREWS

Fig. 12.12 Twin headlamp unit retaining screws (Sec 15)

16.2 Disconnecting the block

16.4 Spring clip (arrowed)

3 Remove the rubber cover.
4 Unhook the spring retaining clip (photo).
5 Lift out the bulb and holder (photo).
6 Refitting is a reversal of removal. Do not handle the units by the bulbs themselves.

17 Headlamp beam – adjusting

1 Headlamp beams should be adjusted by your local dealer to comply with local regulations, and avoid dazzling other road users.
2 Beam adjusting screws are shown in Figs. 12.13 and 12.14.
3 On the single lamp unit types, the adjusters can also be reached through the engine bay crossmember, using a screwdriver (photo).

16.5 Removing the bulb

18 Parking light and side indicator light – removal and refitting

1 Remove the two retaining screws and lift out the unit (photo).
2 Twist out the bulbholder (photo).
3 The bulb is a bayonet fix.

19 Rear lamp cluster – removal, refitting and bulb renewal

1 Remove the luggage compartment rear trim panels.
2 Remove the plastic cover over the lamp unit (photo).
3 The holders and bulbs are a bayonet type fix.

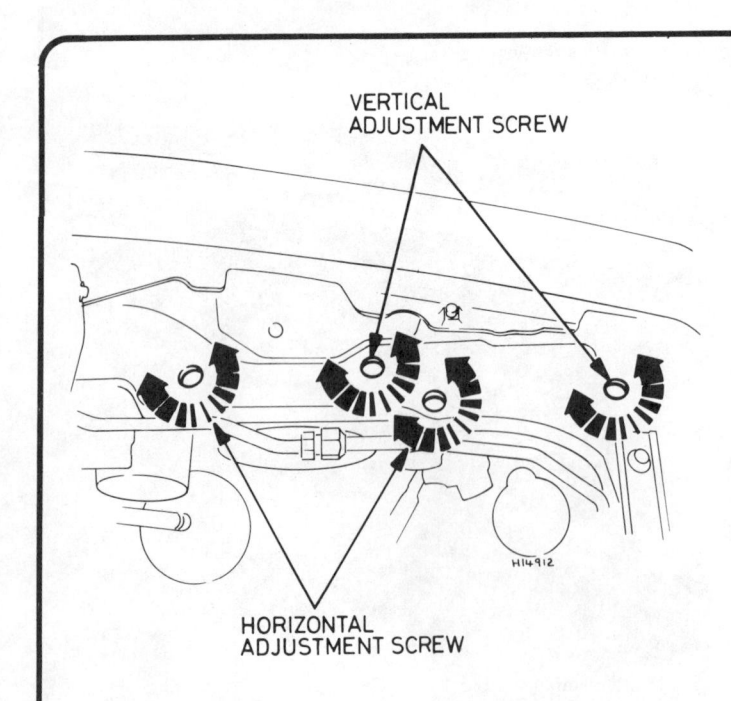

Fig. 12.13 Single unit headlamp beam adjusters (Sec 17)

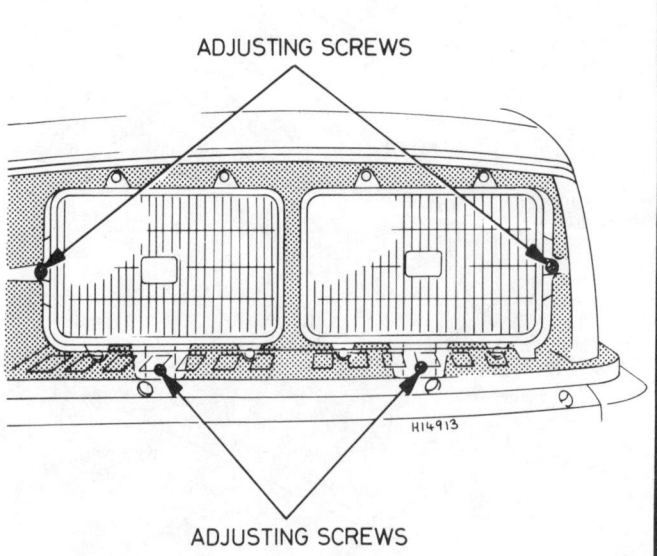

Fig. 12.14 Twin headlamp unit beam adjusters (Sec 17)

Chapter 12 Electrical system

17.3 Adjusting the headlamp beam

18.1 Side indicator light retaining screws (arrowed)

18.2 Bulbholder twists out

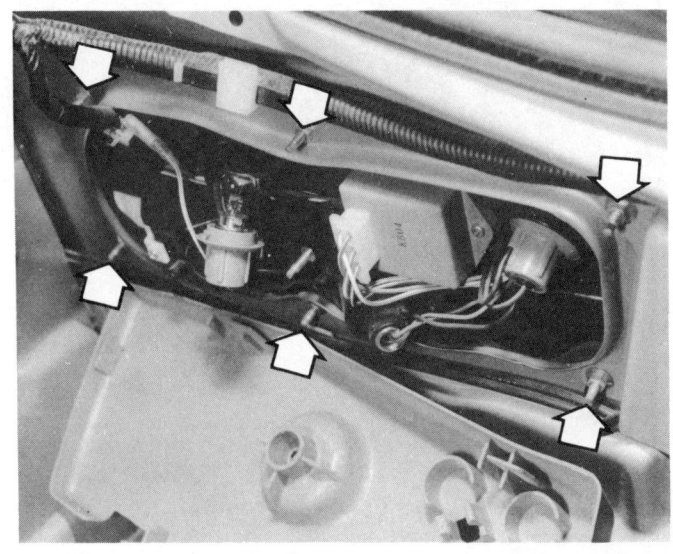

19.4 Plastic cover removed, showing the retaining nuts (arrowed)

4 To remove the lamp unit, disconnect the multi-block connectors and undo the retaining nuts (photo).
5 Lift the unit outward.
6 Refit in the reverse order.

20 Supplementary exterior lamps – removal, refitting and bulb renewal

Front turn indicator light
1 The light unit is mounted in the front bumper.
2 To remove the unit and bulb, undo the two retaining screws (photo).
3 The bulb is a bayonet fit.
4 Refit in the reverse order.

Rear foglight
5 The rear foglight is mounted in the rear bumper in similar fashion to the front indicator light (photos).
6 Remove the screws, take off the lens and remove the bulb, which is a bayonet fix.
7 Refit in the reverse order.

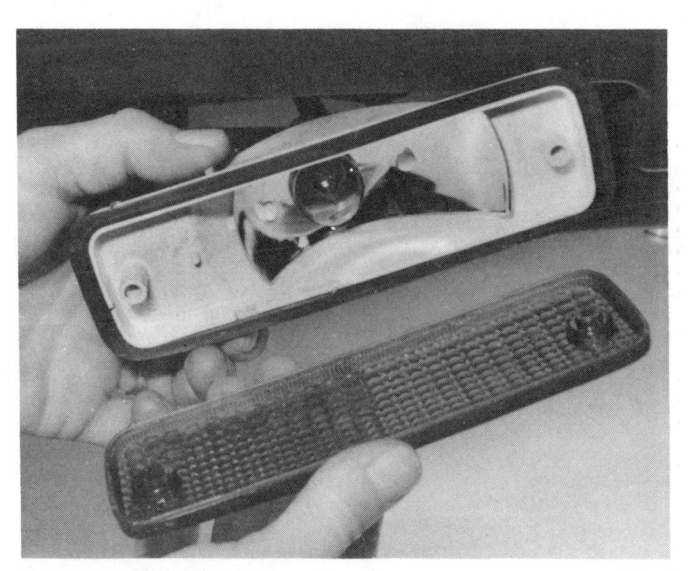

20.2 Front indicator light assembly

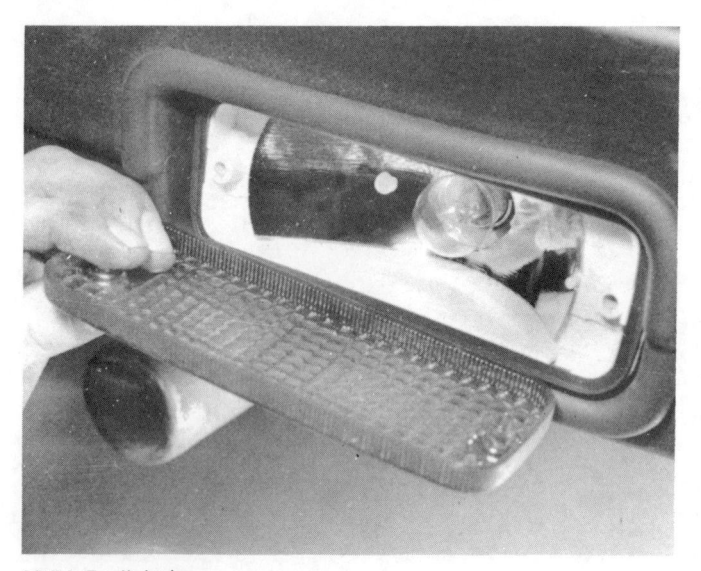

312 Chapter 12 Electrical system

20.5A Foglight lens ...

20.5B ... and body

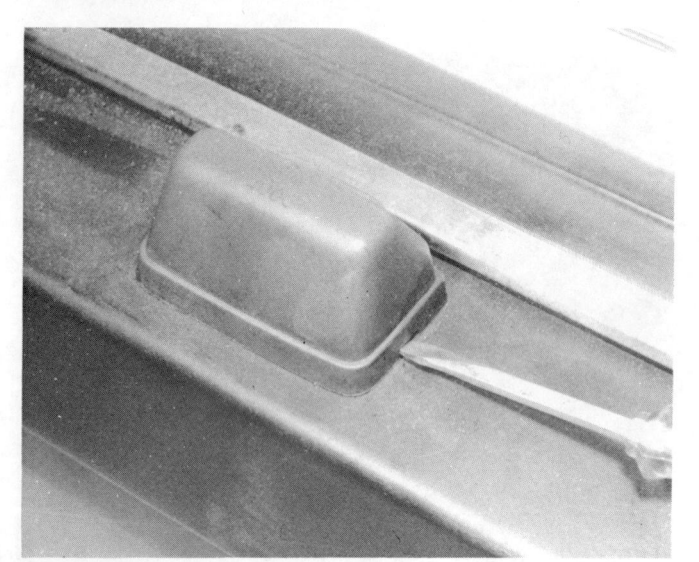

20.10 Lever the unit out ...

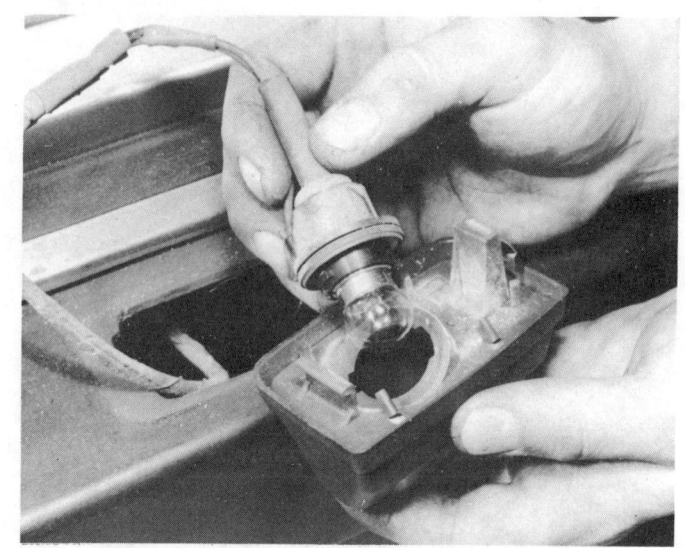

20.11 ... and twist out the bulbholder

Number plate light
8 There are three kinds of number plate light fitted, according to model.
9 One type is mounted in the bumper (type A).
10 To remove it, lever the unit up with a screwdriver (photo).
11 Twist out the bulbholder and the bulb (photo).
12 A second type (type B) has separate, twin units mounted to either side of the number plate.
13 To change the bulbs, remove the screws from the lens cover, remove the lens and take out the bulb.
14 The units themselves are screwed to the rear valance, the retaining bolts being reached from inside the luggage compartment.
15 A third type (type C) is similar to the second except that it is a single unit.

Side turn indicator light
16 To remove the side turn light, release the clip using a screwdriver as shown in Fig. 12.18, being careful not to damage paintwork, and lift the unit out.
17 Twist off the lens (photo).
18 The bulb is a push fit in the holder.
19 Push the unit back into the wing, ensuring the cap snaps into place.

Fig. 12.15 Type B number plate light lens removal (Sec 20)

TO REPLACE BULB ONLY, REMOVE LENS FROM OUTSIDE

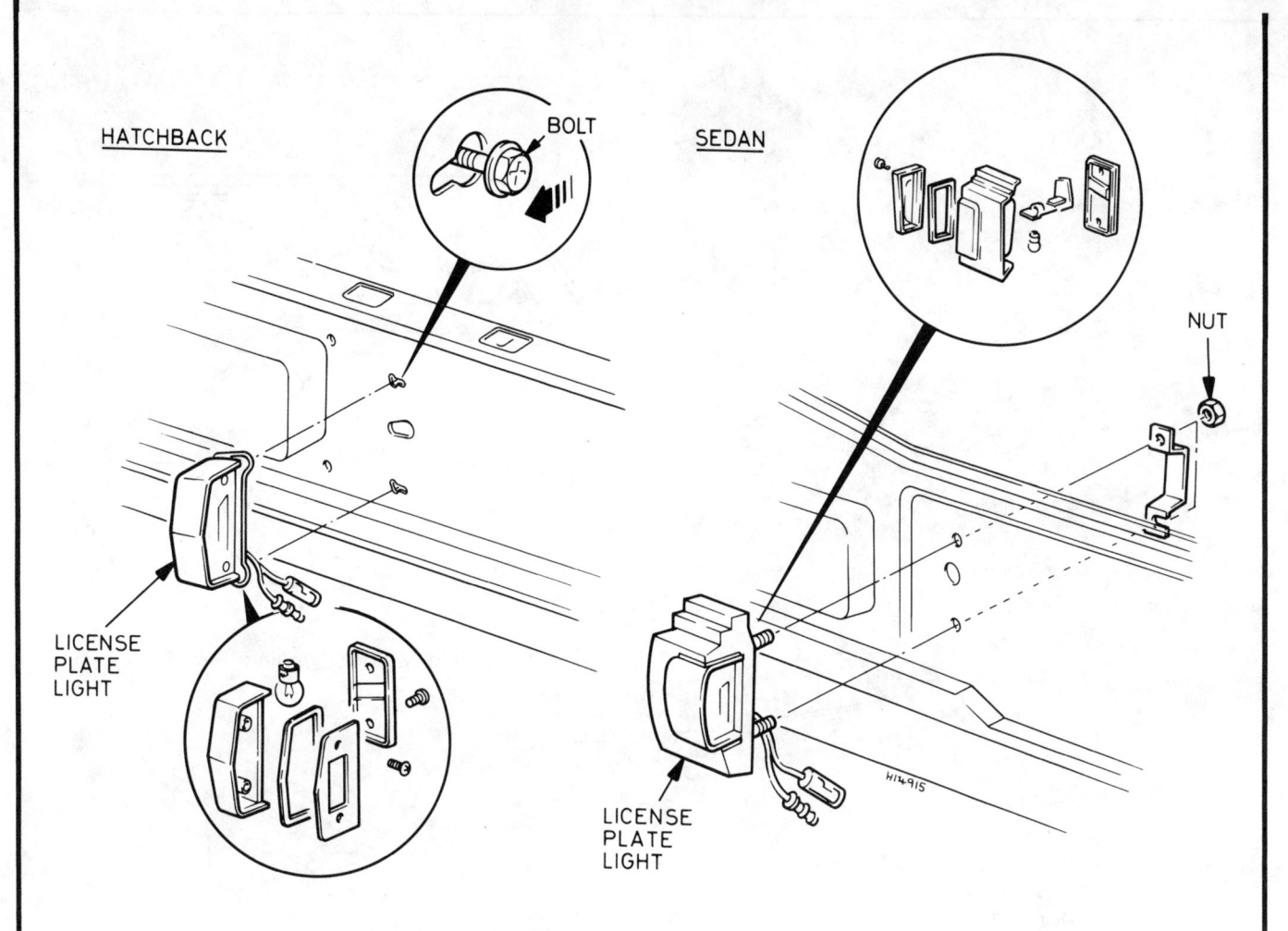

Fig. 12.16 Type B number plate light unit removal (Sec 20)

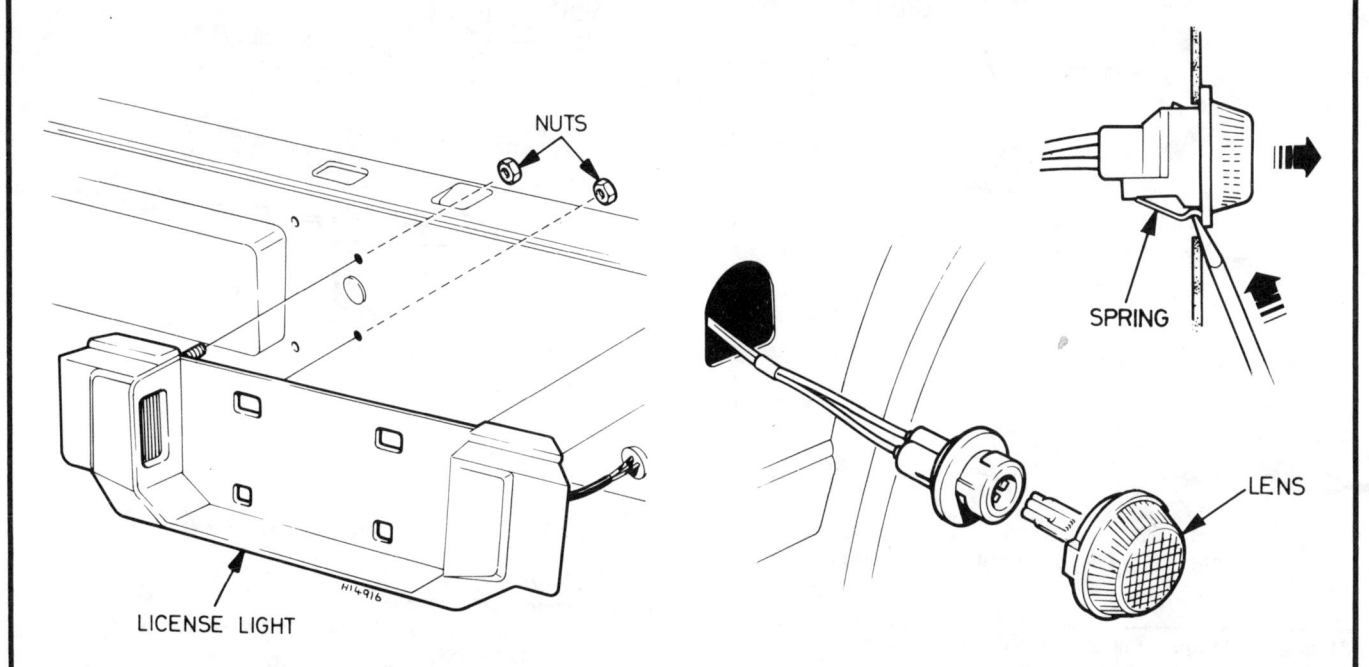

Fig. 12.17 Type C number plate light unit removal (Sec 20)

Fig. 12.18 Side turn indicator light removal (Sec 20)

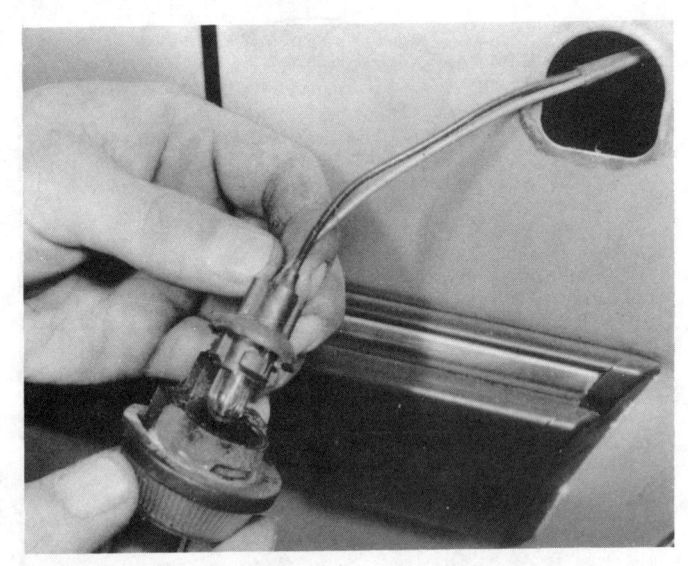

Chapter 12 Electrical system

20.17 Side turn indicator light unit

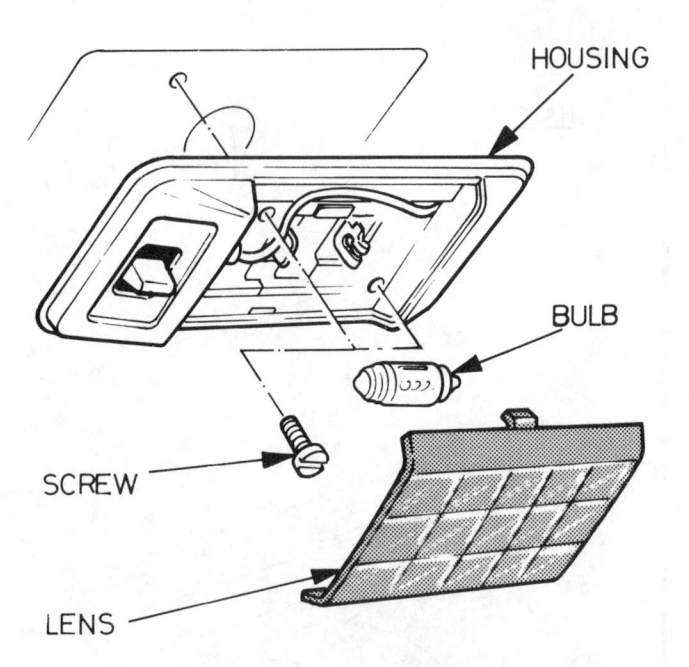

Fig. 12.19 Interior light unit (Sec 21)

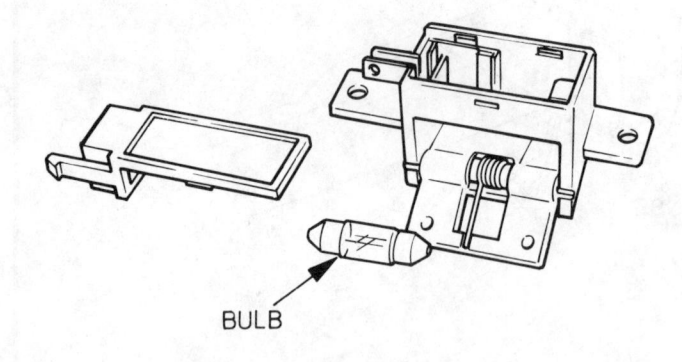

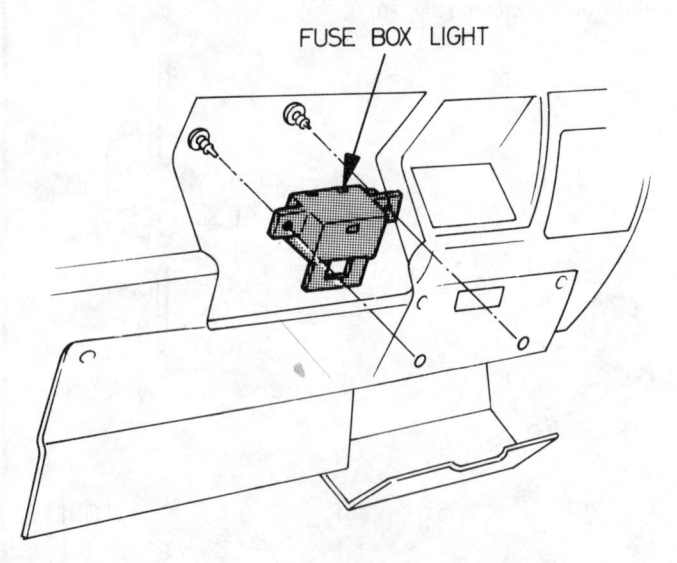

Fig. 12.20 Fusebox light unit (Sec 21)

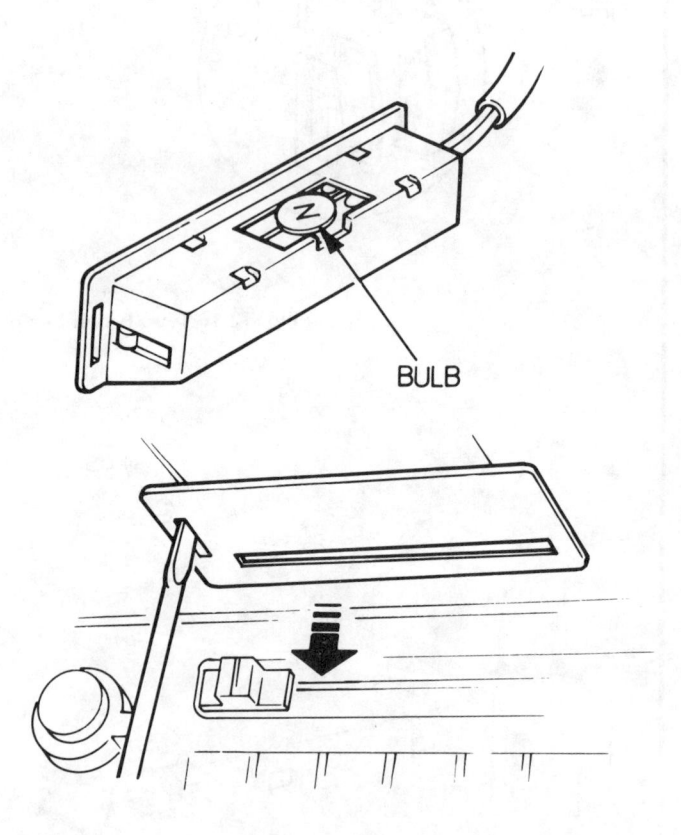

Fig. 12.21 Instrument panel light (Sec 21)

21 Interior lamps – removal, refitting and bulb renewal

Interior light
1. Pry off the lens cover.
2. The bulb is a festoon type and pulls out.
3. To remove the housing, undo the two retaining screws.

Fusebox light
4. Remove the fusebox cover.
5. Remove the screws from the lamp holder and remove the holder.
6. The bulb is of the festoon type.

Instrument panel light bulb
7. Pry out the lamp holder with a screwdriver.
8. Disconnect the cable connector.
9. Remove the case cover and the bulb.

Chapter 12 Electrical system

Ashtray light
10 Remove the ashtray.
11 Remove the three screws from the ashtray carrier and remove the carrier.
12 Turn the socket 90° anti-clockwise to remove the bulb.

Door courtesy lights
13 Pry off the lens cover (photo).
14 The bulb is a bayonet fix.
15 To remove the unit from the door, remove the two retaining screws (photo).

Vanity mirror light
16 Pry off the lens.
17 The bulb is a festoon type fit in the holder.

Luggage compartment light
18 Pry off the lens.
19 The bulb is a push fit in the holder.
20 To remove the holder, pry it from the panel and disconnect the electrical leads.

All lights
21 Refitting of all interior light assemblies is a reversal of removal.

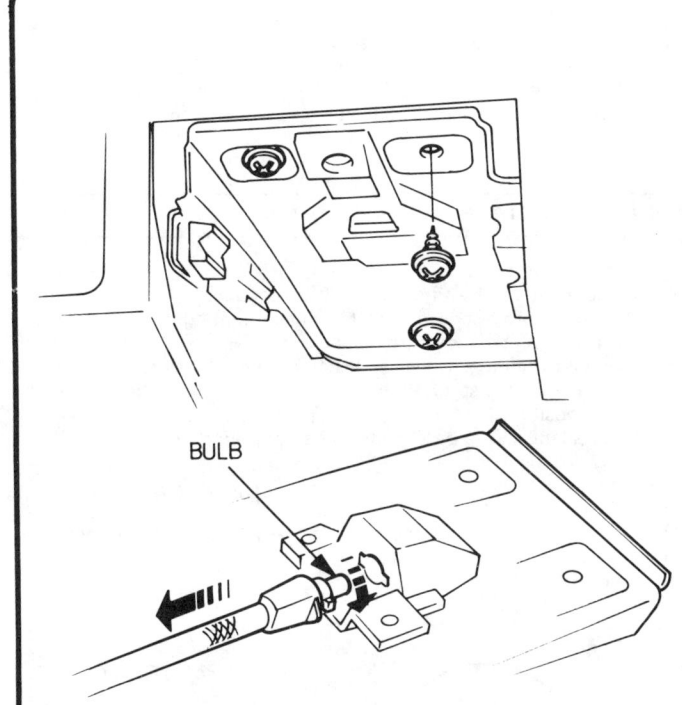

Fig. 12.22 Ashtray light unit (Sec 21)

21.13 Pry off the lens

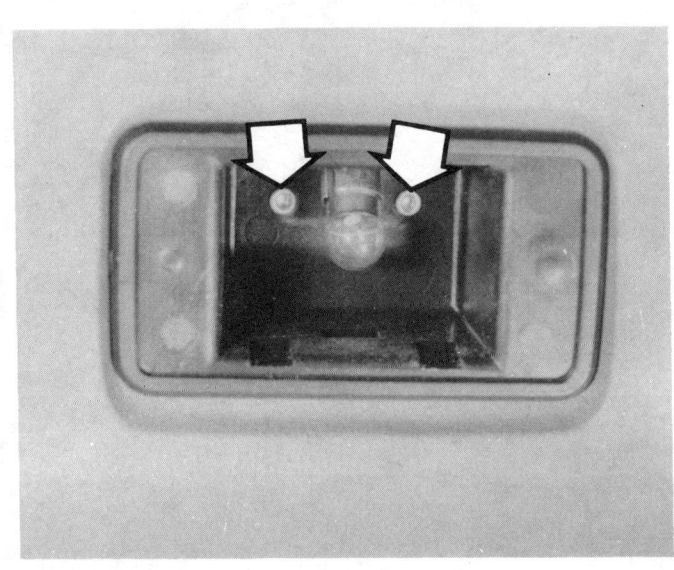

21.15 The bulb, and retaining screws (arrowed)

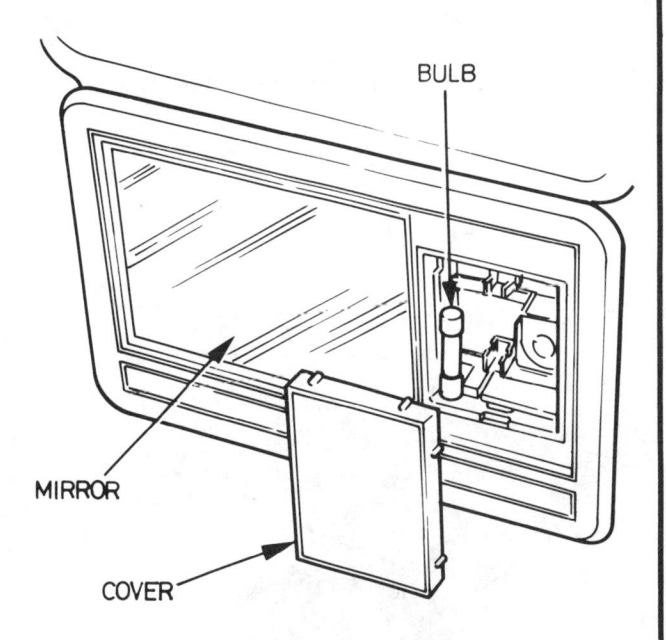

Fig. 12.23 Vanity light unit (Sec 21)

316 Chapter 12 Electrical system

22 Cigar lighter – removal and refitting

1. Disconnect the electrical cables.
2. Remove the ring nut and remove the lighter and protective cover panel.
3. The bulb is contained within the housing on top of the lighter body.
4. Bend up the retaining tags to remove the housing and gain access to the bulb.
5. Refit in the reverse order.

23 Radio – removal and refitting

1. Remove the ashtray.
2. Remove the radio mounting screws.
3. Remove the mounting plate screws.
4. Lower the mounting plate and push the radio out of the dashboard.
5. Disconnect the electrical cable and aerial, and remove the radio.
6. Refit in the reverse order.

24 Radio aerial – removal and refitting

Manual

1. Remove the radio and disconnect the aerial.
2. Remove the two screws from the aerial mounting plate on the roof.
3. Tie a length of stout cord to the aerial lead, and pull the aerial upwards out from the side pillar, leaving the cord inside.
4. Refit in the reverse order, using the cord to guide the new aerial back into position.
5. Use sealant on the base of the aerial mounting plate.

Automatic

6. Extend the aerial.
7. Remove the pins from the cable guide where the cable joins the aerial motor.
8. Remove the motor.

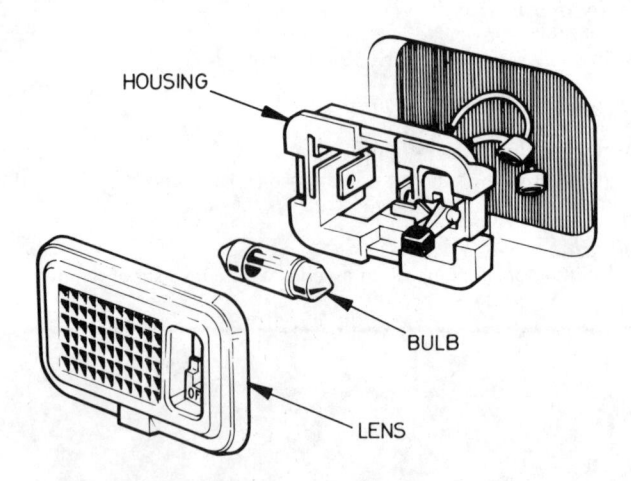

Fig. 12.24 Luggage compartment light (Sec 21)

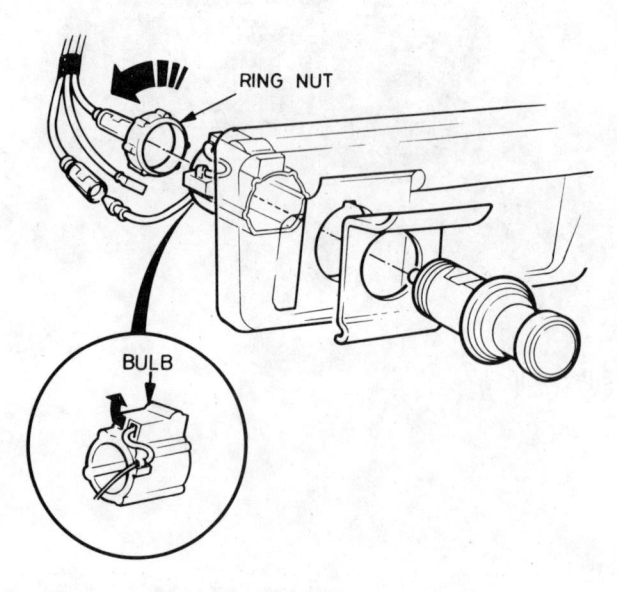

Fig. 12.25 Cigar lighter assembly (Sec 22)

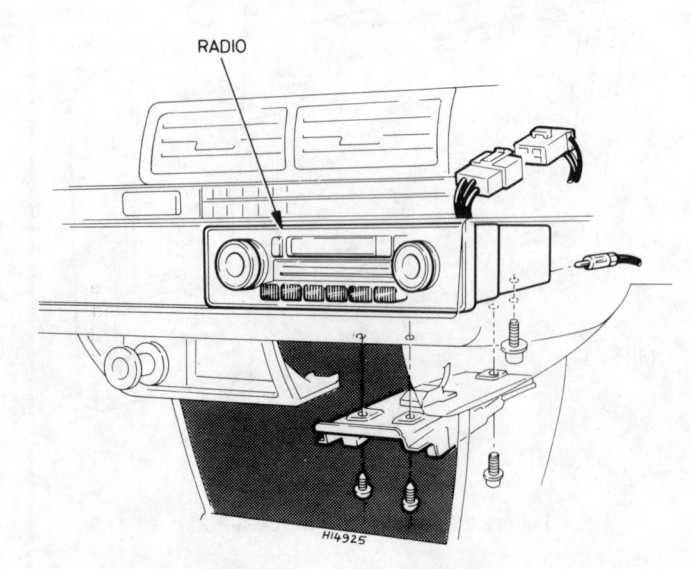

Fig. 12.26 Radio support bracket assembly (Sec 23)

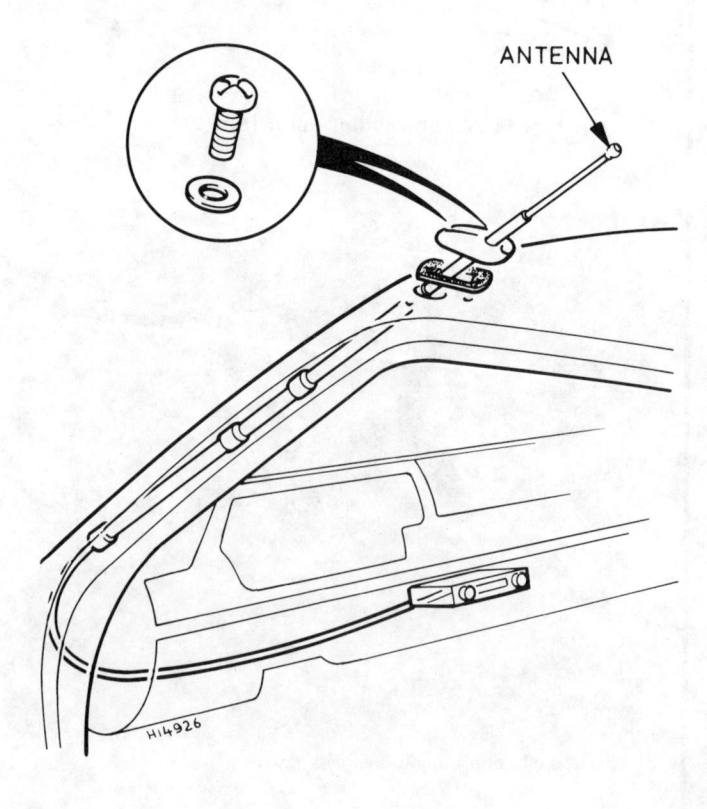

Fig. 12.27 Manual aerial (Sec 24)

Chapter 12 Electrical system

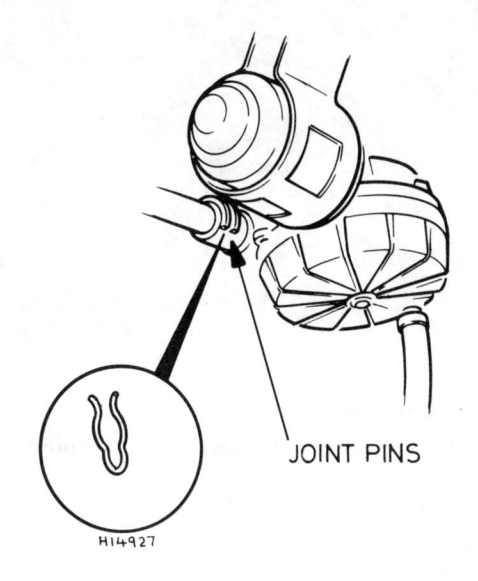

Fig. 12.28 Automatic aerial drive cable joint pins (Sec 24)

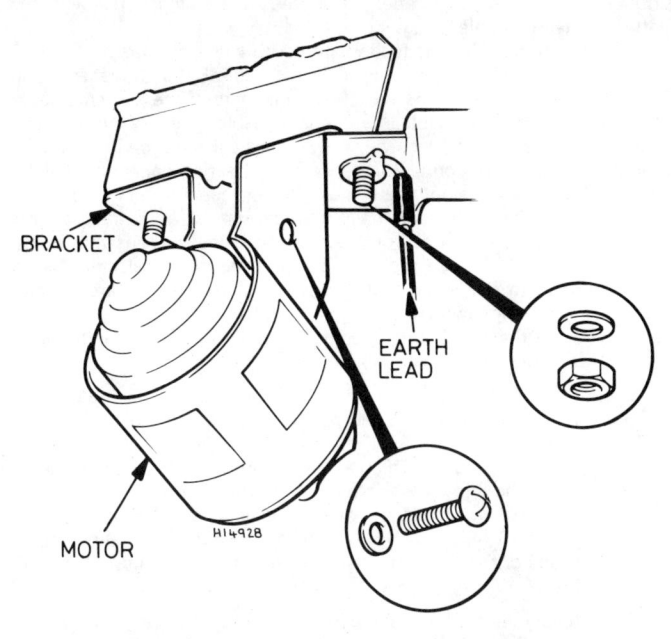

Fig. 12.29 Automatic aerial drive motor and bracket (Sec 24)

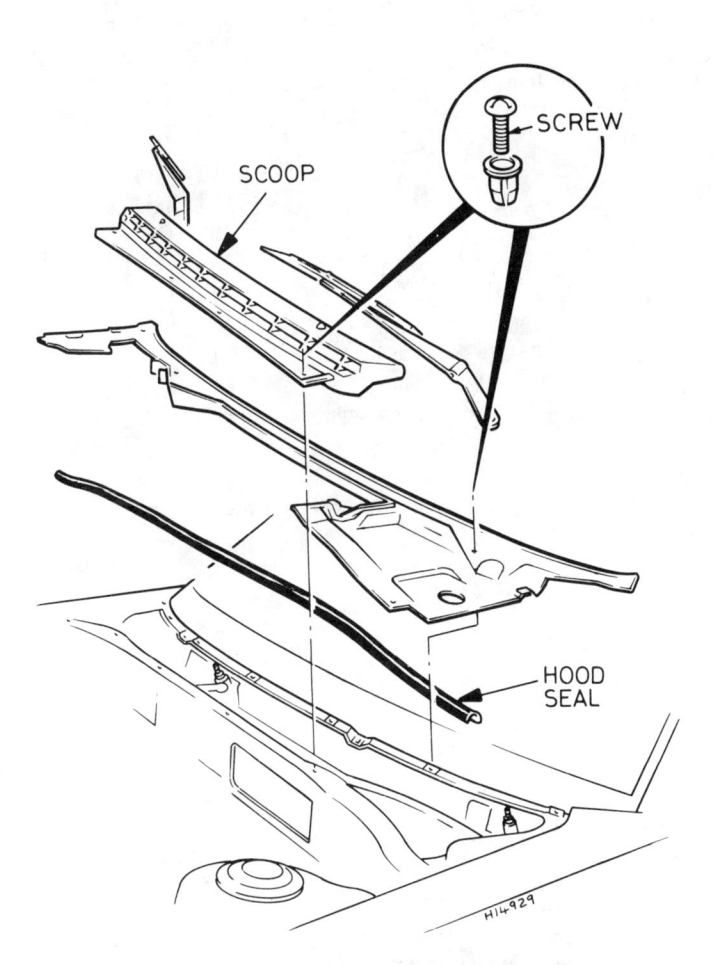

Fig. 12.30 Air scoop, cover and hood seal (Sec 28)

12 Tape the piece of cord to a new aerial, and tape the drive cable to the outer cable.
13 Pull the new aerial into the side pillar.
14 Remove the tapes, expose the cable end and insert it into the motor.
15 Switch the ignition to 'ACC' and retract the aerial.
16 The drive cable will be retracted by the motor.
17 Install the cable guide to the motor and insert the joint pins.
18 Switch off the ignition.
19 Refit the motor and bracket, not forgetting the earth strap.
20 Refit the aerial cover plate on the roof, using sealant on its base.

25 Mobile radio equipment – interference-free installation

Aerials – selection and fitting

The choice of aerials is now very wide. It should be realised that the quality has a profound effect on radio performance, and a poor, inefficient aerial can make suppression difficult.

A wing-mounted aerial is regarded as probably the most efficient for signal collection, but a roof aerial is usually better for suppression purposes because it is away from most interference fields. Stick-on wire aerials are available for attachment to the inside of the windscreen, but are not always free from the interference field of the engine and some accessories.

Motorised automatic aerials rise when the equipment is switched on and retract at switch-off. They require more fitting space and supply leads, and can be a source of trouble.

There is no merit in choosing a very long aerial as, for example, the type about three metres in length which hooks or clips on to the rear of the car, since part of this aerial will inevitably be located in an interference field. For VHF/FM radios the best length of aerial is about one metre. Active aerials have a transistor amplifier mounted at the base and this serves to boost the received signal. The aerial rod is sometimes rather shorter than normal passive types.

A large loss of signal can occur in the aerial feeder cable, especially over the Very High Frequency (VHF) bands. The design of feeder cable is invariably in the co-axial form, ie a centre conductor surrounded by a flexible copper braid forming the outer (earth) conductor. Between the inner and outer conductors is an insulator material which can be in solid or stranded form. Apart from insulation, its purpose is to maintain the correct spacing and concentricity. Loss of signal occurs in this insulator, the loss usually being greater in a poor quality cable. The

9 Lightly pull on the aerial drive cable, and turn the aerial switch to 'extend', with the ignition switch on 'ACC'.
10 The drive cable will be forced out.
11 Feed the aerial out of the side pillar in the same manner as for the manual aerial.

quality of cable used is reflected in the price of the aerial with the attached feeder cable.

The capacitance of the feeder should be within the range 65 to 75 picofarads (pF) approximately (95 to 100 pF for Japanese and American equipment), otherwise the adjustment of the car radio aerial trimmer may not be possible. An extension cable is necessary for a long run between aerial and receiver. If this adds capacitance in excess of the above limits, a connector containing a series capacitor will be required, or an extension which is labelled as 'capacity-compensated'.

Fitting the aerial will normally involve making a 7/8 in (22 mm) diameter hole in the bodywork, but read the instructions that come with the aerial kit. Once the hole position has been selected, use a centre punch to guide the drill. Use sticky masking tape around the area for this helps with marking out and drill location, and gives protection to the paintwork should the drill slip. Three methods of making the hole are in use:

(a) Use a hole saw in the electric drill. This is, in effect, a circular hacksaw blade wrapped round a former with a centre pilot drill.
(b) Use a tank cutter which also has cutting teeth, but is made to shear the metal by tightening with an Allen key.
(c) The hard way of drilling out the circle is using a small drill, say 1/8 in (3 mm), so that the holes overlap. The centre metal drops out and the hole is finished with round and half-round files.

Whichever method is used, the burr is removed from the body metal and paint removed from the underside. The aerial is fitted tightly ensuring that the earth fixing, usually a serrated washer, ring or clamp, is making a solid connection. *This earth connection is important in reducing interference.* Cover any bare metal with primer paint and topcoat, and follow by underseal if desired.

Aerial feeder cable routing should avoid the engine compartment and areas where stress might occur, eg under the carpet where feet will be located. Roof aerials require that the headlining be pulled back and that a path is available down the door pillar. It is wise to check with the vehicle dealer whether roof aerial fitting is recommended.

Loudspeakers

Speakers should be matched to the output stage of the equipment, particularly as regards the recommended impedance. Power transistors used for driving speakers are sensitive to the loading placed on them.

Before choosing a mounting position for speakers, check whether the vehicle manufacturer has provided a location for them. Generally door-mounted speakers give good stereophonic reproduction, but not all doors are able to accept them. The next best position is the rear parcel shelf, and in this case speaker apertures can be cut into the shelf, or pod units may be mounted.

For door mounting, first remove the trim, which is often held on by 'poppers' or press studs, and then select a suitable gap in the inside door assembly. Check that the speaker would not obstruct glass or winder mechanism by winding the window up and down. A template is often provided for marking out the trim panel hole, and then the four fixing holes must be drilled through. Mark out with chalk and cut cleanly with a sharp knife or keyhole saw. Speaker leads are then threaded through the door and door pillar, if necessary drilling 10 mm diameter holes. Fit grommets in the holes and connect to the radio or tape unit correctly. Do not omit a waterproofing cover, usually supplied with door speakers. If the speaker has to be fixed into the metal of the door itself, use self-tapping screws, and if the fixing is to the door trim use self-tapping screws and flat spire nuts.

Rear shelf mounting is somewhat simpler but it is necessary to find gaps in the metalwork underneath the parcel shelf. However, remember that the speakers should be as far apart as possible to give a good stereo effect. Pod-mounted speakers can be screwed into position through the parcel shelf material, but it is worth testing for the best position. Sometimes good results are found by reflecting sound off the rear window.

Unit installation

Many vehicles have a dash panel aperture to take a radio/audio unit, a recognised international standard being 189.5 mm x 60 mm. Alternatively a console may be a feature of the car interior design and this, mounted below the dashboard, gives more room. If neither facility is available a unit may be mounted on the underside of the parcel shelf; these are frequently non-metallic and an earth wire from the case to a

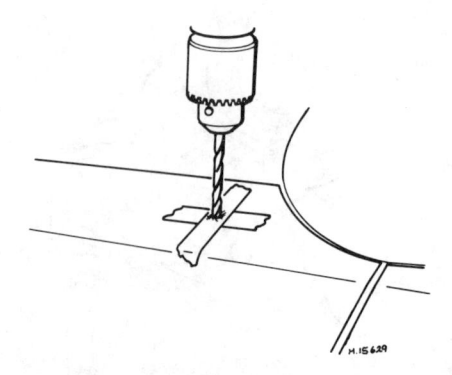

Fig. 12.31 Drilling the bodywork for aerial mounting

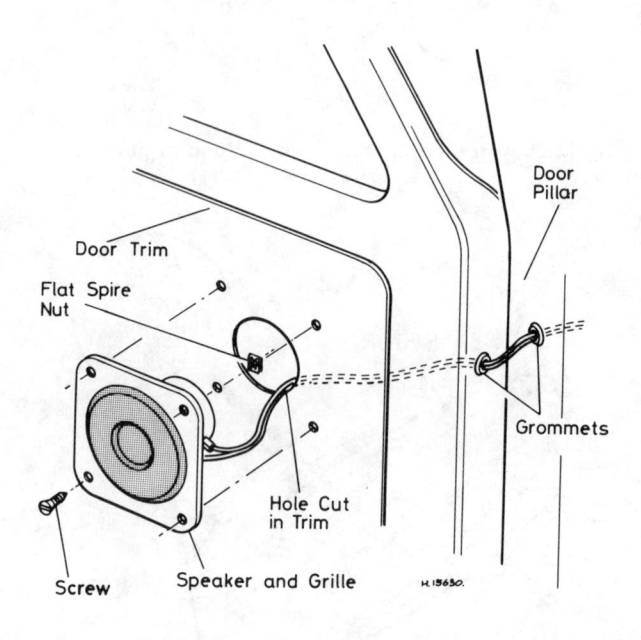

Fig. 12.32 Door-mounted speaker installation

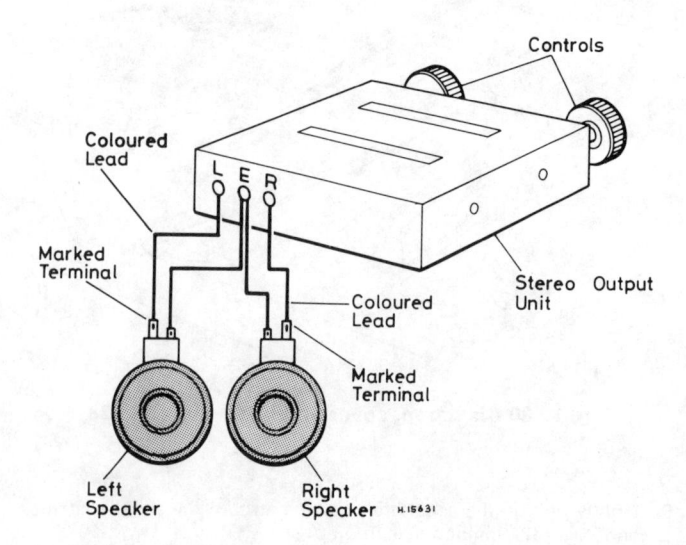

Fig. 12.33 Speaker connections must be correctly made as shown

Chapter 12 Electrical system

good earth point is necessary. A three-sided cover in the form of a cradle is obtainable from car radio dealers and this gives a professional appearance to the installation; in this case choose a position where the controls can be reached by a driver with his seat belt on.

Installation of the radio/audio unit is basically the same in all cases, and consists of offering it into the aperture after removal of the knobs (*not* push buttons) and the trim plate. In some cases a special mounting plate is required to which the unit is attached. It is worthwhile supporting the rear end in cases where sag or strain may occur, and it is usually possible to use a length of perforated metal strip attached between the unit and a good support point nearby. In general it is recommended that tape equipment should be installed at or nearly horizontal.

Connections to the aerial socket are simply by the standard plug terminating the aerial downlead or its extension cable. Speakers for a stereo system must be matched and correctly connected, as outlined previously.

Note: *While all work is carried out on the power side, it is wise to disconnect the battery earth lead.* Before connection is made to the vehicle electrical system, check that the polarity of the unit is correct. Most vehicles use a negative earth system, but radio/audio units often have a reversible plug to convert the set to either + or − earth. *Incorrect connection may cause serious damage.*

The power lead is often permanently connected inside the unit and terminates with one half of an in-line fuse carrier. The other half is fitted with a suitable fuse (3 or 5 amperes) and a wire which should go to a power point in the electrical system. This may be the accessory terminal on the ignition switch, giving the advantage of power feed with ignition or with the ignition key at the 'accessory' position. Power to the unit stops when the ignition key is removed. Alternatively, the lead may be taken to a live point at the fusebox with the consequence of having to remember to switch off at the unit before leaving the vehicle.

Before switching on for initial test, be sure that the speaker connections have been made, for running without load can damage the output transistors. Switch on next and tune through the bands to ensure that all sections are working, and check the tape unit if applicable. The aerial trimmer should be adjusted to give the strongest reception on a weak signal in the medium wave band, at say 200 metres.

Interference

In general, when electric current changes abruptly, unwanted electrical noise is produced. The motor vehicle is filled with electrical devices which change electric current rapidly, the most obvious being the contact breaker.

When the spark plugs operate, the sudden pulse of spark current causes the associated wiring to radiate. Since early radio transmitters used sparks as a basis of operation, it is not surprising that the car radio will pick up ignition spark noise unless steps are taken to reduce it to acceptable levels.

Interference reaches the car radio in two ways:

(a) by conduction through the wiring.
(b) by radiation to the receiving aerial.

Initial checks presuppose that the bonnet is down and fastened, the radio unit has a good earth connection (*not* through the aerial downlead outer), no fluorescent tubes are working near the car, the aerial trimmer has been adjusted, and the vehicle is in a position to receive radio signals, ie not in a metal-clad building.

Switch on the radio and tune it to the middle of the medium wave (MW) band off-station with the volume (gain) control set fairly high. Switch on the ignition (but do not start the engine) and wait to see if irregular clicks or hash noise occurs. Tapping the facia panel may also produce the effects. If so, this will be due to the voltage stabiliser, which is an on-off thermal switch to control instrument voltage. It is located usually on the back of the instrument panel, often attached to the speedometer. Correction is by attachment of a capacitor and, if still troublesome, chokes in the supply wires.

Switch on the engine and listen for interference on the MW band. Depending on the type of interference, the indications are as follows.

A harsh crackle that drops out abruptly at low engine speed or when the headlights are switched on is probably due to a voltage regulator.

A whine varying with engine speed is due to the dynamo or

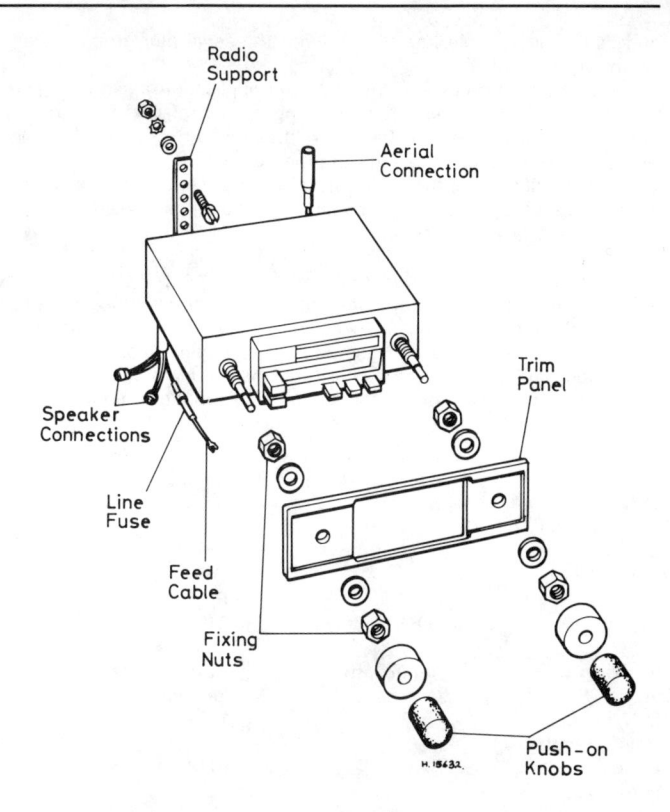

Fig. 12.34 Mounting component details for radio/cassette unit

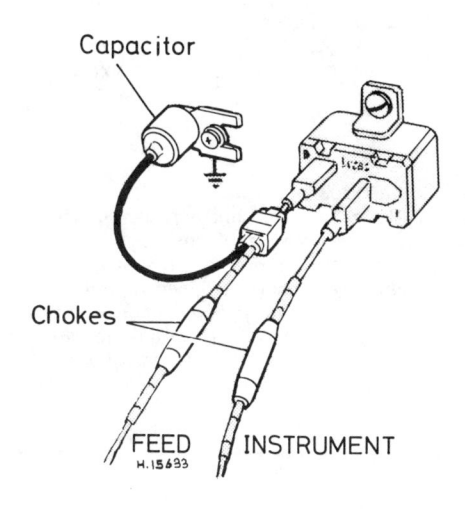

Fig. 12.35 Voltage stabiliser interference suppression

alternator. Try temporarily taking off the fan belt — if the noise goes this is confirmation.

Regular ticking or crackle that varies in rate with the engine speed is due to the ignition system. With this trouble in particular and others in general, check to see if the noise is entering the receiver from the wiring or by radiation. To do this, pull out the aerial plug, (preferably shorting out the input socket or connecting a 62 pF capacitor across it). If the noise disappears it is coming in through the aerial and is *radiation noise.* If the noise persists it is reaching the receiver through the wiring and is said to be *line-borne.*

Interference from wipers, washers, heater blowers, turn-indicators, stop lamps, etc is usually taken to the receiver by wiring, and simple treatment using capacitors and possibly chokes will solve the problem. Switch on each one in turn (wet the screen first for running wipers!)

and listen for possible interference with the aerial plug in place and again when removed.

Electric petrol pumps are now finding application again and give rise to an irregular clicking, often giving a burst of clicks when the ignition is on but the engine has not yet been started. It is also possible to receive whining or crackling from the pump.

Note that if most of the vehicle accessories are found to be creating interference all together, the probability is that poor aerial earthing is to blame.

Component terminal markings

Throughout the following sub-sections reference will be found to various terminal markings. These will vary depending on the manufacturer of the relevant component. If terminal markings differ from those mentioned, reference should be made to the following table, where the most commonly encountered variations are listed.

Alternator	Alternator terminal (thick lead)	Exciting winding terminal
DIN/Bosch	B+	DF
Delco Remy	+	EXC
Ducellier	+	EXC
Ford (US)	+	DF
Lucas	+	F
Marelli	+B	F

Ignition coil	Ignition switch terminal	Contact breaker terminal
DIN/Bosch	15	1
Delco Remy	+	–
Ducellier	BAT	RUP
Ford (US)	B/+	CB/–
Lucas	SW/+	–
Marelli	BAT/+B	D

Voltage regulator	Voltage input terminal	Exciting winding terminal
DIN/Bosch	B+/D+	DF
Delco Remy	BAT/+	EXC
Ducellier	BOB/BAT	EXC
Ford (US)	BAT	DF
Lucas	+/A	F
Marelli		F

Suppression methods – ignition

Suppressed HT cables are supplied as original equipment by manufacturers and will meet regulations as far as interference to neighbouring equipment is concerned. It is illegal to remove such suppression unless an alternative is provided, and this may take the form of resistive spark plug caps in conjunction with plain copper HT cable. For VHF purposes, these and 'in-line' resistors may not be effective, and resistive HT cable is preferred. Check that suppressed cables are actually fitted by observing cable identity lettering, or measuring with an ohmmeter – the value of each plug lead should be 5000 to 10 000 ohms.

A 1 microfarad capacitor connected from the LT supply side of the ignition coil to a good nearby earth point will complete basic ignition interference treatment. *NEVER fit a capacitor to the coil terminal to the contact breaker – the result would be burnt out points in a short time.*

If ignition noise persists despite the treatment above, the following sequence should be followed:

(a) Check the earthing of the ignition coil; remove paint from fixing clamp.

(b) If this does not work, lift the bonnet. Should there be no change in interference level, this may indicate that the bonnet is not electrically connected to the car body. Use a proprietary braided strap across a bonnet hinge ensuring a first class electrical connection. If, however, lifting the bonnet increases the interference, then fit resistive HT cables of a higher ohms-per-metre value.

(c) If all these measures fail, it is probable that re-radiation from metallic components is taking place. Using a braided strap between metallic points, go round the vehicle systematically – try the following: engine to body, exhaust system to body, front suspension to engine and to body, steering column to body (especially French and Italian cars), gear lever to engine and to body (again especially French and Italian cars), Bowden cable to body, metal parcel shelf to body. When an offending component is located it should be bonded with the strap permanently.

(d) As a next step, the fitting of distributor suppressors to each lead at the distributor end may help.

(e) Beyond this point is involved the possible screening of the distributor and fitting resistive spark plugs, but such advanced treatment is not usually required for vehicles with entertainment equipment.

Electronic ignition systems have built-in suppression components, but this does not relieve the need for using suppressed HT leads. In some cases it is permitted to connect a capacitor on the low tension supply side of the ignition coil, but not in every case. Makers' instructions should be followed carefully, otherwise damage to the ignition semiconductors may result.

Suppression methods – generators

For older vehicles with dynamos a 1 microfarad capacitor from the D (larger) terminal to earth will usually cure dynamo whine. Alternators should be fitted with a 3 microfarad capacitor from the B+ main output terminal (thick cable) to earth. Additional suppression may be obtained by the use of a filter in the supply line to the radio receiver.

It is most important that:

(a) Capacitors are never connected to the field terminals of either a dynamo or alternator.

(b) Alternators must not be run without connection to the battery.

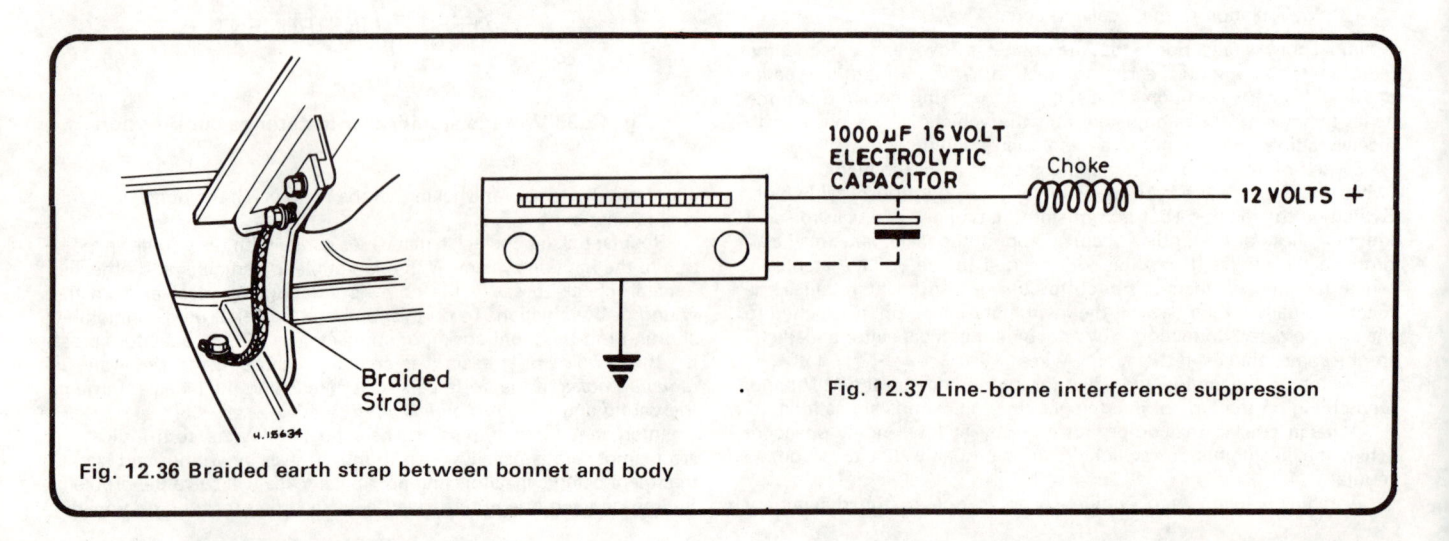

Fig. 12.36 Braided earth strap between bonnet and body

Fig. 12.37 Line-borne interference suppression

Chapter 12 Electrical system

Suppression methods – voltage regulators

Voltage regulators used with DC dynamos should be suppressed by connecting a 1 microfarad capacitor from the control box D terminal to earth.

Alternator regulators come in three types:

(a) Vibrating contact regulators separate from the alternator. Used extensively on continental vehicles.
(b) Electronic regulators separate from the alternator.
(c) Electronic regulators built-in to the alternator.

In case (a) interference may be generated on the AM and FM (VHF) bands. For some cars a replacement suppressed regulator is available. Filter boxes may be used with non-suppressed regulators. But if not available, then for AM equipment a 2 microfarad or 3 microfarad capacitor may be mounted at the voltage terminal marked D+ or B+ of the regulator. FM bands may be treated by a feed-through capacitor of 2 or 3 microfarad.

Electronic voltage regulators are not always troublesome, but where necessary, a 1 microfarad capacitor from the regulator + terminal will help.

Integral electronic voltage regulators do not normally generate much interference, but when encountered this is in combination with alternator noise. A 1 microfarad or 2 microfarad capacitor from the warning lamp (IND) terminal to earth for Lucas ACR alternators and Femsa, Delco and Bosch equivalents should cure the problem.

Suppression methods – other equipment

Wiper motors – Connect the wiper body to earth with a bonding strap. For all motors use a 7 ampere choke assembly inserted in the leads to the motor.

Heater motors – Fit 7 ampere line chokes in both leads, assisted if necessary by a 1 microfarad capacitor to earth from both leads.

Electronic tachometer – The tachometer is a possible source of ignition noise – check by disconnecting at the ignition coil CB terminal. It usually feeds from ignition coil LT pulses at the contact breaker terminal. A 3 ampere line choke should be fitted in the tachometer lead at the coil CB terminal.

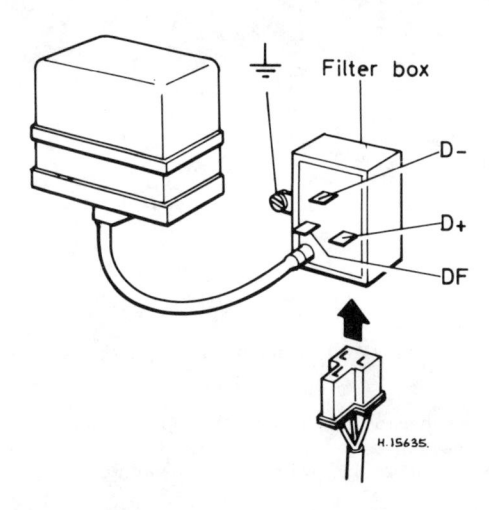

Fig. 12.38 Typical filter box for vibrating contact voltage regulator (alternator equipment)

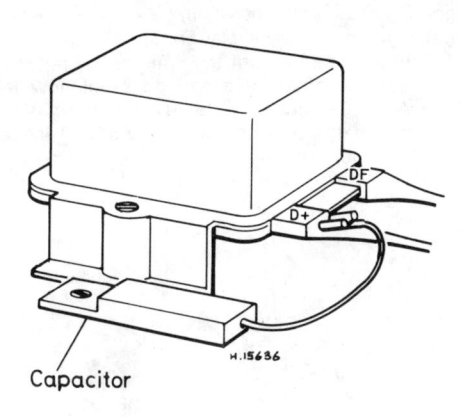

Fig. 12.39 Suppression of AM interference by vibrating contact voltage regulator (alternator equipment)

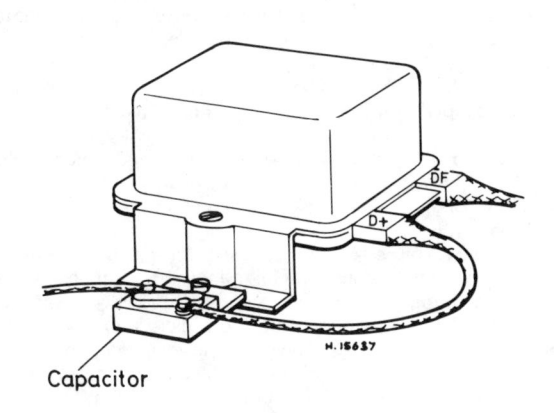

Fig. 12.40 Suppression of FM interference by vibrating contact voltage regulator (alternator equipment)

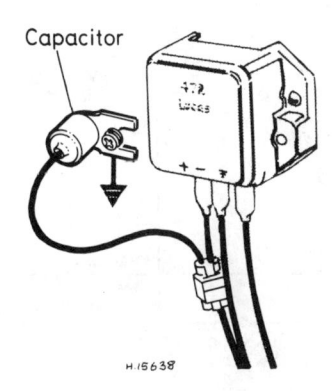

Fig. 12.41 Electronic voltage regulator suppression

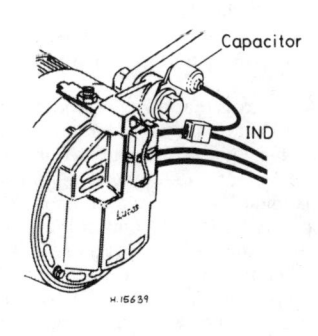

Fig. 12.42 Suppression of interference from electronic voltage regulator when integral with alternator

Chapter 12 Electrical system

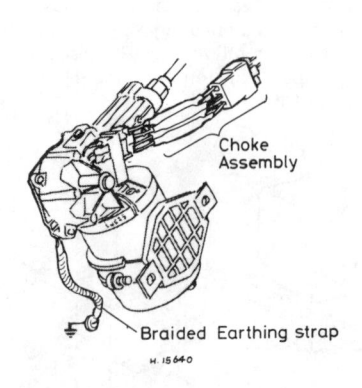

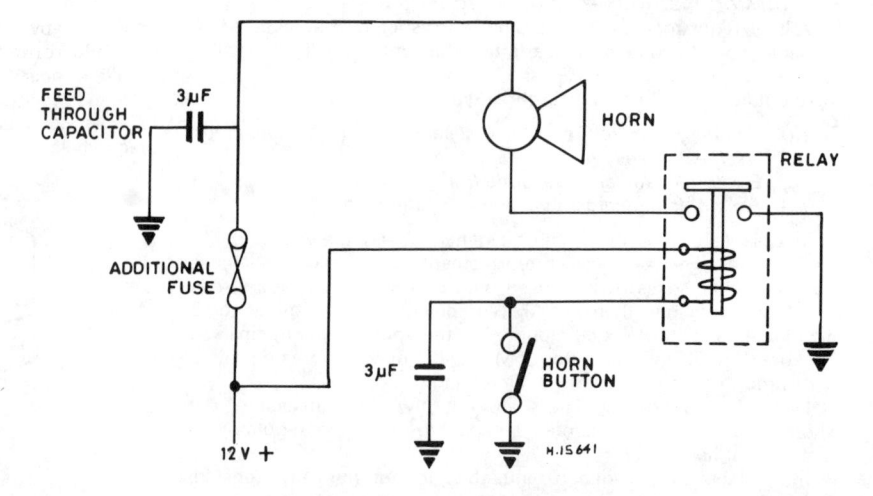

Fig. 12.43 Wiper motor suppression

Fig. 12.44 Use of relay to reduce horn interference

Horn – A capacitor and choke combination is effective if the horn is directly connected to the 12 volt supply. The use of a relay is an alternative remedy, as this will reduce the length of the interference-carrying leads.

Electrostatic noise – Characteristics are erratic crackling at the receiver, with disappearance of symptoms in wet weather. Often shocks may be given when touching bodywork. Part of the problem is the build-up of static electricity in non-driven wheels and the acquisition of charge on the body shell. It is possible to fit spring-loaded contacts at the wheels to give good conduction between the rotary wheel parts and the vehicle frame. Changing a tyre sometimes helps – because of tyres' varying resistances. In difficult cases a trailing flex which touches the ground will cure the problem. If this is not acceptable it is worth trying conductive paint on the tyre walls.

Fuel pump – Suppression requires a 1 microfarad capacitor between the supply wire to the pump and a nearby earth point. If this is insufficient a 7 ampere line choke connected in the supply wire near the pump is required.

Fluorescent tubes – Vehicles used for camping/caravanning frequently have fluorescent tube lighting. These tubes require a relatively high voltage for operation and this is provided by an inverter (a form of oscillator) which steps up the vehicle supply voltage. This can give rise to serious interference to radio reception, and the tubes themselves can contribute to this interference by the pulsating nature of the lamp discharge. In such situations it is important to mount the aerial as far away from a fluorescent tube as possible. The interference problem may be alleviated by screening the tube with fine wire turns spaced an inch (25 mm) apart and earthed to the chassis. Suitable chokes should be fitted in both supply wires close to the inverter.

Radio/cassette case breakthrough

Magnetic radiation from dashboard wiring may be sufficiently intense to break through the metal case of the radio/cassette player. Often this is due to a particular cable routed too close and shows up as ignition interference on AM and cassette play and/or alternator whine on cassette play.

The first point to check is that the clips and/or screws are fixing all parts of the radio/cassette case together properly. Assuming good earthing of the case, see if it is possible to re-route the offending cable – the chances of this are not good, however, in most cars.

Next release the radio/cassette player and locate it in different positions with temporary leads. If a point of low interference is found, then if possible fix the equipment in that area. This also confirms that local radiation is causing the trouble. If re-location is not feasible, fit the radio/cassette player back in the original position.

Alternator interference on cassette play is now caused by radiation from the main charging cable which goes from the battery to the output terminal of the alternator, usually via the + terminal of the starter motor relay. In some vehicles this cable is routed under the dashboard, so the solution is to provide a direct cable route. Detach the original cable from the alternator output terminal and make up a new cable of at least 6 mm² cross-sectional area to go from alternator to battery with the shortest possible route. *Remember – do not run the engine with the alternator disconnected from the battery.*

Ignition breakthrough on AM and/or cassette play can be a difficult problem. It is worth wrapping earthed foil round the offending cable run near the equipment, or making up a deflector plate well screwed down to a good earth. Another possibility is the use of a suitable relay to switch on the ignition coil. The relay should be mounted close to the ignition coil; with this arrangement the ignition coil primary current is not taken into the dashboard area and does not flow through the ignition switch. A suitable diode should be used since it is possible that at ignition switch-off the output from the warning lamp alternator terminal could hold the relay on.

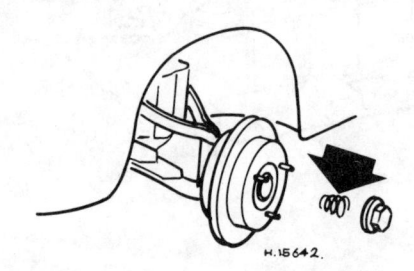

Fig. 12.45 Use of spring contacts at wheels

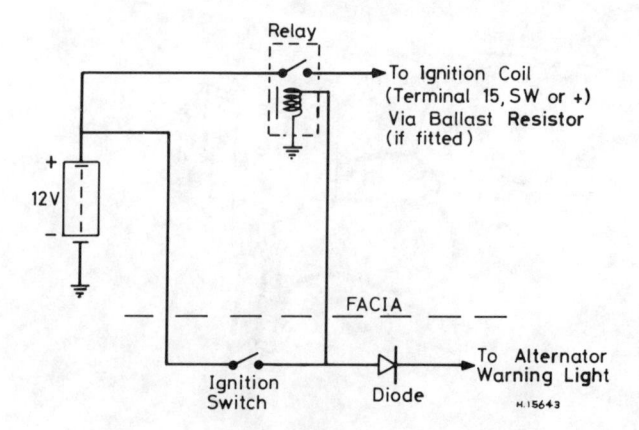

Fig. 12.46 Use of ignition coil relay to suppress case breakthrough

Chapter 12 Electrical system

Connectors for suppression components

Capacitors are usually supplied with tags on the end of the lead, while the capacitor body has a flange with a slot or hole to fit under a nut or screw with washer.

Connections to feed wires are best achieved by self-stripping connectors. These connectors employ a blade which, when squeezed down by pliers, cuts through cable insulation and makes connection to the copper conductors beneath.

Chokes sometimes come with bullet snap-in connectors fitted to the wires, and also with just bare copper wire. With connectors, suitable female cable connectors may be purchased from an auto-accessory shop together with any extra connectors required for the cable ends after being cut for the choke insertion. For chokes with bare wires, similar connectors may be employed together with insulation sleeving as required.

VHF/FM broadcasts

Reception of VHF/FM in an automobile is more prone to problems than the medium and long wavebands. Medium/long wave transmitters are capable of covering considerable distances, but VHF transmitters are restricted to line of sight, meaning ranges of 10 to 50 miles, depending upon the terrain, the effects of buildings and the transmitter power.

Because of the limited range it is necessary to retune on a long journey, and it may be better for those habitually travelling long distances or living in areas of poor provision of transmitters to use an AM radio working on medium/long wavebands.

When conditions are poor, interference can arise, and some of the suppression devices described previously fall off in performance at very high frequencies unless specifically designed for the VHF band. Available suppression devices include reactive HT cable, resistive distributor caps, screened plug caps, screened leads and resistive spark plugs.

For VHF/FM receiver installation the following points should be particularly noted:

(a) Earthing of the receiver chassis and the aerial mounting is important. Use a separate earthing wire at the radio, and scrape paint away at the aerial mounting.
(b) If possible, use a good quality roof aerial to obtain maximum height and distance from interference generating devices on the vehicle.
(c) Use of a high quality aerial downlead is important, since losses in cheap cable can be significant.
(d) The polarisation of FM transmissions may be horizontal, vertical, circular or slanted. Because of this the optimum mounting angle is at 45° to the vehicle roof.

Citizens' Band radio (CB)

In the UK, CB transmitter/receivers work within the 27 MHz and 934 MHz bands, using the FM mode. At present interest is concentrated on 27 MHz where the design and manufacture of equipment is less difficult. Maximum transmitted power is 4 watts, and 40 channels spaced 10 kHz apart within the range 27.60125 to 27.99125 MHz are available.

Aerials are the key to effective transmission and reception. Regulations limit the aerial length to 1.65 metres including the loading coil and any associated circuitry, so tuning the aerial is necessary to obtain optimum results. The choice of a CB aerial is dependent on whether it is to be permanently installed or removable, and the performance will hinge on correct tuning and the location point on the vehicle. Common practice is to clip the aerial to the roof gutter or to employ wing mounting where the aerial can be rapidly unscrewed. An alternative is to use the boot rim to render the aerial theftproof, but a popular solution is to use the 'magmount' – a type of mounting having a strong magnetic base clamping to the vehicle at any point, usually the roof.

Aerial location determines the signal distribution for both transmission and reception, but it is wise to choose a point away from the engine compartment to minimise interference from vehicle electrical equipment.

The aerial is subject to considerable wind and acceleration forces. Cheaper units will whip backwards and forwards and in so doing will alter the relationship with the metal surface of the vehicle with which it forms a ground plane aerial system. The radiation pattern will change correspondingly, giving rise to break-up of both incoming and outgoing signals.

Interference problems on the vehicle carrying CB equipment fall into two categories:

(a) Interference to nearby TV and radio receivers when transmitting.
(b) Interference to CB set reception due to electrical equipment on the vehicle.

Problems of break-through to TV and radio are not frequent, but can be difficult to solve. Mostly trouble is not detected or reported because the vehicle is moving and the symptoms rapidly disappear at the TV/radio receiver, but when the CB set is used as a base station any trouble with nearby receivers will soon result in a complaint.

It must not be assumed by the CB operator that his equipment is faultless, for much depends upon the design. Harmonics (that is, multiples) of 27 MHz may be transmitted unknowingly and these can fall into other user's bands. Where trouble of this nature occurs, low pass filters in the aerial or supply leads can help, and should be fitted in base station aerials as a matter of course. In stubborn cases it may be necessary to call for assistance from the licensing authority, or, if possible, to have the equipment checked by the manufacturers.

Interference received on the CB set from the vehicle equipment is, fortunately, not usually a severe problem. The precautions outlined previously for radio/cassette units apply, but there are some extra points worth noting.

It is common practice to use a slide-mount on CB equipment enabling the set to be easily removed for use as a base station, for example. Care must be taken that the slide mount fittings are properly earthed and that first class connection occurs between the set and slide-mount.

Vehicle manufacturers in the UK are required to provide suppression of electrical equipment to cover 40 to 250 MHz to protect TV and VHF radio bands. Such suppression appears to be adequately effective at 27 MHz, but suppression of individual items such as alternators/dynamos, clocks, stabilisers, flashers, wiper motors, etc, may still be necessary. The suppression capacitors and chokes available from auto-electrical suppliers for entertainment receivers will usually give the required results with CB equipment.

Other vehicle radio transmitters

Besides CB radio already mentioned, a considerable increase in the use of transceivers (ie combined transmitter and receiver units) has taken place in the last decade. Previously this type of equipment was fitted mainly to military, fire, ambulance and police vehicles, but a large business radio and radio telephone usage has developed.

Generally the suppression techniques described previously will suffice, with only a few difficult cases arising. Suppression is carried out to satisfy the 'receive mode', but care must be taken to use heavy duty chokes in the equipment supply cables since the loading on 'transmit' is relatively high.

Glass-fibre bodied vehicles

Such vehicles do not have the advantage of a metal box surrounding the engine as is the case, in effect, of conventional vehicles. It is usually necessary to line the bonnet, bulkhead and wing valances with metal foil, which could well be the aluminium foil available from builders merchants. Bonding of sheets one to another and the whole down to the chassis is essential.

Wiring harness may have to be wrapped in metal foil which again should be earthed to the vehicle chassis. The aerial base and radio chassis must be taken to the vehicle chassis by heavy metal braid. VHF radio suppression in glass-fibre cars may not be a feasible operation.

In addition to all the above, normal suppression components should be employed, but special attention paid to earth bonding. A screen enclosing the entire ignition system usually gives good improvement, and fabrication from fine mesh perforated metal is convenient. Good bonding of the screening boxes to several chassis points is essential.

26 Windscreen wiper blades and arms – removal and refitting

1 The wiper blades should be renewed as soon as they cease to wipe the windscreen efficiently.

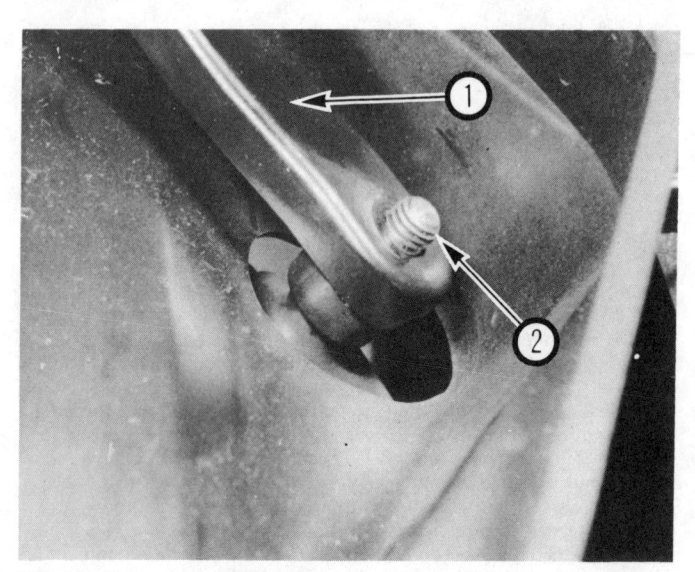

Chapter 12 Electrical system

26.3 Wiper arm and driveshaft (nut removed)
1 Wiper arm   2 Driveshaft

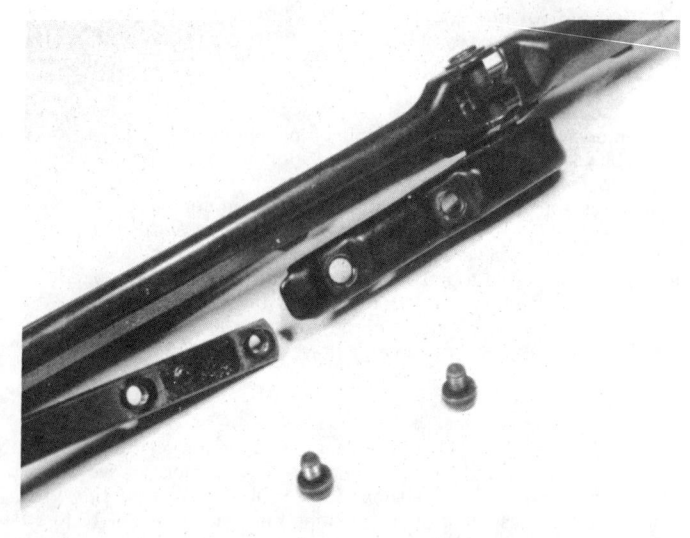

26.4 Wiper blade-to-wiper arm attachment screws

2 Replacement rubber blades are available, but it is better to renew the complete blade arm, as described here.
3 Remove the nut from the wiper arm (photo), and lift the wiper arm and blade arm from the driveshaft.
4 Remove the two screws securing the wiper blade to the wiper arm, and remove the blade (photo).
5 Refitting is a reversal of removal, but ensure that the wiper arm is fitted back on the shaft in its original position, to ensure the blade does not foul the windscreen edge during operation.

27 Windscreen wiper motor – removal and refitting

1 Remove the wiper arms, as described in Section 26.
2 Remove the scoop and bonnet seal, and the plastic cover shield. These are held in place by plastic screw type clips, and it is as well to have a supply of new clips, as they are easily broken.
3 Disconnect the balljoint between the wiper arm linkage and motor drive arm. It pulls off (photo).
4 Remove the clip and pull off the rubber cover (photo).
5 Disconnect the multi-connector and remove the motor mounting bolts (photo).
6 If it is defective, the motor should be renewed as a unit.
7 Refitting is a reversal of removal.
8 Apply a little silicone grease to the wiper linkage balljoints on reassembly.

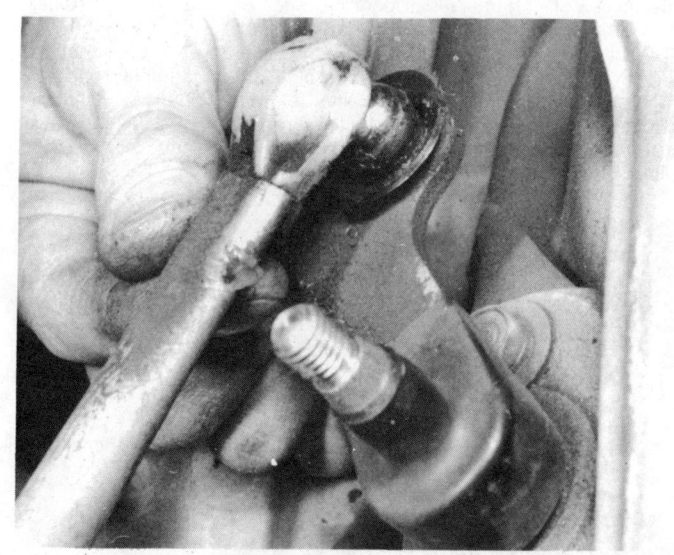

27.3 Disconnecting the wiper linkage balljoint

28 Headlamp wiper unit – removing and refitting

1 Lift the cap, undo the nut and remove the wiper arm (photo).
2 Disconnect the washer tube (photo).
3 Remove the front bumper moulding and grille.
4 Disconnect the electrical coupler.
5 Remove the motor mounting bolts (photo) and remove the motor unit.
6 Renew as a complete unit if defective.
7 Refit in the reverse order.

29 Windscreen wiper and headlamp wash system – general

1 The windscreen, rear window and headlamp washers are all fed by plastic tubing from the reservoir in the engine bay, near the battery.
2 Keep the washer jets aimed at the optimum wash position.

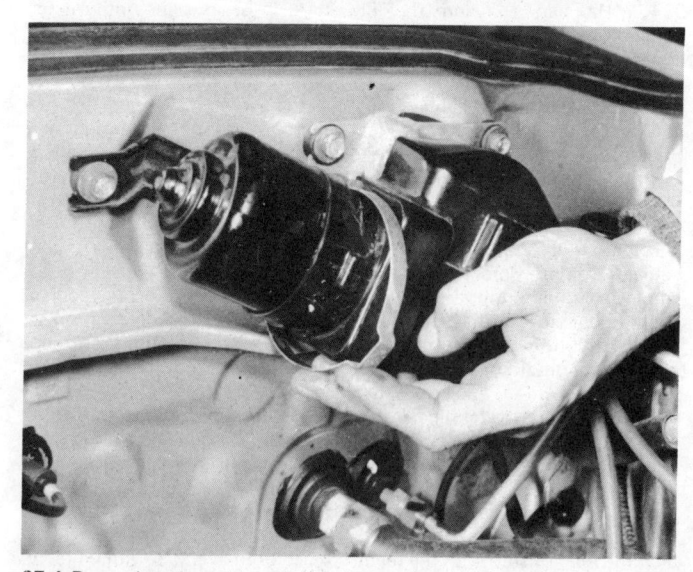

27.4 Removing the rubber cover clip

Chapter 12 Electrical system

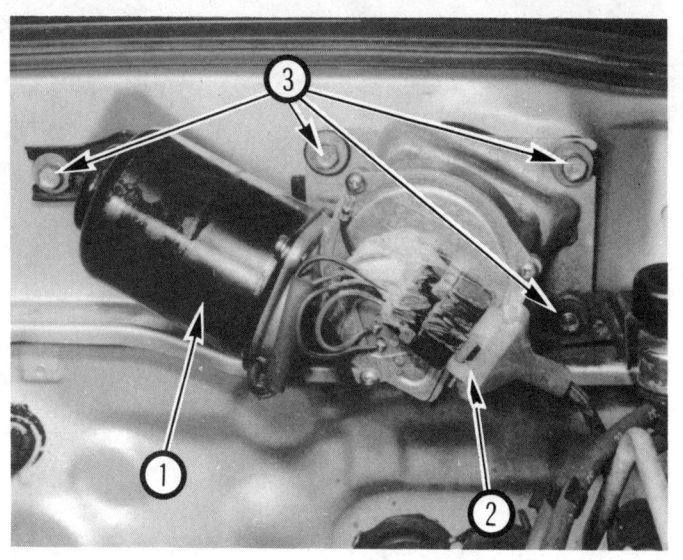

27.5 View of the wiper motor
1 Motor
2 Connector
3 Mounting screws

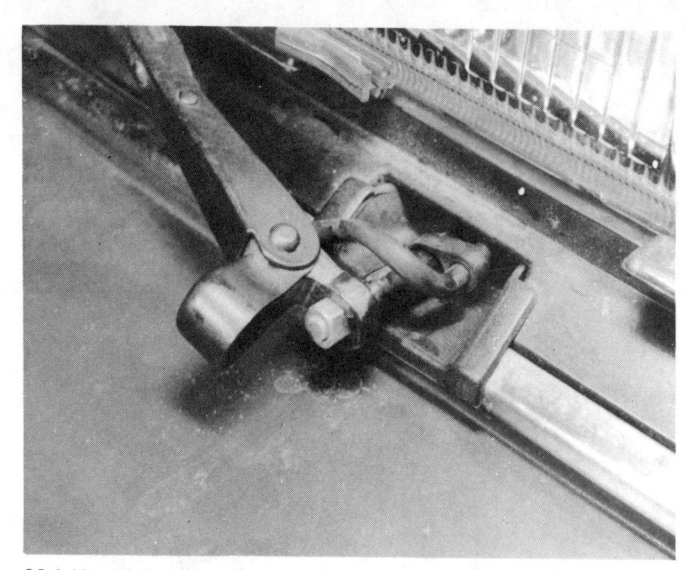

28.1 Headlight wiper retaining nut

28.2 Washer tube connection (arrowed)

28.5 Wiper motor mounting bolts (arrowed)

3 If the jets become blocked, disconnect the plastic tubing and flush the tubes through with fresh water.
4 Blow through the washer jets from the outlet side to clear any obstruction.
5 Remove the washer reservoir and flush it out in clean water.
6 To remove the reservoir, disconnect the electrical leads to the motor, and the plastic tubing, remove the retaining bolts, and lift it out.
7 If the motor becomes defective, renew the complete reservoir assembly.
8 Refit the reservoir, connect the leads and tubing, fill with clean water and retest the system.

30 Cruise control system – general

1 The cruise control system consists of an electronic control unit, mounted in the dashboard, which is fed various signals from sensors mounted in the systems which it controls (photo).
2 A cable connects the accelerator pedal and cruise control actuator, mounted in the engine compartment (photo).

30.1 Cruise control speed sensor

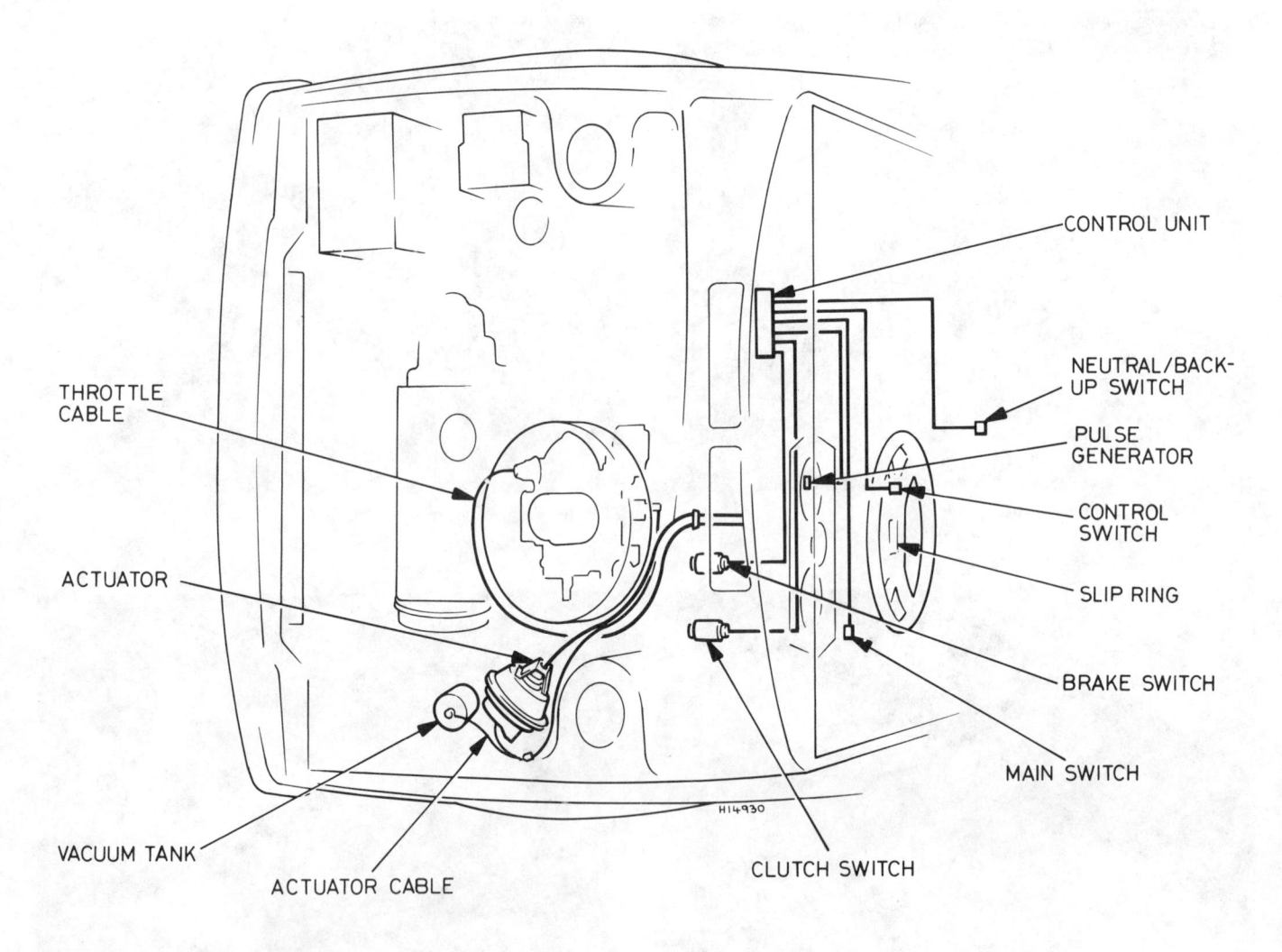

Fig. 12.47 Cruise control system and component location – North American type shown (Sec 30)

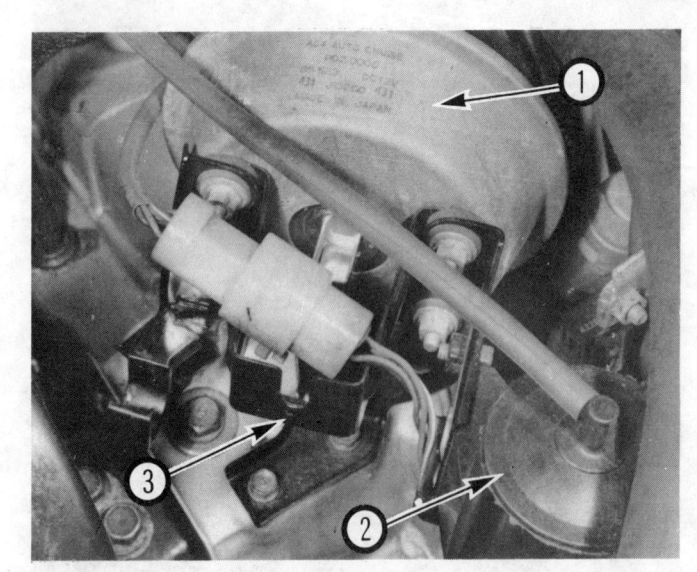

30.2 Cruise control actuator unit
1 Actuator
2 Vacuum tank
3 Cable bracket

3 Testing of the system should be left to your dealer, but the actuator cable can be changed and the actuator unit removed and refitted as follows.
4 To change the cable, loosen the locknut and adjuster at the actuator bracket, and unhook the cable end.
5 Unhook the cable from the accelerator linkage, and feed it through the engine bulkhead.
6 Refit in the reverse order, and adjust the cable to the clearance shown in Fig. 12.48.
7 To remove the actuator unit, disconnect the vacuum tube from the vacuum tank and the electrical connector.
8 Remove the cable as described above.
9 Remove the actuator mounting bolts and remove the actuator.
10 Refit in the reverse order and adjust the cable.

31 Speedometer cable – removal and refitting

1 Remove the transmission end of the cable as described in Chapters 6 and 7 (photo).
2 The instrument panel end is a push-fit into the instrument panel.
3 To remove it from the instrument panel, first detach the panel, as described in Section 13.
4 Depress the clips on the sides of the cable housing and pull the cable out from the instrument panel (photo).
5 Refit in the reverse order.

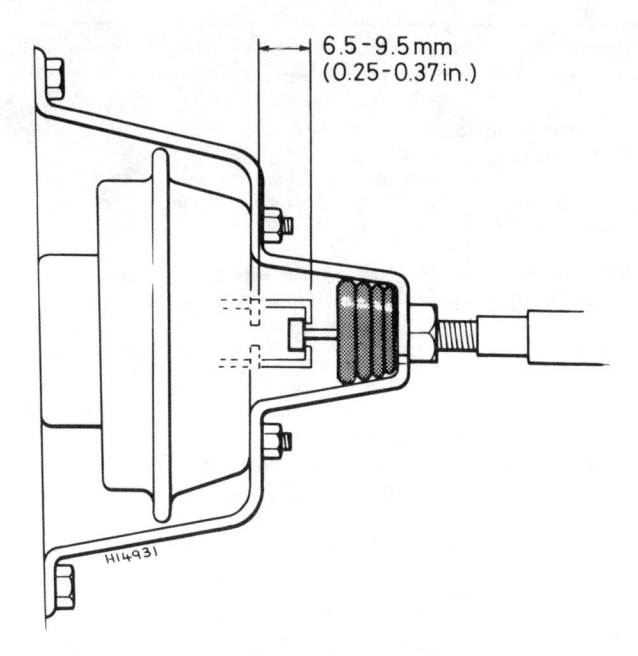

Chapter 12 Electrical system

Fig. 12.48 Actuator cable adjustment (Sec 30)

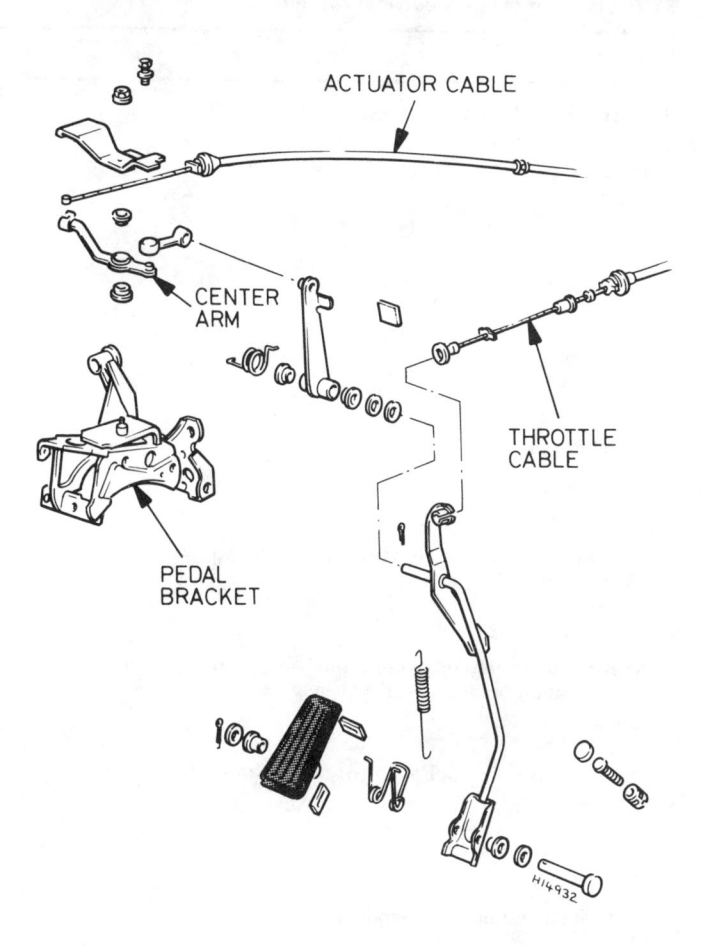

Fig. 12.49 Accelerator pedal linkage assembly (Sec 30)

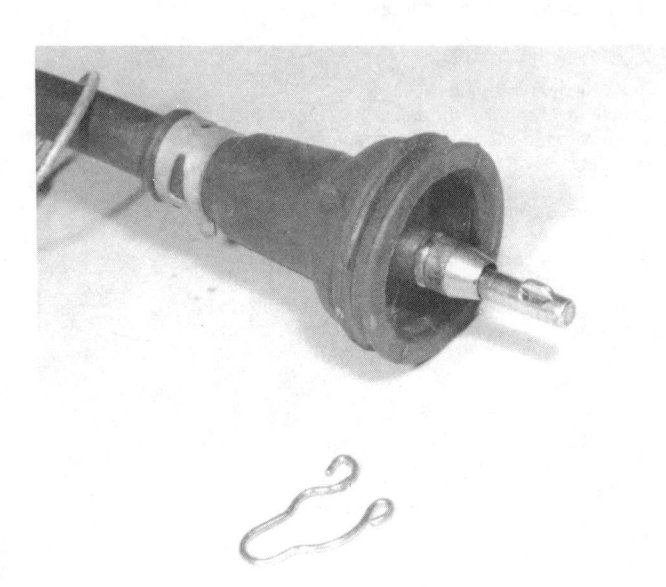

31.1 Speedometer cable transmission end

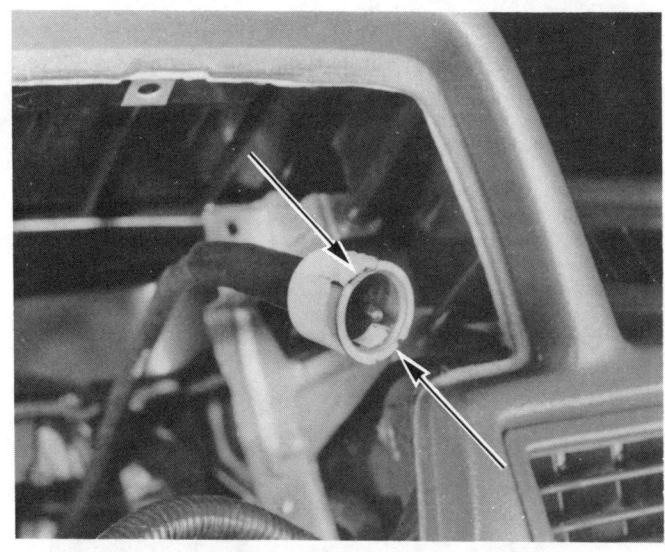

31.4 Instrument panel end of the speedometer cable. Press the clips as shown

32 Warning lights – general

1 Several of the systems fitted to these vehicles employ a warning light of some description.
2 Should a warning light illuminate and not go out after remedial action (like doing up a seat belt), check that all electrical connections are sound and making good contact.
3 Check the fuse and relay for that system if they are employed, and check also the warning light bulbs themselves.
4 If the warning lamp still remains illuminated, have the system checked by your dealer.

33 Optional systems – general

1 Several optional systems are available and may be fitted to your vehicle.
2 These include electronic navigator and compass, heated driver's seat etc.
3 The systems are generally reliable in service, and require little maintenance; being mainly electronic, integrated circuits.
4 Should any fail, have them tested by your dealer.

34 Fault diagnosis – electrical system

Symptom	Reason(s)
Starter motor fails to turn engine	
No electricity at starter motor	Battery discharged
	Battery defective internally
	Battery terminal leads loose or earth lead not securely attached to body
	Loose or broken connections in starter motor circuit
	Starter motor switch or solenoid faulty
Electricity at starter motor: faulty motor	Starter motor pinion jammed in mesh with flywheel gear ring
	Starter brushes badly worn, sticking, or brush wires loose
	Commutator dirty, worn or burnt
	Starter motor armature faulty
	Field coils earthed
Starter motor turns engine very slowly	
Electrical defects	Battery in discharged condition
	Starter brushes badly worn, sticking, or brush wires loose
	Loose wires in starter motor circuit
Starter motor operates without turning engine	
Mechanical damage	Pinion or flywheel gear teeth broken or worn
	Battery in discharged condition
Starter motor noisy or excessively rough engagement	
Lack of attention or mechanical damage	Pinion or flywheel gear teeth broken or worn
	Starter motor retaining bolts loose
Battery will not hold charge for more than a few days	
Wear or damage	Battery defective internally
	Electrolyte level too low or electrolyte too weak due to leakage
	Plate separators no longer fully effective
	Battery plates severely sulphated
Insufficient current flow to keep battery charged	Battery plates severely sulphated
	Drivebelt slipping
	Battery terminal connections loose or corroded
	Alternator not charging
	Short in lighting circuit causing continual battery drain
	Regulator unit nor working correctly
Ignition light fails to go out, battery runs flat in a few days	
Alternator not charging	Drivebelt loose and slipping or broken
	Brushes worn, sticking, broken or dirty
	Brush springs weak or broken
	Commutator dirty, worn or burnt
	Alternator field coils burnt, open, or shorted
	Commutator worn
	Pole pieces very loose
Regulator or cut-out fails to work correctly	Regulator incorrectly set
	Cut-out incorrectly set
	Open circuit in wiring of cut-out and regulator unit
Horn	
Horn operates all the time	Horn push either earthed or stuck down
	Horn cable to horn push earthed
Horn fails to operate	Blown fuse
	Cable or cable connection loose, broken or disconnected
	Horn has an internal fault
Horn emits intermittent or unsatisfactory noise	Cable connections loose
	Horn incorrectly adjusted
Lights	
Lights do not come on	If engine not running, battery discharged
	Sealed beam filament burnt out or bulbs broken
	Wire connections loose, disconnected or broken
	Light switch shorting or otherwise faulty

34 Fault diagnosis – electrical system

Symptom	Reason(s)
Lights come on but fade out	If engine not running battery discharged Light bulb filament burnt out or bulbs or sealed beam units broken Wire connections loose, disconnected or broken Light switch shorting or otherwise faulty
Lights give very poor illumination	Lamp glasses dirty Lamp badly out of adjustment
Lights work erratically – flashing on and off, especially over bumps	Battery terminals or earth connection loose Light not earthing properly Contacts in light switch faulty

Wipers

Symptom	Reason(s)
Wiper motor fails to work	Blown fuse Wire connection loose, disconnected, or broken Brushed badly worn Armature worn or faulty Field coils faulty
Wiper motor works very slowly and takes excessive current	Commutator dirty, greasy or burnt Armature bearings dirty or unaligned Armature badly worn or faulty
Wiper motor works slowly and takes little current	Brushes badly worn Commutator dirty, greasy or burnt Armature badly worn or faulty
Wiper motor works but wiper blades remain static	Wiper motor gearbox parts badly worn
Wipers do not stop when switched off or stop in wrong place	Auto-stop device faulty

35 Wiring diagrams – general

Where a wire in a diagram ends with a number in a triangle, the circuit continues from the same triangle/number combination elsewhere in the diagram.

Where a wire in a diagram ends with a number in a circle, the circuit continues in one of the sub-system diagrams.

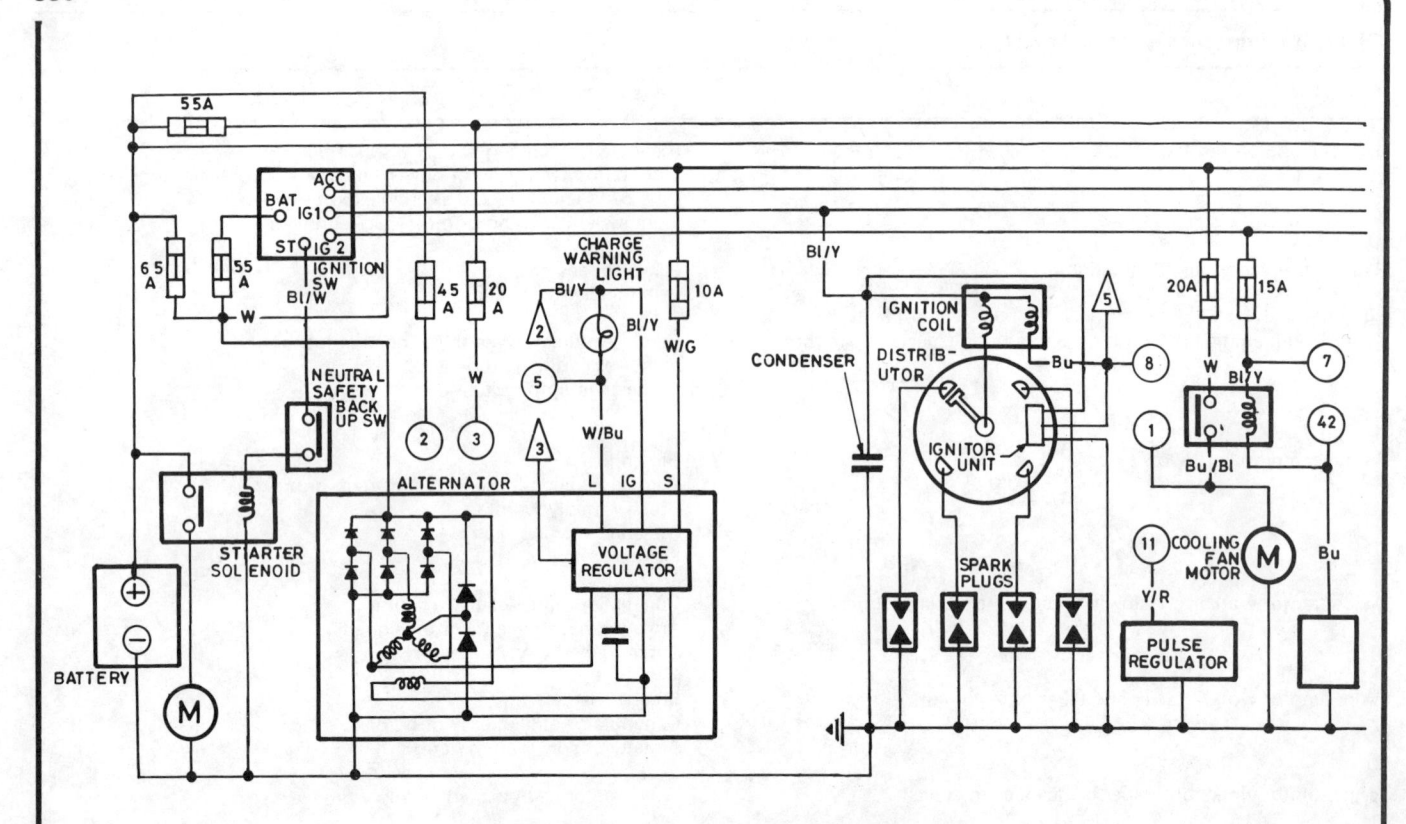

Fig. 12.50 Wiring diagram for 1984/5 UK models

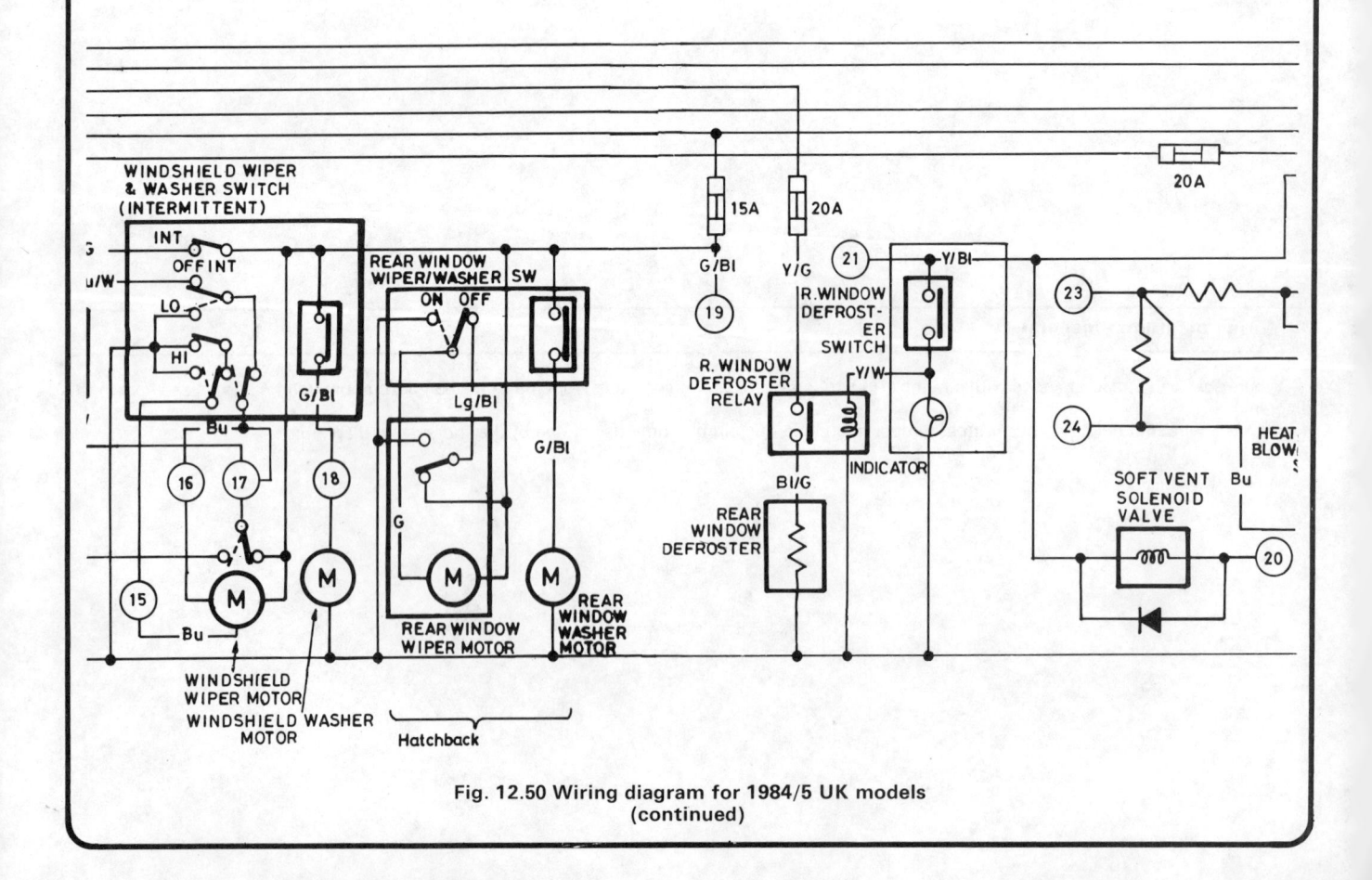

Fig. 12.50 Wiring diagram for 1984/5 UK models (continued)

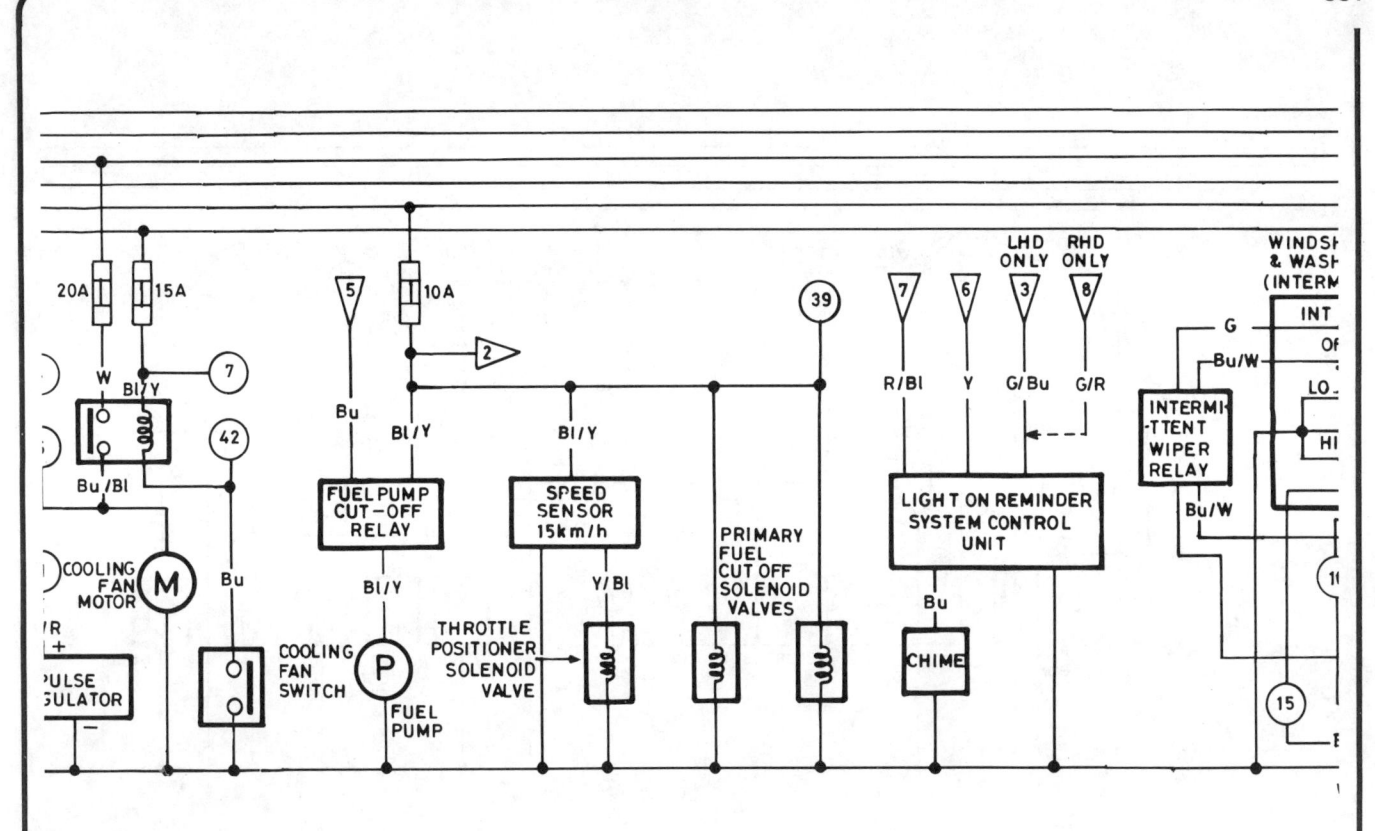

Fig. 12.50 Wiring diagram for 1984/5 UK models (continued)

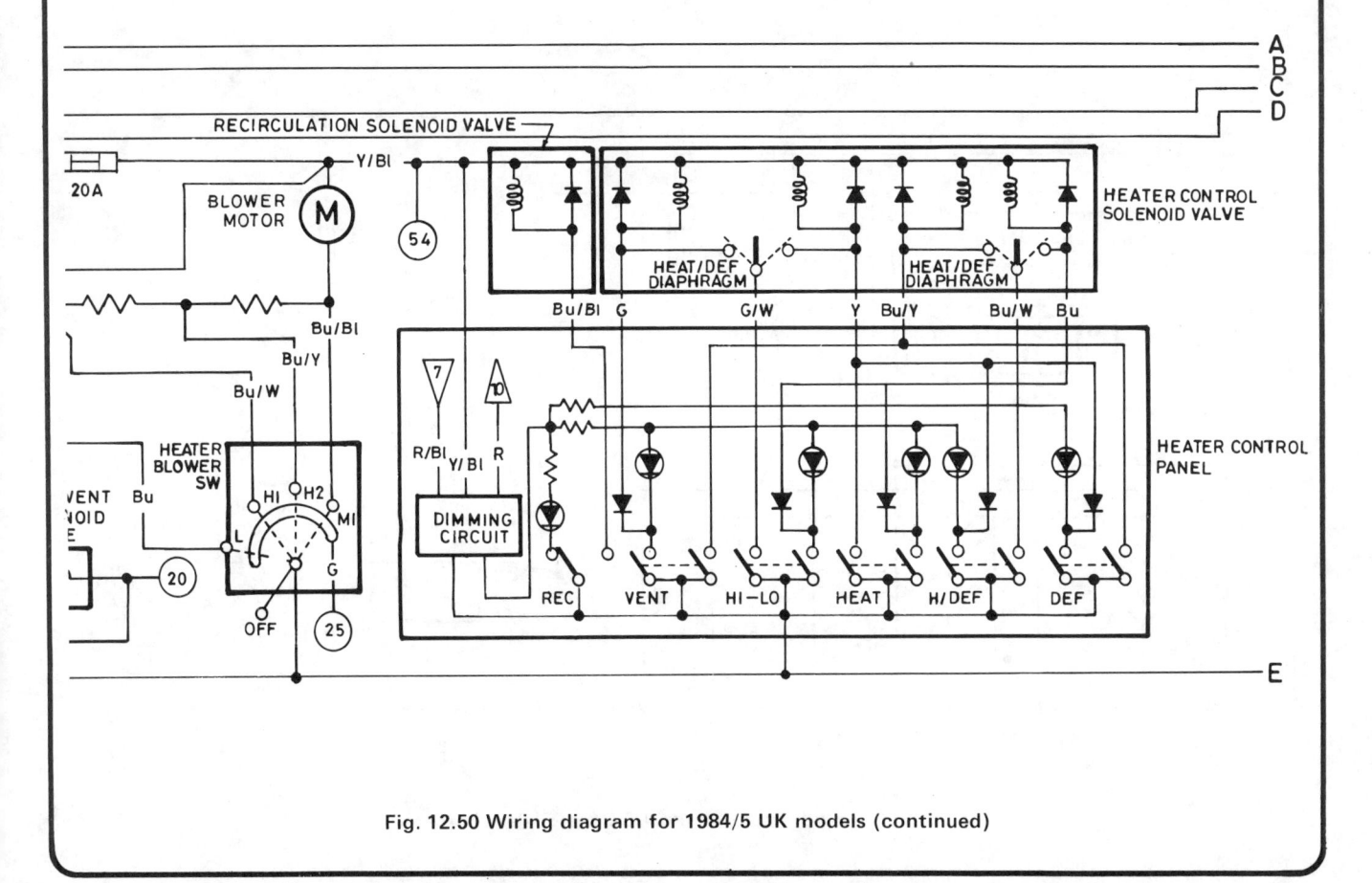

Fig. 12.50 Wiring diagram for 1984/5 UK models (continued)

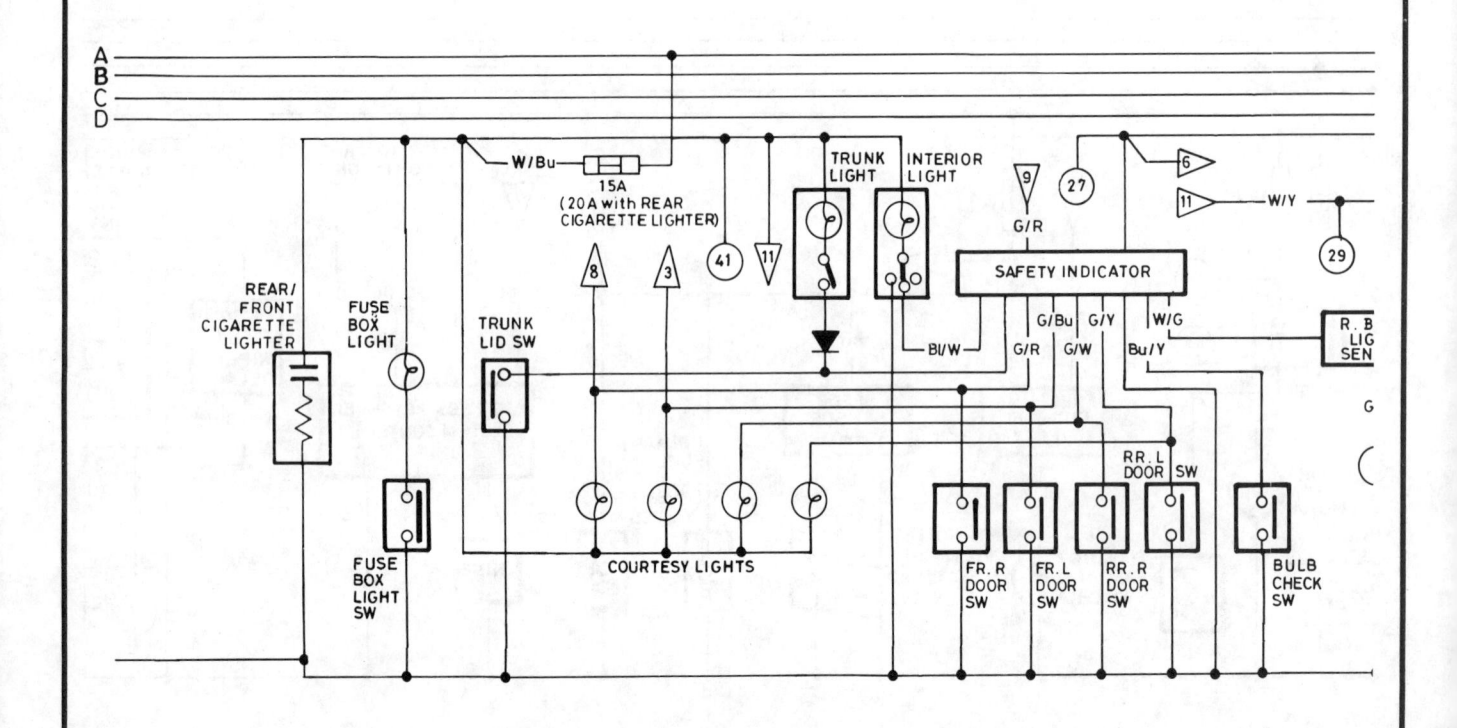

Fig. 12.50 Wiring diagram for 1984/5 UK models (continued)

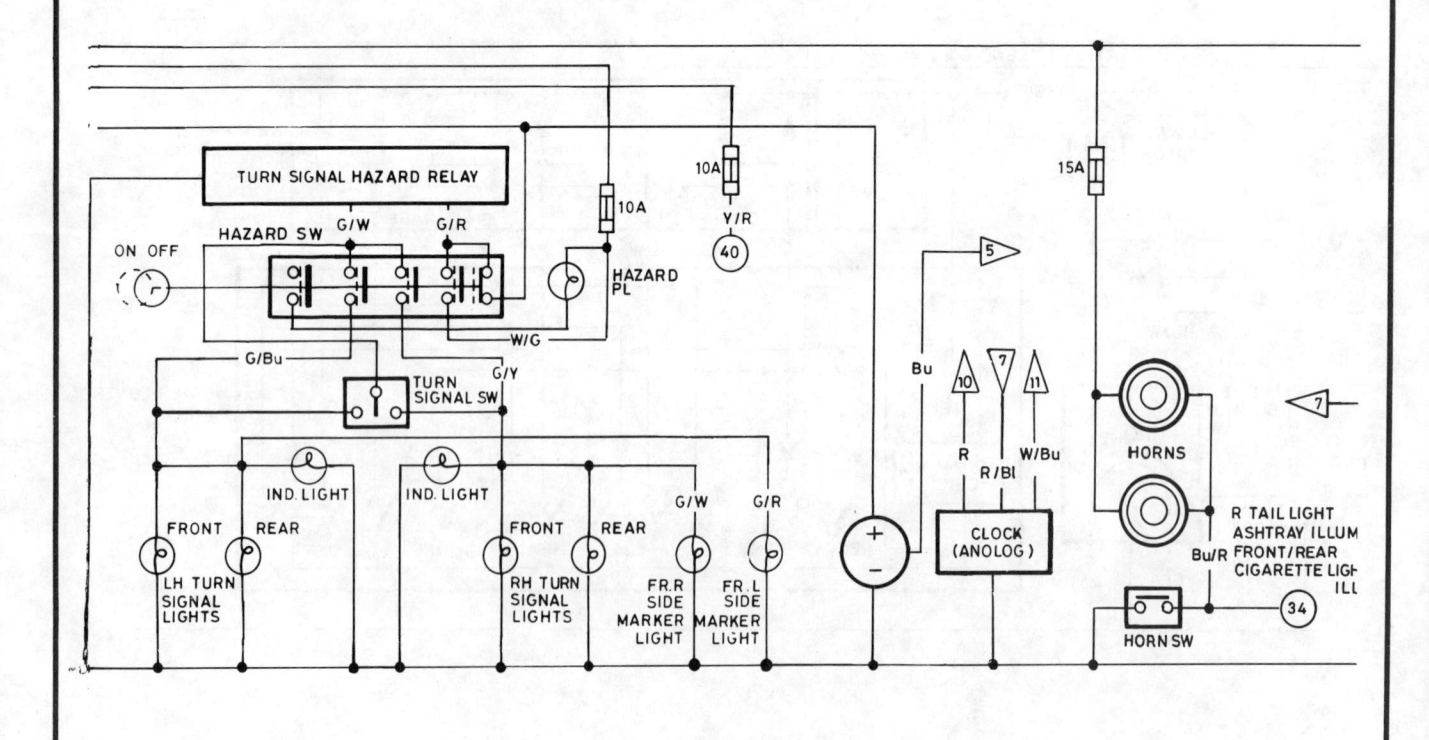

Fig. 12.50 Wiring diagram for 1984/5 UK models (continued)

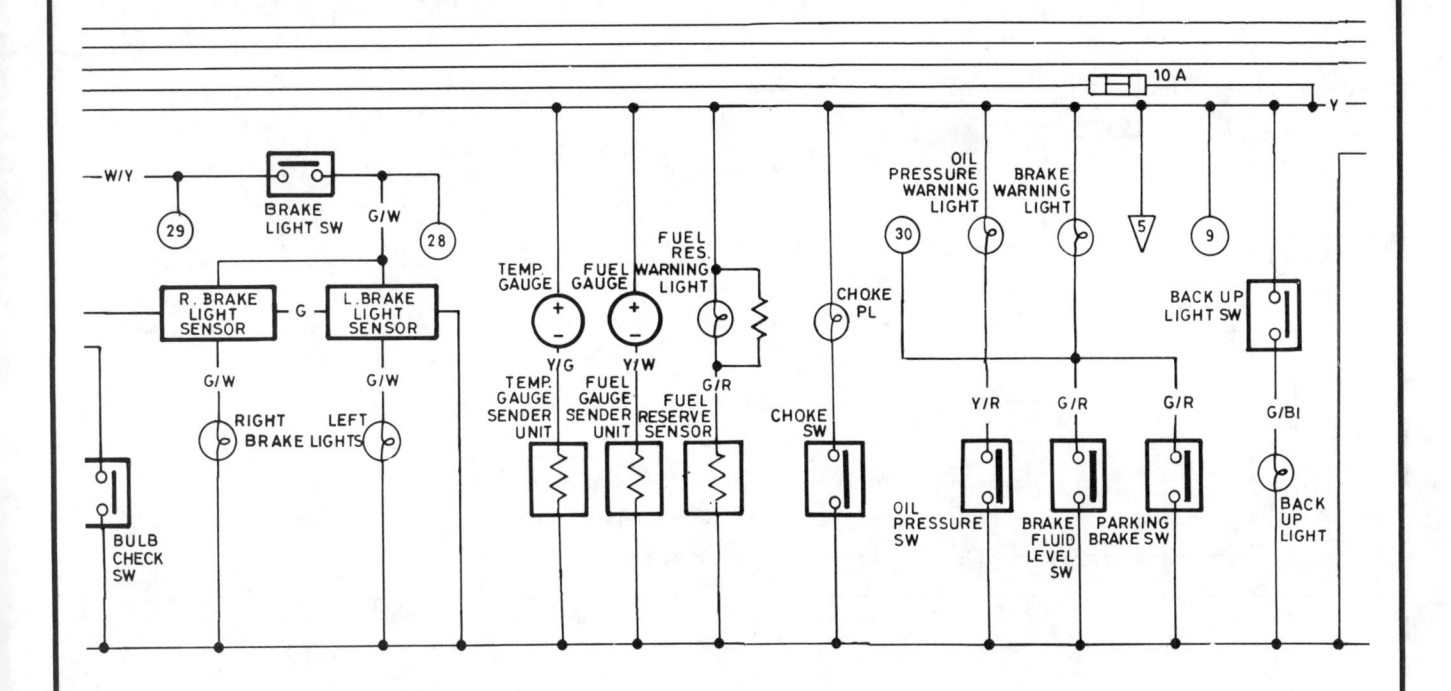

Fig. 12.50 Wiring diagram for 1984/5 UK models (continued)

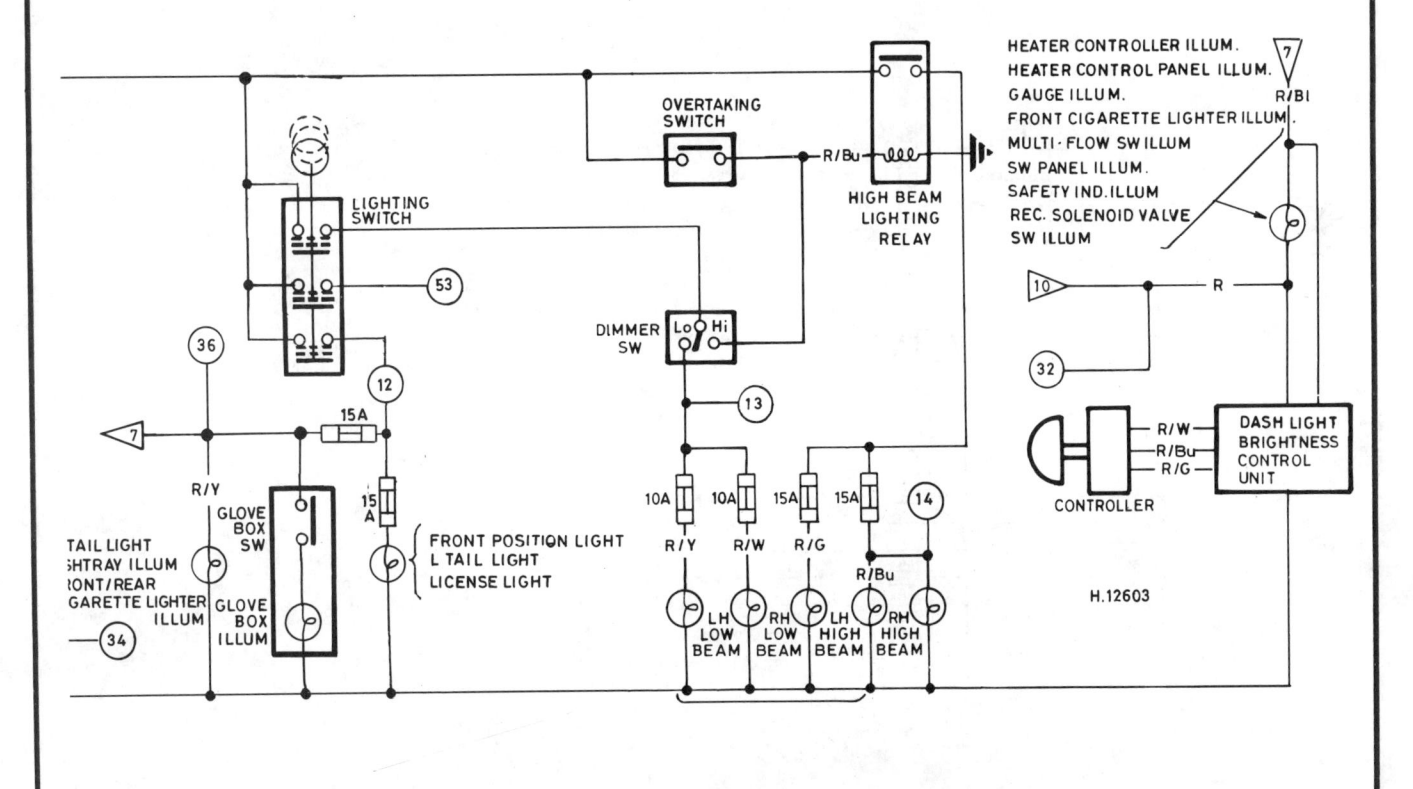

Fig. 12.50 Wiring diagram for 1984/5 UK models (continued)

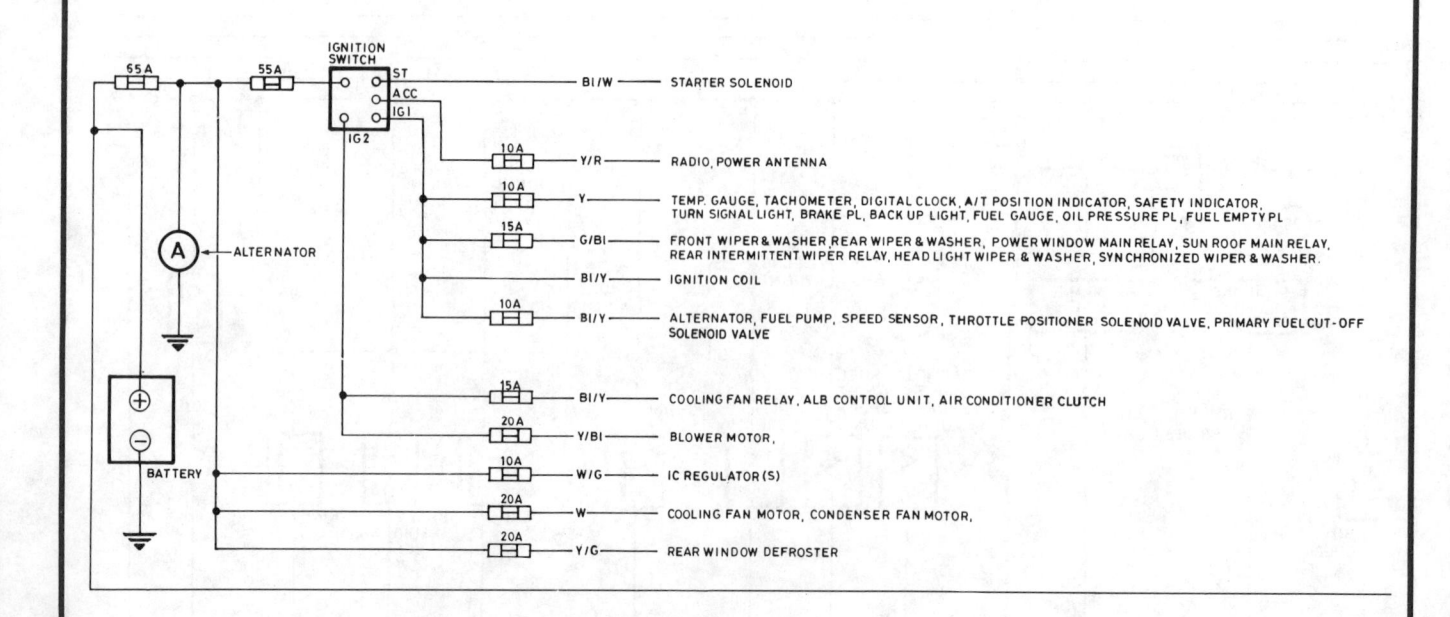

334

```
         65A       55A   IGNITION
                         SWITCH
   ┌──[  ]──┬──[  ]──────○ ST────────── Bl/W ───── STARTER SOLENOID
   │        │            ○ ACC
   │        │         ┌──○ IG1
   │        │     IG2 │
   │        │         ├──[ 10A ]── Y/R ─── RADIO, POWER ANTENNA
   │        │         │
   │        │         ├──[ 10A ]── Y ──── TEMP. GAUGE, TACHOMETER, DIGITAL CLOCK, A/T POSITION INDICATOR, SAFETY INDICATOR,
   │        │         │                   TURN SIGNAL LIGHT, BRAKE PL, BACK UP LIGHT, FUEL GAUGE, OIL PRESSURE PL, FUEL EMPTY PL
   │        │         ├──[ 15A ]── G/Bl ── FRONT WIPER & WASHER, REAR WIPER & WASHER, POWER WINDOW MAIN RELAY, SUN ROOF MAIN RELAY,
   │(A)─ALTERNATOR    │                   REAR INTERMITTENT WIPER RELAY, HEAD LIGHT WIPER & WASHER, SYNCHRONIZED WIPER & WASHER.
   │        │         ├───────── Bl/Y ─── IGNITION COIL
   │        │         │
   │        │         ├──[ 10A ]── Bl/Y ── ALTERNATOR, FUEL PUMP, SPEED SENSOR, THROTTLE POSITIONER SOLENOID VALVE, PRIMARY FUEL CUT-OFF
   │        │                              SOLENOID VALVE
   │        │
   │ ┌──┐   ├──[ 15A ]── Bl/Y ── COOLING FAN RELAY, ALB CONTROL UNIT, AIR CONDITIONER CLUTCH
   │ │ +│   ├──[ 20A ]── Y/Bl ── BLOWER MOTOR
   │ │ −│   │
   │ └──┘   ├──[ 10A ]── W/G ─── IC REGULATOR (S)
   │BATTERY │
   │        ├──[ 20A ]── W ───── COOLING FAN MOTOR, CONDENSER FAN MOTOR
   │        │
   │        └──[ 20A ]── Y/G ─── REAR WINDOW DEFROSTER
   ⏚
```

Fig. 12.50 Wiring diagram for 1984/5 UK models (continued)

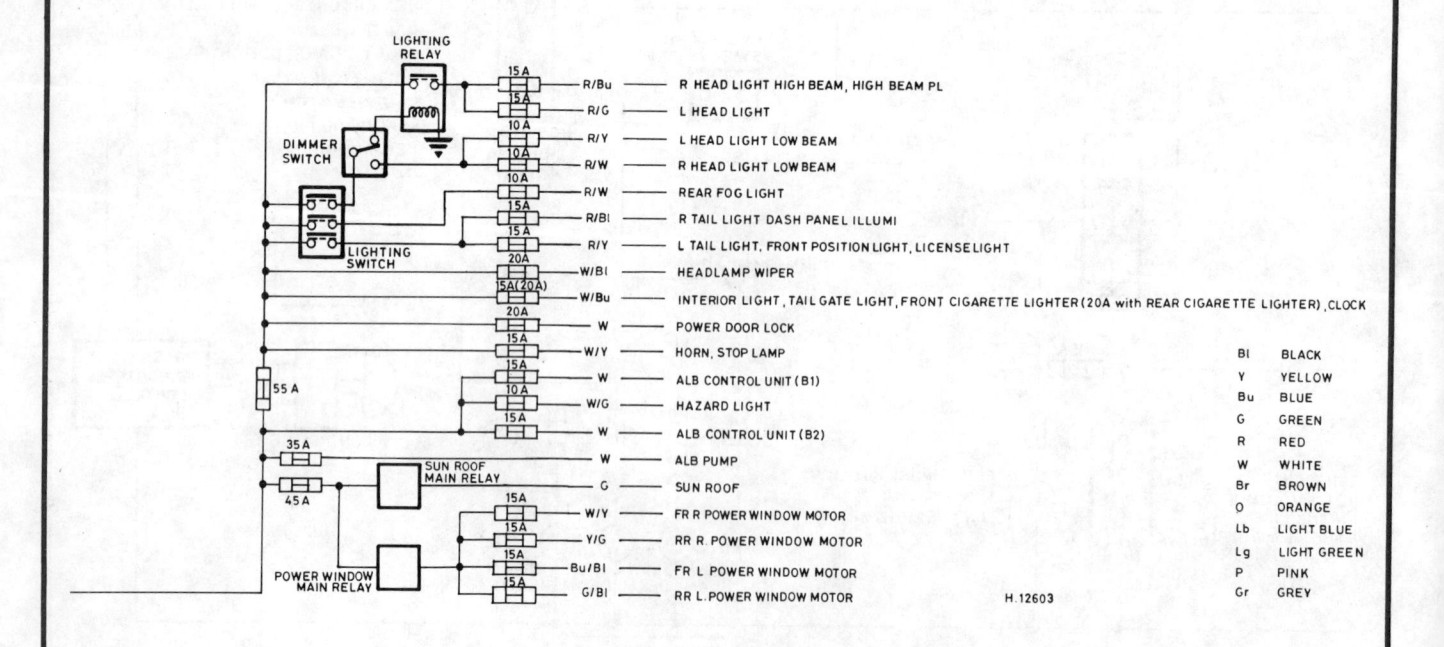

```
                LIGHTING
                RELAY
              ┌────────┬──[ 15A ]── R/Bu ── R HEAD LIGHT HIGH BEAM, HIGH BEAM PL
              │        ├──[ 15A ]── R/G ─── L HEAD LIGHT
   DIMMER     │        ├──[ 10A ]── R/Y ─── L HEAD LIGHT LOW BEAM
   SWITCH     │        ├──[ 10A ]── R/W ─── R HEAD LIGHT LOW BEAM
              │        ├──[ 10A ]── R/W ─── REAR FOG LIGHT
              │        ├──[ 15A ]── R/Bl ── R TAIL LIGHT DASH PANEL ILLUMI
   LIGHTING   │        ├──[ 15A ]── R/Y ─── L TAIL LIGHT, FRONT POSITION LIGHT, LICENSE LIGHT
   SWITCH     │        ├──[ 20A ]── W/Bl ── HEADLAMP WIPER
              │        ├──[5A(20A)] W/Bu ── INTERIOR LIGHT, TAIL GATE LIGHT, FRONT CIGARETTE LIGHTER (20A with REAR CIGARETTE LIGHTER), CLOCK
              │        ├──[ 20A ]── W ───── POWER DOOR LOCK
              │        ├──[ 15A ]── W/Y ─── HORN, STOP LAMP
   55A        │        ├──[ 15A ]── W ───── ALB CONTROL UNIT (B1)
              │        ├──[ 10A ]── W/G ─── HAZARD LIGHT
              │        ├──[ 15A ]── W ───── ALB CONTROL UNIT (B2)
   35A        │        │
              │        └────────── W ───── ALB PUMP
        SUN ROOF
        MAIN RELAY ──── G ───── SUN ROOF
   45A        ├──[ 15A ]── W/Y ─── FR R POWER WINDOW MOTOR
              ├──[ 15A ]── Y/G ─── RR R POWER WINDOW MOTOR
              ├──[ 15A ]── Bu/Bl ── FR L POWER WINDOW MOTOR
   POWER WINDOW
   MAIN RELAY ├──[ 15A ]── G/Bl ── RR L POWER WINDOW MOTOR
```

Bl	BLACK
Y	YELLOW
Bu	BLUE
G	GREEN
R	RED
W	WHITE
Br	BROWN
O	ORANGE
Lb	LIGHT BLUE
Lg	LIGHT GREEN
P	PINK
Gr	GREY

H.12603

Fig. 12.50 Wiring diagram for 1984/5 UK models (continued)

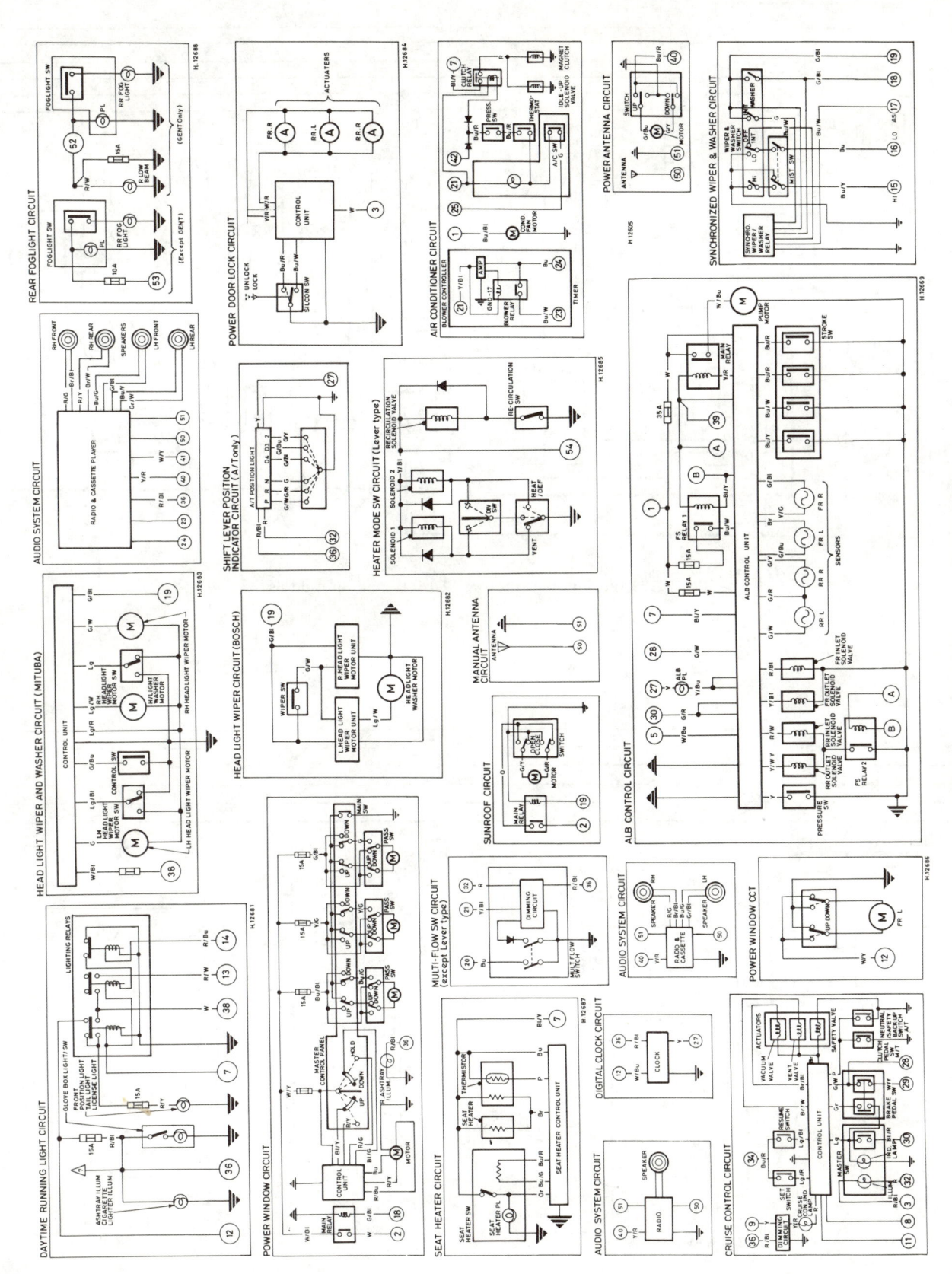

Fig. 12.51 Sub-system wiring diagrams for 1984/5 UK models

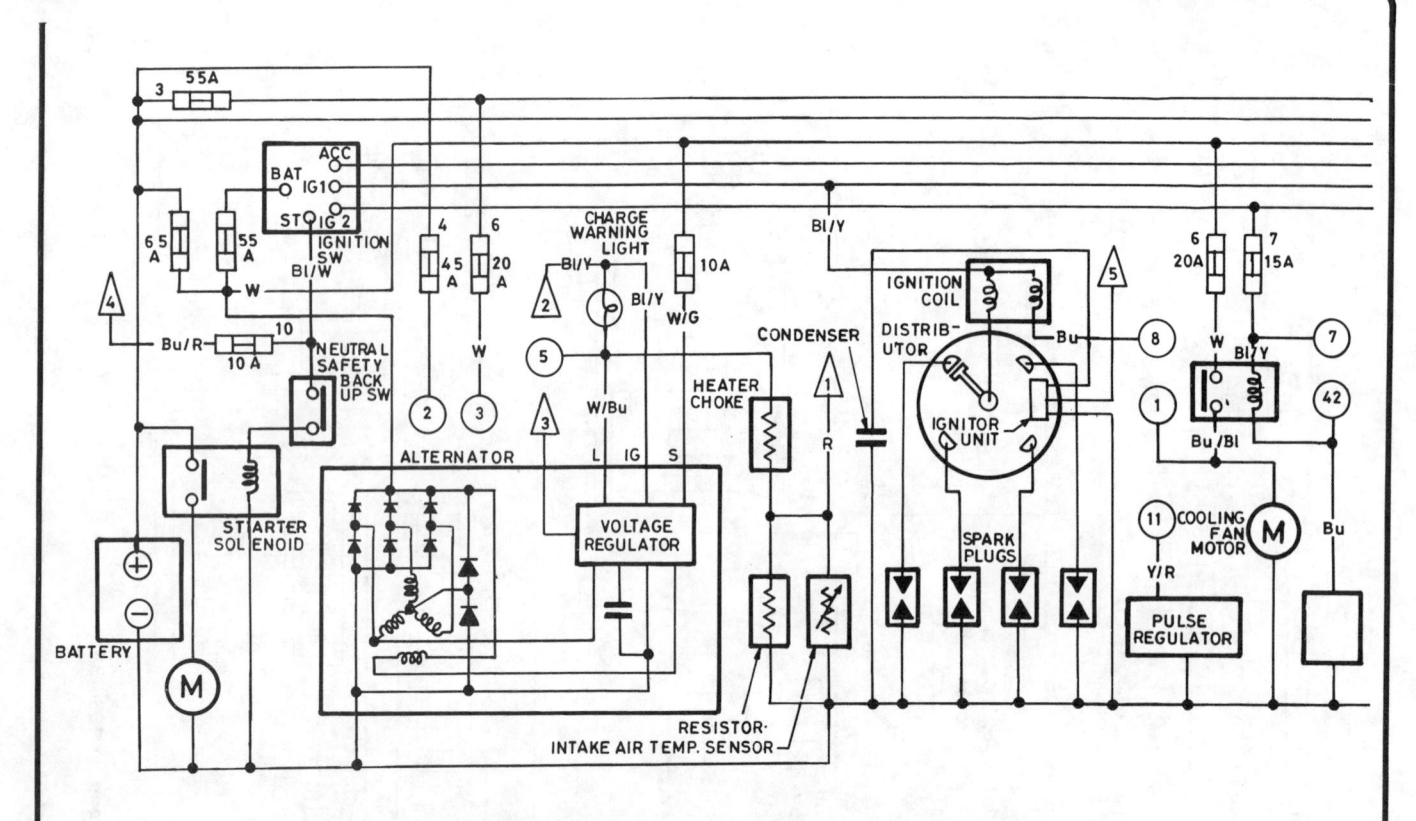

336

Fig. 12.52 Wiring diagram for 1984/5 Canadian models

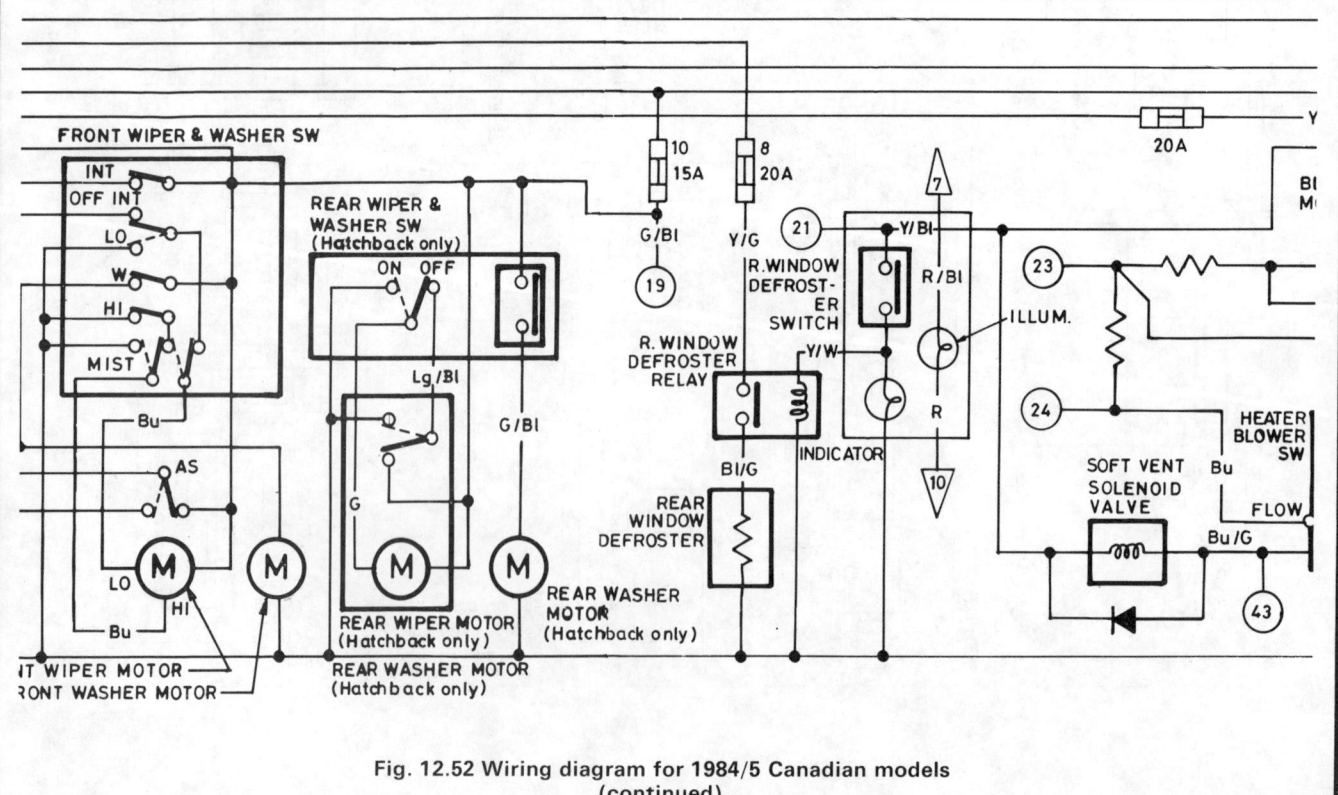

Fig. 12.52 Wiring diagram for 1984/5 Canadian models (continued)

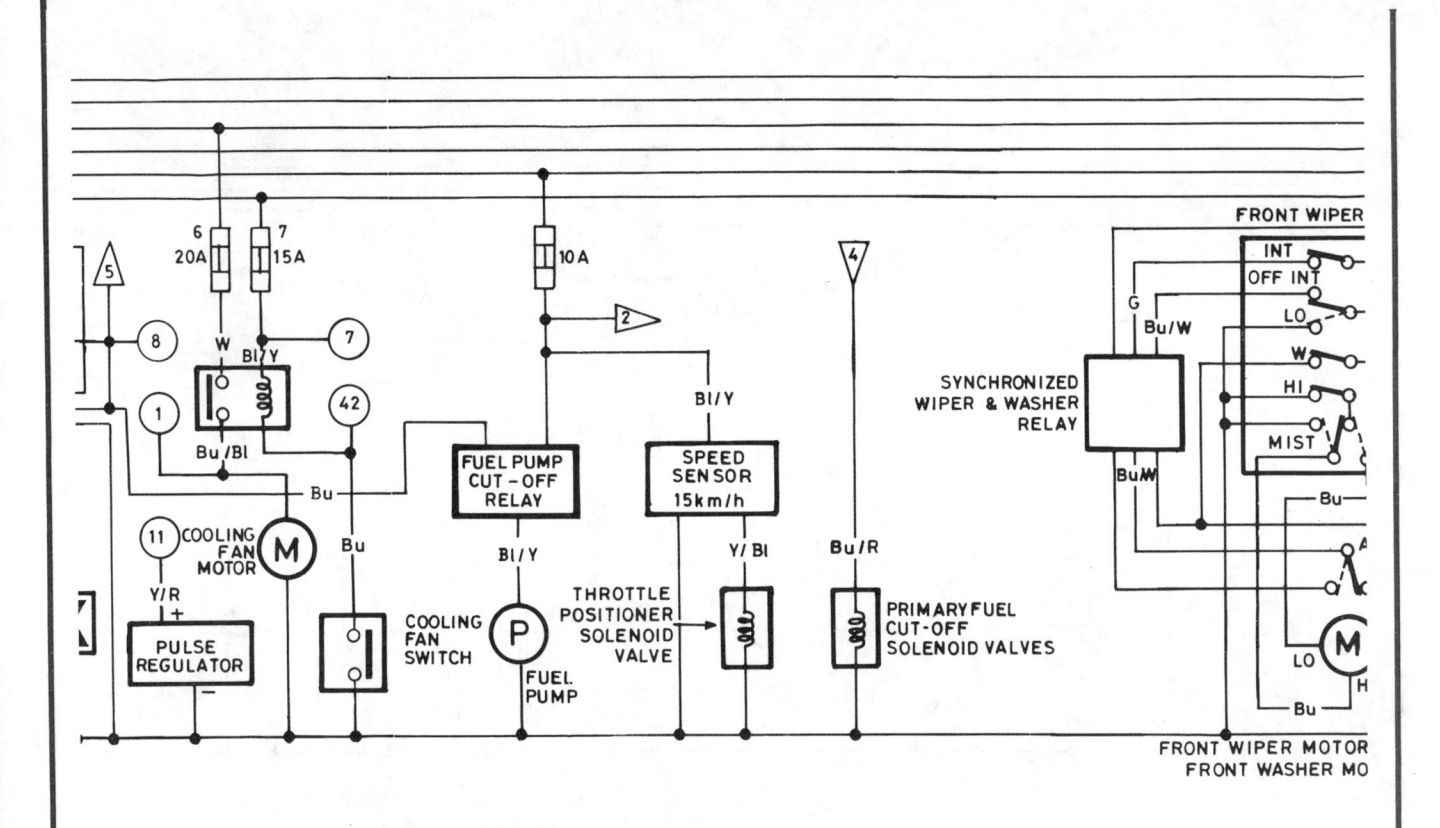

Fig. 12.52 Wiring diagram for 1984/5 Canadian models (continued)

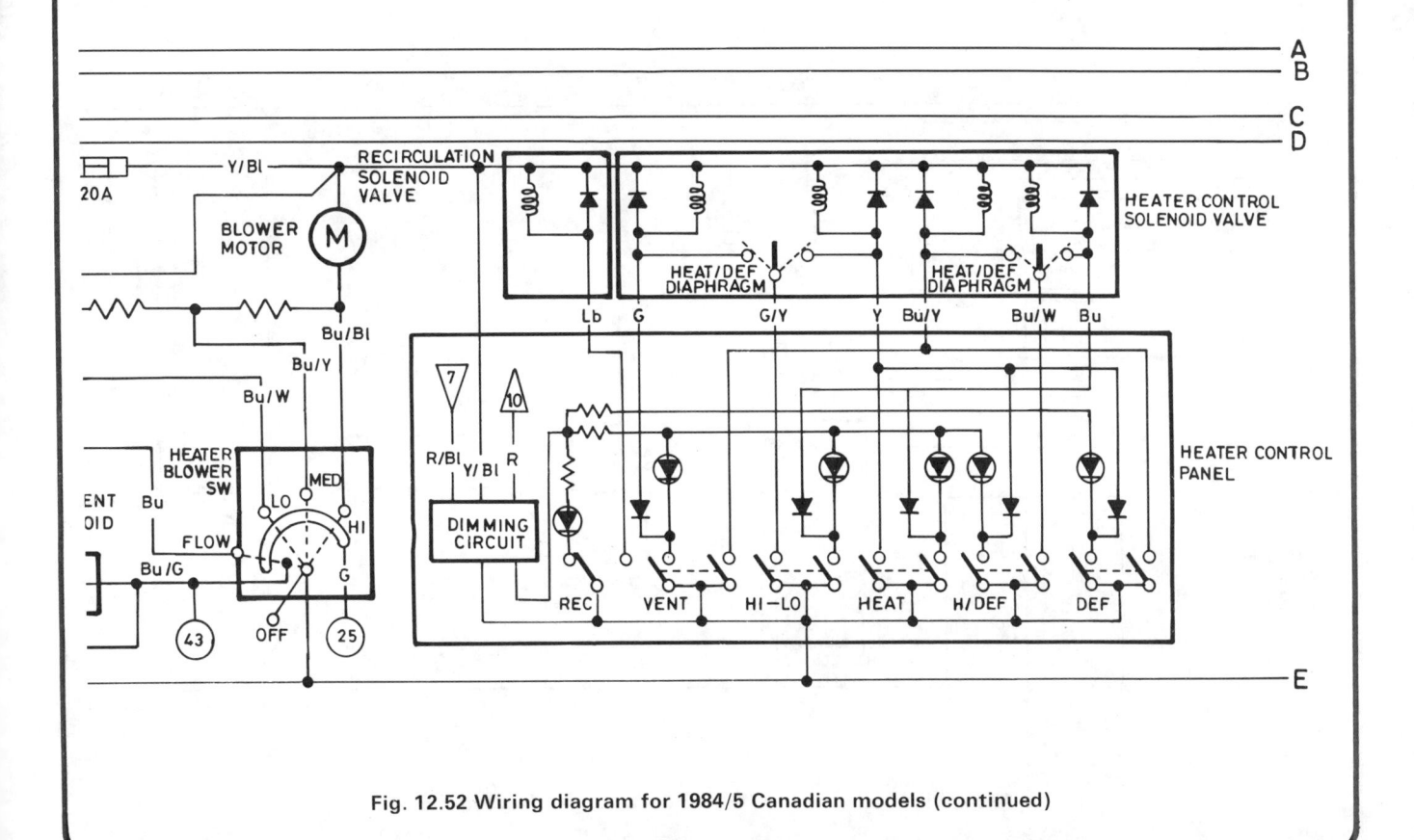

Fig. 12.52 Wiring diagram for 1984/5 Canadian models (continued)

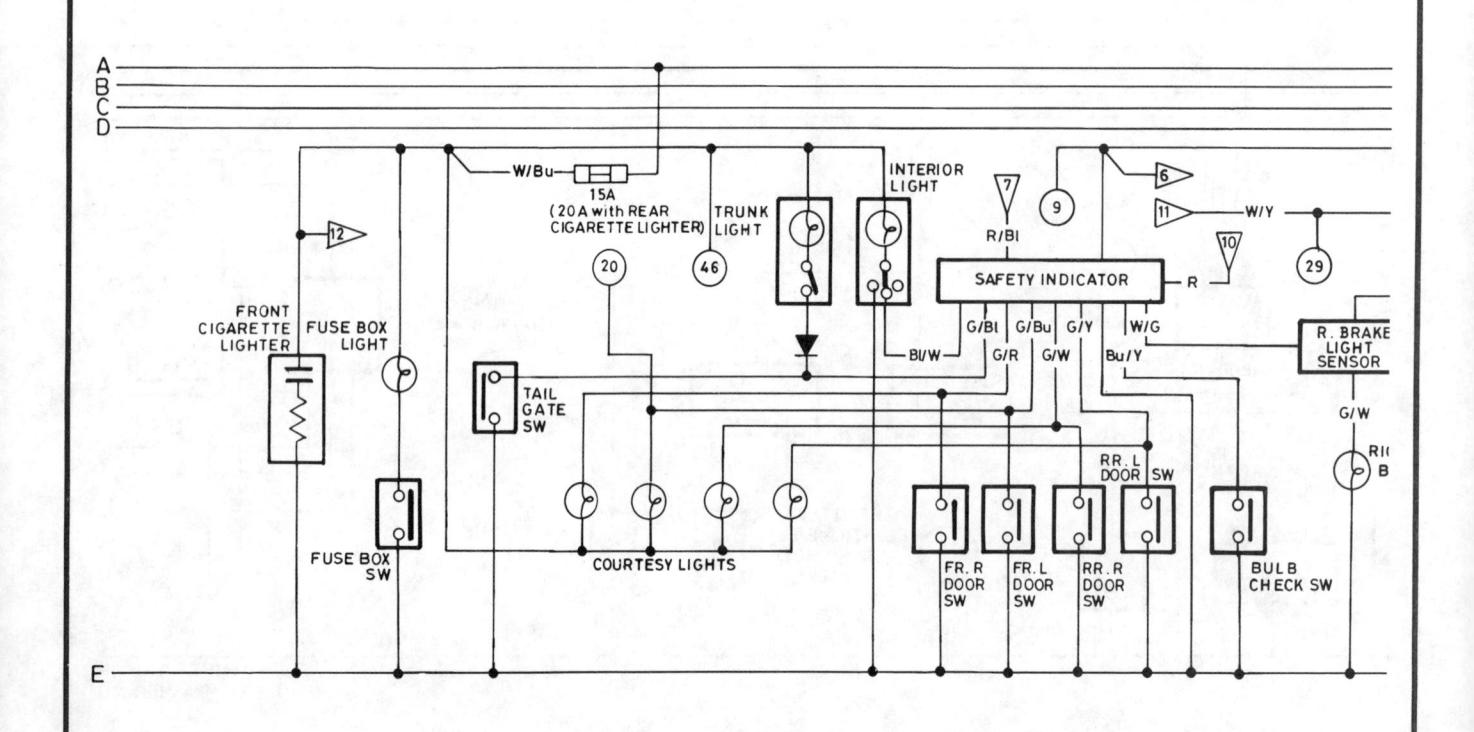

Fig. 12.52 Wiring diagram for 1984/5 Canadian models (continued)

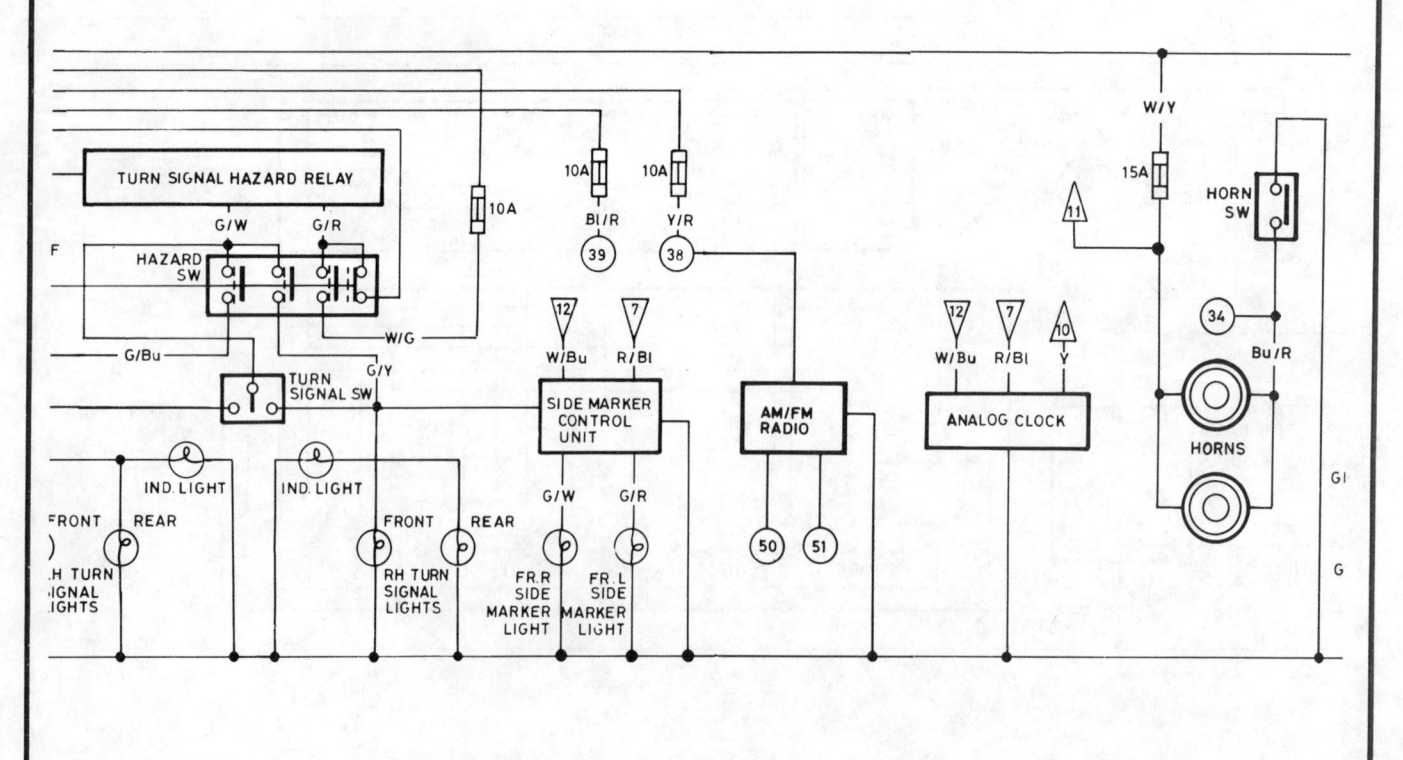

Fig. 12.52 Wiring diagram for 1984/5 Canadian models (continued)

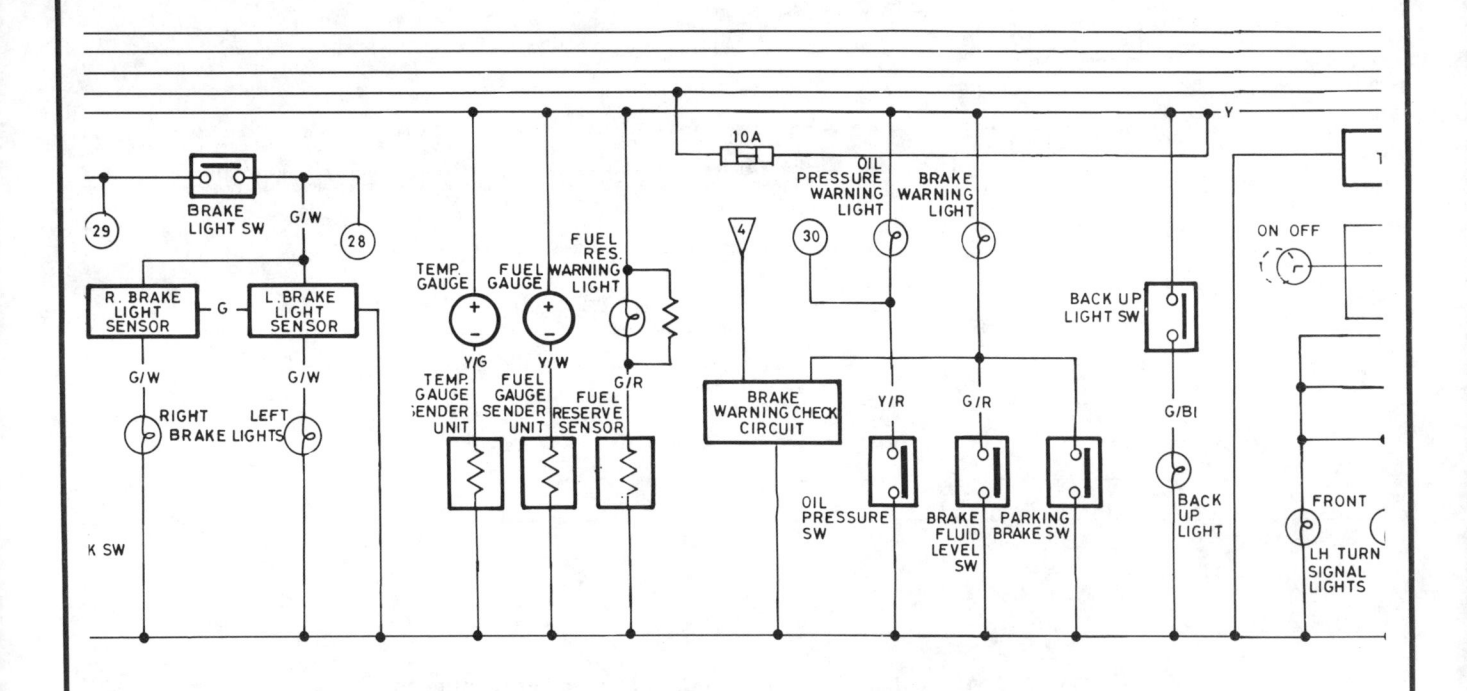

Fig. 12.52 Wiring diagram for 1984/5 Canadian models (continued)

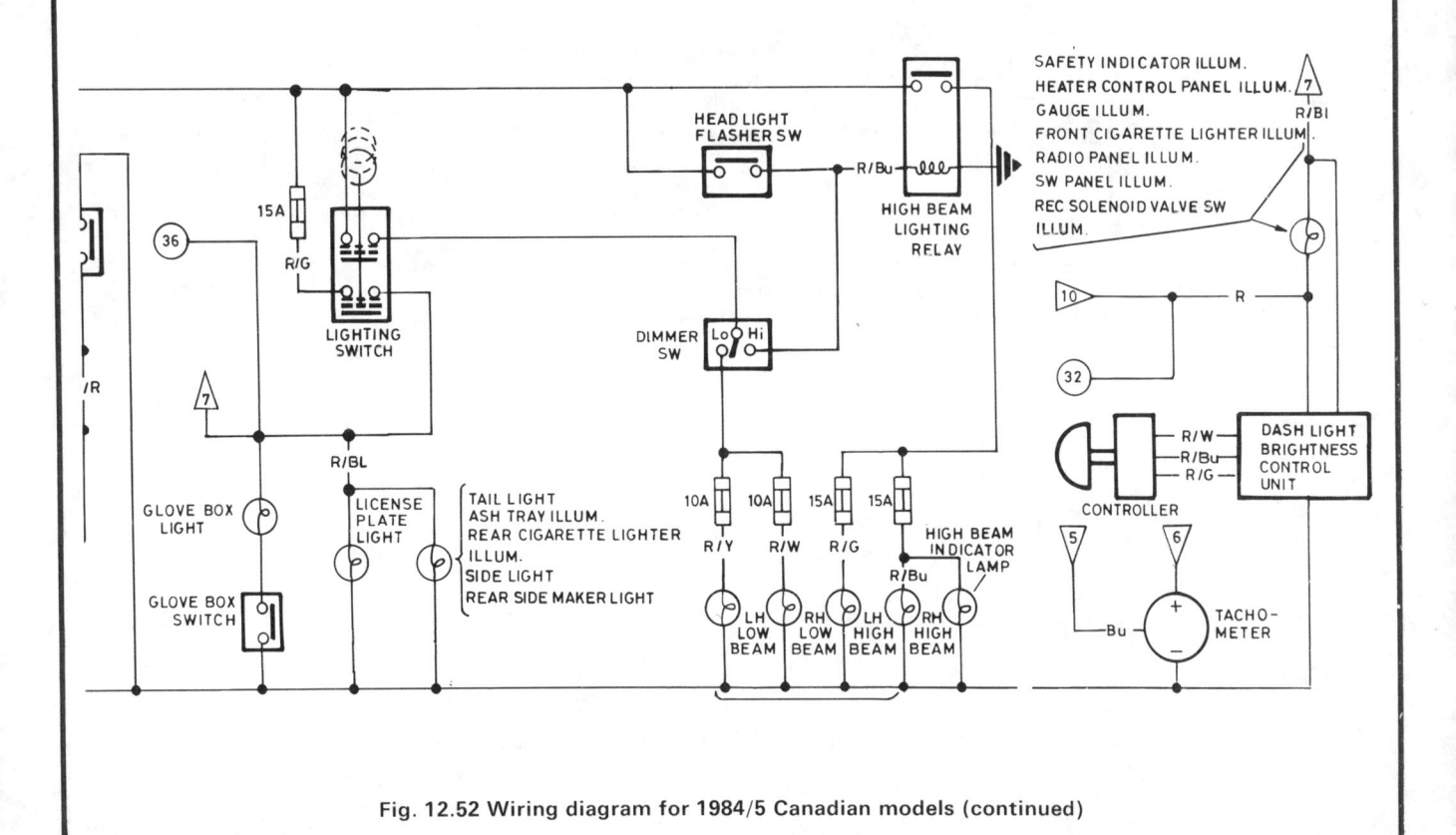

Fig. 12.52 Wiring diagram for 1984/5 Canadian models (continued)

339

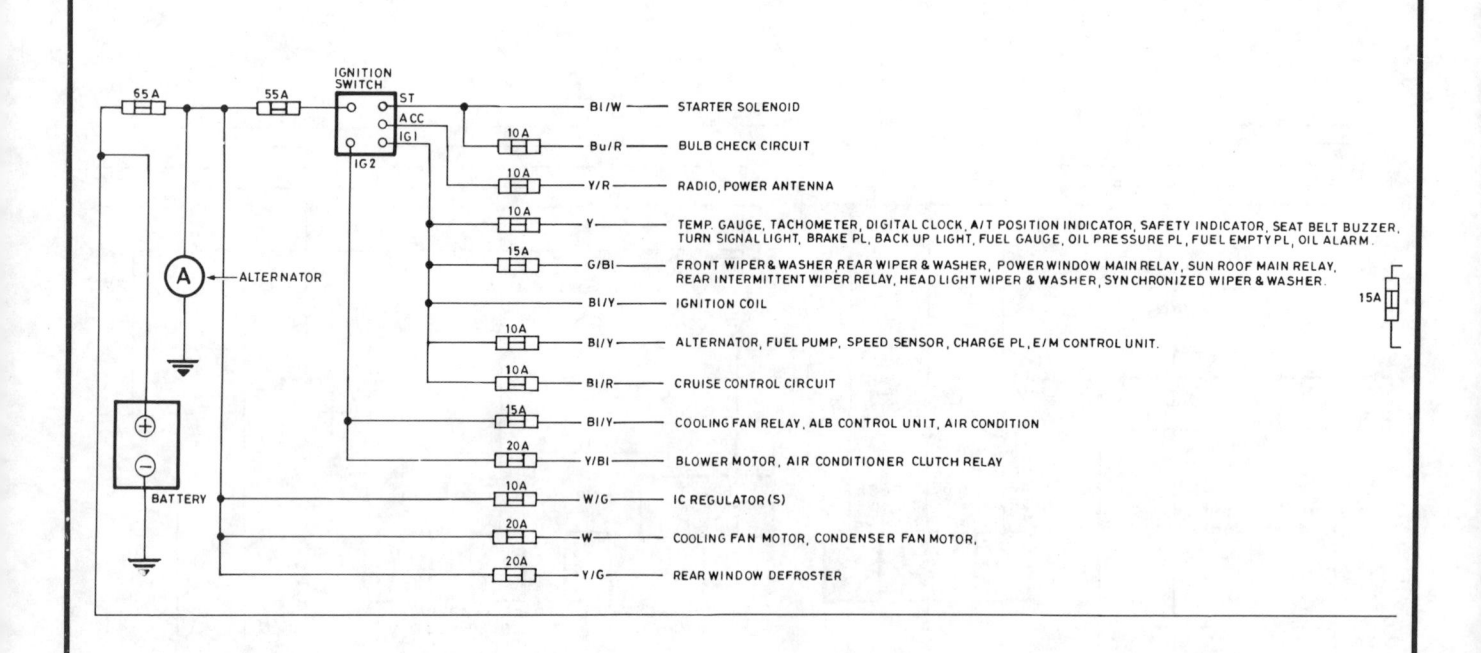

340

Fig. 12.52 Wiring diagram for 1984/5 Canadian models (continued)

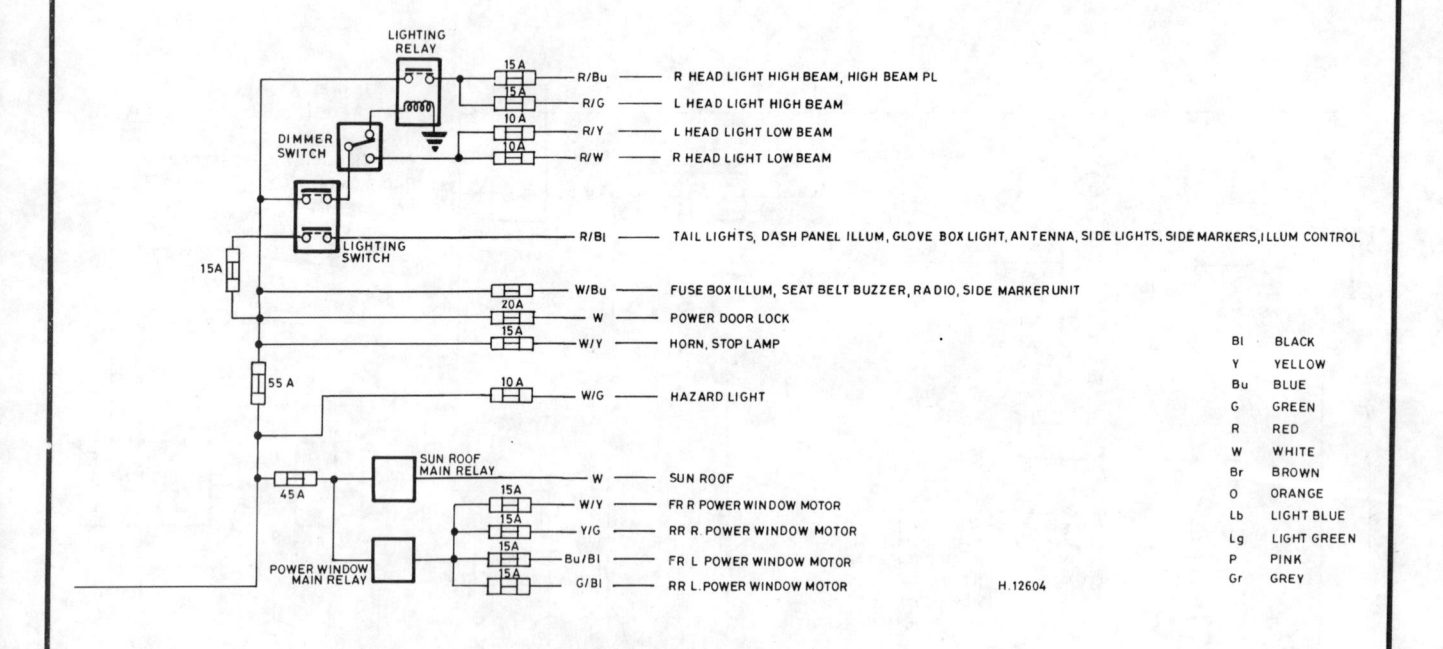

Fig. 12.52 Wiring diagram for 1984/5 Canadian models (continued)

Fig. 12.53 Sub-system wiring diagrams for 1984/5 Canadian models

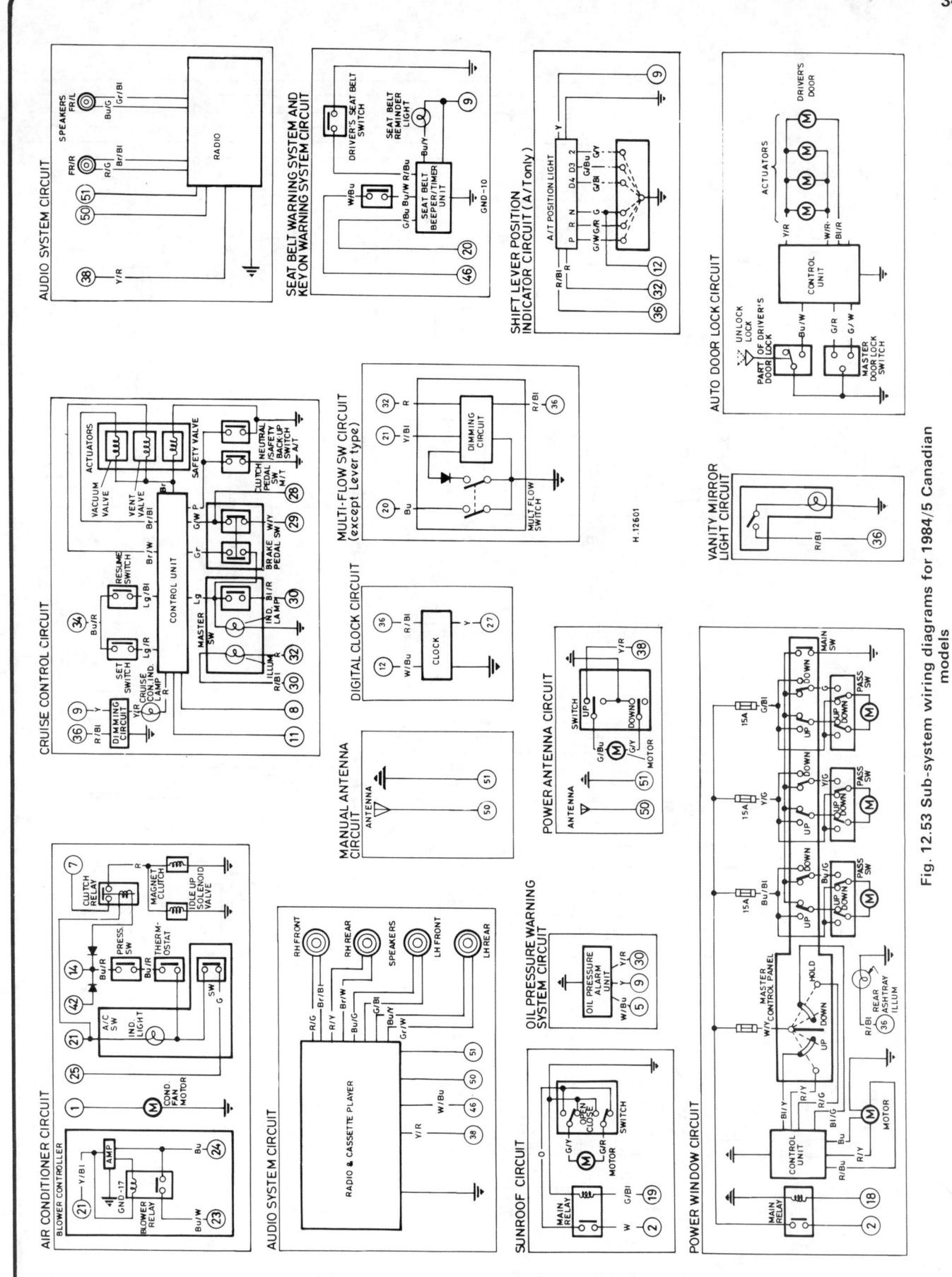

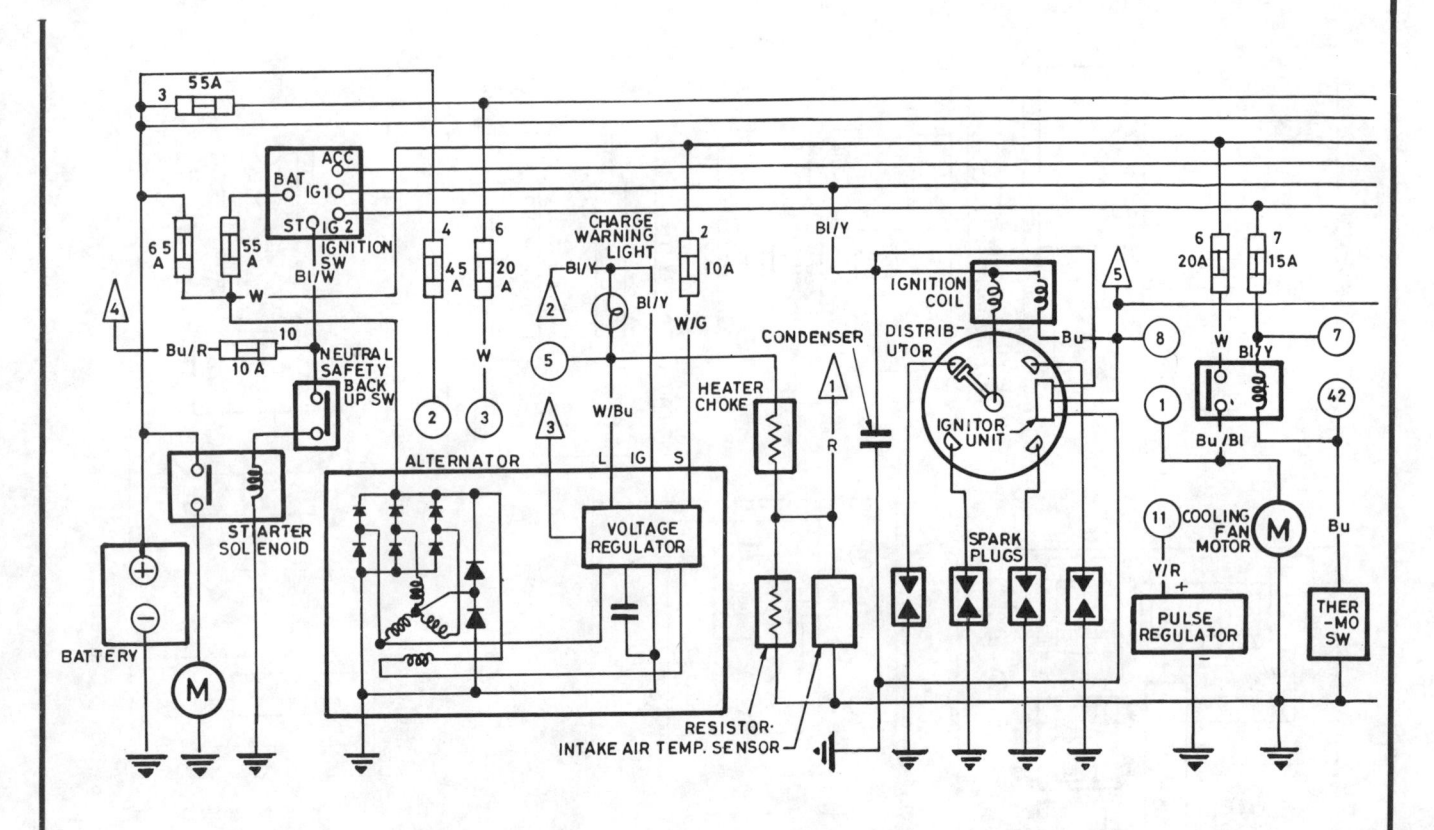

Fig. 12.54 Wiring diagram for 1984/5 North American models with carburettor

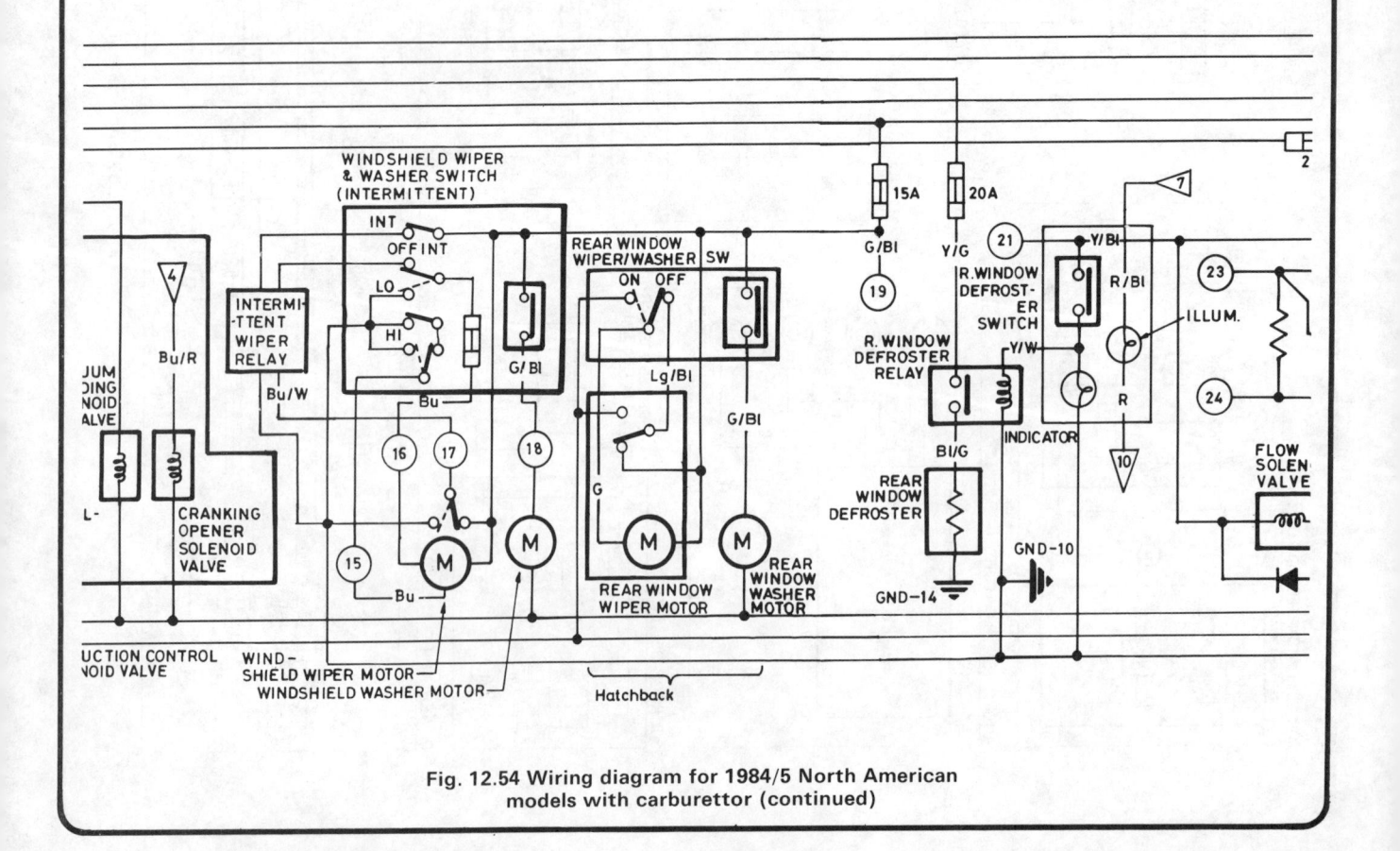

Fig. 12.54 Wiring diagram for 1984/5 North American models with carburettor (continued)

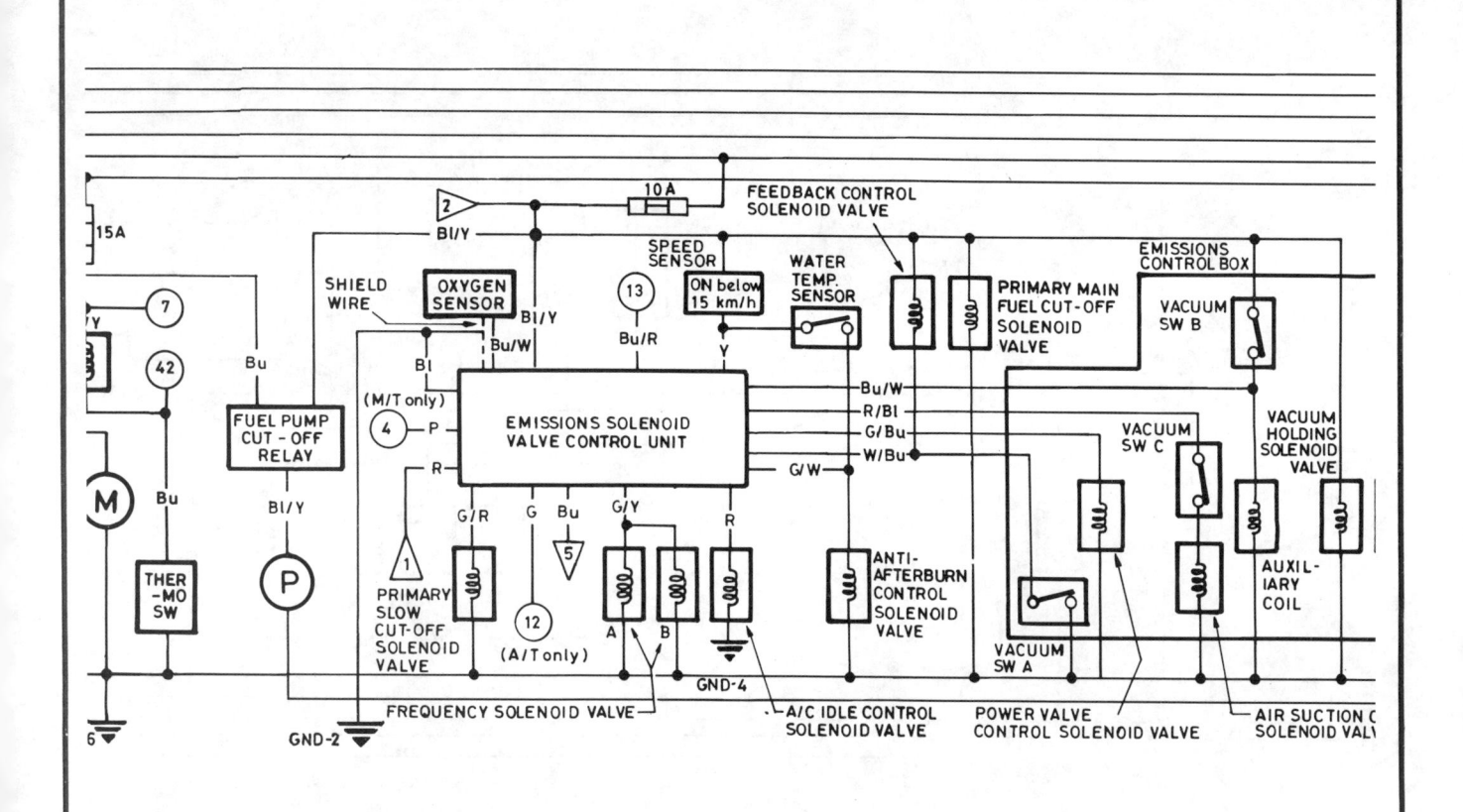

Fig. 12.54 Wiring diagram for 1984/5 North American models with carburettor (continued)

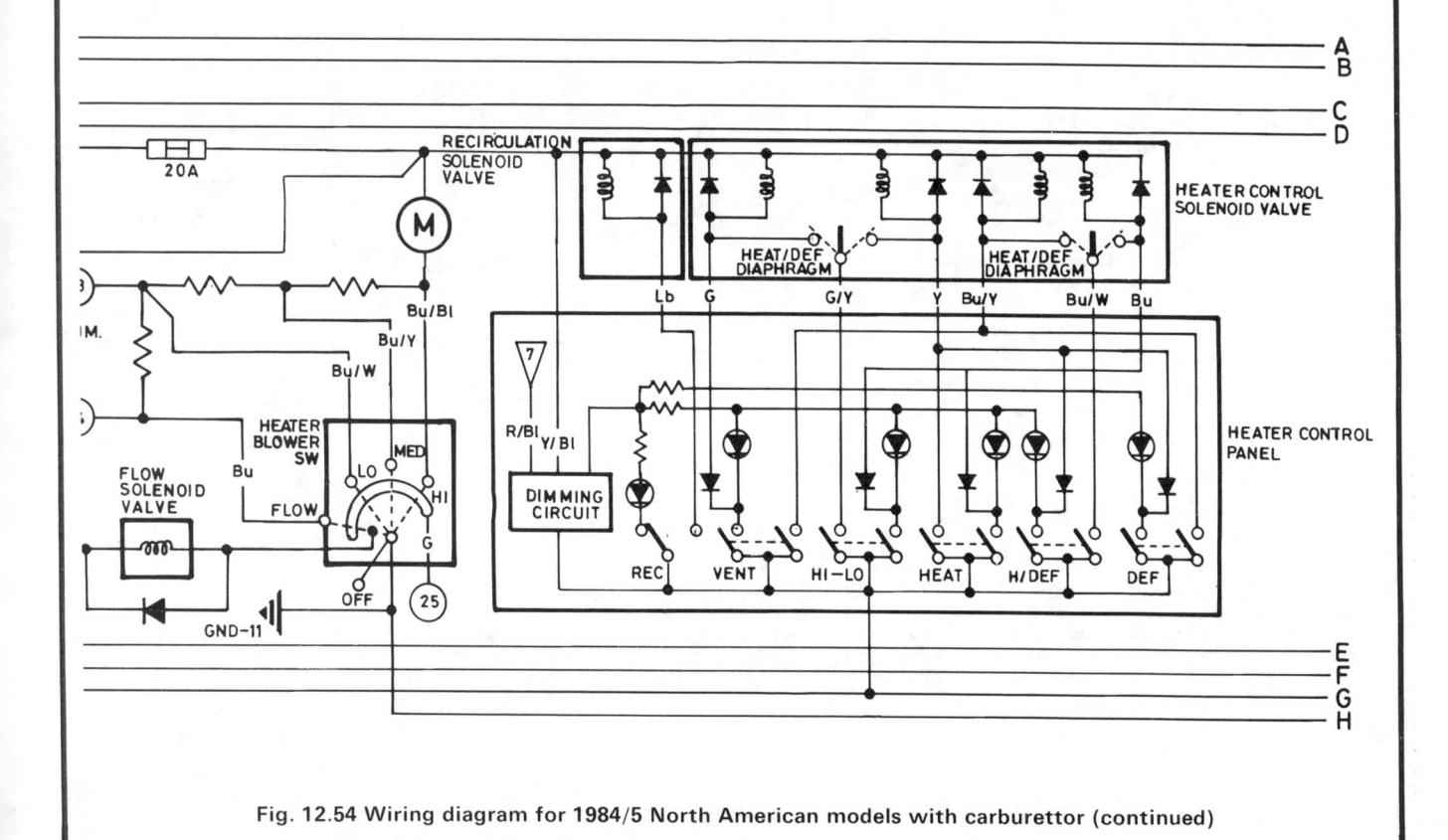

Fig. 12.54 Wiring diagram for 1984/5 North American models with carburettor (continued)

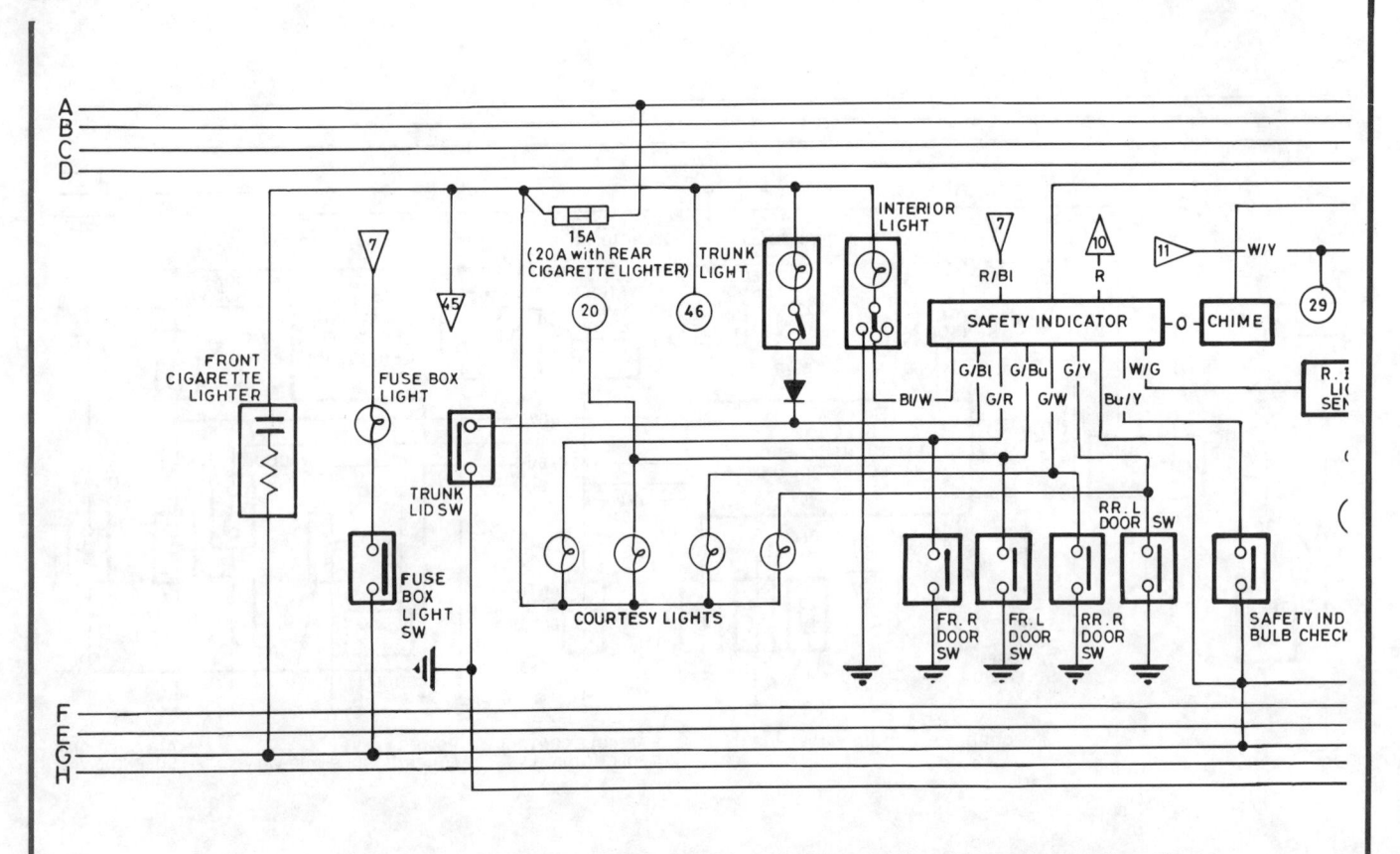

Fig. 12.54 Wiring diagram for 1984/5 North American models with carburettor (continued)

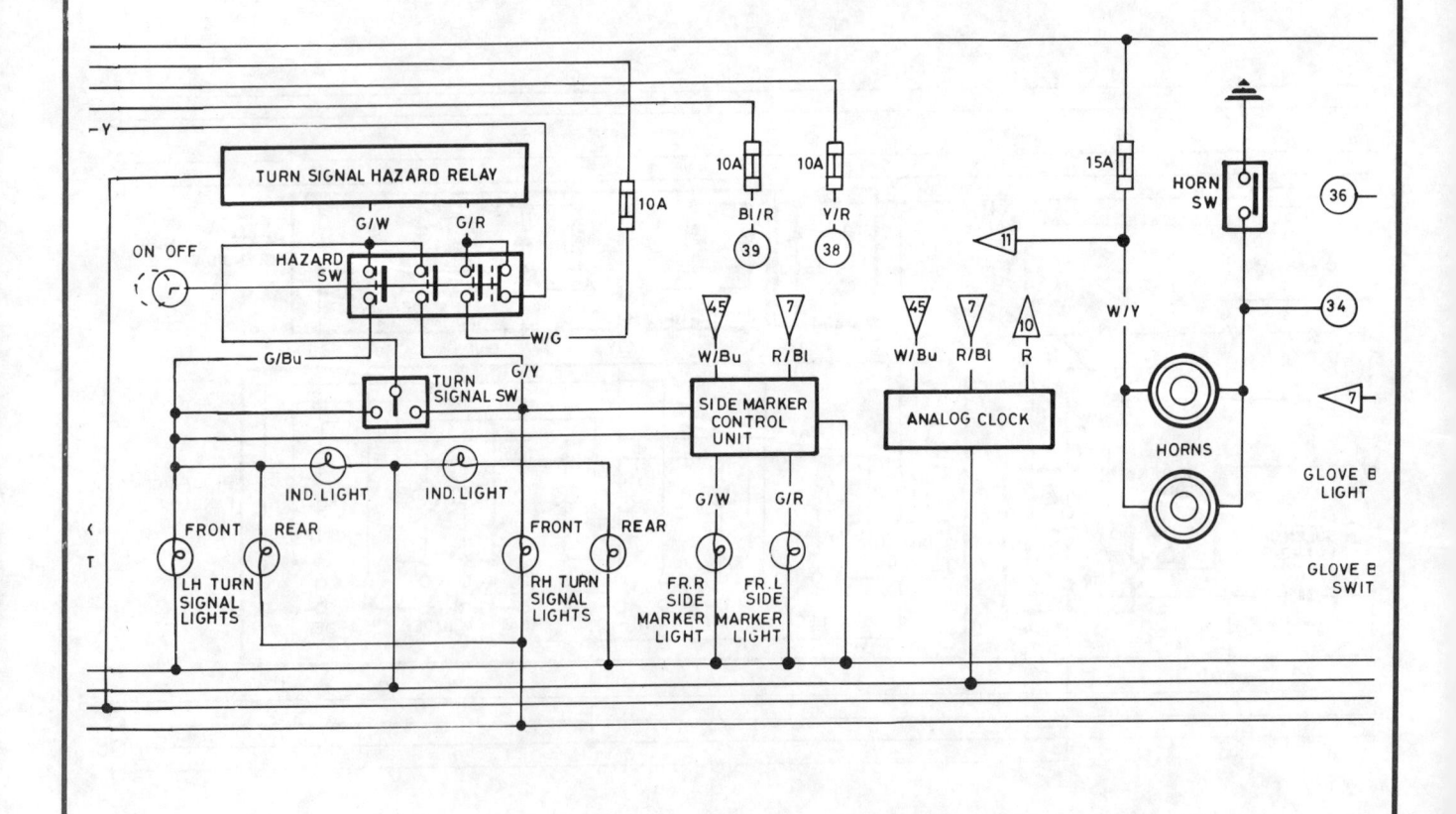

Fig. 12.54 Wiring diagram for 1984/5 North American models with carburettor (continued)

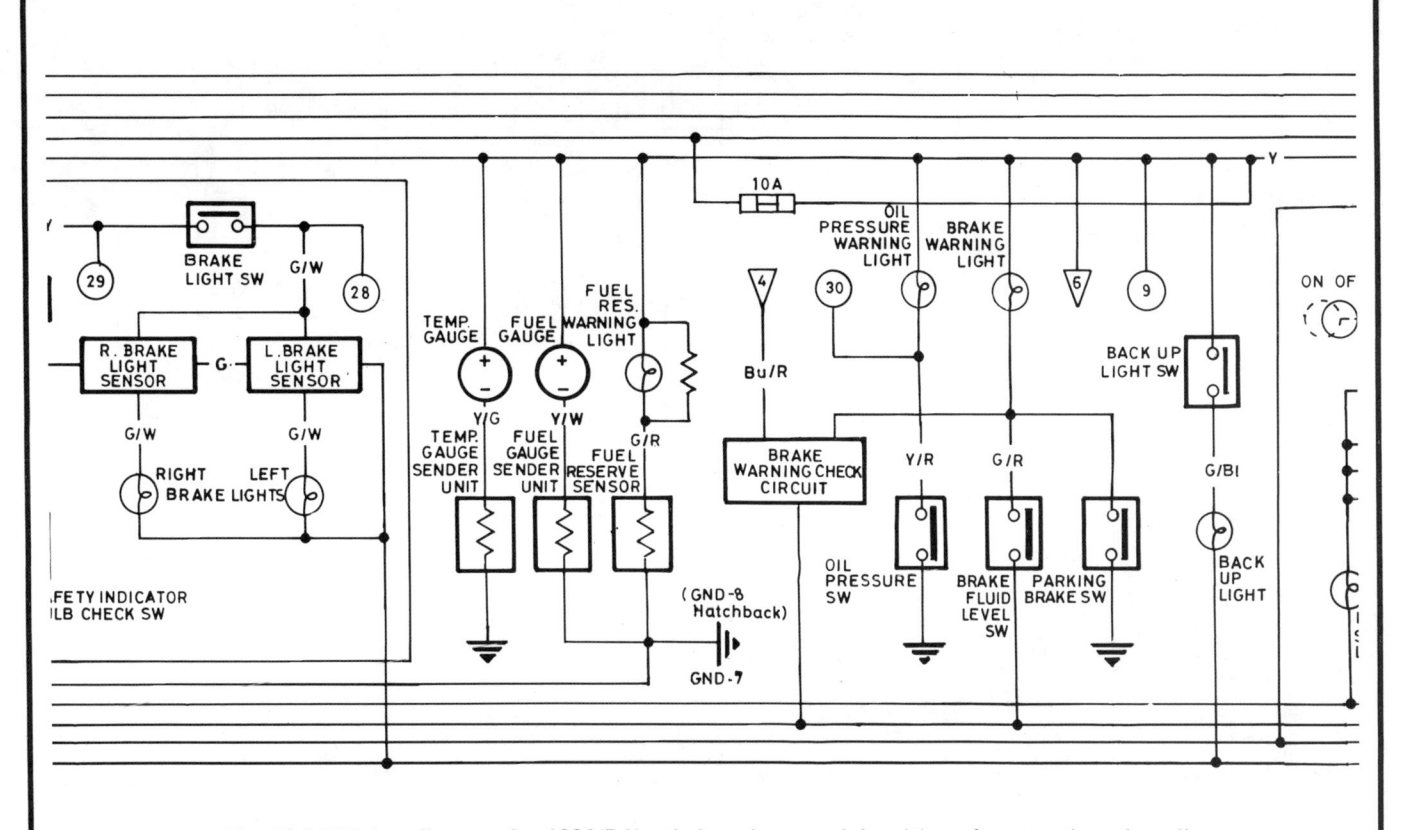

345

Fig. 12.54 Wiring diagram for 1984/5 North American models with carburettor (continued)

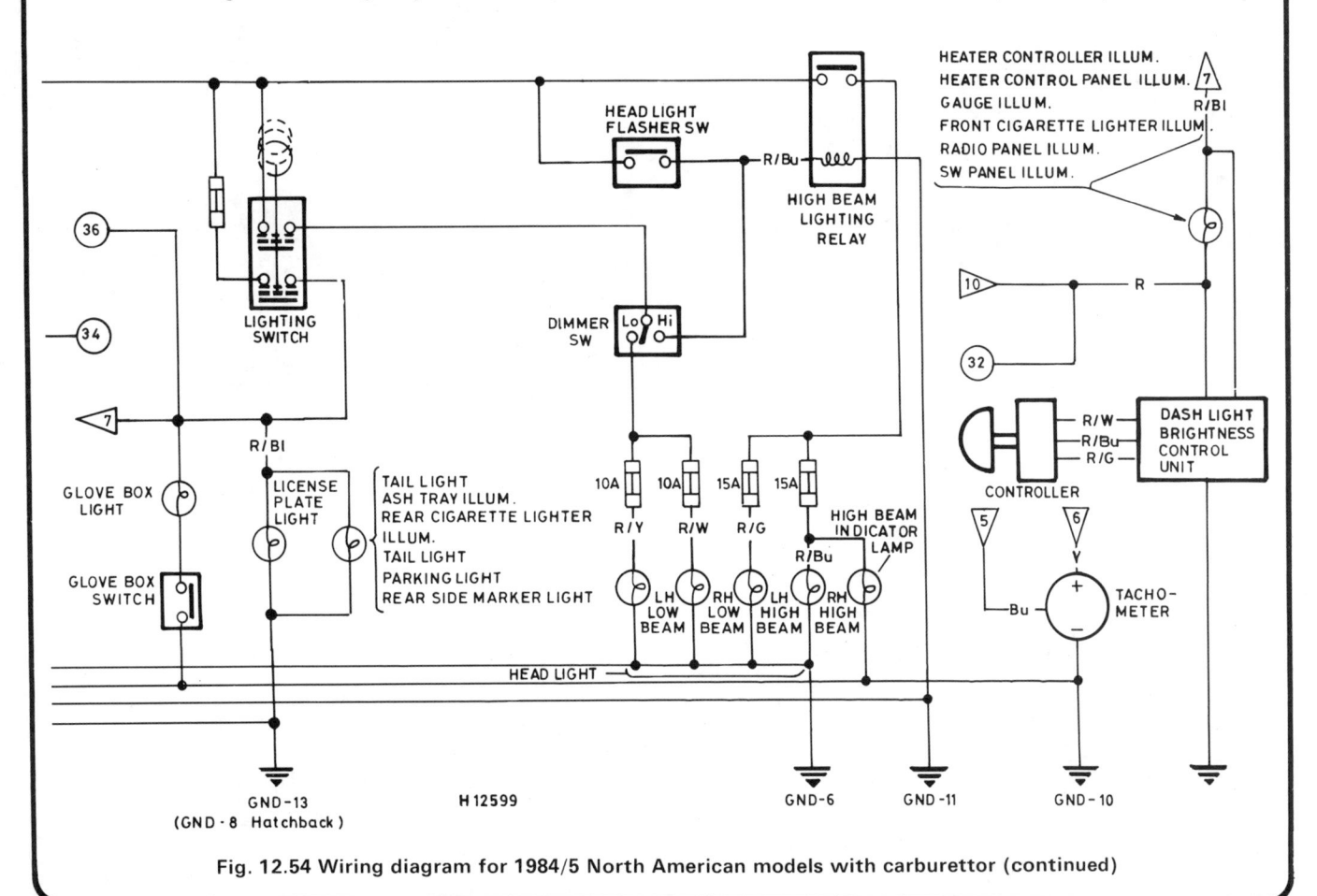

Fig. 12.54 Wiring diagram for 1984/5 North American models with carburettor (continued)

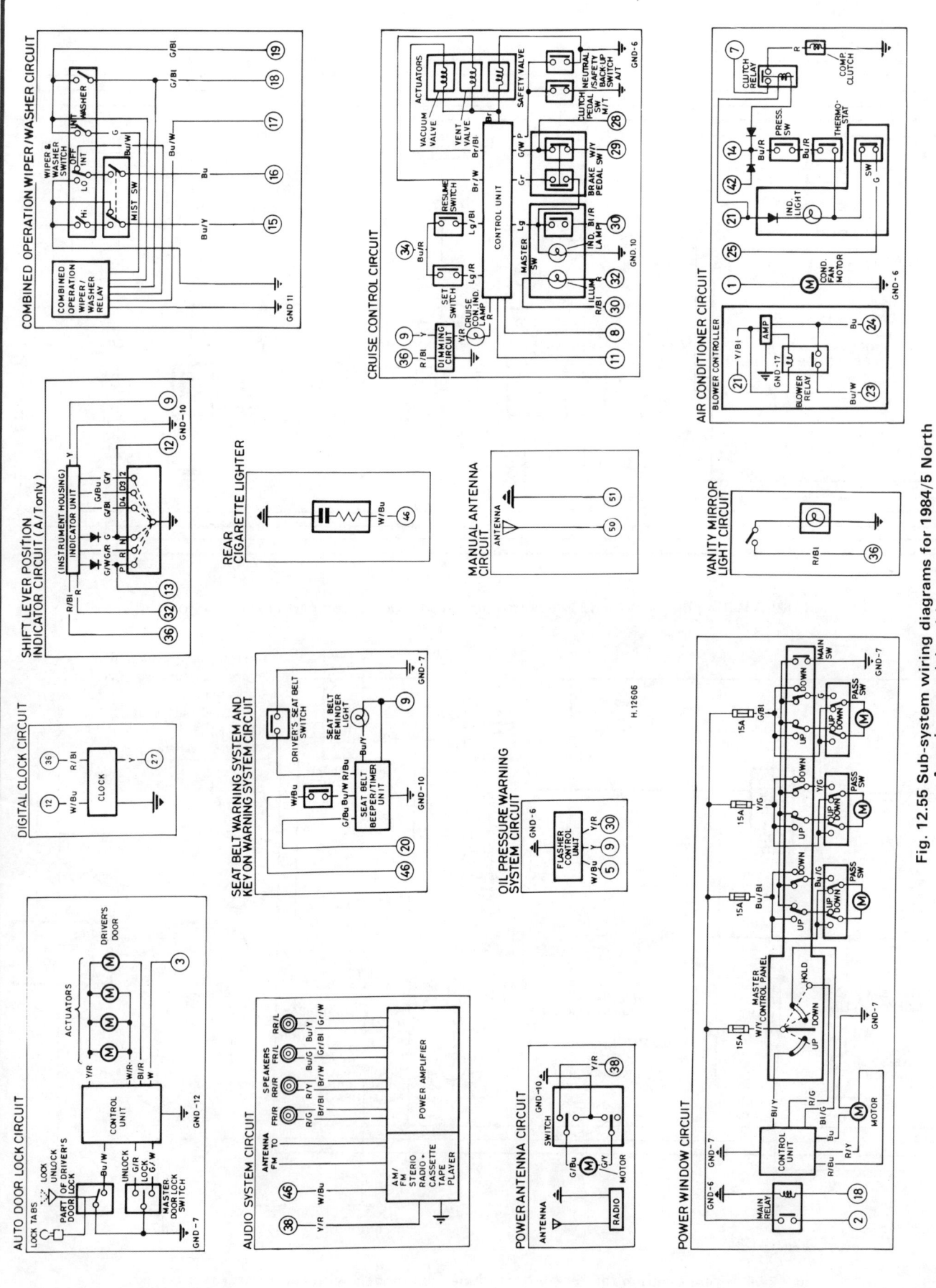

Fig. 12.55 Sub-system wiring diagrams for 1984/5 North American models with carburettor

347

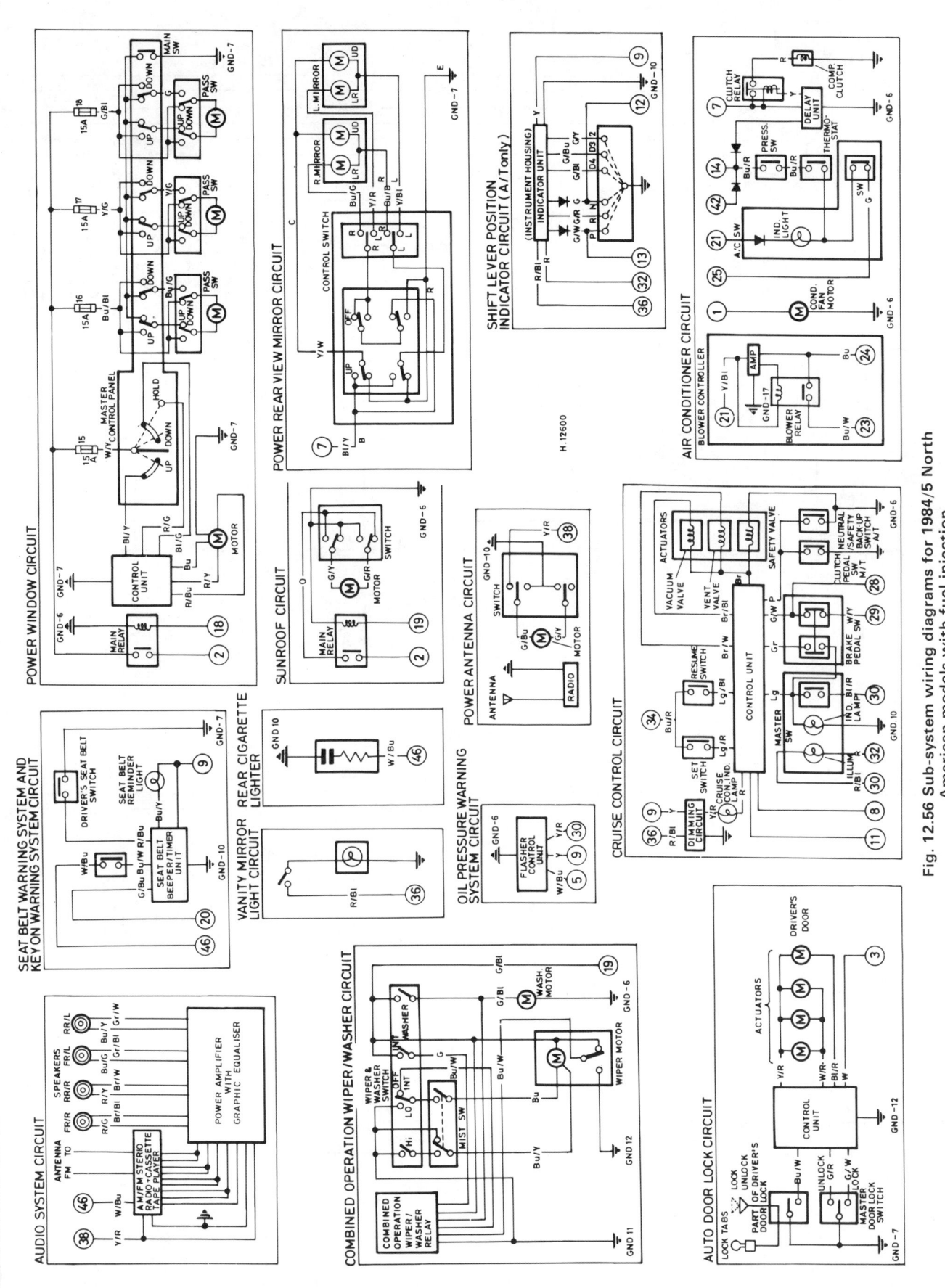

Fig. 12.56 Sub-system wiring diagrams for 1984/5 North American models with fuel injection

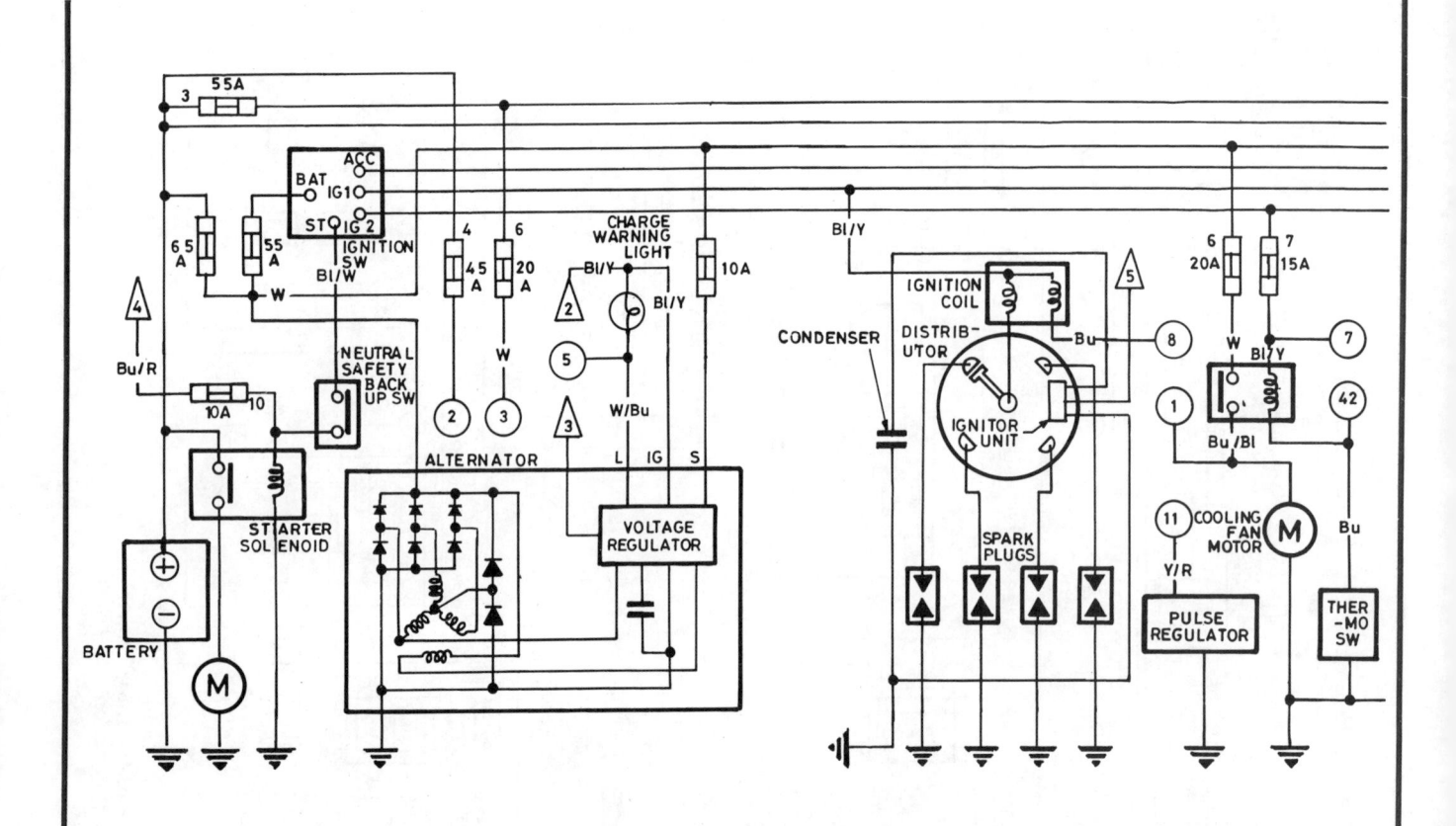

348

Fig. 12.57 Wiring diagram for 1984/5 North American models with fuel injection

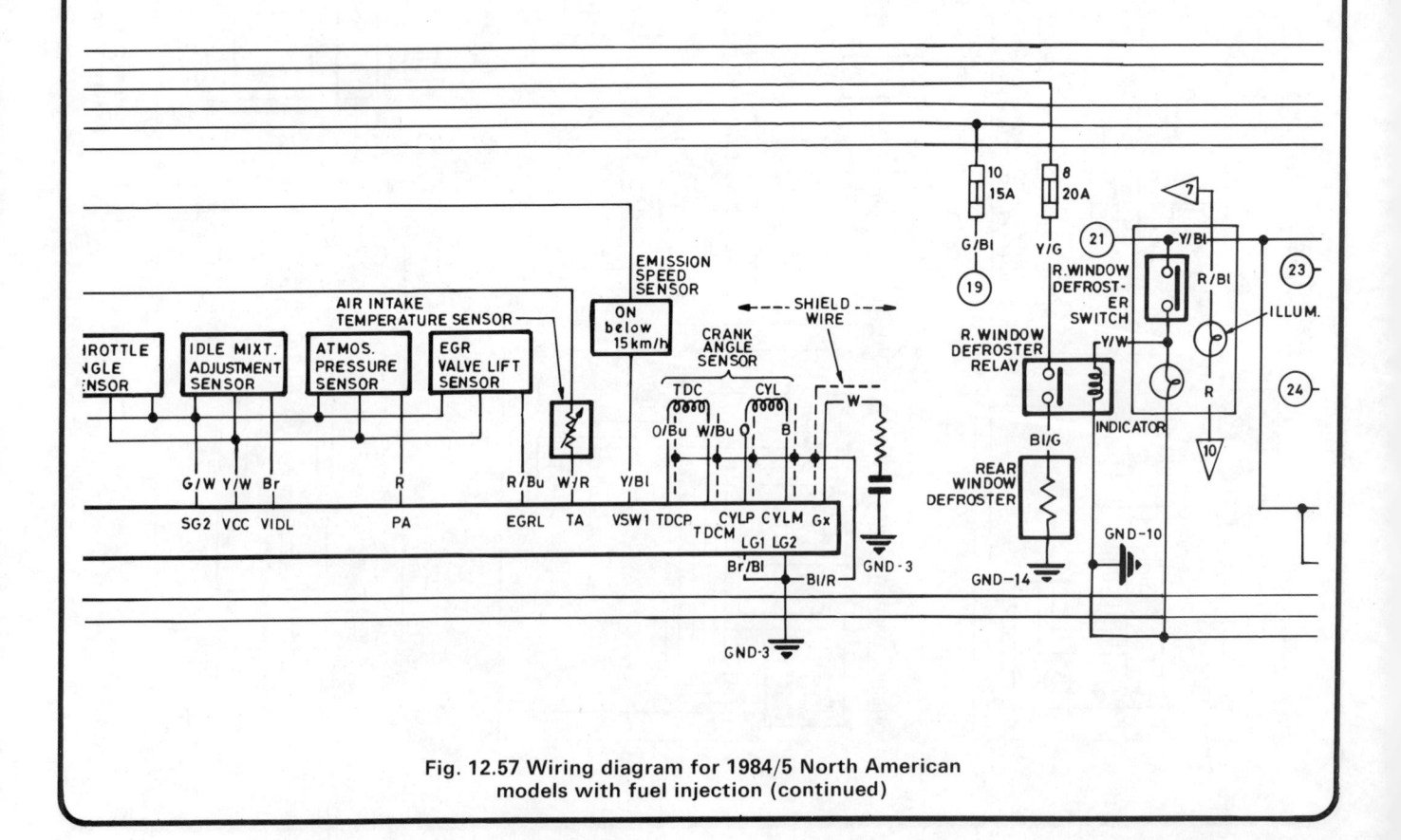

Fig. 12.57 Wiring diagram for 1984/5 North American models with fuel injection (continued)

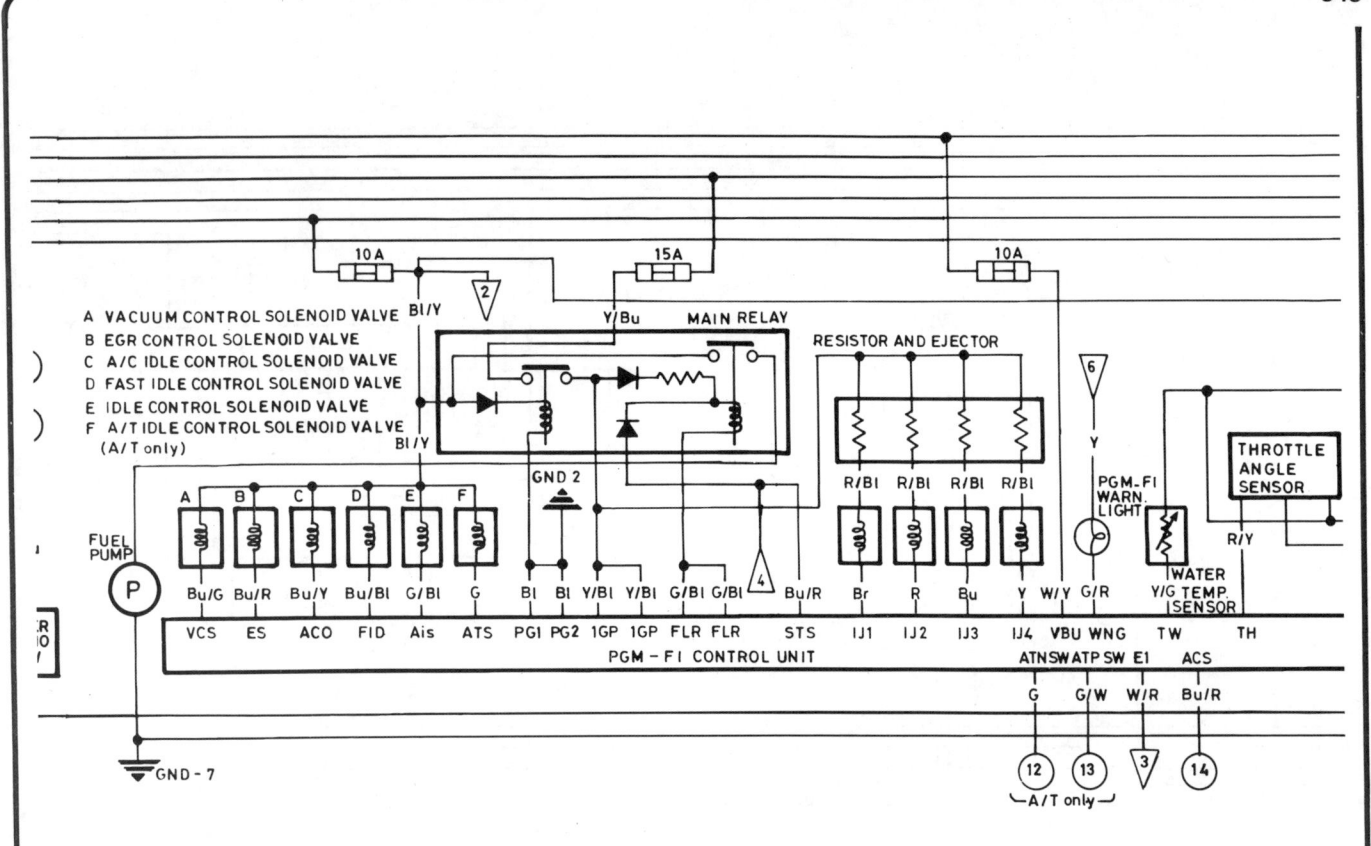

Fig. 12.57 Wiring diagram for 1984/5 North American models with fuel injection (continued)

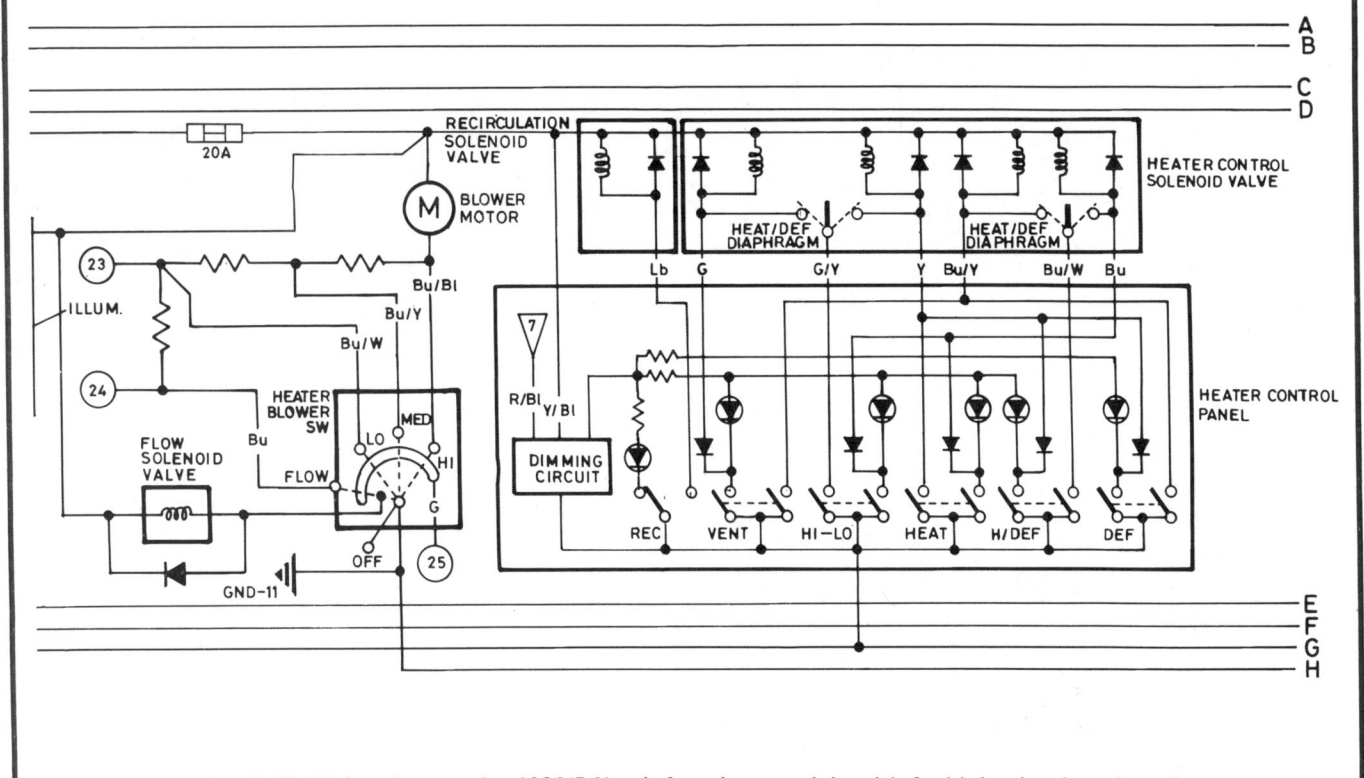

Fig. 12.57 Wiring diagram for 1984/5 North American models with fuel injection (continued)

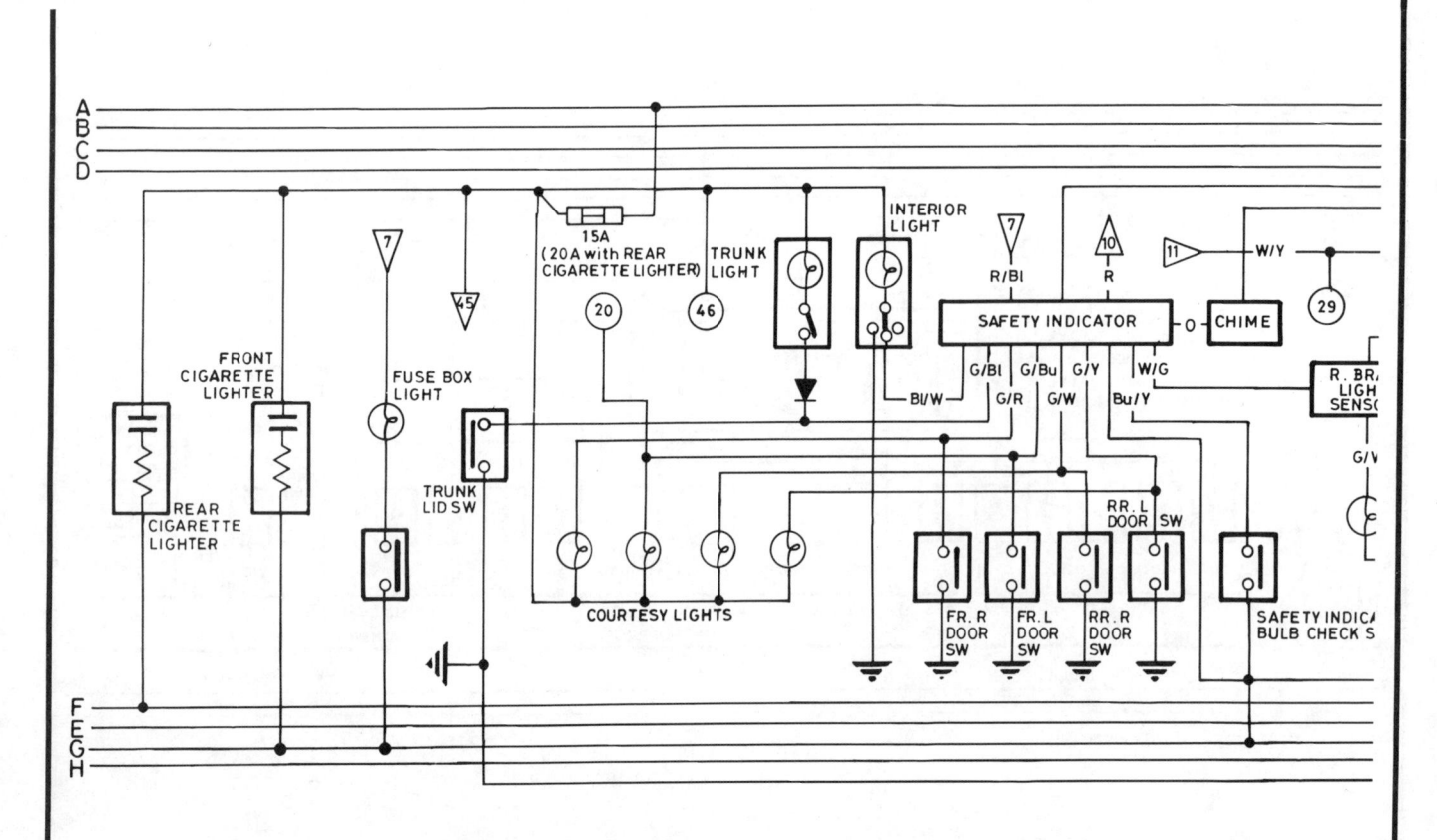

Fig. 12.57 Wiring diagram for 1984/5 North American models with fuel injection (continued)

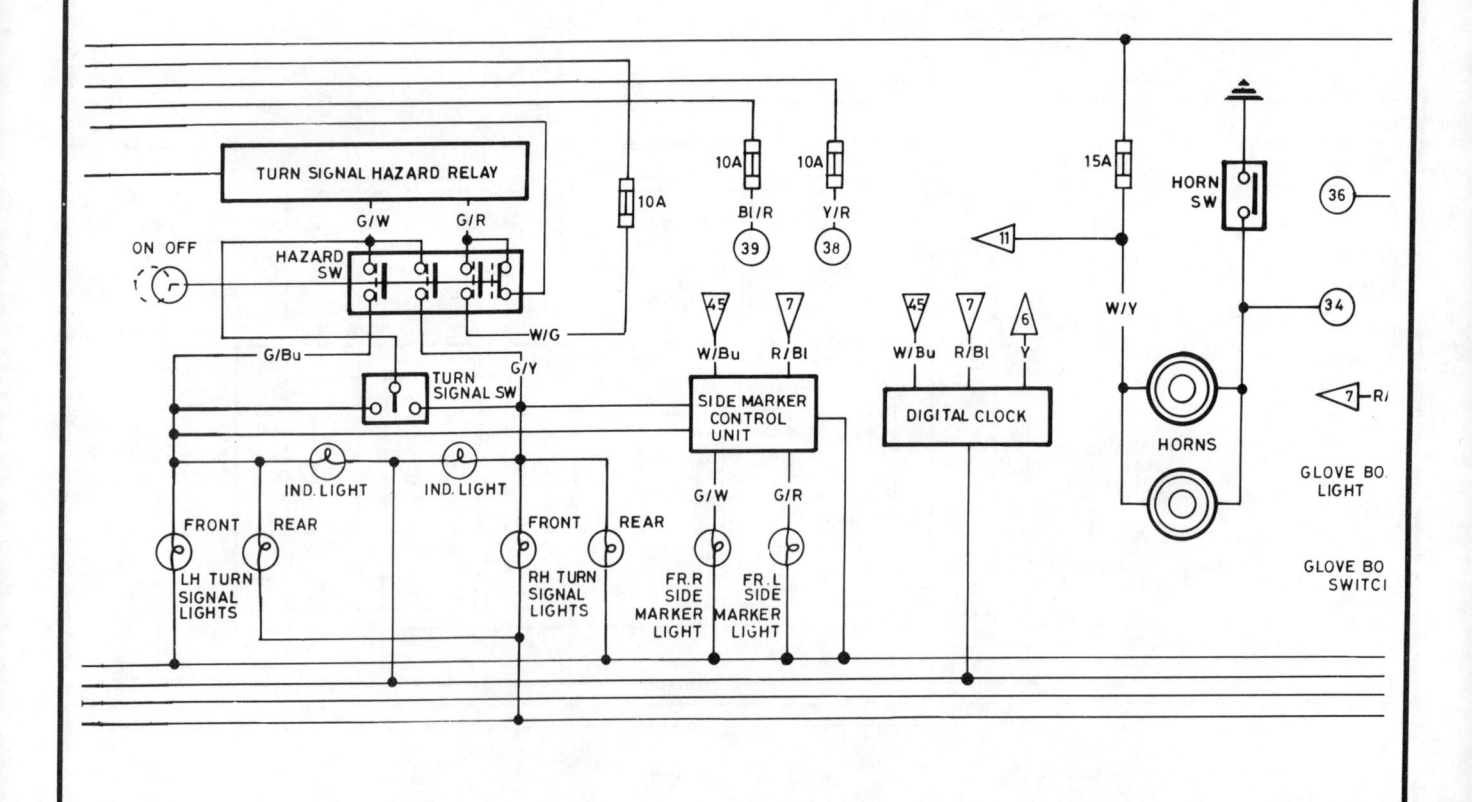

Fig. 12.57 Wiring diagram for 1984/5 North American models with fuel injection (continued)

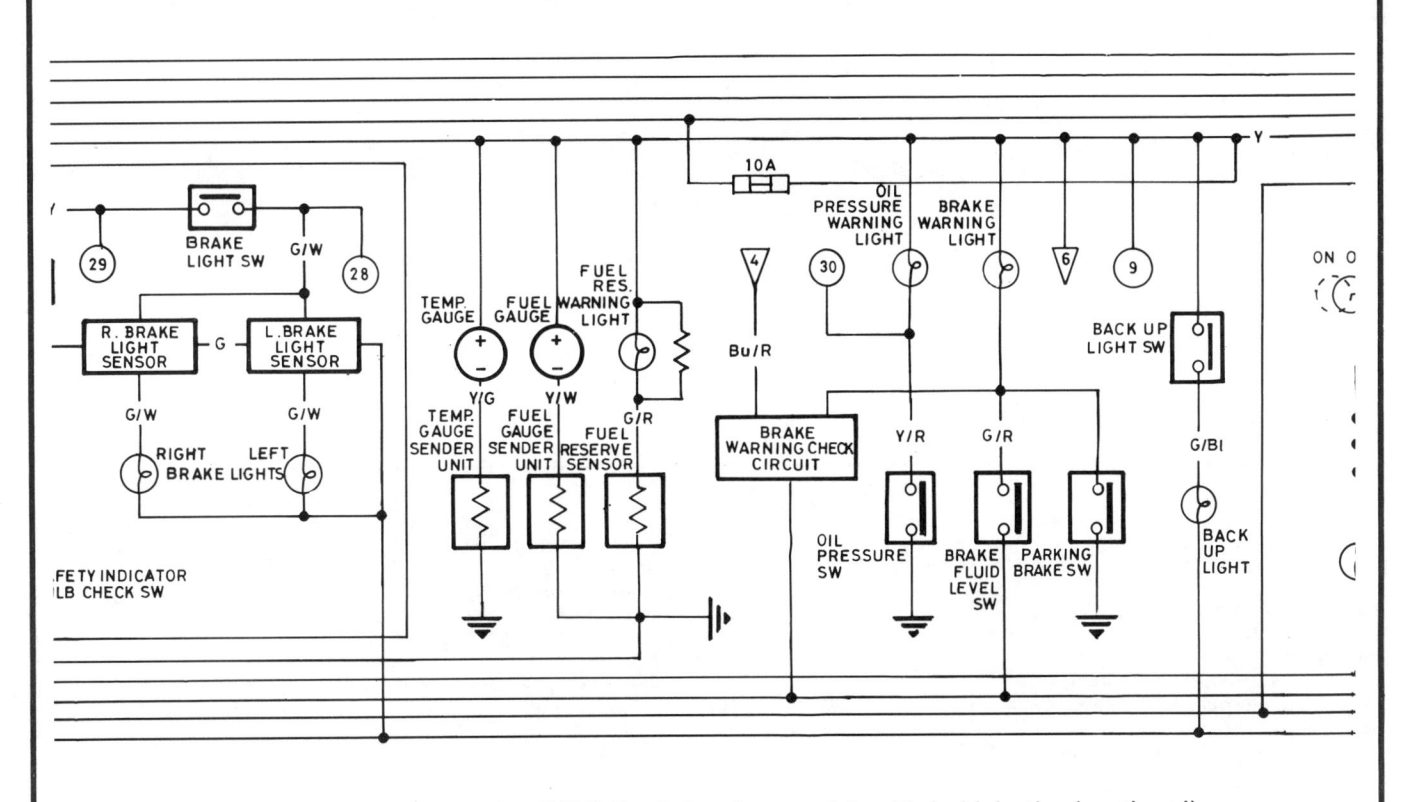

Fig. 12.57 Wiring diagram for 1984/5 North American models with fuel injection (continued)

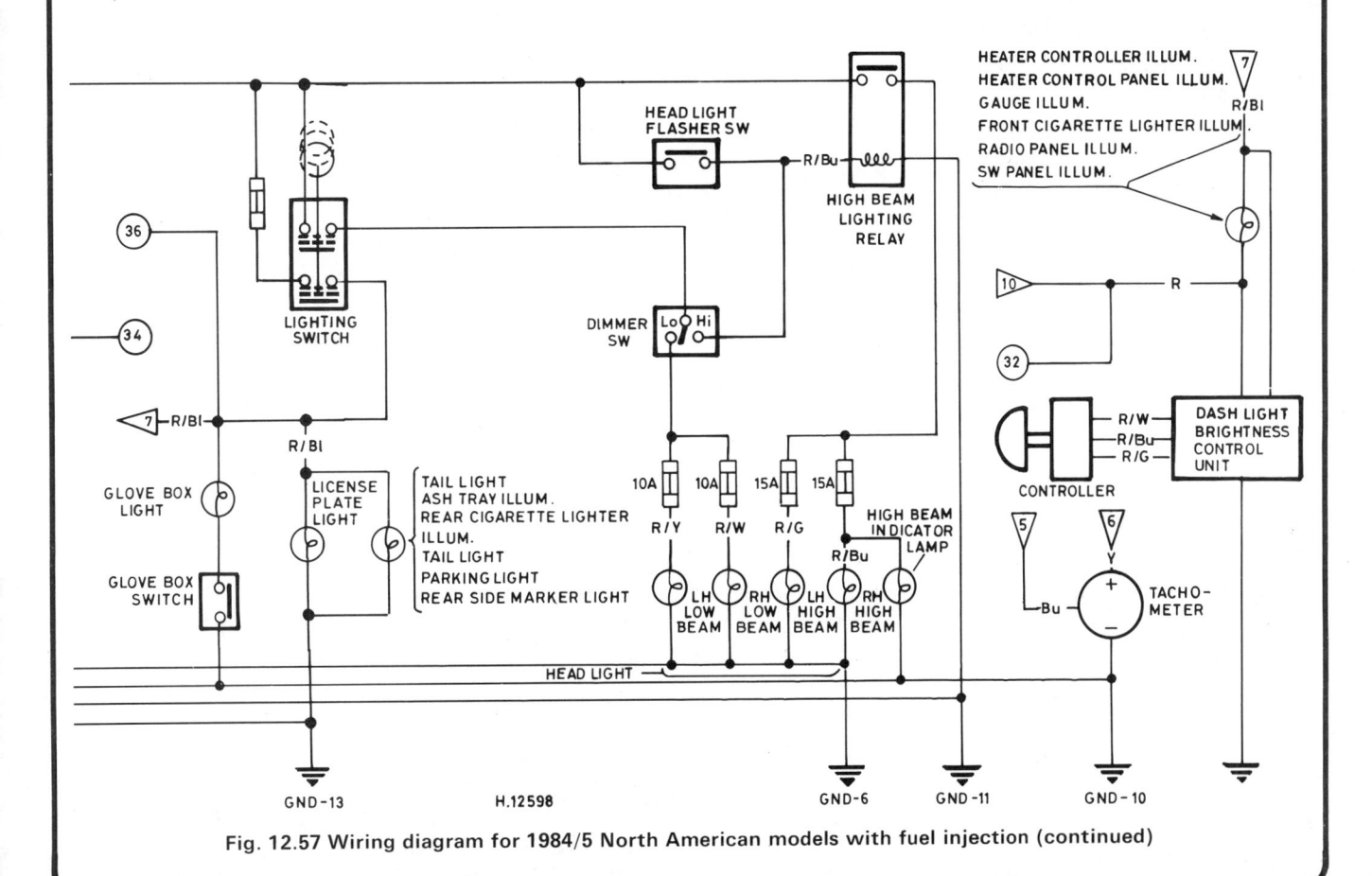

Fig. 12.57 Wiring diagram for 1984/5 North American models with fuel injection (continued)

Index

A

Accelerator pump
  adjustment – 83
Air cleaner
  assembly, removal and refitting – 75
  element removal and refitting – 75
Air conditioner
  compressor removal and refitting – 293
  condenser removal and refitting – 295
  description and maintenance – 292
  evaporator removal and refitting – 295
Alternator
  description, maintenance and precautions – 298
  overhaul – 299
  removal and refitting – 298
  testing *in situ* – 298
Anti-lock braking system (ALB) – 227
Ashtray light – 315
Automatic transmission – 175 *et seq*
Automatic transmission
  description – 177
  differential oil seals renewal – 198
  dismantling – 182
  fault diagnosis – 203
  fluid: level checking, topping-up and renewing – 180
  gearshift indicator adjustment and checks – 201
  gearshift selector removal and refitting – 201
  governor removal and refitting – 184
  mainshaft and countershaft
    bearings and seals removal and refitting – 196
    gear clearance measurement – 197
    removal – 184
  maintenance, routine – 180
  main valve body removal – 184
  major assemblies inspection and overhaul – 187
  reassembly – 198
  removal and refitting – 181
  reverse idler gear removal – 196
  shift cable adjustment – 203
  specifications – 175
  throttle control – 92
  torque converter removal and refitting – 198
  torque wrench settings – 176

B

Battery
  charging – 298
  inspection – 297
  removal and refitting – 297

Bearings, engine
  selection – 54
Bodywork and fittings – 254 *et seq*
  air conditioner – 292, 293, 295
  bonnet – 257, 258
  boot lid – 272
  bumpers – 258
  centre console – 282
  dashboard – 283
  description – 254
  doors – 264, 265, 268, 270
  front console – 282
  heater and ventilation – 287, 288, 289, 292
  interior trim – 280
  maintenance
    bodywork and underframe – 254
    upholstery and carpets – 255
  mirror, rear view
    external – 270
    internal – 287
  radiator grille – 258
  rear hatch – 271
  rear hatch/boot and fuel filler lid
    remote release lever and latches – 272
  rear quarter light – 275
  repair
    major damage – 257
    minor damage – 255
  seat belts – 285
  seats – 284, 285
  sunroof – 275, 277
  torque wrench settings – 251
  wind deflector – 277
  windscreen and rear window – 275
Bodywork repair sequence (colour) – 262, 263
Bonnet
  lock and release assembly
    removal, refitting and adjustment – 258
  removal and refitting – 257
Boot lid
  removal and refitting – 272
Brake booster servo unit
  description and maintenance – 223
  removal and refitting – 225
Brake fluid level switch
  testing – 226
Brake light switch
  removal, testing and refitting – 227
Brake pedal
  removal, refitting and adjustment – 226
Braking system – 208 *et seq*
Braking system
  anti-lock system (ALB) – 227

Index

brake booster servo unit – 223, 225
description – 209
dual proportioning valve – 221
fault diagnosis – 228
fluid level switch – 226
front disc brakes – 210, 214, 215
handbrake – 225, 226
hydraulic system
    bleeding – 222
    pipes and hoses – 222
light switch – 227
load sensing valve – 221
maintenance, routine – 210
master cylinder – 218
pedal – 226
rear brakes
    disc – 210, 214, 215
    drum – 213, 218
specifications – 208
torque wrench settings – 209
wheel hubs, bearings and discs – 215
Bumpers, front and rear
removal and refitting – 258

C

Camshaft
removal and inspection – 40
Camshaft drivebelt
removal, refitting and adjusting – 48
Capacities, general – 6
Carburettor
adjustments general – 81
description – 81
float level inspection and adjustment – 91
idle controller (air conditioning models) adjustment – 82
idle speed and mixture adjustment – 81
overhaul – 88
removal and refitting – 88
settings – 71, 72
Carpets, maintenance – 255
Choke, automatic
choke opener inspection and renewal – 87
coil heater testing and renewal – 87
description – 84
fast idle adjustment – 87
fast idler unloader inspection – 86
linkage adjustment – 84
Choke, automatic (Canadian models)
adjustment – 94
fast idle adjustment – 96
Choke cable
removal, refitting and adjustment – 92
Choke, manual
fast idle adjustment – 83
Choke relief valve
adjustment – 83
Cigar lighter – 316
Clutch – 142 et seq
Clutch
adjustment – 144
cable renewal – 144
description – 142
fault diagnosis – 146
inspection – 144
maintenance, routine – 144
pedal removal and refitting – 144
refitting – 146
release bearing removal, inspection and refitting – 145
removal – 144
specifications – 142
torque wrench settings – 142
Coil, ignition
description and testing – 130
removal and refitting – 130

Connecting rods – 56
Console
removal and refitting
    centre – 282
    front – 282
Conversion factors – 15
Coolant mixture – 65
Cooling system – 63 et seq
Cooling system
coolant mixture – 65
description – 64
draining, flushing and refilling – 65
drivebelts – 69
fault diagnosis – 70
maintenance, routine – 65
radiator – 66
radiator cooling fan – 67
specifications – 63
temperature gauge and sender unit – 67
thermosensor – 67
thermostat – 66
thermoswitch – 67
torque wrench settings – 63
water pump – 68
Crankshaft
reassembly – 58
removal and inspection – 53
Cruise control system – 325
Cut-off valves (fuel system)
testing and renewing – 87
Cylinder block
dismantling – 50
inspection – 55
reassembly – 58
Cylinder head
inspection – 45
reassembly and refitting – 46
removal – 40

D

Dashboard
removal and refitting – 283
Decarbonising – 58
Dimensions, vehicle – 6
Disc brakes, front
caliper removal, overhaul and refitting – 214
disc inspection and renovation – 215
pads inspection and renewal – 210
Disc brakes, rear
caliper removal, overhaul and refitting – 217
disc inspection and renewal – 215
pads inspection and renewal – 212
Distributor
overhaul
    carburettor models – 133
    fuel injection models – 137
removal and refitting – 132
Doors
courtesy lights – 315
interior handle removal and refitting – 264
latch removal and refitting – 265
lock cylinder removal and refitting – 265
outside handle removal and refitting – 264
power lock removal and refitting – 265
removal and refitting – 270
side moulding removal and refitting – 268
striker removal, refitting and adjusting – 270
trim panel removal and refitting – 264
window removal and refitting
    front – 265
    rear – 268
window weatherstrip removal and refitting – 268
window winder regulator scissors assembly
    removal and refitting – 268

Index

Drivebelts
    removal, refitting and adjusting
        air conditioning pump – 69
        alternator/water pump – 69
        power steering pump – 69
Driveshafts – 204 *et seq*
Driveshafts
    description – 204
    fault diagnosis – 207
    inboard joint dismantling, inspection and reassembly – 206
    maintenance, routine – 205
    outboard joint inspection – 207
    removal and refitting – 205
    specifications – 204
    torque wrench settings – 204
Drum brakes, rear
    brake lining inspection and renewal – 213
    drum inspection and renovation – 218
    wheel cylinder removal, overhaul and refitting – 218
Dual proportioning valve (braking system) – 221

E

Electrical system – 296 *et seq*
Electrical system
    alternator – 298, 299
    battery – 297, 298
    cigar lighter – 316
    cruise control system – 325
    description – 297
    fault diagnosis – 23, 328
    fuses – 297, 301
    headlamps – 308, 310, 324
    instrument panel – 305
    lamps and bulbs – 308, 310, 311, 314
    maintenance, routine – 297
    optional systems – 327
    radio – 316, 317
    relays and control units – 301
    specifications – 296
    speedometer cable – 326
    starter motor – 301
    switches – 307
    torque wrench settings – 297
    warning lights – 327
    windscreen wiper – 323, 324
    wiring diagrams – 330 to 351
Emission controls
    air jet controller – 105
    air vent cut-off diaphragm – 101
    anti-afterburn valve – 105
    charcoal cannister – 101
    cranking opener solenoid valve – 103
    dashpot system – 101
    description – 99
    evaporative controls (carburettor models) – 99
    exhaust gas recirculation (EGR) system – 103
    fault diagnosis – 128
    feedback control system – 105
    fuel cut-off relay – 105
    high altitude reduced emission – 105
    ignition control system – 101
    secondary air supply system – 103
    speed sensor – 105
    throttle controller – 103
    two-way valve – 101
Emission controls (fuel injection models)
    catalytic converter – 119
    crankcase controls (FI) – 115
    description – 115
    evaporative controls (FI) – 119
    exhaust gas recirculation system (FI) – 119
    ignition timing controls (FI) – 119
Engine – 27 *et seq*

Engine
    ancillary components removal – 36
    bearing selection – 54
    camshaft – 40
    camshaft drivebelt – 48
    crankshaft – 53, 58
    cylinder block – 50, 55, 58
    cylinder head – 40, 45, 46
    decarbonising – 58
    description – 29
    dismantling – 36
    fault diagnosis – 25, 61
    final assembly and refitting – 60
    firing order – 27
    maintenance, routine – 31
    oil filters and pressure relief valve – 52
    oil pump – 51
    piston rings – 56
    pistons and connecting rods – 56
    removal and refitting with transmission – 31
    removal: general – 31
    rocker shaft – 41
    separation from transmission – 36
    specifications – 27
    torque wrench settings – 29
    valve, auxiliary – 45
    valve clearances – 49
    valves, intake and exhaust – 43
English, use of – 11
Exhaust system – 122

F

Fault diagnosis – 23 *et seq*
Fault diagnosis
    automatic transmission – 203
    braking system – 228
    clutch – 146
    cooling system – 70
    driveshafts – 207
    electrical system – 23, 328
    engine – 25, 61
    fuel and emission control systems – 128
    ignition system – 141
    manual transmission – 174
    suspension and steering – 253
Firing order – 27
Front suspension
    description – 231
    fault diagnosis – 253
    lower arm removal and refitting – 236
    maintenance, routine – 231
    radius rod removal and refitting – 233
    specifications – 229
    spring height inspection – 243
    stabiliser bar removal and refitting – 232
    steering knuckle and wheel hub bearings
        removal and refitting – 236
    strut removal, overhaul and refitting – 234
    torque wrench settings – 230
Front turn indicator light – 311
Fuel, exhaust and emission control systems – 71 *et seq*
Fuel filter
    removal and refitting – 76
Fuel gauge
    testing – 77
Fuel injection system
    air intake system (FI)
        removal, refitting and components
        testing – 108
    components removal, refitting and testing – 112
    description – 107
    electronic control unit (ECU) – 111
    emission controls – 115 to 122
    throttle cable (FI) removal and adjustment – 115

Fuel pump
    cut-off relay – 78
    removal and refitting – 78
Fuel system
    accelerator pump – 83
    air cleaner – 75
    automatic choke – 84, 86, 87, 94, 96
    carburettor – 81, 88, 91
    choke cable – 92
    choke relief valve – 83
    cut-off valves – 87
    description – 73
    emission controls – 99 to 107
    fault diagnosis – 128
    fuel filter – 76
    fuel gauge – 77
    fuel injection system – 107 to 122
    fuel pump – 78
    fuel tank – 76
    fuel tank sender unit – 77
    idle controller (air conditioning models) – 82
    idle speed and mixture – 81
    intake air control system – 80
    intake air temperature sensor – 87
    low fuel level warning light – 78
    maintenance, routine – 74
    manual choke fast idle – 83
    positive crankcase ventilation (PCV)
        filter – 78
        valve – 78
    power valve (North American models) – 97
    specifications – 71
    throttle cable – 92
    throttle control (automatic transmission)
        bracket – 92
        cable – 92
    torque wrench settings – 72
Fuel tank
    removal and refitting – 76
Fuel tank sender unit
    removal, testing and refitting – 77
Fusebox light – 314
Fuses
    general – 301
    specifications – 297

G

Glossary – 11

H

Handbrake
    adjustment – 225
    cable removal and refitting – 226
    warning light switch adjusting and testing – 226
Headlamps
    beam adjusting – 310
    bulbs renewal – 308
    removal and refitting – 308
    wash system – 324
    wiper unit removal and refitting – 324
Heater and ventilation
    description – 287
    heater
        blower removal and refitting – 289
        controls and cables
        removal, refitting and adjusting – 289
        unit removal and refitting – 288
    'mild flow' assembly removal and refitting – 289
    vacuum-operated components – 292
Hydraulic system (brakes)
    bleeding – 222
    pipes and hoses – 222

I

Ignition system – 129 *et seq*
Ignition system
    advance and retard mechanism testing – 139
    coil – 130
    description – 130
    distributor – 132, 133, 137
    fault diagnosis – 141
    HT cables – 130
    igniter unit testing – 139
    maintenance, routine – 130
    reluctor air gap adjustment – 139
    spark plugs – 130, 131
    specifications – 129
    timing adjustment – 140
Instrument panel
    light – 314
    removal and refitting – 305
Intake air
    control system inspection – 80
    temperature sensor testing – 87
Interior light – 314

J

Jacking – 7

L

Lamp units
    removal, refitting and bulb renewal
        headlamps – 308
        interior lamps – 314
        parking light and side indicator light – 310
        rear lamp cluster – 310
        supplementary exterior lights – 311
Load sensing valve (braking system)
    removal, refitting and adjusting – 221
Low fuel warning light
    testing – 78
Lubricants and fluids, recommended – 22
Lubrication chart – 22
Luggage compartment light – 315

M

Maintenance, routine
    bodywork and fittings
        exterior – 254
        interior – 255
    braking system
        ALB system check (if applicable) – 20, 21, 228
        ALB system high pressure hoses renewal – 20, 21
        fluid level check/top up – 20
        fluid renewal (including ALB system) – 20, 21, 222, 228
        front brake discs and calipers inspection – 20, 21
        front brake pads wear check – 20, 210
        handbrake inspection – 20, 21, 225
        hoses and hydraulic lines check – 20, 21, 222
        load sensing valve check – 20, 21, 221
        rear brakes inspection – 20, 21, 212, 213
    clutch release arm travel check – 20, 21, 144
    cooling system
        coolant level check/top up – 20
        coolant renewal – 65
        drivebelts check/adjust – 69
        hoses and connections inspection – 20, 21
    driveshafts constant velocity joints check – 20, 21, 206, 207
    electrical system
        alternator drivebelt inspection – 20, 21
        battery electrolyte level check/top up (if applicable) 20, 297

Index

equipment operation check – 20
windscreen washer fluid level check/top up – 20
engine
    oil and filter change – 20, 21, 31, 52
    oil level check/top up – 20, 21, 31
    valve clearances check/adjust – 20, 21, 49
exhaust system check – 20, 21, 122
fuel and emission control systems
    air cleaner filter element renewal – 20, 21, 74, 75
    blow-by filter renewal – 20
    catalytic converter heat shield inspection (North America) – 21
    choke mechanism inspection – 20, 21, 86, 87, 94
    EGR system inspection (North America) – 21, 103
    evaporative control system inspection (North America) – 21
    fuel filter renewal – 20, 21, 74, 76
    idle speed and idle CO content check – 20, 21, 81
    PCV valve renewal – 20, 21, 74, 78
    throttle control system (manual transmission) inspection – 20, 21
ignition system
    distributor cap, rotor and ignition wiring inspection – 20, 21, 133, 137
    spark plugs renewal – 20, 21, 130
    timing and control systems inspection – 20, 21, 140
safety precautions – 16
service schedules – 20, 21
steering
    front wheel alignment check – 20, 21
    power steering fluid level check/top up (if applicable) – 20, 251
    power steering pump belt inspection – 20, 21, 252
    power steering system check – 20, 21, 252
    tie rod ends, gearbox and rubber boots check – 20, 21
suspension
    mounting bolts check – 20, 21
    rear wheel bearing grease renewal (North America) – 21
transmission
    automatic transmission fluid change – 20, 21, 180
    manual transmission oil change – 20, 21, 149
    oil/fluid level check/top up – 20, 149, 180
tyres
    condition check – 244
    pressures check/adjust – 20, 230
wheels condition check – 244
Manifolds, inlet and exhaust
    removal and refitting – 124
Manual transmission – 147 *et seq*
Manual transmission
    countershaft bearing and mainshaft oil seal
        removal, inspection and refitting – 160
    description – 149
    differential oil seal removal and refitting – 162
    differential unit removal, inspection and refitting – 162
    dismantling into components – 150
    fault diagnosis – 174
    gearchange lever, shift rod and torque rod
        removal and refitting – 171
    gearshift selector mechanism
        removal, inspection and refitting – 165
    mainshaft and countershaft reassembly – 157
    mainshaft, countershaft and components inspection – 153
    maintenance, routine – 149
    reassembly – 168
    removal and refitting – 150
    reversing light switch testing – 173
    specifications – 147
    torque wrench settings – 149
Master cylinder (braking system)
    removal, overhaul and refitting – 218
Mirror, rear view
    removal and refitting
        exterior – 270
        interior – 287

N

Number plate light – 312

O

Oil filters and pressure relief valve
    removal and refitting – 52
Oil pump
    removal, inspection and refitting – 51

P

Parking light and side indicator light – 310
Piston rings
    removal, refitting and inspection – 56
Pistons and connecting rods
    inspection – 56
Positive crankcase ventilation (PCV)
    filter removal and refitting – 78
    valve checking, removal and refitting – 78
Power valve (fuel system North American models) – 97

R

Radiator
    removal, repair and refitting – 66
Radiator cooling fan
    removal, testing and refitting – 67
Radiator grille
    removal and refitting – 258
Radio
    aerial removal and refitting – 316
    mobile equipment: interference-free installation – 317 to 323
    removal and refitting – 316
Rear foglight – 311
Rear hatch
    removal and refitting – 271
Rear hatch/boot and fuel filler lid
    remote release lever and latches – 272
Rear lamp cluster – 310
Rear quarterlight
    removal and refitting – 275
Rear suspension
    description – 231
    fault diagnosis – 253
    hub carrier and wheel bearings – 241
    lower arm removal and refitting – 240
    maintenance, routine – 231
    radius rod removal and refitting – 240
    shock absorber removal and refitting – 238
    specifications – 229
    spring height inspection – 243
    stabiliser bar removal and refitting – 241
    torque wrench settings – 230
Rear window
    removal and refitting – 275
Relays and control units – 301
Repair procedures, general – 12
Rocker shaft assembly
    dismantling, inspection and reassembly – 41
Routine maintenance *see* **Maintenance, routine**

S

Safety precautions – 16
Seat belts – 285
Seats
    removal and refitting
        front – 284
        rear (Hatchback) – 285
        rear (Saloon) – 284
Side turn indicator light – 312
Spare parts
    buying – 9
    to carry in car – 25

Index

Spark plugs
    conditions (colour) – 131
    removal, inspection and refitting – 130
Speedometer cable – 326
Starter motor
    description – 301
    testing *in situ* – 301
Steering
    description – 231
    fault diagnosis – 253
    lock removal and refitting – 250
    maintenance, routine – 231
    manual steering
        column removal, overhaul and refitting – 245
        rack and gearbox overhaul – 248
        rack and gearbox refitting – 250
        rack and gearbox removal – 247
        rack guide adjustment – 244
    power steering
        column removal and refitting – 251
        fluid topping up and renewal – 251
        inspection – 252
        pump belt adjustment – 252
        pump drivebelt removal, refitting and adjusting – 69
        pump overhaul – 252
        pump removal and refitting – 252
        rack guide adjustment – 252
        rack removal and refitting – 252
    specifications – 229
    speed sensor removal and refitting – 252
    tie-rod and balljoint renewal – 251
    torque wrench settings – 230
    wheel
        removal and refitting – 245
        rotational play inspection – 244
Sunroof
    adjustment – 277
    motor, drain tube and frame removal and refitting – 277
    removal and refitting – 275
Suspension and steering – 229 *et seq*
Suspension *see* **Front suspension** *and* **Rear suspension**
Switches
    removal and refitting
        combination – 307
        cruise control – 307
        horn – 307

T

Temperature gauge and sender unit
    removal, testing and refitting – 67
Thermosensor
    removal, testing and refitting – 67
Thermostat
    removal, testing and refitting – 66
Thermoswitch
    removal, testing and refitting – 67
Throttle cable
    removal, refitting and adjusting – 92

Throttle control (automatic transmission)
    bracket adjustment – 92
    cable adjustment – 92
Timing, ignition
    adjustment – 140
Tools
    general – 13
    to carry in car – 25
Towing – 7
Transmission *see* **Automatic transmission** *and* **Manual transmission**
Tyres
    care and maintenance – 244
    pressures – 230
    size – 230

U

Underframe
    maintenance – 254
Upholstery
    maintenance – 255

V

Valve, auxiliary
    removal, inspection and refitting – 45
Valve clearances
    adjustment – 49
Valves, intake and exhaust
    removal, inspection and refitting – 43
Vanity mirror light – 315
Vehicle identification numbers – 9

W

Warning lights – 327
Water pump
    removal and refitting – 68
Weights, kerb – 6
Wheels
    bearings adjustment – 237
    bearings (rear) removal and refitting – 241
    care and maintenance – 244
    changing – 7
    settings – 230
    toe-in adjustment – 238
Wind deflector
    removal and refitting – 277
Windscreen
    removal and refitting – 275
Windscreen wiper
    blades and arms removal and refitting – 323
    general – 324
    motor removal and refitting – 324
Wiring diagrams – 330 to 351
Working facilities – 13